Mu-Ping Nieh, Frederick A. Heberle, John Katsaras (Eds.)
Characterization of Biological Membranes

Also of interest

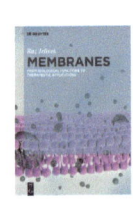

Membranes.
From Biological Functions to Therapeutic Applications
Jelinek, 2018
ISBN 978-3-11-045368-3, e-ISBN 978-3-11-045369-0

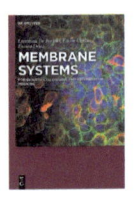

Membrane Systems.
For Bioartificial Organs and Regenerative Medicine
De Bartolo, Curcio, Drioli, 2017
ISBN 978-3-11-026798-3, e-ISBN 978-3-11-026801-0

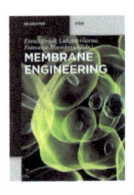

Membrane Engineering.
Drioli, Giorno, Macedonio, 2018
ISBN 978-3-11-028140-8, e-ISBN 978-3-11-028139-2

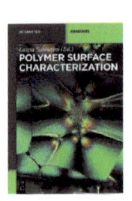

Polymer Surface Characterization
Sabbatini (Eds.), 2018
ISBN 978-3-11-027508-7, e-ISBN 978-3-11-028811-7

Characterization of Biological Membranes

Structure and Dynamics

Edited by
Mu-Ping Nieh
Frederick A. Heberle
John Katsaras

DE GRUYTER

Editors
Prof. Dr. Mu-Ping Nieh
University of Connecticut
Polymer Program, Institute of Materials Science
Department of Chemical and Biomolecular Engineering
97 North Eagleville Road
Storrs, CT 06269-3136, USA
mu-ping.nieh@uconn.edu

Dr. Frederick A. Heberle
Oak Ridge National Laboratory
Shull Wollan Center – A Joint Inst. for Neutron Sciences
2008 Oak Ridge, TN 37831-6453, USA
heberlefa@ornl.gov

Prof. Dr. John Katsaras
Oak Ridge National Laboratory
Shull Wollan Center – A Joint Inst. for Neutron Sciences
2008 Oak Ridge, TN 37831-6453, USA
katsarasj@ornl.gov

ISBN 978-3-11-054464-0
e-ISBN (PDF) 978-3-11-054465-7
e-ISBN (EPUB) 978-3-11-054468-8

Library of Congress Control Number: 2018965309

Bibliographic information published by the Deutsche Nationalbibliothek
The Deutsche Nationalbibliothek lists this publication in the Deutsche Nationalbibliografie;
detailed bibliographic data are available on the Internet at http://dnb.dnb.de.

© 2019 Walter de Gruyter GmbH, Berlin/Boston
Typesetting: Integra Software Services Pvt. Ltd.
Printing and binding: CPI books GmbH, Leck
Cover image: Jill Hemman

www.degruyter.com

Preface

Membranes are functional assemblies found in all living systems. The cell or plasma membrane (PM) – comprised predominantly of lipids (organized in a bilayer), proteins, and carbohydrates – acts as a selective barrier between the cell's interior and its outside environment. It allows for the transport of small molecules, and is involved in processes such as cell signaling and adhesion, as well as endocytosis/exocytosis. However, the structure of the PM, and especially its lateral organization, has been a topic of debate since the fluid mosaic model of membranes was proposed by Singer and Nicolson in 1972. In this compilation we examine the structure and function of biological membranes, the proteins within them, and the techniques for studying them.

Although biomembranes have been extensively studied, in recent years the development of new or revamped biophysical techniques to study their static and dynamic structures at the molecular and atomic levels has been revolutionary. The development of new algorithms and computer hardware technology, in particular, have enabled the study of complex biological membranes and the extraction of information from experimental data not previously possible.

The book is organized in three sections. The first section (Chapters 1–9) is devoted to advanced experimental approaches, including X-ray and neutron scattering, solid-state nuclear magnetic resonance (NMR), atomic force microscopy (AFM), and high-resolution secondary ion mass spectrometry (HR-SIMS). The second (Chapters 10–16) and third (Chapters 17–21) sections describe, respectively, different membrane systems and the theories/simulations used to understand them at the molecular level.

Chapter 1 by Kučerka and Uhríková describes how X-ray and neutron scattering, in combination with molecular dynamics (MD) simulations, can be used to interrogate membrane structure and the interaction of membranes with small molecules like alcohols. Of importance to many of the studies described by Kučerka and Uhríková is the ability to change the neutron contrast of the system by simply by altering the amount of D_2O. In Chapter 2, Salditt, Komorowski, and Frank describe the principles of grazing incidence diffraction (GID), grazing incidence small-angle X-ray scattering (GISAXS), and small-angle x-ray scattering (SAXS), and provide examples for each technique. In contrast to SAXS, GID and GISAXS require the availability of aligned membranes. The authors conclude by providing us with an outlook on X-ray free electron laser sources and the new opportunities they offer regarding biomembrane research. Chapter 3 by Eells et al. introduces yet another scattering technique, namely neutron reflectometry (NR), that allows for high-resolution (fractions of a nanometer) structure determination of membranes and membrane-associated proteins along the bilayer normal. Moreover, the authors point out that the strength of NR is its ability to determine the structure of conformationally flexible peripheral membrane proteins and combined with the neutron's ability to differentiate between protium and deuterium, to resolve the individual constituents of membrane-bound protein-protein complexes. Finally, Eells et al. describe integrative modeling strategies that supplement the low-resolution reflectometry data with additional experimental and computational information resulting in high-resolution, three-dimensional models of membrane-bound protein structure.

The first three chapters dealt with the static structure of membranes. This changes in Chapter 4 where Kelley, Butler, and Nagao describe the collective dynamics of model membrane systems studied by neutron spin echo (NSE). Specifically, they show how NSE can be used to determine the membrane's mechanical properties (e.g., bending modulus), which are thought to control key functions such as bilayer shape, protein binding, and molecular transport. The chapter concludes by considering another important membrane property, namely, membrane viscosity and discusses future challenges regarding NSE and membrane studies. In Chapter 5, Xia and Nieh describe time-resolved small-angle neutron scattering (SANS), differential scanning calorimetry (DSC), and fluorescence correlation spectroscopy to determine the spontaneous lipid transfer rates in two different morphology nanoparticles: i.e., discoidal bicelles and unilamellar vesicles.

Up to this point, the studies described involved either X-ray or neutron scattering techniques. This changes in Chapter 6 where Cheung and Cash describe different NMR techniques and nuclei that can be used to study biological systems. For example, nuclear overhauser spectroscopy (NOESY) identifies correlations between two nuclei near each other and is used to study membrane-bound entities. On the other hand, pulsed gradient spin echo NMR spectroscopy is used to study translational motion directly in biological fluids and under simulated physiological conditions. The NMR theme is continued in Chapter 7 by Brown in which he describes solid-state NMR and how the average structure and dynamics of biomolecules are addressed through a combination of experiment and theory. The different studies described by Brown reveal emergent properties of membrane lipids that are explained through an energy landscape affecting raft-forming mixtures, bilayer fusion, and lipid-protein interactions.

AFM was invented by IBM's Binning and Rohrer in the early 1980s to image materials on the nanoscale. In Chapter 8, Poloni, Won, and Yip describe how AFM offers a unique platform to investigate biomaterials, including the study of spatiotemporal dynamics of biological processes under physiologically relevant conditions and high spatial resolution. They present a detailed protocol for the preparation of supported lipid bilayers (SLBs) and their in situ imaging by AFM, including the experimental design for studies of protein– and peptide–membrane interactions. The book's first section is concluded by the chapter of Gorman et al., which describes high-resolution secondary ion mass spectrometry (SIMS), a technique capable of imaging the distributions of distinct lipid and protein species in the PM of intact mammalian cells. The authors conclude with future uses of high-resolution SIMS to elucidate membrane structure–function relationships.

The second part of the book starts with the chapter by Davis and Schmidt, which gives a broad overview of the contributions of NMR experiments to our understanding of cholesterol in model membrane systems. The use of NMR is then extended to the study of mitochondria by Bozelli and Epand in Chapter 11. Specifically, they shed new insights into the structure of mitochondrial membranes, as well as the thermodynamics of their lipid constituents. The following chapter by DiPasquale et al. describes studies on the location of vitamin E in the bilayer using neutron diffraction and substrate-aligned bilayers whose phospholipids contain polyunsaturated fatty acids (PUFAs). The authors highlight the need for better physical understanding of the oxidation–anitoxidant relationship through studies of vitamin E and its location in different membranes.

In Chapter 13, Steinkühler and Dimova discuss the use of giant unilamellar vesicles (GUVs) as a platform for the study of viscoelastic and mechanical properties of biomembranes. From an experimental point of view, GUVs are very attractive in that they lend themselves ideally to interrogation by light microscopy and can be observed by label-free microscopy techniques using phase contrast or differential interference contrast. The chapter concludes with a discussion of GUVs and a series of biophysical measurements used to determine membrane physical properties. In the following chapter (Chapter 14), London discusses recent developments in the preparation of asymmetric membranes that more closely mimic cellular membranes, concluding with a description of new methods used to alter the lipid composition and asymmetry of living cells. The membrane asymmetry theme is continued in the chapter by Kiessling and Tamm in which they describe the development, characterization, and application of asymmetric supported planar bilayers. Specifically, they highlight their use of asymmetric membranes to address transbilayer coupling of lipid domains, SNARE-mediated membrane fusion, and viral entry into cells. In the final chapter from this section of the book (Chapter 16), Craig et al. describe the use of styrene-maleic acid (SMA) copolymers as an alternative membrane platform to study the structure and function of proteins – SMAs are capable of mimicking lipid bilayers while maintaining protein integrity.

The final section of the book begins with the chapter by Schick describing possible origins for lipid domain formation. Schick argues the case that lateral heterogeneities in membranes are characteristic of a microemulsion, rather than the result of phase separation or critical fluctuations. This is followed by the chapter by Nickels and Katsaras (Chapter 18) describing the combined use of static and dynamic neutron scattering experiments to study model membranes, and the use of coarse-grained and atomistic MD simulations to interpret the dynamic scattering data. In Chapter 19 Cherniavskyi and Tieleman discuss the Martini model for MD simulations – currently the most widely used coarse-grained force field for MD simulations of biological membranes and membrane proteins. They describe the model's advantages and limitations and present recent examples of large-scale mixed membrane simulations. Chapter 20 by Shen et al. describes the different simulations available in terms of length and time scales, including the all-atom CHARMM force field, the Martini model that was described in Chapter 19, and the dissipative particle dynamics (DPD) explicit solvent lipid model. The book ends with the chapter by Chan and Cheng describing recent progress in MD simulations of small molecule–lipid bilayer interactions, concluding with MD simulations using advanced sampling techniques.

The research described in this book is recent and presented by some of the leading researchers in the field. It is our hope that advanced graduate students and researchers in the field will find it informative and inspirational. It is our opinion that biomembranes research using biophysical techniques and computer simulations holds the potential for important and wide-ranging scientific discoveries in the not-too-distant future.

Mu-Ping Nieh
Frederick Heberle
John Katsaras

Contents

Preface —— V

List of contributing authors —— XIII

Part I Structural and dynamic characterization

Norbert Kučerka, Daniela Uhríková
1   Biophysical perspectives of lipid membranes through the optics of neutron and X-ray scattering —— 3

Tim Salditt, Karlo Komorowski, Kilian Frank
2   X-ray structure analysis of lipid membrane systems: solid-supported bilayers, bilayer stacks, and vesicles —— 43

Rebecca Eells, David P. Hoogerheide, Paul A. Kienzle, Mathias Lösche, Charles F. Majkrzak, Frank Heinrich
3   Structural investigations of membrane-associated proteins by neutron reflectometry —— 87

Elizabeth G. Kelley, Paul D. Butler, Michihiro Nagao
4   Collective dynamics in model biological membranes measured by neutron spin echo spectroscopy —— 131

Yan Xia, Mu-Ping Nieh
5   Spontaneous lipid transfer rate constants —— 177

Eugene Cheung, Darian Cash
6   Fundamentals of Nuclear Magnetic Resonance spectroscopy (NMR) and its applications —— 195

Michael F. Brown
7   Collective dynamics in lipid membranes —— 231

Laura N. Poloni, Amy Won, Christopher M. Yip
8   Mapping protein– and peptide–membrane interactions by atomic force microscopy: strategies and opportunities —— 269

Brittney L. Gorman, Ashley N. Yeager, Corryn E. Chini, Mary L. Kraft
9   Imaging the distributions of lipids and proteins in the plasma membrane with high-resolution secondary ion mass spectrometry —— 287

Part II Biomimetic, biorelated, or biological systems

James H. Davis, M. L. Schmidt
10 Cholesterol in model membranes —— 325

José Carlos Bozelli Junior, Richard M. Epand
11 Study of mitochondrial membrane structure and dynamics on the molecular mechanism of mitochondrial membrane processes —— 365

Mitchell DiPasquale, Michael H.L. Nguyen, Thad A. Harroun, Drew Marquardt
12 Monitoring oxygen-sensitive membranes and vitamin E as an antioxidant —— 391

Jan Steinkühler, Rumiana Dimova
13 Giant vesicles: A biomimetic tool for assessing membrane material properties and interactions —— 415

Erwin London
14 Formation and properties of asymmetric lipid vesicles prepared using cyclodextrin-catalyzed lipid exchange —— 441

Volker Kiessling, Lukas K. Tamm
15 Application and characterization of asymmetric-supported membranes —— 465

Andrew F. Craig, Indra D. Sahu, Carole Dabney-Smith, Dominik Konkolewicz, Gary A. Lorigan
16 Styrene-maleic acid copolymers: a new tool for membrane biophysics —— 477

Part III Molecular dynamics – simulation and theory

M. Schick
17 On the origin of "Rafts": The plasma membrane as a microemulsion —— 499

Jonathan D. Nickels, John Katsaras
18 Combining experiment and simulation to study complex biomimetic membranes —— 515

Yevhen Cherniavskyi, D. Peter Tieleman
19 Simulations of biological membranes with the Martini model —— 551

Zhiqiang Shen, Alessandro Fisher, Huilin Ye, Ying Li
20 Multiscale modeling of lipid membrane —— 569

Chun Chan, Xiaolin Cheng
21 Molecular dynamics simulation studies of small molecules interacting with cell membranes —— 603

Index —— 631

List of contributing authors

José Carlos Bozelli Junior
Department of Biochemistry and Biomedical Sciences
McMaster University
1280 Main Street West
Hamilton, Ontario
Canada L8S 4K1

Michael F. Brown
Department of Chemistry and Biochemistry
University of Arizona
Tucson, AZ 85721, USA
and
Department of Physics
University of Arizona
Tucson, AZ 85721, USA

Paul D. Butler
NIST Center for Neutron Research
National Institute of Standards and Technology
Gaithersburg, MD 20899-6102, USA
and
Department of Chemical & Biomolecular Engineering
University of Delaware
Newark, DE 19716, USA
and
Department of Chemistry
University of Tennessee
Knoxville, TN 37996, USA

Darian Cash
Moderna Therapeutics Inc.
200 Technology Square
Cambridge, MA 02139, USA
Darian.Cash@modernatx.com

Chun Chan
College of Pharmacy
Medicinal Chemistry and Pharmacognosy
The Ohio State University
Columbus, OH 43210, USA

Xiaolin Cheng
College of Pharmacy
Medicinal Chemistry and Pharmacognosy
The Ohio State University
Columbus, OH 43210, USA
and
Biophysics Graduate Program
The Ohio State University
Columbus, OH 43210, USA
and
Translational Data Analytics Institute
The Ohio State University
Columbus, OH 43210, USA

Yevhen Cherniavskyi
Department of Biological Sciences and Centre for Molecular Simulation
University of Calgary
2500 University Drive N.W.
Calgary, Alberta
Canada T2N1N4

Eugene Cheung
Moderna Therapeutics Inc.
200 Technology Square
Cambridge, MA 02139, USA
eugene.cheung@modernatx.com

Corryn E. Chini
Department of Chemistry
University of Illinois at Urbana-Champaign
Urbana, IL 61801, USA

Andrew F. Craig
Department of Chemistry and Biochemistry
Miami University
651 E. High Street
Oxford, OH 45056, USA

Carole Dabney-Smith
Department of Chemistry and Biochemistry
Miami University
651 E. High Street
Oxford, OH 45056s, USA

James H. Davis
Department of Physics
University of Guelph
Guelph, Ontario
Canada N1G 2W1

Rumiana Dimova
Max Planck Institute of Colloids and Interfaces
Potsdam, Germany

Mitchell DiPasquale
Department of Chemistry and Biochemistry
University of Windsor
Windsor, Ontario
Canada N9B 3P4

Rebecca Eells
Department of Physics
Carnegie Mellon University
Pittsburgh, PA 15213, USA

Richard M. Epand
Department of Biochemistry and Biomedical Sciences
McMaster University
1280 Main Street West
Hamilton, Ontario
Canada L8S 4K1
epand@mcmaster.ca

Alessandro Fisher
Department of Biomedical Engineering
University of Connecticut
Storrs, CT 06269, USA

Kilian Frank
Institut für Röntgenphrysik
Universität Göttingen
Friedrich-Hund-Platz 1
37077 Göttingen, Germany
www.roentgen.physik.uni-goettingen.de

Brittney L. Gorman
Center for Biophysics and Computational Biology
University of Illinois at Urbana-Champaign
Urbana, IL 61801, USA

Thad A. Harroun
Department of Physics
Brock University
St. Catharines, Ontario
Canada L2S 3A1

Frank Heinrich
Department of Physics
Carnegie Mellon University
Pittsburgh, PA 15213, USA

and

Center for Neutron Research
National Institute of Standards and Technology
Gaithersburg, MD 20899, USA

David P. Hoogerheide
Center for Neutron Research
National Institute of Standards and Technology
Gaithersburg, MD 20899, USA

Paul A. Kienzle
Center for Neutron Research
National Institute of Standards and Technology
Gaithersburg, MD, 20899, USA

Karlo Komorowski
Institut für Röntgenphysik
Universität Göttingen
Friedrich-Hund-Platz 1
37077 Göttingen, Germany
www.roentgen.physik.uni-goettingen.de

Charles F. Majkrzak
Center for Neutron Research
National Institute of Standards and Technology
Gaithersburg, MD, 20899, USA

Drew Marquardt
Department of Chemistry and Biochemistry
University of Windsor
Windsor, Ontario
Canada N9B 3P4

Michihiro Nagao
NIST Center for Neutron Research
National Institute of Standards and
Technology
Gaithersburg, MD 20899-6102, USA

and

Center for Exploration of Energy and Matter
Department of Physics
Indiana University
Bloomington, IN 47408, USA

Michael H.L. Nguyen
Department of Chemistry and Biochemistry
University of Windsor
Windsor, Ontario
Canada N9B 3P4

John Katsaras
Neutron Scattering Directorate
Oak Ridge National Laboratory
Oak Ridge, TN, USA

and

Shull Wollan Center
Oak Ridge National Laboratory
Oak Ridge, TN, USA
katsarasj@ornl.gov

Elizabeth G. Kelley
NIST Center for Neutron Research
National Institute of Standards and
Technology
Gaithersburg, MD 20899-6102, USA

Volker Kiessling
Department of Molecular Physiology and
Biophysics
University of Virginia
Charlottesville VA 22908, USA
vgk3c@virginia.edu

Dominik Konkolewicz
Department of Chemistry and Biochemistry
Miami University
651 E. High Street
Oxford, OH 45056s, USA
d.konkolewicz@miamioh.edu

Mary L. Kraft
Center for Biophysics and Computational
Biology
University of Illinois at Urbana-Champaign
Urbana, IL 61801, USA

and

Department of Chemical and Biomolecular
Engineering
University of Illinois at
Urbana-Champaign
Urbana, IL 61801, USA

and

Department of Chemistry
University of Illinois at
Urbana-Champaign
Urbana, IL 61801, USA

Norbert Kučerka
Department of Physical Chemistry of Drugs
Faculty of Pharmacy
Comenius University in Bratislava
83232 Bratislava, Slovakia

and

Frank Laboratory of Neutron Physics
Joint Institute for Nuclear Research
141980 Dubna – Moscow Region, Russia

Ying Li
Department of Mechanical Engineering
University of Connecticut
Storrs, CT 06269, USA

and

Institute of Materials Science
University of Connecticut
Storrs, CT 06269, USA
yingli@engr.uconn.edu

Erwin London
Dept. of Biochemistry and Cell Biology
Stony Brook University
Stony Brook, NY 11794-5215, USA
erwin.london@stonybrook.edu

Gary A. Lorigan
Department of Chemistry and
Biochemistry
Miami University
651 E. High Street
Oxford, OH 45056s, USA
gary.lorigan@miamioh.edu

Mathias Lösche
Department of Physics
Carnegie Mellon University
Pittsburgh, PA 15213, USA

and

Department of Biomedical Engineering
Carnegie Mellon University
Pittsburgh, PA 15213, USA

and

Center for Neutron Research
National Institute of Standards and
Technology
Gaithersburg, MD 20899, USA

Jonathan D. Nickels
Department of Chemical and Environmental
Engineering
University of Cincinnati
Cincinnati, OH, USA
jonathan.nickels@uc.edu

Mu-Ping Nieh
Department of Chemical and Biomolecular
Engineering
University of Connecticut
Storrs, CT 06269, USA

and

Polymer Program
Institute of Materials Science
University of Connecticut
Storrs, CT 06269, USA

and

Department of Biomedical
Engineering
University of Connecticut
Storrs, CT 06269, USA

Laura N. Poloni
Department of Chemical Engineering and
Applied Chemistry
University of Toronto
Toronto, Ontario
Canada M5S 1A1

and

Donnelly Centre for Cellular and Biomolecular
Research
University of Toronto
Toronto, Ontario
Canada M5S 1A1

Indra D. Sahu
Department of Chemistry and Biochemistry
Miami University
651 E. High Street
Oxford, OH 45056s, USA

Tim Salditt
Institut für Röntgenphysik
Universität Göttingen
Friedrich-Hund-Platz 1
37077 Göttingen, Germany
www.roentgen.physik.uni-goettingen.de

M. Schick
Department of Physics
Box 35160
University of Washington
Seattle WA 98195, USA

M. L. Schmidt
Department of Physics
University of Guelph
Guelph, Ontario
Canada N1G 2W1

Zhiqiang Shen
Department of Mechanical
Engineering
University of Connecticut
Storrs, CT 06269, USA

Jan Steinkühler
Max Planck Institute of Colloids and Interfaces
Potsdam, Germany

Lukas K. Tamm
Department of Molecular Physiology and
Biophysics
University of Virginia
Charlottesville VA 22908, USA
lkt2e@virginia.edu

D. Peter Tieleman
Department of Biological Sciences and
Centre for Molecular Simulation
University of Calgary
2500 University Drive N.W.
Calgary, Alberta
Canada T2N1N4

Daniela Uhríková
Frank Laboratory of Neutron Physics
Joint Institute for Nuclear Research
141980 Dubna – Moscow Region, Russia

Amy Won
Department of Chemical Engineering and
Applied Chemistry
University of Toronto
Toronto, Ontario
Canada M5S 1A1

and

Donnelly Centre for Cellular and Biomolecular
Research
University of Toronto
Toronto, Ontario
Canada M5S 1A1

and

Department of Biochemistry
University of Toronto
Toronto, Ontario
Canada M5S 1A1

Huilin Ye
Department of Mechanical Engineering
University of Connecticut
Storrs, CT 06269, USA

Ashley N. Yeager
Department of Chemical and Biomolecular
Engineering
University of Illinois at Urbana-Champaign
Urbana, IL 61801, USA

Christopher M. Yip
Department of Chemical Engineering and
Applied Chemistry
University of Toronto
Toronto, Ontario
Canada M5S 1A1

and

Donnelly Centre for Cellular and Biomolecular
Research
University of Toronto
Toronto, Ontario
Canada M5S 1A1

and

Department of Biochemistry
University of Toronto
Toronto, Ontario
Canada M5S 1A1
christopher.yip@utoronto.ca

Yan Xia
Department of Chemical and Biomolecular
Engineering
University of Connecticut
Storrs, CT 06269, USA

Part I Structural and dynamic characterization

Norbert Kučerka, Daniela Uhríková
1 Biophysical perspectives of lipid membranes through the optics of neutron and X-ray scattering

Abstract: Throughout the biological world, cell membranes are crucial to life, with lipids being one of their major components. Although the basic notion of the fluid mosaic model still holds true, the plasma membrane has been shown to be considerably more complex, especially with regard to the diversity and function of lipids. Besides proteins playing an active role in carrying out the various functions that take place in a biological membrane, much attention has recently focused on the importance of lipids in membrane function. After all, how better to explain the diversity of lipids found in nature? Biological membrane mimetics, such as liposomes, lipid bilayers, and model membranes, are used in a broad range of scientific and technological applications due to the unique physical properties of these supramolecular aggregates of amphiphilic molecules. They serve as platforms for studying the soft matter physics of membranes and membrane dynamics, interactions of bilayers with drugs or biologically important molecules like DNA or peptides, and effects of various additives or environmental changes. The modern state-of-the-art research takes advantage of joining brilliance of X-ray scattering sources with some peculiar properties of neutrons and combines results with the power of computer simulations. The advances in chemistry, and deuteration possibilities in particular, allow for better experimental spatial resolution and possibility to pinpoint labels within membranes. It is only a matter of time for many biological functions that occur at the membrane interface to be matched with the structural properties of these membranes. Several examples of pursuing correlations between the structural results of model biomembrane and functions taking place therein will be discussed.

Keywords: biological membrane, lipid bilayer, cholesterol, alcohols, metal ions, DNA, neutron and X-ray scattering

Norbert Kučerka, Department of Physical Chemistry of Drugs, Faculty of Pharmacy, Comenius University in Bratislava, Bratislava, Slovakia; Frank Laboratory of Neutron Physics, Joint Institute for Nuclear Research, Dubna – Moscow Region, Russia
Daniela Uhríková, Department of Physical Chemistry of Drugs, Faculty of Pharmacy, Comenius University in Bratislava, Bratislava, Slovakia

https://doi.org/10.1515/9783110544657-001

1.1 Biological membranes

Biological membrane is a main building block in living organism, where it plays a crucial role in encompassing and defining cells and biological tissue. Membranes form a natural hydrophobic barrier that separates the cytosol from the extracellular environment. However, these complex mesoscopic assemblies have been shown repeatedly to be far more elaborate than simple double-layered structures [1] that serve as permeability barriers [2]. Although the fluid mosaic model [3] remains a dominant theme in our understanding of biomembranes [4], it has evolved to include the notions of a crowding [5] and rafts [6] and turned membranes into highly functional dynamic machines that are central to a host of biological processes, including the transport of materials, cell defense, recognition, adhesion, and signaling [7]. Consequently, as structure is often tightly coupled to function, the myriad-specific functions occurring in these membranes are plausibly correlated with the lipidome's size and diversity [8].

Biological membranes consist mainly of lipids and proteins, where it is widely accepted that the membrane's underlying structure is imparted by the lipid bilayer. Due to the compositional complexity of biological membranes, the physical properties and functional roles of individual lipid species are exceedingly difficult to determine. In order to gain insight into the roles of individual components, it is necessary to study model membrane systems that contain the lipid species of interest. For example, in eukaryotic cells, the predominant lipid species are glycerol-based phospholipids, including phosphatidylcholine (PC), phosphatidylethanolamine (PE), phosphatidylserine, phosphatidylglycerol (PG), phosphatidylinositol, and cardiolipin (CL), while major phospholipids observed in prokaryotic membranes constitute PE, PG, and CL [9]. Furthermore to the headgroup diversity, each lipid species exhibits although characteristic yet fairly diverse fatty acid composition.

It is well known that lipid bilayers form spontaneously due to the hydrophobic effect [10], whereby their structure is dictated by the fine balance of the forces that minimize the system's total free energy. This includes both entropic and enthalpic components that are related to the disruption of the hydrogen bonding network between water molecules, van der Waals attractive forces, *trans*-gauche isomerization, and, most likely, other interactions [11,12]. Changes to hydrocarbon chains can affect all the mentioned intrabilayer interactions, resulting in different equilibrium structures. It is thus not surprising that lipids with different length hydrocarbon chains and degree of unsaturation were found to form bilayers with different thicknesses and lateral areas at the bilayer–water interface [13]. The mishap in maintaining the fine balance of membrane unsaturation can then again result in disorder and malfunction, as has been recognized recently in gestational diabetes mellitus [14]. Because the cell membrane is a first line of defense against invading species, it is key to understanding various diseases and pharmaceutical treatments. Small molecules, such as cholesterol, melatonin, vitamins, peptides, and many others, incorporate into the lipid matrix altering the membrane's structure and physical properties. These changes to the membrane in turn affect its functionality and its interactions with biomolecules, while the mechanism of action is often elusive.

There are many experimental techniques suitable for studying biomembranes at the microscopic level. Scattering techniques, on the other hand, have traditionally been used to determine the structure of 3D crystals on atomic levels. However, advances achieved over past decades developed to an extent that scattering approaches can successfully characterize the physical properties of disordered materials such as biomimetic membranes. The X-ray and neutron scattering methods are now applied to elucidate the material properties previously thought to be the domain of other techniques and even provide possibilities not present in any other methods.

Neutron and X-ray scattering are similar in that both techniques are capable of providing dynamical and structural information [15]. However, the principal differences between the two techniques are in their interactions with matter. As X-rays are electromagnetic waves that primarily interact with electrons, the amplitude of X-ray scattering increases in a simple way with atomic number. On the other hand, neutrons are elementary particles that interact with atomic nuclei, and neutron scattering amplitudes depend in a complex manner on the mass, spin, and energy levels of nuclei. Additionally, differences in the interaction of neutrons with the various isotopes of the same element allow for the powerful and commonly used method of contrast variation. This technique, in which hydrogen atoms that are ubiquitous in biologically relevant samples are substituted with deuterium atoms, is commonly used in neutron scattering studies. Neutron diffraction can then determine the distribution of water or individual components through deuterium labeling. The ability to isolate individual molecular groups at atomic level of detail is unique among biophysical techniques, as it does not require model fitting or other interpretation of the data. Furthermore, discovery of the center of mass distribution of a chemical group is information directly comparable to molecular model simulations without the need for additional computations [16]. The recent advances in X-ray and neutron scattering methods in general are increasingly providing us with a unique access to the much-touted structure–function relationship in biomembranes that is universally sought out in biology and pharmacology.

1.2 Structural significance of lipid diversity

Advances in colloid and interface science have stimulated a renewed interest in the study of lipid–water systems. At the same time, much progress has been made in the analysis of small-angle X-ray (SAXS) and neutron scattering (SANS) data [17]. The popularity of small-angle scattering for the study of biologically relevant materials stems from the fact that it provides detailed information on the size, shape, and conformation of molecular assemblies in solution. As a result, structural biophysics has taken advantage of recent developments to accurately determine the structure of lipid bilayers. An example of this is the joint refinement of X-ray and neutron scattering data [18], which has been rethought in terms of improving the values of lipid areas [19].

The models of X-ray scattering length density (XSLD) and neutron scattering length density (NSLD) emphasize different, but complementary features of the bilayer [13] (e.g., compare Figure 1.1a and b). It follows that a combined approach describes the structural features accentuated by each technique, but in a manner that the data are analyzed simultaneously. This is illustrated in the way the lipid molecular area A is determined, a parameter central to bilayer structure. For both X-ray and neutron scattering, A is calculated using the bilayer's thickness and additional volumetric information. However, it should be emphasized that the two scattering

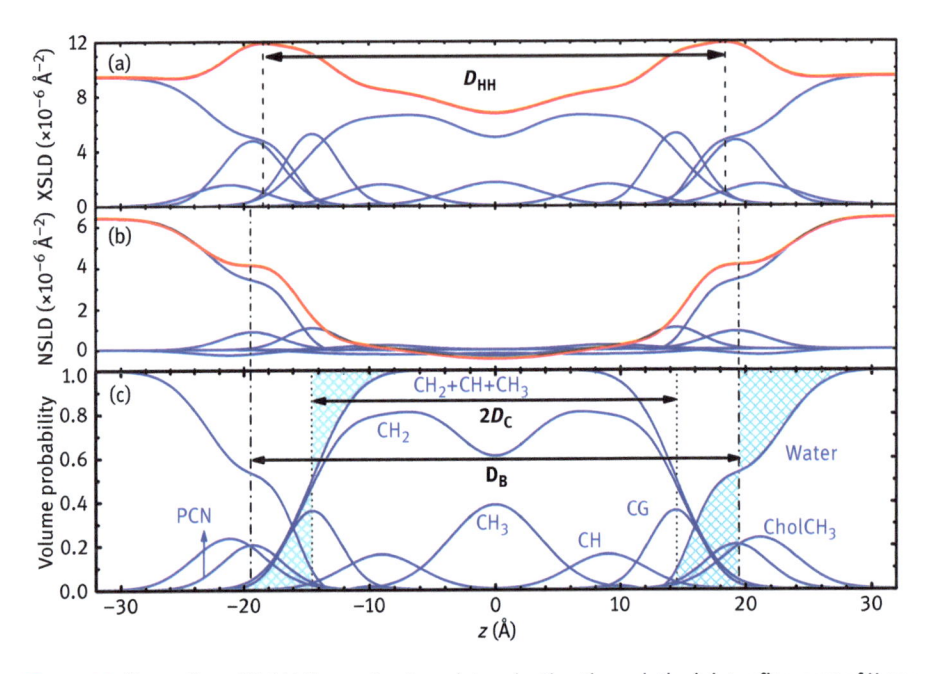

Figure 1.1: Illustration of lipid bilayer structure determination through the joint refinement of X-ray and neutron scattering data. The scattering density profile (SDP) representation of a bilayer in real space is shown, where the top panels show X-ray scattering length density (XSLD) with amplitudes calculated from the number of electrons and electron radius (**a**), and neutron scattering length density (NSLD) based on neutron coherent scattering amplitudes (**b**) of lipid component distributions. The total scattering length densities are denoted by the thick red lines. Panel (**c**) shows volume probability distributions, where the total probability is equal to 1 at each point across the bilayer. The locations where the shaded areas are equal define the Gibbs dividing surface between the bilayer headgroup and hydrocarbon regions (effectively D_C), and that between the lipid bilayer and the water phase (effectively $D_B/2$).

techniques are sensitive to different bilayer thicknesses. The thickness best resolved by X-rays is the distance between the electron density maxima found in the lipid headgroup region, D_{HH}, while in the case of neutron scattering, the contrast variation technique allows finding accurately the total bilayer thickness, D_B. Even though they are the two most robust experimentally determined parameters, D_{HH} and D_B are not directly comparable and neither measure on its own contains all of the desired bilayer structural information.

Area per lipid A follows from the volume probability underlying universally both of the scattering density profiles. The volume probability gives the Gibbs dividing surfaces for the water region and for the hydrocarbon region shown in Figure 1.1 and defined to be at $D_B/2$ and D_C, respectively. The parameter D_B, also known as the Luzzati thickness [20], impacts the model structure through the water distribution. It is defined by the equality of the integrated water probabilities $P_W(z)$ to the left of this surface and the integrated deficit of water probabilities to the right

$$\int_0^{D_B/2} P_W(z)dz = \int_{D_B/2}^{d/2} (1 - P_W(z))dz,$$

where $d/2$ is a point beyond which $P_W(z) = 1$, that is, a point beyond the furthest atom belonging to the lipid. From this, D_B can be expressed in the form

$$D_B = d - 2\int_0^{d/2} P_W(z)dz.$$

Finally, the latter integral is equivalent to the integrated deficit of lipid probability and is equal to $d/2 - V_L/A$, where V_L is a total lipid volume. The equation then yields the first of the following equalities:

$$A = \frac{2V_L}{D_B} = \frac{(V_L - V_{HL})}{D_C}.$$

The second equality in the equation follows from the equivalent derivation applied to the dividing surface between the hydrocarbon and headgroup regions (V_{HL} being a volume of headgroup region). Even though the experimentally obtained scattering data contain information about the bilayer's structure in the z direction (along the bilayer normal), the above derivation allows us to evaluate the structure in the lateral direction, namely A. It should be emphasized that while the latter part of this equation was widely employed in X-ray scattering models, the first equality has important implications in the case of neutron scattering [21]. For protonated lipid bilayers dispersed in D_2O, neutrons are particularly sensitive to the overall bilayer thickness D_B. The previous equation thus directly yields lipid area from highly precise measurements of V_L. The simultaneous analysis of X-ray and neutron scattering data results in robust structural parameters that describe all the key bilayer features.

Biological activities of surface-active compounds are known to depend on lipid acyl chain length [22]. An exciting possibility is that the biological membrane has at its disposal a wide range of lipid lengths to stimulate membrane proteins at different locations. Intuitively, the length of a bilayer's acyl chains affects, to first order, bilayer thickness. Unfortunately, bilayer thickness values reported in literature, even in the case of given lipid bilayers, vary widely depending on the experimental method and analysis model [20] and prevent thus drawing any conclusions. However, the series of bilayer thickness parameters that are placed on a relative scale reveal conveniently such effect of acyl chain length [13]. Figure 1.2a shows a universal increase of bilayer thickness upon the extension of lipid acyl chain for lipids of various chemical compositions in the lipid headgroup region (i.e., PC vs. PG vs. PE lipids) and acyl chain region (i.e., fully saturated vs. monounsaturated vs. mixed chain lipids). In addition, the results showed no thermal effect to the chain length imposed thickness changes [13].

Figure 1.2a supports the notion that a change in the acyl chain length increases bilayer thickness similarly in the different bilayers. However, this is only part of the

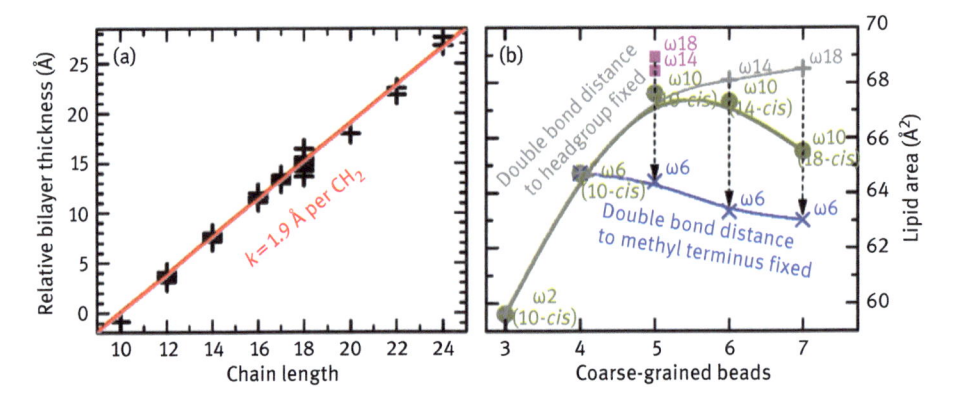

Figure 1.2: (a) Relative bilayer thicknesses as a function of chain length for lipids with different headgroups, acyl chain unsaturation, and temperature taken from the literature (see [13]). The slope of the linear function suggests a universal bilayer thickness increase of 1.9 Å for each CH_2 group (note the distance between the groups corresponds to half this number since the bilayer is spanned by two lipid molecules). (b) Area/Lipid obtained from coarse-grained MD simulations of systems corresponding roughly to the systems of diC12:1PC, diC16:1PC, diC20:1PC, diC24:1PC, and diC28:1PC (dark yellow solid dots) that are commonly utilized for modeling monounsaturated PC lipids. Gray plus signs, on the other hand, follow the results for systems with double bond position fixed at the third bead from the headgroup (i.e., 10-*cis*), and blue crosses those where the double bond was fixed with respect to the methyl terminus (i.e., ω6 position). Dashed arrows indicate points where double bonds were shifted away from the lipid headgroup region, toward the bilayer center. All points are labeled with the double bond position, while due to the coarse-grained nature of the MARTINI model, the mapping between total number of carbons and double bond position is somewhat arbitrary [23]. Results are adapted from Kučerka et al. [13] and Kučerka et al. [24].

story. Another significant change takes place in the plane of the bilayer, namely the change to the area per lipid [25]. As a first approximation, one can estimate the behavior of A from the fact that bilayer thickness increases linearly with each additional carbon and so does lipid volume. However, this turned out to underestimate the values when compared to experimental results [26], demonstrating that it is not possible to simplify changes in the lateral direction in the same way that one does in the transverse direction. It therefore seems that lipid areas are strongly dependent on the acyl chain and headgroup compositions. Indeed, the fine balance between attractive and repulsive intrabilayer forces in the unsaturated lipid bilayers has been explained in terms of double bond position by the coarse-grained molecular dynamics simulations [24]. Figure 1.2b reveals an increase in area when the double bond's distance is fixed relative to the lipid's headgroup, whereas it decreases when the double bond is fixed relative to the bilayer center. As discussed previously [27], increased hydrocarbon chain length results in increased van der Waals attraction, which in turn leads to an ordering of the hydrocarbon chains that effectively reduces the area/lipid. However, lipid chain disorder also depends on double bond position, and it apparently has the largest effect when the double bond is located in the middle of the hydrocarbon chain [28].

The origin of different behavior between bilayer thickness and area per lipid as a function of acyl chain length is born by the forces that are responsible for minimizing the system's total energy. A simple formulation of the free energy for a planar bilayer involves attractive components that are the result of hydrophobic forces within the hydrocarbon chain region, headgroup dipolar interactions, and the repulsive components including steric interactions, hydration forces, and entropic effects due to acyl chain confinement [29]. In this, the interplay between the lateral interactions and headgroup chemical diversity documents why lipid areas likely play a central role. For example, the largest areas are observed for PG lipids even though PG does not possess the largest headgroup. This is however consistent with the observation that the introduction of anionic PG lipids results in decreased membrane stability [30]. The larger areas of PG lipids can also play important roles in regulating protein translocation [31], modulating bacterial membrane permeability [32], and enhancing membrane protein folding [33]. In contrast, structural studies of PE bilayers show a low number of water molecules hydrating their headgroups (between 4 and 7, compared to ~12 for a typical fluid PC bilayer). In fact, the steric exclusion interactions and strong hydrogen bonding between PE headgroups that are responsible for such low levels of hydration are unique among the glycerophospholipids. When compared to PC headgroups with strong repulsive interactions below areas of ~48 $Å^2$ preventing them achieving the minimal packing of their acyl chains (minimal area of an all-*trans* chain is ~20 $Å^2$) [34], the chains of DLPE, with its gel phase area ~41 $Å^2$ [35], appear to achieve such packing. The fluid phase PE area most likely represents the packing limit for fluid chains that is dictated completely by chain interactions, as opposed to the prevailing head–head interactions found in the other classes of lipids

[29]. The discussed structural results support the notion that lipid headgroups govern bilayer packing, while their properties are fine-tuned through the composition of their lipid acyl chains [13].

When scrutinizing the consequences of increased acyl chain length in the fully saturated bilayers, the evidence for an extended linear dependence of areas in Figure 1.3a was observed experimentally [26]. The decreased lipid area as a function of increasing chain length implies a smaller increase of the entropic contribution resulting from rotational isomerization (i.e., chain disorder), compared to the hydrophobic and van der Waals interactions in saturated chain lipids (diCn:0PC). On the other hand, increase in lipid area can be caused by the increased probability of *trans*-gauche isomerization that also happens to increase with chain length and temperature. Figure 1.3a shows the differences between the chain length dependencies for saturated and mixed chain lipids (diCn:0,18:1PC). As discussed above, the presence of a *cis*-double bond perturbs the packing of the hydrocarbon chains, which results in increased chain disorder and a concomitant increase in lipid lateral area. The addition of two methylene groups to mixed chain lipids then results in an increased area/lipid (blue symbols in Figure 1.3a), in contrast to the decrease experienced by saturated hydrocarbon chain lipids (black symbols in Figure 1.3a). This again suggests that rotational isomerization has a much more pronounced effect on lipid areas of unsaturated chain lipid bilayers than attractive van der Waals interactions. Importantly, the theoretically observed dependence on the position of double bond in Figure 1.2b is well reiterated for mono-unsaturated lipid bilayers (diCn:1PC) experimentally (dark yellow symbols) in Figure 1.3a.

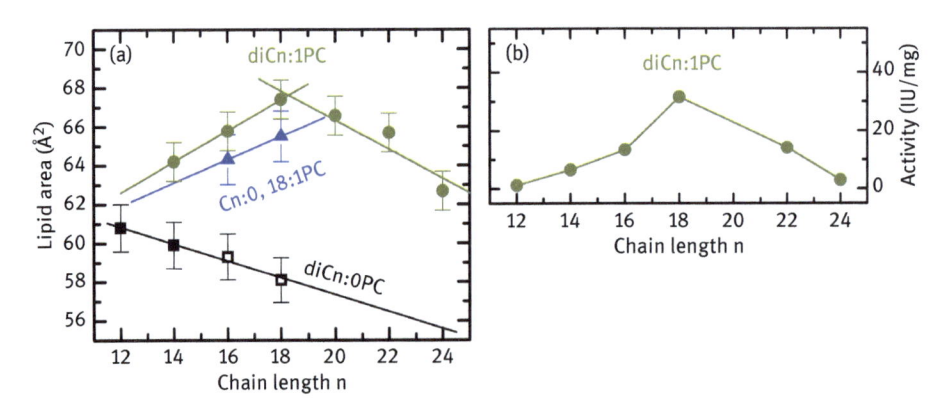

Figure 1.3: (a) Area per lipid as a function of hydrocarbon chain length for the various PC lipids at 30 °C. Saturated fatty acid chains are denoted by black squares, while blue triangles and dark yellow circles denote the presence of one or two mono-unsaturated fatty acid chains, respectively. All areas shown by solid symbols are obtained using 30 °C data, while areas indicated by open squares are calculated from fluid phase data using the appropriate thermal area expansivities (from Kučerka et al. [26]). (b) The dependence of specific Ca^{2+}-ATPase activity at 37 °C as a function of hydrocarbon chain length of mono-unsaturated PC lipids to which it was reconstituted from Karlovská et al.[27].

The above-discussed results also suggest importantly that lipid area is a good indicator of the lateral interactions within the bilayer. This is further corroborated by an activity of Ca^{2+}-ATPase reconstituted into lipid bilayers that was found to depend on various structural parameters such as thermodynamic phase, structure and charges of polar headgroups, and of the utter importance the hydrocarbon chain length [27]. The results displayed in Figure 1.3b indicate almost exactly the same dependence of ATPase activity on the lipid chain length as was observed in the case of lateral area in bilayers made of the same mono-unsaturated lipids (Figure 1.3a). Enzymatic activity was found to be maximal in bilayers composed of diC18:1PC lipids, and decreased, as much as fourfold, in both shorter and longer chain lipid bilayers. The structural data obtained via scattering techniques thus point to a decrease of membrane lateral pressure as a very likely mechanism of stimulating and/or quenching membrane incorporated proteins that provide it with an active function.

Bilayer membranes are characterized by large lateral stresses born by the above-mentioned interactions and depending on the depth within the membrane [36]. The balance of repulsive and attractive forces in the interfacial region is dictated for the most part by the chemical composition of lipid headgroups, as discussed above. The repulsive interactions within the hydrophobic region are then closely related to the chain length and its unsaturation, thus the dynamics of hydrocarbon chains [37]. Interestingly, such interactions may affect not only the proper function of large integral membrane proteins, but the integration and location of other small components of the membrane. For instance, molecular dynamics simulations have shown that the angle of cholesterol with respect to the bilayer normal varies with the number of double bonds present in the lipid fatty acid chains [38] and/or the bilayer thickness [39]. The results showed that the frequency of cholesterol's flip-flop between bilayer leaflets dramatically increases with the increasing disorder of lipid chains. Recently, neutron studies of deuterated cholesterol incorporated into polyunsaturated fatty acid (PUFA) bilayers found cholesterol sequestered inside the membrane, in contrast to its usual position with the hydroxyl group located near the lipid/water interface [40–42].

The average position of cholesterol molecules within the bilayer could be handily deduced from neutron diffraction experiments when labeling the parts of cholesterol with deuterium atoms. The hydrogen/deuterium substitution, for the most part, does not change the chemical and structural properties of the biologically relevant systems, while the differential sensitivity of neutrons to hydrogen and deuterium allows to increase the sensitivity of neutron scattering experiments considerably. The Fourier reconstruction results straight-forwardly in the NSLD calculated from experimental scattering form factors F_h as

$$\mathrm{NSLD}(z) = \left(\frac{1}{d}F_0 + \mathrm{NSLD_W}\right) + \frac{2}{d}\sum_{h=1} F_h \cos\left(\frac{2\pi h z}{d}\right),$$

which consists of nonlabeled parts of hydrated membrane (M) and the parts labeled (L) with deuterium (D) in the case of labeled sample

$$NSLD_D(z) = NSLD_M P_M(z) + NSLD_D P_L(z),$$

or "labeled" with hydrogen (H) in the case of nonlabeled sample

$$NSLD_H(z) = NSLD_M P_M(z) + NSLD_H P_L(z).$$

The subtraction of the two NSLD profiles then includes only the probability distribution of labeled components

$$P_L(z) = \frac{NSLD_D(z) - NSLD_H(z)}{NSLD_D - NSLD_H}.$$

The resulting difference NSLD profile obtained solely from experiment can provide the distribution of the label with nanometer resolution that is directly comparable to molecular modeling utilizing simulations [16]. Figure 1.4a and b portray such approach in the case of cholesterol-loaded PUFA bilayers.

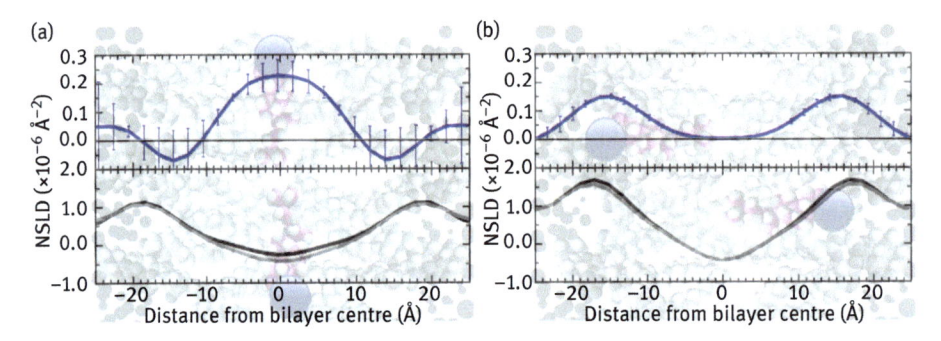

Figure 1.4: Neutron scattering length density (NSLD) profiles in bottom panels and their differences (deuterated minus nondeuterated cholesterol) in top panels show the distribution of cholesterol's deuterium label (depicted by a big blue sphere in the background schematics). It is found in the center of highly disordered bilayers (a), while cholesterol orients in its canonical orientation in the case of more ordered bilayers (b).

The samples in the neutron diffraction experiments are typically prepared on the flat substrate for an unambiguous differentiation between the direction perpendicular and parallel to the membrane plane [43]. This allows, in addition, the very same sample to be reproducibly measured in different contrast conditions when hydrated through water vapor phase. Samples hydrated with 100% D_2O are best suited for the determination of the bilayer's overall structure (i.e., total thickness, area per lipid, etc.) because of the excellent contrast provided between the hydrating medium and the bilayer [19]. On the other hand, fine structural details are more obvious in samples hydrated with 8 mol% D_2O, where the water contribution has a net zero NSLD and the bilayer structure is not obscured by scattering from the solvent. Nevertheless, all the difference profiles obtained from samples hydrated with various percent D_2O solutions should result in the same distribution function. This is due to the fact that both the NSLD associated with the lipid and that of the hydrating medium are subtracted with the end result corresponding to the difference between the labeled and unlabeled sample NSLDs (i.e., six deuterium minus six hydrogen atoms per cholesterol molecule) [44].

The results shown in Figure 1.4 clearly show cholesterol's two very different orientations in different bilayers [45]. The central location for cholesterol is observed in Figure 1.4a for PUFA bilayers doped with up to 30 mol% palmitoyl-oleoyl-phosphatidylcholine (POPC). The increasing amount of more ordered lipid however allows to change the cholesterol's orientation when PUFA bilayers contain more than 50 mol% POPC. The deuterium label in Figure 1.4b appears to be approximately 15 Å from the bilayer center, placing cholesterol's hydroxyl group within the bilayer's hydrophobic/hydrophilic interfacial region. This observation confirms the notion that cholesterol's orientation can be altered by modifying the profile of membrane lateral pressure through changing the ratio of PUFA/saturated chain lipids. Considering the lipid heterogeneity of biological membranes, this result may be viewed as a prerequisite for the mechanism by which cholesterol transports through the cell membrane. It is even possible to imagine the aversion that certain lipids have for each other to drive the formation of functionalized domains with selective profiles of membrane lateral pressure.

1.3 The effect of cholesterol on model membranes

In a further support of the role of membrane lateral pressure, it is interesting to note that activity of Na,K-ATPase discussed previously was shown to be sensitive not only to phospholipid chain length but also to cholesterol content [46]. Maximal protein activation was seen in long-chain phospholipids in the absence of cholesterol, while it shifted toward medium-chain phospholipids in the presence of cholesterol. This observation clearly suggests that not only the lipid species affect the orientation of cholesterol in membranes but that cholesterol alters the structure of bilayers. However, it is not clearly resolved whether the cholesterol affects the lateral packing of lipids [47], or it alters the hydrophobic matching between the lipid and protein [27]. In binary lipid/cholesterol mixtures, cholesterol's addition to fluid phase lipid bilayers results in increased acyl chain order [9–13], while having the opposite effect on lipid headgroups. The consensus of many studies, including NMR [14, 15], EPR [16, 17], and fluorescence [18], is that cholesterol acts as a "spacer" molecule, increasing the separation between lipid headgroups, thereby reducing possible interactions between them. These studies further demonstrate that while the addition of cholesterol decreases the extent of water penetration into the membrane's hydrophobic region, there is a concomitant increase in headgroup hydration.

When it comes to SANS experiments, data analysis has traditionally been based on the Kratky–Porod approximation, or by fitting the data to a simple single-strip model of the bilayer [21, 22, 36–38]. Although this approach is rather popular, it neglects the inner structural details of the bilayer. More accurate models to analyze SANS data [29, 39] were inspired by the results from molecular-dynamics simulations, which show additional substructure within the bilayer. The full form of scattered intensity measured for the dilute system of polydisperse unilamellar vesicles (ULVs) can be calculated as the square of the planar bilayer form factor $F(q)$ (q is the scattering vector $q = 4\pi \sin(\theta/\lambda)$, while 2θ is the scattering angle and λ is the neutron wavelength) and multiplied by the function which includes the particle's sphericity and the system's polydispersity as follows [48]:

$$I(q) = F^2(q) \frac{8\pi^2(z+1)(z+2)}{s^2 q^2} \left\{ 1 - \left(1 + \frac{4q^2}{s^2}\right)^{-(z+3)/2} \cos\left[(z+3)\arctan\left(\frac{2q}{s}\right)\right] \right\},$$

where $s = R_m/\sigma_R^2$ and $z = R_m^2/(\sigma_R^2 - 1)$ are the products of the ULV's mean radius R_m and its polydispersity σ_R (i.e., the width of the size distribution function). The bilayer form factor $F(q)$ is the well-known Fourier transform that for centro-symmetric profile is written as

$$F(q) = 2 \int_0^{d/2} [\text{NSLD}(z) - \text{NSLD}_W] \cos(qz) \, dz,$$

where $NSLD_W$ is the neutron-scattering length density of water solvent that acts as a background in the sample. The bilayer's profile $NSLD(z)$ can be further separated according to various multishell schemes [49]. Of particular importance in this development was the replacement of a sharp water–bilayer interface by a "smooth" one [50]. This function consists of a linear term for the distribution of water penetrating the headgroup (see Figure 1.5a).

There are two sets of parameters that determine the entire SANS curve: in the low-q region, scattering is sensitive to large length scales, that is, the overall size of the ULVs, while the information about bilayer structure is contained in the mid- and high-q regions (see Figure 1.5b). Two parameters describe the size distribution function of ULVs (i.e., mean radius and the width of its distribution), while two others define the bilayer model (i.e., area per lipid and amount of penetrating water). In addition, there are two linear parameters corresponding to a multiplicative scaling coefficient and an additive background constant. In total, there are six fitting parameters, which reach the limitation of standalone SANS data analysis [49] (in contrast to SANS and SAXS-combined approach discussed above). Structural parameters are refined in terms of an iterative model-fitting approach, which result in the bilayer profile.

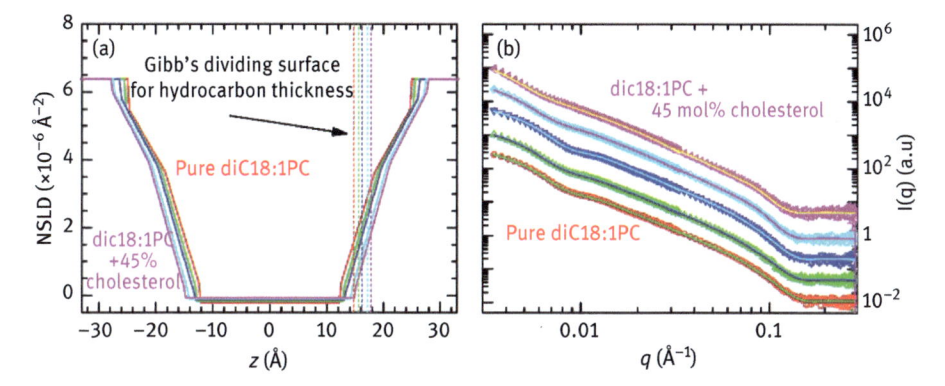

Figure 1.5: (a) Neutron scattering length density (NSLD) profiles of lipid bilayers made of diC18:1PC and various amounts of cholesterol (0, 17, 29, 38, and 45 mol%). The model comprises constant NSLD for hydrocarbon chains in the bilayer center, while linear functions describe the gradual increase of water molecules penetrating the lipid headgroup region. The dashed lines depict the Gibb's dividing surface between headgroup and hydrocarbon regions. (b) Small-angle neutron scattering (SANS) curves showing the experimentally measured intensities (points) for bilayers in the form of unilamellar vesicles (ULVs). Solid curves represent the fitting results of models shown in panel A. Adapted from Kučerka et al. [50].

Figure 1.5a shows the characteristically large contrast observed between the hydrocarbon region in the bilayer center and solvent outside. The contrast gradually decreases within an interfacial region until the boundary between the lipid headgroups and water is reached. All NSLD profiles show this behavior, typical of protonated phospholipid bilayers dispersed in D_2O. In addition, the bilayer properties are continuously changing also as a function of cholesterol content. First, there are increases to the hydrocarbon NSLD with increasing amounts of cholesterol – although this change is very small because of the cholesterol and the lipid hydrocarbon region having similar NSLDs. Second, the monotonically increasing bilayer parameters in Figure 1.6a and b with cholesterol content serve a direct observation of cholesterol-induced increases to the hydration of the lipid headgroup region. Most importantly, however, results in Figure 1.6c and d show definite increases in bilayer thickness as a function of cholesterol concentration [50]. The rigid hydrophobic molecule of cholesterol in an obvious way increases the thickness of the hydrocarbon chain region in the case of shorter lipid bilayers. However, cholesterol may be expected to increase or decrease the thickness of the hydrocarbon chain region in the case of long-chain lipid bilayers. The observations of a thickening of the hydrocarbon chain region, even for diC22:1PC bilayers, imply then that cholesterol prefers to further order the lipid's hydrocarbon chain over the possibility of rectifying the hydrocarbon chain mismatch [39, 50]. Cholesterol promotes the formation of a liquid-ordered phase by condensing the bilayer along its lateral direction.

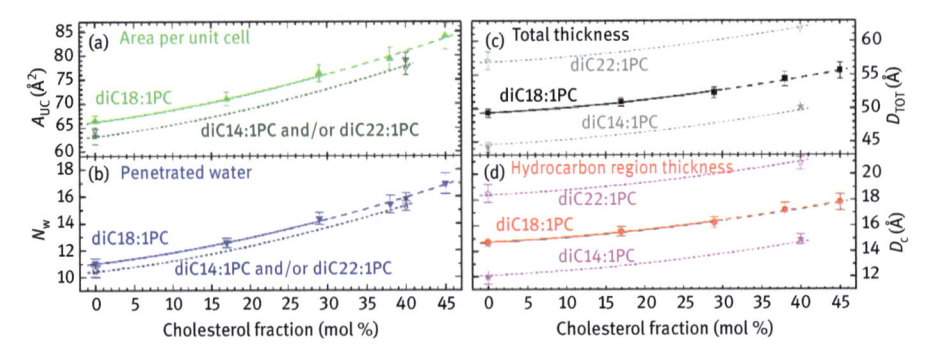

Figure 1.6: Bilayer parameters plotted as a function of cholesterol concentration for three mono-unsaturated phospholipids with different chain length. All the parameters [i.e., area per lipid (a), and amount of water molecules penetrating the lipid headgroup region (b), total bilayer thickness (c), and that corresponding to the hydrocarbon region (d)] exhibit monotonic dependencies as functions of cholesterol concentration. The uniform behavior in the case of different chain length lipids suggests the same mechanism for the cholesterol's effect on these membranes. Adapted from Kučerka et al. [50].

SAXS experiments, although missing the high contrast available in the case of previously discussed SANS, can benefit from highly brilliant modern synchrotron sources. Consequently, SAXS data often provide higher resolution when compared to that of SANS measurements, and they may reveal additional structural results. For example, SAXS experiments can detect an asymmetric distribution of lipid densities across the bilayer. In this case, bilayer asymmetry is readily detectable because of the distinct scattering features associated with the effect. The scattering from symmetric structures is characterized by periodic oscillations that cross zero and result thus in typical zero minima in the scattered form factors [51, 52]. A lack of zero intensity, on the other hand, indicates bilayer asymmetry, as can be deduced from the complete form of the Fourier transform with a complex exponential

$$I(q) \approx \left[\int_{-d/2}^{d/2} [\text{XSLD}(z) - \text{XSLD}_W] \cos(qz) dz \right]^2 + \left[\int_{-d/2}^{d/2} [\text{XSLD}(z) - \text{XSLD}_W] \sin(qz) dz \right]^2.$$

It was shown, based on this experimental approach, that the addition of 44 mol% cholesterol results in some of the bilayers (i.e., $n = 14$, 16, and 18) to become asymmetric, where cholesterol was found to distribute unequally between the bilayer's two leaflets [53]. The subtle changes in the total XSLD shown in Figure 1.7 reveal the asymmetry-related sharpening of the lipid-component distributions in the cholesterol-depleted side, while their broadening in the cholesterol-enriched side [39]. The MD simulations corroborated the effect of bilayer asymmetry on the scattering form factors (namely, the "lift-off" in the first minimum) and provided further details. By

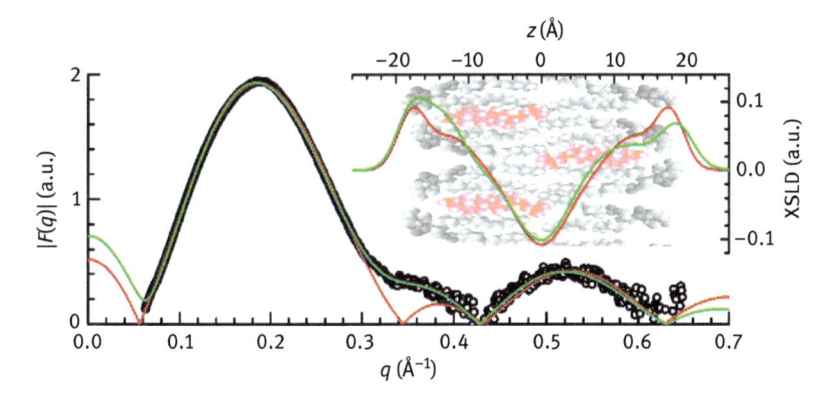

Figure 1.7: X-ray scattering form factors $F(q)$ obtained for diC14:1PC bilayers with 44 mol% cholesterol. The insert shows 1D XSLD profiles in the forefront of bilayer schematics depicting an unequal distribution of cholesterol (magenta) between the two bilayer leaflets. The red curves are best fits to the data using a symmetric bilayer model, while the green curves are best fits to the data using an asymmetric model.

varying the degree of cholesterol asymmetry in each case, the degree of asymmetry was established by matching the magnitude of the experimentally observed "lift-off." The best match with the ULV experimental data provided a final cholesterol distribution corresponding to the ratio of 51/29 [39]. It seems that cholesterol, by distributing itself unequally between the bilayer's two leaflets, induces bilayer asymmetry in symmetric bilayers whose hydrocarbon chains contain up to 18 carbons. It is interesting to note that the magnitude of bilayer asymmetry coincides with the difference in length between the lipid's hydrocarbon chains and cholesterol. Asymmetry is most pronounced in the thinnest bilayer (i.e., diC14:1PC) whose hydrocarbon thickness is significantly shorter than a cholesterol molecule. With decreasing hydrophobic mismatch, however, there is a decrease in cholesterol's asymmetric distribution, eventually distributing itself equally between the two leaflets of diC20:1PC bilayers. This suggests that symmetric bilayers are formed when the bilayer's hydrocarbon thickness is close to the length of cholesterol. It is possible that, in addition to bilayer disorder discussed above, hydrophobic mismatch is after all another important factor when lipids are mixed with other membrane components.

The membrane's composition and the inclusion of small molecules, such as cholesterol discussed above, alter the membrane's structure and physical properties, which in turn affect its various functions. Recent studies have shown the lipid membrane to be extremely important also in enabling amyloid fibril formation and its ensuing toxicity [54, 55]. The formation of insoluble amyloid fibrils composed of proteins in β-sheet conformation found on the surface of neuron plasma membrane is a fingerprint of Alzheimer's disease (AD) and various others related to protein conformational disorders. In this, cholesterol has been shown to enhance the amyloid binding and fibril formation when present in a membrane [56, 57]. In opposite to cholesterol that is mostly hydrophobic, membrane can accommodate also hydrophilic molecules. One example is melatonin – a pineal hormone that is produced in the human brain during sleep and that sets the sleep–wake cycle (and circadian rhythm) [58]. Intriguingly, it has been suggested to have a protective role against amyloid toxicity [59, 60], while the underlying molecular mechanism of this protection is not well understood. It may involve a nonspecific interaction of the molecule with the membrane and decreasing the gel-to-fluid phase transition temperature by increasing the membrane disorder [61–64]. It has been also shown that melatonin may compete with cholesterol for binding to lecithin and that it may even displace cholesterol from the phospholipid bilayer [65]. Melatonin's ability to control cholesterol content, ergo membrane rigidity, may thus reduce the effects of cholesterol on the membrane, as well as cholesterol-mediated processes [66].

In order to better understand how melatonin and cholesterol affect the interaction of biomolecules with the membrane, it is necessary to systematically determine the effects that these molecules have on membrane structure. Small-angle neutron diffraction (SAND) experiments allow for the *in situ* manipulation of sample conditions but, more importantly, provide quantitative data on the distribution of structural moieties, their sizes, shapes, and correlation lengths [67]. Variations in diffraction maxima (see Figure 1.8a) can signal important changes to membrane structure, which,

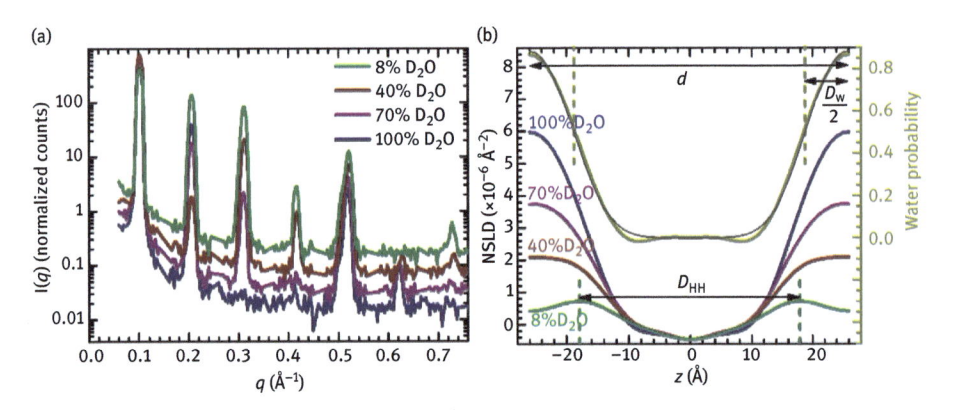

Figure 1.8: (a) Small-angle neutron diffraction curves recorded for lipid bilayers hydrated with different D_2O/H_2O water vapor. Note the vertical shift between the various curves introduced for the clarity of presentation. (b) Neutron scattering length density (NSLD) profiles of lipid bilayers calculated in various D_2O/H_2O contrast conditions scaled according to the left-hand axis. The distance between two peaks corresponding to lipid headgroups is obtained from 8% D_2O profile (green color), and it defines the thickness parameter D_{HH}. The dark yellow curve (scaled according to the right-hand axis) shows the profile of water probability distribution across the bilayer, describing thus the water encroachment (central position is depicted by broken lines). The gray line represents the error function fit to the distribution calculated from experimental data. The total interlamellar spacing is shown by d and the thickness of water layer between adjacent lipid bilayers is depicted by D_W.

in turn, may have biological implications. With regard to structural biology, the various scattering techniques complement crystallographic studies that, in many cases, require hard-to-obtain, high-quality crystals of macromolecules. Due to the intrinsic disorder present in biomimetic systems – disorder that is considered important for the proper function of biological systems – the vast majority of membrane samples do not form perfect, or even near perfect crystals, that are needed to solve structure to atomic resolution. The limited amount of attainable data from such samples is then best described by broad statistical distributions and membrane overall characteristics. For example, in the case of model membrane systems (i.e., positionally correlated structures), the position and amplitude of Bragg reflections reveal the membrane's lamellar periodicity (d) and one-dimensional NSLD profile (Figure 1.8b). Further, by changing the system's "contrast" through the exchange of H_2O for D_2O, it is possible to determine the extent that water penetrates into the bilayer [68]. In fact, such contrast variation approach is the only direct method that allows to solve the infamous scattering phase problem without disturbing the sample structure [69, 70]. Consequently, the distribution profile of water penetration is inferred from difference NSLD profiles obtained at various contrast conditions. It first parses each $NSLD(z)$ into contributions corresponding to the membrane (M) and water (W) molecules

$$NSLD(z) = NSLD_M P_M(z) + NSLD_W P_W(z),$$

while it considers the two probabilities being contrast independent and satisfying the spatial conservation principle (i.e., $P_M(z) + P_W(z) = 1$ at each point z across the membrane). Since $NSLD_M$ does not depend on external contrast conditions, the subtraction of NSLD profiles measured at two different D_2O contents includes only the probability of water, which can be expressed as

$$P_W(z) = \frac{NSLD_1(z) - NSLD_2(z)}{NSLD_{W1} - NSLD_{W2}}.$$

The water probability $P_W(z)$ is between 0 and 1 if $NSLD_1(z)$ and $NSLD_2(z)$ are obtained on an absolute scale [68], and it provides an encroachment of water in a comparison to the membrane profile (see Figure 1.8b). As discussed previously, NSLD distribution corresponds to the membrane profile directly if obtained at 8% D_2O. The resulting profile is then characteristic of two peaks representing the opposing lipid headgroups of the bilayer, and their distance (D_{HH}) provides one of the membrane thickness parameters (see Figure 1.8b).

Membrane thickness is a structural parameter that is directly related to lipid–lipid and lipid–protein interactions in biomembranes. The experimental results in Figure 1.9 confirm the previously reported and well-known bilayer thickening effect induced by cholesterol. This manifestation of cholesterol in bilayer thickness can be interpreted as a disorder–order transition of the lipid's acyl chains. The hydrocarbon chains experience increased order due to their interactions with the rigid cholesterol molecules. Interestingly, the experimental results show also that the addition of melatonin has the

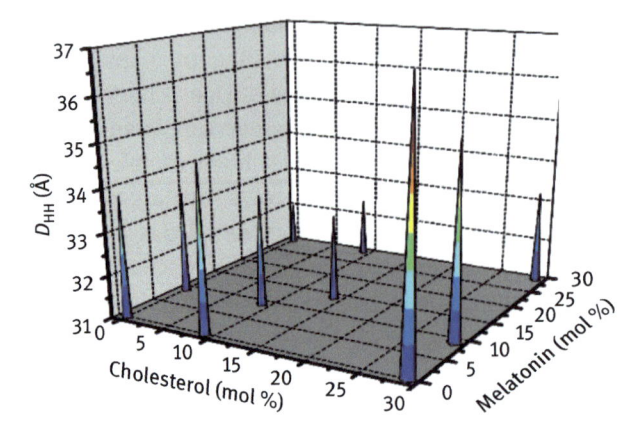

Figure 1.9: Head-to-head distance D_{HH} obtained from NSLDs of diC18:1PC bilayers loaded with various amounts of cholesterol and melatonin. Bilayer thickness clearly increases with an increasing amount of cholesterol, while it decreases upon the addition of melatonin which counteracts the effect of cholesterol.

exact opposite effect, that is, melatonin causes bilayer thinning [71]. This result suggests that melatonin incorporates itself into the bilayer's headgroup region and acts as a spacer therein. According to this scheme, melatonin increases the free volume in the bilayer's hydrocarbon region. This free volume is then readily taken up by the disordered chains at the expense of their effective length (i.e., reduction of bilayer thickness). This notion is also supported by the encroachment of water molecules deeper into the membrane, as readily extracted from contrast varied SAND measurements [71].

The effect of melatonin and cholesterol on model lipid bilayers was corroborated by MD simulations [71]. In agreement with the experimental SAND data, MD simulations show that cholesterol increases the acyl chain order in lipid bilayers, while melatonin decreases this order. In other words, melatonin increases the fluidity of the membrane due to its preferred location just inside the crossover region describing the lipid headgroups and the fatty acid chains. Both experimental and theoretical data are in good agreement and show that the effect of melatonin on bilayer thickness is opposite to that of cholesterol. These observations may prove to be important for other studies on amyloid toxicity, as they may lend some insight into understanding the molecular mechanism of melatonin's protection in AD. For example, melatonin levels in the body have been shown to decrease with age [72]. As AD is more prevalent later in life, the effects of melatonin and cholesterol on lipid membrane become increasingly important as their amounts in membranes also change with age. The conclusions of various investigations can, therefore, provide an understanding for the possible structural changes taking place within biological membranes.

1.4 Alcohol interactions with lipid bilayers

The proper functioning of biological membranes appears to be well correlated to the structural properties, thermodynamic conditions, and composition of lipid matrices. The various additives are known to enable or inhibit these functions provided, for the most part, by membrane proteins. The capacity of a substance to interact with biological membrane and modulate its functionality is defined by biological activity. In the case of molecules with long hydrocarbon chains, such biological activity is known to depend on the chain length. Intriguingly, it increases with increasing chain length up to a maximum, while it decreases again with further chain length increases. Such cutoff behavior has been observed in many cases of various drugs or toxins, suggesting the mechanism for this effect depending on the hydrophobic mismatch between the substance and lipid bilayer in the transmembrane direction, and/or corresponding interactions in the lateral direction [22]. General anesthetics are one of the typical examples displaying the cutoff effect.

Alcohols and other general anesthetics have been known to surgeons for over the century [73]. Despite their successful applications, however, the understanding of anesthesia mechanism is still lacking. According to the general consensus, the place of their action is being recognized either within proteins or lipid membrane [74–76]. However, the hypothesis based on the unspecific interactions between anesthetics and membrane lipids may be more plausible due to, for example, a wide spectrum of membrane proteins that are affected. In any of the two cases, the general anesthetics, including aliphatic alcohols, offer an exciting example on the structure–function correlation that is sought out in biological membranes. Amphiphilic molecules, like long-chain alcohols, intercalate into membranes and change their structural and/or dynamical properties, which in turn might affect membrane-bound protein conformations and result in protein functional changes that are involved in general anesthesia. Although it is not known which structural perturbations are responsible for these effects, it is evident that the exploration of the long-chain alcohol interactions with model membrane can contribute to a better understanding of the cutoff dependencies in various biological activities of homologous series of amphiphilic compounds [77, 78].

Neutron scattering techniques have proven to be valuable tools in structural biology, biophysics, and materials science due to their ability to provide a complex set of parameters. The most straightforward parameter obtained from diffraction measurements (such as shown in Figure 1.8a) corresponds to the lamellar spacing between repeating unit cells, d. It is inversely related to the position of Bragg diffraction peaks (q_h) as $d = 2\pi h/q_h$. Experimental results for oriented lipid multilayers in excess water then suggest the increase of total lamellar spacing upon the addition of the alcohols with their tail length increasing from 10 to 18 carbons [79]. The extension of changes is proportional to the alcohol's tail length, although the changes are relatively small (Figure 1.10a).

The total lamellar d-spacing in the case of hydrated membrane system consists of two components: the total thickness of lipid bilayer D_B and the thickness of water layer in-between the bilayers (see also Figure 1.8b). The water layer comes in as a result of interbilayer interactions that are characteristic to the given lipid composition and thermodynamic state including the hydration conditions. One of its particularly interesting contributions is determined by the membrane viscoelastic properties and it relates to the softness of membrane [80]. Literature results nevertheless show a marginal dependence of water layer thickness on the alcohol tail lengths, suggesting no changes to these characteristics [79]. Figure 1.10a then shows that major changes in these bilayer systems happen to D_B, and accordingly, it displays very similar trend to that of total lamellar d-spacing when examined as a function of tail length of the added alcohol. Interestingly however, there is also a significant difference in the behavior of the two thickness parameters. While the addition of any and all investigated alcohols increases the d-spacing, D_B is always smaller when comparing alcohol-loaded bilayers to neat DOPC bilayers. This can be seen in Figure 1.10a by a broken line being below the points in the case of d-spacing data, and it is above the points in the case of bilayer thickness. Obviously, the water layer increases upon the addition of any alcohols to the neat DOPC bilayers; yet, it stays mostly constant for alcohols with different tail lengths.

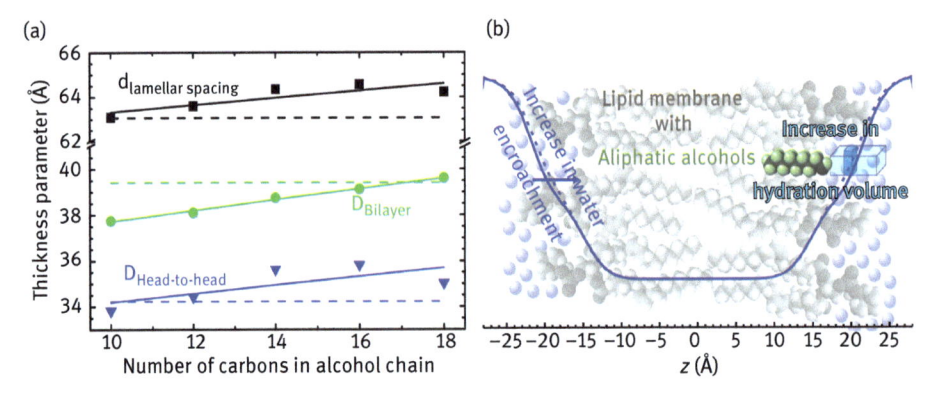

Figure 1.10: (a) Various bilayer thickness parameters obtained for neat dioleoyl-phoshatidylcholine (DOPC) bilayers (dashed lines) and those with the addition of 0.3:1 molar ratio of tail length varied alcohols (solid points). The total d-spacing of lipid multilayers in excess water (black color), the total bilayer thickness without interbilayer water layer (green color), and head-to-head distance obtained from NSLD profiles measured at 8% D_2O contrast conditions. The solid lines are linear approximations of data shown to emphasize the average changes. (b) Illustration of water encroachment shown by blue lines. Fewer water molecules penetrate the bilayer composed of neat DOPC (solid line) compared to the scenario with alcohol (depicted by green color) intercalating the bilayer (broken line). Water molecules (blue spheres) fill an additional space (the dark portion of a blue rectangle) resulting from increased water penetration and enlarged lateral area above the alcohol molecule. Adapted from Kondela et al. [79].

The above-discussed observations suggest the changes may be happening to the lipid headgroups. It is therefore beneficial to look closely at one more parameter extractable easily from SAND experiments. The fine structural details of systems dispersed in water are more obvious in samples hydrated with 8 mol% D_2O, as discussed already. Such obtained NSLD profiles provide directly the features related to lipid bilayer, where carbonyl–glycerol groups are typical of a pronounced maximum. The distance between the opposite maxima, thus in addition to providing yet another thickness parameter D_{HH}, allows to scrutinize the bilayer internal structure. D_{HH} plotted in Figure 1.10a shows a behavior that is very similar to that of total d-spacing. Considering that it is derived from the structure of bilayers themselves, while D_B relates to the water–bilayer interface, the combination of their changes points to differences in the encroachment of water molecules upon the addition of alcohols. The distances of both lipid headgroups and water–bilayer interface from the bilayer center increase accordingly to the tail length of intercalating alcohols. However, changes in D_B, relative to neat DOPC bilayers, are shifted toward the center by about 1 Å at each side of bilayer, when compared to the relative changes of D_{HH}. In other words, the addition of alcohols results in some of the space in the polar headgroup region (note also the expansion of lateral area upon the addition of alcohols [79]) being filled with extra water molecules (Figure 1.10b). This in turn contributes to the decrease of membrane lateral pressure in the region directly above alcohols and supports the modulation of membrane mechanical properties by general anesthetics as a likely mechanism of the anesthesia effect.

1.5 Cation-containing lipid membranes

Amongst the intramembrane interactions lipid–lipid, lipid–protein, and even protein–protein, the significance of the aqueous phase for the proper functioning of biological membranes cannot be overestimated. Plasma membrane properties such as membrane fluidity, bending and compressibility moduli, electrostatics, and aggregation and fusion are in addition associated with ions that are ubiquitous in the cytosol and the extracellular fluid. Not surprisingly, ions such as Na^+, K^+, Ca^{2+}, Mg^{2+}, Zn^{2+}, or Cl^- have been found to play a prominent role regarding bilayer structure. Their functions within cell membranes are understood to influence the gating of ion channels, membrane fusion, and membrane "fluidity" to name but a few [67]. Although a complete understanding of the physicochemical processes taking place in biomembranes has yet to be established, their functionality is known to depend strongly on the type of ion, the chemical composition of the membrane's interface, its thermodynamic state, and degree of hydration [81].

The different effect of monovalent versus divalent cations was observed in bacterial model membranes composed of lipopolysaccharides (LPS) utilizing SAND measurements [82]. Due to a particular sensitivity to the hydration conditions, the diffraction data showed that water penetrates Ca^{2+}-loaded LPS bilayers to a lesser extent than those loaded with Na$^+$. While Ca^{2+} cations make LPS bilayers more compact and less permeable to water, a significant amount of water was found to penetrate deep into Na$^+$-LPS bilayers, including the bilayer's hydrophobic core center. In fact, the disappearance of diffraction peaks when liquid-crystalline Na$^+$-LPS multilayers were subjected to high levels of hydration (see Figure 1.11a) suggested that the deep water penetration increases up to a critical point, beyond which the long-range correlation of bilayer assembly is destroyed [83]. Such a destabilization mechanism could provide a general explanation of how nonlamellar phases are formed and how small molecules penetrate the outer membrane of gram-negative bacteria.

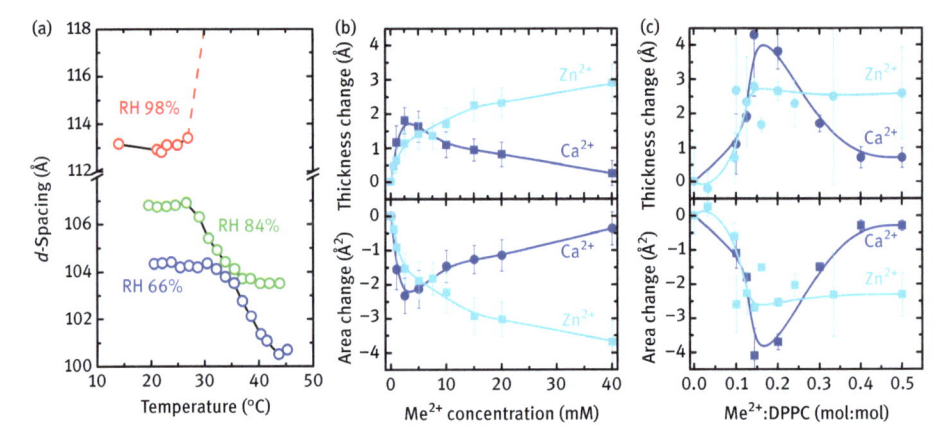

Figure 1.11: (a) Temperature dependence of the lamellar d-spacing at the various levels of relative humidity (RH) controlled externally. The multilamellar structure of Na$^+$-loaded LPS bilayers transitions from gel to liquid-crystalline phase at 66% and 84% RH. However, the repeating nature of the structure disappears when temperature increases above 28 °C at 98% RH. Zn^{2+} and Ca^{2+} cation-induced changes to bilayer structure expressed through the bilayer thickness, and area per unit cell A_{UC} obtained from SANS (b) and SAND (c) experiments. The A_{UC} comprises a DPPC molecule and the appropriate number of cations and is calculated from the inverse relation to the bilayer thickness (i.e., $A_{UC} = V_{UC}/D_{Bilayer}$). Changes are calculated with respect to neat DPPC bilayers. Results are adapted from Abraham et al. [83], Uhríková et al. [84], and Kučerka et al. [85].

The increased levels of hydration are most likely associated with an enhanced biological activity in the bacterial membranes and represent thus an intriguing research issue. The complexity of the system however does not allow for straightforward correlations of structural results. The effect of ions on the other hand can be revealed effectively utilizing model systems. For example, systems of dipalmitoylphosphatidylcholine (DPPC) lipids dispersed in CaCl$_2$ or ZnCl$_2$ solutions have been studied by SANS [84,86], and planar multilayers hydrated through water vapor by SAND [85]. The results confirm that both Ca^{2+} and Zn^{2+} cations bind to DPPC bilayers. This fact is reflected in the increased lamellar d-spacing that is most likely the result of charge-induced repulsion between bilayers. This observation is consistent with other zwitterionic lipid multilayers, which swelled in the presence of CaCl$_2$ at 1–50 mM concentrations to the point that they became unbound (i.e., infinite d-spacing) [87–89]. In the case of vesicular systems, multilamellar DPPC dispersions convert completely into ULVs when the surface charge density is higher than 1–2 μC/cm^2 [90]. The initial swelling is most likely due to an increase in interbilayer electrostatic repulsion that is screened at high salt concentrations. Supposedly however, membrane structure is not affected at these high salt concentrations [91]. Interestingly, the low salt concentration results reveal that even the changes in internal structure (e.g., D_B and A_{UC} shown in Figure 1.11b and c) contribute to the increases in d-spacing. Moreover, Ca^{2+} and Zn^{2+} do not have the same effects.

Initially, both Ca^{2+} and Zn^{2+} cause DPPC bilayers to thicken, while further increases in Ca^{2+} concentration result in the bilayer thinning, eventually reverting to having the same thickness as pure DPPC. In the case of Zn^{2+} however, there is a typical binding isotherm behavior displayed by a monotonic increase of d-spacing and D_B, after which they both seem to plateau. The most pronounced difference is exhibited by D_B whose changes are compensated by the changes in interlamellar water spacing, resulting in a d-spacing that is practically unaffected by changes in cation concentration [85]. One can therefore disregard the effect that adjacent bilayers have on each other in the stacked system. This is supported also by the results, whereby ULVs dispersed in the water solution loaded with either Zn^{2+} or Ca^{2+} ions displayed similar trends in bilayer thickness changes. The data then imply that a small curvature present in the ULVs is not an important factor affecting the bilayer structure of ion containing DPPC membranes and that the effect of cations on bilayer thickness is the result of electrostatic interactions, rather than geometrical constraints due to bilayer curvature.

Finally, the experimental observations supported by MD simulations suggest differences in the binding specificity of Ca^{2+} and Zn^{2+} cations to lipid bilayers [85]. As confirmed by the radial distribution functions calculated from the simulations, both cations interact strongly with the negatively charged oxygen. However, the distributions also show that Ca^{2+} forms a contact pair with any headgroup atom much more favorably than Zn^{2+}. The differences between the structural changes experienced by DPPC bilayers when Ca^{2+} or Zn^{2+} is introduced may be rationalized well by the physical differences between the two cations. Calcium ions have a larger ionic radius (1 Å) than zinc ions (0.75 Å) [81]. In addition, the hydrogen bonding of water molecules extends beyond the ion's primary hydration shell – their crystal arrangements in bulk water have been proposed to be Zn[H$_2$O]$_6^{2+} \cdot$[H$_2$O]$_{12}$ and Ca[H$_2$O]$_{6.1}^{2+} \cdot$[H$_2$O]$_{5.29}$ [92, 93]. Calcium ions as a consequence of their larger size thus require smaller amounts of energy for the removal of hydration shell water. These observations imply that lipid–ion interactions do not only depend on which cation is present but also on their interactions with water, one of the smallest and often neglected biomolecules [94]. Apparently, Ca^{2+} ions create about 1.6 times fewer pairs with surrounding water molecules than do Zn^{2+} ions (Figure 12a), when interacting with DPPC bilayers. This smaller Ca^{2+} hydration shell most likely allows for more proximal and stronger contacts with the

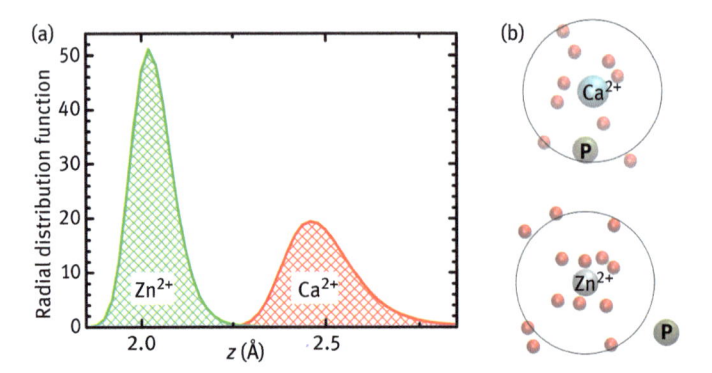

Figure 1.12: (a) Radial distribution functions for Zn^{2+}- and Ca^{2+}-water pairs determined from MD simulations. Areas under the peaks correspond to the number of water molecules present in the hydration shell of each cation interacting with the lipid bilayer (this number is smaller for Ca^{2+}, compared to Zn^{2+}). (b) Schematics of the hydration shells around Ca^{2+} (top) and Zn^{2+} (bottom) suggest fewer hydrating water molecules (red spheres) and closer proximity to the lipid's phosphate (P) in the case of Ca^{2+} cation. Results are adapted from Kučerka et al. [85].

lipid phosphate (see Figure 12b). Intriguingly, all of the mentioned interactions become increasingly important when functionalizing the model membrane systems with specific applications such as drug and/or gene delivery discussed in the following section.

1.6 Structural polymorphism of DNA–PC–Me^{2+} aggregates

Divalent metal cations play an important role as a mediator of many interactions: the binding or insertion of proteins to membranes, membrane fusion, transport of small molecules across membrane, etc. In addition to proteins, also the role of nucleic acid (NA)–lipid interactions in the functioning of cells and formation of a number of cellular structures became of great interest nowadays. For example, the role of NA–lipid–Me^{2+} in nuclear pore assembly [95, 96], signal transduction, and stimulation of DNA and RNA synthesis by endonuclear lipids [97], their presence in chromatin [98], or regulation of cell's lipid composition during cell division [99] received an attention in recent decades. There are many divalent metals of interest, including calcium and magnesium, iron, manganese, copper, zinc, nickel, and cobalt that are essential to life. On the other hand, they may become toxic beyond the level of appropriate concentration. We have shown different effect of calcium and zinc on DPPC bilayers in previous paragraph. Here, we illustrate differences in polymorphic behavior of aggregates formed due to DNA interactions with DPPC bilayers in the presence of Me^{2+}, while selecting Ca^{2+}

as typical earth metal, transition metal Co^{2+}, and Zn^{2+}. Intriguingly, the fluorescence experiments have confirmed the ability of Zn^{2+} to condense DNA in presence of lipid bilayer and to protect it against thermal denaturation up to the level comparable with Ca^{2+} and Mg^{2+} [100]. Cobalt is a trace element, integral part of vitamin B12, cobalamin, and it is involved in the production of erythropoietin, a hormone that stimulates the formation of erythrocytes. This property of cobalt was applied in the past as a therapy for anemia. On the other hand, Co^{2+} ions are genotoxic *in vitro* and *in vivo*, likely due to the involvement of oxidative stress and DNA repair inhibition, and they proved carcinogenic in rodents [101].

Generally, in a system DNA–phospholipid–divalent metal cations, one must consider the following binding events:

$$DNA + PC + Me^{2+} \rightarrow \begin{cases} PC + Me^{2+} \\ DNA + PC + Me^{2+} \\ DNA + Me^{2+} \end{cases}$$

Phospholipid–Me^{2+} interactions: Divalent metal cations bind naturally to negatively charged phospholipids [102, 103] but rather weakly to zwitterionic lipids as PC and PE [104, 105]. The binding mechanism of Me^{2+} on PC bilayers has been studied using many physicochemical methods. As a result of these studies, it is generally accepted that the preference for Me^{2+} binding weakens with an increasing degree of hydrocarbon chain unsaturation, and that it depends on the phospholipid phase (gel > fluid). Due to Me^{2+} binding, the negative charge of phosphate group of the P–N^+ dipole is neutralized and the lipid bilayer becomes positively charged. The electrostatic repulsion between bilayers makes them to swell in excess water up to the level promoting formation of ULVs, as discussed in the previous section. It should be mentioned that the phenomena of "unbound state," that is, spontaneous formation of ULVs, depends on experimental circumstances, such as temperature and concentrations of both lipid and Me^{2+}.

DNA–Me^{2+} interactions: DNA interacts readily with Me^{2+}, which may alter its structure inducing for example right-handed-to-left-handed helical transition [106], denaturation [107, 108], etc.; and thus, its function is modulated. Alkaline earth Me^{2+} (Ca^{2+}, Mg^{2+}) preferentially interacts with the phosphate groups of DNA, stabilizing the polynucleotide molecule [109]. Transition metal cations interact more extensively with DNA bases inducing disruption of base pairing, and destabilization of DNA [108, 109]. The affinity of Me^{2+} to DNA bases was reported to decrease in the order Hg^{2+} > Cu^{2+} > Pb^{2+} > Cd^{2+} > Zn^{2+} > Mn^{2+} > Ni^{2+} > Co^{2+} > Fe^{2+} > Ca^{2+} > Mg^{2+}, Ba^{2+} [110].

DNA–phospholipid–Me^{2+} interactions: Divalent metal cations mediate the interactions between DNA and PC bilayers, as has been shown four decades ago [111]. Microcalorimetry, turbidimetry, infrared spectroscopy, electron spin resonance spectroscopy, and other experimental methods were employed to characterize "a new phase" formed due to DNA interactions with PC vesicles in presence of Me^{2+} [112–116]. Electron freeze fracture micrographs of DNA–PCs–Ca^{2+} mixtures suggested structures with long-range organization [117–119]. Finally, small-angle X-ray diffraction experiments have confirmed the ability of divalent cations to condense DNA into aggregates with internal long-range order of similar topology as it was found in complexes of DNA-cationic lipid studied extensively as nonviral delivery vectors for human gene therapy (Figure 1.13).

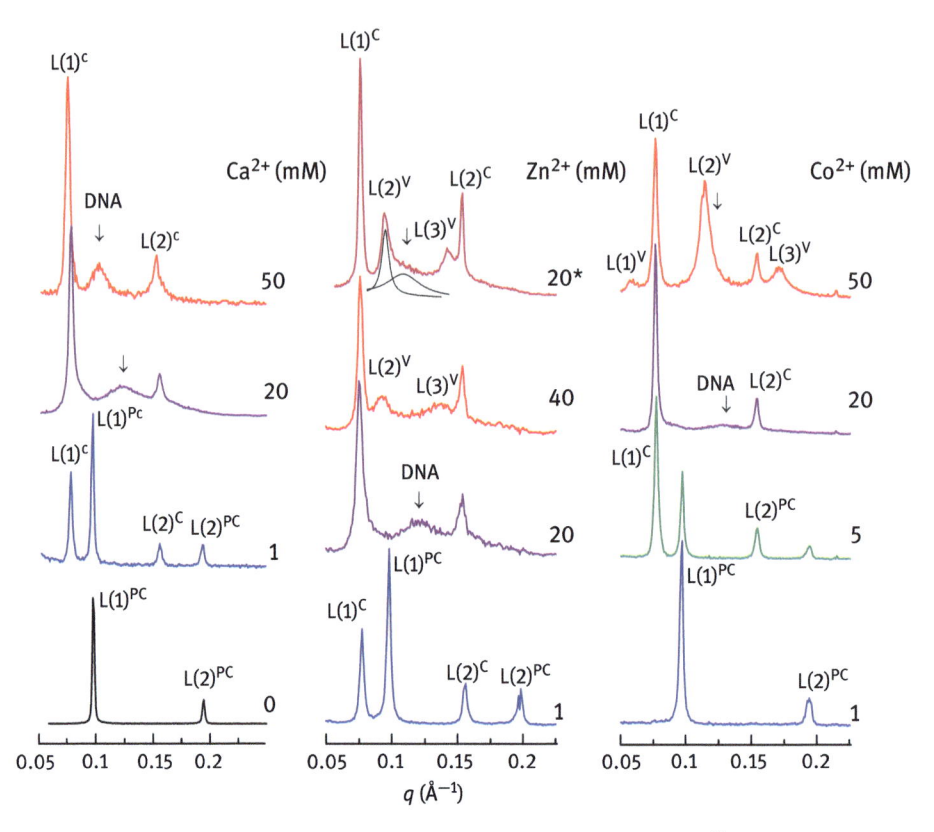

Figure 1.13: Small-angle X-ray diffraction (SAXD) curves of DNA–DPPC–Me^{2+} aggregates (DPPC:DNA = 3:1 mol/base) at selected concentrations of Me^{2+}, and at $T = 20$ °C. The total ionic strength in the diffractogram denoted by asterisk (*) was modulated with NaCl to $Is = 122$ mM.

For easier orientation in the structural polymorphism of aggregates, it is useful to introduce the lipid itself. Fully hydrated DPPC at 20 °C forms a lamellar phase (L) (Figure 1.13). Its temperature behavior is well characterized by a tilted gel phase ($L_{\beta'}$) below 35 °C, rippled gel phase (P_β) below 42 °C, and liquid-crystalline phase (L_α) above 42 °C [120].

About 1 mM solution of CaCl$_2$ or ZnCl$_2$ is sufficient for DNA complexation between DPPC bilayers (Figure 1.13). The diffractograms show the superposition of two one-dimensional structures. Peaks L(1)^{PC} and L(2)^{PC} are attributed to a phase with a periodicity close to that formed by the lipid itself (d ~65 Å). The second lamellar phase with the repeat distance d ~80 Å is identified as a condensed L^C phase. This concentration of Me^{2+} (1 mM) is, however, not sufficient to facilitate regular packing of DNA strands; thus, we do not observe any peak related to DNA–DNA interhelical distance. Sketch of the structure is shown in Figure 1.14c. Such a coexistence of two lamellar phases is frequently reported in aggregates formed using both saturated or monounsaturated PC and DNA also at higher concentrations of Ca^{2+}, Mg^{2+}, and Mn^{2+} [121–124]. Note that the structure of DNA–DPPC in 1 mM solution of CoCl$_2$ is different. We observe only one lamellar phase with d ~65 Å that is comparable to DPPC itself. In this experiment, all samples were prepared at the same conditions. In the case of Co^{2+}, a solution with a higher concentration of salt is necessary for DNA complexation between lipid bilayers.

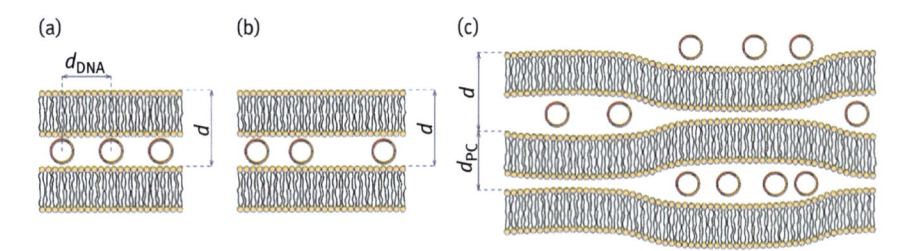

Figure 1.14: Schematic sketch of structures showing condensed lamellar phase (L^C) with (**a**) and without (**b**) regular DNA organization. The panel (**c**) shows a coexistence of two lamellar phases formed in consequence of lateral segregation of DNA strands (with permission Uhríková [125]).

About 20 mM of Me^{2+} (Ca^{2+}, Zn^{2+}, and Co^{2+}) is a sufficient concentration to induce a formation of condensed lamellar phase (L^C) with DNA strands packed regularly between DPPC bilayers in a gel state (T = 20 °C) (Figure 1.13). Two sharp peaks correspond to a lamellar phase with periodicity d ~80 Å, resulting from lipid bilayer stacking. The broad peak of lower intensity is an evidence for DNA–DNA organization. The interhelical DNA–DNA distance, d_{DNA} ~50–60 Å, is typical for DNA–DPPC–Me^{2+} aggregates in L$^C_\beta$ phase with the lipid in a gel state [126, 127]. Similar diffractograms were observed for L^C phase in DNA–cationic lipid complexes [128, 129]. Figure 1.14a shows a schematic sketch of such L^C phase.

Finally, high concentrations of salts ($c^{2+}_{Me} \geq 40$ mM) affect aggregation process further and may change the structures. For example, we observe L$^C_\beta$ phase with a well-defined DNA peak (d_{DNA} ~61 Å) as the only structure of DNA–DPPC–Ca^{2+} prepared at 50 mM of Ca^{2+} and at 20 °C. However, diffraction patterns of DNA–DPPC dispersed either in 40 mM of ZnCl$_2$ or 50 mM of CoCl$_2$ show systems with two phases. An analysis of the pattern with zinc revealed the coexistence of L$^C_\beta$ phase with lattice parameters $d = 85$ Å and $d_{DNA} = 61$ Å, and another lamellar phase, L^V, of a large periodicity $d^V = 135$ Å. Similar lattice parameters were extracted from diffractograms of DNA–DPPC in 50 mM of $CoCl_2$: L_β^C phase with $d = 82$ Å and $d_{DNA} = 55$ Å, and L^V phase with $d^V = 109$ Å. Since, the periodicity of L^V phase (~100–135 Å) is too big to be "accommodated" in a structure as shown in Figure 1.14c, this has indicated rather a destruction of the long-range lamellar structure of the lipid by its swelling into excess water, as discussed above for phospholipid–Me^{2+} interactions. In such a case, the periodicity is dictated by the total number of ions in solution (for DLVO theory, see, e.g., Ref. [130])

Further, we have therefore examined the effect of "a quantity" of ions. Figure 1.13 displays also the diffractogram of DNA–DPPC in 20 mM of $ZnCl_2$, where we modulated ionic strength (Is) by adding of NaCl (total ionic strength $Is = 122$ mM). The observed structural parameters ($d = 82$ Å, $d_{DNA} = 57$ Å, and $d^V = 131$ Å) are close to those found in previously discussed systems (i.e., DNA–DPPC in 40 mM $ZnCl_2$). It has been shown that neither the repeat distance nor the bilayer thickness of neutral phospholipids changed with the solution ionic strength ranging between 1 and 500 mM NaCl [131]. Figure 1.15 shows the effect of ionic strength on structural parameters of both detected phases, L^C and L^V. With increasing Is, we can see two opposite effects: the repeat distance d_L^C of condensed L^C phase formed by DNA–DPPC–Zn^{2+} increases slightly at $c_{Zn}^{2+} > 20$ mM (Figure 1.15, inset), and its total change does not exceed $\Delta d_L^C \sim 3$ Å. However, the repeat distance d_L^V of L^V phase decreases nonlinearly, changing significantly from ~200 Å to ~70 Å at $Is = 330$ mM, reaching almost the periodicity of pure DPPC (~64 Å). L^V phase was identified as being formed by DPPC + Zn^{2+} bilayers, and it is macroscopically separated from the mixture [132]. The mixture of DNA–DPPC–Co^{2+} at 50 mM of $CoCl_2$ has shown similar SAXD pattern. In addition to L^C phase formed by DNA condensed between DPPC and Co^{2+} bilayers, a part of the mixture is structured in a lamellar phase (L^V). Unfortunately, SAXD cannot discriminate between the coexistence of two phases that are, though macroscopically separated, both present within the studied volume of a sample or that coexist within one structure.

The driving force for mutual condensation of DNA by cationic vesicles to form an ordered, composite phase is the gain in electrostatic energy. The electrostatic energy depends on the surface charge densities of separated macroions, the structure and composition of the condensed phases, and the salt concentration in solution [133, 134]. The minima in electrostatic free energy occur when the fixed negative charges on DNA surface are balanced by the same number of positive charges on the bilayer surface, that is, at isoelectric point. Due to the high mobility of Me^{2+}, it is difficult to evaluate the isoelectric point in this system. The mechanism of DNA–phospholipid–Me^{2+} interactions and binding stoichiometry are still under discussion (see, e.g., Refs. [121, 124, 135]). DNA binding onto zwitterionic bilayers in presence of Me^{2+} is not driven by the release of counterions as it was confirmed for DNA–cationic/neutral lipid systems [136].

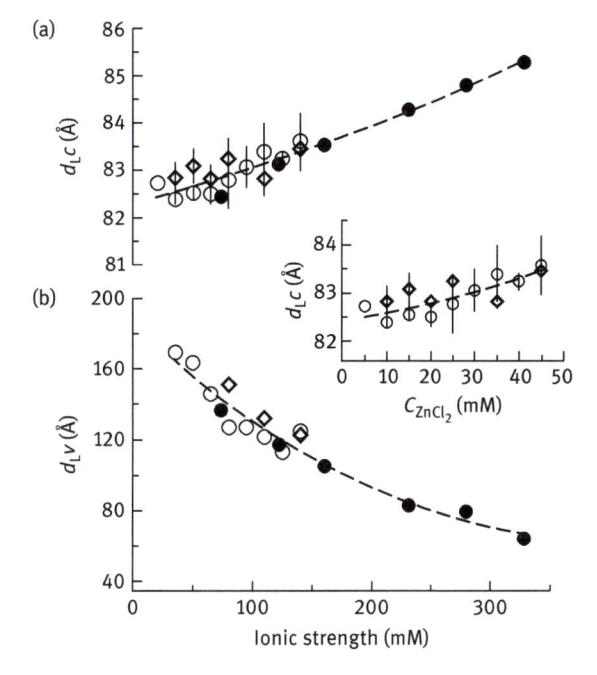

Figure 1.15: Dependence of the repeat distances d on the ionic strength (Is) of solutions (20 °C) for phases L^C (a) and L^V (b). The DNA–DPPC–Zn^{2+} aggregates were prepared either by DNA interacting with DPPC in solutions of Zn^{2+} at various concentrations (prepared in 5 mM of NaCl and represented by empty symbols) or ZnCl$_2$ was kept at constant concentration of 20 mM and ionic strength was modulated by NaCl (full symbols). The circles represent data from SANS, while the same samples measured by SAXD are shown by diamonds. **Inset:** the d of L^C as a function of ZnCl$_2$ concentration (c_{ZnCl_2}). Results are adapted from Uhríková et al. [132].

If salt is added to the system, the mobile salt ions screen electrostatic interactions between fixed charges along DNA and the P–N$^+$ dipole of phospholipid headgroups.

Structural polymorphism is reduced when the lipid is in a liquid-crystalline state (L_α phase). Figure 1.16a shows an evolution of structure with increasing temperature for DNA–DPPC–Ca^{2+} in 20 mM of Ca^{2+}. Note that the peak related to the DNA-DNA organization vanishes gradually into the background when the sample is heated, while it cannot be identified above ~60 °C. The analysis of the system reveals that the disorder in DNA lattice is caused mainly by in-plane fluctuations of DNA strands [126]. Due to the high mobility of metal cations, they do not induce constraints large enough to support DNA strands with the regular packing observed in complexes with cationic amphiphiles. Figure 1.16b illustrates temperature dependence of the repeat distance d_L^C for two different DNA–DPPC–Me^{2+} aggregates, prepared in 20 mM of Ca^{2+} and 40 mM of Co^{2+}, respectively. When comparing with pure DPPC, we see that DNA–DPPC–Me^{2+} aggregates show only one-phase transition: from gel to liquid-

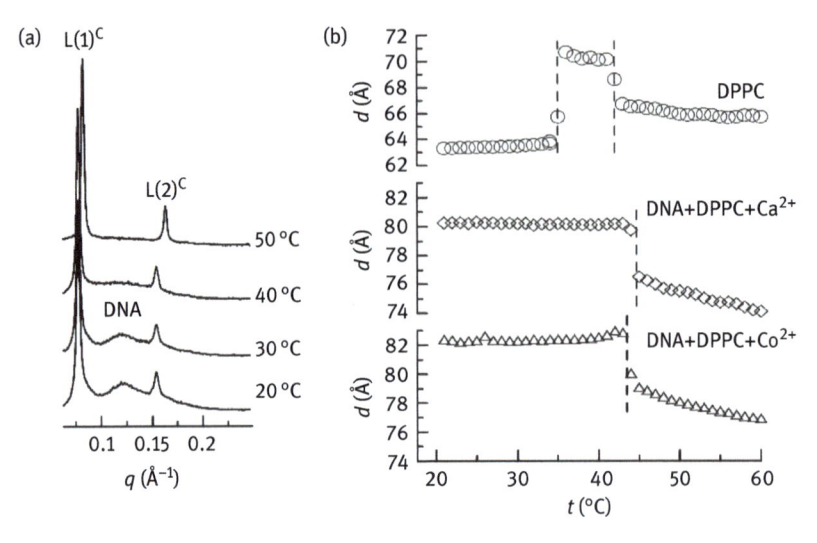

Figure 1.16: (a) SAX diffractograms of DPPC:DNA = 3:1 mol/base aggregates at 20 mM of Ca^{2+}. (b) Temperature dependences of the repeat distances (d) of fully hydrated DPPC bilayers and L^C phase of DNA–DPPC–Me^{2+} aggregates prepared at 20 mM of Ca^{2+} and 40 mM of Co^{2+}.

crystalline phase. DNA strands located in the water layer between bilayers dump the rippling of the DPPC bilayer, typically observed in the range 35–42 °C (P_β phase).

In a fluid lamellar L_α phase, a temperature increase induces an increase of the population of gauche conformers in lipid acyl chains accompanied by the lateral expansion of the bilayer, what is manifested by a decrease of the lipid thickness [137]. Indeed, Figure 1.16b shows d_L^C dropping by ~3–4 Å in the case of L^C phase due to DPPC acyl chains melting at $L_\beta^C \to L_\alpha^C$ phase transition. On the other hand, fluctuations of the lipid bilayer in L_α are enhanced, what is accompanied by a diffusion of water molecules from the bulk water phase in between lipid bilayers, thus an increase of the thickness of water layer [138]. The one of these two processes prevails the change in the repeat distance. As shown in Figure 1.16b, the repeat distances of both aggregates decrease with increasing temperature in L_α phase. In fact, DNA and Me^{2+} dump the bilayer fluctuations and make the bilayers "stiffer" showing higher resistance of lipid bilayers against water diffusion from the bulk aqueous phase [123]. The transversal thermal expansion coefficient at constant pressure π determined from

$$\alpha = \frac{1}{d}\left(\frac{\partial d}{\partial T}\right)_\pi,$$

where T is an absolute temperature and d is a suitable parameter to assess the system. For DPPC, we determined $\alpha = -(0.75 \pm 0.04) \times 10^{-3}\,K^{-1}$ by a linear least square fitting of data in the range 50–60 °C. In triple complexes DNA–DPPC–Me^{2+} at 2–40 mM of Me^{2+}, and the same temperature range, α varies from about -1.6×10^{-3} to -2.1×10^{-3} K^{-1}. The α depends significantly on used Me^{2+}, decreasing in the order $\alpha_{DPPC} > \alpha_{Co^{2+}} > \alpha_{Ca^{2+}} \approx \alpha_{Zn^{2+}}$ [139].

The gel-to-liquid-crystalline phase transition of DPPC happens within the range $T_{mDPPC} = 41.3 \pm 1.8$ °C, depending on the used experimental method [140]. Without Me^{2+}, DNA affects only slightly the thermodynamic parameters of DPPC [132]. On the other hand, DNA complexed with Me^{2+} affects phase transition of DPPC, as indicated in Figure 1.16b. Figure 1.17a summarizes changes to the DPPC's T_m detected in DNA–DPPC–Me^{2+} mixtures at different concentrations of cations using SAXD and differential scanning calorimetry (DSC). ΔT_m values expressed as $\Delta T_m = T_t - T_{mDPPC}$, where T_t is the phase transition temperature either of a condensed phase or a phase formed by DPPC + Me^{2+} in mixtures DNA–DPPC–Me^{2+} derived from both methods, are in a good accord. The effect of Me^{2+} on the melting of DPPC acyl chains bound in L^C phase of DNA–DPPC–Me^{2+} aggregates is evident, where T_m increases in the order: $T_{mDPPC} \approx T_{mCo} < T_{mCa} < T_{mZn}$. DSC profiles of DNA–DPPC–Zn^{2+} mixtures exhibit two well-defined maxima allowing to assess T_m and the enthalpy of both L^C phase of DNA–DPPC–Zn^{2+}, and L^{PC} phase of DPPC + Zn^{2+}. However, DSC does not distinguish between the lipid trapped in the DNA–DPPC–Zn^{2+} aggregate (L^{PC} phase) and that forming L^V phase (Figure 1.13), and thus, we use an abbreviation PC (L^{PC}). The effect of Zn^{2+} on DPPC phase transition without any DNA determined independently is shown in Figure 1.17a as well. In addition to T_m, the enthalpy (ΔH) of transition can be

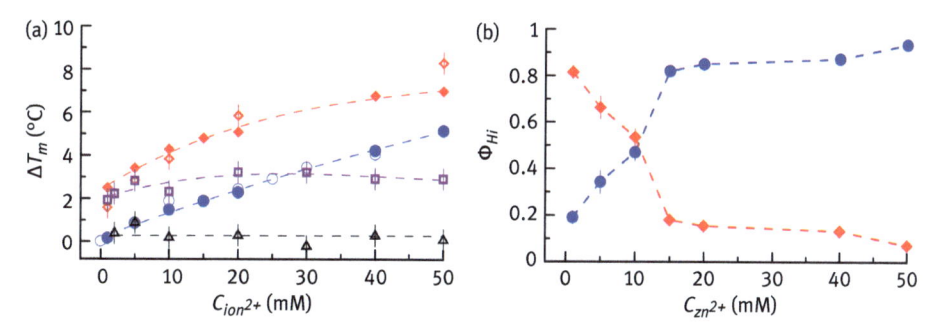

Figure 1.17: (a) Difference between the gel–liquid-crystal phase transition of DPPC (T_{mDPPC}) and that of DNA–DPPC–Me^{2+} as a function of Me^{2+} concentration ($\Delta T_m = T_t - T_{mDPPC}$). The ΔT_m for L^C phase of DNA–DPPC–Co^{2+} is displayed by empty triangles, L^C phase of DNA–DPPC–Ca^{2+} by empty squares, and L^C phase of DNA–DPPC–Zn^{2+} by empty diamonds. Data are derived from the discontinuities of repeat distance d of L^C phase in DNA–DPPC–Me^{2+} aggregates extracted from SAXD data [139]. The complementary data obtained from DSC are shown in the case of DNA–DPPC–Zn^{2+} mixture for L^C phase (full diamonds), L^{PC} phase (full circles), and for DPPC + Zn^{2+} mixture (empty circles). (b) The enthalpy fraction Φ_{H_i} of individual phases L$_i$ (i = PC, C) versus ZnCl$_2$ concentration. The results are adapted from Uhríková et al. [132].

calculated by integrating the heat capacity versus temperature curve of DSC profile. In order to quantify the relative changes in the volume fraction of individual phases, we express the enthalpy fraction Φ_{H_i}:

$$\Phi_{H_i} = \frac{\Delta H_i}{\Delta H_{tot}},$$

where ΔH_i is the enthalpy of i = L^C, L^{PC} phases and ΔH_{tot} is the total enthalpy. The ΔH_i of ith phase as well as the total enthalpy were determined by integration of the calorimetric profile [132]. Relative changes in the enthalpy of individual phases as a function of $ZnCl_2$ concentration are depicted in Figure 1.17b. Note the pronounced changes in the volume fractions of both phases at low concentrations of Zn^{2+} (<20 mM). Volume fraction of lipid involved in the condensed lamellar phase with DNA (L^C phase) decreases significantly with increasing concentration of zinc. High content of the salt has an influence on thermal stability. For example, aggregates prepared with short fragmented DNA in 20 mM of $ZnCl_2$ have lost their long-range order when heated to 60 °C [132]. We did not detect such behavior in aggregates prepared with highly polymerized DNA. As mentioned above, both cations Zn^{2+} and Co^{2+} have shown good affinity for DNA bases. In solutions with higher concentrations of these Me^{2+}, binding sites of both DNA and DPPC are saturated, and cations do not mediate the binding any more. The electrostatic screening of Me^{2+} charge due to ion accumulation and formation of a diffuse double layer at the lipid bilayer surface then leads to a macroscopic phase separation.

To conclude, DNA–phospholipid–Me^{2+} aggregates show higher structural varieties in comparison to complexes of DNA-cationic liposomes prepared with cationic lipid or surfactant. Divalent cations are active in promoting DNA condensation into the ternary complexes. As we demonstrated shortly, using one phospholipid (DPPC) and three cations of Me^{2+} (Ca^{2+}, Zn^{2+}, and Co^{2+}) modified thermodynamic properties and polymorphic behavior of ternary complexes. An attenuation of DNA binding in solutions at higher ionic strength appears as a drawback for their consideration as gene carriers in human therapy. On the other hand, the toxicity of metals results from the formation of complexes with organic compounds. The knowledge of the structure and phase behavior of the formed aggregates can thus contribute to the understanding of their toxicity.

Acknowledgments

The authors thank S.S. Funari for his assistance with SAXD experiments, and A. Lengyel for providing selected data from his PhD thesis. Financial support provided by the EC Program (FP7/2007–2013) under grant agreement no. 226716 (HASYLAB project II-20100372 EC), by the JINR topical theme 04-4-1121-2015/2020, by the CENI/ILL agreement 2014-2018, and by grant VEGA 1/0916/16 is gratefully acknowledged.

References

[1] Gorter, E., & Grendel, F. On bimolecular layers of lipoids on the chromocytes of the blood. J. Exp. Med. 1925, 41, 439–443.
[2] Danielli, J.F., & Davson, H. A contribution to the theory of permeability of thin films. J. Cell. Comp. Physiol. 1935, 5, 495–508.
[3] Singer, S.J., & Nicolson, G.L. The fluid mosaic model of the structure of cell membranes. Science. 1972, 175, 720–731.
[4] Nicolson, G.L. The fluid-mosaic model of membrane structure: still relevant to understanding the structure, function and dynamics of biological membranes after more than 40 years. Biochim. Biophys. Acta. 2014, 1838, 1451–1466.
[5] Engelman, D.M. Membranes are more mosaic than fluid. Nature. 2005, 438, 578–580.
[6] Simons, K., & Ikonen, E. Functional rafts in cell membranes. Nature. 1997, 387, 569–572.
[7] Escriba, P.V., Gonzalez-Ros, J.M., Goni, F.M., et al. Membranes: a meeting point for lipids, proteins and therapies. J. Cell. Mol. Med. 2008, 12, 829–875.
[8] van Meer, G., Voelker, D.R., & Feigenson, G.W. Membrane lipids: where they are and how they behave. Nat. Rev. Mol. Cell. Biol. 2008, 9, 112–124.
[9] Cullis, P.R., & Hope, M.J. Physical properties and structural roles of lipids in membranes. In: D.E. Vance & J.E. Vance, eds. Biochemistry of Lipids and Membranes: The Benjamin/Cumings Publishing Company, Don Mills, Ontario. Inc.; 1985, 25–72.
[10] Tanford, C. The Hydrophobic Effect: Formation of Micelles and Biological Membranes. Second Edition ed. New York: John Wiley and sons; 1980.
[11] Cevc, G., & Marsh, D. Phospholipid Bilayers. Physical Principles and Models. New York: John Wiley & Sons, Inc.; 1987, 157–231.
[12] Yeagle, P. The structure of biological membranes. Boca Raton: CRC Press; 1992. Second Ed. 2005, 201–241.
[13] Kučerka, N., Heberle, F.A., Pan, J., & Katsaras, J. Structural significance of lipid diversity as studied by small angle neutron and X-ray scattering. Membranes (Basel). 2015, 5, 454–472.
[14] Weijers, R.N. Lipid composition of cell membranes and its relevance in type 2 diabetes mellitus. Curr. Diabetes Rev. 2012, 8, 390–400.
[15] Katsaras, J., Pencer, J., Nieh, M.P., Abraham, T., Kučerka, N., & Harroun, T.A. Neutron and X-Ray Scattering from Isotropic and Aligned Membranes. Structure and Dynamics of Membranous Interfaces: John Wiley & Sons, Hoboken, New Jersey. Inc. 2008, 107–134.
[16] Harroun, T.A., Kučerka, N., Nieh, M.P., & Katsaras, J. Neutron and X-ray scattering for biophysics and biotechnology: examples of self-assembled lipid systems. Soft Matter. 2009, 5, 2694–2703.
[17] Heberle, F.A., Pan, J., Standaert, R.F., Drazba, P., Kučerka, N., & Katsaras, J. Model-based approaches for the determination of lipid bilayer structure from small-angle neutron and X-ray scattering data. Eur. Biophys. J. 2012, 41, 875–890.
[18] Wiener, M.C., & White, S.H. Fluid bilayer structure determination by the combined use of x-ray and neutron diffraction. I. fluid bilayer models and the limits of resolution. Biophys. J. 1991, 59, 162–173.
[19] Kučerka, N., Nagle, J.F., Sachs, J.N., et al. Lipid bilayer structure determined by the simultaneous analysis of neutron and X-ray scattering data. Biophys. J. 2008, 95, 2356–2367.
[20] Nagle, J.F., & Tristram-Nagle, S. Structure of lipid bilayers. Biochim. Biophys. Acta. 2000, 1469, 159–195.

[21] Nagle, J.F., Zhang, R., Tristram-Nagle, S., Sun, W., Petrache, H.I., & Suter, R.M. X-ray structure determination of fully hydrated L alpha phase dipalmitoylphosphatidylcholine bilayers. Biophys. J. 1996, 70, 1419–1431.

[22] Balgavý, P., & Devínsky, F. Cut-off effects in biological activities of surfactants. Adv. Colloid. Interface Sci. 1996, 66, 23–63.

[23] Wassenaar, T.A., Ingolfsson, H.I., Bockmann, R.A., Tieleman, D.P., & Marrink, S.J. Computational lipidomics with insane: A versatile tool for generating custom membranes for molecular simulations. J. Chem. Theory Comput. 2015, 11, 2144–2155.

[24] Kučerka, N., Gallová, J., Uhríková, D., et al. Areas of monounsaturated diacylphosphatidylcholines. Biophys. J. 2009, 97, 1926–1932.

[25] Cornell, B.A., & Separovic, F. Membrane thickness and acyl chain length. Biochim. Biophys. Acta. 1983, 733,189–193.

[26] Kučerka, N., Nieh, M.P., & Katsaras, J. Fluid phase lipid areas and bilayer thicknesses of commonly used phosphatidylcholines as a function of temperature. Biochim. Biophys. Acta. 2011, 1808, 2761–2771.

[27] Karlovská, J., Uhríková, D., Kučerka, N., et al. Influence of N-dodecyl-N,N-dimethylamine N-oxide on the activity of sarcoplasmic reticulum Ca(2+)-transporting ATPase reconstituted into diacylphosphatidylcholine vesicles: effects of bilayer physical parameters. Biophys. Chem. 2006, 119, 69–77.

[28] Martinez-Seara, H., Rog, T., Pasenkiewicz-Gierula, M., Vattulainen, I., Karttunen, M., & Reigada, R. Effect of double bond position on lipid bilayer properties: insight through atomistic simulations. J. Phys. Chem. B. 2007, 111, 11162–11168.

[29] Petrache, H.I., Dodd, S.W., & Brown, M.F. Area per lipid and acyl length distributions in fluid phosphatidylcholines determined by (2)H NMR spectroscopy. Biophys. J. 2000, 79, 3172–3192.

[30] Shoemaker, S.D., & Vanderlick, T.K. Intramembrane electrostatic interactions destabilize lipid vesicles. Biophys. J. 2002, 83, 2007–2014.

[31] de Vrije, T., de Swart, R.L., Dowhan, W., Tommassen, J., & de Kruijff, B. Phosphatidylglycerol is involved in protein translocation across Escherichia coli inner membranes. Nature. 1988, 334, 173–175.

[32] Nikaido, H., & Vaara, M. Molecular basis of bacterial outer membrane permeability. Microbiol. Rev. 1985, 49, 1–32.

[33] Seddon, A.M., Lorch, M., Ces, O., Templer, R.H., Macrae, F., & Booth, P.J. Phosphatidylglycerol lipids enhance folding of an alpha helical membrane protein. J. Mol. Biol. 2008, 380, 548–556.

[34] Tristram-Nagle, S., Zhang, R., Suter, R.M., Worthington, C.R., Sun, W.J., & Nagle, J.F. Measurement of chain tilt angle in fully hydrated bilayers of gel phase lecithins. Biophys. J. 1993, 64, 1097–1109.

[35] McIntosh, T.J., & Simon, S.A. Area per molecule and distribution of water in fully hydrated dilauroylphosphatidylethanolamine bilayers. Biochemistry. 1986, 25, 4948–4952.

[36] Cantor, R.S. Lateral pressures in cell membranes: A mechanism for modulation of protein function. J. Phys. Chem. B. 1997, 101, 1723–1725.

[37] Marsh, D. Lateral pressure in membranes. Biochim. Biophys. Acta. 1996, 1286, 183–223.

[38] Marrink, S.J., de Vries, A.H., Harroun, T.A., Katsaras, J., & Wassall, S.R. Cholesterol shows preference for the interior of polyunsaturated lipid membranes. J. Am. Chem. Soc. 2008, 130, 10–11.

[39] Kučerka, N., Perlmutter, J.D., Pan, J., Tristram-Nagle, S., Katsaras, J., & Sachs, J.N. The effect of cholesterol on short- and long-chain monounsaturated lipid bilayers as determined by molecular dynamics simulations and X-ray scattering. Biophys. J. 2008, 95, 2792–2805.

[40] Harroun, T.A., Katsaras, J., & Wassall, S.R. Cholesterol hydroxyl group is found to reside in the center of a polyunsaturated lipid membrane. Biochemistry. 2006, 45, 1227–1233.

[41] Harroun, T.A., Katsaras, J., & Wassall, S.R. Cholesterol is found to reside in the center of a polyunsaturated lipid membrane. Biochemistry. 2008, 47, 7090–7096.
[42] Kučerka, N., Marquardt, D., Harroun, T.A., Nieh, M.P., Wassall, S.R., & Katsaras, J. The functional significance of lipid diversity: orientation of cholesterol in bilayers is determined by lipid species. J. Am. Chem. Soc. 2009, 131, 16358–16359.
[43] Katsaras, J., Kučerka, N., & Nieh, M.P. Structure from substrate supported lipid bilayers (Review). Biointerphases. 2008, 3, FB55.
[44] Wiener, MC, King, GI, & White, SH. Structure of a fluid dioleoylphosphatidylcholine bilayer determined by joint refinement of x-ray and neutron diffraction data. I. scaling of neutron data and the distributions of double bonds and water. Biophys. J. 1991, 60,568–576.
[45] Kučerka, N., Marquardt, D., Harroun, T.A., et al. Cholesterol in bilayers with PUFA chains: doping with DMPC or POPC results in sterol reorientation and membrane-domain formation. Biochemistry. 2010, 49, 7485–7493.
[46] Cornelius, F. Modulation of Na,K-ATPase and Na-ATPase activity by phospholipids and cholesterol. I. Steady-State kinetics. Biochemistry. 2001, 40, 8842–8851.
[47] Lee, A.G. Ca2+ -ATPase structure in the E1 and E2 conformations: mechanism, helix-helix and helix-lipid interactions. Biochim. Biophys. Acta. 2002, 1565, 246–266.
[48] Pencer, J., Krueger, S., Adams, C.P., & Katsaras, J. Method of separated form factors for polydisperse vesicles. J. Appl. Crystallogr. 2006, 39, 293–303.
[49] Kučerka, N., Nagle, J.F., Feller, S.E., & Balgavý, P. Models to analyze small-angle neutron scattering from unilamellar lipid vesicles. Phys. Rev. E. Stat. Nonlin. Soft Matter Phys. 2004, 69, 051903.
[50] Kučerka, N., Pencer, J., Nieh, M.P., & Katsaras, J. Influence of cholesterol on the bilayer properties of monounsaturated phosphatidylcholine unilamellar vesicles. Eur. Phys. J.E. 2007, 23, 247–254.
[51] Brzustowicz, M.R., & Brunger, A.T. X-ray scattering from unilamellar lipid vesicles. J. Appl. Cryst. 2005, 38, 126–131.
[52] Kučerka, N., Pencer, J., Sachs, J.N., Nagle, J.F., & Katsaras, J. Curvature effect on the structure of phospholipid bilayers. Langmuir. 2007, 23, 1292–1299.
[53] Kučerka, N., Nieh, M.P., & Katsaras, J. Asymmetric distribution of cholesterol in unilamellar vesicles of monounsaturated phospholipids. Langmuir. 2009, 25, 13522–13527.
[54] Friedman, R., Pellarin, R., & Caflisch, A. Amyloid aggregation on lipid bilayers and its impact on membrane permeability. J. Mol. Biol. 2009, 387, 407–415.
[55] Gellermann, G.P., Appel, T.R., Tannert, A., et al. Raft lipids as common components of human extracellular amyloid fibrils. Proc. Natl. Acad. Sci. U S A. 2005, 102, 6297–6302.
[56] Puglielli, L., Tanzi, R.E., & Kovacs, D.M. Alzheimer's disease: the cholesterol connection. Nature Neuroscience. 2003, 6, 345–351.
[57] Di Paolo, G., & Kim, T.W. Linking lipids to Alzheimer's disease: cholesterol and beyond. Nat. Rev. Neurosci. 2011, 12, 284–296.
[58] Benloucif, S., Burgess, H.J., Klerman, E.B., et al. Measuring melatonin in humans. J. Clin. Sleep. Med. 2008, 4, 66–69.
[59] Karasek, M. Melatonin, human aging, and age-related diseases. Exp. Gerontol. 2004, 39, 1723–1729.
[60] Olcese, J.M., Cao, C., Mori, T., et al. Protection against cognitive deficits and markers of neurodegeneration by long-term oral administration of melatonin in a transgenic model of Alzheimer disease. J. Pineal. Res. 2009, 47, 82–96.
[61] Severcan, F., Sahin, I., & Kazanci, N. Melatonin strongly interacts with zwitterionic model membranes--evidence from Fourier transform infrared spectroscopy and differential scanning calorimetry. Biochim. Biophys. Acta. 2005, 1668, 215–222.

[62] Garcia, J.J., Reiter, R.J., Guerrero, J.M., et al. Melatonin prevents changes in microsomal membrane fluidity during induced lipid peroxidation. FEBS. Lett. 1997, 408, 297–300.

[63] Saija, A., Tomaino, A., Trombetta, D., et al. Interaction of melatonin with model membranes and possible implications in its photoprotective activity. Eur. J. Pharm. Biopharm. 2002, 53, 209–215.

[64] de Lima, V.R., Caro, M.S., Munford, M.L., et al. Influence of melatonin on the order of phosphatidylcholine-based membranes. J. Pineal. Res. 2010, 49, 169–175.

[65] Bongiorno, D., Ceraulo, L., Ferrugia, M., Filizzola, F., Ruggirello, A., & Liveri, V.T. Localization and interactions of melatonin in dry cholesterol/lecithin mixed reversed micelles used as cell membrane models.J. Pineal. Res. 2005, 38, 292–298.

[66] Rosales-Corral, S.A., Acuna-Castroviejo, D., Coto-Montes, A., et al. Alzheimer's disease: pathological mechanisms and the beneficial role of melatonin. J. Pineal. Res. 2012, 52, 167–202.

[67] Pabst, G., Kučerka, N., Nieh, M.P., Rheinstädter, M.C., & Katsaras, J. Applications of neutron and X-ray scattering to the study of biologically relevant model membranes. Chem. Phys. Lipids. 2010, 163, 460–479.

[68] Kučerka, N., Nieh, M.P., Pencer, J., Sachs, J.N., & Katsaras, J. What determines the thickness of a biological membrane. Gen. Physiol. Biophys. 2009, 28, 117–125.

[69] Worcester, D.L., & Franks, N.P. Structural analysis of hydrated egg lecithin and cholesterol bilayers. II. Neutrol diffraction. J. Mol. Biol. 1976, 100, 359–378.

[70] Leonard, A., Escrive, C., Laguerre, M., et al. Location of cholesterol in DMPC membranes. A comparative study by neutron diffraction and molecular mechanics simulation. Langmuir. 2001, 17, 2019–2030.

[71] Drolle, E., Kučerka, N., Hoopes, M.I., et al. Effect of melatonin and cholesterol on the structure of DOPC and DPPC membranes. Biochim. Biophys. Acta. 2013, 1828, 2247–2254.

[72] Sack, R.L., Lewy, A.J., Erb, D.L., Vollmer, W.M., & Singer, C.M. Human melatonin production decreases with age. J. Pineal. Res. 1986, 3, 379–388.

[73] Robinson, D.H., & Toledo, A.H. Historical development of modern anesthesia. J. Invest. Surg. 2012, 25, 141–149.

[74] Mullins, L.J. Some Physical Mechanisms in Narcosis. Chem. Rev. 1954, 54, 289–323.

[75] Lee, A.G. Model for action of local anaesthetics. Nature. 1976, 262, 545–548.

[76] Franks, N.P., & Lieb, W.R. Molecular mechanisms of general anaesthesia. Nature. 1982, 300, 487–493.

[77] Heimburg, T., & Jackson, A.D. The thermodynamics of general anesthesia. Biophys. J. 2007, 92, 3159-65.

[78] Graesboll, K., Sasse-Middelhoff, H., & Heimburg, T. The thermodynamics of general and local anesthesia. Biophys. J. 2014, 106, 2143–2156.

[79] Kondela, T., Gallová, J., Hauß, T., Barnoud, J., Marrink, S.J., & Kučerka, N. Alcohol Interactions with Lipid Bilayers. Molecules. 2017, 22, 2078.

[80] Chu, N., Kučerka, N., Liu, Y., Tristram-Nagle, S., & Nagle, J.F. Anomalous swelling of lipid bilayer stacks is caused by softening of the bending modulus. Phys. Rev. E. Stat. Nonlin. Soft Matter Phys. 2005, 71, 041904.

[81] Binder, H., & Zschornig, O. The effect of metal cations on the phase behavior and hydration characteristics of phospholipid membranes. Chem. Phys. Lipids. 2002, 115, 39–61.

[82] Kučerka, N., Papp-Szabo, E., Nieh, M.P., et al. Effect of cations on the structure of bilayers formed by lipopolysaccharides isolated from Pseudomonas aeruginosa PAO1. J. Phys. Chem. B. 2008, 112, 8057–8062.

[83] Abraham, T., Schooling, S.R., Nieh, M.P., Kučerka, N., Beveridge, T.J., & Katsaras, J. Neutron diffraction study of Pseudomonas aeruginosa lipopolysaccharide bilayers. J. Phys. Chem. B. 2007, 111, 2477–2483.

[84] Uhríková, D., Kučerka, N., Lengyel, A., et al. Lipid bilayer – DNA interaction mediated by divalent metal cations: SANS and SAXD study. J. Phys.: ConferenceSeries. 2012, 351, 012011-1-9.

[85] Kučerka, N., Dushanov, E., Kholmurodov, K.T., Katsaras, J., & Uhríková, D. Calcium and zinc differentially affect the structure of lipid membranes. Langmuir. 2017, 33, 3134–3141.

[86] Uhríková, D., Kučerka, N., Teixeira, J., Gordeliy, V., & Balgavý, P. Structural changes in dipalmitoylphosphatidylcholine bilayer promoted by Ca2+ ions: a small-angle neutron scattering study. Chem. Phys. Lipids. 2008, 155, 80–89.

[87] Yamada, L., Seto, H., Takeda, T., Nagao, M., Kawabata, Y., & Inoue, K. SAXS, SANS and NSE studies on "Unbound State" in DPPC/Water/CaCl2 system. J. Phys. Soc. Jpn. 2005, 74, 2853–2859.

[88] Inoko, Y., Yamaguchi, T., Furuya, K., & Mitsui, T. Effects of cations on dipalmitoyl phosphatidylcholine/cholesterol/water systems. Biochim. Biophys. Acta. 1975, 413, 24–32.

[89] Uhríková, D., Teixeira, J., Lengyel, A., Almasy, L., & Balgavý, P. Formation of unilamellar dipalmitoylphosphatidylcholine vesicles promoted by Ca2+ ions: A small-angle neutron scattering study. Spectrosc.-An Int. J. 2007, 21, 43–52.

[90] Hauser, H. Phospholipid Vesicles. In: G. Cevc, ed. Phospholipid Handbook. New York: Marcel Dekker, Inc. 1993:603–637.

[91] Petrache, H.I., Tristram-Nagle, S., Harries, D., Kučerka, N., Nagle, J.F., & Parsegian, V.A. Swelling of phospholipids by monovalent salt. J. Lipid. Res. 2006, 47, 302–309.

[92] Bock, C.W., Markham, G.D., Katz, A.K., & Glusker, J.P. The arrangement of first- and second-shell water molecules in trivalent aluminum complexes: results from density functional theory and structural crystallography. Inorg. Chem. 2003, 42, 1538–1548.

[93] David, F., Vokhmin, V., & Ionova, G. Water characteristics depend on the ionic environment. Thermodynamics and modelisation of the aquo ions. J. Mol. Liquids. 2001, 90, 45–62.

[94] Chaplin, M. Do we underestimate the importance of water in cell biology?. Nat. Rev. Mol. Cell Biol. 2006, 7, 861–866.

[95] Kuvichkin, V.V. DNA-lipid interactions in vitro and in vivo. Bioelectrochemistry. 2002, 58, 3–12.

[96] Kuvichkin, V.V. The mechanism of a nuclear pore assembly: A molecular biophysics view. J. Membrane. Biol. 2011, 241, 109–116.

[97] Martelli, A.M., Fala, F., Faenza, I., et al. Metabolism and signaling activities of nuclear lipids. Cell Mol. Life Sci. 2004, 61, 1143–1156.

[98] Albi, E., Lazzarini, R., & Viola Magni, M. Phosphatidylcholine/sphingomyelin metabolism crosstalk inside the nucleus. Biochem. J, 2008, 410, 381–389.

[99] Atilla-Gokcumen, G.E., Muro, E., Relat-Goberna, J., et al. Dividing cells regulate their lipid composition and localization. Cell. 2014, 156, 428–439.

[100] Lengyel, A., Uhríková, D., Klacsová, M., & Balgavý, P. DNA condensation and its thermal stability influenced by phospholipid bilayer and divalent cations. Colloids. Surf. B. 2011, 86, 212–217.

[101] De Boeck, M., Kirsch-Volders, M., & Lison, D. Cobalt and antimony: genotoxicity and carcinogenicity. Mutat Res. 2003, 533, 135–152.

[102] Macdonald, P.M., & Seelig, J. Calcium binding to mixed phosphatidylglycerol-phosphatidylcholine bilayers as studied by deuterium nuclear magnetic resonance. Biochemistry. 1987, 26, 1231–1240.

[103] Sinn, C.G., Antonietti, M., & Dimova, R. Binding of calcium to phosphatidylcholine-phosphatidylserine membranes. Colloids Surf. A-Physicochemical and Engineering Aspects. 2006, 282, 410–419.

[104] Altenbach, C., & Seelig, J. Ca2+ binding to phosphatidylcholine bilayers as studied by deuterium magnetic resonance. Evidence for the formation of a Ca2+ complex with two phospholipid molecules. Biochemistry. 1984, 23, 3913–3920.
[105] Satoh, K. Determination of binding constants of Ca^{2+}, Na^+, and Cl^- ions to liposomal membranes of dipalmitoylphosphatidylcholine at gel phase by particle electrophoresis. Biochim. Biophys. Acta. 1995, 1239, 239–248.
[106] Jovin, T.M., Soumpasis, D.M., & McIntosh, L.P. The transition between B-DNA and Z-DNA. Annu. Rev. Phys. Chem. 1987, 38, 521–560.
[107] Dove, W.F., & Davidson, N. Cation effects on the denaturation of DNA. J. Mol. Biol. 1962, 5, 467–478.
[108] Eichhorn, G.L., & Shin, Y.A. Interaction of metal ions with polynucleotides and related compounds. XII. the relative effect of various metal ions on DNA helicity. J. Am. Chem. Soc. 1968, 90, 7323–7328.
[109] Luck, G., & Zimmer, C. Conformational aspects and reactivity of DNA. effects of manganese and magnesium ions on interaction with DNA. Eur. J. Biochem. 1972, 29, 528–536.
[110] Duguid, J., Bloomfield, V.A., Benevides, J., & Thomas Jr., G.J. Raman spectroscopy of DNA-metal complexes. I. Interactions and conformational effects of the divalent cations: Mg, Ca, Sr, Ba, Mn, Co, Ni, Cu, Pd, and Cd. Biophys. J. 1993, 65, 1916–1928.
[111] Budker, V.G., Kazatchkov, Y.A., & Naumova, L.P. Polynucleotides adsorb on mitochondrial and model lipid membranes in the presence of bivalent cations. FEBS. Letters. 1978, 95, 143–146.
[112] Vojčíková, L., & Balgavý, P. Interaction of DNA with dipalmitoylphosphatidylcholine model membranes: A microcalorimetric study. Studia. Biophysica. 1988, 125, 5–10.
[113] Vojčíková, L., Švajdlenka, E., & Balgavý, P. Spin label and microcalorimetric studies of the interaction of DNA with unilamellar phosphatidylcholine liposomes. Gen. Physiol. Biophys. 1989, 8, 399–406.
[114] Balgavý, P., Vojčíková, L., & Švajdlenka, E. Microcalorimetric study the DNA melting in the presence of phosphatidylcholine liposomes and magnesium ions. Acta Facultatis Pharm. Univ. Comenianae. 2002, 49, 17–25.
[115] Bruni, P., Gobbi, G., Morganti, G., Iacussi, M., Maurelli, E., & Tosi, G. Use and activity of metals in biological systems. I. The interaction of bivalent metal cations with double-stranded polynucleotides and phospholipids. Gazz. Chim. Ital. 1997, 127, 513–517.
[116] Bruni, P., Cingolani, F., Iacussi, M., Pierfederici, F., & Tosi, G. The effect of bivalent metal ions on complexes DNA-liposome: a FT-IR study. J. Mol. Struc. 2001, 565–566.
[117] Tarahovsky, Y.S., Khusainova, A.V., Gorelov, A.V., et al. DNA initiates polymorphic structural transitions in lecithin. FEBS. Letters. 1996, 390, 133–136.
[118] Tarahovsky, Y.S., Deev, A.A., Masulis, I.S., & Ivanitsky, G.R. Structural organization and phase behavior of DNA-calcium- dipalmitoylphosphatidylcholine complex. Biochemistry-Moscow 1998, 63, 1126–1131.
[119] Khusainova, R.S., Dawson, K.A., Rochev, I., Gorelov, A.V., & Ivanitskii, G.R. Structural changes in DNA-Ca2+-dipalmitoylphosphatidylcholine complexes during changes in the molar ratio of nucleotide/lipid. Microcalorimetric study. Dokl. Akad. Nauk. 1999, 367, 553–556.
[120] Stumpel, J., Eibl, H.J., & Nicksch, A. X-ray analysis and calorimetry on phosphatidylcholine model membranes. Biochim. Biophys. Acta. 1983, 727, 246–254.
[121] McManus, J., Radler, J.O., & Dawson, K.A. Phase behaviour of DPPC in a DNA-calcium-zwitterionic lipid complex studied by small angle x-ray scattering. Langmuir. 2003, 19, 9630–9637.
[122] Francescangeli, O., Stanic, V., Gobbi, L., et al. Structure of self-assembled liposome-DNA-metal complexes. Phys. Rev. e. 2003, 67, art-011904.

[123] Uhríková, D., Hanulová, M., Funari, S.S., Khusainova, R.S., Šeršeň, F., & Balgavý, P. The structure of DNA–DOPC aggregates formed in presence of calcium and magnesium ions: a small angle synchrotron X ray diffraction study. Biochim. Biophys. Acta. 2005, 1713, 15–28.

[124] Bruni, P., Francescangeli, O., Marini, M., Mobbili, G., Pisani, M., & Smorlesi, A. Can neutral liposomes be considered as genetic material carriers for human gene therapy?. Mini-Rev. Org Chem. 2011, 8, 38–48.

[125] Uhríková, D. Divalent Metal Cations in DNA–Phospholipid Binding. In: A. Iglič & C.V. Kulkarni, eds. Advances in Planar Lipid Bilayers and Liposomes, vol 20. Burlington: Elsevier 2014:111–135.

[126] Uhríková, D., Lengyel, A., Hanulová, M., Funari, S.S., & Balgavý, P. The structural diversity of DNA-neutral phospholipids-divalent metal cations aggregates: a small-angle synchrotron X-ray diffraction study. Eur. Biophys. J. 2007, 36, 363–375.

[127] Uhríková, D., Rapp, G., & Balgavý, P. Condensation of DNA and phosphatidylcholine bilayers induced by Mg(II) ions – a synchrotron X-ray diffraction study. In: M. Melník & A. Sirota, eds. Challenges for Coordination Chemistry in the new Century. Bratislava: Slovak Technical University Press; 2001, 219–224.

[128] Lasic, D.D., Strey, H., Stuart, M.C.A., Podgornik, R., & Frederik, P.M. The structure of DNA-Liposome complexes. J. Am. Chem. Soc. 1997, 119, 832–833.

[129] Radler, J.O., Koltover, I., Salditt, T., & Safinya, C.R. Structure of DNA-Cationic liposome complexes: DNA Intercalation in multilamellar membranes in distinct interhelical packing regimes. Science. 1997, 275, 810–814.

[130] Izumitani, Y. Cation dipole interaction in the lamellar structure of DPPC bilayers. J. Colloid and Interface Sci. 1994, 166, 143–159.

[131] Pabst, G., Hodzic, A., Štrancar, J., Danner, S., Rappolt, M., & Laggner, P. Rigidification of neutral lipid bilayers in the presence of salts. Biophys. J. 2007, 93, 2688–2696.

[132] Uhríková, D., Pullmannová, P., Bastos, M., Funari, S.S., & Teixeira, J. Interaction of short-fragmented DNA with dipalmitoylphosphatidylcholine bilayers in presence of zinc. Gen. Physiol Biophys. 2009, 28, 146–159.

[133] May, S., Harries, D., & Ben Shaul, A. The phase behavior of cationic lipid-DNA complexes. Biophys. J. 2000, 78, 1681–1697.

[134] Harries, D., May, S., Gelbart, W.M., & Ben Shaul, A. Structure, stability, and thermodynamics of lamellar DNA-lipid complexes. Biophys. J. 1998, 75, 159–173.

[135] Mengistu, D.H., Bohinc, K., & May, S. Binding of DNA to zwitterionic lipid layers mediated by divalent cations. J. Phys. Chem. B. 2009, 113, 12277–12282.

[136] Wagner, K., Harries, D., May, S., Kahl, V., Radler, J.O., & Ben Shaul, A. Direct evidence for counterion release upon cationic lipid-DNA condensation. Langmuir. 2000, 16, 303–306.

[137] Ruocco, M.J., & Shipley, G.G. Characterization of the sub-transition of hydrated dipalmitoylphosphatidylcholine bilayers – kinetic, hydration and structural study. Biochim. Biophys. Acta. 1982, 691, 309–320.

[138] Petrache, H.I., Tristram-Nagle, S., & Nagle, J.F. Fluid phase structure of EPC and DMPC bilayers. Chem. Phys. Lipids. 1998, 95, 83–94.

[139] Lengyel, A. Interakcia DNA s fosfolipidovými lipozómami v prítomnosti dvojmocných katiónov. / DNA interaction with phospholipid bilayer in presence of divalent cations. [PhD thesis]. Faculty of Pharmacy: Comenius University in Bratislava; 2010, 137–142.

[140] Koynova, R., & Caffrey, M. Phases and phase transitions of the phosphatidylcholines. Biochim. Biophys. Acta. 1998, 1376, 91–45.

Tim Salditt, Karlo Komorowski, Kilian Frank
2 X-ray structure analysis of lipid membrane systems: solid-supported bilayers, bilayer stacks, and vesicles

Abstract: In this chapter, we describe X-ray diffraction analysis of lipid model membranes, including fundamentals of experiment and analysis. We start with solid-supported single bilayers and monolayers, then discuss solid-supported multilamellar stacks, and finally vesicles in solution. For oriented membranes, we discuss specular and nonspecular reflectivity, as well as grazing incidence diffraction, and for vesicles we discuss small-angle X-ray scattering. In each case, illustrative examples of current applications are given. The chapter closes with an outlook on free electron laser sources and new opportunities for research on lipid model membranes and biomembranes.

Keywords: solid-supported lipid membranes, lipid vesicles, X-ray reflectivity, small-angle X-ray scattering

2.1 Introduction

Biological membranes are essential for fundamental functions of the eukaryotic cell, such as compartmentalization, vesicular transport, enzymatic activity, or signal transduction. As in other fields of structural biology and biophysics, these functions cannot be explained without resort to molecular structure. However, "structural biology" of lipids and the biophysical principles of the underlying lipid self-assembly cannot be grounded on macromolecular crystallography, which is the basis of the structural biology for proteins and nucleic acids. Instead of a macromolecular crystal, the most important model systems for biological membranes are liquid-crystalline lipid mesophases, lipid bilayers, and lipid vesicles. They all reflect the basic bilayer structure of membranes, as well as their self-assembly properties in the presence of aqueous solution. Structure analysis of lipids must hence be carried out for noncrystallographic states. This is the reason why despite the associated loss in resolution and information, X-ray diffraction of fluid membranes in hydrated environment has become a classic tool of biophysics. This has started in form of the conceptually simple and widely spread approach of small-angle X-ray scattering (SAXS) from isotropic lipid suspensions, in the seminal studies of lipid phase

---

Tim Salditt, Karlo Komorowski, Kilian Frank, Institut für Röntgenphysik, Universität Göttingen, Göttingen, Germany

https://doi.org/10.1515/9783110544657-002

diagrams by Luzzati and others [1, 2]. SAXS of soft matter systems in general, and in particular of lipids has already received much coverage in many reviews and textbooks [3, 4]. In this chapter, we introduce basic aspects of X-ray diffraction from aligned bilayers and membrane stacks. At the same time, we include vesicle SAXS, in view of the opportunities for recent studies of vesicle docking and fusion. More generally, we will focus on the advantages opened up by brilliant third-generation undulator sources for studies of membrane systems by several different X-ray diffraction techniques. Solid-supported membranes, which are highly aligned, are therefore amenable to modern surface sensitive scattering X-ray diffraction techniques. These techniques include specular and nonspecular reflectivity, as well as grazing incidence diffraction (GID) and reciprocal space mapping (RSM). They offer advantages over SAXS for the investigation of the structure of lipid bilayer systems, primarily since they allow for a precise distinction between the scattering vector component normal q_z and parallel $q_\parallel$ to the bilayer, see Figure 2.1, opening up a way to study lateral structure of the bilayers also in weakly ordered systems, typical for fluid membranes. Lateral micro- and nanophase segregation, the partitioning of membrane-active molecules, or the short range order of lipids can be addressed by these techniques, in particular if the signal is amplified by using stacks with many thousands of bilayers. Using lipid self-assembly and careful preparation methods as reviewed in Refs. [5–7], highly aligned membrane stacks can be achieved, where the distribution of the bilayer normal vector, that is, the mosaicity of a membrane stack, is on the order of 0.01°. In this case, very quantitative treatments of specular and

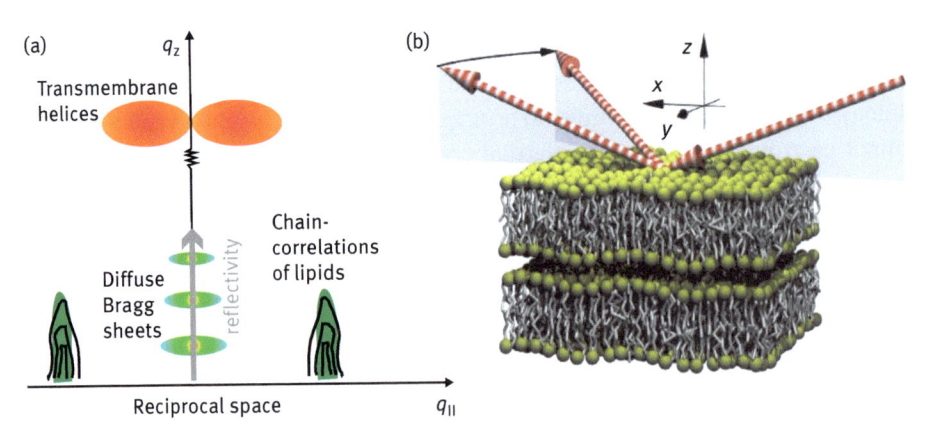

Figure 2.1: (a) Schematic of reciprocal space mapping (RSM) of oriented bilayers, as a function of momentum transfer parallel $q_\parallel$ and perpendicular q_z to the plane of the membranes, and (b) sketch of the multilamellar configuration. In the SAXS range, the lamellar peaks and the super-lattice peaks of the rhombohedral (R) phase (Section 2.3) are recorded. In the WAXS range, the correlation peak of the lipid tails is observed with a characteristic banana shape. In some systems with reconstituted proteins, signals of transmembrane helices can be detected.

nonspecular reflectivity, GID, and RSMs can be employed, which correctly account for total external reflection or refraction phenomena [3, 8–11].

Each of the different X-ray diffraction techniques and experimental approaches to study the structure of lipid model membranes requires different sample preparation, instrumentation, and data analysis. The conceptually simplest technique is scattering from isotropic lipid suspensions in the SAXS regime. Lipid suspensions for SAXS either consist of multilamellar vesicles (MLVs, onions) or of small unilamellar vesicles (SUVs) often also denoted as liposomes. To generate liposome suspensions, lipids have to be sonicated and/or extruded through filter paper with small pores. To account for the SAXS signal of liposomes, the polydispersity has to be incorporated by an integral over the vesicle radius R. For larger liposomes with considerable shape and size fluctuations, interference between opposing bilayer patches averages out, and the simpler bilayer form factor of a quasi-planar but powder-averaged system describes the data sufficiently well [12, 13]. SAXS from unilamellar vesicles is fairly simple and robust, in particular if the radius is large enough and if the vesicles are radially symmetric. MLVs, which are closer to the thermodynamic equilibrium state, can be considered as smectic liquid crystalline (LC) phases. Like their LC counterparts, they exhibit different thermal modes such as undulations and compressional waves, which can prominently affect the structure factor resulting in characteristic lineshapes [14–16]. Therefore, not only structure, but also information on elasticity properties of membranes can be deduced from such experiments. At the same time, nontrivial structure factors can complicate the structure analysis and can make the analysis of the bilayer electron density profile less robust.

For aligned membranes, X-ray reflectivity is the method of choice to extract the vertical bilayer structure, that is, its electron density profile (EDP) $\rho(z)$. Reflectivity can be applied to a single solid-supported bilayer, as well as to multilamellar stacks. Single bilayer reflectivity enables a particularly robust and straightforward measurement with no additional parameters for the structure factor. By ways of total external reflection, the signal is high and sensitive enough for a single membrane, and in contrast to studies of in-plane structure, multilamellar signal amplification is not actually needed. The advantages of single bilayer reflectivity are the following: (1) $\rho(z)$ is obtained on an absolute scale [17]. (2) The structural parameters of the inorganic substrate can be determined from a control experiment without the bilayer. (3) Fitting of the reflectivity signal is fairly robust using for example box models for $\rho(z)$, and in some cases, even unique inversion of reflectivity data becomes possible [18]. The disadvantage is that the substrate may induce structural alterations and even cause denaturing of membrane proteins. In this case, soft polymer cushions [19] can help, but complicate both preparation and data analysis.

Multilamellar stacks of membranes with up to thousands of bilayers clearly circumvent any detrimental effects of a substrate, since the signal is dominated by membranes from within the stack. As for MLVs, thermal fluctuations, elasticity constants, and interaction potentials become important, but due to the alignment, these phenomena themselves can be studied in much greater detail [7]. Thermal fluctuations are particularly pronounced when at full hydration [20], and in this case, full q-range fits can become quite challenging. A proper treatment of smectic elasticity in aligned membrane stacks with appropriate boundary conditions has been presented [21]. Contrarily, at partial hydration, one often neglects these effects and resorts to a simplified Fourier analysis based on the integrated peak intensities. In this case, the structure factor is neglected, and the analysis is based on the assumption that the peak intensities only depend on the form factor $F(q_z)$. In this chapter, we will not consider this simplified approach, but the better justified full q-range fits of reflectivity as presented in Ref. [20]. Nevertheless, in some cases Fourier synthesis is a simple and robust alternative, in particular at partial hydration, and if only relative changes in $\rho(z)$ are of interest.

The main advantage of oriented bilayer stacks is the signal enhancement for the analysis of lateral membrane structure. Like single bilayers, aligned multilamellar lipid membranes deposited on solid surfaces are amenable to modern surface sensitive scattering X-ray diffraction techniques, but with a signal level which is increased by the number of membranes N, that is, up to three or four orders of magnitude. In contrast to SAXS, oriented bilayer systems allow for a precise distinction between the scattering vector component normal q_z and parallel $q_\parallel$ to the bilayer, see Figure 2.1. Figure 2.1 illustrates in a coarse overview the scattering geometry of aligned membranes, including reflectivity, grazing incidence small-angle X-ray scattering (GISAXS), and GID from lipid films, and locates the associated features of membrane components in a RSM.

For each of the techniques addressed in this chapter, we provide the basic principles below, along with the fundamental equations required to model the signal and to deduce structural parameters. We then present illustrative examples for each case from our own recent work, for the simple reason of being most familiar with our own studies. We also keep in mind recent progress in X-ray optics, which not only offers increased signal levels, faster measurement times, and extension to dynamical time-resolved studies but also novel approaches based on coherent diffractive imaging, which will be included in an outlook.

This chapter is organized as follows: After this introduction, we first briefly provide an overview of the techniques addressed and the corresponding requirements of sample preparation. Section 2.2 is then devoted to X-ray reflectivity of single lipid bilayer on solid support, followed by multilamellar stacks addressed in Section 2.3. SAXS on vesicle suspensions is treated in Section 2.4, before we consider a SAXS study of synaptic vesicles (SVs) as a real biological membrane organelle in Section 2.5. We close the chapter with an outlook on X-ray diffractive imaging of vesicles.

2.2 X-ray reflectivity of solid-supported bilayers and monolayers

X-ray reflectivity is an interface-sensitive scattering method, which gives access to the scattering length density profile $\rho(z)$ of a structured interface, based on fitting or inverting [18] the reflectivity signal $R(q_z)$ as a function of momentum transfer perpendicular to the interface [3]. Starting from the plateau of total external reflection, up to eight orders of magnitude in the reflectivity signal are recorded. Alignment as well as background subtraction are critical. The total interface profile comprises a solid support (typically glass or silicon wafer), an oxide and/or polymer layer, and silane or surfactant layers depending on the preparation, as well as the lipid bilayer profile. The signal of the bilayer profile can be isolated, and the electron density along the membrane normal can be determined on absolute scale. The resolution is given by the highest momentum transfer, typically $q_z \simeq 0.6\,\text{Å}^{-1}$ in the fluid phase and $0.7\,\text{Å}^{-1}$ in the gel phase of the bilayers, using synchrotron radiation, or $0.35\,\text{Å}^{-1}$ for in-house sealed tube experiments [17]. There are many excellent treatments of X-ray reflectivity in general [3, 8, 22–24], and many reviews in particular on X-ray reflectivity from Langmuir monolayers at the air–water interface [25, 26], which was historically the first experimental setup used to study lipids by X-ray reflectivity. An early review of bilayer reflectivity is found in Ref. [27]. Here, we concentrate on X-ray reflectivity from solid-supported bilayers and monolayers presenting the technique and data analysis with one example for each case.

2.2.1 Experiment and data analysis for single solid-supported bilayers

The principal deposition method of solid-supported lipid bilayers is vesicle fusion [28–30]. As detailed in Ref. [17], silicon or glass wafers are cleaned and made hydrophilic, for example, by plasma etching. Lipids (with their abbreviated names listed in Table 2.1) are mixed in the desired amount, dissolved, dried in vacuum, and resuspended in buffer solution. For example in Ref. [17], 150 mM NaCl and 10 mM HEPES (see Table 2.1), pH 7.4 was used for zwitterionic and neutral lipids; for charged lipids, 2 mM $CaCl_2$ was added to facilitate fusion. SUVs can be obtained by sonication with a tip sonicator or alternatively by extrusion methods. A typical lipid concentration of 0.1 mg/ml is used when bringing the vesicle suspension into contact with the wafer. Vesicles fuse and rupture at the surface. Within an incubation time of about 30 min at room temperature, and after rinsing away excess vesicles, a homogeneous coverage of a single membrane is obtained.

Table 2.1: Abbreviations and full names of lipids and chemicals used in the text.

Abbreviation	Full name
DOPC	1,2-Dioleoyl-*sn*-glycero-3-phosphocholine
DOPS	1,2-Dioleoyl-*sn*-glycero-3-phospho-L-serine
DOTAP	1,2-Dioleoyl-3-trimethylammonium-propane
DMPC	1,2-Dimyristoyl-*sn*-glycero-3-phosphocholine
OPPC	1,2-Oleoyl-palmitoyl-*sn*-glycero-3-phosphocholine
DOPE	1,2-Dioleoyl-*sn*-glycero-3-phosphoethanolamine
DODAB	Dioctadecyl-dimethylammonium-bromide
HEPES	4-(2-hydroxyethyl)-1-piperazineethanesulfonic acid
OTS	Octadecyl-tri-chloro-silane
PEG	Polyethylene glycol

Once that homogeneous bilayer coverage is achieved, the sample should be kept in buffer at all times to prevent rupture and drying, and is transferred to a buffer-filled X-ray compatible chamber.

The technical aspects of measuring reflectivity are demanding, regarding alignment and diffuse background rejection, in particular for high resolution studies if the signal is to be recorded up to high q_z. Figure 2.2 addresses some of the challenges, such as patching different q_z ranges recorded with different attenuators, correcting for the illumination, background measurement in form of an offset-scan, as well as control of radiation damage if synchrotron radiation is used [17]. Radiation damage is found to depend strongly on the photon energy E_{ph}, making it favorable to use $E_{ph} \simeq 20 - 30$ keV.

Even before molecular changes of the chemical moieties due to free radical generation by photoelectrons lead to structural changes, beam-induced charging of the solid surface can alter the membrane structure [31]. While the latter effect is reversible, the first is not. Various cross-checks can be used to verify that the measurement and raw data treatment is indeed reproducible and correct, for example, by verifying that the plateau of total external reflection is equal to one, or that the maxima of several rocking scans are at the nominal $\alpha_i = \alpha_f$, before further analysis.

In the following, we describe the treatment of X-ray reflectivity from solid-supported membranes, following the presentation in Ref. [17]. We consider an interface with electron density profile (EDP) $\rho(z)$ between a medium 1 with electron density ρ_1 and a medium 2 with density ρ_2. Within the semi-kinematical approximation, the signal is then given by the so-called master formula of reflectivity [3]

$$R(q_z) = R_F(q_z) \left| \frac{1}{\Delta \rho_{12}} \int \frac{d\rho(z)}{dz} e^{iq_z z} dz \right|^2, \qquad (2.1)$$

where $R_F(q_z)$ is the Fresnel reflectivity of the ideal interface between the two media, q_z is the scattering vector (vertical momentum transfer), and $\Delta \rho_{12}$ is the density contrast.

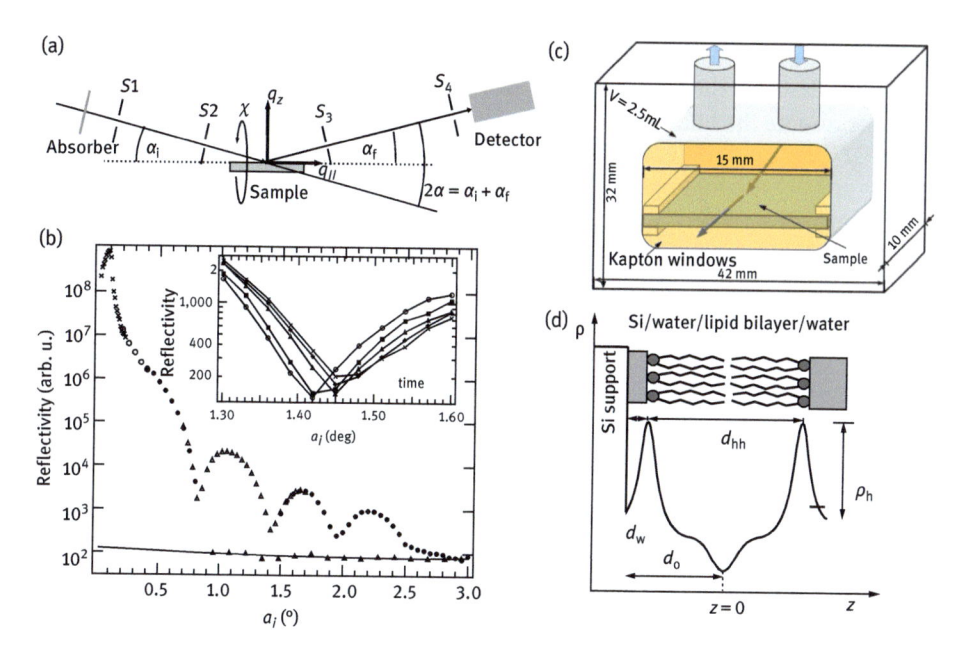

Figure 2.2: Fundamentals of bilayer reflectivity experiments. (a) Schematic of a X-ray reflectivity setup, with the X-ray beam impinging at an incident angle α_i. The signal is recorded at a reflection angle $\alpha_f = \alpha_i$ (specular reflectivity) and is corrected for the varying size of the footprint (illumination correction). Angles are converted to the vertical momentum transfer $q_z = 4\pi/\lambda \sin \alpha_i$. The size of the collimated beam on the sample and background rejection (nonspecular or diffuse reflectivity) are controlled by careful setting of slits (S1–S4). (b) X-ray reflectivity scan (before footprint correction) and offset scan (nonspecular background) of a supported bilayer on a silicon wafer, recorded with undulator radiation at beamline ID01/ESRF (France). To avoid detector saturation and to prevent radiation damage, attenuators are successively removed such that only data points for $q_z \geq 1.6$ Å^{-1} are recorded with the full synchrotron beam. In addition, the sample is translated perpendicular to the optical axis to successively expose pristine sample regions, and a fast shutter is used to prevent X-ray exposure when not recording data, for example, during motor movement. If these measures are not taken, radiation damage is likely to occur and can manifest itself in form of a shift of minima with dose, see the region around the second minimum in the inset, in buffer solution. (c) Schematic of the sample chamber, where the sample is placed in buffer solution, which can be exchanged by external pumps. (d) Schematic of the bilayer profile. Adapted from Ref. [17].

Note that $\rho(z)$ is the laterally averaged density profile. $R_F(q_z)$ can be expressed as $|(q_z - q'_z)/(q_z + q'_z)|^2$ with $q'^2_z = q_z^2 - q_c^2$. The critical momentum transfer, below which total external reflection occurs, is given by $q_c = 4\pi/\lambda \sin(\alpha_c) \simeq 4\sqrt{\pi r_0 \Delta \rho_{12}}$, where r_0 denotes the classical electron radius. It is useful to decompose the total EDP into a substrate step function and a bilayer profile $\rho_0(z)$, assuming a thin water layer between the solid surface and the bilayer

$$\rho(z) = (\rho_s - \rho_{\text{water}}) \cdot \text{erf}\left(\frac{z + d_0}{\sqrt{2}\sigma}\right) + \rho_0(z), \quad (2.2)$$

where $(\rho_s - \rho_{\text{water}})$ is the contrast between substrate and water, and the error function is used to describe the interfacial width with root mean square (rms) roughness σ. $\rho_0(z)$ represents the bilayer electron density, the quantity of principal interest here.

Next, we insert the EDP eq. (2.2) in (2.1) to obtain

$$R(q_z) = R_F(q_z) \left| \int \frac{1}{\sqrt{2\pi\sigma^2}} e^{-(1/2)((z+d_0)/\sigma)^2} e^{iq_z z} dz \right.$$

$$\left. + \frac{1}{\Delta\rho_{12}} \int \frac{\partial \rho_0(z)}{\partial z} e^{iq_z z} dz \right|^2 \quad (2.3)$$

$$= R_F(q_z) \left| e^{-iq_z d_0} e^{-q_z^2 \sigma^2/2} + f(q_z) \right|^2$$

$$= R_F(q_z) \left[e^{-q_z^2 \sigma^2} - 2i e^{-q_z^2 \sigma^2/2} \sin(q_z d_0) f(q_z) + |f(q_z)|^2 \right].$$

The first term in the sum represents the reflectivity of the substrate, the second accounts for interference between solid and the bilayer, and the third is the bilayer form factor $|f(q_z)|^2$, that is, the transform of the bilayer profile centered at $z = 0$

$$f(q_z) = \int_{-d/2}^{d/2} \frac{1}{\Delta\rho_{12}} \frac{\partial \rho_0(z)}{\partial z} e^{iq_z z} dz, \quad (2.4)$$

where $\Delta\rho_{12}$ is the contrast of the water and solid and d_0 is the thickness of the water layer plus half the membrane thickness. The bilayer density profile $\rho_0(z)$ is parameterized in terms of the first N_0 Fourier coefficients f_n [32]

$$\rho_0(z) = \rho_{\text{water}} + \Delta_{\max}\Delta\rho_{12} \sum_{n=1}^{N_0} f_n v_n \cos\left(\frac{2\pi n z}{d}\right), \quad (2.5)$$

with f_n and v_n modulus and phase of coefficient n, respectively, and $\Delta_{\max}\Delta\rho_{12}$ a prefactor describing the amplitude of ρ_0 with respect to the solid-water contrast. Note that the zero of ρ_0 is taken to be equal to water, which is a justified approximation for many bilayer profiles. If mirror symmetry of the bilayer can be assumed, the phases v_n are real and equal to ±1. Inserting eq. (2.5) into (2.4) yields

$$f(q_z) = \sum_{n=1}^{N_0} f_n \Delta_{\max} \left[\frac{i 8\pi^2 n^2 \sin(0.5 q_z d)}{q_z^2 d^2 - 4\pi^2 n^2} \cos(n\pi) \right], \quad (2.6)$$

which contains all the parameters to be used in least-squares fitting of an experimental reflectivity curve. Fixing the electron density of water to $0.334 \, e^-/\text{Å}^3$ and of the solid to, for example, $0.699 \, e^-/\text{Å}^3$ for crystalline silicon, all other parameters are fit parameters. Relevant bilayer parameters, such as the phospholipid head group density ρ_h, head-to-head distance d_{hh}, or the density associated with the central minimum corresponding to the terminal methyl groups of the hydrocarbon chains ρ_c, can be easily derived from f_n and v_n [32]. The water layer thickness between solid and the bilayer is $d_w = (d_0 - d_{hh})/2$.

Figure 2.3 shows the reflectivity curves for DOPC/DOTAP, DOPC/DOPS, and DOPC lipid mixtures in the fluid phase, [17], along with the best least-squares fit (solid line). The intensity of specular reflected X-rays for lipid mixtures in the fluid phase was measured up to $q_z \approx 0.6\,\text{Å}^{-1}$ covering eight orders of magnitude in

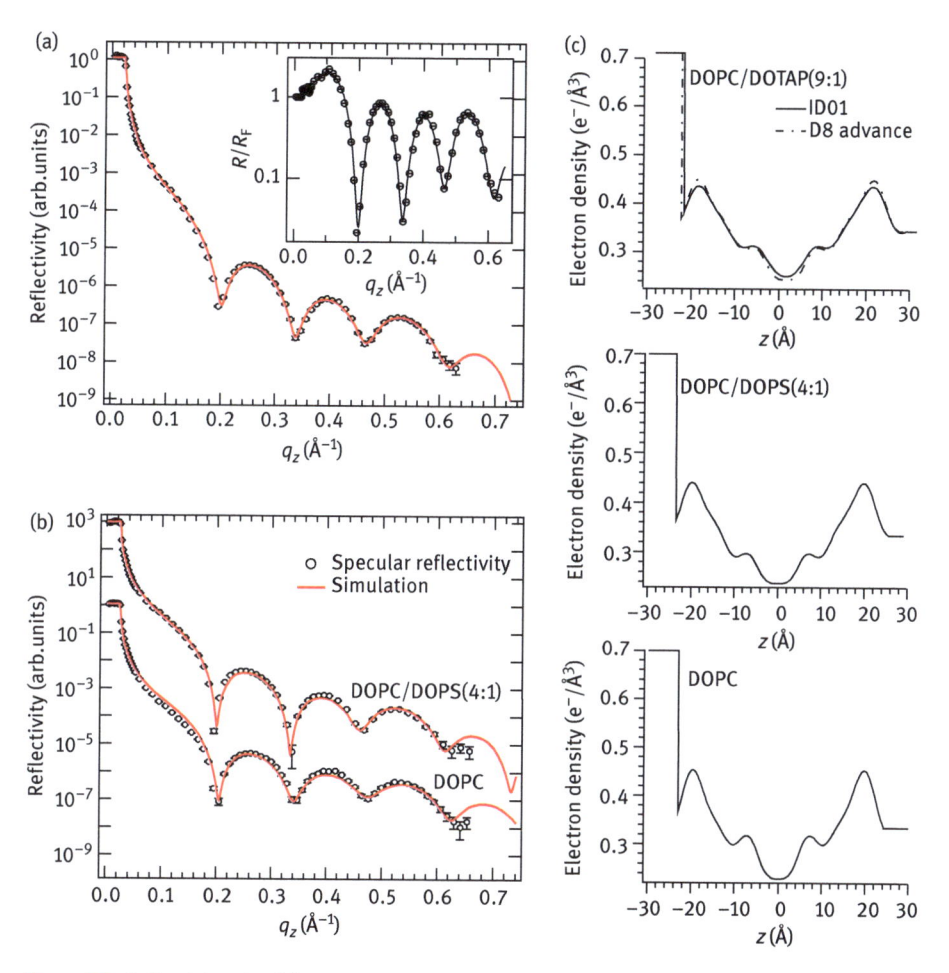

Figure 2.3: Reflectivity of solid-supported lipid bilayers, for zwitterionic, cationic, and anionic mixtures in the fluid phase, recorded with synchrotron radiation (ESRF/ID01) at $T = 23.8\,°C$. The best fit (solid lines) to the reflectivity curves is shown for (a) DOPC/DOTAP(9:1), and (b) DOPC/DOPS(4:1) and DOPC. The inset in (a) shows the reflectivity plotted as R/R_F versus q_z. (c) Electron density profiles corresponding to the fits in (a, b). For DOPC/DOTAP (top), the profile reconstructed from an in-house (sealed-tube) reflectometer is shown for comparison. Adapted from Ref. [17].

intensity, before reaching the background level. The head-to-head distance of the DOPC bilayer was determined to $d_{hh} = 39.6$ Å, which is slightly larger than the value 37.1 Å found for oriented stacks of DOPC [33]. This may be an indication of a substrate induced effect in the headgroup area. The single bilayer result agrees well with the value of 40 Å at 22 °C measured by AFM for solid-supported DOPC [34].

An important asset of X-ray reflectivity is the fact that it gives access to the electron density on absolute scale and thus provides valuable data for comparison with molecular dynamics (MD) simulations [35]. Further illustrative and insightful examples of how X-ray reflectivity is used for structural biophysics are found in Refs. [36–38] for single component lipid bilayers, in Refs. [39, 40] for multi-component bilayers, and in Refs. [41, 42] for solid-supported bilayers with reconstituted membrane proteins. The development of more complex bilayer systems with soft polymer cushions is described in Refs. [43, 44]. Reflectivity from solid-supported membranes can also be used to study the interaction with biomolecules or even organelles such as SVs with lipid membranes [45, 46].

2.2.2 Charged monolayers with counterions as an example for box-model fits

Next, we describe a very powerful and widespread approach to analyze X-ray reflectivity of bilayers and monolayers, or more generally of thin films, by decomposing the density profile $\rho(z)$ in terms of a discrete set of piecewise constant slabs, a so-called box model. We illustrate this approach by means of an example of charged lipid monolayers on silanized solid surfaces in contact with an electrolyte [47]. The goal of this study was to measure the spatial distribution of counterions near a charged surface, for which the classical Poisson–Boltzmann (PB) theory was formulated more than 100 years ago. It describes the interplay between entropy, which favors the delocalization of ions released in solution, and the electrostatic attraction by the surface. It is a cornerstone of modern colloid and biomolecular science. However, a gap has persisted up to quite recently between the very detailed theoretical predictions of the counterion density profile on molecular scales and the experimental capability to observe the actual distribution of ions. Since modern theoretical studies have predicted fundamental deviations from the classical theory, it had become even more relevant to study this problem with many different techniques, as reviewed in Ref. [48], including X-ray reflectivity from charged lipid monolayers. We have selected this example here, since it shows that many interface components have to be correctly captured in terms of corresponding boxes (silanized interface) and free-form distribution (counterion cloud), in order to draw a conclusion on a particular component, here the counterions.

Figure 2.4 illustrates the charged lipid monolayer in contact with its cloud of bromide counterions. The special case studied was at vanishing bulk electrolyte

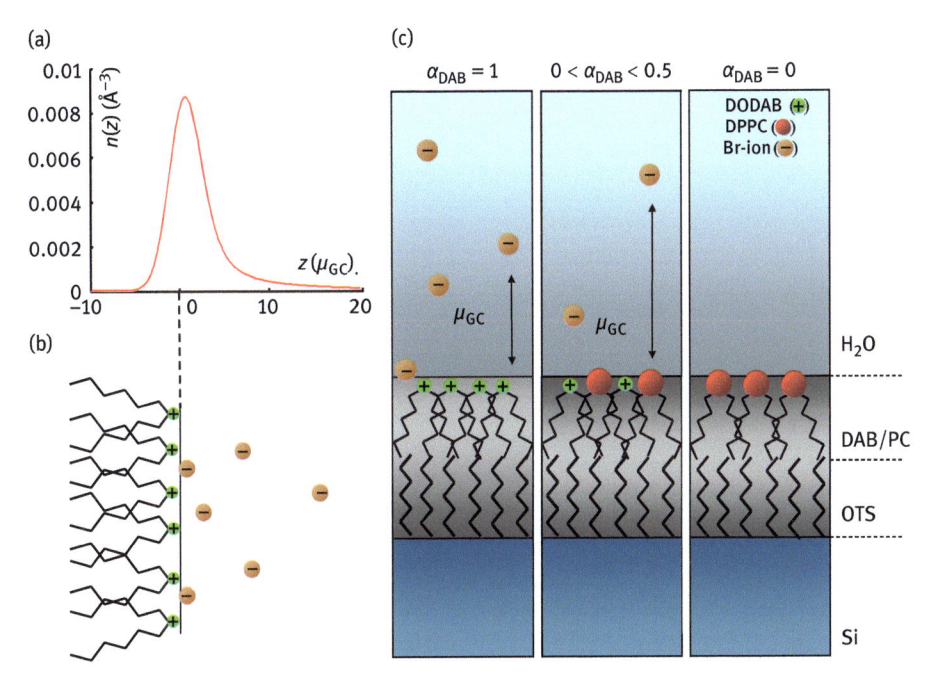

Figure 2.4: (a) Counterion distribution $n(z)$ near a monolayer with surface charge density σ_s. The idealized Poisson–Boltzmann distribution is convolved with a Gaussian of width ξ to account for experimental broadening of the profile. (b) Sketch of a DODAB monolayer with bromide counterions. (c) Sketch of the mixed monolayer on a silanized silicon wafer, for three different charged molar fractions α_{DAB}. The corresponding Gouy–Chapman length $\mu_{GC} = (2\pi l_B \sigma_s)^{-1}$ increases with decreasing surface charge density σ_s and becomes infinite for an uncharged surface. l_B is the Bjerrum length, which is about 7.1 Å in water at room temperature. Adapted from Ref. [47].

concentration, that is, counterions only with no added salt. The interface density profile was probed by X-ray reflectivity, with the surface charge density of the interface as an external control parameter. The silicon wafer was first silanized to render it hydrophobic for subsequent deposition of a lipid monolayer. Lipids were deposited by vesicle fusion with controlled charge density, which was varied by integrating certain amounts of a neutral lipid. The idealized case of a flat, positively charged solid–liquid interface with bromide counterions would result in an algebraic decay of the counterion profile according to the classical PB theory. To properly describe the experimental situation, the PB profile was convolved with a Gaussian to account for interface roughness, thermal fluctuations, and finite experimental resolution.

Figure 2.5a shows the normalized reflectivity curve $\Phi(q_z)$ of a pure DODAB monolayer on silane (octadecyl-tri-silane (OTS)) with bromide counterions. Here, $\Delta \rho_{12}$ corresponds to the density contrast between silicon and water, and ρ is the

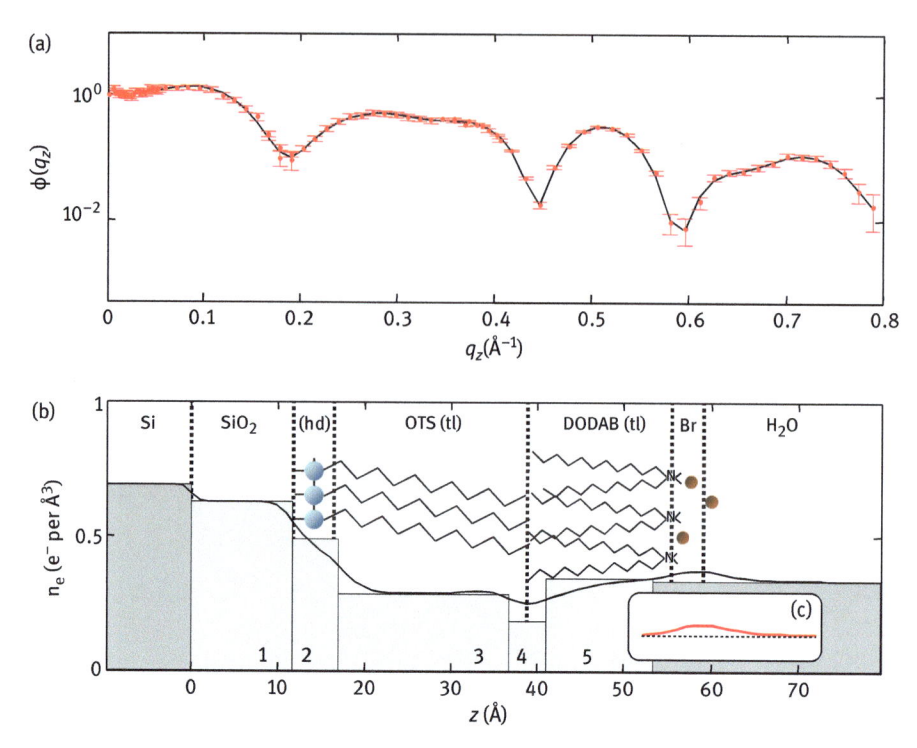

Figure 2.5: (a) Normalized reflectivity curve of a pure DODAB monolayer on OTS, along with the best slab-model fit. (b) Electron density profile along the interface normal z, corresponding to the least-squares fit in (a). The boxes are used to parameterize the silicon, silicon oxide, the silane headgroup, and tails, as well as the DODAB tails and headgroup. The bromide counterion cloud was modeled separately with a Poisson–Boltzmann distribution convolved with a Gaussian, as sketched in (c). Adapted from Ref. [47].

interface density profile of the best box-model fit which is sketched in (b). The reflectivity was measured at the ID01 undulator beamline of the European Synchrotron Radiation Facility (ESRF), Grenoble (France). Reflectivity curves at varied surface charge density were consistently modeled by a classical, convolved PB model. In addition to a model analysis at a single X-ray energy, the photon energy was varied around the bromine K-edge (13.475 keV) to change the scattering power of the bromide counterions, leaving the scattering contributions from all other components of the sample constant [47, 48]. The corresponding small change in $\Phi(q_z)$ confirmed the expected change in the bromide scattering contribution. In summary, no deviation from the classical mean-field theory was observed, down to an accuracy of ca. 3 Å, limited as much by the interfacial roughness as by the experimental resolution. Consequently, no deviations from the classical theory of a counterion cloud were observed within the experimental accuracy.

After this short summary of the monolayer reflectivity experiment, let us consider the underlying description of the density profile $\rho(z)$ for full q-range fits and the data analysis used in Ref. [47] in more detail. In the semi-kinematic framework, the data are modeled in terms of $\Phi(q_z) := R(q_z)/R_f(q_z) = |(1/\Delta\rho_{12})d\rho_e(z)/dz\, e^{iq_z z}dz|^2$. $\Delta\rho_{12}$ is, as in the supported lipid bilayer experiments, the net density contrast of the silicon–water interface. A general box or slab model assumes a density profile consisting of N boxes of densities ρ_n, that is, $N+1$ interfaces at positions z_n of widths (or roughnesses) σ_n. The interface widths ensure smooth transitions between neighboring boxes, since discontinuous steps in the electron density would be unrealistic. The box model parameterization can be written as

$$\rho(z) = \frac{1}{2}\sum_{n=0}^{N}(\rho_{n+1}-\rho_n)\left(\mathrm{erf}\left(\frac{z-z_n}{\sqrt{2\sigma_n^2}}\right)-1\right)+\rho_0, \quad (2.7)$$

where ρ_0 is the substrate density. The derivative entering the reflectivity is thus a sum of Gaussians

$$\frac{\partial\rho}{\partial z} = \sum_{n=0}^{N}\frac{(\rho_{n+1}-\rho_n)}{\sqrt{2\pi\sigma_n^2}}\exp\left(-\frac{(z-z_n)^2}{2\sigma_n^2}\right) \quad (2.8)$$

so that the normalized model reflectivity reads

$$\Phi(q_z) = \left|\frac{1}{\Delta\rho_{12}}\sum_{n=0}^{N}(\rho_{n+1}-\rho_n)\exp\left(-\frac{q_z^2}{2\sigma_n^2}\right)\cdot\exp(-iq_z z_n)\right|^2 \quad (2.9)$$

The parameters entering the model are the substrate roughness σ_0 and for every box the density ρ_j, thickness d_j, and roughness σ_j, so $N_P = 3(N+1)+2$ parameters are needed for N independent boxes. This high number of free parameters complicates the use of conventional least-squares fitting algorithms. Furthermore, this simple model is limited to boxes with the shape of error functions. In particular, the description of the algebraic spatial distribution of counterions requires a more detailed parameterization of the density profile. Therefore, a convolved PB profile was introduced as a modification, added to the box model. For a position z normal to the monolayer, the counterion distribution is

$$n(z) = \frac{\sigma_s}{\mu_{GC}}\frac{1}{(z/\mu_{GC}+1)^2}. \quad (2.10)$$

To take into account roughness, thermal fluctuations, and instrumental resolution effects, it is convolved with a Gaussian of width $\xi \approx 3$ Å [49], leading to

$$n(z) = n'(z-z_0) = \frac{\sigma_s}{\mu_{GC}}\cdot\frac{1}{\sqrt{2\pi\xi^2}}\int_{-\infty}^{z}\frac{\exp(-\tau^2/(2\xi^2))}{(1+(z-\tau)/\mu_{GC})^2}d\tau, \quad (2.11)$$

which enters the box model density profile as a counterion effective electron density $\Delta n_e(z) = Z_{\text{eff}}(E) \cdot n(z)$. Here, $Z_{\text{eff}}(E) = f_{\text{Br}}(E) - v_{\text{Br}}\rho_e^{H_2O}$ is the effective number of scattering bromide counterion electrons. $v_{\text{Br}}\rho_e^{H_2O}$ is the number of bulk water electrons in the bromide ionic volume, using an electron density of water of $\rho_e^{H_2O} = 0.334\,\text{Å}^{-3}$. The scattering factor $f_{\text{Br}}(E) \simeq f'_{\text{Br}}(E) + f_{\text{Br}}^0$ varies by 10–15% upon variation of the X-ray energy around the bromine K-edge from $E_e = 13.475$ keV to $E_o = 13.228$ keV while the scattering of the other components of the sample remains constant. The counterion distribution is described by three fit parameters, the reference position z_0, the Gouy–Chapman length μ_{GC}, and the width ξ. The remaining fit parameters parameterize the boxes for silicon, silicon oxide, the silane headgroup, and tails, as well as the DODAB tails and headgroup.

The box model is "subjective" in the sense that it assumes a certain shape of the density transitions. As an alternative, an "objective" free-form model can be formulated [50, 51], in which the density profile can have virtually any shape. It can be understood as the limit of a box model for $N \to \infty$ and zero interface roughness. In Ref. [47], such an unbiased model was independently compared to the data and confirmed the PB shape of the counterion distribution within the experimental resolution. In the free-form model, the normalized density profile $\bar{\rho} = \rho(z)/\Delta\rho_{12}$ is sampled at $N + 1$ positions $z_j = j\Delta z$ with a sum of Heaviside step functions

$$\bar{\rho} = 1 + \sum_{j=0}^{N} a_j\, H(z - j\Delta z) \tag{2.12}$$

with normalization condition $\sum_{j=0}^{N} a_j = -1$. Typically, $N \approx 100$ and $\Delta z \approx 0.25\, d_r$ are used. $d_r = \pi/q_{\max}$ is the full width at half maximum (FWHM) of the real-space resolution, given by the sampling theorem, which states that the achievable resolution is limited by the maximal scattering vector [38, 51]. The modeled slabs are thus sampled finer than the experimental resolution but the density profile is then convolved with a normalized Gaussian resolution function of rms-width $\varepsilon = 2\sqrt{2\ln 2}\, d_r$ to ensure physically viable smooth transitions. This results in a normalized density profile

$$\bar{\rho}_{\text{real}} = 1 + \frac{1}{2}\sum_{j=0}^{N}\left(a_j \cdot \text{erfc}\left(\frac{j\text{d}z - z}{\sqrt{2\varepsilon^2}}\right)\right). \tag{2.13}$$

The boundary conditions of the density profile $\text{d}\bar{\rho}/\text{d}z(0) = \text{d}\bar{\rho}/\text{d}z(z_N) = 0$ lead to $a_0 = a_N = 0$. The derivative

$$\frac{\text{d}\bar{\rho}_{\text{real}}}{\text{d}z} = \frac{1}{\sqrt{2\pi\varepsilon^2}}\sum_{j=0}^{N} a_j \exp\left(-\frac{(j\text{d}z - z)^2}{2\varepsilon^2}\right) \tag{2.14}$$

enters the normalized model reflectivity

$$\Phi(q_z) = \left| \sum_{j=0}^{N} a_j \exp(-ijdzq) \right|^2 \exp(-q^2\varepsilon^2), \tag{2.15}$$

which is fitted to the experimental curve.

During the least-squares fitting process of the free-form model, the parameter vector $[a_j]_{j=0}^{N}$ is varied with a differential evolution algorithm. This particularly robust algorithm is able to find global minima without getting trapped in local ones by an evolutionary search strategy based on a random starting guess [51, 52]. For multiple runs, one obtains classes of systematically different results due to the intrinsic randomness of the search procedure. A class is defined by the property that averaging all of its profiles yields a reflectivity similar to the ones obtained from all of its parameter vectors. Physically unrealistic classes can be excluded based on comparison with known molecular dimensions or electron densities. The most probable result is then given as the mean of the remaining class and has maximum information entropy [51].

In summary, two models with different assumptions, one being based on slabs on different density, the other on a free-form approach, can be fitted to the experimental data over the full q-range to obtain EDPs on absolute scale. Both approaches have shown independently that the counterion distribution near a lipid monolayer in contact with an electrolyte is in good agreement with a convolved PB distribution.

2.3 Multilamellar membranes

Using lipid self-assembly and careful preparation methods as reviewed in Refs. [5, 6, 32], highly aligned membrane stacks can be achieved, where the distribution of the bilayer normal vector, the so-called mosaicity, is on the order of 0.01°. In this case, very quantitative treatments of specular and nonspecular reflectivity, GID, and RSMs can be employed, which correctly account for total external reflection or refraction phenomena [3, 8, 24]. However, moderate alignment is in general sufficient to distinguish momentum transfer parallel and perpendicular to the bilayer. Aligned multilamellar membranes have even been prepared from biological membranes such as the purple membrane [53, 54]. Multilamellar stacks are particularly well suited to enhance the signal (by a factor of N) for weak scattering signals, resulting for example from in-plane fluid correlations of lipids and proteins [35], peptide pores [41, 55], and stalk structures [56, 57]. Fundamentals of grazing incidence small-angle scattering (GISAXS) in general are found in Ref. [58], and of GID in Ref. [59]. Diffraction from aligned lipid–water systems has been reviewed in Ref. [60]. Apart from phase diagrams and lateral structures in multi-component membranes which can be probed by GISAXS or GID, also thermal fluctuations break lateral translational invariance of the system, resulting in diffuse nonspecular scattering. Aligned membranes can independently yield information on the density profile and the thermal fluctuations, which in the simplest case are described by two elasticity parameters. This is exploited in the fitting approach based on smectic elasticity as presented in Refs. [61, 62]. Independent of any particular model, the correlation functions describing thermal or static membrane displacements can also be directly determined from the diffuse scattering [63]. For thin stacks, however, it becomes important to take into account the boundary effects of the planar surfaces, as shown in Ref. [21], which in our view is the most rigorous treatment of thermal fluctuations in solid-supported bilayers.

Next, we review the fundamentals of multilamellar reflectivity, as illustrated in Figure 2.6, including experimental aspects and analysis. We begin with the case of partial hydration, where fluctuation effects can be neglected. To this end, we follow the development presented in Ref. [41], and use the parameterization of the interface density profile corresponding to a stack of lipid membranes, as defined in Figure 2.6a. Δ_{sub} is the density contrast between the substrate (e.g., silicon or glass) and a microscopic layer of water between substrate and lipid, Δ_{film}, is the contrast between water and the film, that is, the water/lipid mixture. This contrast is very small and its influence on the reflectivity can often be neglected. Finally, Δ_{top} is the contrast between the film and air. All values are given in units of the overall contrast ρ_{12} (solid–air), satisfying $\Delta_{sub} + \Delta_{film} + \Delta_{top} = 1$. For the case of full hydration (bulk water), ρ_{12} is replaced by the water/silicon contrast. Δ_{max} is the prefactor of the bilayer form factor in units of $\Delta\rho_{12}$. With these parameters, the EDP of N bilayers on a solid substrate is written as

$$\rho(z) = \underbrace{(\rho_{Si} - \rho_{water})}_{Si-\text{water interface}} \cdot \text{erf}\left(-\frac{z + d_0}{\sqrt{2}\sigma_0}\right) + \rho_{water} + \sum_{n=0}^{N-1} \rho_0(z - nd)$$
$$+ \underbrace{(\rho_{water} - \rho_{air})}_{\text{water}-\text{air interface}} \cdot \text{erf}\left(-\frac{z - Nd - d_1}{\sqrt{2}\sigma_1}\right). \tag{2.16}$$

The error functions $\text{erf}((z + d_0)/\sqrt{2}\sigma)$ and $\text{erf}((z - Nd - d_1)/\sqrt{2}\sigma_1)$ with roughness parameters σ_0 and σ_1 take into account the intrinsic width and roughness of the solid–water and water–air interface, assuming a thin water layer on top of the stack, facing humid air inside the measurement chamber at partial hydration. Inserting eq. (2.16) in (2.1), we obtain [41]

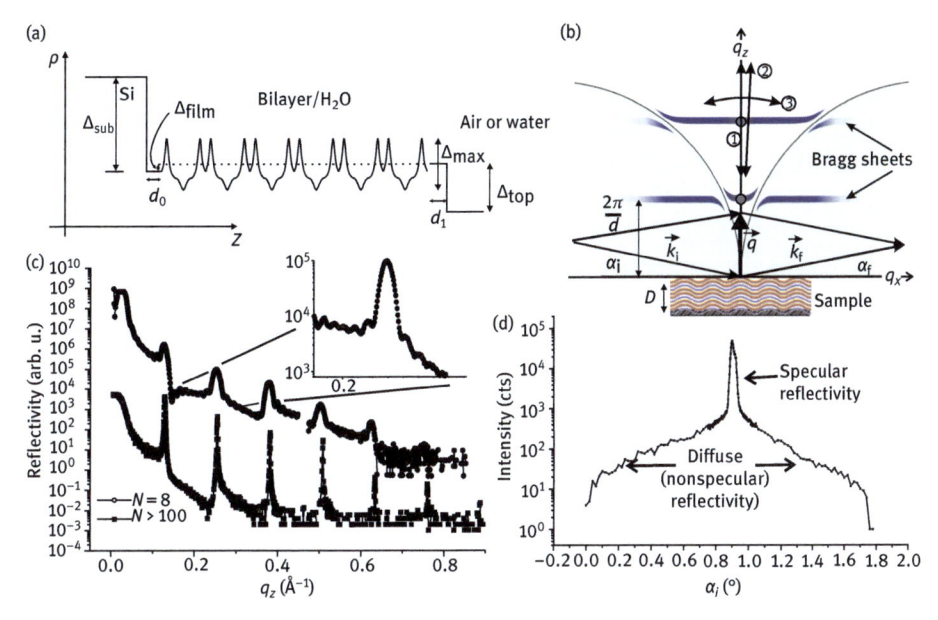

Figure 2.6: Fundamentals of multilamellar reflectivity experiments. (a) Parametrization of the density profile defining the parameters used in the main text. (b) A sketch of reciprocal space with the geometries used to measure specular and diffuse (nonspecular) scattering. The blue stripes represent diffuse Bragg sheets arising from conformal bilayer fluctuations. Conventional scan geometries are indicated as (1) for specular reflectivity, (2) offset scan to subtract the diffuse background, and (3) rocking scan to measure mosaicity and check alignment. Reciprocal space mappings (RSM) are most useful for analysis of the diffuse scattering. In the plane of incidence, refraction effects modify the scattering at the Laue/Bragg boundary where the incidence or exit beam falls below the solid surface [64], and out-of-plane scans are therefore more suitable to cover a large range of parallel momentum transfer [63]. (c) Reflectivity scans of multilamellar DMPC in the fluid L_α-phase, after subtraction of the diffuse background. A thick stack with hundreds of bilayers and a thin so-called oligo-membrane stack with $N = 8$ bilayers are shown for comparison (shifted vertically for clarity). Note the differences in peak width (resolution dominated for the thick sample) and presence of the thickness fringes in the thin sample (see inset). (d) Rocking scan measured at the first Bragg peak of DMPC in the fluid L_α-phase, showing the sharp specular and the broad diffuse scattering component in logarithmic representation. The mosaicity is given by the central width as $0.014°$. If the lateral correlation length of height fluctuations does not diverge as can happen in bulk liquid crystalline phases, the specular and nonspecular contributions are well separated. Adapted from Ref. [32].

$$R(q_z) = R_F(q_z) \left| \Delta_{\text{sub}} \int \frac{1}{\sqrt{2\pi\sigma_0^2}} e^{-(1/2)((z+d_0)/\sigma_0)^2} \cdot e^{iq_z z} dz + \frac{1}{\rho_{12}} \int \sum_{n=0}^{N-1} \frac{d\rho_0(z-nd)}{dz} \cdot e^{iq_z z} dz \right.$$

$$\left. + \Delta_{\text{top}} \int \frac{1}{\sqrt{2\pi\sigma_1^2}} e^{-(1/2)((z-Nd-d_1)/\sigma_1)^2} \cdot e^{iq_z z} dz \right|^2$$

$$= R_F(q_z) \left| \underbrace{\Delta_{\text{sub}} e^{-iq_z d_0} e^{-q_z^2 \sigma_0^2/2}}_{\text{Si}-\text{water}} + \underbrace{f(q_z) s(q_z)}_{\text{lipid bilayer}} + \underbrace{\Delta_{\text{top}} e^{iq_z(Nd+d_1)} e^{-q_z^2 \sigma_1^2/2}}_{\text{water}-\text{air}} \right|^2 , \tag{2.17}$$

with

$$f(q_z) = \int_{-d/2}^{d/2} \frac{d\rho_0(z)}{dz} e^{iq_z z} dz \quad \text{and} \quad s(q_z) = \sum_{n=0}^{N-1} e^{iq_z nd} . \tag{2.18}$$

Here, $f(q_z)$ is the form factor of the bilayer and $s(q_z)$ is the structure factor of the lipid bilayer stack. If thermal fluctuations, which become important at full hydration, are taken into account, the structure factor has to be replaced by

$$s(q_z) = \sum_{m=0}^{N-1} e^{iq_z md} \cdot e^{iq_z u_m}, \tag{2.19}$$

where u_m denote the amplitude of thermal fluctuations for the mth bilayer, followed by ensemble averages in thermal equilibrium [21]. Next, we include an angle dependent absorption term $a(q_z)$ taking into account the path length of the beam in the sample (both incident and exit beams) with the total sample thickness D according to

$$a(q_z, z) = \exp\left(\frac{16\pi^2 \beta(z-D)}{\lambda^2 q_z}\right), \tag{2.20}$$

where β is the imaginary component of the index of refraction and D is the total film thickness.

As in the example of the single membrane above, the bilayer density profile $\rho_{bl}(z)$ can be conveniently parameterized in terms of the first N_o Fourier coefficients f_n,

$$\rho_0(z) = \langle \rho_0 \rangle + \Delta_{\max} \rho_{12} \sum_{n=1}^{N_o} f_n \cos\left(\frac{2\pi n z}{d}\right), \tag{2.21}$$

where $\Delta_{\max}$ is again an amplitude prefactor introduced for convenience to scale the f_n to $f_1 = 1$. Inserting eq. (2.21) into (2.18) again yields

$$f(q_z) = \sum_{n=1}^{N_0} f_n \Delta_{\max} \left[\frac{i\, 8\pi^2 n^2\, \sin(0.5 q_z d)}{q_z^2 d^2 - 4\pi^2 n^2} \cos(n\pi) \right]. \tag{2.22}$$

As described in Refs. [32, 41], it is difficult to describe the width of the Bragg peaks quantitatively. Using the structure factor as above, the simulations always yield sharper peaks than observed experimentally. This can be rescued by modifying the structure factor to account for domain size, for example, as done in Ref. [41]:

$$s(q_z, L) = \sum_{n=0}^{N-1} \exp(inq_z d - nd/L) = \frac{e^{iq_z Nd - Nd/L} - 1}{e^{-iq_z d + d/L} - 1}, \tag{2.23}$$

where L is an effective (mean) domain size. Further, one takes into account small modifications, accounting for inhomogeneous coverage, as described in Ref. [41], $R(q_z) = (1 - x)R_{\text{blank}}(q_z) + xR_{\text{film}}$, where R_{blank} and R_{film} denote the reflectivity of the wafer without and with the lipid film, respectively.

$$R(q_z) = (1-x)R_{\text{blank}}(q_z) + xR_{\text{lipid}}(q_z)$$

$$= (1-x)R_{F_{\text{blank}}}(q_z) e^{-q_z^2 \sigma^2} + xR_{F_{\text{lipid}}}(q_z)$$

$$\cdot \left| \Delta_{\text{sub}} e^{-iq_z d_0} e^{-q_z^2 \sigma_0^2 - q_z^2 \sigma_0^2/2} a(q_z, 0) + f(q_z)s(q_z) + \Delta_{\text{top}} e^{iq_z(Nd+d_1)} e^{-q_z^2 \sigma_1^2/2} \right|^2. \tag{2.24}$$

To account for instrumental resolution, the model reflectivity is convolved with a normalized Gaussian

$$R'(q_z) = R(q_z) \otimes G(q_z) = \int_{-\infty}^{\infty} R(q_z - t) \sqrt{\frac{C}{\pi}} e^{-Ct^2} dt, \tag{2.25}$$

yielding Bragg peaks with a characteristic width $1/\sqrt{2C}$. This full q-range fitting model was used to determine the bilayer density profile $\rho(z)$ of several different multi-component membrane systems, in particular of lipid bilayers containing varied molar ratios of different lipids and the antimicrobial peptides magainin and alamethicin [41]. However, it is only valid for partial hydration, when thermal fluctuations are suppressed, since it does not take into account undulation or compression modes, which become pronounced at full hydration.

In fact, it is long known from bulk systems that multilamellar stacks with smectic symmetry are subject to pronounced undulations and compressional modes [65] which are typically described by classical smectic elasticity theory [66], based on the Hamiltonian

$$\frac{H}{V} = \frac{1}{2} B \left(\frac{\partial u}{\partial z} \right)^2 + \frac{1}{2} K \left(\nabla_{xy}^2 u \right)^2, \tag{2.26}$$

where $u(x, y, z)$ is a continuum displacement field of the membranes with respect to a perfect lattice. B [J/m³] and K [J/m] are the bulk moduli for compression and curvature, respectively. K is related to the bending modulus of a single membrane K_s by $K = K_s/d$. Thermal fluctuations are very pronounced at full hydration, when the compressional modulus B of the stack is small, and less important at partial hydration, when B is high. Furthermore, the solid surface effectively reduces thermal fluctuations (in particular long range undulations), making it possible to get higher resolution profiles $\rho(z)$ than in the bulk, in particular for partially hydrated states [67]. Initially, smectic elasticity was formulated as a continuum model, often with periodic boundary effects, as in the famous Caillé theory [14]. Later, the displacements of individual layers were treated as N discrete functions [21, 68–70].

A suitable way to incorporate the thermal fluctuations was presented by Ref. [21], where the boundary condition at the flat substrate was taken into account by choosing for the fluctuation modes an orthogonal set of eigenfunctions which vanish at the substrate, while the boundary condition at the top was treated by taking the associated surface tension to zero. A complete description of the fluctuation spectrum was obtained, including the dependence of the correlation function on z and on the in-plane distance r.

For specular reflectivity, only the rms fluctuation amplitude σ_n for each bilayer is needed, as pointed out in an earlier review [71], which we follow here. The values of the fluctuation amplitudes are simply determined as [21]

$$\sigma_n = \eta \left(\frac{d}{\pi}\right)^2 \sum_{n=1}^{N} \frac{1}{2n-1} \sin^2\left(\frac{2n-1}{2}\pi\frac{n}{N}\right), \qquad (2.27)$$

where $\eta = \pi\, k_B T 2\, d^2 \sqrt{KB}$ is a dimensionless parameter first introduced by Caillé [14], which quantifies the importance of fluctuations. The most important kind of imperfection in lipid films is often the inhomogeneous coverage, that is, the distribution of the total number of bilayers N on lateral length scales of several micrometers, deriving either from the nonequilibrium deposition process or from an equilibrium dewetting instability [72]. The effect can be modeled by a coverage function for which a convenient analytical form can be chosen as

$$c(n) = \left[1 - \left(\frac{n}{N}\right)^\alpha\right]^2, \qquad (2.28)$$

where α is an empirical parameter controlling the degree of coverage. Dewetted patches are observed in thin oligo-membrane films hydrated from water vapor, as well as in thick films in excess water at high temperatures, where the multilamellar stack partially unbinds from the substrate [73]. In order to determine correct density profiles $\rho(z)$ from the least-squares fits, the effects of thermal and static fluctuations have to be incorporated in the structure factor. Full details on this approach can be found in Refs. [20, 21], where this theory was used to extract precise interaction potentials of oligo-membranes as a function of osmotic pressure. Figure 2.7 illustrates

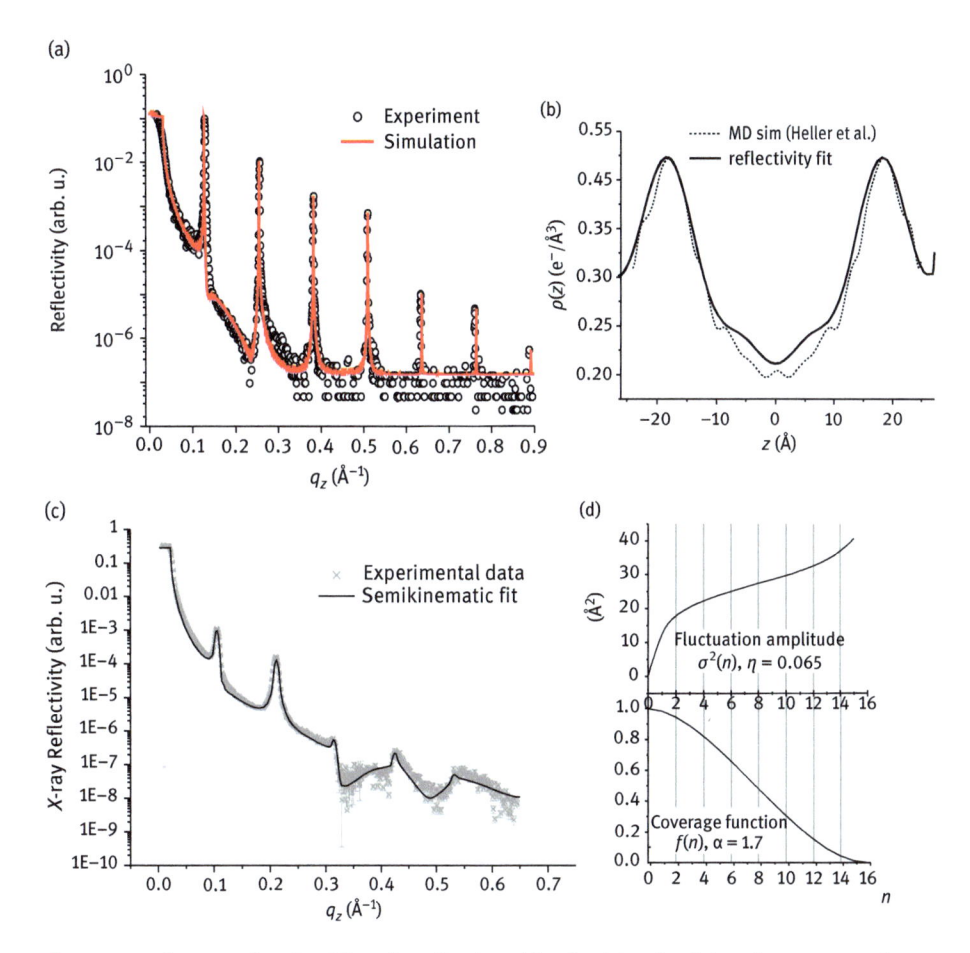

Figure 2.7: Full q-range fits of multilamellar reflectivity. (a) Reflectivity of multilamellar DMPC membranes in the fluid L_α phase at partial hydration ($d = 55$ Å), fitted to a model with five free Fourier components which define the electron density profile on an absolute scale [32]. (b) Full q-range fitting enables a comparison of the density profile (solid line), here of OPPC bilayers, with MD simulations (dotted line) without free scaling parameters [32]. (c) X-ray reflectivity of 16 DMPC bilayers on a silicon substrate measured in an aqueous polyethylene glycol (PEG) solution of 3.6 wt% concentration, at 40 °C in the fluid L_α phase. The continuous line shows a full q-range fit to the data using the smectic structure factor with appropriate boundary conditions [21]. The Bragg peak positions indicate a lamellar periodicity of 59.5 Å. The thermal fluctuations are already quite pronounced at this hydration, which results in a 5 Å thicker water layer. This is already sufficient for thermal fluctuations to suppress the higher order Bragg peaks. This effect can be well understood when evaluating the inter-bilayer interaction potentials [20]. (d) Fluctuation amplitudes $\sigma^2(n)$ (Caillé parameter $\eta = 0.065$), and coverage function $f(n)$ corresponding to the fit in (c). (a,b) Adapted from Ref. [32], and (c, d) from Ref. [21].

the transition from partial hydration to full hydration, that is, from the case where thermal fluctuations can be largely neglected to the case where the reflectivity curve becomes heavily affected by fluctuations. In practice, a lot of applications of multilamellar reflectivity do not even use either of the two approaches and refrain from full q-range fitting. Instead, they use a very old method of determining the density profile by Fourier synthesis from only the lamellar Bragg peak intensities. This should, however, be regarded with caution and only carried out at partial hydration, where the structure factor is less important. This approach incorporates all effects related to the structure factor (fluctuations, defects, and coverage) in an effective form factor, yielding qualitative density profiles, which can be sufficient to monitor relative changes in structure, but are subject to strong systematic aberrations. However, these can to some extent be "calibrated" with full q-range fits [71]. Note that the Fourier synthesis method does offer the advantage of determining the phase by the swelling method, and that it can faithfully reconstruct the bilayer thickness and water layer [57].

Next, we address the nonspecular (diffuse) scattering by presenting an example in which out-of-equilibrium membrane fluctuations have been studied after photoexcitation [74]. This represents a situation where we do not have a proper fluctuation model at hand, and where the analysis therefore has to be based on a model-independent approach as in Ref. [63]. In this pump-probe diffraction study, outlined in Figure 2.8, the nonspecular X-ray scattering from fluorescently labeled phospholipid multi-bilayers was recorded after optical excitation in a time-resolved manner. Bilayer shape fluctuations were monitored in a stack of DOPC bilayers, deposited on a quartz surface in solution. The membranes were excited by a nanosecond laser pulse, matched in wavelength to the absorption band of Texas-red labeled lipids, which were mixed into the membranes, as routinely done for optical fluorescence microscopy. After energy uptake, the system response was probed by well-controlled picosecond X-ray pulses, covering a broad range of time and length scales from the near-molecular to the mesoscopic range. The characteristic diffraction pattern of the membrane stack (nonspecular diffuse scattering) was then recorded in GISAXS geometry as a function of time delay τ, from a few picoseconds to several microseconds. The laser excitation with 1 kHz repetition rate was synchronized to the synchrotron pulses selected by a high speed chopper system at beamline ID09 of the ESRF. From the diffuse (nonspecular) Bragg peaks of the multilamellar system, the evolution of membrane height–height correlation functions was monitored. Pronounced deviations of the collective undulation spectra from thermal equilibrium were observed. In particular, it was found that pulsed laser illumination even at quite moderate peak intensities of about 10^5 W/cm^2 leads to significant changes of the in-plane membrane correlation length by up to 50%, as well as to the excitation of transient conformal undulation modes of a well-defined lateral wavelength. The observed phenomena evolved on nano-to-microsecond timescales

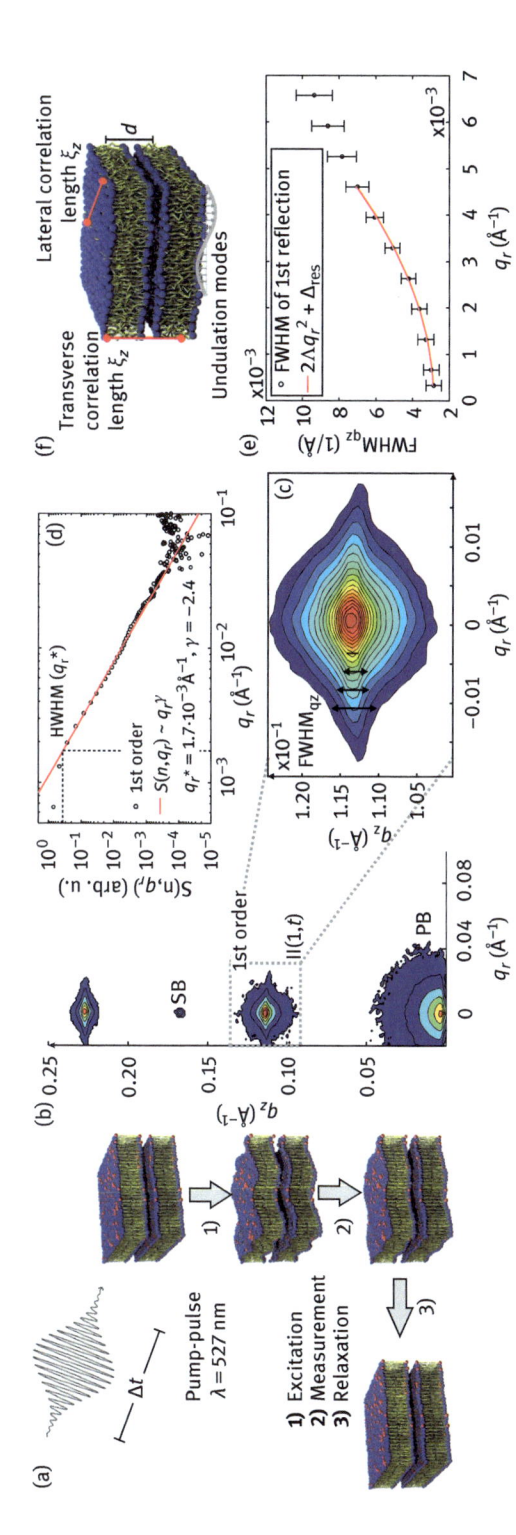

Figure 2.8: Diffuse (nonspecular) reflectivity. (a) Schematic of a laser pump/X-ray probe experiment probing out-of-equilibrium fluctuations of multilamellar lipid membranes. The time evolution of diffuse scattering reflecting membrane undulations is studied in response to a short pulse excitation [74]. (b) Typical lamellar diffraction pattern as recorded by a two-dimensional detector, with primary beam (PB), specular beam (SB), and the two first lamellar diffraction orders. Without any lateral translational symmetry breaking by fluctuations, SB and PB would be the only scattering contributions. The fact that the diffuse scatter forms lamellar Bragg sheets indicates conformal (i.e., highly correlated) undulations. (c) The lineshape analysis of the diffuse lamellar reflections gives access to the lateral and vertical correlation functions, by analysis of (d) the algebraic decay of the scattering intensity with lateral momentum transfer, and by (e) the increase of the vertical Bragg sheet FWHM [7, 63, 73]. (f) Relaxation after photo-excitation was found to involve a characteristic undulation mode (modulation instability), with a lateral wavelength of about 130 nm reaching its maximum about 1 μs after excitation. Adapted from Ref. [74].

after optical excitation and could be modeled in terms of a modulation instability in the lipid multilamellar stack. The energy uptake at molecular level first leads to nonthermalized local vibrations ("local heat"), while the long range undulation modes have not yet taken up any heat. In such a transient state, the lipid molecules tend to expand their intermolecular distances, but on short time scales, this motion is quenched. For compensation, the membranes buckle, opening up a faster relaxation mechanism than the thermal expansion of the entire system. The system hence relaxes by a particular "buckling" pathway, which involves specific undulation modes. The experiments thus showed that local molecular energy uptake by photon absorption can lead to peculiar long range mesoscopic changes in the membrane. It takes microseconds to reach equilibrium again, with an energy dissipation pathway involving a characteristic sequence of modes at length scales of a few hundred nanometers. While this setting of optical excitation is rather artificial, it shows a generic aspect of nonequilibrium effects in membranes: The out-of-equilibrium response to molecular heat uptake need not directly manifest itself on molecular length scales, but rather on mesoscopic scales coupled by collective dynamics.

Finally, we include an example of GID. As we mention above, transitional symmetry is broken not only by height and density fluctuations on small spatial frequencies as probed by GISAXS but also on larger spatial frequencies in the membrane plane, resulting from structure modifications by membrane additives, peptides, pores, or simply the molecular structure of pure lipid membranes, for example, the liquid structure of local acyl chain packing. In aligned bilayers, such signals can be cleanly resolved in reciprocal space, that is, in the $q_z/q_\parallel$ plane without any powder averaging. Figure 2.9 illustrates the experimental aspects of the scattering geometry as well as the basic chain correlation signals, as observed by mapping the reciprocal space for three different bilayer phases, lamellar, hexagonal, and rhombohedral [75]. Note that in the rhombohedral phase as discovered in Ref. [56], lipid stalks form a super-lattice in a matrix of liquid-phase multilamellar membranes. Electron density reconstructions from such phases have provided insight into stalks as important intermediate structure of membrane fusion, which before were postulated by analytical theory and observed by numerical simulations. A wide range of lipids was subsequently screened to study the influence of different lipids on the structure and energetics of stalks [57]. As a result, one can now distinguish stalk promoting and stalk inhibiting lipid mixtures [40]. Figure 2.9 shows how the orientational distribution of acyl chains changes between lamellar, inverse hexagonal, and rhombohedral phases, as measured and quantitatively analyzed in Ref. [75].

The treatment of the 2d intensity patterns described below follows the supporting information of Ref. [75]. The image is transformed to pixel coordinates (u, v) relative to the primary beam, followed by a transformation to angles Ψ and α_f, as defined in Figure 2.9, by

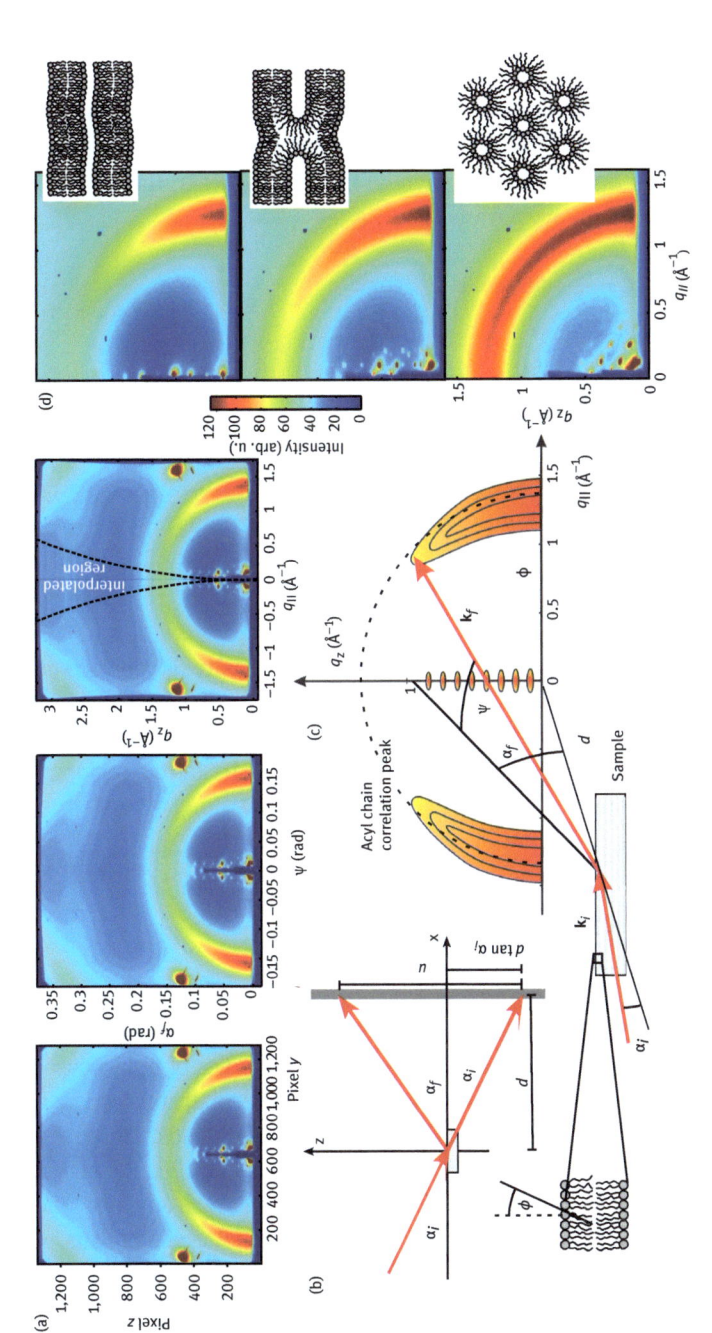

Figure 2.9: Example for reciprocal space maps (RSM) recorded in the scattering geometry of GID from aligned multilamellar membranes in different phase states. (a) The diffraction pattern as measured on the 2d detector (left) is converted to diffraction angles (center), and then to $q_z/q_\parallel$ (right). (b) Parametrization of angles and distances used for the variable transformation. (c) Schematic of a RSM recorded from oriented bilayers, as a function of momentum parallel $q_\parallel$ and perpendicular q_z. (d) Typical intensity distribution for the lamellar (L), the rhombohedral (R), and the inverse hexagonal (H$_{II}$) phase, along with a schematic for each phase. In the SAXS regions, the lamellar peaks and the super-lattice peaks of the R-phase and H$_{II}$ phase are recorded. In the WAXS range, the correlation peak of the lipid tails is observed with a characteristic banana shape. Adapted from Ref. [75].

$$\Psi = \arctan\left(\frac{v}{d}\right) \tag{2.29}$$

$$\alpha_f = \frac{u - d \tan \alpha_i}{\sqrt{d^2 + v^2}} \tag{2.30}$$

α_i is the incident angle of the primary beam on the sample and d is the sample-to-detector distance. A useful tool to determine the position of the 2d detector experimentally from a scattering pattern is provided in Ref. [76]. Next, the image is transformed from angles to momentum transfer ($q_\parallel$, q_z):

$$q_x = \frac{2\pi}{\lambda}(\cos\alpha_f \cos\psi - \cos\alpha_i), \tag{2.31}$$

$$q_y = \frac{2\pi}{\lambda}(\cos\alpha_f \sin\psi), \tag{2.32}$$

$$q_z = \frac{2\pi}{\lambda}(\sin\alpha_f + \sin\alpha_i), \tag{2.33}$$

$$q_\parallel = \sqrt{q_x^2 + q_y^2} \tag{2.34}$$

The coordinate transformation and re-gridding of the intensity is performed numerically. Note that the accessible reciprocal space is restricted due to the scattering geometry with a fixed angle of incidence. In particular, scattering on and close to the q_z-axis is only partially observable. The intensity in this region of the detector images is interpolated from the neighboring regions along a line.

2.4 SAXS from lipid vesicles

SAXS as a technique in general [77–80], and SAXS analysis of lipid vesicles in particular [4], is well covered in literature. Historically, SAXS is the oldest X-ray scattering technique used to analyze lipid phases. In fact, the field began with studies of lipid phases used for phase diagram studies and only later for structural analysis of the bilayer itself, phase diagram studies, see (citation: Cevc, G. (Ed.) Phospholipidshandbook, CRC Press, Boca Raton 1993) for a review, and only later. Here, we focus solely on SUVs. For MLVs, we refer to the work by Pabst *et al.* as well as by Kucerka, who have much advanced the X-ray analysis of MLVs both with X-rays and neutrons as well as joint refinements. A good starting point is Ref. [81], a recent review is provided by Ref. [82].

For SUVs, where the resolution on the bilayer EDP, that is, the equivalent of the profile function $\rho(z)$ introduced in the reflectivity section, is not quite as high as in MLVs, a simple approach based on three Gaussians is often sufficient. More sophisticated and more accurate decompositions of the EDP by several extensions, for example, to five Gaussians, were applied in the analysis [83]. Beyond the Gaussian models, better constrained models have been formulated which parse the EDP of lipids into different molecular components, each with a volume probability function redundant. For phosphatidylcholines, the following decomposition is common, as introduced by Kucerka and Nagle [84]: choline methyl (CholCH$_3$); phosphate + CH$_2$CH$_2$N (PCN); carbonyl + glycerol (CG4); hydrocarbon methylene (CH$_2$); and hydrocarbon terminal methyl (CH$_3$), see also Ref. [85] and references therein. Further, generalizations to asymmetric bilayer profiles have been addressed in Refs. [86, 87]. For a state-of-the-art presentation of least-squares fitting to vesicle data, along with a range of useful models, we refer to Refs. [4, 13].

Figure 2.10 illustrates the fundamentals of SAXS from lipid vesicle suspensions, along with the experimental workflow, including data acquisition and data processing. As sketched in (a), an incident X-ray beam with wave vector $\vec{k}_i$ and wave number $|\vec{k}| = 2\pi/\lambda$ for wavelength λ is scattered from an isotropic suspension of vesicles. The scattered X-rays with wave vector $\vec{k}_j$ and momentum transfer $\vec{q} = \vec{k}_j - \vec{k}_i$ are recorded on the 2d detector. Due to the (incoherent) powder average over the entire distribution, the diffraction pattern is isotropic even if the scattering particles were not and depends only on the scattering angle 2θ, or correspondingly the modulus of the momentum transfer $q = |\vec{q}| = 4\pi/\lambda \sin\theta$. The scattering intensity can be written as

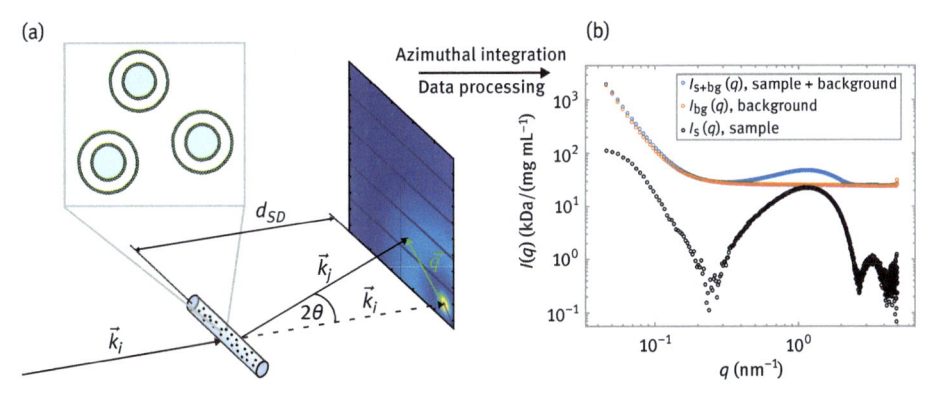

Figure 2.10: Principle of a SAXS experiment on lipid vesicles with exemplary data of a DOPC:DOPE (1:1) vesicle suspension. (a) Schematic of the SAXS geometry and the corresponding powder-averaged diffraction pattern, recorded by a 2d detector at a distance d_{SD} behind the sample. $\vec{q} = \vec{k}_j - \vec{k}_i$ is the scattering vector, where $\vec{k}_i$ and $\vec{k}_j$ are the wave vectors of the incident and the scattered X-ray beam, respectively. (b) Background subtraction $I_s(q) = I_{s+bg}(q) - I_{bg}(q)$ is performed on the azimuthally averaged curves. Adapted from Ref. [88].

$$I(q) = \Delta\rho^2 V_p^2 \langle |f(\vec{q})|^2 |s(\vec{q})|^2 \rangle \sim F(q)S(q), \quad (2.35)$$

where $\langle \cdots \rangle$ denotes the powder average, and we use the factorization of structure factor $S(q) := |s(q)|^2$ and form factor $F(q) := |f(q)|^2$. Form and structure factor amplitudes are denoted in small letters, and the squared quantities which are proportional to intensity in capital letters. $\Delta\rho$ is the difference between the scattering length density of the solvent and the average scattering length density of the particle, and V_p is the volume of the particle. For a sufficiently dilute system of noninteracting particles, one can approximate $S(q) \to 1$. The intensity for a dilute, polydisperse system of particles of radius R with the number size distribution $p(R)$ follows from an incoherent polydispersity integration, assuming a reasonable model for the polydispersity function $p(R)$, and is given by

$$I(q) = \Delta\rho^2 \int_0^\infty p(R) V_p(R)^2 \langle |f(\vec{q}, R)|^2 \rangle dR. \quad (2.36)$$

In general, there are several ways to analyze SAXS data to obtain structural information on the objects, ranging from bead models suitable to solution scattering from proteins [89], decomposition into a pair correlation function, to analytical models, see also Ref. [90] for a review of SAXS analysis. Here, we focus on analytical form and structure factor models derived by appropriate assumptions of the electron density as a common approach in the SAXS analysis of lipid vesicles. To this end, we largely follow approaches introduced in Refs. [12, 81].

2.4.1 SAXS analysis of lipid vesicles

The bilayer EDP, which enters into the form factor amplitude $f(q)$, is described by three Gaussians

$$\rho(z) = \sum_{i=1}^{3} \rho_i \exp\left[-\frac{(z-z_i)^2}{2\sigma_i^2}\right], \quad (2.37)$$

where z is the coordinate normal to the bilayer. The headgroup regions and the hydrophobic chain region are parameterized by an amplitude ρ_i, peak position z_i, and width σ_i of the respective Gaussian.

2.4.1.1 The flat bilayer and the spherical vesicle model

For noninteracting vesicles, we assume $S(q) = 1$. We use two models based on Ref. [12], namely the flat bilayer model, which assumes that the SAXS signal is dominated by the powder-averaged bilayer structure, and a spherical vesicle model, which takes the overall spherical shape of the vesicle into account. In addition to the bilayer structure parameters, the spherical vesicle model contains the mean vesicle radius R_0 and polydispersity σ_R as parameters.

The flat bilayer model assumes that interference between different bilayer patches averages out in polydisperse ensembles of vesicles. The sample can thus be considered a "perfect powder" of flat bilayer patches with random orientations. It is therefore sufficient to calculate the one-dimensional Fourier transform of the electron density $\rho(z)$

$$f_{\text{flat}}(q) = \int p(z)\exp(iqz)dz, \tag{2.38}$$

as the form factor amplitude of the bilayer. As interactions between vesicles are neglected, the scattering intensity is given by $\langle |f_{\text{flat}}(q)|^2 \rangle$ [12]

$$I_{\text{flat}}(q) \propto \frac{1}{q^2} \sum_{i=1}^{N} \sum_{j=1}^{N} p_i p_j \sigma_i \sigma_j \exp\left[-\frac{q^2(\sigma_i^2 + \sigma_j^2)}{2}\right] \cos[q(z_i - z_j)], \tag{2.39}$$

Powder averaging of the one-dimensional Fourier transform is taken into account by a factor q^{-2}. In the spherical vesicle model, the form factor amplitude is calculated as the radially symmetric Fourier transform of the electron density $\rho(z)$

$$f_{\text{sphere}(q)} = \int_0^\infty \rho(r) r^2 \frac{\sin(qr)}{qr} dr. \tag{2.40}$$

To account for the polydispersity of the vesicle suspension, the SAXS intensity is integrated over different radii weighted with a Gaussian distribution around R_0 with variance σ_R^2. As shown by Ref. [12], the result can be written as

$$I_{\text{sphere}}(q) \propto \frac{1}{q^2} \sum_{i=1}^{N} \sum_{j=1}^{N} p_i p_j \sigma_i \sigma_j \exp\left[-\frac{q^2(\sigma_i^2 + \sigma_j^2)}{2}\right] [A_{ij}(q) - B_{ij}(q) + C_{ij}(q)], \tag{2.41}$$

where

$$A_{ij}(q) = [(R_0 + z_i)(R_0 + z_j) + \sigma_R^2] \cos[q(z_i - z_j)], \tag{2.42}$$

$$B_{ij}(q) = \exp(-2q^2\sigma_R^2)[(R_0 + z_i)(R_0 + z_j) + \sigma_R^2 - 4q^2\sigma_R^4] \cos[q(2R_0 + z_i + z_j)], \tag{2.43}$$

and

$$C_{ij}(q) = 2q\sigma_R^2 \exp(-2q^2\sigma_R^2)(2R_0 + z_i + z_j) \sin[q(2R_0 + z_i + z_j)]. \tag{2.44}$$

2.4.1.2 The docking model

As an example of which further structural information can be obtained from vesicle SAXS, beyond the simple EDP, we consider docking of vesicles, which is a relevant intermediate state in membrane fusion. To this end, we briefly review results from Ref. [88], where docking was induced by addition of divalent salts, and the resulting bilayer apposition was studied by SAXS. In addition, soluble N-ethylmaleimide-sensitive-factor attachment receptor (SNARE) proteins were reconstituted into vesicles and studied in view of docking [91] and fusion. According to the flat bilayer model, the form factor amplitude of the electron density of a single bilayer is calculated by the Fourier transform in eq. (2.38), yielding [81]

$$f(q) = \sqrt{2\pi} \left[2\sigma_h \rho_h \exp\left(-\frac{\sigma_h^2 q^2}{2}\right) \cos(qz_h) + \sigma_c \rho_c \exp\left(-\frac{\sigma_c^2 q^2}{2}\right) \right]. \tag{2.45}$$

The first term corresponds to the form factor amplitude of the bilayer headgroups, the second term to the one of the hydrophobic chain regions. At the docking site, two bilayers come in close apposition, which in real space can be described by a convolution of the electron density $\rho(z)$ with the sum of two Dirac delta functions with distance d.

$$s(z) = \delta(z) + \delta(z-d). \tag{2.46}$$

The Fourier transform of the delta function is simply a complex phase shift, thus

$$s(q) = 1 + \exp(-iqd). \tag{2.47}$$

The modulus square of eq. (2.47) yields the structure factor

$$S(q) = 2 + 2\cos(qd). \tag{2.48}$$

The same structure factor would be obtained from a MLV model with N bilayers with periodicity d [81, 85], setting $N = 2$. The scattering intensity $I(q)$ of a mixture of single (undocked) and docked bilayers is obtained by combining eqs. (2.45) and (2.48) and taking the powder average as q^{-2}

$$I(q) \propto \frac{1}{q^2} \left[v_d |f(q)|^2 S(q) + (1 - v_d)|f(q)|^2 \right]. \tag{2.49}$$

Equation (2.49) takes into account a fraction $(1 - v_d)|f(q)|^2$ of undocked bilayers. Note that even if all vesicles dock, $v_d < 1$, corresponding to the fraction of vesicle surface involved in adhesion.

With this simple model curve $I_{\text{mod}}(q)$ at hand, the experimental scattering intensities $I_{\text{exp}}(q_i)$ at sampling points q_i with $i = 1, \ldots, N$ were fitted, after accounting for scaling factor and background as

$$I_{tot}(q) = c_1 \cdot I_{mod}(q) + I_{bg}(q) \,. \tag{2.50}$$

Standard least-squares fitting was used to minimize the reduced χ^2-function $\chi^2_{red} = 1/(N-p-1) \sum_{i=1}^{N} (I_{exp}(q_i) - I_{tot}(q_i))^2 / \sigma_i^2$, where p is the number of free model parameters and σ_i^2 is the variance of the intensity $I_{exp}(q_i)$. In Ref. [88], the following parametrization and notation of the EDP was used, both for symmetric and asymmetric EDPs. The Gaussian parameters representing the headgroups are denoted as $\sigma_h = \sigma_{h1} = \sigma_{h2}$ and $\rho_h = \rho_{h1} = \rho_{h2}$, or as $\sigma_{h1} \neq \sigma_{h2}$ and $\rho_{h1} \neq \rho_{h2}$, constrained to a symmetric, or an asymmetric EDP model, respectively. The width σ_c of the Gaussian representing the chain region was a free parameter; the amplitude and the position were set to $\rho_c = -1$ (arb. units) and $z_c = 0$ (nm), respectively. Additional fit parameters were the positions of the two outer Gaussians denoted as $z_{h1,2} = \pm z_h$. As shown in Figure 2.11, an asymmetric bilayer model (with increased number of parameters) results in a decrease of χ^2, but already small uncertainties for example in the experimental background subtraction can lead to a situation where asymmetric EDPs are falsely favored. Therefore, it remains a challenge to disentangle these effects and to elaborate the intrinsic curvature induced asymmetry expected at small R_0. Contrarily, evidence for docking from SAXS curves is quite robust (Figure 2.12), as it manifests itself by intensity modulations of the bilayer form factor.

2.4.2 SAXS analysis of synaptic vesicles

In this section, we address SAXS analysis of synaptic vesicles (SVs) which are cell organelles involved in synaptic neurotransmission. This example demonstrates the potential of SAXS to investigate more complex systems beyond pure lipid vesicles. SVs are heterogeneous organelles varying in size with densely packed proteins facing both toward the interior and the exterior of the vesicles (Figure 2.13a). The SAXS model for SVs sketched in Figure 2.13b is built based on a radial EDP $\rho(r)$ for the lipid bilayer, modeled by three Gaussians, and the inner and outer protein layers, modeled as Gaussian chains. Importantly, the Gaussian chains break the spherical symmetry as they effectively model distinct protein patches characterized by the radius of gyration R_g and the effective number of protein patches N_c. Transmembrane proteins and amino acid residues associated with the headgroups are included in the bilayer contribution. Here, we briefly review the SAXS analysis of SVs, following [92, 93].

The excess scattering length density of the bilayer profile is given by

$$\rho(r) = \sum_i \rho_i \exp\left(-\frac{(r-R_i)^2}{2t_i^2}\right), \tag{2.51}$$

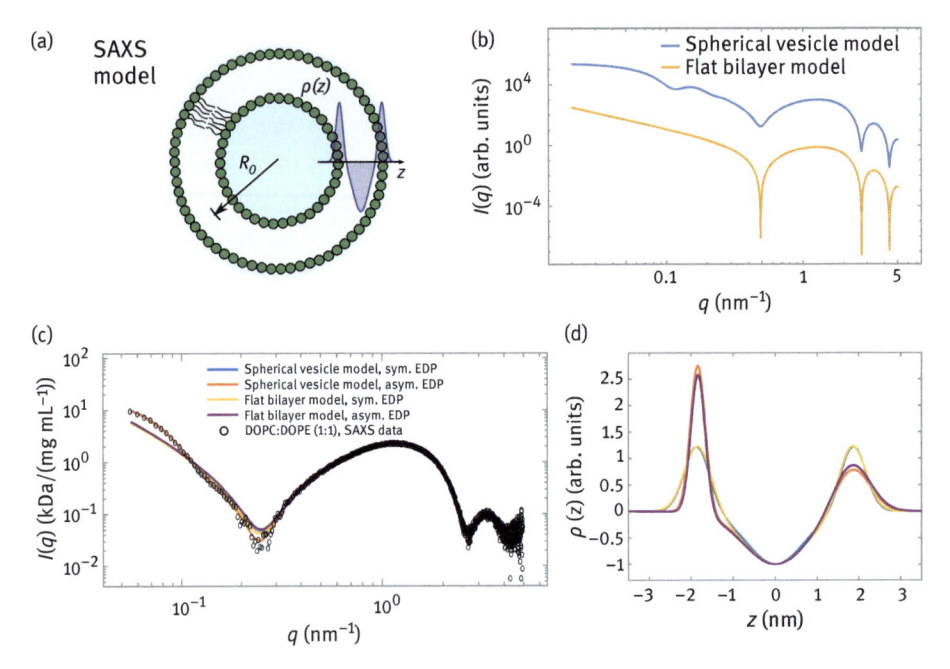

Figure 2.11: SAXS analysis of DOPC:DOPE (1:1) vesicles extruded through 30 nm membranes. (a) Sketch of the spherical vesicle model for a lipid vesicle with radius R_0. The model uses a radial EDP $\rho(r)$ modeled by three Gaussians. (b) Comparison of the spherical vesicle model (blue line) and the flat bilayer model (orange line). (c) Spherical vesicle and flat bilayer model fits to SAXS data of DOPC:DOPE (1:1) vesicles, both for a symmetric and an asymmetric EDP. (d) Reconstructed EDPs from the least-squares fits in (c). Further, from the spherical vesicle model fit, the mean radius $R_0 = 14.35$ nm and the width of the size distribution $\sigma_R = 7.4$ nm were obtained, given here in the case of a symmetric EDP model. The results show that the parameter values deduced from SAXS fitting are often found to be highly model dependent and that unreasonably high asymmetries in the EDP are easily encountered, in case the EDP is not constrained to a symmetric case. This difficulty can screen the intrinsic (small) asymmetry between inner and outer leaflet expected for high curvature. Therefore, careful assessment of the models used, and the experimental workflow, in particular the background subtraction, is required. Adapted from Ref. [88].

with peak position R_i, amplitude ρ_i, and width t_i, $i \in \{\text{in, out, tail}\}$, for each of the Gaussians representing the headgroups and the tail region. $t_{\text{in}} = t_{\text{out}}$ is chosen to describe a symmetric bilayer. The characteristic radius R of the outer bilayer surface is defined as $R = R_{\text{out}} + t_{\text{out}}\sqrt{2\pi}/2$. $R_{\text{tail}} = R - ((t_{\text{out}} + t_{\text{tail}})/2)\sqrt{2\pi}$ and $R_{\text{in}} = R - ((t_{\text{out}} + t_{\text{tail}} + t_{\text{in}})/2)\sqrt{2\pi}$ are expressed in terms of the bilayer characteristics to reduce the number of model parameters. The thickness of the bilayer is characterized by $D = \sqrt{2\pi}(t_{\text{in}} + t_{\text{tail}} + t_{\text{out}})$. Its total excess scattering length with respect to the aqueous buffer is β_b. There are N_c^{in} and N_c^{out} Gaussian chains distributed randomly and without

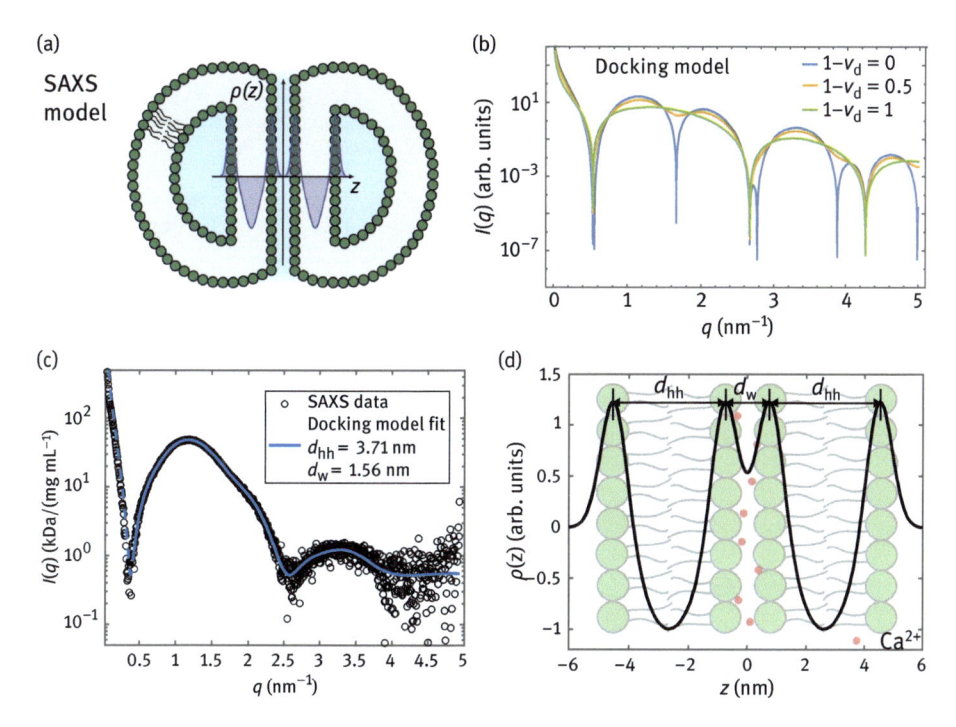

Figure 2.12: SAXS analysis of docked DOPC:DOPS (1:1) vesicles induced by Ca^{2+}. (a) Sketch of the docking model. (b) Simulations of the docking model for different values of v_d. The limiting cases are given by $1-v_d = 1$ and $1-v_d = 0$, corresponding to no docking or (unphysical) docking of the entire vesicle surface. (c) Least-squares fit to SAXS data using the docking model. (d) Reconstructed EDP from the fit in (c). The most important structural parameter deduced from these fits is the width of the water layer in between the two bilayers. The values were found to be in quantitative agreement with predictions of the strong coupling regime beyond mean-field Poisson–Boltzmann theory. Adapted from Ref. [88].

correlations forming the inner and outer protein shell. The individual Gaussian chains are characterized by their radii of gyration, R_g^{in} and R_g^{out}, and their common average excess scattering length density ρ_c. The distance between the inner headgroup and the center of mass of the Gaussian chains facing the lumen is $t_{in}\sqrt{2\pi}/2 + R_g^{in}$, and the distance between the outer headgroup and the center of mass of the Gaussian chains facing outward is $t_{out}\sqrt{2\pi}/2 + R_g^{out}$. The Gaussian chains do therefore not fully penetrate the bilayer although they partly overlap with the extending tails of the bilayer profile [95]. A combination of these results leads to the following form factor:

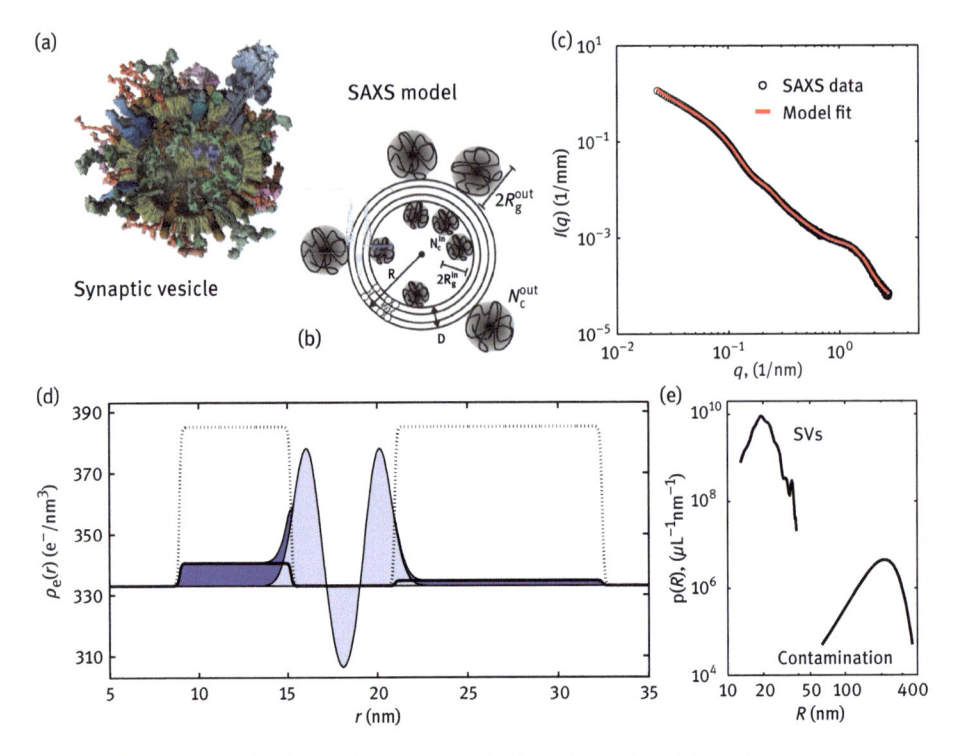

Figure 2.13: SAXS analysis of synaptic vesicles (SVs). (a) Biochemical model of SVs based on the stoichiometry of lipids and proteins [94]. (b) SAXS model with bilayer and Gaussian chains, describing the protein shells. (c) SAXS curve (black line) and least-squares model fit (red line). (d) Resulting electron density profile of the bilayer, with mean (solid) and local (dashed) densities of the protein layers. (e) The polydispersity curve of SVs was measured by cryo-electron microscopy. A fraction of larger aggregates had to be included to obtain agreement between data and model fit. Note that the SAXS curve, the EDP, and the size distributions are given in absolute units. Adapted from Ref. [92].

$$F(q, R) = \frac{1}{M^2} \times \left[\beta_b^2 F_b^2(q, R) + \sum_{i=\text{in, out}} N_c^i \beta_c^{i\,2} P_c^i(q) \right.$$

$$+ \sum_{i=\text{in, out}} 2N_c^{i\,2} \beta_b \beta_c^i S_{b\,c}^i(q, R) + \sum_{i=\text{in, out}} N_c^i (N_c^i - 1) \beta_c^{i\,2} S_c^i(q, R) \quad (2.52)$$

$$\left. + S_c^{\text{in out}}(q, R) \prod_{i=\text{in, out}} N_c^i \beta_c^i \right].$$

The five terms are now described one by one: $M = \beta_b + N_c^{\text{in}} \beta_c^{\text{in}} + N_c^{\text{out}} \beta_c^{\text{out}}$ denotes the excess scattering length, with $\beta_c^i = (4\pi/3) R_g^{i\,3} \rho_c$ the total excess scattering length of a single Gaussian chain in the modeled protein layer and with $i =$ in, out, as in all following equations. The first term contains the normalized amplitude of the self-correlation of the bilayer profile, given by

$$F_b(q, R) = \sum_{i = \text{in, tail, out}} \frac{F_{bi}(q, R_i)}{M_{bi}}, \qquad (2.53)$$

with

$$F_{bi}(q, R_i) = 4\sqrt{2} t_i \rho_i \exp\left(-\frac{t_i^2 q^2}{2}\right) q^{-1} \left[t_i^2 q \cos(qR_i) + R_i \sin(qR_i)\right]. \qquad (2.54)$$

$M_{bi} = \rho_i(4\pi/3)\left((R_i + t_i\sqrt{2\pi}/2)^3 - (R_i - t_i\sqrt{2\pi}/2)^3\right)$ is the excess scattering length of one peak of the bilayer profile [96]. The second term in the form factor describes the self-correlation terms of the Gaussian chains:

$$P_c^i(q) = \frac{2\left[\exp(-x^i) - 1 + x^i\right]}{x^{i2}}, \qquad (2.55)$$

with $x^i = q^2 R_g^{i2}$. The third term accounts for the interference cross-terms $S_{bc}^{\text{in}}(q, R)$ and $S_{bc}^{\text{out}}(q, R)$ between the bilayer and the Gaussian chains, given by

$$S_{bc}^i(q, R) = F_b(q, R)\psi^i(x^i) \frac{\sin\left(q\left[R_{\text{tail}} \mp \left(D/2 + R_g^i\right)\right]\right)}{q\left[R_{\text{tail}} \mp \left(D/2 + R_g^i\right)\right]}, \qquad (2.56)$$

where $\psi^i(x^i) = [1 - \exp(-x^i)]/x^i$ the effective form factor amplitude of the Gaussian chains [103]. Finally, the interference of chains inside and outside the bilayer is described by the fourth term

$$S_c^i(q, R) = \left[\psi^i(x^i) \frac{\sin\left(q\left[R_{\text{tail}} \mp \left(D/2 + R_g^i\right)\right]\right)}{q\left[R_{\text{tail}} \mp \left(D/2 + R_g^i\right)\right]}\right]^2, \qquad (2.57)$$

and the interference between the chains of the inner and outer shells across the bilayer is taken into account by the fifth term

$$S_c^{\text{in, out}}(q, R) = \prod_{i = \text{in, out}} \psi^i(x^i) \frac{\sin\left(q\left[R_{\text{tail}} \mp \left(D/2 + R_g^i\right)\right]\right)}{q\left[R_{\text{tail}} \mp \left(D/2 + R_g^i\right)\right]}. \qquad (2.58)$$

Rather strong approximations were made in the derivation of the last two terms, see Ref. [93], and have to be re-checked. However, the model was found to agree well with the experimental SAXS data, as shown in Figure 2.13c. The reconstructed radial EDP and the bimodal size distribution are given in absolute units Figure 2.13d, e. The size distribution of the SVs was fixed by data obtained from cryo-electron microscopy. A second size distribution describing a fraction of larger membranous particles was included.

2.5 Outlook

We have started this chapter by considering X-ray reflectivity from a single model bilayer and have ended with SAXS analysis of an entire membrane organelle. Different experimental approaches appear to be in opposition: model bilayer versus biological membrane, solid-supported membranes versus isotropic suspension, full hydration versus partial hydration, and single bilayer versus multilamellar stack. Distinct choices in the model system and techniques are made with regard to the specific question. We do not want to close the chapter without pointing out the interest of hybrid approaches. Combining different elements of the above, it can be advantageous, for example, to prepare a model bilayer system with full stoichiometric control and to use it in probing the interaction with biological membranes, for example, SVs in interaction with monolayers or bilayers [45, 46], or fusion experiments involving carefully designed proteo-liposomes and SVs. Many more developments are ongoing and further progress in X-ray methods may find their applications in membrane research, see a list of keywords in Table 2.2, and a few remarks below.

Table 2.2: *Brainstorming* list with keywords of experimental innovations and possible applications, where they may gain impact.

Experimental innovation	Possible impact
X-ray focusing	Biological membranes in cells and tissues
Highly brilliant undulator sources	Diffractive imaging of membranes
FEL peak brilliance	Single vesicle imaging
	Diffract and destroy principle
FEL pump-probe	Ultra-fast structural dynamics of membranes
XPCS	Slow dynamics of membranes
X-ray compatible microfluidics	Reduction in sample volume
	Mitigation of radiation damage

XPCS: X-ray photon correlation spectroscopy.

2.5.1 Diffraction and imaging

All of the X-ray techniques discussed here are based on macroscopic beams. These are well suited for diffraction experiments, which typically are used to probe the averaged structure of a large ensemble of identical scattering objects, for example, $\simeq 10^3$ membranes in a stack or $\simeq 10^{12}$ vesicles in a capillary. These requirements are not very well suited for studies of biological membranes in cells or tissues, which are characterized by local heterogeneities and hierarchical structures. Novel capabilities in X-ray optics and focusing have now overcome these long standing limitations and can complement reciprocal space with real space resolution. By scanning SAXS, for example, myelin structures have been mapped in nerve tissue, and isolated phases of hexatic order were observed in the predominantly lamellar structure [97]. X-ray propagation imaging and phase retrieval have been used to locally resolve thickness, density, and more generally the density profile of membranes bulging over an aperture [98, 99]. In this way, local deviations in the density profiles became accessible.

Finally, we want to briefly consider the opportunities for research on membrane structure and dynamics opened up by the advent of coherent X-ray imaging. Progress in X-ray sources provides highly brilliant X-ray radiation which enables entirely new approaches in X-ray structure analysis. The use of ultra-short fully coherent X-ray pulses at X-ray free electron lasers (FELs), for example, has been proposed for single particle imaging [100] and has already been demonstrated for viruses. In future, FEL sources could also find applications in the analysis of vesicles and membrane assemblies, injected sequentially into an FEL beam by a liquid jet system. In Figure 2.14, we show simulations and a schematic of possible future experiments for the investigation of single membrane organelles and docking or fusion events of lipid vesicles.

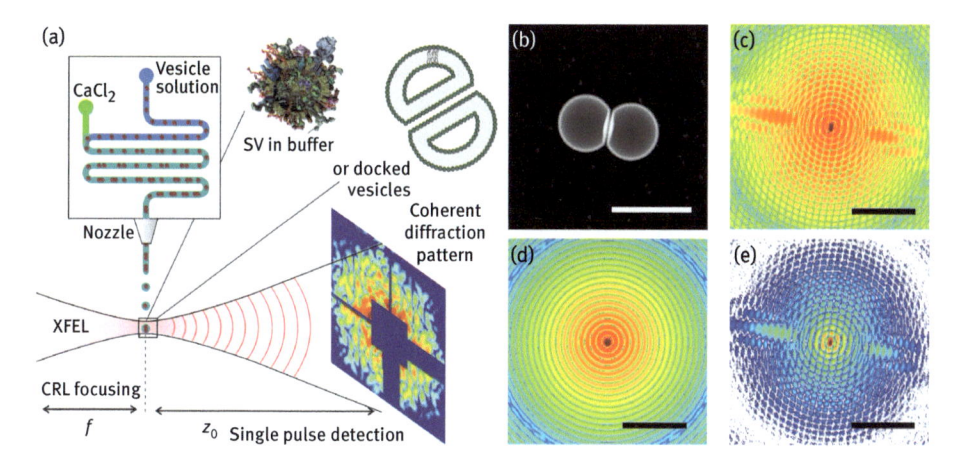

Figure 2.14: (a) Schematic overview of an XFEL vesicle imaging experiment. Synaptic vesicles or pure lipid vesicles with salt solution, to induce docking or fusion, are injected sequentially into a focused XFEL beam, composed of a pulse train of femtosecond pulses. Single pulse coherent diffraction patterns of individual objects are recorded on a detector. (b) Simulated projection of the electron density of two docked vesicles. Scale bar: 100 nm. (c) The corresponding simulated diffraction pattern shows periodic modulations perpendicular to the docking site, potentially allowing an in-silico selection of vesicle docking or fusion events from the detector images. Simulated with the Condor software package [101] for 6 keV photon energy, 1 mJ pulse energy, 100 nm focus size. (d) Simulated diffraction pattern of a single vesicle before docking. The scattering characteristic features of docking are absent. (e) Same pattern as in (c), simulated including Poisson noise. Recording thousands of such shots, complete histograms of structural quantities could be generated. Note that simulations assumed the scattering length density of lipid vesicles but did not take into account the solvent (excess scattering length) so that the real expected signal will be correspondingly weaker. Scale bars: 1 nm^{-1}.

Further, short FEL pulses can be used to not only probe the structure but also dynamics of membranes by pump-probe techniques, in particular of membranes driven out-of-equilibrium effects. However, first FEL data from oriented multi-bilayers have also indicated considerable technical challenges associated with the stochastic nature of single FEL pulses generated by self-amplified stimulated emission [102]. Finally, with all due respect for X-ray methods, there is always much more to be learned from the combination with other techniques. From elastic and inelastic neutron scattering to NMR spectroscopy, from fluorescence correlation spectroscopy to super-resolution optical microscopy, from cryo-electron microscopy to electron tomography, impressive progress is being made, and correlative and complementary use of techniques seems increasingly promising and important. Beyond other experiments, it seems, however, that the combination of X-ray diffraction with numerical simulations and especially MD simulations [35] is particularly synergistic.

Acknowledgments

We thank all of our collaborators on the original work reviewed here, in particular Sebastian Aeffner, Simon Castorph, Doru Constantin, Sajal Ghosh, Klaus Giewekemeyer, Chenghao Li, Ulrike Mennicke, Eva Nováková, Tobias Reusch, Annalena Salditt, Britta Weinhausen, and Yihui Xu (in alphabetical order). We are thankful to the ESRF (Grenoble, France), DESY Photon Science (Hamburg, Germany), and the Swiss Light Source (SLS, Villigen, Switzerland) for generous beam time allocation. Finally, we are grateful for very enjoyable collaborations with Reinhard Jahn and Marcus Müller. Financial support by the German Science Foundation through grants SFB 937/Project A7 and SFB 803/Project B1 is gratefully acknowledged.

References

[1] Luzzati, V., Reiss-Husson, F., Rivas, E., and Gulik-Krzywicki, T., Structure and polymorphism in lipid-water systems, and their possible biological implications. Ann. N.Y. Acad. Sci. 137, 409–413 (1966).
[2] Luzzati, V., Tardieu, A., Gulik-Krzywicki, T., Rivas, E., and Reiss-Husson, F., Structure of the cubic phases of lipid-water systems. Nature 220, 485– (1968).
[3] Als-Nielsen, J., *Elements of Modern X-Ray Physics* (Wiley, Chichester 2011).
[4] Heberle, F. A., Pan, J., Standaert, R. F., Drazba, P., Kucerka, N., and Katsaras, J., Model-based approaches for the determination of lipid bilayer structure from small-angle neutron and x-ray scattering data. Eur. Biophys. J. 41, 875–890 (2012).
[5] Salditt, T., Lipid-peptide interaction in oriented bilayers probed by interface-sensitive scattering methods . Curr. Opin. Struct. Biol. 13, 467 (2003).
[6] Salditt, T. and Brotons, G., Biomolecular and amphiphilic films probed by surface sensitive x-ray and neutron scattering. Anal. Bioanal. Chem. 379, 960–973 (2004).

[7] Salditt, T., Thermal fluctuations and stability of solid-supported lipid membranes. J. Phys.: Condens. Matter 17, R287–R314 (2005).
[8] Tolan, M., *X-ray scattering from soft-matter thin films: Materials science and basic research* (Springer-Verlag: Berlin/Heidelberg/New York, 1999).
[9] Daillant, J. and Alba, M., High-resolution x-ray scattering measurements: I. surfaces. Reports on Progress in Physics 63, 1725 (2000).
[10] Renaud, G., Lazzari, R., and Leroy, F., Probing surface and interface morphology with grazing incidence small angle x-ray scattering. Surface Science Reports 64, 255–380 (2009).
[11] Miller, C., Majewski, J., Watkins, E., Mulder, D., Gog, T., and Kuhl, T., Probing the local order of single phospholipid membranes using grazing incidence x-ray diffraction. Phys. Rev. Lett. 100, 058103 (2008).
[12] Brzustowicz, M.R. and Brunger, A.T., X-ray scattering from unilamellar lipid vesicles. J. Appl. Cryst. 38, 126–131 (2005).
[13] Székely, P., Ginsburg, A., Ben-Nun, T., and Raviv, U., Solution x-ray scattering form factors of supramolecular self-assembled structures. Langmuir26, 13110–13129 (2010).
[14] Caille, A., Remarques sur la diffusion des rayons x dans les smectiques. CR Acad. Sci. Serie B. 274, 891–893 (1972).
[15] Safinya, C., Roux, D., Smith, G., Sinha, S., Dimon, P., Clark, N., and Bellocq, A., Steric interactions in a model multimembrane system: A synchrotron study. Phys. Rev. Lett. 57, 2718 (1986).
[16] Zhang, R., Tristram-Nagle, S., Sun, W., Headrick, R., Irving, T., Suter, R., and Nagle, J., Small-angle x-ray scattering from lipid bilayers is well described by modified Caillé theory but not by paracrystalline theory. Biophys. J. 70, 349–357 (1996).
[17] Novakova, E., Giewekemeyer, K., and Salditt, T., Structure of two-component lipid membranes on solid support: An x-ray reflectivity study. Phys. Rev. E. 74, 051911 (2006).
[18] Hohage, T., Giewekemeyer, K., and Salditt, T., Iterative reconstruction of a refractive-index profile from x-ray or neutron reflectivity measurements. Phys. Rev. E. 77, 051604 (2008).
[19] Sackmann, E., Supported membranes: Scientific and practical applications. Science. 271, 43 (1996).
[20] Mennicke, U., Constantin, D., and Salditt, T., Structure and interaction potentials in solid-supported lipid membranes studied by x-ray reflectivity at varied osmotic pressure. Eur Phys. J. E. 20, 221–230 (2006).
[21] Constantin, D., Mennicke, U., Li, C., and Salditt, T., Solid-supported lipid multilayers: Structure factor and fluctuations. Eur. Phys. J. E. 12, 283–290 (2003).
[22] Russell, T., X-ray and neutron reflectivity for the investigation of polymers. Materials Science Reports. 5, 171–271 (1990).
[23] Pietsch, U., Holy, V., and Baumbach, T., *High-resolution x-ray scattering: From thin films to lateral nanostructures* (Springer-Verlag New York, LLC, 2004).
[24] Daillant, J. and Gibaud, A., eds., *X-ray and neutron reflectivity: Principles and applications* (Springer-Verlag: Berlin/Heidelberg, 2009).
[25] Als-Nielsen, J., Jacquemain, D., Kjaer, K., Leveiller, F., Lahav, M., and Leiserowitz, L., Principles and applications of grazing incidence x-ray and neutron scattering from ordered molecular monolayers at the air-water interface. Physics Reports. 246, 251–313 (1994).
[26] Kaganer, V. M., Möhwald, H., and Dutta, P., Structure and phase transitions in Langmuir monolayers. Rev. Mod. Phys. 71, 779–819 (1999).
[27] Smith, G. S. and Majewski, J., X-ray and neutron scattering studies of lipid monolayers and single bilayers," in *Lipid bilayers: Structure and interactions*, Katsaras, J. and Gutberlet, T., eds. (Springer: Berlin/Heidelberg, 2001), pp. 127–147.

[28] Watts, T. H., Brian, A. A., Kappler, J. W., Marrack, P., and McConnell, H. M., Antigen presentation by supported planar membranes containing affinity-purified I-A. Proc Natl. Acad. Sci. U.S.A. 81, 7564 (1984).
[29] Richter, R. P., Bérat, R., and Brisson, A. R., Formation of solid-supported lipid bilayers: An integrated view Langmuir. 22, 3497–3505 (2006).
[30] Reviakine, I. and Brisson, A., Formation of supported phospholipid bilayers from unilamellar vesicles investigated by atomic force microscopy. Langmuir. 16, 1806–1815 (2000).
[31] Ghosh, S., Salgin, B., Pontoni, D., Reusch, T., Keil, P., Vogel, D., Rohwerder, M., Reichert, H., and Salditt, T., Structure and volta potential of lipid multilayers: Effect of x-ray irradiation. Langmuir (2012).
[32] Salditt, T., Li, C., Spaar, A., and Mennicke, U., X-ray reflectivity of solid-supported multilamellar membranes. Eur. Phys. J. E. 7, 105 (2002).
[33] Liu, Y. and Nagle, J. F., Diffuse scattering provides material parameters and electron density profiles of biomembranes. Phys. Rev. E. 69, 040901– (2004).
[34] Leonenko, Z., Finot, E., Ma, H., Dahms, T. S., and Cramb, D., Investigation of temperature-induced phase transitions in DOPC and DPPC phospholipid bilayers using temperature-controlled scanning force microscopy. Biophys. J. 86, 3783–3793 (2004).
[35] Hub, J. S., Salditt, T., Rheinstädter, M. C., and De Groot, B. L., Short-range order and collective dynamics of DMPC bilayers: a comparison between molecular dynamics simulations, x-ray, and neutron scattering experiments. Biophys. J.. 93, 3156–3168 (2007).
[36] Miller, C. E., Majewski, J., Gog, T., and Kuhl, T. L., Characterization of biological thin films at the solid-liquid interface by x-ray reflectivity. Phys. Rev. Lett. 94, 2381041 (2005).
[37] Watkins, E., Miller, C., Mulder, D., Kuhl, T., and Majewski, J., Structure and orientational texture of self-organizing lipid bilayers. Phys. Rev. Lett. 102, 238101 (2009).
[38] Nickel, B., Nanostructure of supported lipid bilayers in water. Biointerphases 3, FC40–FC46 (2008).
[39] Ghosh, S. K., Aeffner, S., and Salditt, T., Effect of PIP2 on bilayer structure and phase behavior of DOPC: An X-ray Scattering Study. ChemPhysChem. 12, 2633–2640 (2011).
[40] Khattari, Z., Köhler, S., Xu, Y., Aeffner, S., and Salditt, T., Stalk formation as a function of lipid composition studied by x-ray reflectivity. BBA – Biomembranes 1848, 41–50 (2015).
[41] Li, C. and Salditt, T., Structure of magainin and alamethicin in model membranes studied by x-ray reflectivity. J. Biophys. 91, 3285–3300 (2006).
[42] Xu, Y., Kuhlmann, J., Brennich, M., Komorowski, K., Jahn, R., Steinem, C., and Salditt, T., Reconstitution of snare proteins into solid-supported lipid bilayer stacks and x-ray structure analysis. Biochim Biophys. Acta (BBA)-Biomembranes 1860, 566–578(2018).
[43] Tanaka, M. and Sackmann, E., Polymer-supported membranes as models of the cell surface. Nature. 437, 656–663 (2005).
[44] Sackmann, E. and Tanaka, M., Supported membranes on soft polymer cushions: fabrication, characterization and applications. Trends Biotechnol. 18, 58–64 (2000).
[45] Ghosh, S. K., Castorph, S., Konovalov, O., Jahn, R., Holt, M., and Salditt, T., In vitro study of interaction of synaptic vesicles with lipid membranes. NJP 12, 105004 (2010).
[46] Ghosh, S. K., Castorph, S., Konovalov, O., Salditt, T., Jahn, R., and Holt, M., Measuring ca^{2+} induced structural changes in lipid monolayers: Implications for synaptic vesicle exocytosis. Biophys. J. 102, 1394–1402 (2012).
[47] Giewekemeyer, K. and Salditt, T., Counterion distribution near a monolayer of variable charge density. Europhys. Lett.. 79, 18003 (2007).

[48] Koelsch, P., Viswanath, P., Motschmann, H., Shapovalov, V., Brezesinski, G., Möhwald, H., Horinek, D., Netz, R. R., Giewekemeyer, K., Saldit, T., Schollmeyer, H., von Klitzing, R., Daillant, J., and Guenoun, P., Specific ion effects in physicochemical and biological systems: Simulations, theory and experiments. Colloids and Surf. A: Physicochem. Eng. Asp. 303, 110–136 (2007).
[49] Fleck, C. C. and Netz, R. R., Counterion density profiles at charged flexible membranes. Phys. Rev. Lett. 95, 128101 (2005).
[50] Lovell, M. and Richardson, R., Analysis methods in neutron and x-ray reflectometry. Curr. Opin. Colloid & Interface Sci. 4, 197–204 (1999).
[51] Politsch, E. and Cevc, G., Unbiased analysis of neutron and X-ray reflectivity data by an evolution strategy. J. Appl Cryst. 35, 347–355 (2002).
[52] Storn, R. and Price, K., Differential evolution–a simple and efficient heuristic for global optimization over continuous spaces. J. Global Optim. 11, 341–359 (1997).
[53] Koltover, I., Salditt, T., Rigaud, J.-L., and Safinya, C. R., Stacked 2d crystalline sheets of the membrane-protein bacteriorhodopsin: A specular and diffuse reflectivity study. Phys. Rev. Lett. 81, 2494–2497 (1998).
[54] Koltover, I., Raedler, J. O., Salditt, T., Rothschild, K. J., and C. R. Safinya, Phase behavior and interactions of the membrane-protein bacteriorhodopsin. Phys. Rev. Lett. 82, 3184–3187 (1999).
[55] Constantin, D., Membrane-mediated repulsion between gramicidin pores. Biochim. Biophys. Acta (BBA) – Biomembranes 1788, 1782–1789 (2009).
[56] Yang, L. and Huang, H. W., Observation of a membrane fusion intermediate structure. Sci. 297, 1877–1879 (2002).
[57] Aeffner, S., Reusch, T., Weinhausen, B., and Salditt, T., Energetics of stalk intermediates in membrane fusion are controlled by lipid composition. Proc. Natl. Acad. Sci. U.S.A. 109, E1609–E1618 (2012).
[58] Rauscher, M., Salditt, T., and Spohn, H., Small-angle x-ray scattering under grazing incidence: The cross section in the distorted-wave born approximation. Phys. Rev. B 52, 16855 (1995).
[59] Dosch, H., *Critical phenomena at surfaces and interfaces: Evanescent X-ray and neutron scattering*, vol. 126 (Springer Tracts in Modern Physics, Berlin 1992).
[60] Katsaras, J. and Gutberlet, T., *Lipid bilayers: Structure and interactions* (Springer, Berlin, London 2001).
[61] de Jeu, W. H., Ostrovskii, B. I., and Shalaginov, A. N., Structure and fluctuations of smectic membranes. Rev. Mod. Phys. 75, 181 (2003).
[62] Nagle, J. F. and Tristram-Nagle, S., Structure of lipid bilayers. BBA – Reviews on Biomembranes 1469, 159–195 (2000).
[63] Salditt, T., Vogel, M., and Fenzl, W., Thermal fluctuations and positional correlations in oriented lipid membranes. Phys. Rev. Lett. 90, 178101– (2003).
[64] Münster, C., Salditt, T., Vogel, M., Siebrecht, R., and Peisl, J., Nonspecular neutron scattering from highly aligned phospholipid membranes. EPL (Europhysics Letters) 46, 486– (1999).
[65] Safinya, C. R., Sirota, E. B., Roux, D., and Smith, G. S., Universality in interacting membranes: The effect of cosurfactants on the interfacial rigidity. Phys Rev. Lett. 62, 1134–1137 (1989).
[66] de Gennes, J. P. P.-G., *The physics of liquid crystals* (Clarendon Press, Oxford, 1993), 2nd ed.
[67] Salditt, T., Münster, C., Lu, J., Vogel, M., Fenzl, W., and Souvorov, A., Specular and diffuse scattering of highly aligned phospholipid membranes. Phys. Rev. E. 60, 7285– (1999).
[68] Hołyst, R., Landau-Peierls instability, x-ray-diffraction patterns, and surface freezing in thin smectic films. Phys. Rev. A. 44, 3692–3709 (1991).
[69] Lei, N., Safinya, C., and Bruinsma, R., Discrete harmonic model for stacked membranes: Theory and experiment. J. Phys. II France 5, 1155–1163 (1995).

[70] Romanov, V. P. and Ul'yanov, S. V., Dynamic and correlation properties of solid supported smectic-a films. Phys. Rev. E. 66, 061701 (2002).
[71] Li, C., Constantin, D., and Salditt, T., Biomimetic membranes of lipid-peptide model systems prepared on solid support. J. Phys.: Condens. Matter 16, S2439–S2453 (2004).
[72] Perino-Gallice, L., Fragneto, G., Mennicke, U., Salditt, T., and Rieutord, F., Dewetting of solid-supported multilamellar lipid layers. Eur Phys J E 8, 275–282 (2002).
[73] Vogel, M., Münster, C., Fenzl, W., and Salditt, T., Thermal unbinding of highly oriented phospholipid membranes. Phys. Rev. Lett. 84, 390 (2000).
[74] Reusch, T., Mai, D., Osterhoff, M., Khakhulin, D., Wulff, M., and Salditt, T., Non-equilibrium collective dynamics in photo-excited lipid multilayers by time resolved diffuse x-ray scattering. Phys. Rev. Lett. 111.26 (2013) 268101.
[75] Weinhausen, B., Aeffner, S., Reusch, T., and Salditt, T., Acyl-chain correlation in membrane fusion intermediates: X-ray diffraction from the rhombohedral lipid phase. Biophys. J.. 102, 2121–2129 (2012).
[76] Kieffer, J. and Karkoulis, D., Pyfai, a versatile library for azimuthal regrouping. J. Phys.: Conf. Ser., vol. 425 (IOP Publishing, 2013), vol. 425, p. 202012.
[77] Guinier, A., *X-ray diffraction in crystals, imperfect crystals, and amorphous bodies, 1994* (New York: Dover Publications Inc, 1994).
[78] Glatter, O. and Kratky, O., eds., *Small angle x-ray scattering* (Academic Press Inc, London LTD, 1982).
[79] Feigin, L. A. and Svergun, D. I., *X-ray and neutron small-angle scattering* (Springer, Plenum Press, New York 1987).
[80] Roe, R.-J., *Methods of X-ray and neutron scattering in polymer science*, vol. 739 (Oxford University Press, New York 2000).
[81] Pabst, G., Rappolt, M., Amenitsch, H., and Laggner, P., Structural information from multilamellar liposomes at full hydration: Full q-range fitting with high quality x-ray data. Phys. Rev. E. 62, 4000–4009 (2000).
[82] Pabst, G., Kučerka, N., Nieh, M.-P., Rheinstädter, M., and Katsaras, J., Applications of neutron and x-ray scattering to the study of biologically relevant model membranes. Chem. Phys. Lipids 163, 460–479 (2010).
[83] Klauda, J. B., Kucerka, N., Brooks, B. R., Pastor, R. W., and Nagle, J. F., Simulation-based methods for interpreting x-ray data from lipid bilayers. Biophys. J. 90, 2796–2807 (2006).
[84] Kučerka, N., Nagle, J. F., Sachs, J. N., Feller, S. E., Pencer, J., Jackson, A., and Katsaras, J., Lipid bilayer structure determined by the simultaneous analysis of neutron and x-ray scattering data. Biophys. J. 95, 2356–2367 (2008).
[85] Heftberger, P., Kollmitzer, B., Heberle, F. A., Pan, J., Rappolt, M., Amenitsch, H., Kucerka, N., Katsaras, J., and Pabst, G., Global small-angle x-ray scattering data analysis for multilamellar vesicles: The evolution of the scattering density profile model . J. Appl. Cryst. 47, 173–180 (2014).
[86] Kucerka, N., Nieh, M.-P., and Katsaras, J., Asymmetric distribution of cholesterol in unilamellar vesicles of monounsaturated phospholipids. Langmuir 25, 13522–13527 (2009).
[87] Eicher, B., Heberle, F.A., Marquardt, D., Rechberger, G. N., Katsaras, J., and Pabst, G., Joint small-angle X-ray and neutron scattering data analysis of asymmetric lipid vesicles. J. Appl. Cryst. 50, 419–429 (2017).
[88] Komorowski, K., Salditt, A., Xu, Y., Yavuz, H., Brennich, M., Jahn, R., and Salditt, T., Vesicle adhesion and fusion studied by small-angle x-ray scattering. Biophys. J. 114.8, 1908–1920 (2018).
[89] Franke, D., Petoukhov, M., Konarev, P., Panjkovich, A., Tuukkanen, A., Mertens, H., Kikhney, A., Hajizadeh, N., Franklin, J., Jeffries, C. et al., Atsas 2.8: a comprehensive data analysis suite for small-angle scattering from macromolecular solutions. J. Appl. Cryst. 50, 1212–1225 (2017).

[90] Lipfert, J. and Doniach, S., Small-angle x-ray scattering from RNA, proteins, and protein complexes. Annu. Rev. Biophys. Biomol. Struct. 36, 307–327 (2007).
[91] Hernandez, J. M., Stein, A., Behrmann, E., Riedel, D., Cypionka, A., Farsi, Z., Walla, P. J., Raunser, S., and Jahn, R., Membrane fusion intermediates via directional and full assembly of the snare complex. Sci. 336, 1581–1584 (2012).
[92] Castorph, S., Riedel, D., Arleth, L., Sztucki, M., Jahn, R., Holt, M., and Salditt, T., Structure parameters of synaptic vesicles quantified by small-angle x-ray scattering. Biophys. J.. 98, 1200–1208 (2010).
[93] Castorph, S., Arleth, L., Sztucki, M., Vainio, U., Ghosh, S. K. , Holt, M., Jahn, R., and Salditt, T., Synaptic vesicles studied by SAXS: Derivation and validation of a model form factor. J. Phys: Conf. Ser. 247, 012015 (2010).
[94] Takamori, S., Holt, M., Stenius, K., Lemke, E. A., Grønborg, M., Riedel, D., Urlaub, H., Schenck, S., Brügger, B., Ringler, P., Müller, S. A., Rammner, B., Gräter, F., Hub, J. S., Groot, B. L. D., Mieskes, G., Moriyama, Y., Klingauf, J., Grubmüller, H., Heuser, J., Wieland, F., and Jahn, R., Molecular anatomy of a trafficking organelle. Cell 127, 831–846 (2006).
[95] Pedersen, J. S. and Gerstenberg, M. C., Scattering form factor of block copolymer micelles. Macromolecules. 29, 1363–1365 (1996).
[96] Gradzielski, M., Langevin, D., Magid, L., and Strey, R., Small-angle neutron scattering from diffuse interfaces. J. Phys. Chem. 99, 13232–13238 (1995).
[97] Carboni, E., Nicolas, J.-D., Töpperwien, M., Stadelmann-Nessler, C., Lingor, P., and Salditt, T., Imaging of neuronal tissues by x-ray diffraction and x-ray fluorescence microscopy: evaluation of contrast and biomarkers for neurodegenerative diseases. Biomed. Opt. Express. 8, 4331–4347 (2017).
[98] Beerlink, A., Mell, M., Tolkiehn, M., and Salditt, T., Hard x-ray phase contrast imaging of black lipid membranes. Appl. Phys. Lett. 95, 203703 (2009).
[99] Beerlink, A., Thutupalli, S., Mell, M., Bartels, M., Cloetens, P., Herminghaus, S., and Salditt, T., X-ray propagation imaging of a lipid bilayer in solution. Soft Matter. 8, 4595–4601 (2012).
[100] Gaffney, K. J. and Chapman, H. N., "Imaging atomic structure and dynamics with ultrafast X-ray scattering," Science ,316 1444–1448 (2007).
[101] Hantke, M. F., Ekeberg, T., and Maia, F. R., Condor: A simulation tool for flash x-ray imaging. J. Appl. Cryst. 49, 1356–1362 (2016).
[102] Mai, D. D., Hallmann, J., Reusch, T., Osterhoff, M., Düsterer, S., Treusch, R., Singer, A., Beckers, M., Gorniak, T., and Senkbeil, T., Single pulse coherence measurements in the water window at the free-electron laser FLASH. Opt. Express. 21, 13005–13017 (2013).
[103] Hammouda, B. (1992). Structure factor for starburst dendrimers. Journal of Polymer Science Part B: Polymer Physics, 30(12), 1387–1390.

Rebecca Eells, David P. Hoogerheide, Paul A. Kienzle, Mathias Lösche, Charles F. Majkrzak, Frank Heinrich

3 Structural investigations of membrane-associated proteins by neutron reflectometry

Abstract: Neutron reflectometry (NR) is a powerful technique for probing the structure of lipid bilayer membranes and membrane-associated proteins. Measurements of the specular neutron reflectivity as a function of momentum transfer can be performed in aqueous environments, and inversion of the resulting reflectivity data yields structural profiles along the membrane normal with a spatial resolution approaching a fraction of a nanometer. With the inherent ability of the neutron to penetrate macroscopic distances through surrounding material, neutron reflectivity measurements provide unique structural information on biomimetic, fully hydrated model membranes and associated proteins under physiological conditions. A particular strength of NR is in the characterization of structurally and conformationally flexible peripheral membrane proteins. The unique ability of neutron scattering to differentiate protium from selectively substituted deuterium enables the resolution of individual constituents of membrane-bound protein–protein complexes. Integrative modeling strategies that supplement the low-resolution reflectometry data with complementary experimental and computational information yield high-resolution three-dimensional models of membrane-bound protein structures.

3.1 Introduction

With both researchers in the life sciences and interested students in mind, this chapter provides an overview of theoretical and practical aspects of neutron reflectometry (NR) from biomimetic lipid membranes that enables the structural characterization of membrane-bound proteins. We focus on peripheral membrane proteins,

---

Rebecca Eells, Department of Physics, Carnegie Mellon University, Pittsburgh, USA
David P. Hoogerheide, Paul A. Kienzle, Center for Neutron Research, National Institute of Standards and Technology, Gaithersburg, USA
Mathias Lösche, Department of Physics, Carnegie Mellon University, Pittsburgh, USA; Center for Neutron Research, National Institute of Standards and Technology, Gaithersburg, USA; Department of Biomedical Engineering, Carnegie Mellon University, Pittsburgh, USA
Charles F. Majkrzak, Center for Neutron Research, National Institute of Standards and Technology, Gaithersburg, USA
Frank Heinrich, Department of Physics, Carnegie Mellon University, Pittsburgh, USA; Center for Neutron Research, National Institute of Standards and Technology, Gaithersburg, USA

https://doi.org/10.1515/9783110544657-003

which are challenging systems to investigate by conventional structural biology methods due to the thermodynamic nature of their interaction with the lipid membrane and their structural and conformational flexibility. At the same time, they are particularly well amenable to the characterization with neutrons. While the theory of NR and many of the discussed protocols are directly applicable to integral membrane proteins, we do not cover the difficult problem of integral membrane protein reconstitution into lipid membranes.

3.1.1 Cellular membranes and proteins

In their most fundamental role, membranes constitute the boundaries of a cell that separate the cytoplasm from the extracellular environment and partition the cell internally into regions of different functionality, i.e., the organelles. However, this function is far from being a passive one, as cellular membranes play active and essential roles in processes such as cell signaling, selective transport between compartments they separate, information transduction and processing, and cellular morphogenesis. To perform these functions, cellular membranes consist of complex mixtures of phospholipids and proteins [1], which can be membrane-integral or membrane-peripheral [2]. Integral membrane proteins often span the entire bilayer, that is, they are transmembrane proteins [3, 4]. Peripheral proteins do not predominantly interact with the hydrophobic core of the lipid bilayer, but rather associate with the lipid headgroups and with other membrane proteins via a multitude of hydrophobic, electrostatic, and bio-specific contacts [5]. Peripheral membrane association of these proteins is often reversible, such that there is an equilibrium between membrane-associated and cytosolic states that may depend on the status of the cell. Peripheral membrane proteins often assume multiple conformations depending on their environment, and interactions between the protein and the membrane, even if merely transient, can induce structural rearrangements or function-related conformational changes [6]. Furthermore, stochastic motions of disordered protein regions can play important functional roles.

Considering their importance for cellular function, it is not surprising that membrane protein malfunction is implicated in a wide variety of diseases, including cystic fibrosis [7], type-2 diabetes [8], heart disease [9], neurological disorders [10], and cancer [11]. The penetration of toxins and pathogens into the cell also involve protein–membrane interactions [12, 13]. Acting as an efficient, selective barrier between the cell and its environment, membranes manage mostly to keep such incursions at bay. However, in an evolutionary tug o' war, many pathogens develop specificity for binding to proteins on their host target cell membranes to gain entry, thereby circumventing or hijacking the cell's defense mechanisms. As a result, about half of current drug targets are membrane proteins [2], and certain classes of membrane proteins, such as G-protein coupled receptors (GPCRs), are among the most actively investigated targets of current drug development [14]. The ability to identify new drug targets and develop new therapeutics has been limited by a lack of high resolution, three-dimensional structures. While membrane proteins comprise ≈30% of mammalian proteomes, their structures determined so far provide only ≈1% of the entries in the Protein Data Bank (PDB) [15].

3.1.2 Challenges in structural biology of membrane proteins

The three major techniques to obtain high-resolution (atomic-level) protein structures are cryo-electron microscopy (cryo-EM), nuclear magnetic resonance (NMR), and X-ray diffraction (XRD). The majority of membrane protein structures has been determined with XRD from protein crystals grown in detergent [15]. In such crystals, detergent molecules associate with protein surfaces that are natively embedded in a lipid bilayer. However, membrane protein structures are, at least in part, determined by the proteins' interactions with these environments and may depend on the biophysical properties of the native membrane [16]. Therefore, it is often difficult to assess how well detergent contacts approximate the native lipid environment.

As an alternative to the use of detergents in XRD, membrane proteins can be embedded in lipid bicelles [17, 18] or in lipidic cubic phases (LCP) [19]. In particular, recent developments in LCP technologies ranging from precrystallization assays to data collection, as well as the commercial availability of tools to utilize these technologies, have the potential to drastically increase the number of structures obtained from detergent-free samples [20, 21]. For proteins with large extra-membrane domains, a liquid analog of LCP, known as the sponge phase, has been used [22]. Two major technical challenges for the LCP crystallization approach remain in harvesting the crystals and data collection due to the small crystal size [20].

NMR has proven invaluable for the determination of high-resolution structures and dynamics of soluble proteins and solubilized membrane proteins [23, 24]. Both solution and solid-state NMR can be applied to study integral and peripheral membrane proteins. For solution NMR, micelles, bicelles, or nanodiscs can provide a lipid membrane environment. However, large-weight complexes of membrane proteins with micelles or bicelles often tumble too slowly to permit the fast rotational diffusion needed to obtain well-resolved spectra [24]. Solid-state NMR (ssNMR), in principle, does not have the same molecular weight constraint as solution NMR, and well-resolved spectra are achieved through magic angle spinning or by oriented sample spectroscopy [16, 25]. With ssNMR, proteins can be characterized within a bilayer environment mimicking the native membrane such as liposomes [25] or oriented membranes [26]. However, most membrane protein structures solved with this technique so far have been limited to molecular weights less than 10 kDa due to the complexity of the ssNMR spectra [24, 27]. Sample heterogeneity and dynamics can also impose limits on ssNMR measurements since they lead to broadened resonances and spectral overlap. Sensitivity enhancement techniques such as ^{1}H detection, dynamic nuclear polarization, and nonuniform sampling algorithms are recent developments that will extend the applicability of ssNMR toward larger proteins [25].

Cryo-EM is the third major structural method for membrane proteins. While cryo-EM was limited to the investigation of large complexes in the past, recent improvements in instrumentation, such as the direct electron detection camera, and image processing resulted in increased resolution, lower size limits, and better classification of heterogeneous samples [15, 28]. Thus, cryo-EM allows for the characterization of membrane proteins in a lipid bilayer environment at resolutions that, in some cases, are comparable to resolutions achieved by XRD.

3.1.3 Neutron reflectometry

NR for biological systems is a well-established technique [29–35], and all major neutron scattering facilities worldwide have biological NR capability [36–38]. In comparison with the high-resolution techniques in structural biology discussed above, NR provides distinct advantages for characterizing in-plane disordered, fluid lipid membranes and the association of membrane proteins in fully buffer-immersed biomimetic environments. While NR is intrinsically a low-resolution technique, when complemented with high-resolution methods and MD simulation, atomistic detail of biomimetic protein–membrane complexes is obtained. NR has been successfully applied to study intrinsically disordered proteins and peptides [39–44], and peripheral membrane proteins that may only associate with the membrane transiently (see Figure 3.1) [45–57]. The ability of neutron scattering to use selective deuteration to highlight parts of the structure allows for the characterization of individual constituents of membrane-bound protein–protein complexes [58]. Since NR is nondestructive, the sample can be manipulated *in situ* during a series of measurements. By changing environmental conditions, biological processes can be simulated, for example, through the introduction of cofactors or by applying external cues [52], and the evolution of the system can be monitored.

NR yields one-dimensional structural profiles along the normal direction of the interface from which the beam is reflected. In investigations of membranes deposited at such an interface, the compositional profile represents in-plane averages parallel to the bilayer at each position along the normal direction, thus yielding a temporal and spatial ensemble over all molecular configurations (see Section 3.2). NR requires planar substrates of low surface roughness that carry the interfacial structure of interest, as well as samples that are homogeneous and stable over typically hour-long measurements. The development of biomimetic, in-plane fluid sparsely tethered lipid membranes (stBLMs) for NR [59–63] has addressed this requirement (see Section 3.3). Conducting the NR experiment requires specialized sample environment that allows for in situ sample manipulation and maximizes the signal-to-noise ratio of the

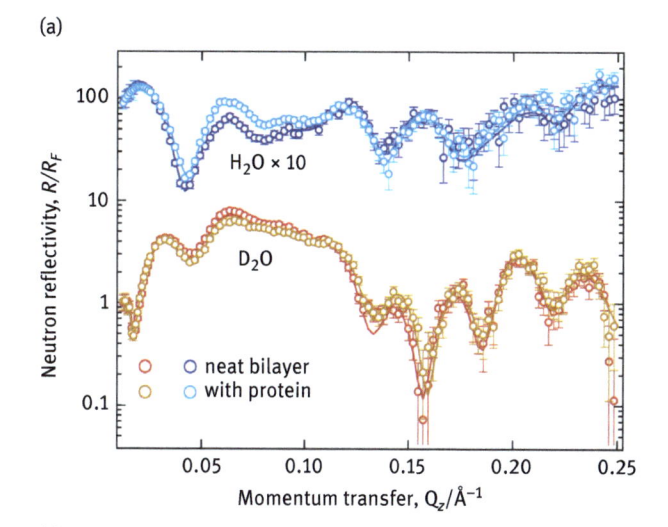

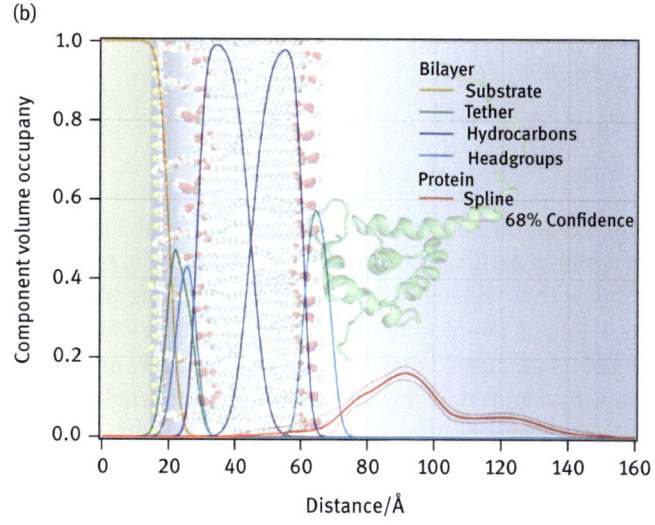

Figure 3.1: Neutron reflection measurements of the peripheral membrane protein HIV-I Gag matrix on a sparsely tethered bilayer lipid membrane. (a) NR reflectivities of the membrane, immersed in two isotopically distinct (H₂O and D₂O-based) buffers before and after incubation with protein. All data were measured on one membrane sample, such that all four measurements could be used to refine one general model structure. (b) Component volume occupancy profiles along the bilayer normal of the membrane–protein complex from a composition-space model. The profile of the protein was determined using a free-form spline (red traces) and localizes the compact core and flexible tail of the protein with respect to the lipid bilayer. In a more refined model, the orientation of the protein was subsequently determined by rigid body modeling using an ensemble of NMR structures (PDB: 2H3F). The background image is a visual representation of the protein on the membrane surface that derives from this refined analysis. (Figure adapted from Eells et al. [46].)

measurement (see Section 3.5). The implementation of molecular modeling strategies [64, 65] in the recent decade has transformed the technique [33]. With a current NR setup, structural features with a thickness of 10 Å can be resolved with a spatial resolution of a fraction of an nm, and volume occupancies of components as low as 5–10% at any position along the surface normal can be reliably determined. A fully atomistic interpretation can be achieved with integrated modeling strategies that utilize complementary experimental data and molecular dynamics (MD) simulations [49, 56, 66, 67] (see Section 3.5).

3.2 Fundamentals of neutron reflectometry

We begin by reviewing the remarkably accurate physical and mathematical description of specular reflection which, in practice, reduces to a one-dimensional quantum mechanical scattering process. This theory is described without detailed derivations of the relevant mathematical equations since comprehensive expositions can be readily found elsewhere [30, 68–80].

3.2.1 Wave-like behavior of neutron quantum objects

It so happens that to obtain structural information about soft condensed matter in general – and biological material in particular – on a subnanometer scale, scattering methods employing neutrons with wavelengths between 1 and 10 Å are ideal. The neutron is a quantum particle and its interactions with matter on such a length scale must be described in terms of its associated wave behavior, specifically as characterized by the neutron's wave function. How this wave function is affected through interaction with a material object in a scattering process can be predicted by the Schrödinger equation of motion in a probabilistic manner. The neutron interaction with matter is predominantly via the nuclear potential in nonmagnetic materials. Experimentally obtained scattering patterns must be analyzed mathematically to deduce the corresponding structure of the scattering object. In essence, this is common to all wave diffraction methods, including X-ray and neutron crystallography. In contrast, methods that image objects in real space, such as optical and electron microscopy, use particles with wavelengths that are far shorter than the objects' spatial dimensions so that wave diffraction effects are negligible. The length scale probed by neutron reflectivity is about an order of magnitude larger than the neutron wavelength, that is, of the order of a nanometer rather than an Angstrom. This means that the potential representing the interaction between a neutron and the scattering structure, such as a lipid membrane with embedded proteins, is effectively continuous, varying smoothly as a function of distance, except at well-defined boundaries, for example, between material and vacuum or aqueous reservoir.

The wave function associated with a freely propagating neutron quantum particle, as it might be found in a beam on a scattering instrument, is localized and occupies a finite volume of space (at least for the practical purposes of interest here). This localized wave function is referred to as a wave packet and typically has dimensions of the order of a fraction to tens of microns, depending on how the particular state was prepared after emission from a source, for example, by monochromating crystals and angle-defining pairs of apertures [81]. This neutron wave packet can be described mathematically by a coherent superposition of basis states, often taken to be a normal distribution of plane waves each having a given value of wave vector $\mathbf{k}$, whose magnitude $k = 2\pi/\lambda$ depends on the neutron wavelength λ of the state. The width of the distribution of the packet's basis wave vectors along a given direction in space is inversely proportional to the width of the probability amplitude distribution of where the neutron quantum particle is located along that same direction. This is a fundamental expression of the well-known uncertainty principle of quantum mechanics [82]. Each packet representing an individual neutron in the beam has an associated mean wave vector $\bar{\mathbf{k}}$. In practice, the neutrons composing a beam can be taken to be an ensemble of independent individual members which are completely noninteracting with one another. Incident beam intensities are oftentimes so low that a single given neutron incident on a sample object is reflected and captured in a detector before the next following neutron in the beam is even incident. The Schrödinger wave equation which describes the scattering of an incident neutron from a material object is explicitly time-dependent, specifically in terms of what happens to the neutron's associated wave packet before and after. However, since we are interested in the structure, and not the dynamics of biological interfaces, only elastic scattering processes are relevant. Here, the total energy of the neutron is conserved and it is possible to solve a simpler, time-independent wave equation using single plane wave (steady-state) representations of the neutron. Nonetheless, it is important to consider the finite spatial extent of the neutron wave packet when interpreting experimentally measured reflectivity data in terms of the steady-state solutions, as will be discussed in the following section.

3.2.2 Specular neutron reflectivity and the scattering length density profile

The neutron reflectivity R is defined as the number of neutrons reflected from a nominally flat interface, divided by the number of neutrons incident at a glancing angle (typically several degrees) from that interface (Figure 3.2). Specular reflection occurs when the glancing angles of incidence (I) and reflection (F) are equal to one another so that the momentum and wave vector transfer, $\mathbf{Q} = \mathbf{k}_F - \mathbf{k}_I$, are along the surface normal. If the interface is perfectly smooth and the materials at the interface are characterized by a homogeneous in-plane density distribution, only specular scattering can occur. If, on the other hand, in-plane inhomogeneity does exist, then

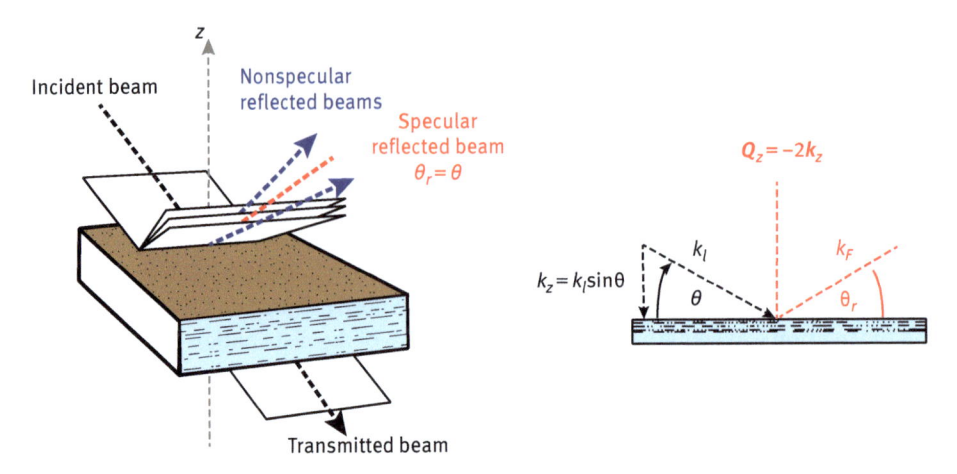

Figure 3.2: Scattering configuration for reflection at glancing angles of incidence. Specular reflection occurs when the angles of incidence and reflection are equal to one another so that the momentum and wave vector transfers are along the nominal surface normal. The neutron reflectivity $R(\theta)$ is defined as the ratio of the number of incident neutrons over that of the specularly reflected neutrons as a function of incident angle θ.

the specular signal gives the in-plane average over an area equal to the projected transverse coherent extent of the neutron wave packet wave fronts [81], and some incident neutrons will contribute to nonspecular scattering. If this nonspecular component remains small, it can be neglected to a good approximation; however, if it is large, a more elaborate analysis is required. For the systems of interest here, we limit our discussion to specular reflection and corresponding in-plane-averaged compositional depth profiles along the interface normal.

The complex reflection amplitude r is the specularly reflected neutron wave function related to the neutron scattering length density (nSLD) profile along the nominal normal direction to the planar film structure. The nSLD for a given material is the sum of the numbers, per unit volume, of nuclei of each specific constituent element multiplied by its corresponding scattering power, as characterized by a scalar coherent scattering length, which has been determined and tabulated for most isotopes [83].

However, only the reflected intensity, that is, the complex square of the reflection amplitude $R = |r|^2$ expressed as a function of wave vector transfer, $R(Q_z)$, with $Q_z = 4\pi \sin\theta/\lambda$, can be directly measured. R corresponds to the probability of finding a neutron quantum particle scattered into a particular angle. In turn, the reflectivity $R(\theta)$ for a model nSLD profile can be obtained via solution of the one-dimensional Schrödinger wave equation. This allows for fitting a parameterized nSLD model to a measured reflectivity as a function of glancing angle, $R(\theta)$ [30]. Any solution for the nSLD profile thereby obtained is not guaranteed to be unique because of the lack of phase information in R. To remove such ambiguity, either additional, independent information about the sample, or multiple measurements with composite systems that include known structural parts such as a reference layers of known scattering properties are required (see Section 3.2.3).

An arbitrary nSLD profile can be subdivided into n layers, each of a specified thickness d over which the nSLD value is approximated to be constant. The smaller the d becomes, the larger is the range of wave vector transfer Q_z over which the reflectivity must be measured to resolve differences in nSLD on that spatial length scale. This follows from a straightforward, semiquantitative consideration of the Fourier relationship between real and scattering (reciprocal) space, which exists in the Born approximation solution for the reflectivity [84]. The same approximate relationship between r and the nSLD profile also affords a means to approximate the uncertainty in the values of nSLD obtained for a given statistical accuracy in the reflectivity data measured [84]. Figure 3.3 illustrates schematically the relationship between the specular reflectivity and nSLD for a measurement performed on α-hemolysin [85] embedded in a lipid bilayer membrane [66].

In practice, the smallest resolvable length scale in NR is roughly determined by the maximum momentum transfer, Q_z^{max}, of the measurement, or the value of Q_z at which the specular reflection becomes comparable to the measurement background, whichever is smaller. The spatial resolution of a measurement with such a Q_z^{max} is approximately π/Q_z^{max}, or about 13 Å for the data shown in Figure 3.3a ($Q_z^{max} = 0.25\,\text{Å}^{-1}$). This should be compared to the intrinsic resolution limit determined by the root mean square roughness of the substrate (typically $\approx 3\,\text{Å}$ for a silicon substrate, but potentially much larger for additional deposited layers). Rough surfaces also reduce the Q_z at which the reflected intensity drops below the background intensity [86]. Thus, smooth substrates and minimal background are central to obtaining high-quality density profiles.

Knowing the sizes of each distinct area of different SLD in the plane of the film is important. As discussed in Section 3.2.1, it is necessary to know over what in-plane area the neutron packet wave front is capable of coherently averaging. This is determined by the transverse extent of the wave front – at a sufficiently uniform phase – that is projected onto the sample surface. In one limit, at which each individual neutron averages over all different types of in-plane areas, a single coherent specular reflectivity signal will be measured. In the opposite limit, if each neutron can only view one of a number of distinct regions of a given SLD at a time, then the measured reflectivity will correspond to an area-weighted incoherent sum of the reflectivities for each type of in-plane region. In either case, an appropriate analysis of the data can be performed with sufficient knowledge of the in-plane SLD distribution. Figure 3.4 illustrates in-plane averaging pictorially. It is preferable to prepare a sample with the greatest in-plane homogeneity possible on a length scale comparable to the projected transverse extent of a neutron packet wave front: a contiguous film has advantages over one with islands.

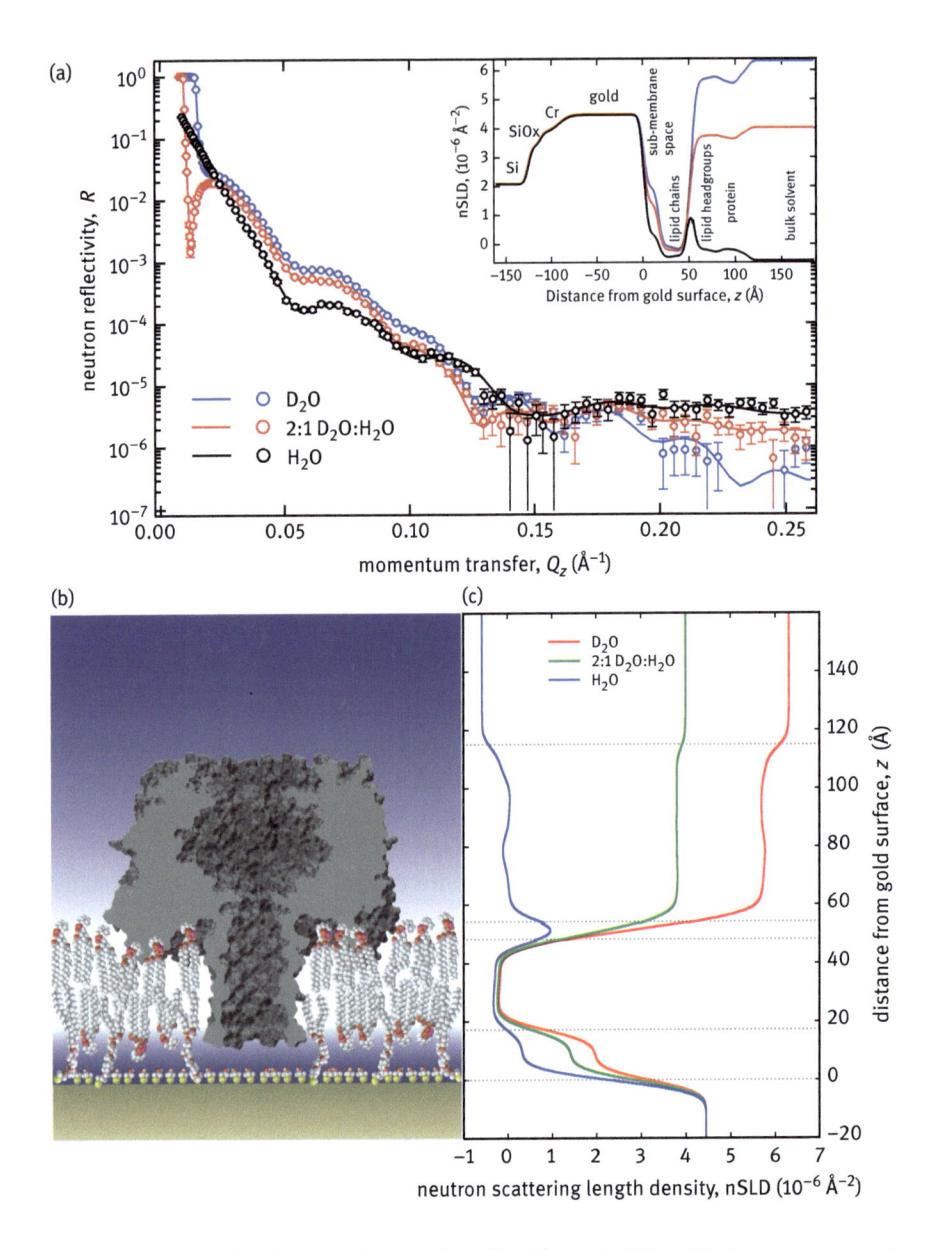

Figure 3.3: Relationship between the specular reflectivity and nSLD profile for a measurement performed on α-hemolysin in a tethered lipid bilayer membrane (see Section 3.3.1) prepared on a solid (silicon / silicon oxide / chromium / gold) substrate. (a) The measured neutron specular reflectivity for various mixtures of H_2O and D_2O in the aqueous reservoir adjacent to the lipid bilayer. Inset: nSLD profiles corresponding to the three measurements with isotopically distinct buffers, across the entire sample obtained by a simultaneous fit of all measured reflectivity curves. (b) Rendering of the bilayer and protein structure based on the nSLD profiles. (c) Magnification of the nSLD profiles across the lipid bilayer region, scaled to the depiction in (b). (Figure adapted from McGillivray et al. [66].)

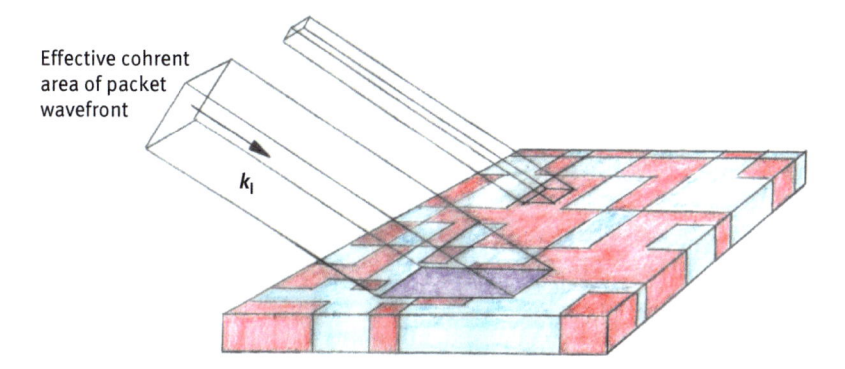

Figure 3.4: The effective transverse coherence area perpendicular to the propagation direction of the neutron wave packet is projected onto the film surface defining an area over which in-plane variations in nSLD are averaged in the specular process (note that the glancing angle of incidence enhances the projection along one in-plane direction). The length scale of the nSLD variations must be small enough for averaging to occur within the projected area. The purple shaded area represents the coherent average of separate areas of "red" and "blue" nSLD. If the length scale of the nSLD variations are not small enough, then the net measured reflected intensity $|r|^2$ is an area-weighted sum of reflectivities, each corresponding to an in-plane averaged SLD within a particular respective area, as depicted schematically for the case of two distinct areas of different SLD (red and blue areas). For a typical reflectometer, the neutron is prepared such that the transverse coherent extent of the wave front is on the order of a micrometer.

3.2.3 Uniqueness and phase information

The inherent loss of explicit phase information is a problem common to all diffraction measurements, but often additional, independent knowledge about the structure and composition of the system is sufficient to eliminate ambiguities. In other cases, a unique solution can be obtained through isomorphic substitution of constituent atoms, for example, by varying the SLD of a surrounding fluid reservoir or solid supporting substrate, or by implementing external references such as neighboring, distinct layers in a multilayered film structure [84, 87–92]. This entails collecting multiple reflectivity data sets, each corresponding to a composite system consisting of a common layered part of interest, the structure or composition of which is unknown, plus a reference part that is completely known. The real and imaginary parts of the reflection amplitude r for the unknown part of interest alone can be mathematically determined, uniquely and independently at each given value of Q_z. Once the amplitude r is obtained, a direct inversion to obtain the corresponding unique nSLD profile can be performed in a model-independent way through the solution of an integral or differential equation, to the extent that sufficiently accurate data up to an appropriate value of Q_z can be measured. Alternatively, the reflectivity data sets from the composite system can be simultaneously fit [92]. Both approaches yield the same unique nSLD profile within numerical and statistical accuracy since both are sensitive to the essential phase information contained within the composite system reflectivity data.

Figure 3.5 shows an example of direct inversion whereby the nSLD profile was extracted from two composite system reflectivity data sets, each having a common unknown Cr/Au and lipid multilayer component. However, one composite system had a silicon substrate as reference, whereas the other had a substrate of Al_2O_3 [90]. The SLD profile was obtained by direct inversion. First, the real part of the reflection amplitude, $\Re(r)$, corresponding to the multilayer alone was determined algebraically, independently at each and every Q_z value of the two composite reflectivity data sets. This $\Re(r)$ for the membrane possesses a one-to-one correspondence with a single nSLD profile. The direct inversion of $\Re(r)$ yields the corresponding nSLD profile independent of any model and without any adjustable parameters, that is, it was retrieved ab initio. The profile obtained by this inversion is unique to the extent allowed by the truncation of the reflectivity data at Q_z^{max} and the degree of statistical uncertainty in the collected data. For comparison, Figure 3.5 also shows the nSLD profile predicted for this system by a MD simulation.

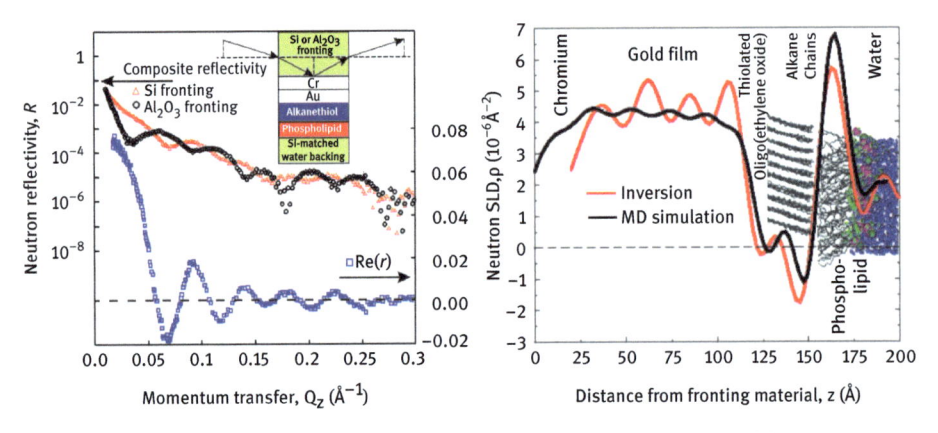

Figure 3.5: An example of direct inversion whereby the nSLD profile was extracted from two composite system reflectivity data sets, shown on the left, each containing a common unknown Cr/Au and lipid multilayer component, but each on a different reference substrate with known scattering properties. $\Re(r)$ for the multilayer system alone, extracted from the two composite system reflectivity data sets, is also plotted. The direct inversion of $\Re(r)$ yields the corresponding nSLD profile shown on the right, independent of any model and without any adjustable parameters. The oscillations are an artifact of the truncation of the composite system reflectivity data at Q_z^{max}. For comparison, the nSLD predicted for this system by a MD simulation is also shown. (Figure adapted from Majkrzak et al. [90].)

3.3 Planning a neutron reflectometry experiment

To determine the structure of a particular lipid membrane–protein complex with NR, the experimenter must identify a suitable model membrane system and experimental conditions – the aqueous membrane environment, bilayer composition and optimized protocol for protein addition – that in combination yield stably bound protein at sufficiently high interfacial density. Complementary surface sensitive techniques aid the characterization of the system of interest and are indispensable tools to optimize experimental conditions for structural characterization. In this section, we discuss in detail sparsely tethered lipid bilayer membranes (stBLMs) optimized for NR and two complementary surface-sensitive techniques for sample precharacterization: electrochemical impedance spectroscopy (EIS) [66, 93, 94] and surface plasmon resonance (SPR) [45, 95].

3.3.1 Sparsely tethered lipid membranes

Lipid bilayer membranes for biological applications of NR must meet several criteria that exclude investigations of natural membranes in vivo, but analogous studies can be performed using model membranes [63, 73, 96, 97]. For the best experimental resolution, the model membrane must be planar and of low interfacial roughness [86]. The relatively low flux at current neutron sources requires long NR measurement times, typically several hours per condition. The membrane must, therefore, be long-term stable. Large sample sizes of several cm^2 are advantageous as they allow for a larger incident beam cross-section and, therefore, higher flux on the sample. To ensure proper analysis of the NR data, the sample must be as homogenous in-plane as possible. Bilayers that are inhomogeneous, even on length scales below the coherence length of the neutron (see Section 3.2.2), require more complex modeling and lower the confidence with which structural features can be determined. Membrane defects in the interfacial bilayer may present nucleation points for nonspecific protein interactions that obstruct structural characterization of the biomimetic protein–membrane complex. Further, the interfacial bilayer should be representative of a lipid membrane in vivo, and be accessible for buffer exchange and protein addition in a series of NR measurements. As such, the model membrane system needs to be flexible in terms of its lipid composition and maintain lipid diffusion rates comparable to those of biological membranes. Finally, to focus on the structural changes induced by protein association with the membrane, the interfacial bilayer should be structurally inert with regard to changes in environmental conditions such as pH, temperature, and ionic strength.

A variety of model membranes have been developed for NR with those criteria in mind. A widely used platform is Langmuir monolayers, which are monomolecular films that resemble one leaflet of a lipid bilayer floating on a fluid (aqueous) surface. Proteins bound to Langmuir monolayers have been extensively studied with X-ray [98–100] and neutron reflection [101–103]. X-ray reflectivity measurements in particular have a large accessible momentum transfer range [104], and consequently high resolution. However, Langmuir monolayers are fragile, making it difficult to inject proteins into the subphase without distorting the model membrane, and measurements usually require large amounts of protein because of large aqueous volume underneath the lipid layer.

A different class of experimental model systems suited for reflectometry investigations are double-layer lipid membranes supported by planar solid substrates – typically silicon wafers – either prepared as membrane stacks in a controlled humidity environment or as fluid-immersed single bilayers. Stacked membranes provide the advantage of periodicity, resulting in reflected intensities that form one-dimensional diffraction patterns [105, 106] with intrinsically higher spatial resolution than reflection experiments from single bilayers. In addition, increased sensitivity due to the large amount of membrane material in the sample permits the determination of in-plane structure and membrane fluctuations [107]. While stacked membranes can readily host peptides to reveal their association with lipids [108, 109], they are not well suited for the study of proteins, nor do these systems permit the in situ manipulation of protein–membrane complexes.

Fluid-immersed single lipid bilayer membranes on a solid support are arguably the most versatile class of model membrane systems for studying lipid-protein complexes with NR. This class includes purely solid-supported membranes [110, 111], hybrid membranes [112–115], tethered membranes [66, 93, 94, 116–119], polymer-cushioned membranes [120–124], and floating membranes [125, 126]. To keep bilayers on the solid support in-plane fluid, they can be deposited by various preparation protocols that create a nanometer-thin water layer or a molecular layer of a hydrated polymer between the membrane and the substrate.

We routinely use a hybrid of these two stabilization schemes in which an oligomeric, hydrophilic ethylene oxide ($EO_6 \ldots EO_9$) tether ligates the bilayer covalently to a gold-coated substrate to form a sparsely tethered lipid bilayer membrane (stBLM, see Figure 3.6) [59–61, 127, 128]. The tether provides ≈15 Å of hydrated submembrane space between bilayer and solid support. Spacing of the tethers is achieved by coadsorption with β-mercaptoethanol (βME), which also passivates exposed areas of the gold surface. While the narrow submembrane reservoir prevents studies of membrane proteins with large extra-membrane domains on both sides of the lipid bilayer, it ensures conformity of the membrane with the ultra-flat substrate, and thereby low interfacial roughness. stBLMs can be prepared virtually defect-free from a wide range of lipid mixtures. The high resilience of this model system [59, 129] permits the exchange of the adjacent buffer to make use of nSLD contrast variation (see Section 3.3.6) and to introduce or remove protein-containing solutions [52]. Complementary surface-sensitive techniques such as EIS [66, 93, 94] and SPR [45, 95] aid the characterization of the membrane systems and are indispensable tools to identify optimal conditions for structure determination by NR. The assessment of fluorescent probe dynamics within the membrane, for example by fluorescence correlation spectroscopy (FCS), confirms the in-plane fluidity of the model systems [130].

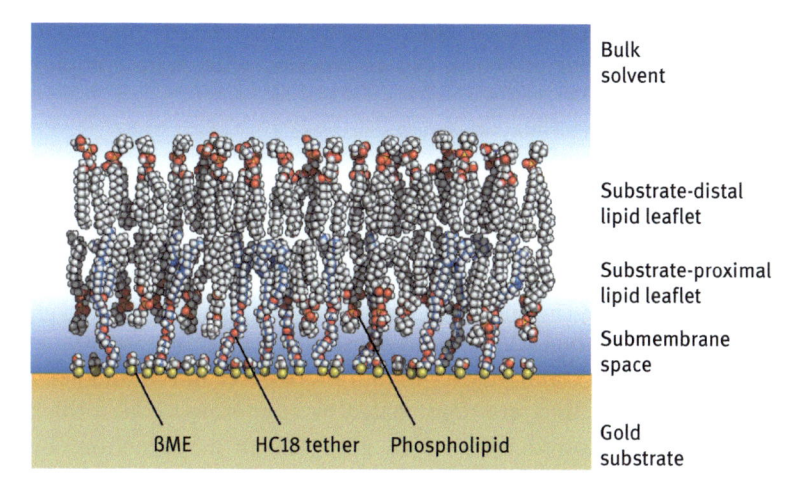

Figure 3.6: Cartoon of a sparsely tethered bilayer lipid membrane (stBLM) on a gold-coated support, composed of a tether lipid (designated as HC18 (Budvytyte et al. [61]); drawn with blue carbon atoms) that form membrane anchors for a POPC/POPS bilayer. Coadsorbed β-mercaptoethanol (βME) prevents the formation of a dense tether layer by occupying void substrate areas between tether molecules, thus passivating the gold surface. The ≈15 Å-thick water-filled submembrane space decouples the lipid bilayer membrane from the substrate and enables lipid diffusion in both lipid leaflets at a rate similar to that in giant unilamellar vesicles (Shenoy et al. [130]). The proximity of the membrane to the substrate quenches bilayer undulations, thus increasing the effective resolution with which bilayer components and bilayer-associated proteins can be resolved by NR.

3.3.2 Choosing a lipid composition

Synthetic model membranes reduce the chemical and functional complexity of cellular membranes such that a limited number of components still represent the molecular interactions that underlie a biological function of interest [96, 97]. In many cases, it is the selective interaction of a particular protein with the membrane and its specific organization in the lipidic environment that holds the key for understanding its function in a complex machinery. A first step is then to identify the cellular target membrane for the protein of interest. Since each of the internal membranes that form the boundaries of organelles or of the cell itself is composed of a unique mixture of phospholipids, sphingolipids, and sterols, the composition of the target membrane is an important feature to reproduce when designing a model membrane [131, 132]. Common molecular mechanisms for targeting peripheral proteins to membranes include electrostatic attraction to charged lipids, biochemical specificity to lipids such as phosphoinositides, and hydrophobic interactions via protein lipidation, for example, through posttranslational fatty acid or isoprenyl ligations – or a mix of all of the above [133]. To characterize the interactions that drive binding and find a lipid composition of optimal complexity, it can be useful to probe each interaction individually using only a subset of lipids, before working with more complex mixtures [95].

3.3.3 Preparation of sparsely tethered lipid membranes

NR investigations require large, flat and atomically smooth substrates to maximize resolution and the signal-to-noise ratio in a measurement. Gold-coated silicon wafers of 2″ or 3″ diameter are typically used as substrates for stBLMs. Substrates are cleaned in sulfuric acid for 15 min, rinsed with ultrapure water and ethanol, and dried in a stream of nitrogen. They are then coated by sputtering a ≈ 40 Å-thick chromium adhesion layer followed by a ≈150 Å gold film in a magnetron. By optimizing the sputtering process, a root mean square roughness of <5 Å can be routinely achieved on silicon wafers polished to a surface roughness <3 Å, as confirmed by atomic force microscopy or X-ray reflection. The most critical parameter during deposition to achieve a low surface roughness is the pressure of the sputter gas, which should be <1 mTorr. Deposition power and voltage are parameters that require additional optimization. After sputtering, the substrates can be vacuum-sealed and kept for several days until use. Optimal gold layer thicknesses for complementary characterization techniques are ≈ 450 Å for SPR, and ≈ 2,000 Å for EIS. SPR and EIS have more relaxed requirements on surface roughness and require sample areas of 1 cm^2 or less. Therefore, glass slides can be conveniently used instead of silicon wafers. Due to the sensitivity of EIS with respect to even a small number of membrane defects, it is advisable to form stBLMs immediately after gold deposition for this technique.

The lipid composition of the membrane has to be matched with one of the available stBLM tether chemistries [59–61, 127, 128]. In our experience, biomimetic lipid membranes with unsaturated lipids are best supported by the HC18 tether, which has an EO$_6$ spacer between the terminal thiol and a glycerol backbone that branches into two unsaturated oleoyl chains [61]. Gold-coated substrates are then incubated in an ethanolic solution of a thiol-terminated tether lipid and βME, typically at a molar ratio of 30:70 and a total concentration of 0.2 mM. Upon incubation, the organic molecules form a self-assembled monolayer (SAM) via thiol bonds to the gold surface, and the ratio of tether lipid and βME controls the tether surface density. However, the ability to form insulating, defect-free lipid bilayer membranes is impaired if the tether density falls below a certain threshold, and a 30:70 (tether lipid:βME) ratio constitutes a compromise between low tether density and membrane integrity [59].

A standard technique for completing the stBLM after SAM formation is rapid solvent exchange (RSE) [59, 134]. Here, the SAM of tether molecules and βME is exposed to a ≈ 10 mg/mL solution of lipids in organic solvent. Most zwitterionic lipids readily dissolve in ethanol, whereas anionic lipids require more polar solvents such as mixtures of methanol and chloroform with small amounts of water. The RSE technique leads to membranes of superior quality with standard lipids [59] but may lead to poor results with mixtures of lipids that have significantly different solvent requirements [60]. Vesicle fusion is an alternative that provides better results for lipid mixtures. To aid the formation of defect-free bilayers via vesicle fusion, osmotic shock can assist membrane completion [135, 136]. A standard protocol involves mixing the lipids at the desired molar ratios in chloroform, then evaporating the solvent under vacuum for 12 h. The dried lipid films are re-hydrated in aqueous buffer or pure water with 1–2 M NaCl to a lipid concentration of 5 mg/mL, sonicated until clear, and allowed to incubate the SAM for 1 h. Thereafter, the vesicle solution is slowly replaced by a low salt buffer (50 mM NaCl) to promote vesicle rupture and bilayer completion, followed by a vigorous rinse to remove any remaining vesicles from the bilayer surface. Parameters in this generic procedure that can be optimized for a particular lipid composition are ionic strength, vesicle size, vesicle concentration, temperature, and buffer pH [135].

3.3.4 Assessing membrane quality

Defect-free lipid membranes are a prerequisite for a successful structure determination of membrane-associated proteins with NR. However, the size of membrane defects often is below the resolution limit of direct imaging techniques such as fluorescence microscopy or atomic force microscopy [94, 137]. Therefore, the propensity of a desired lipid composition for forming complete lipid membranes of low defect-density is most easily quantified by measuring the electrical properties of the membrane. On solid substrates such as those used for stBLMs, EIS is an ideal tool to assess bilayer quality [59, 94, 117] and is also well suited to monitor the effect of protein adsorption on membrane integrity [66, 93]. EIS measures the impedance of the membrane as a function of alternating current (AC) frequency [138], from which the density and size distribution of membrane defects can be determined [94, 139]. To characterize stBLMs, a three-electrode configuration is typically used: the gold coated substrate serves as the working electrode, a saturated silver-silver chloride reference electrode is placed in solution on the opposite side of the membrane, and a 0.25-mm-diameter platinum wire coiled around the barrel of the reference acts as an auxiliary electrode. Equivalent circuit models (ECMs) describe the electrochemical system in terms of circuit components and are used to extract physical parameters of the sample from the EI spectra.

Figure 3.7 shows a typical EIS spectrum of a stBLM represented as a Cole-Cole plot (real part vs imaginary part of the complex capacitance $C = (2\pi jfZ)^{-1}$,

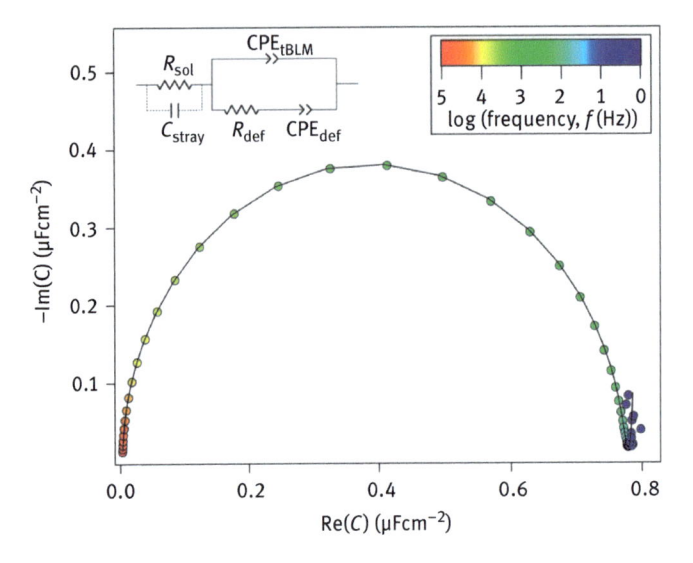

Figure 3.7: Cole–Cole plot of the electrochemical impedance spectra of a stBLM with a complex lipid mixture (30 mol% POPE, 19.5 mol% POPC, 0.5 mol% DMPC, 20 mol% POPS, 30 mol% cholesterol). The solid black line is the best-fit to a model defined by the equivalent circuit shown in the inset. The capacitance of the bilayer is $C_{tBLM} \approx 0.79\ \mu Fcm^{-2}$ (purely capacitive CPE, $\alpha = 1$) and the bilayer resistance is $R_{def} \approx 2\ M\Omega cm^2$.

where f is the AC excitation frequency and Z is the complex impedance). The combined capacitance of the membrane and Helmholtz layer dominates the semicircular shape of the spectrum. The membrane capacitance, inversely related to the thickness of the interfacial film, is approximately equal to the semicircle diameter. The example shown here represents a high-quality bilayer with low defect density. Signatures of membrane defects will show in the low-frequency region of the spectrum at the right-hand side of the figure: the minimum of the curve is a measure of DC defect current (i.e., the curve approaches the $\Re e(C)$ axis closely for highly resistant bilayers); a low-frequency tail develops beyond the minimum (blue-magenta points in the plot) for defective membranes. In membranes that incorporate protein pores, this tail can easily become larger than the semicircle that represents the membrane capacitance.

The solid line is the calculated impedance of the ECM shown in the inset of Figure 3.7 after fitting to the experimental data. This model uses two constant phase elements (CPEs), defined through their impedance $Z_{CPE} = Q_{CPE}/(j\omega)^\alpha$ to describe the electrical properties of the bilayer and its defects. Q_{CPE} is the CPE coefficient, measured in (farad cm^{-2} s$^{\alpha-1}$). The CPE exponent α can vary between 0 (purely resistive) and 1 (purely capacitive) and is typically close to 1 for the unperturbed bilayer in the measured frequency range, as expected for a nonconducting capacitor. In distinction, the CPE exponent for defect areas is close to 0.5, due to ion mobility limitations in the submembrane space and, therefore, a frequency dependence of the relaxation of the potential across the bilayer via ion current through the defects. Such a frequency-dependent impedance is frequently encountered at electrode surfaces as well [94]. The other ECM elements account for the solution resistance R_{sol}, capacitances associated with the measurement setup, C_{stray}, and the frequency-independent resistance of bilayer defects.

3.3.5 Quantification of protein binding to the membrane

Measuring membrane-binding affinities is an essential prerequisite for NR measurements, in particular for peripheral membrane proteins. The lipid composition of the membrane and buffer conditions are the most frequently optimized parameters of the model membrane system with respect to binding affinity, affecting the protein concentration to which a membrane needs to be exposed during an NR experiment. Currently, NR requires 5–10% protein volume occupancy at the membrane to reliably determine the structural profile of a protein–membrane complex. Not in all cases, this coverage can be achieved under conditions that mimic the situation *in vivo*. Whether the binding affinity under physiological conditions is indeed too weak to achieve a sufficient surface coverage for NR, or whether the knowledge about the *in vivo* system is insufficient can often not be determined. As a consequence, a protein's binding affinity is often increased by expanding the fraction of lipids in the membrane that promote a particular binding mechanism. An increased anionic lipid content, for example, will enhance electrostatic interactions of a basic protein with the membrane. Optimizing the ionic strength of the aqueous buffer is another way of tuning this particular interaction, as long as this does not affect protein solubility. In such optimization, often a balance has to be achieved between mimicking *in vivo* conditions and achieving optimal conditions for structure determination. In addition, peripheral membrane proteins often adopt distinct binding conformations under different environmental conditions, which significantly complicate this optimization process [46] but may also provide valuable mechanistic insight.

Membrane affinities of proteins can be measured with direct or indirect assays. In direct assays, the protein–membrane interaction is continuously monitored, while indirect assays physically separate and then quantify membrane-bound and free protein fractions. These assays employ a variety of model membranes such as unilamellar vesicles, membranes tethered to solid supports and free-standing planar bilayers. Indirect assays often invoke centrifugation and chromatographic separation methods [140] while direct quantification methods frequently utilize intrinsic fluorescence or resonance energy transfer assays, isothermal titration calorimetry [140], quartz crystal microbalance with dissipation (QCM/D) [141, 142], bilayer overtone analysis [56], and SPR [56, 140, 143, 144]. The advantages and disadvantages of each method will not be discussed here, but methods that utilize planar supported bilayers are well-suited for a direct application to NR experiments. SPR is one such method and it is routinely applied to measure protein binding to stBLMs.

SPR is an optical technique, sensitive to refractive index changes near a metal–solute interface, used to measure the membrane association of biomolecules in real-time [144]. A common optical setup, the Kretschmann configuration [145], is shown in Figure 3.8. Collective oscillations of free electrons (surface plasmons) are excited at the metal–solute interface under a narrow range of optical conditions. When polarized light strikes the metal film under total internal reflection, and energy and in-plane momentum of the incident monochromatic light simultaneously match that of the surface plasmon mode, photons are converted into surface plasmons [143]. This occurs at a specific angle of incidence, the resonance angle, and is observed as a reduction in the intensity of the reflected beam. When the refractive index near the metal–solute interface changes as protein binds to the membrane, the resonance angle changes. This shift is monitored as a function of time over a range of protein solution concentrations. Equilibrium values of the resonance angle for each protein concentration yield the binding curve from which thermodynamic parameters, such as the equilibrium dissociation constant K_d and the saturation SPR response, can be determined (see Figure 3.8).

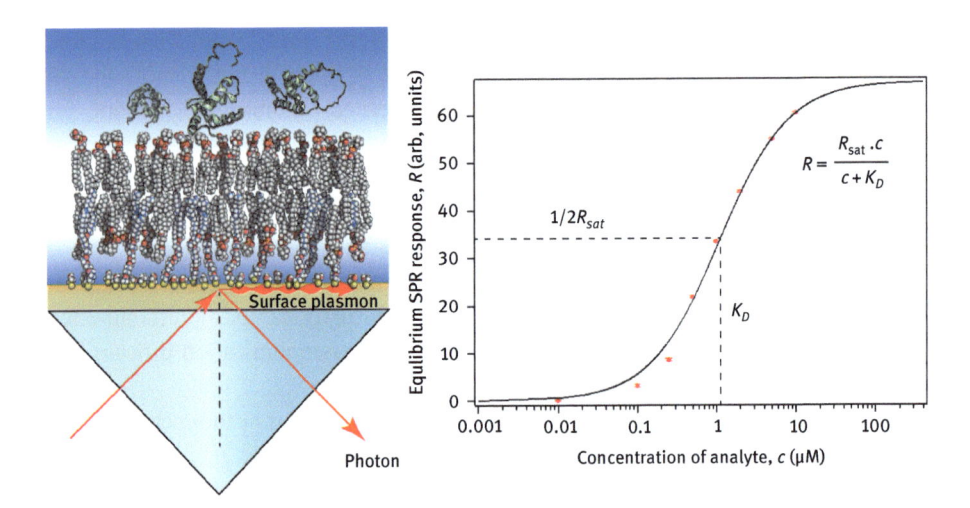

Figure 3.8: Left: Surface plasmon resonance in the Kretschmann configuration for the characterization of protein binding to an stBLM on a gold film. Surface plasmons at the metal–solute interface are excited by incident polarized light under conditions of total internal reflection. The resonance angle at which the conversion occurs (SPR response, R) depends on the refractive index of the material at the metal–solute interface and is monitored as a function of protein concentration in the solute. Shifts in resonance angle are proportional to a change in the amount of membrane-associated protein. Right: Binding curve modeled with a Langmuir isotherm. The Langmuir model yields the dissociation constant K_D of the binding process which is related to the free energy of binding per protein. K_D has the units of a concentration and represents the protein concentration at which one half of the membrane surface sites are occupied by protein (one half of the saturation response R_{sat}).

A standard model for analyzing protein–membrane binding is the Langmuir adsorption model [146]. It assumes a single-step binding event of a ligand (protein) to a receptor (membrane surface site). Each surface site typically contains several lipid molecules.

$$P + \text{Mem} \underset{k_{\text{off}}}{\overset{k_{\text{on}}}{\rightleftharpoons}} P\text{Mem}$$

In this kinetic representation of the Langmuir model, the dissociation constant is defined by the ratio of the rate constants of the association and dissociation processes at equilibrium, $K_D = k_{\text{on}}/k_{\text{off}}$. K_D is also directly related to the free energy of binding $\Delta G = RT \ln(K_D/c^o)$, in which R is the ideal gas constant, T is the temperature, and c^o the standard reference concentration. The Langmuir model does not account for protein–protein interactions in solution or at the membrane, nor does it take into account contributions to the free energy of binding from lipid reorganization or protein crowding at the membrane. Advanced models consider more complex scenarios such as multistate binding events in which conformational changes of the protein occur following membrane binding [146–153].

3.3.6 Optimizing the scattering length density contrast by sample deuteration

The ability to isotopically label, and thus highlight, parts of the interfacial structure is an advantage of neutron scattering that allows specific substructures to be distinguished from other sample components [154]. The exchange of protium for deuterium in biological structures enables large changes in the nSLD of sample components because the scattering cross-sections of ^{1}H and ^{2}H are extremely different, and hydrogen atoms are ubiquitous in biological matter. Therefore, ^{1}H/^{2}H exchange is the most facile and most frequently utilized method to create contrast for neutron scattering in biological samples. There are three major strategies of taking advantage of isotopic labeling in NR from membranes and membrane-bound proteins. Most easily, variation of the bulk solvent nSLD provides scattering contrast, in particular for solvent-containing elements of the interfacial structure (see Figure 3.9). Second, isotopic labeling of a particular protein or specific regions in a protein highlights those regions with respect to its molecular surrounding. This is particularly useful for resolving the structure of protein complexes at the membrane. Finally, lipid deuteration can be used to tune the contrast between the lipid membrane and embedded molecules, such as proteins with trans-membrane domains.

Finding the best deuteration scheme is an optimization problem that requires experience and computational strategies. It is generally good practice to simulate NR data based on anticipated experimental outcomes to assess the sensitivity of the experiment toward the structural features of interest, and it is convenient to combine this exercise with a systematic exploration of deuteration schemes. The use of advanced methods such as the calculation of the information content of the reflectivity data anticipated from different experimental setups puts this optimization process on a quantitative footing [155].

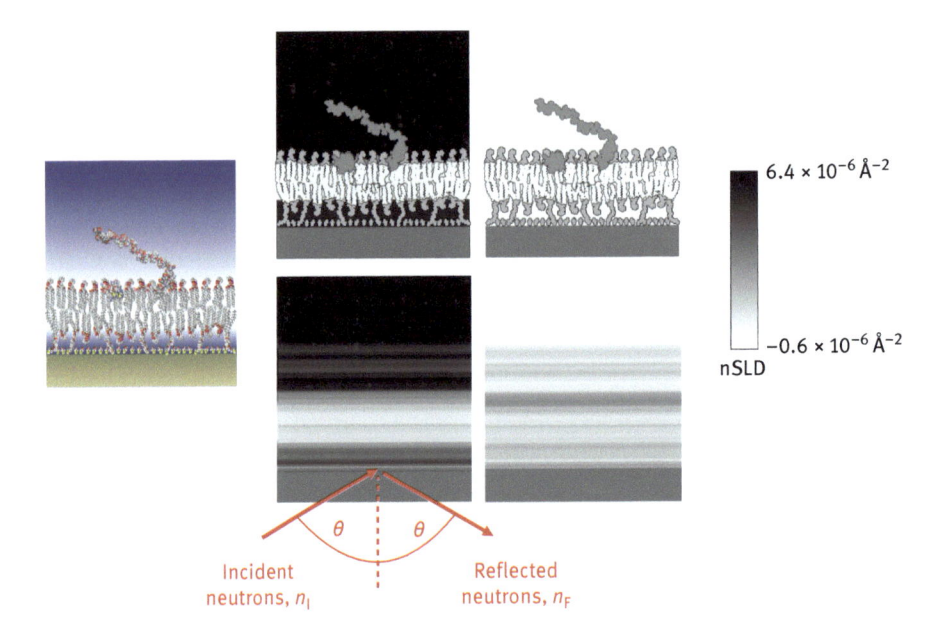

Figure 3.9: Bulk solvent contrast in two related NR experiments highlights different structural features of the surface architecture. Subsequent measurements of the same sample with bulk solvents of different nSLD and simultaneous analysis of all measured contrasts constrains model parameters strongly and increases the confidence, in particular in regions that contain solvent. Left: The colored cartoon shows the structure of a tethered membrane composed of a lipid bilayer and associated protein (α-synuclein (from Jiang et al. [156])). Center and Right: The grayscale images visualize the same cartoon with color-coded nSLD (center: D$_2$O-based bulk solvent (nSLD = 6.34 × 10^{-6} Å^{-2}), right: H$_2$O-based bulk solvent (nSLD = −0.56 × 10^{-6} Å^{-2})). The bottom row shows the in-plane averaged nSLD profiles that give rise to two distinct reflectivity curves. The directions of the incident and reflected neutron beam relative to the sample orientation are indicated in red.

3.4 Conducting the neutron reflectometry experiment

Figure 3.10 shows a schematic view of a typical neutron reflectometer located at a reactor neutron source where a single-crystal monochromator is used to obtain a quasi-monochromatic beam of neutrons. In contrast, reflectometers at pulsed neutron sources distinguish neutrons of different wavelengths via time-of-flight methods, but collect essentially the same reflectivity data as a function of Q_z. Included in the instrument diagram of Figure 3.10 are polarizing and spin flipping devices for measurements utilizing polarized neutrons (the neutron is a Fermion with spin 1/2). The use of polarized neutron beams, particularly in the study of magnetic materials, is reviewed in ref [157]. Polarized neutron beams are of relevance for NR studies of nonmagnetic materials when magnetic reference layers are used for phase-sensitive measurements. In this section, we discuss some of the practical considerations involved in carrying out NR measurements.

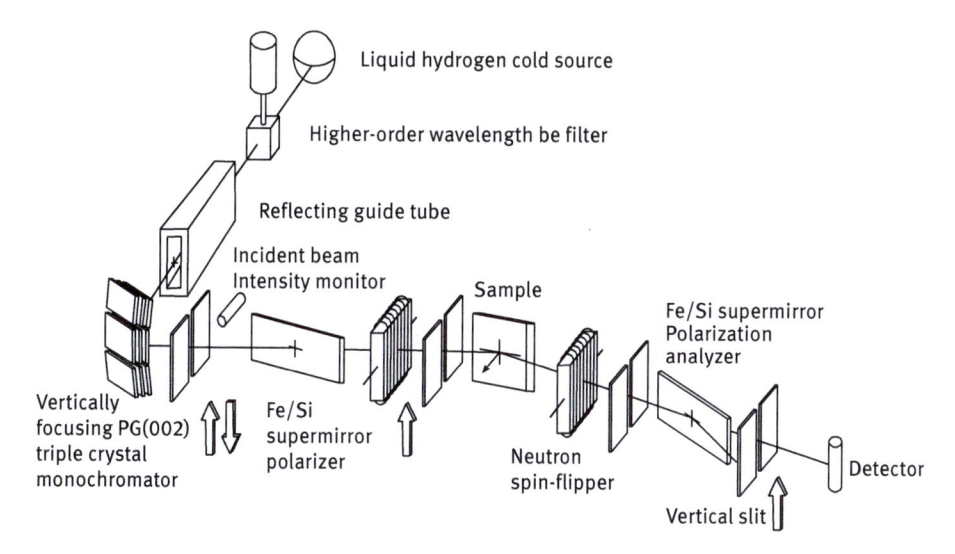

Figure 3.10: Schematic illustration of a typical neutron reflectometer at a reactor source. Thermal neutrons from the reactor core are moderated in the liquid hydrogen cold source to obtain a beam with wavelengths larger than 2 Å. A Be filter eliminates neutrons with a wavelength below 4 Å. A single wavelength of typically 5 Å is selected by a pyrolytic graphite triple crystal monochromator via Bragg reflection and diverted into the instrument at an angle of $\approx 90°$. A pair of presample apertures defines the neutron beam incident on the sample whose intensity is determined in a beam monitor. A pair of postsample apertures reduce background scattering at the detector position while allowing the specularly reflected beam to be registered in the detector. The angle of incidence and the scattering angle can be independently varied. Included in the schematic are polarizing and spin-flipping devices used with polarized beams if the sample contains magnetic reference structures.

3.4.1 Data acquisition and instrumental resolution

Independent from whether a reflectometer uses a monochromatic neutron beam (see Figure 3.10) or whether neutrons of a range of wavelengths are incident on the sample, data collection typically involves measuring the specular reflectivity (see Section 3.2.2), additional nonspecular neutron intensities from which the background contained in the specular signal can be determined, and the incident beam intensity for every instrument configuration used during the measurement. Obtaining a single reflectivity curve for data analysis from this set of raw NR data is called data reduction. It typically involves the subtraction of the background from the specular signal and the division of the background-corrected specular signal by the incident beam intensity. Details vary for different instruments. Scattering facilities often provide remote online data reduction and analysis services that allow for immediate data processing during the measurement [158]. This is particularly useful for NR studies of systems such as those discussed here, as the accuracy of the measurement is a key factor for a successful protein structure determination and remote data reduction software allows for an early assessment of all separately measured raw data.

To obtain accurate SLD depth profiles from specular NR data, it is essential that the measurement is performed with commensurate precision and accuracy. It is common to measure the reflectivity with uncertainties as low as a fraction of a percentage. To do so it requires that systematic errors be minimized by proper alignment of the reflecting surface of the sample relative to the incident beam, exact definition and calibration of slit apertures, and proper compensation of backlash in gear drive mechanisms that control rotation angles. Proper accounting for the propagation of systematic uncertainties in signal, background, and incident beam intensity measurements is necessary as well.

As discussed in Section 3.2.1, each individual neutron wave packet has a coherent distribution of component basis states, each with a different wave vector $\mathbf{k}_j$ and a mean wave vector $\mathbf{k}_M$. Therefore, the beam of independent packets has an incoherent distribution of these mean wave vectors, specified by their angular divergence as well as (a narrow) distribution of wavelengths. A measured reflectivity curve, $R(Q_z) = |r_{\text{meas}}(Q_z)|^2$, is then a convolution of the actual reflectivity of the sample, $|r_{\text{sample}}(Q_z)|^2$, and the instrumental resolution characterized by the effective ΔQ associated with the beam angular divergence and wavelength spread. For monochromatic reflectometers at reactor sources, $\Delta\lambda/\lambda$ is typically of the order of 1–5% and $\Delta\theta$ is of the order of a few minutes of arc.

3.4.2 Low-background fluids sample cells

The ability of neutrons to be transmitted through long distances with negligible scattering or absorption loss in certain materials such as single-crystal silicon makes it possible to measure the reflectivity from lipid membranes on solid substrates adjacent to fluid aqueous reservoirs. NR experiments of membrane-associated proteins on stBLMs typically utilize a fluids cell that allows for in situ buffer exchange (see Figure 3.11) [90]. This enables the study of the evolution of the interfacial structure after addition of small molecules or proteins, or after changing the ionic strength, pH, or temperature of the bulk solvent. Each such condition is typically measured using two isotopically distinct bulk solvents, for example H_2O and D_2O-based buffers (see Section 3.3.6). To verify that the sample is not changing during the measurement of any of those sequential conditions, every individual reflectivity measurement consists of two scans over the entire Q_z-range that can be later combined.

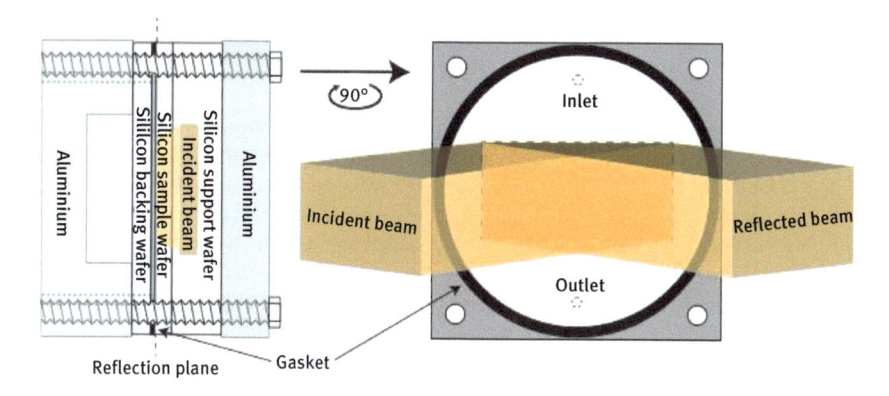

Figure 3.11: The NCNR fluids sample cell for NR measurements. Left: The fluids cell consists of a stack of three 3″ diameter silicon wafers clamped into an aluminum frame that can be mounted at the sample position of the NR instrument. A supported membrane is prepared on a 5–mm-thick Si wafer ("sample wafer") that faces a ≈ 100–μm-thick exchangeable aqueous reservoir (blue layer) defined by a Viton gasket (black). Right: View from the sample wafer facing the aqueous reservoir. The fluid inlet and outlet are drilled into the backing wafer. The footprint of the incident beam on the sample wafer is kept constant during the NR measurement by progressively opening the beam-defining slit apertures as the incident angle θ increases. Note that the incident and reflected neutron beams (orange) traverse macroscopic distances in the 20 mm thick Si "fronting" support wafer and the sample wafer with small attenuation. The refractive bending of the beam upon entry to and exit from the rectangular Si crystal support (Majkrzak et al. [30], Maranville et al. [159]) is not shown. Numerical data reduction and fitting programs perform the appropriate corrections (Kirby et al. [38], Maranville [158]).

The main limitation to the accessible Q_z-range of a neutron measurement using a fluids cell, and thus, the main limitation on the spatial resolution of the experiment, is the amount of background scattering originating from the sample. Therefore, in the remainder of this section, strategies to minimize this scattering are discussed. The beam geometry at the fluids cell sample position of an instrument such as that shown in Figure 3.10 is presented in Figure 3.12a. The monochromatic neutron beam is

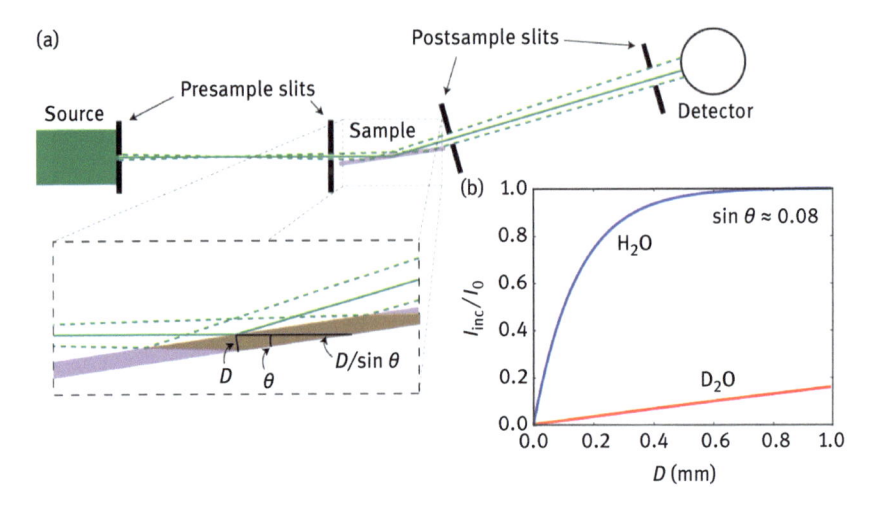

Figure 3.12: Design considerations for an NR experiment. (a) Schematic of the sample geometry at a neutron reflectometer. Pre- and postsample slits are chosen to provide the tightest possible collimation while still illuminating the sample and capturing the entire reflected beam. Maximally divergent neutron paths are shown as dashed lines. The inset shows the beam paths incident on the sample; the brown shading shows the area of the sample illuminated by the beam that generates isotropic background. (b) Fraction of incident intensity incoherently scattered from H_2O and D_2O sample reservoirs as a function of reservoir thickness D shown in the inset in (a). Multiplying this fraction by the fractional solid angle observed by the detector gives an estimate for the expected background (incident angle 4.56°, $Q_z = 0.2$ Å^{-1} for 5 Å neutrons; $\sin\theta \approx 0.08$).

collimated by two presample slits. Background scattering arises primarily from isotropic incoherent scattering from the liquid in the sample cell. Therefore, minimizing the volume of liquid that is both illuminated by the beam and observed by the detector is the most direct approach for background reduction. This requires the narrowest possible (most highly collimated) incident and reflected beams. A ratio of the presample slit openings of 1:1 is optimal, with the value of the slit opening chosen in an angle-dependent manner to illuminate the same total area of the sample over the entire Q_z-range. Two postsample slit openings are chosen to admit the entire reflected specular beam while admitting a minimum of ancillary scattering.

For membrane protein structure determination, a typical number of (nonbackground) neutrons observed in a single NR spectrum with $Q_z^{max} = 0.25$ Å^{-1} are 200,000. The counting time required to observe this number of neutrons varies considerably with the source intensity but is usually several hours. To improve the measurement speed, a divergent beam can also be utilized, where the first presample slit is open wider than the second presample slit. This configuration increases the incident flux on the sample, decreasing count times but increasing the background observed by the detector, and reduces the angular resolution of the measurement. Optimal slit conditions are often chosen to balance experiment speed and background reduction.

The liquid reservoir dimensions of the fluids cell are chosen to minimize the background generated by isotropic incoherent scattering. Neglecting scattering from the cell support materials, a simple estimate of the background level as a fraction of the incident intensity for a given flow cell design is as follows. A beam scattered by the liquid reservoir of thickness D at an angle of incidence θ has a path length through the material $D/\sin\theta$ (see Figure 3.12a). The incoherently scattered fraction of the incident beam is given by:

$$\frac{I_{inc}}{I_0} = 1 - e^{-\frac{\epsilon D}{\sin\theta}}$$

Here, ϵ is a materials property of the reservoir liquid representing the inverse extinction length from incoherent scattering. For 5 Å neutrons incident on D_2O, $\epsilon_{D_2O} = (72.2\,\text{mm})^{-1}$; for H_2O, in which hydrogen has a much larger incoherent cross-section than deuterium, $\epsilon_{H_2O} = (1.84\,\text{mm})^{-1}$ [83]. In Figure 3.12b, the background level is plotted as a function of reservoir thickness for an incident angle of 4.56°. The actual background observed in a measurement is approximated by:

$$\frac{I_{bkg}}{I_0} \approx \frac{\Omega_d I_{inc}}{I_0}$$

where $\Omega_d \approx A_d/4\pi L_{ds}^2$, the fractional solid angle subtended by a detector of area A_d located a distance L_{ds} from the sample, is typically ~10^{-5}.

From Figure 3.12b, it is inferred that an optimal signal-to-noise ratio requires the reservoir thickness to be even smaller than 100 µm. In addition, if the Si support media that surrounds the sample and the reservoir can be decreased to thicknesses of ½ mm or less, inelastic contributions to the background scattering from Si are diminished. It has been shown that such a combination of improvements in sample cell design that minimizes incoherent scattering from both the buffer and the sample support structure allows measurements of reflectivities as low as 10^{-8} at $Q_z^{max} = 0.7$ Å^{-1} [113]. In practice, it remains difficult to routinely prepare such thin samples and mechanically stabilize them during the measurement.

3.4.3 Measuring membrane-associated proteins

Sample preparation, in particular that of stBLMs, is detailed in Section 3.3.3. The last step in this procedure, lipid bilayer completion by osmotic shock-aided vesicle fusion, is typically carried out directly in the fluids cell (see Section 3.4.2). After bilayer completion, the cell is mounted and aligned at the sample stage of the NR instrument. Data are typically collected for the as-prepared bilayer first, using at least two isotopically distinct bulk solvents, such as H_2O- and D_2O-based buffer. A buffer exchange is accomplished by flushing the cell with at least six times its volume. A complete set of reflectivity data consists of measurements of the specular reflection and the scattering background over the entire range of incident angles, as well as a beam-normalization scan that determines the intensity of the incident neutron beam at every angle. Details of those measurements vary between different instruments.

Before adding protein to the sample and at the earliest opportunity, the NR data of the as-prepared lipid bilayer should be analyzed. A structural characterization of a membrane-bound protein at the highest possible resolution requires a complete lipid bilayer, as bilayer defects at the very least complicate data analysis – if they do not lead to nonspecific protein association, which obliterates the determination of the biologically relevant protein–membrane complex structure altogether. As a rule of thumb, lipid bilayers that are less than 90% complete should be discarded, and the best data are obtained from bilayers that are well over 95% complete.

Preparing the protein for measurement is straightforward if the protein stock buffer already matches the desired experimental buffer. In this case, protein at the desired concentration is prepared by diluting an aliquot from the stock solution with the working buffer to the final concentration. If the buffers do not match, dialysis cassettes or spin columns are used for an exchange of the protein buffer. Dialysis is the gentler method but requires large volumes, which is often not cost-effective when exchanging into a D_2O-based buffer. After the buffer exchange, the protein concentration needs to be reassessed such that an aliquot from the new stock can be diluted with working buffer to achieve the desired concentration for the NR experiment based upon the independently determined protein dissociation constant K_D (see Section 3.3.5). A concentration moderately higher than the value of K_D is often a good compromise between sufficient protein coverage for structure determination and a too dense protein coverage that might lead to undesirable protein–protein interactions at the membrane. Thin reservoirs such as those used in NR sample cells can be depleted of protein by membrane binding. For a 100 µm reservoir, this typically occurs when protein binds to the membrane with high affinity ($K_D < 1\,\mu M$). Repeated injections of multiple cell volumes, or a continuous supply of protein, are then required to achieve an optimal protein surface coverage.

Different incubation protocols can be useful for measuring the structure of a membrane-associated protein: (1) The NR measurement is performed while protein is present in solution for the entire time of the measurement, which can be several hours. (2) The lipid membrane is incubated with protein for a set period based on binding kinetics observed with SPR and then replaced by pure buffer before the NR measurement is initiated. The choice between the two strategies is determined by factors such as the rate at which the protein dissociates from the membrane, and the long-term stability of the protein in solution. When an NR measurement takes place with protein in solution, it is useful to measure another set of reflectivity curves after rinsing with pure buffer. Differences of the interfacial structure before and after rinsing then indicate that a fraction of the membrane-bound protein assumes an alternate conformation that dissociates fast from the membrane. A comparison of the measurements before and after buffer rinse can help delineate both protein structures [41].

Automation of buffer exchange reduces manual effort. Various pumping schemes are available for buffer exchange: syringe pumps inject premixed solutions and are convenient to use; peristaltic pumps are ideal for circulating flow and in situ dialysis procedures; and chromatography pumps and valves can be used to automate both mixing and flow. When using such devices, experimenters should be cognizant of the temperature requirements to maintain the stability of any protein-containing solutions. Protein adhesion to tubing walls and denaturation due to shear in microtubing are additional concerns.

3.5 Data analysis

As direct inversion of NR data from biomimetic systems to obtain the structural profile of the sample is in most cases not practical (see Section 3.2), data modeling becomes the prevalent data analysis strategy in biological NR. The ultimate goal of NR data modeling is to identify a parametric molecular model of the biomimetic interface that uniquely corresponds to the measured reflectivity. Intrinsically, NR only yields one-dimensional profiles of neutron scattering length densities along the interface normal while averaging in-plane within the coherence area of the neutron beam (see Section 3.2.2 and Figure 3.3) [30]. Since different materials in the interfacial architecture can have similar nSLDs, this profile may not be uniquely related to a structural profile. With the option of selective deuteration, however, it can usually be ascertained that the structural features of interest bear nSLDs sufficiently different from their immediate molecular surroundings and can, therefore, be resolved (see Section 3.3.6) [154].

NR data modeling makes use of the circumstance that the calculation of a unique reflectivity from a particular interfacial structure is possible, given that the nSLDs of all materials in the structure are known. A typical modeling strategy is then to iteratively find a realization of a suitably parameterized structural model that gives rise to the measured reflectivity. NR modeling is, therefore, a task of finding an appropriate model and a global solution within the parameter space of the model (see Figure 3.13). This procedure does not solve the missing phase problem, discussed in Section 3.2.3, however. Unique solutions are ensured by integrating additional information provided by bounds on model parameters, constraints on the model – such as volumetric data or the chemical connectivity of submolecular moieties – and multiple reflectivity measurements of isomorphic structures at distinct isotopic contrasts. The latter possibility is unique to neutron scattering and usually leads to very stringent constraints on the model when different sets of reflectivity data from one sample are corefined while sharing parameter values for immutable parts of the structure [38]. It has been demonstrated that such an integration of various sources of information leads to unique solutions to the scattering problem [92], in particular, if an approximate structure is already known.

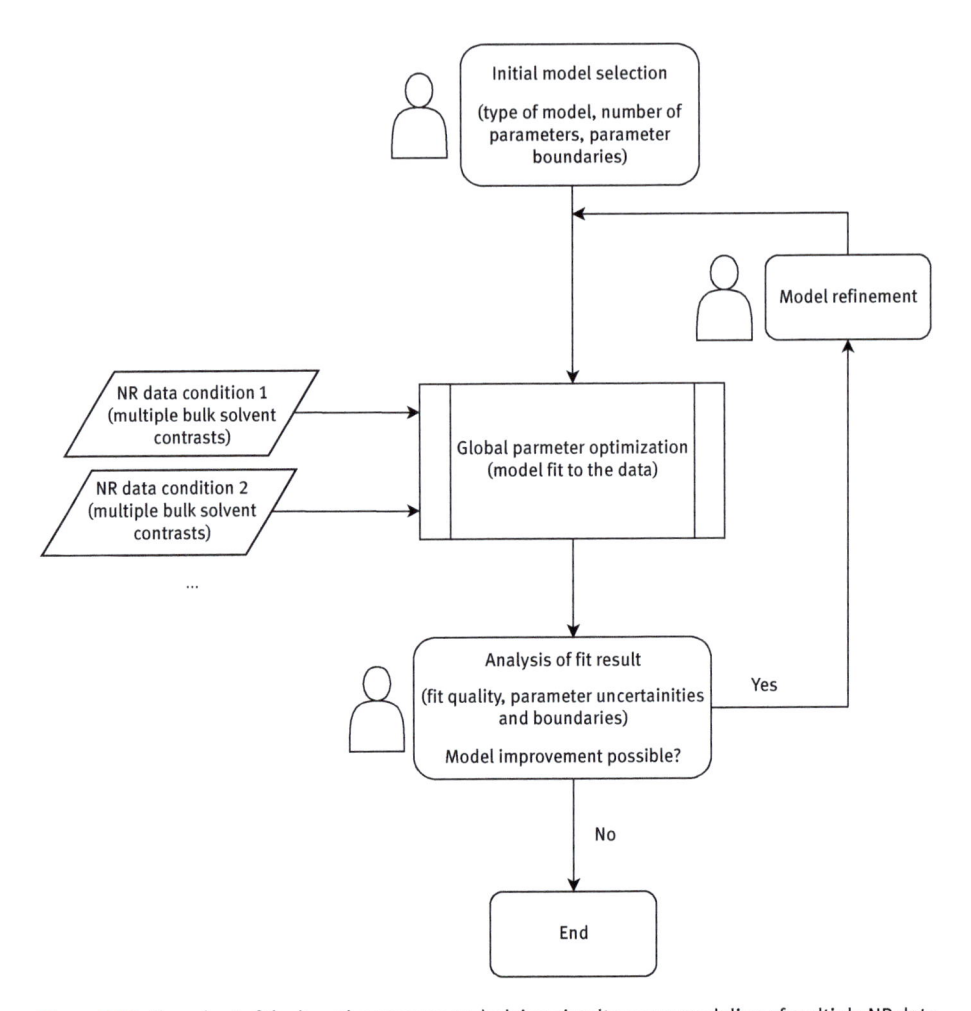

Figure 3.13: Flow chart of the iterative process underlying simultaneous modeling of multiple NR data sets. Steps that require user interaction because they are currently not automated are indicated. Global parameter optimization for the modeling of biomimetic systems requires a robust optimizer that searches the large parameter space efficiently. Solutions are often provided by implementing genetic algorithms (de Haan and Drijkoningen [160]) or a Monte Carlo Markov Chain-based optimizer (Kirby et al. [38], Braak [161], Braak, and Vrugt [162]).

Interfacial structures of proteins on bilayer membranes easily reach a degree of complexity that requires a large number of parameters in a realistic model. Therefore, rigorous methods are needed to determine parameter uncertainties that reveal over-parameterization. We typically use a Monte Carlo Markov Chain (MCMC) global optimizer [38, 161, 162] that yields realistic parameter confidence intervals and provides access to full posterior parameter distributions from which parameter correlations can be identified. Such correlations are useful information for further model optimization. From the posterior, confidence limits on modeling results that are functions of multiple parameters, such as the area per lipid for a lipid bilayer or free-form spline profiles, can be readily calculated. The process of finding an adequately complex model supported by the data still requires empirical testing of models of various complexity (Figure 3.13), as algorithmic model selection [155, 163] has not yet been robustly implemented for NR. While each scattering facility provides software packages for NR data analysis to their beam users [38, 164–166], software modifications that support complex models for biomimetic systems are often required.

3.5.1 Composition-space modeling

A particularly powerful strategy for the analysis of NR data from biomimetic membrane systems is composition-space modeling (see Figure 3.14) [33, 64, 167]. This approach leads to component volume occupancy profiles along the bilayer normal that account for the solid substrate and spatial distributions of all molecular components of the interfacial architecture such as the lipid bilayer and membrane-associated protein. Each such distribution is associated with the scattering length of the particular component, which then allows for a facile computation of the nSLD profile of the entire structure that is needed for the calculation of the model reflectivity. The advantage of composition-space models over other approaches, such as conventional slab models [69], stems from a parameterization of the model that is directly tied to molecular structure. Furthermore, composition-space models readily integrate auxiliary information such as molecular volumes and chemical connectivity, which reduces the number of fit parameters and increases the confidence on unknown parts of the structure. As an example, volume occupancies of the headgroups in a stBLM are tied to their respective hydrocarbon chains since individual volumes for these two components are known from auxiliary methods such as X-ray diffraction [106]. Importantly, composition-space models allow for realistic representations of spatially overlapping molecular distributions, such as that of a protein that penetrates the lipid bilayer. Complete volume filling is achieved without overfilling the available volume or leaving void volumes in the structure. As a result, composition-space modeling has been successfully applied in a large number of studies [49, 58, 67, 95].

Volume occupancy profiles of components with unknown internal structures, such as a membrane-associated protein with unstructured regions, require representations by free-form models. Free-form Hermite splines can accurately describe arbitrary protein profiles (Figure 3.14) and join consistently with lipid bilayer profiles

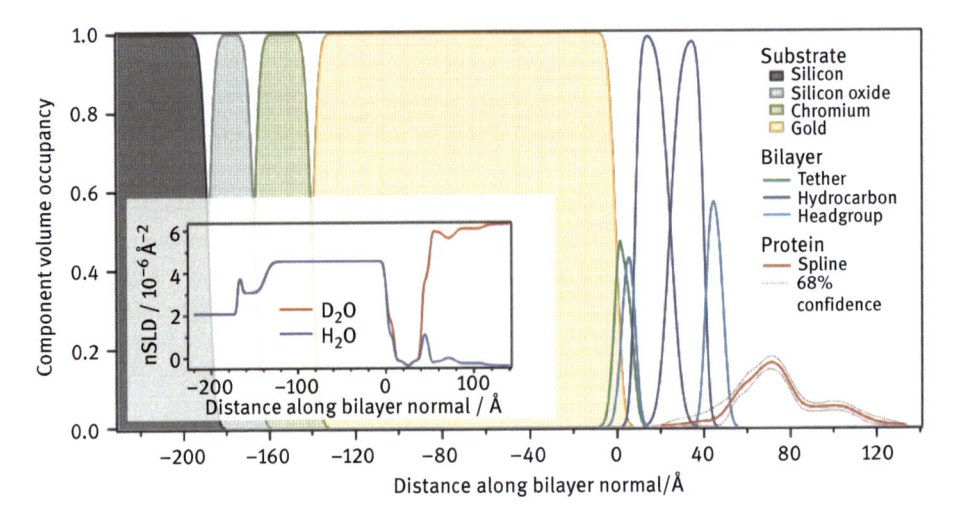

Figure 3.14: Schematic representation of a composition-space model that combines a traditional slab model for the substrate layers with a continuous distribution model for the lipid bilayer and an associated protein. The lipid bilayer is parameterized using volume occupancy distributions of its constituents (Shekhar et al. [64]) and the protein is modeled using a free-form Hermite spline (Heinrich and Lösche [33]). Volume not occupied by either substrate layers or molecular components is taken up by bulk solvent. The inset shows the nSLD profiles calculated from this model for H2O and D2O-based bulk solvent. In combination with similar profiles that fit NR data sets obtained for the as-prepared bilayer (not shown), the resulting general model restricts the parameter space so that the protein structure is determined with high confidence, as shown by the 68% confidence intervals associated with the red profile in the main panel.

to form protein–membrane complexes [33]. In this approach, the protein nSLD has a constant value, which is the average nSLD of the protein given its sequence. It accounts for partial deuteration due to proton exchange with D_2O- and H_2O-based buffers such that the protein nSLD values differ in isotopically distinct solvents. The number of control points that define the Hermite spline is determined by the spatial extension of the protein along the bilayer normal and iteratively refined. By allowing the control point positions to deviate from an equidistant separation, the flexibility of the Hermite spline is increased. Each control point, therefore, carries fit parameters for the volume occupancy of the protein envelope and the separation along the membrane normal. The MCMC optimizer discussed above is an essential tool to avoid over-parameterization as it provides an unbiased determination of the parameter uncertainties [38]. Volume occupancy profiles of constituents of protein–protein complexes at the membrane can be individually resolved if there is scattering contrast between them. This is most easily achieved by deuteration of one of the constituents. The model then includes two separately parameterized spline functions with different average nSLD values of its constituents [58].

3.5.2 Integrative modeling of interfacial structures

To obtain high resolution, three-dimensional information of membrane-bound proteins, integrative modeling strategies are required to supplement the NR data. NR data from surface-associated membrane proteins are routinely refined using crystallographic and NMR structural data within a rigid body modeling approach [33]. Within this approach, a volume occupancy protein profile and a scattering length profile of the high-resolution structure is obtained by projecting the solvent accessible volume and the coherent cross-sections [83] of all atoms of the protein onto the bilayer normal (see Figure 3.15). These two distributions are binned into microslabs of typically 0.5 Å thickness. The solvent accessible volume of the protein is calculated using Connolly's method [168, 169], but can also be derived from experimentally determined average volumes per amino acid, such as from SANS contrast-matching experiments [170]. Both profiles are then directly used within the composition-space model discussed above. The calculated volume occupancy profile can be freely placed at any distance along the bilayer normal (see Figure 3.14), thus allowing the

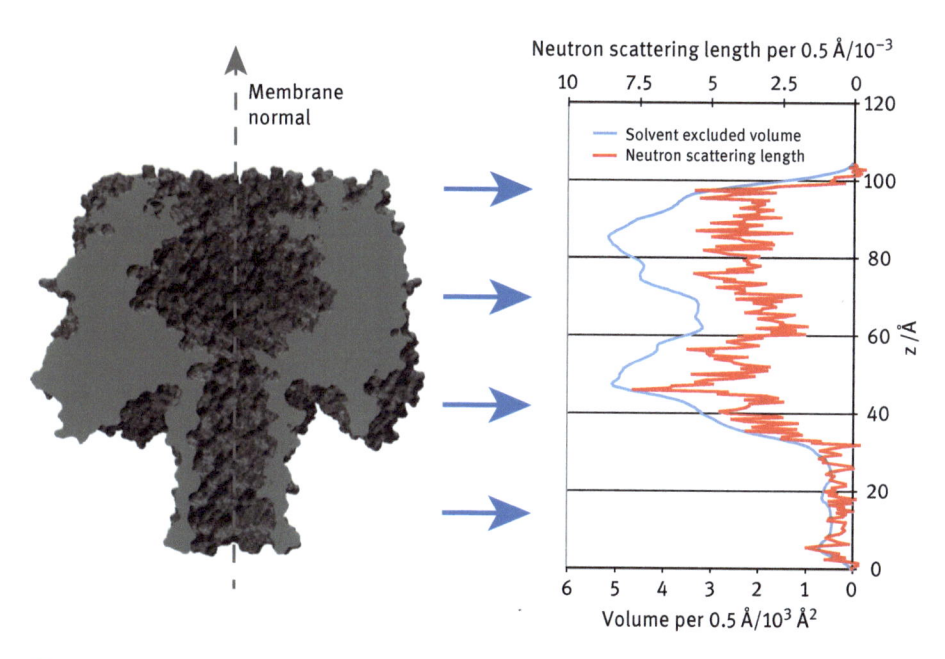

Figure 3.15: Projected, sliced solvent excluding volume and neutron scattering length profiles for the α-hemolysin membrane-pore (Brouette et al. [50]) based on the protein crystal structure (Song et al. [85]). The smooth volume profile was obtained by Connolly's method of rolling a sphere with the diameter of a water molecule over the protein's surface (Pattnaik [145]). The scattering length profile is a projection of discrete scattering centers onto the membrane normal and will be smoothed in successive steps of the data analysis with regard to thermal fluctuations of the atomic positions at room temperature.

protein to penetrate the lipid bilayer or to be strictly peripheral. The height of the protein profile can be scaled to represent different surface coverages. Different to the Hermite spline that is used for model-free protein profiles, the volume occupancy profiles derived from high-resolution structures have variable and potentially more accurate nSLD values associated with every position.

The rigid-body modeling approach underlies the assumption that the membrane-associated structure of the protein is not significantly different from the high-resolution structure. Under this condition, the position and orientation of the protein with respect to the lipid membrane can be determined with high accuracy [49, 66]. To determine the orientation of the protein, the high-resolution protein structure is rotated and projected in discrete steps of typically 5° about two of three Euler angles that are required to define a particular protein orientation. The third Euler angle is irrelevant since NR is invariant against rotational symmetry about the membrane normal [49]. All protein orientations are, thus, parameterized using two Euler angles that are optimized during NR data analysis using interpolation between the precalculated discrete orientations (see Figure 3.16c).

If no satisfactory fit to the data can be determined using a particular high-resolution structure, significant structural changes of the protein upon membrane-interaction, or significant rotational freedom of the rigid protein structure, are likely. A comparison of the volume occupancy profiles obtained from rigid body rotation and free-form modeling can pinpoint the underlying discrepancies. In cases that rigid-body modeling is not viable or ambiguities from the NR data remain, MD or Monte–Carlo simulations are powerful options to gain atomistic information about the membrane-bound protein. Several approaches utilize simulation-based integrative modeling [39, 49, 67, 171, 172], and efforts are underway to establish a complete integrative modeling framework for NR.

A recent example of integrative modeling is shown in Figure 3.16 and is illustrative of both the power and limitations of the technique [56]. The structure of cytoskeletal tubulin heterodimers on biomimetic mitochondrial membranes was interrogated with NR and optimized to the known high-resolution structure [173] using the integrative modeling procedure outlined in this section. The results are presented as volume occupancy profiles in Figure 3.16a and show good agreement between a model-free Hermite spline (black curve) and the profile calculated from a high-resolution structure after optimizing the protein orientation and bilayer penetration depth (red curve). The oriented high-resolution structure is shown in the upper panel of Figure 3.16b and corresponds well to a MD simulation of this system (lower panel). The rotation scheme and Euler angle definitions are shown in Figure 3.16c, while Figure 3.16d shows a probability plot of the tubulin orientation obtained from data analysis. Note, that while the orientation is well constrained in the tilt angle β (which changes the extension of the protein along the bilayer normal and is thus well constrained by NR), it is not as well constrained in rotation α around the heterodimer axis, because of the cylindrical shape of the protein. Likewise, Figure 3.16e shows a significant probability that the

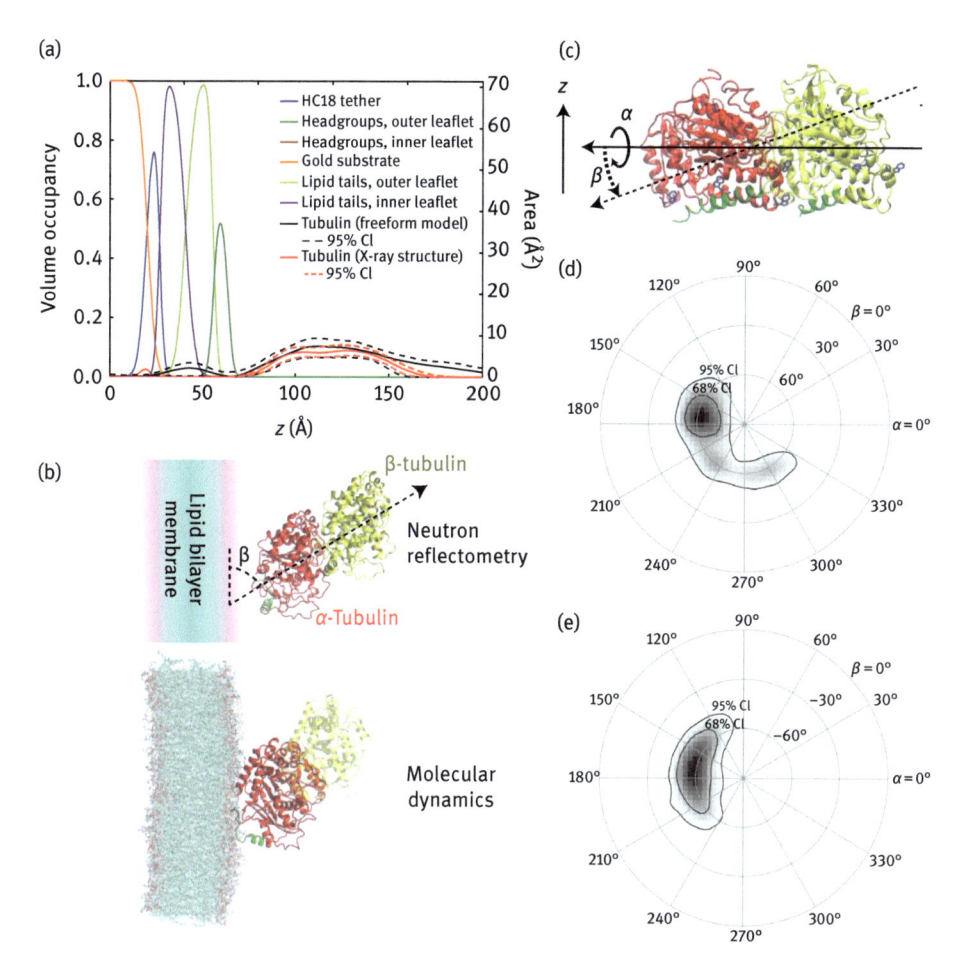

Figure 3.16: Orientation analysis of tubulin on biomimetic mitochondrial membranes. (a) Volume occupancy profiles of bilayer and protein, represented as a freeform model (black curve) and an atomistic model obtained by rigid-body rotation of the high-resolution structure (red curve). (b) Optimized structure of tubulin bound to the membrane through its α subunit obtained from NR (top) and MD simulation (bottom). (c) Definition of Euler rotations for the tubulin heterodimer. (d–e) Polar probability plots of tubulin orientations consistent with the NR data, in which tubulin is bound to the membrane through its α subunit (d) or β subunit (e). Due to the approximate symmetry of the low-resolution tubulin heterodimer structure, NR cannot distinguish between the two possibilities and is also not very sensitive to rotations about the heterodimer axis (Figure adapted from [56]).

tubulin heterodimer is inverted, that is, bound by the other component of the heterodimer. This example shows that NR is sensitive to transformations of a protein that are not about symmetry axes or planes, but cannot distinguish between configurations that produce similar low-resolution density profiles normal to the surface. In these cases, auxiliary techniques (in this example, MD simulations) are useful to distinguish among the structures allowed by the NR data.

3.6 Conclusion

NR from biomimetic lipid model membranes and membrane-associated proteins has seen tremendous improvements over the past decade. Those were triggered by the development of robust biomimetic model membrane systems and by major advances in NR data analysis. Using integrative modeling, it is now possible to obtain high-resolution structural information of membrane-bound proteins that finds direct application in biomedical research and biophysics. The immediate future of the field will be shaped by upcoming major improvements in neutron scattering instrumentation and the installation of new and more powerful neutron sources worldwide, increasing scientific throughput and allowing the field to expand into studying time-dependent processes at the membrane. Integrative modeling will further increase its footprint and is on its way to become the essential data analysis framework for NR from biomimetic systems.

Disclaimer

Certain commercial materials, equipment, and instruments are identified in this work to describe the experimental procedure as completely as possible. In no case does such an identification imply a recommendation or endorsement by NIST, nor does it imply that the materials, equipment, or instrument identified are necessarily the best available for the purpose.

References

[1] Alberts, B., Johnson, A., Lewis, J., et al. Molecular Biology of the Cell, 6 ed., Garland Science, Taylor & Francis Group, LLC, New York 2014, pp. 565–596.
[2] Heijne von, G. The membrane protein universe: what's out there and why bother? J. Intern. Med. 2007, 261, 543–557.
[3] Lodish, H.F., Berk, A., Zipursky, S.L., Matsudaira, P., Baltimore, D., & Darnell, J. Molecular Cell Biology, 4 ed., W.H. Freeman, New York 2000.
[4] van Geest, M., & Lolkema, J.S. Membrane topology and insertion of membrane proteins: search for topogenic signals. Microbiol. Mol. Biol. Rev. 2000, 64, 13–33.
[5] Nastou, K.C., Tsaousis, G.N., Kremizas, K.E., Litou, Z.I., & Hamodrakas, S.J. The human plasma membrane peripherome: visualization and analysis of interactions. Biomed. Res. Int. 2014, 2014, 397145–397112.
[6] Tatulian, S.A. Interfacial enzymes: membrane binding, orientation, membrane insertion, and activity. Methods Enzymol. 2017, 583, 197–230.
[7] Zielenski, J., & Tsui, L.C. Cystic fibrosis: genotypic and phenotypic variations. Annu. Rev. Genet. 1995, 29, 777–807.
[8] Jiang, M., Jia, L., Jiang, W., et al. Protein disregulation in red blood cell membranes of type 2 diabetic patients. Biochem. Biophys. Res. Commun. 2003, 309, 196–200.

[9] Xiao, S., & Shaw, R.M. Cardiomyocyte protein trafficking: relevance to heart disease and opportunities for therapeutic intervention. Trends Cardiovasc. Med. 2015, 25, 379–389.
[10] Meisler, M.H., & Kearney, J.A. Sodium channel mutations in epilepsy and other neurological disorders. J. Clin. Invest. 2005, 115, 2010–2017.
[11] Wymann, M.P., & Schneiter, R. Lipid signalling in disease. Nat. Rev. Mol. Cell. Biol. 2008, 9, 162–176.
[12] de Haan, L., & Hirst, T.R. Cholera toxin: a paradigm for multi-functional engagement of cellular mechanisms (Review). Mol. Membr. Biol. 2004, 21, 77–92.
[13] Ladokhin, A.S. pH-triggered conformational switching along the membrane insertion pathway of the diphtheria toxin T-Domain. Toxins (Basel). 2013, 5, 1362–1380.
[14] Yin, H., & Flynn, A.D. Drugging membrane protein interactions. Annu. Rev. Biomed. Eng. 2016, 18, 51–76.
[15] Hendrickson, W.A. Atomic-level analysis of membrane-protein structure. Nat. Struct. Mol. Biol. 2016, 23, 464–467.
[16] Zhou, H.-X., & Cross, T.A. Modeling the membrane environment has implications for membrane protein structure and function: influenza A M2 protein. Protein Sci. 2013, 22, 381–394.
[17] Ujwal, R., & Bowie, J.U. Crystallizing membrane proteins using lipidic bicelles. Methods. 2011, 55, 337–341.
[18] Ishchenko, A., Abola, E.E., & Cherezov, V. Crystallization of membrane proteins: an overview. Methods Mol. Biol. 2017, 1607, 117–141.
[19] Pebay-Peyroula, E., Rummel, G., Rosenbusch, J.P., & Landau, E.M. X-ray structure of bacteriorhodopsin at 2.5 angstroms from microcrystals grown in lipidic cubic phases. Science. 1997, 277, 1676–1681.
[20] Caffrey, M. A comprehensive review of the lipid cubic phase or in meso method for crystallizing membrane and soluble proteins and complexes. Acta. Crystallogr. F. Struct. Biol. Commun. 2015, 71, 3–18.
[21] Cherezov, V. Lipidic cubic phase technologies for membrane protein structural studies. Curr. Opin. Struct. Biol. 2011, 21, 559–566.
[22] Wadsten, P., Wöhri, A.B., Snijder, A., et al. Lipidic sponge phase crystallization of membrane proteins. J. Mol. Biol. 2006, 364, 44–53.
[23] Jiang, Y., & Kalodimos, C.G. NMR studies of large proteins. J. Mol. Biol. 2017, 429, 2667–2676.
[24] Liang, B., & Tamm, L.K. NMR as a tool to investigate the structure, dynamics and function of membrane proteins. Nat. Struct. Mol. Biol. 2016, 23, 468–474.
[25] Wylie, B.J., Do, H.Q., Borcik, C.G., & Hardy, E.P. Advances in solid-state NMR of membrane proteins. Mol. Phys. 2016, 114, 3598–3609.
[26] Opella, S.J., & Marassi, F.M. Structure determination of membrane proteins by NMR spectroscopy. Chem. Rev. 2004, 104, 3587–3606.
[27] Radoicic, J., Lu, G.J., & Opella, S.J. NMR structures of membrane proteins in phospholipid bilayers. Quart. Rev. Biophys. 2014, 47, 249–283.
[28] Nogales, E. The development of cryo-EM into a mainstream structural biology technique. Nat. Methods. 2016, 13, 24–27.
[29] Brun, A., Darwish, T.A., & James, M. Studies of biomimetic cellular membranes using neutron reflection. J. Chem. Biol. Interfaces. 2013, 1, 3–24.
[30] Majkrzak, C.F., Berk, N.F., Krueger, S., & Perez-Salas, U. Structural Investigations of Membranes in Biology by Neutron Reflectometry, in: F.J.T. Gutberlet, & J. Katsaras. (Eds.), Neutron Scattering in Biology, Springer, Berlin, Heidelberg, 2008, pp. 225–263.
[31] Lakey, J.H. Neutrons for biologists: a beginner's guide, or why you should consider using neutrons. J. R. Soc. Interface. 2009, 6, S567–S573.

[32] Fragneto, G., & Gabel, F. Editorial on the topical issue "neutron biological physics". Eur. Phys. J. E, Soft Matter. 2013, 36, 81.
[33] Heinrich, F., & Lösche, M. Zooming in on disordered systems: neutron reflection studies of proteins associated with fluid membranes. Biochim. Biophys. Acta. 2014, 1838, 2341–2349.
[34] Nanda, H., Resolving Membrane-Bound Protein Orientation and Conformation by Neutron Reflectivity, in: J. M. Ruso, & Piñeiro. Á. (Eds.), Proteins in Solution and at Interfaces. Methods and Application in Biotechnology and Materials Science., Wiley, 2013, pp. 99–111.
[35] Junghans, A., Watkins, E.B., Barker, R.D., et al. Analysis of biosurfaces by neutron reflectometry: from simple to complex interfaces. Biointerphases. 2015, 10, 019014.
[36] Teixeira, S.C.M., Ankner, J., Bellissent-Funel, M.C., et al. New sources and instrumentation for neutrons in biology. Chem. Phys. 2008, 345, 133–151.
[37] Dura, J.A., Pierce, D.J., Majkrzak, C.F., et al. AND/R: advanced neutron diffractometer/reflectometer for investigation of thin films and multilayers for the life sciences. Rev. Sci. Instrum. 2006, 77, 74301–7430111.
[38] Kirby, B.J., Kienzle, P.A., Maranville, B.B., et al. Phase-sensitive specular neutron reflectometry for imaging the nanometer scale composition depth profile of thin-film materials. Curr. Opin. Colloid & Interface Sci. 2012, 17, 44–53.
[39] Pfefferkorn, C.M., Heinrich, F., Sodt, A.J., Maltsev, A.S., Pastor, R.W., & Lee, J.C. Depth of α-synuclein in a bilayer determined by fluorescence, neutron reflectometry, and computation. Biophys. J. 2012, 102, 613–621.
[40] Jiang, Z., Heinrich, F., McGlinchey, R.P., Gruschus, J.M., & Lee, J.C. Segmental deuteration of α-Synuclein for neutron reflectometry on tethered bilayers. J. Phys. Chem. Lett. 2017, 8, 29–34.
[41] Zimmermann, K., Eells, R., Heinrich, F., et al. The cytosolic domain of T-cell receptor ζ associates with membranes in a dynamic equilibrium and deeply penetrates the bilayer. J. Biol. Chem. 2017, 292, 17746–17759.
[42] Hellstrand, E., Grey, M., Ainalem, M.-L., et al. Adsorption of α-Synuclein to supported lipid bilayers: positioning and role of electrostatics. ACS. Chem. Neurosci. 2013, 4, 1339–1351.
[43] Jones, E.M., Dubey, M., Camp, P.J., et al. Interaction of tau protein with model lipid membranes induces tau structural compaction and membrane disruption. Biochemistry. 2012, 51, 2539–2550.
[44] Junghans, A., Watkins, E.B., Majewski, J., Miranker, A., & Stroe, I. Influence of the human and rat islet amyloid polypeptides on structure of phospholipid bilayers: neutron reflectometry and fluorescence microscopy studies. Langmuir. 2016, 32, 4382–4391.
[45] Shenoy, S., Shekhar, P., Heinrich, F., et al. Membrane association of the PTEN tumor suppressor: molecular details of the protein-membrane complex from SPR binding studies and neutron reflection. PLoS ONE. 2012, 7, e32591.
[46] Eells, R., Barros, M., Scott, K.M., Karageorgos, I., Heinrich, F., & Lösche, M., Structural characterization of membrane-bound human immunodeficiency virus-1 Gag matrix with neutron reflectometry. Biointerphases. 2017, 12, 02D408.
[47] Kent, M.S., Murton, J.K., Sasaki, D.Y., et al. Neutron reflectometry study of the conformation of HIV Nef bound to lipid membranes. Biophys. J. 2010, 99, 1940–1948.
[48] Fritz, K., Fritz, G., Windschiegl, B., Steinem, C., & Nickel, B. Arrangement of Annexin A2 tetramer and its impact on the structure and diffusivity of supported lipid bilayers. Soft Matter. 2010, 6, 4084–4094.
[49] Heinrich, F., Nanda, H., Goh, H.Z., Bachert, C., Lösche, M., & Linstedt, A.D. Myristoylation restricts orientation of the GRASP domain on membranes and promotes membrane tethering. J. Biol. Chem. 2014, 289, 9683–9691.

[50] Brouette, N., Fragneto, G., Cousin, F., Moulin, M., Haertlein, M., & Sferrazza, M. A neutron reflection study of adsorbed deuterated myoglobin layers on hydrophobic surfaces. J. Colloid Interface Sci. 2013, 390, 114–120.
[51] Nanda, H., Datta, S.A.K., Heinrich, F., et al. Electrostatic interactions and binding orientation of HIV-1 matrix studied by neutron reflectivity. Biophys. J. 2010, 99, 2516–2524.
[52] Datta, S.A.K., Heinrich, F., Raghunandan, S., et al. HIV-1 Gag extension: conformational changes require simultaneous interaction with membrane and nucleic acid. J. Mol. Biol. 2011, 406, 205–214.
[53] Le Brun, A.P., Holt, S.A., Shah, D.S., Majkrzak, C.F., & Lakey, J.H. Monitoring the assembly of antibody-binding membrane protein arrays using polarised neutron reflection. Eur. Biophys. J. 2008, 37, 639–645.
[54] Kreiner, M., Chillakuri, C.R., Pereira, P., et al. Orientation and surface coverage of adsorbed fibronectin cell binding domains and bound integrin alpha α5β1 receptors. Soft Matter. 2009, 5, 3954–3962.
[55] Le Brun, A.P., Holt, S.A., Shah, D.S.H., Majkrzak, C.F., & Lakey, J.H. The structural orientation of antibody layers bound to engineered biosensor surfaces. Biomaterials. 2011, 32, 3303–3311.
[56] Hoogerheide, D.P., Noskov, S.Y., Jacobs, D., et al. Structural features and lipid binding domain of tubulin on biomimetic mitochondrial membranes. Proc. Natl. Acad. Sci. U.S.A. 2017, 114, E3622–E3631.
[57] Hossain, K.R., Holt, S.A., Le Brun, A.P., Khamici Al, H., Valenzuela, S.M. X-ray and neutron reflectivity study shows that CLIC1 undergoes cholesterol-dependent structural reorganization in lipid monolayers. Langmuir. 2017, 33, 12497–12509.
[58] Yap, T.L., Jiang, Z., Heinrich, F., et al. Structural features of membrane-bound glucocerebrosidase and α-synuclein probed by neutron reflectometry and fluorescence spectroscopy. J. Biol. Chem. 2015, 290, 744–754.
[59] McGillivray, D.J., Valincius, G., Vanderah, D.J., et al. Molecular-scale structural and functional characterization of sparsely tethered bilayer lipid membranes. Biointerphases. 2007, 2, 21–33.
[60] Heinrich, F., Ng, T., Vanderah, D.J., et al. A new lipid anchor for sparsely tethered bilayer lipid membranes. Langmuir. 2009, 25, 4219–4229.
[61] Budvytyte, R., Valincius, G., Niaura, G., et al. Structure and properties of tethered bilayer lipid membranes with unsaturated anchor molecules. Langmuir. 2013, 29, 8645–8656.
[62] Vockenroth, I.K., Ohm, C., Robertson, J.W.F., McGillivray, D.J., Lösche, M., & Köper, I. Stable insulating tethered bilayer lipid membranes. Biointerphases. 3, 2008, FA68.
[63] Fragneto, G. Neutrons and model membranes. Eur. Phys. J-Spec. Top. 2012, 213, 327–342.
[64] Shekhar, P., Nanda, H., Lösche, M., & Heinrich, F. Continuous distribution model for the investigation of complex molecular architectures near interfaces with scattering techniques. J. Appl. Phys. 2011, 110, 102216-1–102216-12.
[65] Belicka, M., Gerelli, Y., Kučerka, N., & Fragneto, G. The component group structure of DPPC bilayers obtained by specular neutron reflectometry. Soft Matter. 2015, 11, 6275–6283.
[66] McGillivray, D.J., Valincius, G., Heinrich, F., et al. Structure of functional Staphylococcus aureus α-hemolysin channels in tethered bilayer lipid membranes. Biophys. J. 2009, 96, 1547–1553.
[67] Shenoy, S.S., Nanda, H., & Lösche, M. Membrane association of the PTEN tumor suppressor: electrostatic interaction with phosphatidylserine-containing bilayers and regulatory role of the C-terminal tail. J. Struct. Biol. 2012, 180, 394–408.
[68] Smith, G.S., & Majkrzak, C.F. Neutron reflectometry, in: E. Prince, H. Fuess, T. Hahn, et al. (Eds.), International Tables for Crystallography, 1st ed., International Union of Crystallography, Chester, England 2006, pp. 126–146.

[69] Ankner, J.F., & Majkrzak, C.F. Subsurface profile refinement for neutron specular reflectivity. SPIE. Proc. 1992, 1738, 260–269.
[70] Penfold, J., Richardson, R.M., Zarbakhsh, A., et al. Recent advances in the study of chemical surfaces and interfaces by specular neutron reflection. J. Chem. Soc. Faraday Trans. 1997, 93, 3899–3917.
[71] Russell, T.P. X-ray and neutron reflectivity for the investigation of polymers. Mater. Sci. Rep. 1990, 5, 171–271.
[72] Fragneto, G., Thomas, R.K., Rennie, A.R., & Penfold, J. Neutron reflection study of bovine β-casein adsorbed on OTS self-assembled monolayers. Science. 1995, 267, 657–660.
[73] Wacklin, H.P. Neutron reflection from supported lipid membranes. Curr. Opin. Colloid & Interface Sc. 2010, 15, 445–454.
[74] Daillant, J. Recent developments and applications of grazing incidence scattering. Curr. Opin. Colloid & Interface Sci. 2009, 14, 396–401.
[75] Kučerka, N., Nieh, M.-P., Pencer, J., Harroun, T., & Katsaras, J. The study of liposomes, lamellae and membranes using neutrons and X-rays. Curr. Opin. Colloid & Interface Sci. 2007, 12, 17–22.
[76] Lovell, M.R., & Richardson, R.M. Analysis methods in neutron and X-ray reflectometry. Curr. Opin. Colloid & Interface Sci. 1999, 4, 197–204.
[77] Lu, J.R., Zhao, X., & Yaseen, M. Protein adsorption studied by neutron reflection. Curr. Opin. Colloid & Interface Sci. 2007, 12, 9–16.
[78] Schlossman, M.L. Liquid-liquid interfaces: studied by X-ray and neutron scattering. Curr. Opin. Colloid & Interface Sci. 2002, 7, 235–243.
[79] Thomas, R.K., & Penfold, J. Neutron and X-ray reflectometry of interfacial systems in colloid and polymer chemistry. Curr. Opin. Colloid & Interface Sci. 1996, 1, 23–33.
[80] Als-Nielsen, J., & Kjær, K. X-Ray Reflectivity and Diffraction Studies of Liquid Surfaces and Surfactant Monolayers, in: Phase Transitions in Soft Condensed Matter, Springer, Boston, MA, New York 1989, pp. 113–138.
[81] Majkrzak, C.F., Metting, C., Maranville, B.B., et al. Determination of the effective transverse coherence of the neutron wave packet as employed in reflectivity investigations of condensed-matter structures. I. Measurements. Phys. Rev. A. 2014, 89, 033851.
[82] Cohen-Tannoudji, C., Diu, B., & Laloe, F. Quantum Mechanics, Wiley-VCH, New York 1992.
[83] Sears, V.F. Neutron scattering lengths and cross sections. Neutron News. 1992, 3, 26–37.
[84] Majkrzak, C.F., Berk, N.F., & Perez-Salas, U. Phase-sensitive neutron reflectometry. Langmuir. 2003, 19, 7796–7810.
[85] Song, L., Hobaugh, M.R., Shustak, C., Cheley, S., Bayley, H., & Gouaux, J.E. Structure of staphylococcal α-hemolysin, a heptameric transmembrane pore. Science. 1996, 274, 1859–1866.
[86] Majkrzak, C.F., Carpenter, E., Heinrich, F., & Berk, N.F. When beauty is only skin deep; optimizing the sensitivity of specular neutron reflectivity for probing structure beneath the surface of thin films. J. Appl. Phys. 2011, 110, 102212.
[87] Majkrzak, C.F., & Berk, N.F. Exact determination of the phase in neutron reflectometry. Phys. Rev. B. 1995, 52, 10827–10830.
[88] Dehaan, V.O., van Well, A.A., Adenwalla, S., & Felcher, G.P. Retrieval of phase information in neutron reflectometry. Phys. Rev. B. 1995, 52, 10831–10833.
[89] Majkrzak, C.F., & Berk, N.F. Exact determination of the phase in neutron reflectometry by variation of the surrounding media. Phys. Rev. B. 1998, 58, 15416–15418.
[90] Majkrzak, C.F., Berk, N.F., Krueger, S., et al. First-principles determination of hybrid bilayer membrane structure by phase-sensitive neutron reflectometry. Biophys. J. 2000, 79, 3330–3340.

[91] Majkrzak, C.F., & Berk, N.F. Phase sensitive reflectometry and the unambiguous determination of scattering length density profiles. Physica. B. 2003, 336, 27–38.
[92] Majkrzak, C.F., Berk, N.F., Kienzle, P., & Perez-Salas, U. Progress in the development of phase-sensitive neutron reflectometry methods. Langmuir. 2009, 25, 4154–4161.
[93] Valincius, G., Heinrich, F., Budvytyte, R., et al. Soluble amyloid β-oligomers affect dielectric membrane properties by bilayer insertion and domain formation: implications for cell toxicity. Biophys. J. 2008, 95, 4845–4861.
[94] Valincius, G., Meškauskas, T., & Ivanauskas, F. Electrochemical impedance spectroscopy of tethered bilayer membranes. Langmuir. 2012, 28, 977–990.
[95] Barros, M., Heinrich, F., Datta, S.A.K., et al. Membrane binding of HIV-1 matrix protein: dependence on bilayer composition and protein lipidation. J. Virol. 2016, 90, 4544–4555.
[96] Chan, Y.-H.M., & Boxer, S.G. Model membrane systems and their applications. Curr. Opin. Chem. Biol. 2007, 11, 581–587.
[97] Khan, M.S., Dosoky, N.S., & Williams, J.D. Engineering lipid bilayer membranes for protein studies. Int. J. Mol. Sci. 2013, 14, 21561–21597.
[98] Weygand, M., Wetzer, B., Pum, D., et al. Bacterial S-layer protein coupling to lipids: x-ray reflectivity and grazing incidence diffraction studies. Biophys. J. 1999, 76, 458–468.
[99] Tietjen, G.T., Gong, Z., Chen, C.-H., et al. Molecular mechanism for differential recognition of membrane phosphatidylserine by the immune regulatory receptor Tim4. Proc. Natl. Acad. Sci. USA. 2014, 111, E1463–14672.
[100] Tietjen, G.T., Baylon, J.L., Kerr, D., et al. Coupling X-Ray reflectivity and in silico binding to yield dynamics of membrane recognition by Tim1. Biophys. J. 2017, 113, 1505–1519.
[101] Vaknin, D., Als-Nielsen, J., Piepenstock, M., & Lösche, M. Recognition processes at a functionalized lipid surface observed with molecular resolution. Biophys. J. 1991, 60, 1545–1552.
[102] Lösche, M., Piepenstock, M.M., Diederich, A.A., Grünewald, T.T., Kjær, K., & Vaknin, D. Influence of surface chemistry on the structural organization of monomolecular protein layers adsorbed to functionalized aqueous interfaces. Biophys. J. 1993, 65, 2160–2177.
[103] Akgun, B., Satija, S., Nanda, H., et al. Conformational transition of membrane-associated terminally acylated HIV-1 nef. Structure. 2013, 21, 1822–1833.
[104] Schalke, M., & Lösche, M. Structural models of lipid surface monolayers from X-ray and neutron reflectivity measurements. Adv. Colloid Interface. 2000, 88, 243–274.
[105] Wiener, M.C., & White, S.H. Structure of a fluid dioleoylphosphatidylcholine bilayer determined by joint refinement of x-ray and neutron diffraction data. III. Complete structure. Biophys. J. 1992, 61, 434–447.
[106] Nagle, J.F., & Tristram-Nagle, S. Structure of lipid bilayers. Biochim. Biophys. Acta. 2000, 1469, 159–195.
[107] Liu, Y., & Nagle, J.F. Diffuse scattering provides material parameters and electron density profiles of biomembranes. Phys. Rev. E. 2004, 69, 40901.
[108] Cruz, J., Mihailescu, M., Wiedman, G., et al. A membrane-translocating peptide penetrates into bilayers without significant bilayer perturbations. Biophys. J. 2013, 104, 2419–2428.
[109] Tristram-Nagle, S.A., Chan, R., Kooijman, E., et al. HIV fusion peptide penetrates, disorders, and softens T-cell membrane mimics. J. Mol. Biol. 2010, 402, 139–153.
[110] Hardy, G.J., Nayak, R., Munir Alam, S., Shapter, J.G., Heinrich, F., & Zauscher, S. Biomimetic supported lipid bilayers with high cholesterol content. J. Mater. Chem. 2012, 22, 19506–19513.
[111] Benedetto, A., Heinrich, F., Gonzalez, M.A., Fragneto, G., Watkins, E., & Ballone, P. Structure and stability of phospholipid bilayers hydrated by a room-temperature ionic liquid/water solution: a neutron reflectometry study. J. Phys. Chem. B. 2014, 118, 12192–12206.

[112] Meuse, C.W., Krueger, S., Majkrzak, C.F., et al. Hybrid bilayer membranes in air and water: infrared spectroscopy and neutron reflectivity studies. Biophys. J. 1998, 74, 1388–1398.
[113] Krueger, S., Meuse, C.W., Majkrzak, C.F., et al. Investigation of hybrid bilayer membranes with neutron reflectometry: probing the interactions of melittin. Langmuir. 2001, 17, 511–521.
[114] Anderson, N.A., Richter, L.J., Stephenson, J.C., & Briggman, K.A. Determination of lipid phase transition temperatures in hybrid bilayer membranes. Langmuir. 2006, 22, 8333–8336.
[115] Anderson, N.A., Richter, L.J., Stephenson, J.C., & Briggman, K.A. Characterization and control of lipid layer fluidity in hybrid bilayer membranes. J. Am. Chem. Soc. 2007, 129, 2094–2100.
[116] Köper, I. Insulating tethered bilayer lipid membranes to study membrane proteins. Mol. Biosyst. 2007, 3, 651–657.
[117] Junghans, A., & Köper, I. Structural analysis of tethered bilayer lipid membranes. Langmuir. 2010, 26, 11035–11040.
[118] Rossi, C., & Chopineau, J. Biomimetic tethered lipid membranes designed for membrane-protein interaction studies. Eur. Biophys. J. 2007, 36, 955–965.
[119] Andersson, J., Knobloch, J.J., Perkins, M.V., Holt, S.A., & Köper, I. Synthesis and characterization of novel anchorlipids for tethered bilayer lipid membranes. Langmuir. 2017, 33, 4444–4451.
[120] Wagner, M.L., & Tamm, L.K. Tethered polymer-supported planar lipid bilayers for reconstitution of integral membrane proteins: silane-polyethyleneglycol-lipid as a cushion and covalent linker. Biophys. J. 2000, 79, 1400–1414.
[121] Naumann, C.A., Prucker, O., Lehmann, T., Ruhe, J., Knoll, W., & Frank, C.W. The polymer-supported phospholipid bilayer: tethering as a new approach to substrate-membrane stabilization. Biomacromolecules. 2002, 3, 27–35.
[122] Tanaka, M., & Sackmann, E. Polymer-supported membranes as models of the cell surface. Nature. 2005, 437, 656–663.
[123] Jablin, M.S., Zhernenkov, M. Toperverg, B.P., et al. In-plane correlations in a polymer-supported lipid membrane measured by off-specular neutron scattering. Phys. Rev. Lett. 2011, 106, 138101.
[124] Singh, S., Junghans, A., Tian, J., et al. Polyelectrolyte multilayers as a platform for pH-responsive lipid bilayers. Soft Matter. 2013, 9, 8938–8948.
[125] Fragneto, G., Charitat, T., Graner, F., Mecke, K., Perino-Gallice, L., & Bellet-Amalric, E. A fluid floating bilayer. Epl-Europhys. Lett. 2001, 53, 100–106.
[126] Hughes, A.V., Howse, J.R., Dabkowska, A., Jones, R., Lawrence, M.J., & Roser, S.J. Floating lipid bilayers deposited on chemically grafted phosphatidylcholine surfaces. Langmuir. 2008, 24, 1989–1999.
[127] Schiller, S.M., Naumann, R., Lovejoy, K., Kunz, H., & Knoll, W. Archaea analogue thiolipids for tethered bilayer lipid membranes on ultrasmooth gold surfaces. Angew. Chem. Int. Ed. Engl. 2003, 42, 208–211.
[128] Vockenroth, I.K., Atanasova, P.P., Jenkins, A.T.A., & Köper, I. Incorporation of alpha-hemolysin in different tethered bilayer lipid membrane architectures. Langmuir. 2008, 24, 496–502.
[129] Vockenroth, I.K., Rossi, C., Shah, M.R., & Köper, I. Formation of tethered bilayer lipid membranes probed by various surface sensitive techniques. Biointerphases. 2009, 4, 19–26.
[130] Shenoy, S., Moldovan, R., Fitzpatrick, J., Vanderah, D.J., Deserno, M., & Lösche, M. In-plane homogeneity and lipid dynamics in tethered bilayer lipid membranes (tBLMs). Soft Matter. 2010, 2010, 1263–1274.
[131] van Meer, G., Voelker, D.R., & Feigenson, G.W. Membrane lipids: where they are and how they behave. Nat. Rev. Mol. Cell. Biol. 2008, 9, 112–124.

[132] Lomize, M.A., Pogozheva, I.D., Joo, H., Mosberg, H.I., & Lomize, A.L. OPM database and PPM web server: resources for positioning of proteins in membranes. Nucleic Acids Res. 2012, 40, D370–D376.
[133] Whited, A.M., & Johs, A. The interactions of peripheral membrane proteins with biological membranes. Chem. Phys. Lipids. 2015, 192, 51–59.
[134] Cornell, B.A., Braach-Maksvytis, V., King, L.G., et al. A biosensor that uses ion-channel switches. Nature. 1997, 387, 580–583.
[135] Hardy, G.J., Nayak, R., & Zauscher, S. Model cell membranes: techniques to form complex biomimetic supported lipid bilayers via vesicle fusion. Curr. Opin. Colloid & Interface Science. 2013, 18, 448–458.
[136] Seitz, M., Ter-Ovanesyan, E., Hausch, M., et al. Formation of tethered supported bilayers by vesicle fusion onto lipopolymer monolayers promoted by osmotic stress. Langmuir. 2000, 16, 6067–6070.
[137] Kwak, K.J., Valincius, G., Liao, W.-C., et al. Formation and finite element analysis of tethered bilayer lipid structures. Langmuir. 2010, 26, 18199–18208.
[138] Barsoukov, E., & Macdonald, J.R. (Eds.). Impedance Spectroscopy, John Wiley & Sons, Inc, New Jersey 2005.
[139] Valincius, G., Mickevicius, M., Penkauskas, T., & Jankunec, M. Electrochemical impedance spectroscopy of tethered bilayer membranes: an effect of heterogeneous distribution of defects in membranes. Electrochim. Acta. 2016, 222, 904–913.
[140] Cho, W.W., Bittova, L., & Stahelin, R.V. Membrane binding assays for peripheral proteins. Anal. Biochem. 2001, 296, 153–161.
[141] Richter, R.P., Him, J.L.K., Tessier, B., Tessier, C., & Brisson, A.R. On the kinetics of adsorption and two-dimensional self-assembly of annexin A5 on supported lipid bilayers. Biophys. J. 2005, 89, 3372–3385.
[142] Rodahl, M., Höök, F., Fredriksson, C., et al. Simultaneous frequency and dissipation factor QCM measurements of biomolecular adsorption and cell adhesion. Faraday Discuss. 1997, 107, 229–246.
[143] Knoll, W. Interfaces and thin films as seen by bound electromagnetic waves. Annu. Rev. Phys. Chem. 1998, 49, 569–638.
[144] Löfås, S., Malmqvist, M., Rönnberg, I., Stenberg, E., Liedberg, B., & Lundström, I. Bioanalysis with surface plasmon resonance. Sens. Actuators B: Chem. 1991, 5, 79–84.
[145] Pattnaik, P. Surface plasmon resonance - Applications in understanding receptor-ligand interaction. Appl. Biochem. Biotechnol. 2005, 126, 79–92.
[146] Wei, Y., & Latour, R.A. Determination of the adsorption free energy for peptide-surface interactions by SPR spectroscopy. Langmuir. 2008, 24, 6721–6729.
[147] Hinderliter, A., & May, S. Cooperative adsorption of proteins onto lipid membranes. J. Phys.: Condens. Matter 2006, 18, S1257–S1270.
[148] Minton, A.P. Effects of excluded surface area and adsorbate clustering on surface adsorption of proteins. II. Kinetic models. Biophys. J. 2001, 80, 1641–1648.
[149] May, S., Harries, D., & Ben-Shaul, A. Lipid demixing and protein-protein interactions in the adsorption of charged proteins on mixed membranes. Biophys. J. 2000, 79, 1747–1760.
[150] Heimburg, T., Angerstein, B., & Marsh, D. Binding of peripheral proteins to mixed lipid membranes: effect of lipid demixing upon binding. Biophys. J. 1999, 76, 2575–2586.
[151] Weiss, J.N. The Hill equation revisited: uses and misuses. Faseb J. 1997, 11, 835–841.
[152] Murray, D., Ben-Tal, N., Honig, B., & McLaughlin, S. Electrostatic interaction of myristoylated proteins with membranes: simple physics, complicated biology. Struct./Folding Des. 1997, 5, 985–989.

[153] Ben-Shaul, A., Ben-Tal, N., & Honig, B. Statistical thermodynamic analysis of peptide and protein insertion into lipid membranes. Biophys. J. 1996, 71, 130–137.
[154] Heinrich, F. Deuteration in biological neutron reflectometry. Methods in Enzymol. 2016, 566, 211–230.
[155] Sivia, D.S., & Webster, J. The Bayesian approach to reflectivity data. Physica B. 1998, 248, 327–337.
[156] Jiang, Z., Hess, S.K., Heinrich, F., & Lee, J.C. Molecular details of alpha-synuclein membrane association revealed by neutrons and photons. J. Phys. Chem. B. 2015, 119, 4812–4823.
[157] Majkrzak, C.F., O'Donovan, K.V., & Berk, N.F. (2006). Polarized Neutron Reflectometry. In T. Chatterji (Ed.), *Neutron Scattering from Magnetic Materials* (pp. 397–471). Amsterdam, Boston, Heidelberg, London, New York, Oxford, Paris, San Diego, San francisco, Singapore, Syney, Tokyo: Elsevier. http://doi.org/10.1016/B978-044451050-1/50010-0
[158] Maranville, B.B. Interactive, web-based calculator of neutron and X-ray reflectivity. J. Res. Natl. Inst. Stand. Technol. 2017, 122, 34.
[159] Maranville, B.B., Kirby, B.J., Grutter, A.J., et al. Measurement and modeling of polarized specular neutron reflectivity in large magnetic fields. J. Appl. Crystallogr. 2016, 49, 1121–1129.
[160] de Haan, V., & Drijkoningen, G. Genetic algorithms used in model finding and fitting for neutron reflection experiments. Physica B. 1994, 198, 24–26.
[161] Braak, T. Cajo, J.F. A Markov Chain Monte Carlo version of the genetic algorithm differential evolution: easy Bayesian computing for real parameter spaces. Stat. Comput. 2006, 16, 239–249.
[162] Braak, T., Vrugt, J.A., & Cajo, J.F. Differential Evolution Markov Chain with snooker updater and fewer chains. Stat. Comput. 2008, 18, 435–446.
[163] Samoilenko, I., Feigin, L., Shchedrin, B., & Antolini, R. Selection of a model in reflectometry: use of the linear statistical inference. Physica B. 2000, 283, 262–267.
[164] Nelson, A. Co-refinement of multiple-contrast neutron/X-ray reflectivity data using MOTOFIT. J. Appl. Crystallogr. 2006, 39, 273–276.
[165] van der Lee, A., Salah, F., & Harzallah, B. A comparison of modern data analysis methods for X-ray and neutron specular reflectivity data. J. Appl. Crystallogr. 2007, 40, 820–833.
[166] Gerelli, Y. Aurore: new software for neutron reflectivity data analysis. J. Appl. Crystallogr. 2016, 49, 330–339.
[167] Wiener, M.C., & White, S.H. Fluid bilayer structure determination by the combined use of x-ray and neutron diffraction. II. "Composition-space" refinement method. Biophys. J. 1991, 59, 174–185.
[168] Connolly, M.L. Solvent-accessible surfaces of proteins and nucleic acids. Science. 1983, 221, 709–713.
[169] Connolly, M.L. Analytical molecular surface calculation. J. Appl. Crystallogr. 1983, 16, 548–558.
[170] Perkins, S.J. Protein volumes and hydration effects. The calculations of partial specific volumes, neutron scattering matchpoints and 280 nm absorption coefficients for proteins and glycoproteins from amino acid sequences. Eur. J. Biochem. 1986, 157, 169–180.
[171] Curtis, J.E., Zhang, H., & Nanda, H. SLDMOL: a tool for the structural characterization of thermally disordered membrane proteins. Comput. Phys. Commun. 2014, 185, 3010–3015.
[172] Hughes, A.V., Ciesielski, F., Kalli, A.C., et al. On the interpretation of reflectivity data from lipid bilayers in terms of molecular-dynamics models. Acta. Crystallogr. D. Struct. Biol. 2016, 72, 1227–1240.
[173] Nettles, J.H., Li, H.L., Cornett, B., Krahn, J.M., Snyder, J.P., & Downing, K.H. The binding mode of epothilone A on α,β-tubulin by electron crystallography. Science. 2004, 305, 866–869.

Elizabeth G. Kelley, Paul D. Butler, Michihiro Nagao

4 Collective dynamics in model biological membranes measured by neutron spin echo spectroscopy

Thermal fluctuations and mechanical properties in lipid bilayers

Abstract: Cell membranes are extraordinarily heterogeneous environments composed of many thousands of chemically distinct lipids, sterols, and proteins. It is this very complexity and diversity in membrane composition that allows for its many varied and critical biological functions. These membranes are rather thin, only 3 to 5 nm thick, and present both structural and dynamic features on a wide variety of length and timescales. Within the hierarchy of length and timescales, the membrane's mechanical properties control many of the key functions such as bilayer shape transformations, protein binding, budding, and molecular transport which in turn are related to such things as apoptosis, endocytosis, protein signaling, and drug delivery. In this chapter, we will review how the elastic properties of membranes control the membranes' dynamics by presenting experimental results obtained by means of neutron spin echo spectroscopy. Toward the end of the chapter, we will consider another interesting property, membrane viscosity, and discuss some future aspects and challenges.

Keywords: lipid, membrane, dynamics, neutron scattering, neutron spin echo, bending modulus, area compressibility modulus, membrane viscosity

4.1 Introduction

The biological functions of lipid membranes require that they be highly dynamic. The hydrophobic tails rapidly flex and kink while the individual lipid molecules rotate, protrude, and diffuse on picosecond to nanosecond timescales [1, 2]. Biomembranes are also strikingly fluid with lipid diffusion coefficients on the order of 10^{-8} cm^2/s,

---

Elizabeth G. Kelley, NIST Center for Neutron Research, National Institute of Standards and Technology, Gaithersburg, USA
Paul D. Butler, NIST Center for Neutron Research, National Institute of Standards and Technology, Gaithersburg, USA; Department of Chemical & Biomolecular Engineering, University of Delaware, Newark, USA; Department of Chemistry, University of Tennessee, Knoxville, USA
Michihiro Nagao, NIST Center for Neutron Research, National Institute of Standards and Technology, Gaithersburg, USA; Center for Exploration of Energy and Matter, Department of Physics, Indiana University, Bloomington, USA

https://doi.org/10.1515/9783110544657-004

meaning a lipid molecule can diffuse across a typical cell length in less than 10 s [3, 4]. The fast local dynamics of the molecules allow the cell to easily manipulate the lateral membrane organization necessary for protein–protein interactions and cell signaling [5–7]. At the same time, the membrane must bend, bud, and fuse on much larger length scales as cells take in nutrients, send chemical signals, and even grow and divide [8, 9]. There is a large gap in length scale and timescale between the molecular and macroscopic processes that is bridged by the collective membrane dynamics shown in Figure 4.1. These mesoscale fluctuations involve tens to hundreds of lipids on the length scale of nm to μm and have their own role to play in the membrane's biological functions.

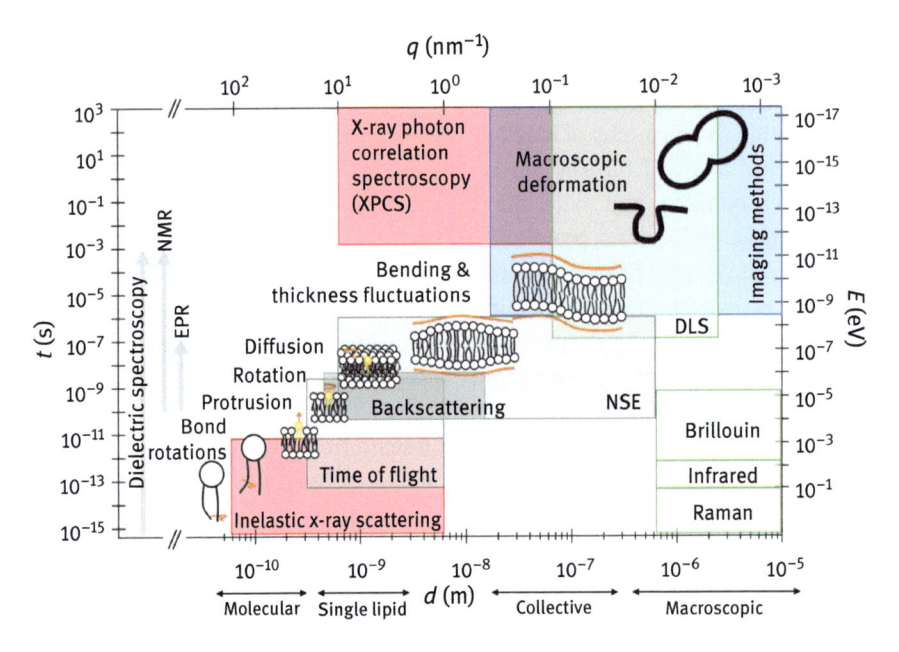

Figure 4.1: The length (d) and timescales (t), and corresponding momentum (Q) and energy (E) transfers, covering the hierarchy of membrane dynamics. The dynamic ranges of the spectroscopic techniques available to measure the different dynamics are also shown, with neutron techniques as gray squares, X-ray in red, light in green, and fluorescence and optical imaging techniques in blue. Spectroscopic techniques such as dielectric spectroscopy, nuclear magnetic resonance (NMR), and electron paramagnetic resonance (EPR) access a broad range of timescales without any specific associated length scales.

Stochastic membrane fluctuations, such as collective bending and thickness fluctuations, lead to small local changes in the membrane shape. These fluctuations were first observed in red blood cell plasma membranes as early as 1890 [10], and more than 125 years of research has revealed that the membrane shape changes are the result of both active and equilibrium processes [11]. At high frequencies, the fluctuations are dominated by thermal excitations and are thought to play a role in preventing cell–cell adhesion and facilitate the diffusion of membrane-embedded proteins [12–17]. These same fluctuations affect membrane protein folding, channel formation, and function within the membrane [18–23], and play a significant role in cell adhesion [24–28], facilitating vesicle budding and trafficking [8, 29], and influencing cell spreading and motility [30–32]. Moreover, changes in membrane fluctuations have been directly linked to cell function and disease. Studies of macrophages show that the membranes are softer and have a greater fluctuation amplitude when the cells are activated, which likely aides in phagocytosis as the cells engulf foreign bodies [33, 34]. Changes in the fluctuation spectrum of red blood cell membranes also are seen after malaria and parasite infections [35, 36], as well as in sickle cell diseases [37–39], and studies suggest that cancer cells are softer than nonmalignant cells, which may contribute to the blebbing and migration of cancerous cells [40–42]. As such, measuring and quantifying the membrane fluctuations have important implications for understanding both biomembranes and cell functions.

This chapter focuses on measuring the equilibrium undulations in model lipid membranes, namely the mesoscale collective bending and thickness fluctuations, and how the measurements can be used to quantify the membrane elastic and viscous properties. Both of these dynamic processes are a direct consequence of the membrane being soft. As illustrated in Figure 4.1, the bending fluctuations are undulations normal to the membrane plane at a constant membrane thickness (often referred to as height fluctuations) [12, 43–45], while thickness fluctuations are undulations of the two membrane leaflets in opposite directions (sometimes referred to as peristaltic or breathing modes) [46–50]. The fluctuation length scales are dictated by the membrane elastic properties, or how resistant the membrane is to bending or compressing in solvent [51–53], the same properties that determine the energy required for the large-scale membrane deformations that occur in cell processes with macroscopic membrane remodeling. Meanwhile, the thickness fluctuation timescale will be determined by the membrane compressibility and how long it takes the lipids to flow as the membrane relaxes – or the membrane viscosity [53] – the same viscosity that dictates the timescales of molecular lipid and protein diffusion in the membrane that we talked about at the beginning of this chapter [54–57]. In other words, the collective fluctuations are governed by the same properties that influence the microscopic and macroscopic membrane functions, and measuring these dynamics gives us a way to quantify both the elastic and viscous properties that are essential to biological membrane functions.

While the mesoscale dynamics are an important bridge between the microscopic and macroscopic scales, accessing the length and timescales necessary to study the collective fluctuations can be experimentally challenging. Biological membranes are on the order of 5 nm thick, meaning we need to be able to resolve bending fluctuations on sub-nanometer scales. Similarly, theory and experiment suggest that thickness fluctuation amplitudes are on the order of 20% of the membrane thickness [47, 52,

58, 59] – or ≈ 1 nm – and estimates from the membrane viscosity tell us that these fluctuations should relax on nanosecond timescales [53]. These dynamics are too slow for traditional spectroscopic techniques such as NMR [2, 60–62], Raman [63–65], infrared [64, 66], dielectric [67], and EPR [68–70] that are sensitive to the motions of fatty acyl tails and the individual lipid molecules. On the other hand, the dynamics are too small and too fast for most light scattering and microscopy methods that are used to study large-scale dynamics and shape changes [34, 71–74]. As illustrated in Figure 4.1, neutron spectroscopy techniques cover a range of length scales and timescales that are hard to access with other characterization techniques and are therefore uniquely suited for characterizing the collective membrane fluctuations. This chapter will focus on neutron spin echo (NSE) spectroscopy and how this technique can be used to measure the collective bending and thickness fluctuations to determine the membrane elastic and viscous properties.

The remainder of this chapter is structured as follows: Section 4.2 presents the basics of neutron scattering and how NSE works. Section 4.3 covers the theoretical background necessary to correlate membrane dynamics with experimental NSE data. In Section 4.4, example NSE measurements of collective bending and thickness fluctuations are presented. A summary and current and future challenges close out the chapter in Section 4.5.

4.2 The NSE Technique

Neutrons are noncharged particles (neutron mass $m_n = 1.67 \times 10^{-27}$ kg) [75] that have both particle and wavelike properties. A free neutron has momentum $\vec{p}_n = \hbar\vec{q}$, where $\vec{q}$ is the wavevector of the associated wave function and $\hbar = h/2\pi$ with h as Planck's constant. $\vec{q}$ describes the propagation direction and speed of the wave with a magnitude $|\vec{q}| = q = m_n v_n/\hbar = 2\pi/\lambda$, where v_n is the velocity of the neutron and λ is its wavelength. Typical neutron wavelengths range from 0.1 to 2 nm, which allows neutron scattering techniques to provide detailed structural information on the nanometer scale. Several neutron scattering methods covered in other chapters illustrate the capability of neutron scattering to provide important information about the lipid bilayer thickness, composition, and asymmetry as well as protein binding and orientation in the membrane.

Neutron scattering also is an important tool for understanding nanoscale dynamics. The neutron energy is given by $E_n = \left|\frac{p_n^2}{2m_n}\right|$ and is on the order of the thermal energy, $k_B T \approx 10$ meV (1 meV = 1.6×10^{-22} J = 2.42×10^{11} Hz). Neutron backscattering and time of flight instruments are capable of resolving 10^{-3} meV changes in neutron energy, providing dynamic information on picosecond to nanosecond timescales [76, 77]. These techniques mostly measure the incoherent or "self" dynamics – the correlations between the relative positions of a given atom at different times – and are very useful for determining the diffusion of hydration water, lipids, and other small molecules embedded in lipid membranes [78–84].

Measuring collective membrane dynamics on the tens or hundreds of nanosecond timescales requires an instrument capable of measuring even smaller changes in neutron energy, also referred to as having a higher energy resolution. Measuring a relaxation time on the order of 100 ns requires resolving a change in neutron energy of $\approx$ 10 neV (10^{-8} eV!), which is currently only achievable using NSE, the highest energy resolution neutron spectroscopy technique [85]. NSE is capable of measuring these really small energy changes because it operates using fundamentally different principles from other spectrometers as we discuss in more detail in Section 4.2.3. Another important difference is that NSE is most suited to measuring the coherent dynamics – dynamics that originate from correlations between relative positions of different atoms at different times – or the exact collective dynamics we are interested in measuring in lipid membranes.

4.2.1 Basics of neutron scattering

A basic scheme of a scattering experiment is shown in Figure 4.2. Consider a neutron beam characterized by its wavevector $\vec{q}$. If the initial wave $\vec{q}_i$ interacts with a nucleus and scatters, then the final wave will have a different wavevector $\vec{q}_f$. Here, we can define the momentum transfer as

$$\hbar\vec{Q} = \hbar(\vec{q}_i - \vec{q}_f) \tag{4.1}$$

where $\vec{Q} = \vec{q}_i - \vec{q}_f$ is the scattering vector, often also referred to as the momentum transfer vector, while the corresponding energy transfer, ω, is defined as the difference between the initial, E_i, and final, E_f, waves

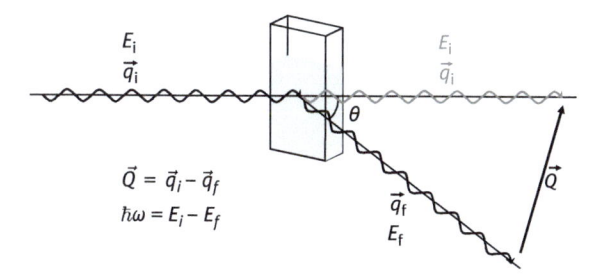

Figure 4.2: Schematic of the geometry of a scattering experiment. The initial neutron wave is characterized by its wavevector $\vec{q}_i$ and energy E_i. The initial wave interacts with the object and is scattered into scattering angle θ. The final neutron is characterized by its wavevector $\vec{q}_f$ and energy E_f. The scattering vector $\vec{Q}$ is expressed as $\vec{Q} = \vec{q}_i - \vec{q}_f$, and the energy exchanged is $\hbar\omega = E_i - E_f$.

$$\hbar\omega = E_i - E_f = \frac{\hbar^2}{2m_n}(q_i^2 - q_f^2) \tag{4.2}$$

Equations (4.1) and (4.2) express the momentum and energy conservation of the scattering process, respectively. Here, we consider the magnitude of $\vec{Q}$ by defining the angle between $\vec{q}_i$ and $\vec{q}_f$ as the scattering angle, θ, and applying $q_i \approx q_f$ ($\hbar\omega \approx 0$ i.e., small energy transfer) as

$$|\vec{Q}| = Q = 2q_i \sin\left(\frac{\theta}{2}\right) = \frac{4\pi}{\lambda}\sin\left(\frac{\theta}{2}\right) \tag{4.3}$$

In the case of a perfectly ordered system, Bragg's law states that constructive interference will occur when $2d\sin\left(\frac{\theta}{2}\right) = n\lambda$, where d is the spacing between scattering planes and n is an integer. For such systems, Q is simply given as $Q = 2\pi/d$, demonstrating the inverse relationship between the length scales in the sample and the Q at which those length scales significantly impact the scattering. In other words, large length scales are associated with small Q values and vice versa. The primary aim of neutron scattering is to determine the probability of neutrons being scattered in $\vec{Q}$ with energy transfer ω, known as the dynamic structure factor $S(\vec{Q}, \omega)$. The Fourier transform of $S(\vec{Q}, \omega)$ with respect to ω is called the intermediate scattering function (ISF), $I(\vec{Q}, t)$, which can be written as

$$S(\vec{Q}, \omega) = \text{FT}\left\{I(\vec{Q}, t)\right\} \tag{4.4}$$

$$I(\vec{Q}, t) = \left\langle \exp\left\{-i\vec{Q} \cdot [\vec{r}(t) - \vec{r}'(0)]\right\} \right\rangle \tag{4.5}$$

where $\langle \cdots \rangle$ indicates an ensemble average over all pairs of atoms, t is the time, and $\vec{r}$ and $\vec{r}'$ are position vectors for the atoms, respectively.

In a static elastic scattering experiment such as small-angle neutron scattering (SANS), we ignore the energy exchange between the neutrons and our samples and integrate the scattering over all neutron energies,

$$S(\vec{Q}) = \int S(\vec{Q}, \omega) d\omega \tag{4.6}$$

which corresponds to the Fourier transform of the instantaneous spatial atomic correlations in the system, that is, the structure of our sample. For the rest of the chapter, we only consider isotropic scattering cases for simplicity and treat the vector $\vec{Q}$ as a scalar Q. Measuring the membrane structure requires counting the number of neutrons scattered at angle θ or corresponding Q.

4.2.2 Inelastic/quasi-elastic scattering techniques

Measuring the dynamic structure factor, $S(Q, \omega)$, requires keeping track of the energy exchanged between the neutrons and the sample (ω) as these are scattered at a defined angle (θ or corresponding Q). This type of measurement is called inelastic or quasi-elastic neutron scattering. Information on the dynamic length scales is given by Q while information on the energy scales is given by ω. Most spectrometers determine the energy exchanged by analyzing the initial and final neutron energies. Therefore, in order to achieve higher energy resolution, narrow $\Delta\lambda/\lambda$ (or $\Delta v_n/v_n$) of the initial neutron beam is required. The need for a narrow wavelength distribution drastically limits the number of useful neutrons, and a scattering experiment with the high energy resolution needed to measure relaxation processes on the order of 100 ns would be impossible because of poor counting statistics. Ferenc Mezei, who invented the NSE principle in 1972 [85], illustrates the severity of this limitation in the first text book on NSE [86]. According to his estimates, achieving an energy resolution of 50 neV on an ideal hypothetical time of flight instrument would not count more than one neutron a day in the detector, which means it would take literally years to get any useful information on the sample dynamics. As explained below, his proposed NSE technique broke through this limitation and significantly improved the energy resolution by eliminating the requirement for a narrow $\Delta\lambda/\lambda$.

4.2.3 Characteristics of the NSE technique

Currently, the best NSE instrument in the world (IN15 in Grenoble, France) has an energy resolution of a few neV (or a few µs) while using a wavelength distribution $\Delta\lambda/\lambda$ on the order of 10% [87], ensuring sufficient neutron flux to perform the scattering experiments. There are a handful of NSE instruments in operation around the world [86–95], with more under construction [96]. There are two spectrometers in North America: the SNS-NSE at Oak Ridge National Laboratory [95] and the NGA-NSE (formerly NG5-NSE) at the National Institute of Standards and Technology [91]. The NGA-NSE is a reactor-based instrument while the SNS-NSE is at a spallation neutron source. The instrument design and operation for reactor versus spallation sources are slightly different due to the differences in the neutron source; however, both instruments operate on the same basic principles outlined below. We note that the quantum mechanical description and details of NSE instrumentation are beyond the scope of this Chapter, and we instead refer the interested reader to a textbook on NSE [92] as well as publications on specific spectrometers [97, 98] for more information.

NSE measures dynamics by taking advantage of the fact that neutrons have a spin. Although a neutron has no net charge, it has a spin degree of freedom of 1/2, which gives the neutron a magnetic moment. The NSE technique uses polarized neutrons, meaning only one state of the neutron spin is selected from the initial neutron beam.

That polarized beam then passes through a variety of magnetic fields as it traverses the instrument, through the sample and onto the detector. The trick to NSE's high energy resolution is to make use of the Larmor precession of the neutron spin in a magnetic field to provide each neutron with an "internal" clock with which to track any change in neutron velocity (or equivalently energy, $E_i = m_n v_i^2/2$) [85, 86, 92]. When the neutron spin direction is parallel to a magnetic field, the polarization of the spin state is maintained. On the other hand, when the neutron spin is perpendicular to the magnetic field, the spin starts to rotate around the magnetic field (called Larmor precession), shown schematically in Figure 4.3b. The neutron spins will precess with a frequency, $\omega_L = -\gamma_L B$ in which γ_L is the Larmor constant for a neutron and B is the average field strength, and the precession angle is proportional to the time the neutron spends in the magnetic field: $\varphi_L = \omega_L t = \omega_L l/v \propto B\lambda$ where l is the length of the field. Larmor precession is the same process that is key to NMR (for nuclear spin of various kinds of nuclei) and EPR (for electron spin) spectroscopic techniques that are also used to measure membrane dynamics [61, 70].

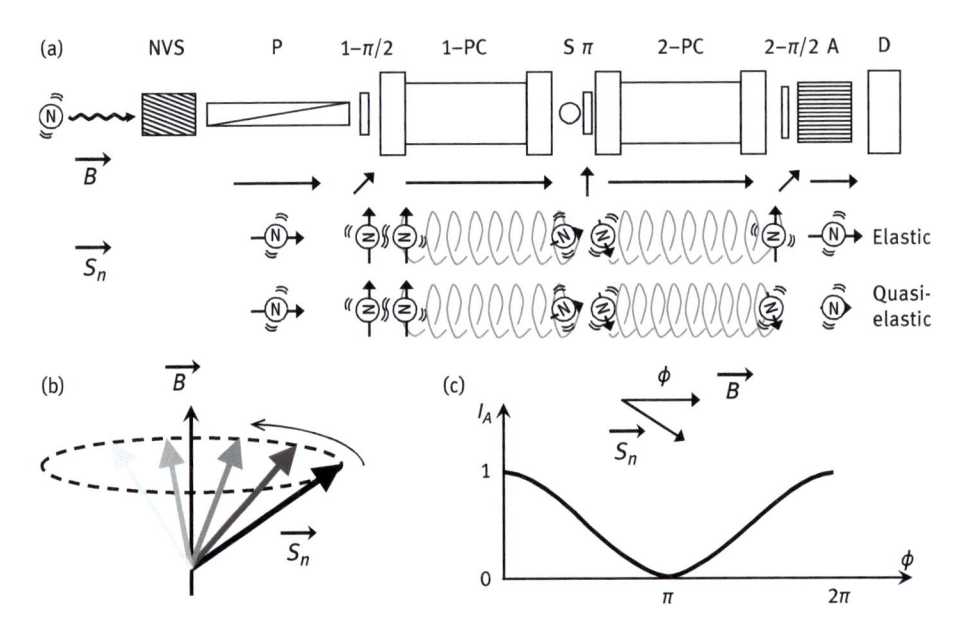

Figure 4.3: (a) A schematic illustration of a reactor-based NSE instrument, which consists of a neutron velocity selector (NVS), a neutron polarizer (P), the first and second π/2 flippers (1-π/2 and 2-π/2), precession fields (1-PC and 2-PC) before and after the sample (S), a π flipper, a neutron spin analyzer (A), and a neutron detector (D). Also the instrumental magnetic field direction $\vec{B}$ and neutron spin direction $\vec{S}_n$ for elastic and quasi-elastic scattering cases are shown. (b) A neutron spin $\vec{S}_n$ is subject to the Larmor precession in a magnetic field $\vec{B}$. (c) The spin analyzer allows neutrons to pass through with cosine probability of $\vec{S}_n$ along $\vec{B}$, namely $I_A = (1 + \cos\phi)/2$.

In NSE, information about the initial velocity is encoded within the neutron itself through Larmor precession, and the initial velocity can be compared with the final velocity for the same neutron. Figure 4.3a shows a layout of a NSE instrument. The basic idea is the following: The incident neutron wavelength and wavelength distribution are defined using a neutron velocity selector (NVS). The beam is then polarized by passing the neutrons through a polarizer (P) to select one state of neutron spins. The polarized neutrons are then flipped perpendicular to the magnetic field by a so called $\pi/2$-flipper (1-$\pi/2$). The precession frequency is controlled by a magnetic field in the main precession coil along the flight path from the $\pi/2$-flipper to the sample (1-PC). After the sample, a π-flipper reverses the precession angle and the neutrons pass through a second precession field (2-PC). A second $\pi/2$-flipper (2-$\pi/2$) stops the precession and the beam passes through a spin analyzer (A) and hits the detector (D). The spin analyzer only allows the cosine probability of the neutron spin direction parallel to the magnetic field to pass through to the detector. Figure 4.3c shows the neutron intensity I_A transmitted through the analyzer with the angle between the magnetic field and the neutron spin defined as ϕ, the net change in precession angle after passing through the two precession fields. In this case, I_A can be written as $I_A = (1 + \cos\phi)/2$. If the sample scatters elastically, that is, $\hbar\omega = 0$, then there is no net change in neutron spin after passing through both precession fields and the initial polarization is recovered, therefore $\phi = 0$ and $I_A = 1$ (Figure 4.3a). The precession angles in the primary (1-PC) and secondary (2-PC) paths are only different when the symmetry is distorted because of a change in the neutron velocity after interacting with the sample, that is, the neutrons scatter quasi-elastically, and both $\phi \neq 0$ and $\hbar\omega \neq 0$. In this case, the final spin polarization is rotated by an angle ϕ from the initial polarization leading to a decrease in the measured intensity, I_A, at the detector. Using this set-up unique to NSE, it is possible to make extremely accurate measurements of the energy change during the scattering process, and therefore to design a spectrometer with high resolution.

The reason NSE can use a relatively broad wavelength distribution and still provide a high energy resolution is because the energy resolution of the neutrons (determined by the distribution of velocity v_n or wavelength λ) is decoupled from the resolution of the energy transfer with the sample (determined by measuring the change in each neutron's polarization). This characteristic allows NSE to be the highest energy resolution technique among neutron spectroscopies with good counting statistics within a realistic experimental time. Furthermore, the measured neutron intensity can be written as $I \propto \int S(Q, \omega) \cos(\omega t) d\omega$ because the spin analyzer only allows the cosine probability of the neutron spin direction (I_A) to pass through to the detector. This equation is a cosine Fourier transformation of $S(Q, \omega)$, equivalent to $I(Q, t)$, meaning NSE automatically provides results in the time domain, while all the other neutron spectrometers work in the energy domain (see eqs. (4.4) and (4.5)). Because NSE works in the time domain, it is best suited to measuring relaxation processes (quasi-elastic scattering) rather than excitation processes (inelastic scattering). In NSE, we define the Fourier time as $t \equiv 2\pi\gamma_L \left(\frac{m_n}{\hbar}\right)^2 \lambda^3 J$, where J is the magnetic field (B) integral along the neutron trajectory, $J = \int B dl$.

Figure 4.4 shows a schematic image of $I(Q, t)$ measured in a scattering experiment. Figure 4.4a shows the correlation function (scattering intensity $I(Q, t)$) with respect to space (Q) and time (t). NSE measures the normalized ISF $I(Q, t)/I(Q, 0)$, which is used to describe the Q-dependent time correlation function as shown in Figure 4.4b.

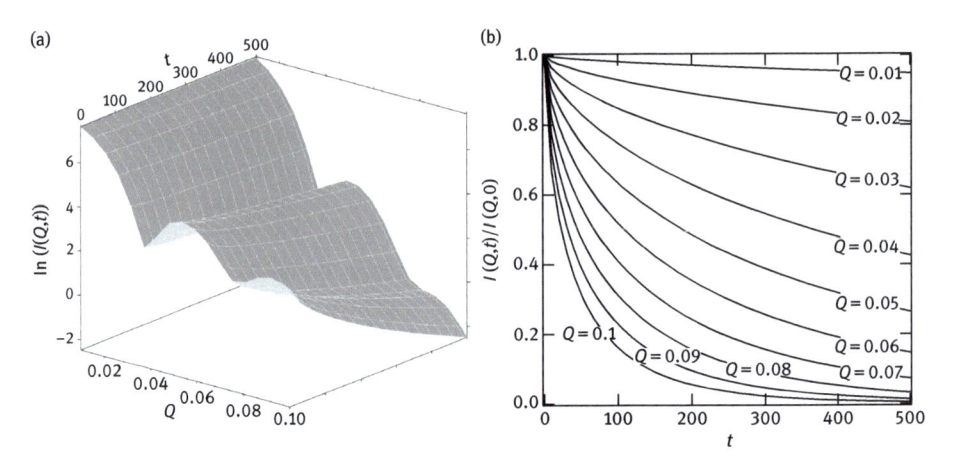

Figure 4.4: A typical example of $I(Q, t)$ in (a) three-dimensional representation as Q and t dependence of $\ln[I(Q, t)]$ and (b) representation of the normalized intermediate scattering function $I(Q, t)/I(Q, 0)$ against t. $I(Q, t = 0)$ in (a) corresponds to the Fourier transform of the instantaneous spatial atomic correlations, that is, structure, and the time decaying $I(Q, t)$ corresponds to how fast the structural correlation is lost with t due to the motions at the corresponding length scale Q.

4.3 Membrane dynamics theory

Membranes in solutions have very large interfacial area. When 0.01 mol of surfactant are dispersed in a liter of water (corresponding to a mole fraction of 10 mmol/L), the total interfacial area of the self-assembled bilayers is on the order of 10^3 m^2, or roughly the size of a football pitch, in just 1 L of water. (A typical head area per molecule is ≈ 0.7 nm^2 while the bilayer thickness is on the order of nm). Therefore, interfacial energy plays a decisive role in determining the shape of the membranes.

The framework for understanding the curvature elasticity of lipid bilayers was established in a seminal paper by Helfrich in 1973 [12]. Because the bilayer is thin and governed by the interfacial energy, he assumed that the bilayer was infinitely thin and wrote the bending energy as what has become known as the Helfrich Hamiltonian

$$F = \int dA \left[\frac{\kappa}{2}(C_1 + C_2 - C_s)^2 + \bar{\kappa} C_1 C_2 \right] \quad (4.7)$$

where C_1 and C_2 are the two principal curvatures of a membrane, C_s is the spontaneous curvature, κ is the bilayer bending modulus, and $\bar{\kappa}$ is the saddle-splay modulus, respectively, and the integration is over the area A. The Gauss–Bonnet theorem can be applied to the second term of eq. (4.7) as $\int C_1 C_2 dA = 4\pi$. The saddle-splay modulus $\bar{\kappa}$ is sensitive to topological changes of the membrane. However, in NSE experiments, the membrane topology does not change and this term is almost constant. Also, if the lipid membrane is fluid and there are no long-range interactions present, a special surface within the bilayer can be defined such that $\bar{\kappa} = 0$ [43, 45], which further supports that the second term in the Helfrich Hamiltonian can be neglected in certain cases.

The interfacial energy also contains a contribution from the elastic energy required to stretch the membrane (which originate from the membrane tension) and is written as a function of the relative area change $\Delta A/A$. However, an area change requires a large compression energy, particularly for short wavelengths, which implies that the surface tension is zero for free unstressed membranes [44, 99]. Further, in order to assume that the interfacial tension is finite, the area change in the membrane must be compensated for by either adding or removing lipids from the bilayer, meaning there would need to be a reservoir of lipid molecules [100]. Technically, there is a very low concentration of free lipids present in the solution and the bilayer can exchange lipid molecules with the solvent; however, the timescales for lipid exchange are orders of magnitude slower than the timescales of the membrane fluctuations and the lipid exchange cannot compensate for the area changes due to the membrane bending [12]. Early work by Faucon et al. included a stretching contribution when analyzing the thermal undulations of giant unilamellar vesicles (GUV, typical sizes on the order of μm) that were measured using an imaging technique [71]. From their analysis, the estimated surface tension was on the order 10^{-5} N/m. However, later work by Yeung and Evans suggested that the dynamics measured by Faucon et al. could also be explained by taking an internal membrane friction into account [101], which will be explained later in Section 4.3.2.

The description of the membrane interfacial energy according to the Helfrich Hamiltonian is an essential starting point to understanding membrane bending dynamics, but the model neglects any change in membrane thickness or internal membrane structure. As we have learned more about the structure and dynamics of lipid bilayers, the models for the membrane deformation energy have also evolved to take into consideration many other contributions. As suggested above, Yeung and Evans and several others have proposed that lipid membrane undulations are affected by an internal membrane friction – a result of the membrane being composed of two monolayers as will be discussed more in Section 4.3.2 [53, 101–103]. In some membranes, the lipid molecules also are tilted with respect to the membrane normal [104, 105]. This molecular tilt is neglected in Helfrich's original free energy expression [12], but more recent studies have shown that the membrane internal structure does contribute to the membrane energy through a molecular tilt modulus [106–112].

In the following subsections, we describe the theory for the membrane dynamics as it applies to NSE measurements. The models for NSE data build in complexity with each subsequent subsection as the theory evolves and more details of the membrane structure are considered.

4.3.1 Single membrane fluctuation dynamics and intermediate scattering function

Because this chapter is about studying lipid bilayer dynamics with NSE, we start with a model for the $I(Q, t)$ (which we directly measure with NSE, see Section 4.2.3) based on the Helfrich bending Hamiltonian that was derived by Zilman and Granek (ZG) [99, 113]. At short time, t, and length, l, scales ($t \leq 1$ μs and $l \leq L$ with L a long length scale cutoff), most membranes can be treated as an assembly of independent and nearly flat membrane sheets. The length scale L depends on the experimental system studied and is defined as the plaquette linear size, which is considered as the pore size in a sponge phase, the Helfrich–Servuss patch size (corresponding to the intermembrane collision length) in a lamellar phase, or the vesicle radius.

4.3.1.1 Single membrane fluctuation dynamics – Zilman and Granek approach

When a nearly flat thin elastic sheet is thermally undulating at a height, $h(\vec{r})$, from a mean surface as illustrated in Figure 4.5, the Helfrich bending Hamiltonian is proportional to the square of the curvature and takes the following form [99, 113]

$$F = \frac{\kappa}{2}\int d^2r\left[\nabla^2 h(\vec{r})\right]^2 = \frac{\kappa}{2L^2}\sum_{\vec{k}} k^4 h_{\vec{k}} h_{-\vec{k}} \tag{4.8}$$

Here, $\vec{r}$ is a two-dimensional vector (x, y) on the planar surface and $h(\vec{r})$ is the local height of the surface (by definition $\langle h(\vec{r})\rangle = 0$). As discussed earlier, the spontaneous curvature of the bilayer is assumed to be zero. The second equality expresses the Hamiltonian in terms of Fourier modes, where $h_{\vec{k}} = \int d^2r h(r) e^{i\vec{k}\cdot\vec{r}}$ is the two-dimensional Fourier transform of $h(\vec{r})$. We note here that different communities have different notations. In this chapter, we are using $\vec{k}$ rather than $\vec{q}$ to denote the Fourier component of the membrane fluctuations and use $\vec{q}$ and $\vec{Q}$ to denote the wavevector and the scattering (or momentum transfer) vector, respectively, as defined in the previous section on scattering. L is the membrane patch size, where we consider the dynamics of a membrane plaquette. The membrane is suspended in

Figure 4.5: A schematic representation of undulating thin sheet around a flat reference surface (shown in light gray) in a plaquette size $L \times L$. The coordination axes are shown in the left corner. The undulations occur with height $h(\vec{r})$ about a nearly flat reference surface in the x, y plane where $\vec{r} = (x, y)$. Also defined are the membrane fluctuation wavevector $\vec{k}$ and the initial and final neutron wavevectors $\vec{q}_i$ and $\vec{q}_f$. The scattering vector is defined as $\vec{Q} = \vec{q}_i - \vec{q}_f$. The vector $\vec{Q}$ in the coordination axes is represented to define the angles for the orientational average (see eq. (4.19)).

aqueous solvent with solvent viscosity, η, and performs thermal undulations that are coupled to the hydrodynamic flow of the solvent. Applying the equipartition theorem to eq. (4.8) leads to the following expression for the equilibrium spectrum of undulations [114]

$$\langle h_{\vec{k}} h_{-\vec{k}} \rangle = \frac{k_B T L^2}{\kappa k^4} \qquad (4.9)$$

The time-dependent correlation function of $h_{\vec{k}}(t)$ thus follows an exponential decay from its equilibrium value [114]

$$\langle h_{\vec{k}}(t) h_{-\vec{k}}(0) \rangle = \frac{k_B T L^2}{\kappa k^4} e^{-\omega(k)t} \qquad (4.10)$$

The relaxation frequency $\omega(k)$ can be determined from a standard hydrodynamic mode analysis [113–116]

$$\omega(k) = \frac{\kappa k^3}{4\eta}. \qquad (4.11)$$

In order to calculate $I(Q, t)$, it is important to first calculate the two-point correlation function $\langle (h(\vec{r}, t) - h(\vec{r}', 0))^2 \rangle$. The two-point correlation function can be determined using either a Langevin equation for the undulating bilayers in viscous solvent [99, 113] or the stochastic field approach [114]. Both approaches give the correlation function as

$$\langle (h(\vec{r}, t) - h(\vec{r}', 0))^2 \rangle = \frac{k_B T}{2\pi^2 \kappa} \int_{\pi/L}^{\pi/d_0} \frac{d^2 k}{k^4} \left(1 - e^{-\omega(k)t}\right) \qquad (4.12)$$

where d_0 represents the size of a lipid molecule (molecular length). At $t \lesssim \eta d_0^3/\kappa$ (very short timescales), the calculation yields the simple diffusion of the monomers.

For $t \gtrsim \eta L^3/\kappa$ (long times), the integral saturates to a constant and the correlation has decayed. In the intermediate time range $\eta d_0^3/\kappa \ll t \ll \eta L^3/\kappa$, the lower and upper limits of the integration are replaced by zero and infinity, and the integration evaluates to [99, 113, 114]

$$\langle (h(\vec{r},t) - h(\vec{r},0))^2 \rangle \simeq 0.17 \left[\sqrt{\frac{k_B T}{\kappa}} \frac{k_B T}{\eta} t \right]^{2/3} \quad (4.13)$$

The above equation indicates that the membrane plaquette dynamics exhibit an anomalous $t^{2/3}$ time dependence and that the mean square displacement $\langle h(\vec{r})^2 \rangle$ follows $\kappa^{-1/3}$.

4.3.1.2 Intermediate scattering function

Now that we have a theoretical expression for the equilibrium height fluctuations – or correlations between the membrane heights as a function of time – that we can measure with NSE, we need to incorporate eq. (4.13) into $I(Q,t)$ to be able to fit experimental data. Considering a system containing a membrane plaquette, p, with a bilayer size of $L \times L$, the $I(Q,t)$ of the system is given by [99]

$$I(\vec{Q},t) = \frac{N_p}{V} \sum_p \left\langle e^{i\vec{Q} \cdot (\vec{R}_p(t) - \vec{R}_p(0))} \right\rangle I_p(\vec{Q},t) \quad (4.14)$$

where $\vec{R}_p$ is the center of mass position of the plaquette p, N_p is the number of molecules in a single plaquette, and V is the macroscopic system volume. The single plaquette ISF is [99, 113]

$$I_p(\vec{Q},t) = \frac{1}{N_p} \left\langle \sum_{i,j} e^{i\vec{Q} \cdot (\vec{R}_i(t) - \vec{R}_j(0))} \right\rangle \quad (4.15)$$

where $\vec{R}_i$ is the position vector of the ith lipid molecule in the center of mass coordinate frame and the sum runs over all molecules in a single plaquette.

The plaquette can also undergo simple diffusion, and in that case, the normalized ISF decays by contributions from both the diffusion and the membrane undulations as [117]

$$\frac{I(Q,t)}{I(Q,0)} = e^{-DQ^2 t} \frac{\langle I_p(\vec{Q},t) \rangle_{\vec{Q}}}{\langle I_p(\vec{Q},0) \rangle_Q} \quad (4.16)$$

where D is the center of mass diffusion constant of the patch, and the bracket indicates the average over all scattering angles (orientational average). For vesicles with ≈ 50 nm radius, the diffusion constant from the Stokes–Einstein equation is on the order of 10^{-12} m^2/s. Therefore, for $t \ll \eta L^3/\kappa$, translational diffusion has a negligible effect ($e^{-DQ^2 t} \approx 1$) and $I(\vec{Q}, t)$ will be determined solely by individual membrane undulations. For the rest of this section, we will focus our attention on the plaquette undulations, neglecting contributions from translational diffusion. However, it is important to note that this simplification is not always possible. Depending on the vesicle size and concentration in the sample as well as Q and the Fourier times accessed in a particular experiment, the translational diffusion contribution to $I(\vec{Q}, t)$ may need to be taken into account.

The internal position vector $\vec{R}_i(t)$ is now expressed in terms of the longitudinal two-dimensional vector $\vec{r}$ and the transverse component $h(\vec{r}, t)$, and therefore, we can rewrite eq. (4.15) using a double integral as follows [99, 113]

$$I_p(\vec{Q}, t) = \frac{1}{N_p d_0^4} \int d^2 r \int d^2 r' \left\langle e^{i \vec{Q}_\parallel \cdot (\vec{r} - \vec{r}')} e^{i Q_\perp [h(\vec{r}, t) - h(\vec{r}', 0)]} \right\rangle \tag{4.17}$$

where the scattering vector $\vec{Q}$ is decomposed into two component vectors: a longitudinal in-plane component, $\vec{Q}_\parallel$, and a perpendicular component $Q_\perp$. The membrane undulations are implicitly assumed to be small. Since $h(\vec{r}, t)$ and $\vec{r}$ are weakly correlated, we assume that their averages can be decoupled. In the present treatment, the white noise appearing in the linear Langevin equation follows Gaussian statistics [99, 113, 114], which supports that we can assume that the statistics of $h(\vec{r}, t)$ are also Gaussian [118]. This assumption relates the height fluctuations with the two-point correlation function, eq. (4.13), as follows [99, 117]

$$\left\langle e^{i Q_\perp [h(\vec{r}, t) - h(\vec{r}', 0)]} \right\rangle = e^{-\frac{Q_\perp^2}{2} \left\langle (h(\vec{r}, t) - h(\vec{r}', 0))^2 \right\rangle} \tag{4.18}$$

Note this equation is for a defined orientation while most lipid membrane samples are randomly orientated in solution. In order to apply the equation to isotropic structures, such as sponge, nonoriented lamellar, and vesicle structures, one needs to take the orientational average over all values of the angle between $\vec{Q}$ and the membrane normal, as [117]

$$I(Q, t) \equiv \langle I_p(\vec{Q}, t) \rangle_{\hat{Q}} = \frac{1}{4\pi} \int_0^{2\pi} d\varphi \int_0^{\pi} \sin\alpha \, d\alpha I_p(\vec{Q}, t) \tag{4.19}$$

The angles α and φ are defined in Figure 4.5. This average is assumed to be dominated by the region at $\varphi \approx 0$ ($Q_\parallel \approx 0$ and $Q_\perp \approx Q$) [113, 117]. At the limit of $\kappa \gg k_B T$, the solution to the integral is approximated as a pure stretched exponential decay [99, 113, 117, 119]

$$\frac{I(Q, t)}{I(Q, 0)} \simeq e^{-(\Gamma_{ZG} t)^{2/3}} \tag{4.20}$$

where the decay rate Γ_{ZG} is

$$\Gamma_{ZG} = 0.025 \sqrt{\frac{k_B T}{\kappa_{ZG}} \frac{k_B T}{\eta}} Q^3 \qquad (4.21)$$

We now have a series of equations that relates the membrane bending modulus to the collective height fluctuations that are experimentally measured with $I(Q, t)/I(Q, 0)$ – meaning we can quantify a membrane elastic property from our NSE measurement. In the original work, eq. (4.21) contained the intrinsic bending modulus, κ. Here, we replace κ with κ_{ZG} because, as we will see in Section 4.3.2, eq. (4.21) will only describe the intrinsic bending modulus in certain cases. Nevertheless, eq. (4.21) establishes that the measured decay rate is inversely related to the bending modulus and the expression gives the trends that you may intuitively expect – a stiff membrane with a large κ will have a lower decay rate and the measured correlations in $I(Q, t)$ will decay slower than a softer membrane with a lower κ.

It is noted here that both the transverse and longitudinal contributions to the scattering have been calculated by Zilman and Granek [99]. These calculations support that the longitudinal contribution is weak and the above description for only the transverse component is a good approximation for isotropically distributed bilayers as long as $\kappa \gg k_B T$. For works that consider the longitudinal, in-plane contribution to the dynamics in aligned lipid bilayers, we refer the reader to works by Rheinstädter and colleagues [81, 120].

4.3.2 Internal membrane dissipation – two coupled monolayers

We just saw how the Helfrich model of the membrane as a thin, structureless sheet can be applied to NSE data analysis. However, we know that the lipid membrane is not structureless, but is in fact composed of two lipid monolayers. If there were no interactions between the monolayers, then they would be free to slide past each other. However, for the bilayer to be stable, the monolayers *need* to interact with each other through van der Waals interactions between the lipid tails. Yeung and Evans suggest that this interaction could lead to "hidden" degrees of freedom in the bilayer bending dynamics due to an internal viscous contribution [101].

A similar idea was also proposed by Seifert and Langer (SL), where by incorporating the effects of potential density variations in each leaflet due to the bending deformations, they suggested that a viscous mode contributed to the membrane fluctuations [102, 103]. For long wavelengths, as represented schematically in Figure 4.6a, bending fluctuations do not significantly perturb the lipid monolayer densities and the standard hydrodynamic theory holds. On the other hand, short fluctuation wavelengths, as depicted in Figure 4.6b, create defects due to the fact that the molecular redistribution required by the perturbation cannot happen quickly enough. As a result, the short wavelength fluctuations are governed by an effective

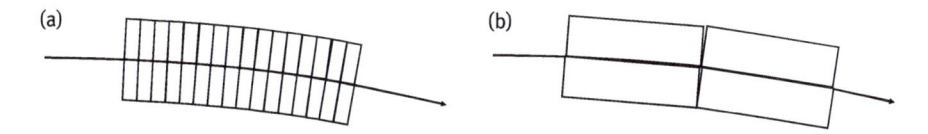

Figure 4.6: Schematic representation of a slightly bent membrane formed by multiple rectangles for (a) long wavelength undulation and (b) short wavelength undulation relative to the size of the lipid molecules. The bending imposes a radius of curvature that spreads the outer part of the rectangles out while compressing the inner portion. This requires the rectangular elements to distort to try to fill the voids and release the compression. For short time short wavelength modes the rectangles cannot accommodate the shape change fast enough, leading to an increase in the bending modulus of the membrane [102].

bending rigidity that is higher than the intrinsic bending modulus that controls the long wavelength fluctuations (i.e., the short wavelength fluctuations require more energy than the long wavelength fluctuations). In order to mathematically express these physical effects, Seifert and Langer included a contribution from density modes in each monolayer to the free energy as [102, 103]

$$F = \int dA \left[\frac{\kappa}{2}(2H)^2 + \frac{K_A}{2}\left\{(\rho_m^+ + 2d_nH)^2 + (\rho_m^- - 2d_nH)^2\right\}\right] \quad (4.22)$$

where the mean curvature H has the relation $2H = C_1 + C_2$, K_A denotes the monolayer area compressibility modulus, $\rho_m^\pm$ is the scaled deviation of the projected density from its equilibrium value for a flat membrane, and d_n is defined as the distance between the mid-plane of the bilayer and the neutral surface of a monolayer. The neutral surface in eq. (4.22) is a special dividing surface where the monolayer neither stretches nor compresses as it is bent [45]. This mathematical definition greatly simplifies the expressions for the bilayer free energy; however, it cannot be measured experimentally making it difficult to assign a numerical value for d_n as we will revisit later in Section 4.4.

The relaxation frequencies for this free energy model are given as follows [102, 103]

$$\omega_1(k) \approx \begin{cases} \frac{\kappa}{4\eta}k^3, & k \ll k_1 \\ \frac{K_A}{2b}\frac{\tilde{\kappa}}{\kappa}k^2, & k_1 \ll k \ll k_2 \\ \frac{K_A}{\eta_m}\frac{\tilde{\kappa}}{\kappa}, & k_2 \ll k \end{cases} \quad (4.23)$$

$$\omega_2(k) \approx \begin{cases} \frac{K_A}{2b}k^2, & k \ll k_1 \\ \frac{\tilde{\kappa}}{4\eta}k^3, & k_1 \ll k \end{cases} \quad (4.24)$$

where $\tilde{\kappa} = \kappa + 2d_n^2 K_A$ is a renormalized bending rigidity including the effect of the elastic stretching and compression, b denotes a friction coefficient for a phenomenological internal dissipation, and η_m is the monolayer surface viscosity. $\tilde{\kappa}$ now accounts for the extra energy required to bend the membrane at short wavelengths. The crossover wavevectors $k_1 \equiv 2\eta K_A/b\tilde{\kappa}$ and $k_2 \equiv \sqrt{2b/\eta_m}$ separate the range of the characteristic dynamics into three regimes, shown in the plot of dispersion relations in Figure 4.7.

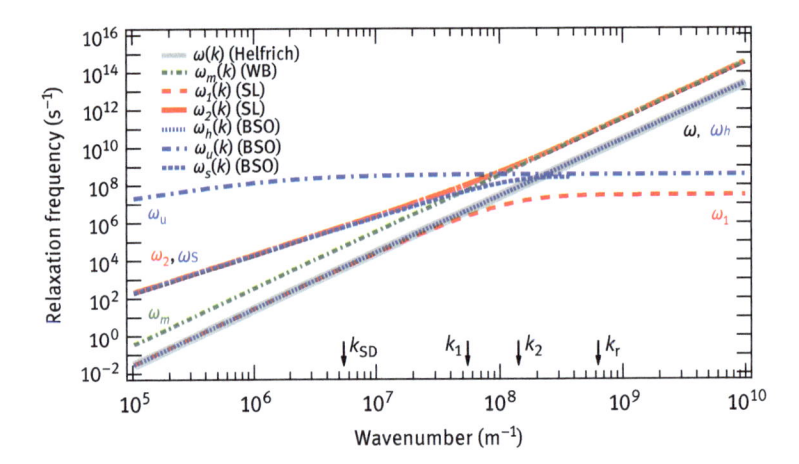

Figure 4.7: Comparison of the dispersion relations from the theories explained in this Section 4.3. The result for Helfrich is the solution of a standard hydrodynamic mode analysis (eq. (4.11)) shown as $\omega(k)$, and the modification of Watson and Brown (WB) gives $\omega_m(k)$. The theory by Seifert and Langer (SL) has two eigenvalues, ω_1 and ω_2 (eqs. (4.23) and (4.24)). The model by Bingham, Smye, and Olmsted (BSO) gives three independent eigenvalues, ω_h, ω_u and ω_s which are expressed in eqs. (4.34) to (4.36). $\omega_s(k)$ is not shown at high k values where the Stokes approximation breaks down, $k \gtrsim k_r$ ($k_r \equiv 2\pi/l_r$) [53]. The arrows at the bottom of the graph show the relevant length scales for the different theories. The parameters used in the calculation are given in the main text (see Section 4.3.4).

For long wavelength undulations (small k), $\omega_1(k)$ corresponds to the usual hydrodynamically damped bending mode and thus overlaps with the $\omega(k)$ predicted by the Helfrich theory. $\omega_2(k)$ is the damping rate of the slipping mode (density difference fluctuations between the monolayers that is damped by the inter-monolayer friction). For $k_1 \ll k$ (short wavelength undulations), $\omega_2(k)$ becomes the damping rate of the bending mode which is affected by the density mode because the lipid densities cannot respond quickly enough to changes in the membrane shape. Thus, an effectively larger bending rigidity $\tilde{\kappa}$ dominates the fluctuations. In this regime, the rate $\omega_1(k)$ is determined by the slipping mode. At the second crossover k_2, the main dissipative mechanism changes from inter-monolayer friction ($k \ll k_2$; slipping) to monolayer surface viscosity ($k_2 \ll k$; membrane hydrodynamics).

The dynamic undulation correlation function can be expressed as the sum of two decaying components as [101, 121–123]

$$\langle h_{\vec{k}}(t)h_{-\vec{k}}(0)\rangle = \frac{k_B T}{\kappa k^4}\left(A_1(k)e^{-\omega_1(k)t} + A_2(k)e^{-\omega_2(k)t}\right) \tag{4.25}$$

where $A_1(k) + A_2(k) = 1$ and

$$A_1(k) = \frac{\omega_2(k) - \kappa k^3/4\eta}{\omega_2(k) - \omega_1(k)} \tag{4.26}$$

A molecular dynamics simulation of a coarse-grained bilayer model has shown good agreement with this theory [123], and NSE data have provided evidence of the slipping mode in lipid bilayers [124, 125].

The ZG theory covered in Section 4.3.1.2 [99, 113] does not include these internal dissipation mechanisms, and as suggested earlier, eq. (4.21) will work for measurements of long wavelength fluctuations measured with dynamic light scattering (DLS, at small scattering vector Q) that are not affected by the internal membrane friction [126]. NSE measures the short wavelength fluctuations (large Q), where theory predicts that the inter-monolayer friction affects the dynamics. Early NSE measurements of lipid membrane bending fluctuations reported a much higher bending modulus than measured with other techniques [127–130]. These works used an effectively larger solvent viscosity (three to four times the actual solvent viscosity) as an additional dissipation mechanism to get κ values that were comparable to other experimental techniques [127–130]; however, from the theories described above, we now know that the additional dissipation mechanism comes not from the solvent, but from the inter-monolayer friction within the bilayer itself.

There are two ways to include the contribution from the density mode into NSE data analyses; one proposed by Watson and Brown (WB) [117, 119] and the other by the group of Monroy [124, 125]. We note here that, as alluded to in the previous section, the ZG theory has another limitation in that it does not properly account for the orientational averaging in the case when $\kappa \gg k_B T$ does not hold. However, this point is beyond the scope of this chapter and we encourage interested readers to refer to [131–134].

In the WB theory, they proposed a modification of the ZG theory to include contributions from the internal membrane friction into the NSE data analysis [117, 119]. Since the NSE timescale is relatively short, the energy dissipation from the surface viscosity is too slow to observe ($\omega_1(k)$ from the SL theory). Therefore, one can assume that the decay in the NSE time window is solely due to the bending modes and dominated by $\omega_2(k)$ [117], and eq. (4.25) becomes

$$\langle h_{\vec{k}}(t)h_{\vec{k}}(0)\rangle \approx \frac{k_B T}{\tilde{\kappa} k^4}\left(\frac{\tilde{\kappa} - \kappa}{\kappa} + e^{-\frac{\tilde{\kappa} k^3}{4\eta}t}\right) \tag{4.27}$$

We see that this correlation function is almost the same as the one presented in Section 4.3.1 for the ZG theory (eqs. (4.10) and (4.12)) except that we need to replace κ

with $\tilde{\kappa}$. Therefore, all the derivations described in Section 4.3.1 still hold, and the decay follows the stretched exponential function derived by Zilman and Granek. The only modification is to the expression for the relaxation rate, which now includes the effective bending modulus and becomes

$$\Gamma_{ZG_{mod}} = 0.025 \sqrt{\frac{k_B T}{\tilde{\kappa}} \frac{k_B T}{\eta}} Q^3 \qquad (4.28)$$

again with $\tilde{\kappa} = \kappa + 2d_n^2 K_A$ [117, 119]. This modification by Watson and Brown [117, 119], to include the internal membrane friction due to the inter-monolayer coupling, thus describes short wavelength membrane dynamics more accurately. This model was first applied to NSE data by Choi and colleagues [135], and gave a realistic number for κ, which supports the validity of the theory to explain short wavelength bending fluctuations without needing to use an effective solvent viscosity.

On the other hand, Arriaga et al. observed a deviation from the predicted Q^3 behavior of the relaxation rates measured by NSE (see e.g., eq. (4.28)). Instead, their data suggested a diffusive type Q^2 behavior for palmitoyl-oleoyl-phosphocholine (POPC) large unilamellar vesicles [124]. They ascribed the origin of this deviation to hybrid curvature-compression modes, equivalent to the models of inter-monolayer friction discussed in this section. They observed a crossover of Γ from Q^3 to Q^2 at $Q \approx 0.4$ nm^{-1}, where at low Q the usual bending type Q^3 relation is seen, while at high Q the dynamics scale with Q^2. Their observation yielded $K_A \approx 80$ mN/m, $b \approx 2 \times 10^9$ Pa·s/m, and $\kappa \approx 15 k_B T$. Interestingly, the inter-monolayer friction constant b increased upon addition of cholesterol [124]. Mell and colleagues developed an expression for the ISF that includes the hybrid mode as [125]

$$\frac{I_{hyb}(Q, t)}{I(Q, 0)} = e^{-0.3 \frac{R k_B T}{2 d_0 K_A} Q^2 t^0} \qquad (4.29)$$

where R is the radius of the vesicles. They showed that both the bending and hybrid modes contribute to the measured NSE data, while theory predicts that the hybrid mode dominates at high Q. As $I_{hyb}(Q, t)$ goes as $\exp(t^0)$, the hybrid mode is a non-decaying contribution seen at longer times. These examples suggest that NSE is sensitive to the internal membrane dynamical contributions and a careful evaluation of the theoretical underpinnings is essential to correctly interpret the data.

4.3.3 Dynamics of "thick" membranes

The original theory by Helfrich ignored any energetic contribution from the membrane structure [12]. While Yeung and Evans [101, 136] and Seifert and Langer [102, 103] successfully included inter-monolayer coupling as an additional dissipation mechanism in membranes, the models still neglect potential changes in the bilayer thickness, even though thickness fluctuations in lipid bilayers have been theoretically predicted since the 1980s [19, 20, 46–48]. Computer simulations also have captured membrane thickness fluctuations [52, 108, 137–141]. Experimentally, the first direct observation of thickness fluctuations in surfactant membranes was made by Farago and colleagues in a stacked lamellar membrane [142, 143], and later more detailed experiments and analyses were performed by Nagao et al. [141, 144, 145] Thickness fluctuations in single component lipid bilayers were measured by Woodka and colleagues using NSE [58], and the method was then extended to a homogeneously mixed lipid membranes [59]. The experimental thickness fluctuation observations were also reproduced in recent computer simulations [141, 146]. Despite the growing experimental and computational data on the thickness fluctuations, a theoretical model that explicitly accounts for the dynamic contributions from the membrane thickness in the dispersion relation has only recently been developed by Bingham, Smye and Olmsted (BSO) [53].

The membrane geometry considered in the BSO model is schematically represented in Figure 4.8. The upper (+) and lower (−) monolayer surfaces are described by the height functions $h_+(\vec{r})$ and $h_-(\vec{r})$ from a planar reference plane (dashed straight line in Figure 4.8) as

$$h_+(\vec{r}) = d_+(\vec{r}) + s(\vec{r}) \tag{4.30}$$

$$h_-(\vec{r}) = -d_-(\vec{r}) + s(\vec{r}) \tag{4.31}$$

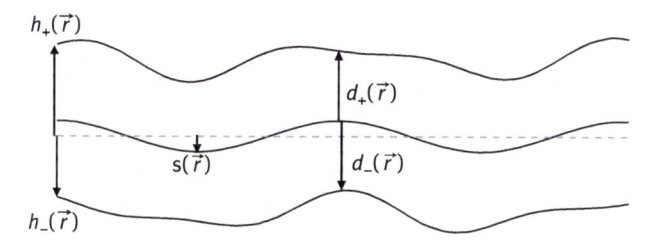

Figure 4.8: The membrane geometry considered in the Bingham, Smye and Olmsted model [53]. The monolayer height $h_\pm(\vec{r})$ is the surface height from the reference surface (shown as a dashed line). The monolayer thickness $d_\pm(\vec{r})$ is defined as the distance between the outer surface and the undulating mid-surface, $s(\vec{r})$.

where $d_\pm(\vec{r})$ is the monolayer thickness for each leaflet, and $s(\vec{r})$ is the height of the inter-monolayer surface from the reference plane. The free energy of each monolayer has been calculated in the framework of the Helfrich Hamiltonian [53]

$$F_\pm = \frac{1}{2}\int dA \left[\kappa_{m\pm}(\nabla^2 h_\pm)^2 + \gamma_{s\pm}(\nabla h_\pm)^2 + K_{A\pm}\left(\frac{d_\pm - d_{0\pm}}{d_{0\pm}} - d_{0\pm}\nabla^2 s\right)^2 \right] \tag{4.32}$$

where $\kappa_{m\pm}$ is the monolayer bending modulus for each leaflet, $\gamma_{s\pm}$ is the surface tension which restricts variations in the monolayer/water interfacial area, and $d_{0\pm}$ is the unperturbed monolayer thickness, respectively. The total free energy was set as

$$F = F_+ + F_- + F_{frame} \tag{4.33}$$

where $F_{frame} = \frac{\gamma_{fr}}{2}\{\nabla(h_+ + h_-)\}^2$ is the contribution from a tension that restricts changes in the total membrane area with the frame tension, γ_{fr}. In this framework, Bingham et al. found three types of relaxation frequencies in a "thick" membrane [53].

The first mode is equivalent to the ripple type dynamic (bending) mode of Brochard and Lennon [116]. This mode appears in the model as a coupled bilayer and internal surface mode. The internal surface undulates in phase with the bilayer surface in order to preserve the thickness of each monolayer, and the relaxation is driven by the membrane bending rigidity and tension and damped to the surrounding viscous medium. The relaxation frequency is written as

$$\omega_h(k) = \frac{(\gamma_s + 2\gamma_{fr} + \kappa_m k^2)k e^{-kd_0}}{2\eta} \tag{4.34}$$

When we ignore the contribution from the tension and the correction term of e^{-kd_0}, $\omega_h(k)$ is equivalent to the standard hydrodynamic model of a thin elastic sheet (eq. (4.11)), which is plotted in Figure 4.7.

The second mode is a pure peristaltic dynamic mode describing undulations in the bilayer where the thickness of the monolayers undulates in phase (thickness fluctuations). The relaxation frequency is calculated as

$$\omega_u(k) = \frac{kK_A}{\eta_m k + 2\eta} = \frac{kK_A}{\eta(l_{SD}k + 2)} \tag{4.35}$$

The thickness fluctuation mode is driven by the area compressibility, K_A, and damped by the solvent and membrane monolayer viscosity, η and η_m, respectively. The mode is split into two regimes based on the Saffman–Delbrück length $l_{SD} = \eta_m/\eta$ [55]. The dissipation of the long wavelength modes, $kl_{SD} \ll 1$, is dominated by the solvent viscosity, while the short wavelength modes, $kl_{SD} \gg 1$, dissipate through the membrane viscosity. The relaxation frequency for the short wavelength modes can be written as $\omega_u \simeq K_A/\eta_m$. This expression suggests that the peristaltic mode can be seen as a local mode in the short wavelength limit, which is similar to the short wavelength behavior of the slipping mode $\omega_1(k)$ in the case where it is damped by the monolayer surface viscosity (see Figure 4.7) [102].

Finally, a dispersion relation for the movement of the internal membrane surface ($s(\vec{r})$ in Figure 4.8) gives the relaxation frequency

$$\omega_s(k) = \frac{K_A k^2 (k^2 d_0^2 - 1)}{\eta(2k + k^2 l_{SD} + 2l_{SD}/l_r^2)} \tag{4.36}$$

The internal surface mode also is driven by the area compressibility; however, it dissipates through the inter-monolayer friction as well as through the solvent and membrane viscosities. The new length scale $l_r = \sqrt{\eta_m/b}$ represents the dimension at which the forces from the membrane viscosity and the inter-monolayer friction balance. This length scale was estimated to be $l_r \sim 10$ nm [53], which is much shorter than l_{SD}.

While the dispersion relations developed by Bingham, Smye, and Olmsted have not yet been incorporated into the expressions for $I(Q,t)$, they nevertheless provide important insights into the dynamics measured with NSE. As we will see in Section 4.4.2, these theories have begun to help link the phenomenologically measured thickness fluctuation timescale back to the inherent membrane elastic and viscous properties.

4.3.4 Comparing the dynamics predicted by different membrane models

In the past three subsections, the description of a biomembrane has evolved from a thin, structureless sheet (Section 4.3.1) to two coupled monolayers (Section 4.3.2) to a "thick" membrane that can undergo fluctuations in thickness as well as bending undulations (Section 4.3.3). With each section, both the illustrations and equations to describe the membrane dynamics have become more complicated as more terms are needed to describe the added contributions. To summarize the different theories, we compare the predicted dispersion relations in Figure 4.7. The following parameters were used to calculate the dispersion relations: $\kappa = 8.5 \times 10^{-20}$ J, $\eta = 8.7 \times 10^{-3}$ Pa·s, $\eta_m = 1 \times 10^{-9}$ Pa·s·m, $b = 1 \times 10^{7}$ Pa·s/m, $K_A = 12\kappa_m/d_0^2$, $\kappa = 2\kappa_m$, $d_t = 2d_0$, $d_t = 2.42 \times 10^{-9}$ m, and $\beta = 24$, where d_t and β are the hydrocarbon thickness of the bilayer and intermonolayer coupling constant, respectively, which will be introduced in Section 4.4.1.

Single membrane fluctuations: The solid gray line plots $\omega(k)$ versus the undulation mode wavenumber predicted from the standard hydrodynamic mode analysis based on the Helfrich bending Hamiltonian (eq. (4.11)). This model predicts a single dispersion relation across the entire fluctuation spectrum.

Two coupled monolayers: Incorporating the effects of inter-monolayer friction introduces a second dissipation mechanism. The modification of the Zilman and Granek (ZG) theory by Watson and Brown (WB) considered the change in the dispersion relation given by eq. (4.11) to incorporate the effects of inter-monolayer friction by replacing κ with $\tilde{\kappa}$. The dispersion relation predicted by WB is shown as the green dash-dotted line marked as $\omega_m(k) = \tilde{\kappa}k^3/4\eta$. $\omega_m(k)$ relaxes faster than $\omega(k)$ because the membrane is effectively stiffer with $\tilde{\kappa} > \kappa$.

WB applied the model developed by Seifert and Langer (SL) to the short wavelength fluctuations measured by NSE. The SL model gives two eigenvalues for the dynamic modes of coupled monolayers, which are shown in red-dashed (w_1) and dashed-solid (w_2) lines. The main dissipation mechanism changes from $w_1(k)$ at long wavelengths ($k \ll k_1$) to $w_2(k)$ at short wavelengths ($k_1 \ll k$). At short wavelengths, $w_2(k)$ corresponds to the bending modes that are affected by the inter-monolayer friction and the fluctuations are governed by the effective bending modulus $\tilde{\kappa}$ also used by WB, so $w_2(k) = w_m(k)$. Meanwhile at long wavelengths, $w_1(k)$ corresponds to the hydrodynamically damped bending modes also predicted by Helfrich and $w_1(k) = w(k)$ at $k \ll k_1$.

Between the crossover wavenumbers, $k_1 < k < k_2$, $w_1(k)$ is determined by the slipping mode and then plateaus at high $k \gg k_2$ where the mode dissipates through the surface viscosity and is independent of wavenumber. However, for the chosen parameters, $k_1 < k < k_2$ range is quite narrow and the slipping mode does not contribute significantly despite its predicted importance by SL. The minor contribution of the slipping mode is due to the choice of η_m as well as b values used to calculate the dispersion relations in Figure 4.7. The experiments by Arriaga and colleagues found a b value that is much larger than the typical value used here ($b \sim 10^9$ Pa·s/m) [124], suggesting more thorough studies, combining experiment, simulation and theory, are needed to better understand inter-monolayer friction and how it affects the membrane dispersion relations.

"Thick" bilayers: The dispersion relation predicted by the model for "thick" membranes by Bingham, Smye, and Olmsted (BSO) are summarized by the blue-dashed (w_h and w_s) and dash-dotted (w_u) lines. In the calculation, we neglected the contributions from the membrane tension ($\gamma_s = \gamma_{fr} = 0$) and plot only the contribution from the height fluctuation as $w_h(k)$. As such $w_h(k)$ completely overlaps with $w(k)$ from the Helfrich description. The thickness fluctuations ($w_u(k)$) are the fastest mode at low k, but plateau to an almost constant value above $k_{SD} \equiv 2\pi/l_{SD}$, the same behavior seen in the slipping mode of the SL model $w_1(k)$. The internal surface mode $w_s(k)$ also converges to the $w_u(k)$ in the high k range.

Now that we have covered the theory needed to describe the membrane undulations, we will see in the next section how it is applied to actual NSE data. As discussed in Section 4.2, NSE measures the spatial and time correlation function as $I(Q, t)$ on the nanometer and nanosecond scales and is thus uniquely suited to probing the small wavelength dynamics predicted to occur from 0.1 nm^{-1} to ~10 nm^{-1} in Q and from 1 ps to sub µs in t (10^8 m$^{-1} < Q < 10^{10}$ m^{-1} and 10^{-12} s $< t < 10^{-6}$ s). As we move into the next section, we remind the reader that NSE measures the scatterer's Q, which corresponds to the scattering vector and not the dynamic mode number which we call k here, though the experimentally measured length scales should be weighted in the corresponding k ranges.

4.4 Dynamics measurements

Having covered the theory on membrane dynamics and how NSE works, we now turn our attention to the design of an appropriate experimental system to which these methods and theories can be applied. While there are many model systems for lipid bilayers, most NSE experiments use large unilamellar vesicles (LUVs). Multilamellar vesicles (MLVs) have a strong correlation peak at $Q \approx 1\,\text{nm}^{-1}$ from the inter-layer spacing [147, 148]. The width of the Bragg peak measured in an elastic scattering technique such as small angle X-ray or neutron scattering (SAXS or SANS) can be used to determine the so-called Caillé parameter that is related to the membrane bending rigidity [149, 150]. However, in a quasi-elastic scattering technique like NSE, Bragg peaks result in a phenomenon known as de Gennes narrowing, in which the measured dynamics markedly slow down at the peak position [151]. Indeed, while Seto et al. were able to obtain an estimate of the bending modulus of multilamellar vesicles using NSE, the data deviated from the predicted Q^3 scaling due to the close proximity of the Bragg reflection [129]. It is therefore best to use unilamellar vesicles to measure membrane fluctuations and determine the membrane's mechanical properties. While MLVs are not ideal for NSE experiments, measurements can be made using aligned multilayers [81, 120, 152]. Note that these experiments will probe the dynamics parallel to the membrane plane in eq. (4.17), which we explicitly do not cover in this chapter.

Giant unilamellar vesicles (GUVs) are widely used as model systems to determine the membrane elastic properties by other methods, either through fluctuation analysis or micro-manipulation techniques [153, 154]. GUVs are microns in diameter, making them ideal for studying by light microscopy methods, but it also means they are quite dilute in solution. If we assume a GUV diameter of roughly 10 μm and the highest achievable packing fraction in solution to be 74% (assuming face-centered cubic, FCC, packing), the lipid concentration in solution would only be 0.17% by volume or about 2 mg/mL lipid in solution. Unfortunately this concentration range is much too small for neutron scattering experiments (especially for measuring dynamics) and we need to increase the amount of lipid in the sample to gain sufficient scattering intensities. Using LUVs with a diameter of ≈ 100 nm, the lipid concentration can be as high as 17%, assuming an FCC packing of vesicle in solution (about 200 mg/mL lipid in the solution), which gives enough scattering intensities for our purposes. Therefore, most NSE experiments use LUVs with diameters ranging from 30 to 200 nm. These vesicles are usually obtained by the extrusion method [153], where aqueous lipid solutions are pushed through a polycarbonate filter yielding relatively monodisperse vesicles whose diameter are determined by the filter pore size.

4.4.1 Bending fluctuations

Now that we have an experimental system, we will cover actual NSE data. Figure 4.9 shows a typical $I(Q,t)/I(Q,0)$ measured using NSE. The solid lines in Figure 4.9 are fits to a single membrane fluctuation model proposed by Zilman and Granek: eq. (4.20) [99, 113]. The first application of ZG theory to NSE data appeared in Takeda's paper in which they studied dipalmitoyl-phosphocholine (DPPC) lipid dispersed in D_2O with $CaCl_2$ [127]. While this lipid/salt mixture is multilamellar, no strong correlation peak was seen in the NSE data and the relaxation rate followed the Q^3 scaling predicted by Zilman and Granek [127]. Subsequent NSE studies on the same system were used to probe the bending dynamics in the fluid and gel phases to estimate the steric interactions between the bilayers in the unbound state [156].

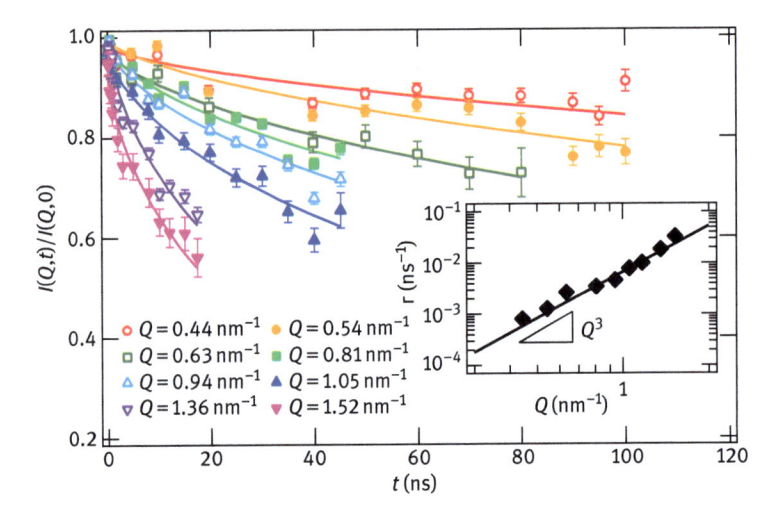

Figure 4.9: Normalized intermediate scattering function $I(Q,t)/I(Q,0)$ measured by NSE for protiated DMPC in D_2O at $T = 35\,°C$. Inset indicates the Q-dependence of the relaxation rate Γ ($\propto Q^3$). Error bars represent ±1 standard deviation throughout the chapter.

Following the early studies of large unilamellar vesicles by Yi et al. [130], NSE has been used to study a wide range of lipid systems including the effects of fatty acyl tail structure [130, 157, 158], inclusion of small molecules such as cholesterol [159, 160], peptides [135, 161], drugs [83, 162–164], and nanoparticles [165], as well as the internal membrane fluctuations [58, 59, 166, 167]. Recently, NSE has also been used to measure more complex systems such as membrane domains [168] and more biologically relevant systems [169, 170].

The inset in Figure 4.9 shows Γ following a Q^3 dependence as expected from the expression for $\Gamma_{ZG_{mod}}$ given in eq. (4.28) with a slope that is proportional to $\tilde{\kappa}$. In Section 4.3.2, the effective bending modulus is given by $\tilde{\kappa} = \kappa + 2d_n^2 K_A$. Again, K_A is the *monolayer* area compressibility modulus and can be calculated as $K_A = 12\kappa_m/d_c^2$, where κ_m is the monolayer bending modulus and d_c is the monolayer hydrocarbon thickness [171]. The monolayer bending modulus is half the bilayer bending modulus, $\kappa_m = \kappa/2$, which means we can rewrite the expression for $\tilde{\kappa}$ in terms of the bilayer properties as $\tilde{\kappa} = \{1 + 48(d_n/2d_c)^2\}\kappa$. The remaining unknown in the expression for $\tilde{\kappa}$ is $d_n/2d_c$, the ratio of the neutral surface to the thickness of the hydrocarbon tails in the bilayer. Remember from Section 4.3.2, the neutral surface is defined as the surface at which the monolayer bending and stretching energies are decoupled and cannot be measured experimentally [45]. Generally, d_n is assumed to lie somewhere between half and the full monolayer thickness with corresponding values of $d_n/2d_c$ ranging from 0.25 to 0.6 [108, 135, 172–175]. The first NSE studies to use the Watson and Brown's equations to analyze NSE data used a value of $d_n/2d_c \approx 0.6$; which puts the neutral surface within the headgroup region of the bilayer [58, 59, 135]. d_n is generally thought to be closer to the interface between the lipid headgroups and tail, with $d_n/2d_c = 0.5$ [176–183]. Putting this value into the expression for $\tilde{\kappa}$ gives $\tilde{\kappa} = 13\kappa$, which shows how much more energy is required to bend the membrane at the NSE scales.

Having an expression for $\tilde{\kappa}$ in terms of the intrinsic bending modulus, $\Gamma_{ZG_{mod}}$ in eq. (4.28) can now be rewritten as [184]

$$\frac{\Gamma_{ZG_{mod}}}{Q^3} = 0.0069\sqrt{\frac{k_B T}{\kappa}\frac{k_B T}{\eta}}. \tag{4.37}$$

Therefore, we can determine the bilayer bending modulus κ from a plot of $\Gamma_{ZG_{mod}}$ versus Q measured in an NSE experiment. Note that changing the value of d_n only changes the numerical prefactor in eq. (4.37) and scales the value of κ by a constant.

Table 4.1 compares values of the bending modulus for DMPC bilayers at a temperature of about 30 °C measured with different experimental techniques. A similar comparison for POPC bilayers is presented in [73] and several other lipids in [185]. All values are on the order of 10 $k_B T$, but can vary significantly depending on the experimental technique. The NSE result is comparable to values determined from fluctuation analysis by Méléard et al. [186] and some computer simulations [175, 185], but differs by almost a factor of 3 with results obtained from other methods such as micropipette aspiration. Nagle and coworkers discussed the discrepancy between κ values measured by different techniques in detail in their recent work [187]. One important difference is that the different techniques measure the bending modulus on different length and timescales, so it is possible that there are other contributions

Table 4.1: Comparison of the bending modulus κ of DMPC bilayers at $T \approx 30\,°C$ for various experimental and simulation techniques in literature.

Technique	T (°C)	κ ($k_B T$)
Neutron spin echo	30	35.6 ± 2.0 [184]
Fluctuation analysis	30	31.1 ± 1.9 [186]
Micropipette aspiration	29	13.4 ± 1.4 [153]
Diffuse x-ray scattering		
with tilt	30	24.6 ± 1 [112]
without tilt		15.6 ± 0.6 [112]
Computer Simulation	30	34.7 ± 1.2 [185]
		29.3 [175]

to the intrinsic membrane rigidity that are not captured by current theories described in Section 4.3.

As the bilayer bends, one leaflet must stretch while the other compresses. This physical relationship is expressed mathematically in a model for thin elastic sheets as $K_{A_b} = \beta \kappa / (2 d_c)^2$ in which K_{A_b} is the *bilayer* area compressibility modulus [171]. The constant β depends on the degree of coupling between the two leaflets: $\beta = 12$ when the two monolayers are fully coupled and the bilayer behaves as a single slab, while $\beta = 48$ when the monolayers are completely uncoupled. Rawicz et al. proposed an intermediate value for lipid bilayers of $\beta = 24$ based on a polymer brush model [188]. Their model assumes that K_{A_b} is related to the bilayer surface pressure, Π, through a constant factor, $K_{A_b} = 6\Pi$, and that the surface pressure is dominated by the entropic contributions from the hydrocarbon tails which are modeled as idealized freely jointed polymer chains (hence the name polymer-brush model) [188]. Any contributions from interactions between the hydrocarbon tails and/or lipid headgroups are neglected, yet despite its simplicity, this model has been shown to hold true for a number of different lipid systems [189]. Because of its success in describing fluid lipid membranes, the polymer brush model is used regularly to relate a measured K_{A_b} to κ and vice versa.

Combining the κ values measured with NSE and d_c values that are measured with an elastic scattering technique such as SANS, we can now use the polymer brush model to calculate values for K_{A_b}. Figure 4.10 shows the resulting values for three phospholipids with different tail lengths and the same headgroup, dimyristoyl-, dipalmitoyl-, and distearoyl-phosphocholine (DMPC, DPPC, and DSPC) [184]. The values of K_{A_b} are independent of the lipid tail length as expected [188] and decrease with increasing temperature.

Here, we should note that the value of β from the polymer brush model may have some connection with the inter-monolayer friction constant b that appeared in the SL and BSO models. However, b did not contribute significantly to the

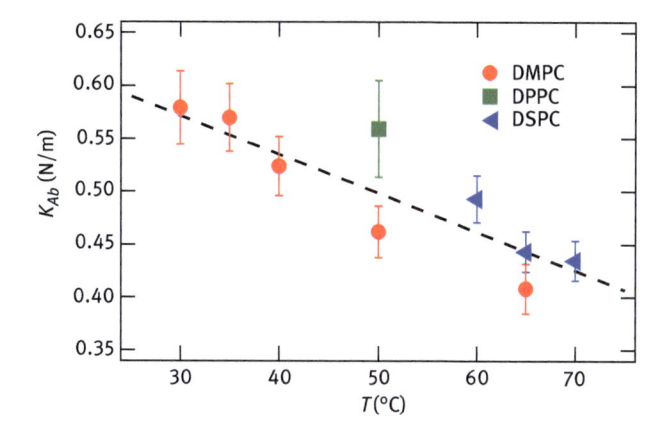

Figure 4.10: Temperature dependence of the bilayer area compressibility modulus K_{A_b} calculated from the values of κ estimated by NSE by adopting the polymer brush theory [188] to relate κ and K_{A_b} by $K_{A_b} = 24\kappa/(2d_c)^2$. The data are adapted with permission from [184]. Copyright 2017 American Chemical Society.

bending mode in these theories. Instead, b appeared in the slipping mode ($\omega_2(k)$) at small k. These small k values are not easy to access using NSE and are likely too fast to observe for DLS. Therefore, it is hard to determine if and how b and β are related. But, if we can independently measure κ and K_{A_b}, then, we can start to calculate the value of β and improve our understanding of both the dispersion relation and b.

4.4.2 Thickness fluctuations

We saw in the BSO model that, in addition to the bending fluctuations we measured in Section 4.4.1, membranes can also undergo thickness fluctuations. To measure the thickness fluctuations with NSE, we take advantage of a unique feature of neutron scattering that the scattering power of different isotopes can be vastly different. In particular, by judiciously replacing H with D, we can carefully control or even eliminate contrast, a technique known as contrast matching, without significantly modifying the properties of the sample. Thus by exchanging H for D in the lipid tails, the scattering contrast of the tails can be "matched" to the surrounding solvent, making the tails "invisible" to the neutrons and thus highlighting the scattering from the lipid headgroups. With this contrast condition, we now emphasize the dynamics from the headgroups and their relative motions.

Recalling the discussion in the previous section, we note that eqs. (4.21) and (4.28) indicate that the undulation fluctuations of a membrane follow a simple

scaling law. Thus, plotting the natural logarithm of the normalized ISF, $\ln[I(Q,t)/I(Q,0)]$ against $(Q^3 t)^{2/3}$, as done in Figure 4.11, should collapse all the data onto a single master curve whose slope is given by $P = \left(0.0069\sqrt{\frac{k_B T}{\kappa}\frac{k_B T}{\eta}}\right)^{2/3}$.
Figure 4.11 shows such a plot of NSE data for tail-deuterated DPPC vesicles in which the hydrocarbon tails are contrast-matched to the D$_2$O solvent [58]. While most of the data do follow the expected scaling, the data at $Q \approx 1.0$ nm^{-1} do not, suggesting that there is another, faster relaxation process contributing at this Q.

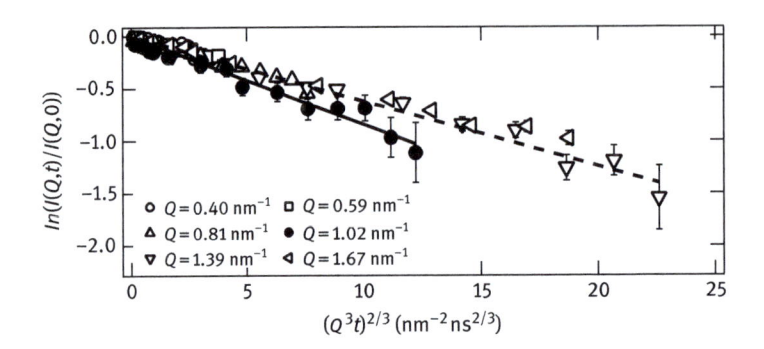

Figure 4.11: A single membrane undulation scaling plot of $\ln[I(Q,t)/I(Q,0)]$ vs $(Q^3 t)^{2/3}$ for tail deuterated DPPC at $T = 50\,°C$ suggested by eq. (4.28) [113]. All the data, except that for $Q = 1.02$ nm^{-1}, collapse onto a single line. Deviation of the data at $Q = 1.02$ nm^{-1} from the undulation fluctuation scaling behavior supports the idea of an additional dynamic contribution at this specific Q.

As suggested by eq. (4.37), the enhanced dynamics can also be emphasized by plotting the data as Γ/Q^3 versus Q which, following the same scaling arguments above for pure undulation dynamics, leads to a horizontal line whose intercept is related to the bending modulus. The data for DMPC, DPPC, and DSPC bilayers plotted in this fashion in Figure 4.12 all show a distinct peak [58]. The deviation for all of the bilayers is very localized at Q around 1 nm^{-1} (length scales of ≈ 3 nm) which we assign to the membrane thickness fluctuations. The Q-value of the peak corresponds to the minimum in the scattered intensity of the membrane form factor measured in SANS, supporting the idea that the excess dynamics are occurring at the length scale of the membrane thickness. The same signature of membrane thickness fluctuations was also seen in NSE data for oil-swollen surfactant bilayers [141, 144, 145, 190]. The peak shift to lower Q from DMPC to DPPC to DSPC in Figure 4.12 then reflects the increase in bilayer thickness with increasing tail length. Note that these thickness fluctuations are only seen in the fluid phase of the membrane [58]. The peak in the data disappears at low temperatures below T_m (see Figure 4.12) when the lipids are in the gel phase and the membrane is more rigid and less dynamic.

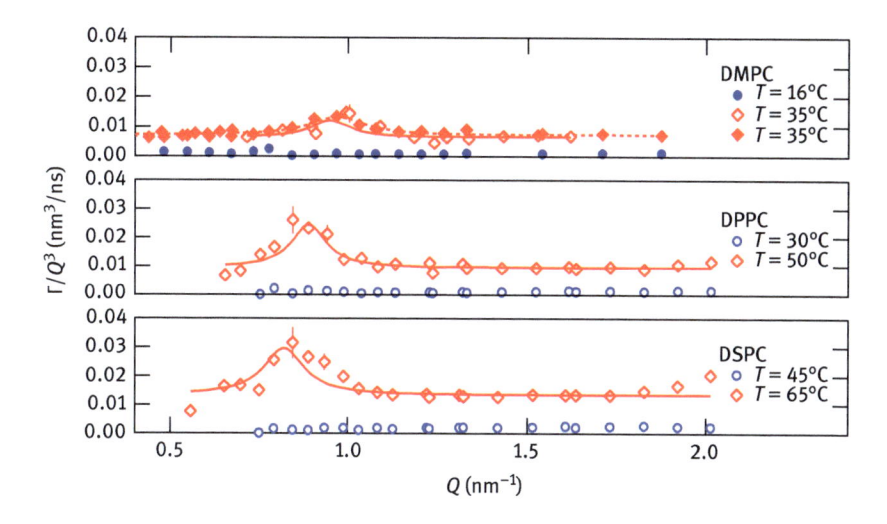

Figure 4.12: The relaxation rate measured by NSE is plotted in terms of Γ/Q^3 against Q. If the dynamics are solely undulation fluctuations, the value of Γ/Q^3 is flat as it is observed below the transition temperature T_m for each lipid. On the other hand, at $T > T_m$ a peak like enhancement is observed indicative of thickness fluctuations. The solid lines are the fit results according to eq. (4.38). Open and full symbols indicate the data collected on the NG5-NSE and IN15, respectively. The data are adapted with permission from [58]. Copyright 2012 The American Physical Society.

In order to further characterize the excess dynamics at the membrane thickness length scale, we assume that they can be captured by a simple additive term to the bending fluctuation decay rate given by eq. (4.37) and can be expressed by the following equation [144]

$$\frac{\Gamma}{Q^3} = \frac{\Gamma_{ZG_{mod}}}{Q^3} + \frac{\Gamma_{TF}}{Q_0^3} \frac{1}{1+(Q-Q_0)^2 \xi^2}, \qquad (4.38)$$

in which the first term describes the underlying bending dynamics, quantified using protiated lipid bilayers, and the second term empirically fits the thickness fluctuation peak. The two important parameters extracted from the Lorentz function are the relaxation time given by the decay constant $\tau_{TF} = 1/\Gamma_{TF}$ and the fluctuation amplitude $\Delta d_c = 2d_c(Q_0\xi)^{-1}$, which is given by the half width at half maximum of the Lorentzian, ξ^{-1} [141, 146, 191].

Earlier in the chapter, we noted that the characteristic length and timescales of the thickness fluctuations were related to the membrane's elastic and viscous properties, suggesting eq. (4.38) can be rewritten in terms of the intrinsic membrane properties.

Statistical mechanics predicts a relationship between the bilayer area compressibility modulus K_{A_b} and the fractional change in area $\sigma_A = \Delta A/A$ as [52, 192]

$$KA_b = \frac{k_B T}{\sigma_A^2 A_0} \tag{4.39}$$

Where A and A_0 are the unit area of the membrane and the area per molecule, respectively. Assuming the bilayer volume compressibility is negligible ($V = Ad_c$, $\Delta V/V = \Delta A/A + \Delta d_c/d_c \approx 0$), σ_A is compensated for by a corresponding change in thickness $\sigma_d = \Delta d_c/d_c$, that is, $\sigma_d^2 = \sigma_A^2$. (Both simulation and experiment suggest that the volume compressibility is at least an order of magnitude smaller than the area compressibility, supporting that the assumption of $\Delta V/V \approx 0$ is reasonable to a first approximation [52, 193].) Therefore, the relation $\sigma_d^2 = (\Delta d/d)^2 = k_B T/(K_{A_b} A_0) = 4(Q_0 \xi)^{-2}$ leads to an expression for the peak width in terms of the bilayer compressibility modulus $\xi^2 = 4K_{A_b} A_0/(Q_0^2 k_B T)$.

The BSO model covered in Section 4.3.3 [53] developed a dispersion relation for the peristaltic mode (thickness fluctuations), $\omega_u(k)$ in eq. (4.35). $\omega_u(k)$ is related to K_{A_b} and damped by the viscosities of the solvent, η, and membrane, η_m. When the wavelengths of the mode are shorter than the Saffman–Delbrück length [55], $l_{SD} = \eta_m/\eta$, the in-plane monolayer viscosity dominates and the damping is independent of the fluctuation wavelength. In general, this condition is satisfied in pure lipid bilayers as can be seen in Figure 4.7, and we can use the dispersion relation from the BSO model to relate the experimentally measured τ_{TF} to the membrane viscosity as [53]

$$\tau_{TF} \approx \frac{\eta_m}{K_{A_b}}. \tag{4.40}$$

These refinements lead to an expression for the Lorentzian based on the bilayer's elastic and viscous properties as [184]

$$\frac{\Gamma}{Q^3} = \frac{\Gamma_{ZG_{mod}}}{Q^3} + \frac{K_{A_b} k_B T}{\eta_m Q_0^3 k_B T + 4\eta_m Q_0 K_{A_b} A_0 (Q - Q_0)^2}. \tag{4.41}$$

In this equation, K_{A_b} is calculated from κ determined by NSE experiments with protiated lipid bilayers using the polymer brush model, $K_{A_b} = 24\kappa/(2d_c)^2$ and assuming that K_{A_b} is the same for protiated and tail-deuterated lipids within the experimental uncertainty. The area per lipid is given by $A_0 = V_L/(d_c + d_h)$, where V_L and d_h represent the lipid volume and the headgroup thickness, respectively. V_L can be determined from density measurement to find the specific molecular volume [194] or calculated from equations for the lipid volume available in literature [195]. The tail thickness can be determined from an elastic small angle scattering experiment (such as SANS or SAXS), and the headgroup thickness, while not yet well defined, is generally assumed to be on the order of $d_h = 1$ nm [196–198]. Q_0 can also be determined from a SANS or diffraction measurement. Thus, all the parameters in eq. (4.41) are known except for the membrane viscosity η_m.

The refined analysis method now defines the peak shape in Γ/Q^3 in terms of the structural and elastic parameters and in the process reduced the number of fit parameters from two (τ_{TF} and ξ^{-1} in eq. (4.38)) to one (η_m in eq. (4.41)). Based on eq. (4.41), the peak width is related to K_{A_b}; however, experimentally the data also contain effects from the instrumental Q resolution. The Q resolution smearing can be taken into account using the expressions developed for SANS instruments [199], and the resolution function can be convoluted with eq. (4.41) to fit the NSE data. The fit result is plotted in Figure 4.13 for the temperature variation of Γ/Q^3 in the fluid phase of the DMPC vesicles in which the instrumental Q resolution was expressed as a Gaussian function with a full width at half maximum of $\approx 13\%$ [184]. Note that the baseline at $Q > Q_0$ does not capture the experimental data, which suggests that additional intramembrane dynamics may be present that are not currently considered.

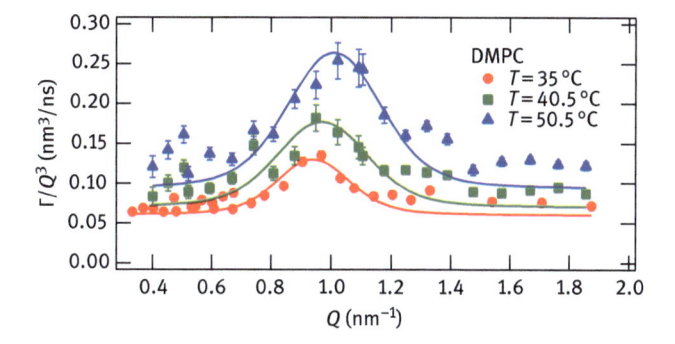

Figure 4.13: Temperature variation of Γ/Q^3 against Q for DMPC membranes is shown together with the fit results according to eq. (4.41) using a single fit parameter: the membrane viscosity η_m. The instrumental Q resolution is also accounted for in the fit. Reprinted with permission from [184]. Copyright 2017 American Chemical Society.

Reformulating the expression for the bilayer thickness fluctuations in terms of the membrane properties now allows us to determine the membrane viscosity η_m from the NSE data. The temperature dependence of η_m determined from NSE measurements is shown in Figure 4.14 for DMPC, DPPC, and DSPC [184]. The estimated values of η_m are on the order of 10 nPa·s·m, and increase with increasing tail length and decreasing temperature, which qualitatively matches the trends seen in linear alkanes [200, 201]. Interestingly, the measured η_m values for these three lipids are about the same ($\eta_m \approx 100$ nPa·s·m) at $T \approx T_m$ and decrease by as much as a factor of 2 with increasing temperature, which suggests a strong coupling between the bilayer phase behavior and the lipid motion.

Reported values for the membrane viscosity in literature vary widely. While some of the discrepancies may be due to differences in studied temperatures, the temperature

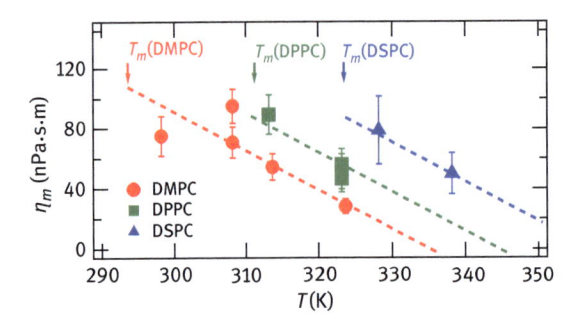

Figure 4.14: The membrane viscosity η_m calculated from the NSE result. The dashed straight lines are guide the eyes. Reprinted with permission from [184]. Copyright 2017 American Chemical Society.

variation seen in Figure 4.14 over a 50 K range does not account for the two orders of magnitude variations seen in literature with different experimental methods. Fluorescent probe measurements, tracking domain motions in phase separated membranes, and falling ball viscometry measurements give membrane viscosity values on the order of ($\approx$ 2 to 3) nPa·s·m [183, 202–206] while measurements based on the diffusion of tracer particles give viscosities on the order of $\approx$ 10 nPa·s·m [207]. On the other hand, membrane viscosities estimated based on red blood cell recovery times in micropipette aspiration measurements are $\approx$ 700 nPa·s·m [208, 209]. The η_m values from NSE fall within the wide range of reported values for lipid membranes, and perhaps more importantly, were determined without incorporating fluorescent probes or tracer particles. Instead, the viscosity values from NSE are determined from the relaxation timescales of the naturally occurring membrane fluctuations. Similar NSE experiments were also used to determine the viscosity of oil-swollen surfactant membranes [190]. In this case, the η_m determined from NSE could be compared with the measured bulk viscosity of the oil-surfactant mixtures η_s as $\eta_m = d_m \eta_s$ with d_m as the membrane thickness. Both measurements gave similar values of η_m further supporting the utility of NSE to also determine the viscosity of membrane systems.

4.5 Summary and future perspective

In this chapter, we presented the observation of collective membrane fluctuations with NSE and the theory needed to relate the measured dynamics to the membrane elastic and viscous properties. By measuring the bending and thickness fluctuations on the nanoscale, we can independently determine the bilayer bending modulus, κ, area compressibility modulus, K_{A_b}, and monolayer viscosity, η_m. These same properties that govern the equilibrium thermal fluctuations will also influence the membrane's viscoelastic response to nonequilibrium deformations that are an essential part of cell function as they package cargo, shape organelles and respond to external forces that we talked about at the beginning of the chapter [210]. Measuring both the bending and thickness fluctuations with NSE also opens the possibility of determining the inter-monolayer coupling constant β. While extracting an independent determination of K_{A_b} from the thickness fluctuation measurements is currently limited by the quality of the experimental data, continued improvement in NSE instrumentation and data analysis may make these measurements more robust in the future. Alternatively, β could be determined by combining NSE with another technique to measure κ or K_{A_b}, such as those outlined in other chapters of this book. Gelation [211], degree of saturation of the lipid tail [212], and mixing lipids with cholesterol [189] are all known to modulate inter-monolayer coupling and the NSE measurements described here can be used to not only gain insights into a wide variety of lipid bilayers, but help advance our understanding of the complex dissipation mechanisms in these systems.

The development of theory and experiment go hand-in-hand, and extracting more information from the NSE measurements requires a solid theoretical basis. Currently, for example, the thickness fluctuations are treated as an excess to the bending fluctuations and assumed to be an additive term in the decay constant expression (eq. (4.41)), and the $I(Q, t)$ data are fit using the stretched exponential function of the single membrane fluctuation model. In principle though, these two independent modes should be treated as two individual relaxations in $I(Q, t)$. Just as Mell and colleagues have developed a relaxation function incorporating the hybrid modes (Section 4.3.2) [125], proper data treatment for the thickness fluctuations requires the correct relaxation function for $I(Q, t)$ to understand the relaxation behavior and dispersion relation of this dynamic mode.

Recent theory, simulation and experimental works have also begun to incorporate tilt degrees of freedom, described by a tilt modulus, into the analysis of the nanoscale dynamics. The tilt modulus introduces another parameter, in addition to the bending and thickness fluctuation parameters, and these models now have literally too many parameters (see, e.g., Table II in Ref. [108]) to reliably extract from a single experimental measurement. Quantifying all of the lipid membrane properties will require combining multiple experimental techniques with the appropriate theories. For example, recent development of diffuse X-ray scattering allows one to determine both the bending and tilt moduli [109, 112, 213, 214], which could be combined with NSE measurements. The length and timescales probed by scattering methods in particular are highly complementary to ranges accessible by computer simulation, even for thickness fluctuations [141, 146], and combining simulation and experiment is a powerful opportunity to probe the complexities of membrane dynamics.

The theoretical and experimental framework outlined in this chapter provides a foundation for understanding the elastic and viscous properties of lipid membranes, and we hope the examples provided here inform and inspire ongoing investigations into more complex and biologically relevant systems. All of the example NSE data shown in the chapter were for simple saturated lipid bilayers, yet a single cell membrane can contain hundreds of chemically distinct lipids. While we know this remarkable complexity is essential to cell function, our understanding of how it fundamentally affects the membrane properties is still developing. For example, decades of research have shown that the effects of cholesterol on the membrane properties are highly dependent on the lipid species and our picture of cholesterol-containing membranes is still evolving. Moreover, cell membranes are not only made of lipids, but can contain upwards of 50% by mass of proteins. Simply altering the thickness of model membranes has been shown to affect the biological activity of several proteins while incorporating peptides into lipid membranes also is known to affect the bilayer structural properties. Clearly, there is a synergy in lipid–protein interactions in determining the membrane properties; however, the nature of these interactions are not well understood. All of these intricacies relate to the synergy among structure, dynamics, and functions of biological membranes, and NSE is providing unique insights into the interdependence of the membrane collective dynamics and the elastic and viscous properties.

Acknowledgments

Access to the NGA-NSE was provided by the Center for High Resolution Neutron Scattering, a partnership between the National Institute of Standards and Technology and the National Science Foundation under Agreement No. DMR-1508249. The access to the IN15 was granted through the proposal No. 9-13-353. M.N. acknowledges funding support of cooperative agreement 70NANB15H259 from NIST, US Department of Commerce.

References

[1] Goetz, R., Gompper, G., & Lipowsky, R. Mobility and elasticity of self-assembled membranes. *Phys. Rev. Lett.*, 82:221–224, 1999.

[2] Klauda, J.B., Roberts, M.F., Redfield, A.G., Brooks, B.R., & Pastor, R.W. Rotation of lipids in membranes: Molecular dynamics simulation, ^{31}P spin-lattice relaxation, and rigid-body dynamics. *Biophys. J.*, 94:3074–3083, 2008.

[3] Vaz, W.L.C., Goodsaid-Zalduondo, F., & Jacobson, K. Lateral diffusion of lipids and proteins in bilayer membranes. *FEBS. Lett.*, 174:199–207, 1984.

[4] Almeida, P.F.F., & Vaz, W.L.C. *Handbook of Biological Physics, Volume I*, chapter 6 Lateral Diffusion in Membranes. Elsevier Science, Amsterdam, pp. 305–357, 1995.

[5] Singer, S.J., & Nicolson, G.L. The fluid mosaic model of the structure of cell membranes. *Science*, 175:720–731, 1972.

[6] Brown, D.A., & London, E. Functions of lipid rafts in biological membranes. *Annu. Rev. Cell Dev. Biol.*, 14:111–136, 1998.
[7] Engelman, D.M. Membranes are more mosaic than fluid. *Nature*, 438:578–580, 2005.
[8] McMahon, H.T., & Gallop, J.L. Membrane curvature and mechanisms of dynamic cell membrane remodelling. *Nature*, 438:590–596, 2005.
[9] McMahon, H.T., & Boucrot, E. Membrane curvature at a glance. *J. Cell. Sci.*, 128:1065–1070, 2015.
[10] Browicz, T. Further observation of motion phenomena on red blood cells in pathological states. *Zbl. Med. Wiss.*, 28:625–627, 1890.
[11] Turlier, H., Fedosov, D.A., Audoly, B., Auth, T., Gov, N.S., Sykes, C., Joanny, J.-F., Gompper, G., & Betz, T. Equilibrium physics breakdown reveals the active nature of red blood cell flickering. *Nat. Phys.*, 12:513–519, 2016.
[12] Helfrich, W. Elastic properties of lipid bilayers: Theory and possible experiments. *Z. Naturforsch.*, 28 c:693–703, 1973.
[13] Netz, R.R. Complete unbinding of fluid membranes in the presence of short-ranged forces. *Phys. Rev. E.*, 51:2286–2294, 1995.
[14] Marx, S., Schilling, J., Sackmann, E., & Bruinsma, R. Helfrich repulsions and dynamical phase separation of multicomponent lipid bilayers. *Phys. Rev. Lett.*, 88:138102, 2002.
[15] Dan, N., Pincus, P., & Safran, S.A. Membrane-induced interactions between inclusions. *Langmuir*, 9:2768–2771, 1993.
[16] Netz, R.R. Inclusions in fluctuating membranes: Exact results. *J. Phys. I France*, 7:833–852, 1997.
[17] Dean, D.S., Parsegian, V.A., & Podgornika, R. Fluctuation mediated interactions due to rigidity mismatch and their effect on miscibility of lipid mixtures in multicomponent membranes. *J. Phys.: Condens. Matter*, 27:214004, 2015.
[18] Maurya, S.R., Chaturvedi, D., & Mahalakshmi, R. Modulating lipid dynamics and membrane fluidity to drive rapid folding of a transmembrane barrel. *Sci. Rep.*, 3:1989, 2013.
[19] Huang, H.W. Deformation free energy of bilayer membrane and its effect on gramicidin channel lifetime. *Biophys. J.*, 50:1061–1070, 1986.
[20] Helfrich, P., & Jakobsson, E. Calculation of deformation energies and conformations in lipid membranes containing gramicidin channels. *Biophys. J.*, 57:1075–1084, 1990.
[21] Andersen, O.S., & Koeppe, R.E., II Bilayer thickness and membrane protein function: An energetic perspective. *Annu. Rev. Biophys. Biomol. Strut.*, 36:107–130, 2007.
[22] Phillips, R., Ursell T., Wiggins P., and Sens P. Emerging roles for lipids in shaping membrane-protein function. *Nature*, 459:379–385, 2009.
[23] Stevenson, P., & Tokmakoff, A. Time-resolved measurements of an ion channel conformational change driven by a membrane phase transition. *Proc. Natl. Acad. Sci.*, 114:10840–10845, 2017.
[24] Raudino, A., & Pannuzzo, M. Adhesion kinetics between a membrane and a flat substrate. An ideal upper bound to the spreading rate of an adhesive patch. *J. Phys. Chem. B.*, 114:15495–15505, 2010.
[25] Fenz, S.F., Bihr, T., Merkel, R., Seifert, U., Sengupta, K., & Smith, A.-S. Switching from ultraweak to strong adhesion. *Adv. Mater.*, 23:2622–2626, 2011.
[26] Bihr, T., Seifert, U., & Smith, A.-S. Nucleation of ligand-receptor domains in membrane adhesion. *Phys. Rev. Lett.*, 109:258101, 2012.
[27] Bihr, T., Seifert, U., & Smith, A.-S. Multiscale approaches to protein-mediated interactions between membranes - relating microscopic and macroscopic dynamics in radially growing adhesions. *New J. Phys.*, 17:083016, 2015.
[28] Fenz, S.F., Bihr, T., Schmidt, D., Merkel, R., Seifert, U., Sengupta, K., & Smith, A.-S. Membrane fluctuations mediate lateral interaction between cadherin bonds. *Nat. Phys.*, 13:906–913, 2017.

[29] Stachowiak, J.C., Brodsky, F.M., & Miller, E.A. A cost-benefit analysis of the physical mechanisms of membrane curvature. *Nat. Cell. Bio.*, 15:1019–1027, 2013.
[30] Zidovska, A., & Sackmann, E. Brownian motion of nucleated cell envelopes impedes adhesion. *Phys. Rev. Lett.*, 94:048103, 2006.
[31] Pierres, A., Benoliel, A.-M., Touchard, D., & Bongrand, P. How cells tiptoe on adhesive surfaces before sticking. *Biophys. J.*, 94:4114–4122, 2008.
[32] Biswas, A., Alex, A., & Sinha, B. Mapping cell membrane fluctuations reveals their active regulation and transient heterogeneities. *Biophys. J.*, 113:1768–1781, 2017.
[33] Pontes, B., Ayala, Y., Fonseca, A.C.C., Romão, L.F., Amaral, R.F., Salgado, L.T., Lima, F.R., Farina, M., Viana, N.B., Moura-Neto, V., & Nussenzveig, H.M. Membrane elastic properties and cell function. *PLOS One.*, 8:e67708, 2013.
[34] Monzel, C., Schmidt, D., Kleusch, C., Kirchenbücler, D., Seifert, U., Smith, A.-S., Sengupta, K., & Merkel, R. Measuring fast stochastic displacements of bio-membranes with dynamic optical displacement spectroscopy. *Nat. Commun.*, 6:8162, 2015.
[35] Cojoc, D., Finaurini, S., Livshits, P., Gur, E., Shapira, A., Mico, V., & Zalevsky, Z. Toward fast malaria detection by secondary speckle sensing microscopy. *Biomed. Opt. Express*, 3: 991–1005, 2012.
[36] Park, Y.-K., Diez-Silva, M., Popescu, G., Lykotrafitis, G., Choi, W., Feld, M.S., & Suresh, S. Refractive index maps and membrane dynamics of human red blood cells parasitized by *Plasmodium falciparum*. *Proc. Natl. Acad. Sci.*, 105:13730–13735, 2008.
[37] Shaked, N.T., Satterwhite, L.L., Telen, M.J., Truskey, G.A., & Wax, A. Quantitative microscopy and nanoscopy of sickle red blood cells performed by wide field digital interferometry. *J. Biomed. Opt.*, 16:030506, 2011.
[38] Yoon, Y.-Z., Hong, H., Brown, A., Kim, D.C., Kang, D.J., Lew, V.L., & Cicuta, P. Flickering analysis of erythrocyte mechanical properties: Dependence on oxygenation level, cell shape, and hydration level. *Biophys. J.*, 97:1606–1615, 2009.
[39] Byun, H., Hillman, T.R., Higgins, J.M., Diez-Silva, M., Peng, Z., Dao, M., Dasari, R.R., Suresh, S., & Park, Y. Optical measurements of biomechanical properties of individual erythrocytes from a sickle cell patient. *Acta Biomater.*, 8:4130–4138, 2012.
[40] Partin, A.W., Schoeniger, J.S., Mohler, J.L., & Coffey, D.S. Fourier analysis of cell motility: Correlation of motility with metastatic potential. *Proc. Natl. Acad. Sci.*, 86:1254–1258, 1989.
[41] Braig, S., Schmidt, B.U.S., Stoiber, K., Händel, C., Möhn, T., Werz, O., Müller, R., Zahler, S., Koeberle, A., Käs, J.A., & Vollmar, A.M. Pharmacological targeting of membrane rigidity: Implications on cancer cell migration and invasion. *New. J. Phys.*, 17:083007, 2015.
[42] Handel, C., Schmidt, B.U.S., Schiller, J., Dietrich, U., Mohn, T., Kiebling, T.R., Pawlizak, S., Fritsch, A.W., Horn, L.-R., Briest, S., Hockel, M., Zink, M., & Kas, J.A. Cell membrane softening in human breast and cervical cancer cells. *New J. Phys.*, 17:083007, 2015.
[43] Suezaki, Y., & Ichinose, H. A theory on the bending moduli of thin membranes by the use of a simple molecular model. *J. Phys. I France*, 5:1469–1480, 1995.
[44] Safran, S.A. *Statistical Thermodynamics of Surfaces, Interfaces, and Membranes*. Addison-Wesley, Massachusetts, 1994.
[45] Safran, S.A. Curvature elasticity of thin films. *Adv. Phys.*, 48(4):395–448, 1999.
[46] Hladky, S.B., & Gruen, D.W.R. Thickness fluctuations in black lipid membranes. *Biophys. J.*, 38:251–258, 1982.
[47] Miller, I.R. Energetics of fluctuation in lipid bilayer thickness. *Biophys. J.*, 45:643–644, 1984.
[48] Hladky, S.B., & Gruen, D.W.R. Response to energetics of fluctuation in lipid bilayer thickness by I. R. Miller. *Biophys. J.*, 45:645–646, 1984.

[49] Popescu, D., Ion, S., Popescu, A., & Movileanu, L. Elastic properties of bilayer lipid membranes and pore formation. In H.T. Tien and A. Ottova-Leitmannova, editors, *Planar Lipid Bilayers (BLMs) and Their Applications*, pages 173-204. Elsevier, Amsterdam 2003.

[50] Movileanu, L., Popescu, D., Ion, S., & Popescu, A.I. Transbilayer pores induced by thickness fluctuations. *Bull. Math. Biol.*, 68:1231–1255, 2006.

[51] Kaufmann, K., Hanke, W., & Corcia, A. *Ion Channel Fluctuations in Pure Lipid Bilayer Membranes: Control by Voltage*. Caruaru, Brazil, 1989.

[52] Lindahl, E., & Edholm, O. Mesoscopic undulations and thickness fluctuations in lipid bilayers from molecular dynamics simulations. *Biophys. J.*, 79:426–433, 2000.

[53] Bingham, R.J., Smye, S.W., & Olmsted, P.D. Dynamics of an asymmetric bilayer lipid membrane in a viscous solvent. *Europhys. Lett.*, 111:18004, 2015.

[54] Seifert, U. Configurations of fluid membranes and vesicles. *Adv. Phys.*, 46:13–137, 1997.

[55] Saffman, P.G., & Delbrück, M. Brownian motion in biological membranes. *Proc. Natl. Acad. Sci. USA.*, 72:3111–3113, 1975.

[56] Hughes, B.D., Pailthorpe, B.A., & White, L.R. The translational and rotational drag on a cylinder moving in a membrane. *J. Fluid Mech.*, 110:349–372, 1981.

[57] Petrov, E.P., & Schwille, P. Translational diffusion in lipid membranes beyond the Saffman-Delbrück approximation. *Biophys. J.*, 107:L41–L43, 2008.

[58] Woodka, A.C., Butler, P.D., Porcar, L., Farago, B., & Nagao, M. Lipid bilayers and membrane dynamics: Insight into thickness fluctuations. *Phys. Rev. Lett.*, 109:058102, 2012.

[59] Ashkar, R., Nagao, M., Butler, P.D., Woodka, A.C., Sen, M.K., & Koga, T. Tuning membrane thickness fluctuations in model lipid bilayers. *Biophys. J.*, 109:106–112, 2015.

[60] Seeling, A., & Seelig, J. Dynamic structure of fatty acyl chains in a phospholipid bilayer measured by deuterium magnetic resonance. *Biochemistry*, 13:4839–4845, 1974.

[61] Brown, M.F., Ribeiro, A.A., & Williams, G.D. New view of lipid bilayer dynamics from ^{2}H and ^{13}C NMR relaxation time measurements. *Proc. Natl. Acad. Sci. USA.*, 80: 4325–4329, 1983.

[62] Leftin, A., & Brown, M.F. An NMR database for simulations of membrane dynamics. *Biochim. Biophys. Acta Biomembr.*, 1808:818–839, 2011.

[63] Yellin, N., & Levin, I.W. Hydrocarbon chain trans-gauche isomerization in phospholipid bilayer gel assemblies. *Biochemistry*, 16:642–647, 1977.

[64] Wallach, D.F.H., Verma, S.P., & Fookson, J. Application of laser Raman and infrared spectroscopy to the analysis of membrane structure. *Biochim. Biophys. Acta*, 559:153–208, 1979.

[65] Pink, D.A., Green, T.J., & Chapman, D. Raman scattering in bilayers of saturated phosphatidylcholines. Experiment and theory. *Biochemistry*, 19:349–356, 1980.

[66] Mendelsohn, R., Davies, M.A., Brauner, J.W., Schuster, H.F., & Dluhy, R.A.. Quantitative determination of conformational disorder in the acyl chains of phospholipid bilayers by infrared spectroscopy. *Biochemistry*, 28:8934–8939, 1989.

[67] Klösgen, B., Reichle, C., Kohlsmann, S., & Kramer, K.D.. Dielectric spectroscopy as a sensor of membrane headgroup mobility and hydration. *Biophys. J.*, 71:3251–3260, 1996.

[68] Lange, A., Marsh, D., Wassmer, K.H., Meier, P., & Kothe, G. Electron spin resonance study of phospholipid membranes employing a comprehensive line-shape model. *Biochemistry*, 24:4383–4392, 1985.

[69] Marsh, D., & Horváth, L.I. Structure, dynamics and composition of the lipid-protein interface. Perspectives from spin-labelling. *Biochim. Biophys. Acta Rev. Biomembr.*, 1376:267–296, 1998.

[70] Subczynski, W.K., & Kusumi, A. Dynamics of raft molecules in the cell and artificial membranes: approaches by pulse EPR spin labeling and single molecule optical microscopy. *Biochim. Biophys. Acta*, 1610:231–243, 2003.

[71] Faucon, J.F., Mitov, M.D., Méléard, P., Bivas, I., & Bothorel, P. Bending elasticity and thermal fluctuations of lipid membranes. Theoretical and experimental requirements. *J. Phys. France*, 50:2389–2414, 1989.

[72] Mutz, M., & Helfrich, W. Bending rigidities of some biological model membranes as obtained from the Fourier analysis of contour sections. *J. Phys. France*, 51:991–1002, 1990.

[73] Dimova, R. Recent developments in the field of bending rigidity measurements on membranes. *Adv. Coll. Int. Sci.*, 208:225–234, 2014.

[74] Monzel, C., & Sengupta, K. Measuring shape fluctuations in biological membranes. *J. Phys. D.: Appl. Phys.*, 49:243002, 2016.

[75] Chadwick, J. Possible existence of a neutron. *Nature*, 129:312, 1932.

[76] Meyer, A., Dimeo, R.M., Gehring, P.M., & Neumann, D.A. The high flux backscattering spectrometer at the NIST Center for Neutron Research. *Rev. Sci. Instrum.*, 74:2759–2777, 2003.

[77] Copley, J.R.D., & Cook, J.C. The disk chopper spectrometer at NIST: a new instrument for quasielastic neutron scattering studies. *Chem. Phys.*, 292:477–485, 2003.

[78] Pfeiffer, W., Henkel, T., Sackmann, E., Knoll, W., & Richter, D. Local dynamics of lipid bilayers studied by incoherent quasi-elastic neutron scattering. *Europhys. Lett.*, 8:201–206, 1989.

[79] Armstrong, C.L., Kaye, M.D., Zamponi, M., Mamontov, E., Tyagi, M., Jenkins, T., & Rheinstädter, M.C. Diffusion in single supported lipid bilayers studied by quasi-elastic neutron scattering. *Soft Matter*, 6:5864–5867, 2010.

[80] Armstrong, C.L., Trapp, M., Peters, J., Seydel, T., & Rheinstädter, M.C. Short range ballistic motion in fluid lipid bilayers studied by quasi-elastic neutron scattering. *Soft Matter*, 7:8358–8362, 2011.

[81] Armstrong, C.L., Häußler, W., Seydel, T., Katsaras, J., & Rheinstädter, M.C. Nanosecond lipid dynamics in membranes containing cholesterol. *Soft Matter*, 10:2600–2611, 2014.

[82] Sharma, V.K., Mamontov, E., Tyagi, M., Qian, S., Rai, D.K., & Urban, V.S. Dynamical and phase behavior of a phospholipid membrane altered by an antimicrobial peptide at low concentration. *J. Phys. Chem. Lett.*, 7:2394–2401, 2016.

[83] Sharma, V.K., Mamontov, E., Ohl, M., & Tyagi, M. Incorporation of aspirin modulates the dynamical and phase behavior of the phospholipid membrane. *Phys. Chem. Chem. Phys.*, 19:2514–2524, 2017.

[84] Yamada, T., Takahashi, N., Tominaga, T., Takata, S., & Seto, H. Dynamical behavior of hydration water molecules between phospholipid membranes. *J. Phys. Chem. B.*, 121:8322–8329, 2017.

[85] Mezei, F. Neutron spin echo: A new concept in polarized thermal neutron techniques. *Z. Physik.*, 255:146–160, 1972.

[86] Mezei, F. *Lecture Notes in Physics*, volume 128. Springer-Verlag, 1979.

[87] Farago, B., Falus, P., Hoffmann, I., Gradzielski, M., Thomas, F., & Gomez, C. The IN15 upgrade. *Neutron News.*, 26:15–17, 2015.

[88] Takeda, T., Komura, S., Seto, H., Nagai, M., Kobayashi, H., Yokoi, E., Zeyen, C.M.E., Ebisawa, T., Tasaki, S., Ito, Y., Takahashi, S., & Yoshizawa, H. A neutron spin echo spectrometer with two optimal field shape coils for neutron spin precession. *Nucl. Instrum. Meth. A.*, 364:186–192, 1995.

[89] Monkenbusch, M., Schätzler, R., & Richter, D. The Jülich neutron spin-echo spectrometer — Design and performance. *Nucl. Instrum. Meth. A.*, 399:301–323, 1997.

[90] Farago, B. Recent neutron spin-echo developments at the ILL (IN11 and IN15). *Physica B.*, 267–268:270–276, 1999.

[91] Rosov, N., Rathgeber, S., & Monkenbusch, M. Neutron spin echo spectroscopy at the NIST Center for Neutron Research. In P. Cebe, B. S. Hsaio, & D. J. Lohse, editors, *Scattering from polymers: characterization by x-rays, neutrons, and light*, pages 103–116. American Chemical Society: Washington D. C., 2000.

[92] Mezei, F. *Lecture Notes in Physics*, volume 601. Springer-Verlag Berlin Heidelberg, 2003.
[93] Nagao, M., Yamada, N.L., Kawabata, Y., Seto, H., Yoshizawa, H., & Takeda, T. Relocation and upgrade of neutron spin echo spectrometer, iNSE. *Physica B.*, 385–386:1118–1121, 2006.
[94] Holderer, O., Monkenbusch, M., Schätzler, R., Kleines, H., Westerhausen, W., & Richter, D. The JCNS neutron spin-echo spectrometer J-NSE at the FRM II. *Meas. Sci. Technol.*, 19:034022, 2008.
[95] Ohl, M., Monkenbusch, M., Arend, N., Kozielewski, T., Vehres, G., Tiemann, C., Butzek, M., Soltner, H., Giesen, U., Achten, R., Stelzer, H., Lindenau, B., Budwig, A., Kleines, H., Drochner, M., Kaemmerling, P., Wagener, M., Möller, R., Iverson, E.B., Sharp, M., & Richter, D. The spin-echo spectrometer at the spallation neutron source (SNS). *Nucl. Instrum. Meth. A.*, 696:85–99, 2012.
[96] Fouquet, P., Ehlers, G., Farago, B., Pappas, C., & Mezei, F. The wide-angle neutron spin echo spectrometer project WASP. *J. Neutron Res.*, 15:39–47, 2007.
[97] Häussler, W., Gohla-Neudecker, B., Schwikowski, R., Streibl, D., & Böni, P. RESEDA — The new resonance spin echo spectrometer using cold neutrons at the FRM-II. *Physica B.*, 397:112–114, 2007.
[98] Hino, M., Oda, T., Kitaguchi, M., Yamada, N.L., Sagehashi, H., Kawabata, Y., & Seto, H. Current status of BL06 beam line for VIN ROSE at J-PARC/MLF. *Phys. Procedia.*, 42:136–141, 2013.
[99] Zilman, A.G., & Granek, R. Membrane dynamics and structure factor. *Chem. Phys.*, 284: 195–204, 2002.
[100] Suezaki, Y. *Physics of Lipid Membranes [in Japanese]*. Kyushu University Publisher, Fukuoka, 2007.
[101] Yeung, A., & Evans, E. Unexpected dynamics in shape fluctuations of bilayer vesicles. *J. Phys. II France*, 5:1501–1523, 1995.
[102] Seifert, U., & Langer, S.A. Viscous modes of fluid bilayer membranes. *Europhys. Lett.*, 23: 71–76, 1993.
[103] Seifert, U., & Langer, S.A. Hydrodynamics of membranes: The bilayer aspect and adhesion. *Biophys. Chem.*, 49:13–22, 1994.
[104] Evans, E., & Needham, D. Physical properties of surfactant bilayer membranes: Thermal transitions, elasticity, rigidity, cohesion, and colloidal interactions. *J. Phys. Chem.*, 91: 4219–4228, 1987.
[105] Akabori, K., & Nagle, J.F. Structure of the DMPC lipid bilayer ripple phase. *Soft Matter*, 11: 918–926, 2015.
[106] Hamm, M., & Kozlov, M.M. Elastic energy of tilt and bending of fluid membranes. *Eur. Phys. J. E.*, 3:323–335, 2000.
[107] May, S., Kozlovsky, Y., Ben-Shaul, A., & Kozlov, M.M.. Tilt modulus of a lipd monolayer. *Eur. Phys. J. E.*, 14:299–308, 2004.
[108] Watson, M.C., Penev, E.S., Welch, P.M., & Brown, F.L.H. Thermal fluctuations in shape, thickness, and molecular orientation in lipid bilayers. *J. Chem. Phys.*, 135:244701, 2011.
[109] Jablin, M.S., Akabori, K., & Nagle, J.F. Experimental support for tilt-dependent theory of biomembrane mechanics. *Phys. Rev. Lett.*, 113:248102, 2014.
[110] Wang, X., & Deserno, M. Determining the lipid tilt modulus by simulating membrane buckles. *J. Phys. Chem. B.*, 120:6061–6073, 2016.
[111] Terzi, M.M., & Deserno, M. Novel tilt-curvature coupling in lipid membranes. *J. Chem. Phys.*, 147:084702, 2017.
[112] Nagle, J.F. Experimentally determined tilt and bending moduli of single-component lipid bilayers. *Chem. Phys. Lipids*, 205:18–24, 2017.
[113] Zilman, A.G., & Granek, R. Undulations and dynamic structure factor of membranes. *Phys. Rev. Lett.*, 77:4788–4791, 1996.
[114] Granek, R. From semi-flexible polymers to membranes: Anomalous diffusion and reptation. *J. Phys. II France*, 7:1761–1788, 1997.

[115] Messager, R., Bassereau, P., & Porte, G. Dynamics of the undulation mode in swollen lamellar phases. *J. Phys. France*, 51:1329–1340, 1990.
[116] Brochard, F., & Lennon, J.F. Frequency spectrum of the flicker phenomenon in erythrocytes. *J. Phys. France*, 36:1035–1047, 1975.
[117] Watson, M.C., Peng, Y., Zheng, Y., & Brown, F.L.H. The intermediate scattering function for lipid bilayer membranes: From nanometers to microns. *J. Chem. Phys.*, 135:194701, 2011.
[118] Van Kampen, N.G. *Stochastic Processes in Physics and Chemistry*. Elsevier, Amsterdam, 1992.
[119] Watson, M.C., & Brown, F.L.H. Interpreting membrane scattering experiments at the mesoscale: The contribution of dissipation within the bilayer. *Biophys. J.*, 98:L09–L11, 2010.
[120] Rheinstädter, M.C., Häußler, W., & Salditt, T. Dispersion relation of lipid membrane shape fluctuations by neutron spin-echo spectrometry. *Phys. Rev. Lett.*, 97:048103, 2006.
[121] Bivas, I., Méléard, P., Mircheva, I., & Bothorel, P. Thermal shape fluctuations of a quasi spherical lipid vesicle when the mutual displacements of its monolayers are taken into account. *Coll. Surf. A.*, 157:21–33, 1999.
[122] Miao, L., Lomholt, M.A., & Kleis, J. Dynamics of shape fluctuations of quasi-spherical vesicles revisited. *Eur. Phys. J. E.*, 9:143–160, 2002.
[123] Shkulipa, S.A., Den Otter, W.K., & Briels, W.J. Thermal undulations of lipid bilayers relax by intermonolayer friction at submicrometer length scales. *Phys. Rev. Lett.*, 96:178302, 2006.
[124] Arriaga, L.R., Rodríguez-García, R., López-Montero, I., Farago, B., Hellweg, T., & Monroy, F. Dissipative curvature fluctuations in bilayer vesicles: Coexistence of pure- bending and hybrid curvature-compression modes. *Eur. Phys. J. E.*, 31:105–113, 2010.
[125] Mell, M., Moleiro, L.H., Hertle, Y., López-Montero, I., Cao, F.J., Fouquet, P., Hellweg, T., & Monroy, F. Fluctuation dynamics of bilayer vesicles with intermonolayer sliding: Experiment and theory. *Chem. Phys. Lipids*, 185:61–77, 2015.
[126] Iñiguez-Palomares, R., Acuña-Campa, H., & Maldonado, A. Effect of polymer on the elasticity of surfactant membranes: A light scattering study. *Phys. Rev. E.*, 84:011604, 2011.
[127] Takeda, T., Kawabata, Y., Seto, H., Komura, S., Ghosh, S.K., Nagao, M., & Okuhara, D. Neutron spin-echo investigations of membrane undulations in complex fluids involving amphiphiles. *J. Phys. Chem. Solids*, 60:1375–1377, 1999.
[128] Komura, S., Takeda, T., Kawabata, Y., Ghosh, S.K., Seto, H., & Nagao, M. Dynamical fluctuation of the mesoscopic structure in ternary $C_{12}E_5$-water-n-octane amphiphilic system. *Phys. Rev. E.*, 63:041402, 2001.
[129] Seto, H., Yamada, N.L., Nagao, M., Hishida, M., & Takeda, T. Bending modulus of lipid bilayers in a liquid-crystalline phase including an anomalous swelling regime estimated by neutron spin echo experiments. *Eur. Phys. J. E.*, 26:217–223, 2008.
[130] Yi, Z., Nagao, M., & Bossev, D.P. Bending elasticity of saturated and monounsaturated phospholipid membranes studied by the neutron spin echo technique. *J. Phys.: Condens. Matter*, 21:155104, 2009.
[131] Mihailescu, M., Monkenbusch, M., Endo, H., Allgaier, J., Gompper, G., Stellbrink, J., Richter, D., Jakobs, B., Sottmann, T., & Farago, B. Dynamics of bicontinuous microemulsion phases with and without amphiphilic block-copolymers. *J. Chem. Phys.*, 115:9563–9577, 2001.
[132] Mihailescu, M., Monkenbusch, M., Allgaier, J., Frielinghaus, H., Richter, D., Jakobs, B., & Sottmann, T. Neutron scattering study on the structure and dynamics of oriented lamellar phase microemulsions. *Phys. Rev. E.*, 66:041504, 2002.
[133] Monkenbusch, M., Holderer, O., Frielinghaus, H., Byelov, D., Allgaier, J., & Richter, D. Bending moduli of microemulsions; comparison of results from small angle neutron scattering and neutron spin-echo spectroscopy. *J. Phys.: Condens. Matter*, 17:S2903–S2909, 2005.

[134] Holderer, O., Frielinghaus, H., Byelov, D., Monkenbusch, M., Allgaier, J., & Richter, D. Dynamic properties of microemulsions modified with homopolymers and diblock copolymers: The determination of bending moduli and renormalization effects. *J. Chem. Phys.*, 122:094908, 2005.

[135] Lee, J.-H., Choi, S.-M., Doe, C., Faraone, A., Pincus, P.A., & Kline, S.R. Thermal fluctuation and elasticity of lipid vesicles interacting with pore-forming peptides. *Phys. Rev. Lett.*, 105:038101, 2010.

[136] Evans, E., Yeung, A., Waugh, R., & Song, J. Dynamic coupling and nonlocal curvature elasticity in bilayer membranes. In R. Lipowsky, D. Richter, & K. Kremer, editors, *The Structure and Conformation of Amphiphilic Membranes*, pages 148–153, Berlin, 1992. Springer.

[137] Marrink, S.J., & Mark, A.E. Effect of undulations on surface tension in simulated bilayers. *J. Phys. Chem. B.*, 105:6122–6127, 2001.

[138] Brannigan, G., & Brown, F.L.H. A consistent model for thermal fluctuations and protein-induced deformations in lipid bilayers. *Biophys. J.*, 90:1501–1520, 2006.

[139] West, B., Brown, F.L.H., & Schmid, F. Membrane-protein interactions in a generic coarse-grained model for lipid bilayers. *Biophys. J.*, 96:101–115, 2009.

[140] Brandt, E.G., Braun, A.R., Sachs, J.N., Nagle, J.F., & Edholm, O. Interpretation of fluctuation spectra in lipid bilayer simulations. *Biophys. J.*, 100:2104–2111, 2011.

[141] Nagao, M., Chawang, S., & Hawa, T. Interlayer distance dependence of thickness fluctuations in a swollen lamellar phase. *Soft Matter*, 7:6598–6605, 2011.

[142] Farago, B., Monkenbusch, M., Goecking, K.D., Richter, D., & Huang, J.S. Dynamics of microemulsions as seen by neutron spin echo. *Physica B.*, 213 & 214:712–717, 1995.

[143] Farago, B. Spin echo studies of microemulsions. *Physica B.*, 226:51–55, 1996.

[144] Nagao, M. Observation of local thickness fluctuations in surfactant membranes using neutron spin echo. *Phys. Rev. E.*, 80:031606, 2009.

[145] Nagao, M. Temperature and scattering contrast dependencies of thickness fluctuations in surfactant membranes. *J. Chem. Phys.*, 135:074704, 2011.

[146] Carrillo, J.-M.Y., Katsaras, J., Sumpter, B.G., & Ashkar, R. A computational approach for modeling neutron scattering data from lipid bilayers. *J. Chem. Theory Comput.*, 13:916–925, 2017.

[147] Lemmich, J., Mortensen, K., Ipsen, J.H., Hønger, T., Bauer, R., & Mouritsen, O.G. Small-angle neutron scattering from multilamellar lipid bilayers: Theory, model, and experiment. *Phys. Rev. E.*, 53:5169–5180, 1996.

[148] Frielinghaus, H. Small-angle scattering model for multilamellar vesicles. *Phys. Rev. E.*, 76:051603, 2007.

[149] Caillé, A. Remarques sur la diffusion des rayons X dans les smectiques A. *C. R. Acad. Sci. Paris, Ser. B.*, 274:891–893, 1972.

[150] Nallet, F., Laversanne, R., & Roux, D. Modelling X-ray or neutron scattering spectra of lyotropic lamellar phases: Interplay between form and structure factors. *J. Phys. II.*, 3:487–502, 1993.

[151] de Gennes, P.G. Liquid dynamics and inelastic scattering of neutrons. *Physica.*, 25:825–839, 1959.

[152] Pfeiffer, W., König, S., Legrand, J.F., Bayerl, T., Richter, D., & Sackmann, E. Neutron spin echo study of membrane undulations in lipid multibilayers. *Europhys. Lett.*, 23:457–462, 1993.

[153] Evans, E., & Rawicz, W. Entropy-driven tension and bending elasticity in condensed- fluid membranes. *Phys. Rev. Lett.*, 64:2094–2097, 1990.

[154] Dimova, R., Aranda, S., Bezlyepkina, N., Nikolov, V., Riske, K.A., & Lipowsky, R. A practical guide to giant vesicles. Probing the membrane nanoregime via optical microscopy. *J. Phys.: Condens. Matter.*, 18:S1151–S1176, 2006.

[155] Olson, F., Hunt, C.A., Szoka, F.C., Vail, W.J., & Papahadjopoulos, D. Preparation of liposomes of defined size distribution by extrusion through polycarbonate membranes. *Biochim. Biophys. Acta*, 557:9–23, 1979.

[156] Yamada, N.L., Seto, H., Takeda, T., Nagao, M., Kawabata, Y., & Inoue, K. SAXS, SANS and NSE studies on "unbound state" in DPPC/water/$CaCl_2$ system. *J. Phys. Soc. Jpn.*, 74:2853–2859, 2005.

[157] Brüning, B., Stehle, R., Falus, P., & Farago, B. Influence of charge density on bilayer bending rigidity in lipid vesicles: A combined dynamic light scattering and neutron spin-echo study. *Eur. Phys. J. E.*, 36:77, 2013.

[158] Pan, J., Cheng, X., Sharp, M., Ho, C.-S., Khadka, N., & Katsaras, J. Structural and mechanical properties of cardiolipin lipid bilayer determined using neutron spin echo, small angle neutron and X-ray scattering, and molecular dynamics simulations. *Soft Matter*, 11:130–138, 2015.

[159] Arriaga, L.R., López-Montero, I., Monroy, F., Orts-Gil, G., Farago, B., & Hellweg, T. Stiffening effect of cholesterol on disordered lipid phases: A combined neutron spin echo + dynamic light scattering analysis of the bending elasticity of large unilamellar vesicles. *Biophys. J.*, 96:3629–3637, 2009.

[160] Brüning, B.A., Prévost, S., Stehle, R., Steitz, R., Falus, P., Farago, B., & Hellweg, T. Bilayer undulation dynamics in unilamellar phospholipid vesicles: Effect of temperature, cholesterol and trehalose. *Biochim. Biophys. Acta*, 1838:2412–2419, 2014.

[161] Hirai, M., Kimura, R., Takeuchi, K., Sugiyama, M., Kasahara, K., Ohta, N., Farago, B., Stadler, A., & Zaccai, G. Change of dynamics of raft-model membrane induced by amyloid-β protein binding. *Eur. Phys. J. E.*, 36:74, 2013.

[162] Brüning, B., & Farago, B. Perfluorooctanoic acid rigidifies a model lipid membrane. *Phys. Rev. E.*, 89:040702(R), 2014.

[163] Boggara, M.B., Faraone, A., & Krishnamoorti, R. Effect of pH and ibuprofen on phospholipid bilayer bending modulus. *J. Phys. Chem. B.*, 114:8061–8066, 2010.

[164] Yi, Z., Nagao, M., & Bossev, D.P. Effect of charged lidocaine on static and dynamic properties of model bio-membranes. *Biophys. Chem.*, 160:20–27, 2012.

[165] Hoffmann, I., Michel, R., Sharp, M., Holderer, O., Appavou, M.S., Polzer, F., Farago, B., & Gradzielski, M. Softening of phospholipid membranes by the adhesion of silica nanoparticles – as seen by neutron spin-echo (NSE). *Nanoscale*, 6:6945–6952, 2014.

[166] Mell, M., Moleiro, L.H., Hertle, Y., Fouquet, P., Schweins, R., López-Montero, I., Hellweg, T., & Monroy, F. Bending stiffness of biological membranes: What can be measured by neutron spin echo? *Eur. Phys. J. E.*, 36:75, 2013.

[167] Moleiro, L.H., Mell, M., Bocanegra, R., López-Montero, I., Fouquet, P., Hellweg, T., Carrascosa, J.L., & Monroy, F. Permeability modes in fluctuating lipid membranes with DNA-translocating pores. *Adv. Coll. Int. Sci.*, 247:543–554, 2017.

[168] Nickels, J.D., Cheng, X., Mostofian, B., Stanley, C., Lindner, B., Heberle, F.A., Perticaroli, S., Feygenson, M., Egami, T., Standaert, R.F., Smith, J.C., Myles, D.A.A., Ohl, M., & Katsaras, J. Mechanical properties of nanoscopic lipid domains. *J. Am. Chem. Soc.*, 137:15772–15780, 2015.

[169] Stingaciu, L.R., O'Neill, H., Liberton, M., Urban, V.S., Pakrasi, H.B., & Ohl, M. Revealing the dynamics of thylakoid membranes in living cyanobacterial cells. *Sci. Rep.*, 6:19627, 2016.

[170] Nickels, J.D., Chatterjee, S., Mostofian, B., Stanley, C.B., Ohl, M., Zolnierczuk, P., Schulz, R., Myles, D.A.A., Standaert, R.F., Elkins, J.G., Cheng, X., & Katsaras, J. Bacillus subtilis lipid extract, a branched-chain fatty acid model membrane. *J. Phys. Chem. Lett.*, 8:4214–4217, 2017.

[171] Boal, D. *Mechanics of the Cell*. Cambridge University Press, 2nd edition, Cambridge, UK, 2002.

[172] Baumgart, T., Das, S., Webb, W.W., & Jenkins, J.T. Membrane elasticity in giant vesicles with fluid phase coexistence. *Biophys. J.*, 89:1067–1080, 2005.

[173] Bitbol, A.-F., Constantin, D., & Fournier, J.-B. Bilayer elasticity at the nanoscale: The need for new terms. *PLOS One.*, 7:e48306, 2012.

[174] Hu, M., de Jong, D.H., Marrink, S.J., & Deserno, M. Gaussian curvature elasticity determined from global shape transformations and local stress distributions: A comparative study using the MARTINI model. *Faraday Discuss.*, 161:365–382, 2013.

[175] Venable, R.M., Brown, F.L.H., & Pastor, R.W. Mechanical properties of lipid bilayers from molecular dynamics simulation. *Chem. Phys. Lipids*, 192:60–74, 2015.

[176] Rand, R.P., & Fuller, N.L. Structural dimensions and their changes in a reentrant hexagonal-lamellar transition of phospholipids. *Biophys. J.*, 66:2127–2138, 1994.

[177] Leikin, S., Kozlov, M.M., Fuller, N.L., & Rand, R.P. Measured effects of diacylglycerol on structural and elastic properties of phospholipid membranes. *Biophys. J.*, 71:2623–2632, 1996.

[178] Winterhalter, M., & Helfrich, W. Bending elasticity of electrically charged bilayers: Coupled monolayers, neutral surfaces, and balancing stresses. *J. Phys. Chem.*, 96:327–330, 1992.

[179] Templer, R.H., Khoo, B.J., & Seddon, J.M. Gaussian curvature modulus of an amphiphilic monolayer. *Langmuir*, 14:7427–7434, 1998.

[180] Campelo, F., McMahon, H.T., & Kozlov, M.M. The hydrophobic insertion mechanism of membrane curvature generation by proteins. *Biophys. J.*, 95:2325–2339, 2008.

[181] Campelo, F., Arnarez, C., Marrink, S.J., & Kozlov, M.M. Helfrich model of membrane bending: From Gibbs theory of liquid interfaces to membranes as thick anisotropic elastic layers. *Adv. Colloid Interface Sci.*, 208:25–33, 2014.

[182] Kollmitzer, B., Heftberger, P., Rappolt, M., & Pabst, G. Monolayer spontaneous curvature of raft-forming membrane lipids. *Soft Matter*, 9:10877–10884, 2013.

[183] Wu, Y., Štefl, M., Olzyńska, A., Hof, M., Yahioglu, G., Yip, P., Casey, D.R., Ces, O., Humpolíčková, J., & Kuimova, M.K. Molecular rheometry: Direct determination of viscosity in L_O and L_D lipid phases via fluorescence lifetime imaging. *Phys. Chem. Chem. Phys.*, 15:14986–14993, 2013.

[184] Nagao, M., Kelley, E.G., Ashkar, R., Bradbury, R., & Butler, P.D. Probing elastic and viscous properties of phospholipid bilayers using neutron spin echo spectroscopy. *J. Phys. Chem. Lett.*, 8:4679–4684, 2017.

[185] Doktorova, M., Harries, D., & Khelashvili, G. Determination of bending rigidity and tilt modulus of lipid membranes from real-space fluctuation analysis of molecular dynamics simulations. *Phys. Chem. Chem. Phys.*, 19:16806–16818, 2017.

[186] Méléard, P., Gerbeaud, C., Pott, T., Fernandez-Puente, L., Bivas, I., Mitov, M.D., Dufourcq, J., & Bothorel, P. Bending elasticities of model membranes: Influences of temperature and sterol content. *Biophys. J.*, 72:2616–2629, 1997.

[187] Nagle, J.F., Jablin, M.S., Tristram-Nagle, S., & Akabori, K. What are the true values of the bending modulus of simple lipid bilayers? *Chem. Phys. Lipids*, 185:3–10, 2015.

[188] Rawicz, W., Olbrich, K.C., McIntosh, T., Needham, D., & Evans, E. Effect of chain length and unsaturation on elasticity of lipid bilayers. *Biophys. J.*, 79:328–339, 2000.

[189] Pan, J., Tristram-Nagle, S., & Nagle, J.F. Effect of cholesterol on structural and mechanical properties of membranes depends on lipid chain saturation. *Phys. Rev. E*, 80:021931, 2009.

[190] Bradbury, R., & Nagao, M. Effect of charge on the mechanical properties of surfactant bilayers. *Soft Matter*, 12:9383–9390, 2016.

[191] Lee, V., & Hawa, T. Investigation of the effect of bilayer membrane structures and fluctuation amplitudes on SANS/SAXS profile for short membrane wavelength. *J. Chem. Phys.*, 139:124905, 2013.

[192] Allen, M.P., & Tildesley, D.J. *Computer Simulation of Liquids*. Oxford Science Publications, Oxford, 1987.

[193] Braganza, L.F., & Worcester, D.L. Structural changes in lipid bilayers and biological membranes caused by hydrostatic pressure. *Biochemistry*, 25:7484–7488, 1986.

[194] Nagle, J.F., & Wilkinson, D.A. Lecithin bilayers. Density measurements and molecular interactions. *Biophys. J.*, 23:159–175, 1978.

[195] Koenig, B.W., & Gawrisch, K. Specific volumes of unsaturated phosphatidylcholines in the liquid crystalline lamellar phase. *Biochim. Biophys. Acta*, 1715:65–70, 2005.

[196] Nagle, J.F., & Tristram-Nagle, S. Structure of lipid bilayers. *Biochim. Biophys. Acta*, 1469:159–195, 2000.
[197] Kučerka, N., Nieh, M.P., & Katsaras, J. Fluid phase lipid areas and bilayer thicknesses of commonly used phosphatidylcholines as a function of temperature. *Biochim. Biophys. Acta*, 1808:2761–2771, 2011.
[198] Kučerka, N., Heberle, F.A., Pan, J., & Katsaras, J. Structural significance of lipid diversity as studied by small angle neutron and X-ray scattering. *Membranes*, 5:454–472, 2015.
[199] Barker, J.G., & Pedersen, J.S. Instrumental smearing effects in radially symmetric small-angle neutron scattering by numerical and analytical methods. *J. Appl. Crystallogr.*, 28:105–114, 1995.
[200] Dymond, J.H., & Øye, H.A. Viscosity of selected liquid n-alkanes. *J. Phys. Chem. Ref. Data*, 23:41–53, 1994.
[201] Doolittle, A.K. Studies of Newtonian flow. I. The dependence of the viscosity of liquids on temperature. *J. Appl. Phys.*, 22:1031–1035, 1951.
[202] Dimova, R., Dietrich C., Hadjisky A., Danov K., and Pouligny B. Falling ball viscosimetry of giant vesicle membranes: Finite-size effects. *Eur. Phys. J. B.*, 12:589–598, 1999.
[203] Petrov, E.P., Petrosyan, R., & Schwille, P. Translational and rotational diffusion of micrometer-sized solid domains in lipid membranes. *Soft Matter*, 8:7552–7555, 2012.
[204] Camley, B.A., Esposito, C., Baumgart, T., & Brown, F.L.H. Lipid bilayer domain fluctuations as a probe of membrane viscosity. *Biophys. J.*, 99:L44–L46, 2010.
[205] Honerkamp-Smith, A.R., Woodhouse, F.G., Kantsler, V., & Goldstein, R.E. Membrane viscosity determined from shear-driven flow in giant vesicles. *Phys. Rev. Lett.*, 111:038103, 2013.
[206] Stanich, C.A., Honerkamp-Smith, A.R., Putzel, G.G., Warth, C.S., Lamprecht, A.K., Mandal, P., Mann, E., Hua, T.-A.D., & Keller, S.L. Coarsening dynamics of domains in lipid membranes. *Biophys. J.*, 105:444–454, 2013.
[207] Hormel, T.T., Kurihara, S.Q., Brennan, M.K., Wozniak, M.C., & Parthasarathy, R. Measuring lipid membrane viscosity using rotational and translational probe diffusion. *Phys. Rev. Lett.*, 112:188101, 2014.
[208] Waugh, R., & Evans, E.A. Thermoelasticity of red blood cell membrane. *Biophys. J.*, 26:115–131, 1979.
[209] Hochmuth, R.M., Buxbaum, K.L., & Evans, E.A. Temperature dependence of the viscoelastic recovery of red cell membrane. *Biophys. J.*, 29:177–182, 1980.
[210] Rahimi, M., & Arroyo, M. Shape dynamics, lipid hydrodynamics, and the complex viscoelasticity of bilayer membranes. *Phys. Rev. E*, 86:011932, 2012.
[211] Tristram-Nagle, S., Liu, Y., Legleiter, J., & Nagle, J.F. Structure of gel phase DMPC determined by X-ray diffraction. *Biophys. J.*, 83:3324–3335, 2002.
[212] Chiantia, S., & London, E. Acyl chain length and saturation modulate interleaflet coupling in asymmetric bilayers: Effects on dynamics and structural order. *Biophys. J.*, 103:2311–2319, 2012.
[213] Liu, Y., & Nagle, J.F. Diffuse scattering provides material parameters and electron density profiles of biomembranes. *Phys. Rev. E*, 69:040901(R), 2004.
[214] Nagle, J.F. X-ray scattering reveals molecular tilt is an order parameter for the main phase transition in a model biomembrane. *Phys. Rev. E*, 96:030401(R), 2017.

Yan Xia, Mu-Ping Nieh

5 Spontaneous lipid transfer rate constants

Abstract: This chapter will introduce three different experimental approaches to measure spontaneous lipid transfer rate constants between lipid bilayers, including time-resolved small angle neutron scattering (TR-SANS), differential scanning calorimetry (DSC), and fluorescence correlation spectroscopy (FCS). The basic principle of TR-SANS and DSC is applicable if the neutron scattering length density (NSLD) (in the case of TR-SANS) or melting transition temperature (in the case of DSC) of the lipid in study is sufficiently different from that of its deuterated counterpart. With minimal disturbance of the chemical properties of the system using the isotope molecules, these two approaches presumably yield more accurate lipid transfer rate constant than the outcome obtained from FCS, which requires the use of fluorophores having a similar chemical structure of the lipid in study. Examples of measuring lipid transfer rate constants will be demonstrated in two well-defined nanoparticles (NPs), discoidal bicelles and unilamellar vesicles (ULVs). At the end of this chapter, we summarize the pros and cons of each method in order to provide the researchers with the principles of selecting the appropriate method for the systems of their interest.

Keywords: lipid transfer, SANS, DSC, FCS

5.1 Introduction

Lipids are the building blocks of biological membranes, serving as the barrier between cells (organelles) and their surrounding environments. Lipid transfer is ubiquitous in cells and plays a significant role in different cellular activities [1], such as endocytotic and exocytotic processes [2, 3]. Evaluation of the lipid dynamics in cell membranes is challenging due to the chemical complexity of biomembranes. Therefore, model membranes have been widely employed for studying the lipid dynamics. Two examples of model membranes are used in this chapter to illustrate the methodologies which are applied, but not limited, to determining the lipid transfer rate constants for discoidal bicelles (Figure 5.1(a)) and unilamellar vesicles (ULVs) (Figure 5.1(b)). Bicelle is composed of a mixture of long- and short-chain phosphatidylcholines (PC), with the long-

---

Yan Xia, Department of Chemical and Biomolecular Engineering, University of Connecticut, Connecticut, United States
Mu-Ping Nieh, Department of Chemical and Biomolecular Engineering, University of Connecticut, Connecticut, United States; Polymer Program, Institute of Materials Science, University of Connecticut, Connecticut, United States; Department of Biomedical Engineering, University of Connecticut, Connecticut, United States

https://doi.org/10.1515/9783110544657-005

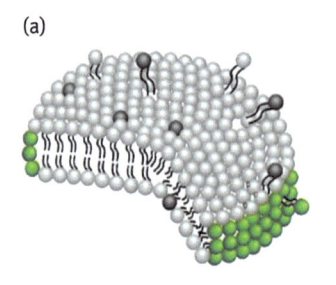

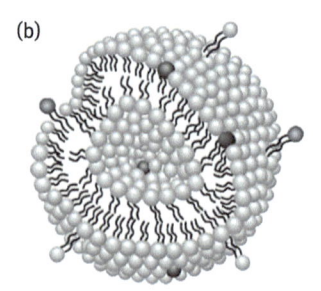

Figure 5.1: (a): Bicelles; (b): Unilamellar vesicles (ULVs). Lipids with gray or white headgroups are long-chain lipids. Lipids with green headgroups are short-chain lipids.

chain lipids constituting the planar bilayer disk while the short-chain lipids sequestering the disk rim [4, 5]. ULV has an aqueous core surrounded by a single bilayer shell. Molecular transfer kinetics has also been shown to affect the formation and stabilities of lipid-based, polymeric, surfactant micelles, and so on. [6–10].

It is crucial to apply appropriate methods to the systems in order to accurately determine the molecular transfer rate constants. In this chapter, we discuss three approaches including small-angle neutron scattering (SANS), differential scanning calorimetry (DSC), and fluorescence correlation spectroscopy (FCS). FCS requires fluorescence-labeled lipid which is assumed to have the same transfer property as that of the lipid of interest. On the other hand, SANS and DSC require isotope substitution (i.e., the use of deuterated lipids to replace protiated lipids), which imposes the minimum perturbation to the systems, resulting in more reliable determination of the rate constants; however, the data collection time is relatively long. Besides these three methods that are highlighted in this chapter, other methods have also been applied for studying lipid transfer, such as the fluorescence spectroscopy [11–13] and electron spin resonance [14, 15]. Audience should search for the references of their own interest.

5.2 The spontaneous lipid transfer model

Lipid transfer can be protein-mediated or spontaneous. The spontaneous lipid transfer is defined as the lipid exchange between two bilayers without the aid of associated proteins [16]. There are several mechanisms associated with lipid transfer in addition to monomeric lipid transfer, which involves only dissociation from and association with the nanoparticles (NPs) of a single lipid. First, lipids can exchange between inner and outer leaflets, so-called "flip-flop," which is normally slow in gel phase as it involves the translocation of the hydrophilic lipid headgroups through the crystalline hydrophobic tail region [17–19]. Another mechanism is the collision-induced lipid transfer, which takes place in a highly interactive system, where the lipid transfer is fast, involving multiple lipids. For instance, a rapid exchange of charged lipids was observed between two oppositely charged bicelles, which formed stacked bicelle aggregates owing to the strong coulombic attractions [20]. In this chapter, the lipid transfer rate constant is based on the assumption of monomeric lipid transfer. To ensure monomeric lipid transfer, charged or PEG lipids can be incorporated into the NPs to inhibit the particles from coalescing [21].

5.3 Time-resolved small angle neutron scattering

Time-resolved small angle neutron scattering (TR-SANS) has been employed widely to investigate the molecular transfer in polymer and lipid systems [10, 18, 22–24], as the neutron scattering length density (NSLD) of protiated components is drastically different from their deuterated counterparts. For example, the NSLD of 1,2-dipalmitoyl-sn-glycero-3-phosphocholine (DPPC) is 2.67×10^{-7} Å^{-2}, while the NSLD of 1,2-dipalmitoyl-d_{62}-sn-glycero-3-phosphocholine (d_{62}-DPPC) is 5.67×10^{-6} Å^{-2}. The scattered intensity is proportional to the NSLD contrast $(\Delta\rho)^2$ as shown in eq. 5.1,

$$I(q) = \varphi V P(q) S(q) (\Delta\rho)^2 + I_{bkg} \tag{5.1}$$

where q is the scattering vector $\frac{4\pi}{\lambda}\sin\left(\frac{\theta}{2}\right)$, λ and θ being the wavelength and the scattering angle, respectively, and φ, V, $P(q)$, $S(q)$, and I_{bkg} are the volume fractions of NPs in the solution, NP molecular volume, form factor which accounts for the nanoparticle morphology, structure factor that takes into account the interparticle interactions, and incoherent background, respectively. $\Delta\rho$ is the difference in NSLD between the NP, ρ_{NP} and the solvent, ρ_s. As shown in eq. (5.1), $I(q)$ decreases as ρ_{NP} approaches to ρ_s. Comparing to other techniques that require labeling (e.g., fluorescence and electron spin resonance), isotope substitution introduces minimum perturbation to the systems.

To conduct TR-SANS experiments, protiated and deuterated samples are prepared separately [25] in a buffer whose NSLD (ρ_s) is determined by the ratio of h-NP to d-NP in the post-mixture shown as below:

$$\rho_s = \rho_h \varphi_h + \rho_d (1 - \varphi_h) \tag{5.2}$$

where φ_h is the molar percentage of h-NPs with respect to the total number of NPs. The buffer is typically a mixture of D$_2$O and H$_2$O, with ρ_s ranging from -5.61×10^{-7} Å^{-2} (i.e., 100% H$_2$O) to 6.40×10^{-6} Å^{-2} (i.e., 100% D$_2$O). The h-NPs and d-NPs are then mixed and subjected to TR-SANS measurements. As the lipid exchange takes place between h and d-NPs, the NSLD of the NPs approaches to that of the solvent, which results in a decay of scattered intensity. Figure 5.2 depicts the principle of using TR-SANS to measure the lipid transfer, presenting that the particles become less "visible" as the lipid transfer proceeds, and eventually become

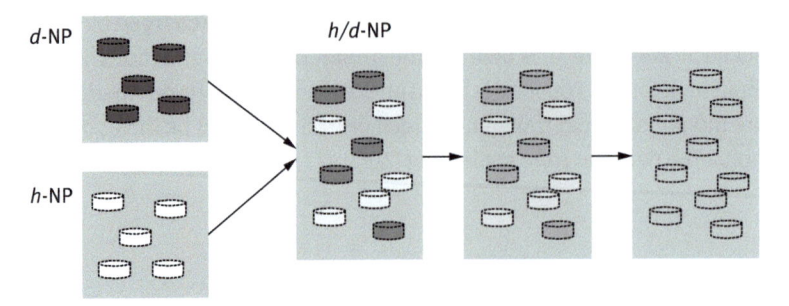

Figure 5.2: Cartoons showing the contrast between NPs (black and white) and solvent (gray). Neutron contrast decays over time, due to the exchange between protiated and deuterated lipids, resulting in less visible NPs in the solvent.

"invisible." Minoru Nakano etc. used TR-SANS to determine the interparticle and flip-flop rate constants in LUVs [23]. As shown in Figure 5.3(b), the intensity dropped as the incubation time increased due to the exchange of lipids between D-LUV and H-LUV.

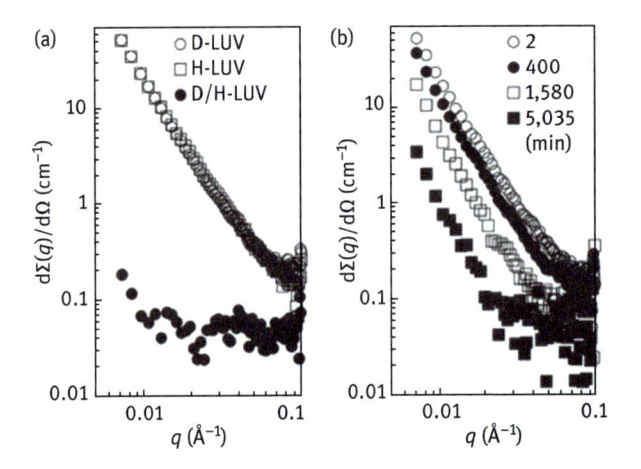

Figure 5.3: (a) Scattering curves of deuterated large unilamellar vesicle (D-LUV), protiated LUV (H-LUV), and LUVs containing equal molar d- and h- DMPC in 50% D_2O buffer at 27.1 °C. (b) Scattering curves of equal molar D-LUV and H-LUV mixtures at different incubation times. Figure reprinted from Phys. Rev. Lett. 2007, 98, 238101.

5.3.1 Kinetics models for lipid transfer

In order to build the correlation between the decay of scattered intensities observed in TR-SANS and lipid transfer rate constants, we provide the kinetics models of monomeric lipid transfer in bicelles and ULVs.

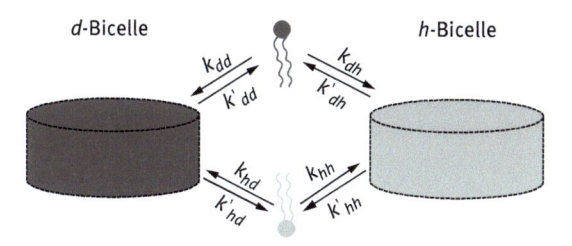

Figure 5.4: The schematics of lipid transfer between bicelles. The black and grey discs represent the deuterated and protiated bicelles, respectively. The black and grey lipids represent the deuterated and protiated lipids, respectively.

In bicelles, only interparticle lipid transfer takes place. According to Figure 5.4, we can write down six differential equations as follows:

$$\frac{d[h]_h}{dt} = k_{hh}[h]_w - k'_{hh}[h]_h \tag{5.3}$$

$$\frac{d[d]_h}{dt} = k_{dh}[d]_w - k'_{dh}[d]_h \tag{5.4}$$

$$\frac{d[d]_d}{dt} = k_{dd}[d]_w - k'_{dd}[d]_d \tag{5.5}$$

$$\frac{d[h]_d}{dt} = k_{hd}[h]_w - k'_{hd}[h]_d \tag{5.6}$$

$$\frac{d[h]_w}{dt} = k'_{hd}[h]_d - k_{hd}[h]_w + k'_{hh}[h]_h - k_{hh}[h]_w \tag{5.7}$$

$$\frac{d[d]_w}{dt} = k'_{dd}[d]_d - k_{dd}[d]_w + k'_{dh}[d]_h - k_{dh}[d]_w \tag{5.8}$$

where $[h]_h$, $[h]_d$, and $[h]_w$ are the concentrations of protiated lipids in h-NP, d-NP, and water phase, respectively; $[d]_d$, $[d]_h$, and $[d]_w$ are the concentrations of deuterated lipids in d-NP, h-NP, and water phase, respectively. The first letter, "x" of the subscript of the association rate constant, k_{xy} or the dissociation rate constant, k'_{xy}, represents the transferred lipid species (either protiated or deuterated), while the second letter, "y" represents the recipient NPs that are initially rich of y species. For instance, k_{hd} is the association rate constant of protiated lipids from water to d-bicelles. To simplify the kinetics model, it is assumed that deuterated lipids have the same transport behavior as the protiated lipids, leading to a symmetric lipid transfer. Eqs. (5.3) to (5.8) are then reduced to eq. (5.9)

$$\frac{d([h]_h - [h]_d)}{dt} = k'_{hh}([h]_h - [h]_d) \tag{5.9}$$

which indicates that the lipid transfer in bicelles can be tracked by the concentration difference between protiated lipids and deuterated lipids in one bicelle. We can easily correlate $([h]_h - [h]_d)$ to NSLD differences between h- or d-NP and the solvent, that is $(\rho_h - \rho_s)$ or $(\rho_d - \rho_s)$, which results in the following governing equation

$$|\Delta\rho(t)| = |\Delta\rho_0| e^{-k_{in}t} \tag{5.10}$$

where we replaced k'_{hh} using k_{in} to generalize the interparticle lipid transfer rate constant. Here, $|\Delta\rho(t)|$ is the absolute NSLD difference between NPs and solvent, and $\Delta\rho_0$ is $\Delta\rho(t)$ at $t = 0$. Based on eq. (5.10), the lipid transfer in bicelles follows a single exponential decay. In nanodiscs whose rims are stabilized by lipoproteins rather than short-chain PCs, the lipid transfer following a single exponential decay was also observed [26]. For a more detailed derivation of the kinetics model, audience should refer to the literature report [18]. The experimental outcome based on eq. (5.10) shows that k_{in} of DMPC bicelles is two orders of magnitude higher than that of DPPC bicelles [18]. This is anticipated as the van der Waals interactions among the acyl chains are stronger in DPPC bicelles than DMPC bicelles; therefore, more energy is required for the dissociation of DPPC from the bilayer. Interestingly, we further found that the DMPC transfer between bicelles was significantly faster than that between ULVs. The drastic decrease of k_{in} in ULVs is not due to the slow flip-flop which will be discussed later in this chapter. Instead, it is explained by the presence of interfacial defects originated from the acyl chain mismatch, which consequently accelerated the lipid transfer. The lipid transfer in the case of lipoprotein-constituted nanodiscs is also found faster than that of ULVs [26].

The NSLD of particles, ρ_{NP}, can be obtained through fitting the scattering curve using appropriate SANS models. Importantly, the structure factor S(q) for describing interparticle interactions becomes negligible for h/d- NPs in CM solvent condition [27], which allows a more accurate determination of ρ_{NP} even for a highly interactive system. However, in order to obtain $\Delta\rho_0$, good fittings of the scattering curves for h- or d-NP in CM solvent are required, where $S(q)$ is no longer negligible if there is strong interparticle interaction and the fitting of scattering curves can be difficult. So, we introduce an alternative way to analyze the scattering data.

Since the scattered intensity is proportional to the neutron contrast, $(\Delta\rho)^2$, eq. (5.10) can be rewritten into eq. (5.11) as follow

$$\left|\sqrt{I(t)} - \sqrt{I(\infty)}\right| = \left|\sqrt{I(0)} - \sqrt{I(\infty)}\right| e^{-k_{in}t} \tag{5.11}$$

where $\sqrt{I(t)}$, $\sqrt{I(0)}$, and $\sqrt{I(\infty)}$ are the square root of the scattered intensity at t, $t = 0$, and $t = \infty$, respectively. $I(t)$, $I(0)$ and $I(\infty)$ are obtained through integrating the whole scattering curve. $I(\infty)$ corresponds to the intensity of h/d-NPs which represents the NPs at the equilibrium state. $I(0)$ is calculated as the average intensity of d-NP/CM solvent and h-NP/CM solvent.

Figure 5.5: The schematics of lipid transfer in ULVs. The black and gray circles represent the cross sections of deuterated and protiated ULVs, respectively. The black and grey lipids represent the deuterated and protiated lipids, respectively.

In the case of ULVs, we take into account both interparticle lipid transfer and flip-flop as shown in Figure 5.5. Ten differential equations are needed in order to describe the lipid transfer process in ULVs as shown below.

$$\frac{d[h]_{in}}{dt} = k'_{h,ff}[h]_{in} - k_{h,ff}[h]_{out} \tag{5.12}$$

$$\frac{d[h]_{out}}{dt} = k_{hh}[h]_w - k'_{hh}[h]_{out} + k_{h,ff}[h]_{in} - k'_{h,ff}[h]_{out} \tag{5.13}$$

$$\frac{d[d]_{in}}{dt} = k'_{d,ff}[d]_{out} - k_{d,ff}[d]_{in} \tag{5.14}$$

$$\frac{d[d]_{out}}{dt} = k_{d,ff}[d]_{in} - k'_{d,ff}[d]_{out} + k_{dd}[d]_w - k'_{dd}[d]_{out} \tag{5.15}$$

$$\frac{d[h]_{in-d}}{dt} = k'_{d,ff}[h]_{out-d} - k_{d,ff}[h]_{in-d} \tag{5.16}$$

$$\frac{d[h]_{out-d}}{dt} = k_{d,ff}[h]_{in-d} - k'_{d,ff}[h]_{out-d} + k_{hd}[h]_w - k'_{hd}[h]_{out-d} \tag{5.17}$$

$$\frac{d[d]_{in-h}}{dt} = k'_{h,ff}[d]_{out-h} - k_{h,ff}[d]_{in-h} \tag{5.18}$$

$$\frac{d[d]_{out-h}}{dt} = k_{h,ff}[d]_{in-h} - k'_{h,ff}[d]_{out-h} + k_{dh}[d]_w - k'_{dh}[d]_{out-h} \tag{5.19}$$

$$\frac{d[h]_w}{dt} = k'_{hh}[h]_{out} - k_{hh}[h]_w + k'_{hd}[h]_{out-d} - k_{hd}[h]_w \tag{5.20}$$

$$\frac{d[d]_w}{dt} = k'_{dd}[d]_{out} - k_{dd}[d]_w + k'_{dh}[d]_{out-h} - k_{dh}[d]_w \tag{5.21}$$

We denote the interparticle lipid transfer rate constants in ULVs the same as those in bicelles. The $k_{x,ff}$ and $k'_{x,ff}$ are the flip-flop rate constants for inner-to-outer-leaflet and outer-to-inner-leaflet, respectively, where "x" represents the NPs which are initially rich of the species, x. For instance, $k_{d,ff}$ is the flip-flop rate constant for lipid to transfer from inner to outer leaflet in d-NPs. In addition to the symmetric interparticle lipid transfer assumption, we also assume that the flip-flop is the same regardless of the transfer direction, which leads to $k_{d,ff} = k'_{d,ff} = k_{h,ff} = k'_{h,ff}$. Eventually, ten differential equations were reduced to eqs. (5.22) and (5.23),

$$|\Delta\rho(t)| = |\Delta\rho_0| \left[\frac{2k_{ff} + k_{in} + \sqrt{4k_{ff}^2 + k_{in}^2 - 2rk_{in}}}{2\sqrt{4k_{ff}^2 + k_{in}^2}} \right] e^{\lambda_1 t}$$

$$- |\Delta\rho_0| \left[\frac{2k_{ff} + k_{in} - \sqrt{4k_{ff}^2 + k_{in}^2 - 2rk_{in}}}{2\sqrt{4k_{ff}^2 + k_{in}^2}} \right] e^{\lambda_2 t} \tag{5.22}$$

$$\lambda_{1,2} = \frac{-(k_{in} + 2k_{ff}) \pm \sqrt{k_{in}^2 + 4k_{ff}^2}}{2} \tag{5.23}$$

where k_{ff} represents the flip-flop rate constant, and r is the number ratio of lipids in the inner leaflet to the total number of lipids in one ULV. Again, we can replace $|\Delta\rho_0|$ and $|\Delta\rho(t)|$ using $|\sqrt{I(0)} - \sqrt{I(\infty)}|$ and $|\sqrt{I(t)} - \sqrt{I(\infty)}|$, respectively. This leads to eq. (24)

$$\left|\sqrt{I(t)} - \sqrt{I(\infty)}\right| = \left|\sqrt{I(0)} - \sqrt{I(\infty)}\right|$$

$$\left\{ \left[\frac{2k_{ff} + k_{in} + \sqrt{4k_{ff}^2 + k_{in}^2 - 2rk_{in}}}{2\sqrt{4k_{ff}^2 + k_{in}^2}} \right] e^{\lambda_1 t} - \left[\frac{2k_{ff} + k_{in} - \sqrt{4k_{ff}^2 + k_{in}^2 - 2rk_{in}}}{2\sqrt{4k_{ff}^2 + k_{in}^2}} \right] e^{\lambda_2 t} \right\} \tag{5.24}$$

We can easily calculate r using the known ULV dimensions, diameter D and bilayer thickness t, resulting in $r = \frac{(D-t)^2}{(D)^2 + (D-t)^2}$. If we assume equal number of lipids in the inner and outer leaflet, r becomes 0.5. We then further reduce the equation to the one reported by Nakado [23]. Eqs. (5.22) to (5.24) indicate that the lipid transfer in ULV follows a double exponential decay with the decay rate constants of λ_1 and λ_2. The pre-factors of both decays contain k_{ff} and k_{in}. Therefore, we cannot simply designate either decay to the interparticle lipid transfer or the flip-flop mechanism. However, we can simplify the eq. (5.22) or eq. (5.24) based on certain assumptions. For instance, if flip-flop is significantly slower than the interparticle lipid transfer, these equations can be reduced to be a single exponential decay, similarly to the lipid transfer in bicelles. On the other hand, if the interparticle lipid transfer is extremely slow, then we will not see any change of scattered intensity.

So far, we have discussed the strategy of mixing d-NPs containing deuterated lipids of interest and h-NPs containing protiated lipids of interest in CM solvent to probe the lipid transfer through essentially monitoring the exchange between

d-lipids and h-lipids. Similar contrast variation strategy can be applied to investigate the transfer of protiated molecules associated with lipid bilayer which is contrast matched by the solvent. Figure 5.6 (a) shows the schematics of the contrast matching. The matrix composed of deuterated 1,2-dioleoyl-sn-glycero-3-phosphocholine (d-DOPC) in this specific study conducted by S. Garg et al. [28] was perfectly matched by the solvent. Protiated cholesterols were incorporated in the donor while the acceptor contained no cholesterols. With the transferring of cholesterols from the donor to the acceptor, the scattered intensity from the donor decreases while it increases for the acceptor, leading to the decrease of overall intensity Fig. 5.6 (b). This method is only valid if the transport behavior of the molecules does not strongly depend on the bilayer composition which changes constantly before the molecular transfer reaches equilibrium.

Other than TR-SANS, neutron reflectometry has also been used to measure the lipid exchange and flip-flop in solid supported bilayers [29]. The solid lipid bilayers were exposed to a solution of isotopically labeled vesicles. As the protiated lipids from the vesicles move into the bilayers, the NSLD of the supported bilayers changes. Time-resolved reflectivity is acquired from the supported bilayers, from which the time-resolved NSLD of supported bilayers can be obtained.

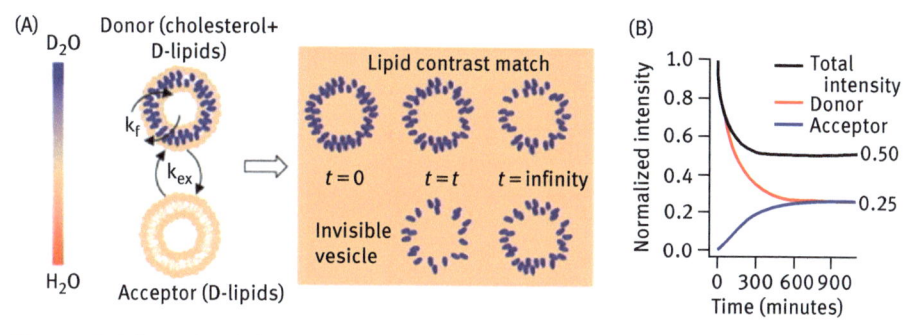

Figure 5.6: Schematics of the contrast matching: (a) donor and acceptor vesicles made of DOPC are contrast-matched by the solvent. The protiated cholesterols are only incorporated in the donor. As cholesterol transfer from donor to acceptor, the acceptor vesicle gradually becomes visible. (b) As the cholesterol exchange proceeds, the scattered intensity from donor decreases while the scattered intensity from acceptor increases, both of which contribute the total intensity change. Figure reprinted from Biophy. J. 2011, 101 (2), 370–377.

5.4 Time-resolved differential scanning calorimetry

DSC is a highly sensitive technique to study the thermotropic properties of many different biological macromolecules. For lipid-based systems, DSC has been widely used to determine thermally induced phase transitions. Figure 5.7 shows the DSC

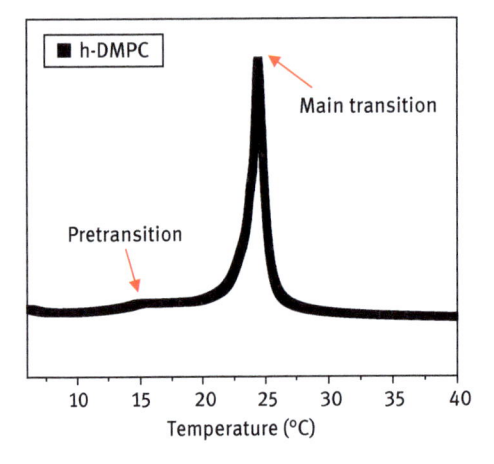

Figure 5.7: DSC thermogram of DMPC ULV.

thermogram of DMPC ULVs, where we can identify the pre-transition from gel phase to ripple phase at ~15 °C and the main transition from ripple phase to liquid crystalline phase at ~24 °C. The structures of each phase (gel, ripple and liquid crystalline) are sketched in Figure 5.8. In order to utilize DSC to measure the transfer rate constants, the melting transition temperature (T_M) of the donor must be differentiable from the acceptor in a thermogram. For instance, the T_M of DMPC is 24 °C, which is 17 degrees lower than that of DPPC (i.e., T_M = 41 °C). The equimolar mixture of DMPC and DPPC yielded two distinct peaks as shown in Figure 5.9, corresponding to the main transition peaks of DMPC located at 29 °C and DPPC located at 48 °C [30]. Note that the melting transition temperatures of DMPC and DPPC were elevated by approximately 4 °C than the commonly reported values due to the presence of polyethylene glycols (PEG).

To conduct TR-DSC experiments, NPs with different T_M are prepared separately in the aqueous buffer of interest and post-mixed before subjecting to the tests. In a study that was conducted to investigate the effect of PEG on the aggregation of sonicated vesicles, Tilcock and Fisher observed a new population formed after DMPC and DPPC vesicles were mixed, whose T_M (i.e., 38 °C) was the average of those of DMPC and DPPC (Figure 5.9(b) to (f)) [30]. This indicates that a new population composed of equimolar of DMPC and DPPC was formed. At the same

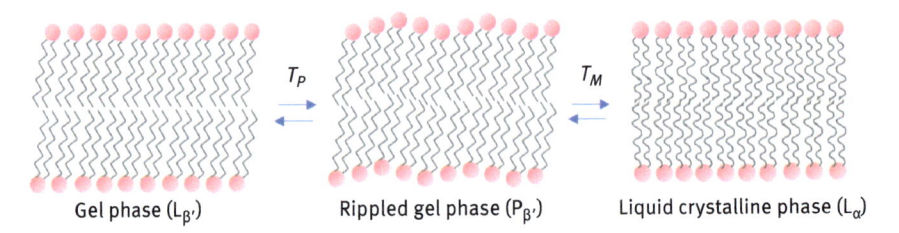

Figure 5.8: Phase transition of lipid bilayers.

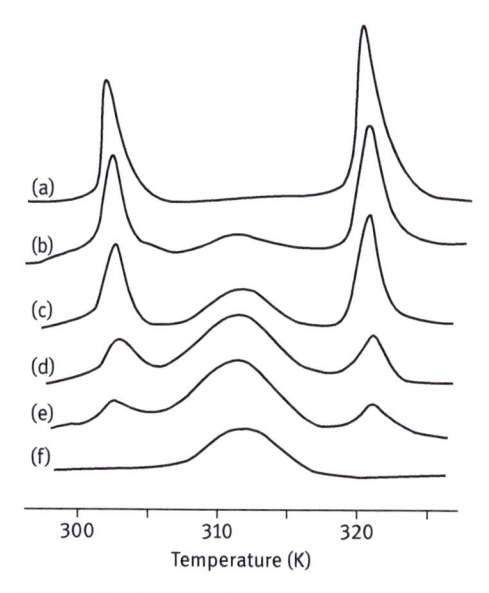

Figure 5.9: Equimolar mixtures of dimyristoylphosphatyidylcholine and dipalmitoylphophatidylchone in 5 mM EDTA were mixed at room temperature (a) and incubated at 60 °C for (b), 7 min; (c), 10 min; (d), 16 min; (e), 20 min; (f), 25 min. Figure was reprinted from Biochim. Biophys. Acta 1982, 688, 645–652.

time, the endothermic peaks of DMPC and DPPC vesicles reduced because of the decreased population, and eventually disappeared. In this case ($T = 60\ °C$), the lipid exchange was rather rapid and induced by the aggregation of DMPC and DPPC vesicles, which leads to the formation of new composite vesicles.

Isotope substitution has been commonly adopted in TR-DSC for measuring lipid transfer rate constants. The T_M of a deuterated lipid is found to be lower compared to its protiated counterpart [31, 32]. Deuterated lipids have a less-ordered configuration in gel phase compared protiated lipids, therefore resulting in a lower T_M [32]. For instance, the T_M of 1,2-dimyristoyl-d54-sn-glycero-3-phosphocholine (d_{54}-DMPC) ULVs is 20.2 °C, while the T_M of h-DMPC ULVs is 24 °C (Figure 5.10a). The difference in the T_M between d- and h-bicelles was also observed [24]. When ULVs are composed of equal molar d- and h-NPs (e.g., bicelles or ULVs), DSC only shows one endothermic peak, indicating that h- and d-lipids are miscible. Importantly, the T_M of h/d-NPs, $T_{M,h/d}$ is the average of d-NPs, $T_{M,d}$ and h-NPs, $T_{M,h}$ (22.1 °C for h/d-DMPC ULV), suggesting a linear relationship between T_M and the composition of h- and d-lipids [33]. In the case of monomeric lipid transfer, as the lipid transfer proceeds, the two endothermic peaks move toward each other and eventually merge into one peak whose position depends on the composition of h- and d-NPs in the mixture, $T_{M,h/d} = T_{M,h}\varphi_h + T_{M,d}(1-\varphi_h)$ (Figure 5.10b) [31]. The difference in T_M (ΔT_M) between h-NP and d-NP can then be calculated, which should show a decrease if there is lipid exchange between h-NP and d-NP.

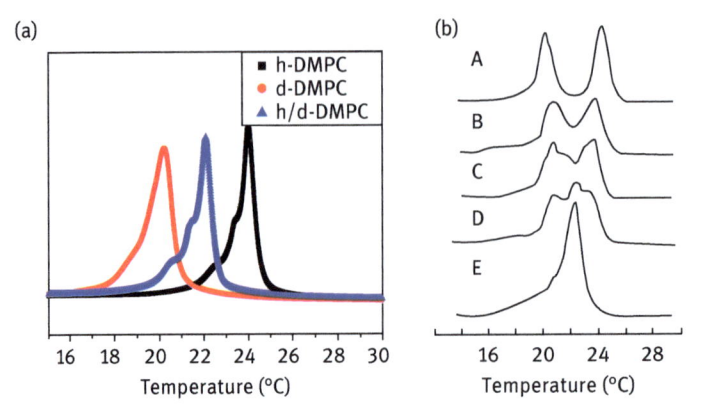

Figure 5.10: (a) DSC thermograms of h-DMPC ULV (black), d54-DMPC ULV (red) and h/d-DMPC ULV containing equal molar of d54 and h-DMPC; (b): DSC thermograms of a 1:1 (mol) mixtures of two vesicles populations of DMPC and d54-DMPC for various incubation times at 35 C incubation temperature: (A) 0.15 h; (B) 3.0 h; (C) 5.0 h; (D) 12.5 h; (E) 25h. Figure 5.10b was reprinted from Biochemistry 1988, 27, 6078.

The same kinetics models used for the TR-SANS are applied here to analyze TR-DSC data. Through correlating $([h]_h - [h]_d)$ in one NP to its T_M in eq. (5.9), the final kinetics equations for bicelles and ULVs are derived and presented using eqs. (5.25) and (5.26), respectively [18].

$$\Delta T_M(t) = \Delta T_{M,0} * e^{-k_{in}t} \tag{5.25}$$

$$\Delta T_M(t) = \Delta T_{M,0} \left[\frac{2k_{ff} + k_{in} + \sqrt{4k_{ff}^2 + k_{in}^2 - 2rk_{in}}}{2\sqrt{4ff + k_{in}^2}} \right] e^{\lambda_1 t}$$

$$- \Delta T_{M,0} \left[\frac{2k_{FF} + k_{in} - \sqrt{4k_{FF}^2 + k_{in}^2 - 2rk_{in}}}{2\sqrt{4k_{FF}^2 + k_{in}^2}} \right] e^{\lambda_2 t} \tag{5.26}$$

where $\Delta T_M(t) = T_{M,h}(t) - T_{M,d}(t)$ and $\Delta T_{M,0} = \Delta T_M(0)$.

We have shown that TR-SANS and TR-DSC data were in good agreement in measuring the DMPC transfer in ULVs, a relatively slow transfer process [18]. Comparing to TR-SANS, DSC is more accessible, yet more invasive. Since obtaining DSC data requires a heating cycle, the material is typically discarded after each data collection. Therefore, the quantity of materials has to be enough to allow multiple measurements. Moreover, inaccuracy may be introduced if lipid transfer is faster than DSC data collection time at the studied temperature range. It is worth noting that bicelles in general show a broader transition peak compared to ULVs due to a loss of co-operativity [34, 35]. Finally, the TR-DSC is not suitable for systems containing multiple endothermic peaks such as DMPC bicelles whose

T_M is difficult to identify. The deconvolution of each peak can be very complicated in the data analysis.

5.5 Fluorescence correlation spectroscopy

FCS uses statistical analysis of the fluctuations of fluorescence in a system to probe the dynamics of molecular events, such as diffusion or conformational fluctuations of biomolecules [36]. It has been used to measure the diffusivity of both small and big molecules, such as the diffusion of fluorescent lipids and proteins in cellular plasma membranes [37, 38] or to probe lipid domain in model membranes [39–41]. Like most of the fluorescence techniques, it requires the use of a fluorophore attached to the molecule of interest, which bears the risk of altering physical properties of the native molecules. For instance, it has been reported that the orientation of BODIPY analogs in the membrane shifted from parallel to normal to the membrane plane as the surface pressure of the membrane increases [42]. The DMPC transfer rate constant measured by FCS was ~30% larger than the one obtained from SANS [18]. Therefore, caution is needed when selecting the appropriate fluorophores.

FCS measures the fluctuations of the time-averaged fluorescence intensity (Figure 5.11), which can be expressed using the autocorrelation function as below eq. (5.27)

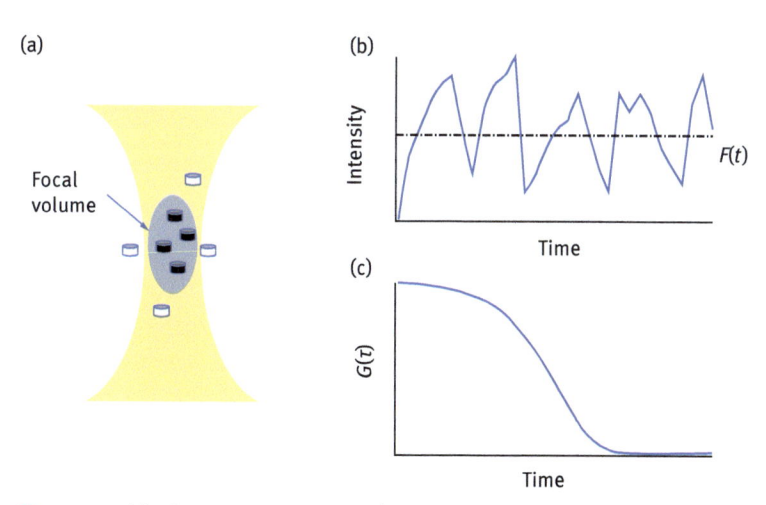

$$G(\tau) = \frac{\delta F(t)\delta F(t+\tau)}{F(t)^2} \tag{5.27}$$

Figure 5.11: (a): observation area of FCS (focal volume); (b): The fluctuation of fluorescence intensity; (c): Autocorrelation function derived from the fluctuation of fluorescence intensity.

where $\delta F(t) = F(t) - \langle F(t) \rangle$ and $F(t)$, $F(t+\tau)$ are the fluorescence intensity at time t and $(t+\tau)$, respectively, and τ is the correlation time. The autocorrelation function is related to the concentration c and diffusion coefficient D of the fluorescence species in the focal volume as shown in the three-dimensional Gaussian equation as below [43],

$$G(\tau) = \left(c\pi^{1.5}\omega_z\omega_{xy}^2\right)^{-1}\left(1+\frac{4D\tau}{\omega_{xy}^2}\right)^{-1}\left(1+\frac{4D\tau}{\omega_z^2}\right)^{-0.5} \quad (5.28)$$

where ω_z and ω_{xy} are the axial and radial radii of the laser focal volume.

To monitor the lipid transfer using FCS, fluorescent particles are mixed with non-fluorescent particles (Figure 5.12). As the lipid transfer proceeds, the concentration of fluorescent particles increases, leading to a decrease of $G(0)$ that is the y-intercept of $G(\tau)$ based on eq. (5.28). Note that FCS only requires a small number of fluorophores to be present in the laser focal volume in order to enhance the measured signal (Figure 5.11(a)). Therefore, the perturbation due to the introduction of fluorophores to the systems can be greatly reduced compared to traditional fluorescence spectroscopy method. In one example of using FCS to measure the DMPC transfer, we incorporated two fluorescent lipids into one DMPC bicelle, so each bicelle should only contain one fluorophore upon mixing with equal mole of non-fluorescent pristine bicelle after lipid transfer reaches equilibrium [18] (Figure 5.12). As the lipid transfer takes place, the number of fluorescent NPs increases in the focal volume, which is then reflected through the decrease of $G(\tau)$ (Figure 5.11). The interparticle lipid transfer rate constant can be calculated using the following equation,

$$c(t) = c_0\left(2 - e^{-k_{in}t}\right) \quad (5.29)$$

where c_0 is the concentration of fluorescent NPs at $t = 0$.

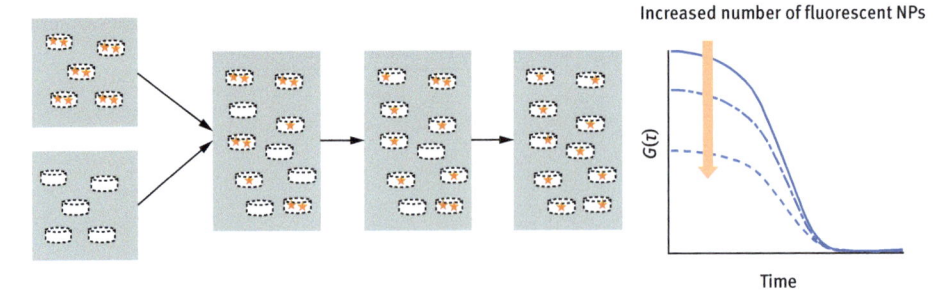

Figure 5.12: Schematics of using FCS to measuring lipid transfer. Fluorescent NPs containing two fluorescent lipids are mixed with non-fluorescent NPs. As the fluorescent lipid moves to the non-fluorescent NP, the number of fluorescent NPs increases.

Despite the disadvantage of using a fluorophore-modified lipid, FCS can be used to measure relatively fast lipid transfers, since it takes just a few minutes to collect a complete data set. On the other hand, each DSC and SANS measurement takes at least 20 min therefore they are only suitable for systems that have a lipid transfer half-life >1 h. However, the measurement of a slow lipid transfer using FCS is often limited by photobleaching of fluorescence probe as it moves through the observation volume. Thus, the half-life of optimal samples for characterization using FCS is approximately from 20 min to 5 h.

5.6 Pros and cons of different techniques

The pros and cons of each technique have been discussed in previous sections in this chapter, which are listed in Table 5.1.

Table 5.1: Advantages and limitations of SANS, DSC, and FCS for the study of molecular transfer.

	SANS	DSC	FCS
Requirement	– Sufficient NSLD contrast between nanoparticles and solvent	– The phase transition peaks must be assignable and well-resolved. Systems need to be thermally stable in the temperature range of DSC measurements. – The transfer rate is slower than the data collection time	– Fluorophore-modified lipid behaves similarly to the molecules of interest and minimally perturbs the system.
Possible Weakness	– isotope substitution may affect physical properties of the system of interest	– Same as SANS – The heating process during measurements may affect the transfer process	– The molecular transfer rate constant is monitored by the fluorophores which may behave differently from the molecules being studied.
Appropriate half-life	– Slower transfer (>1 hour)	– Slower transfer (>1 hour)	– Faster transfer (20 min to 5 hours)
Accessibility	– limited availability of SANS instruments	– Readily available	– Readily available
Materials cost	– < 0.5 mL	– 5 ~ 10 mL	– < 0.5 mL

Source: Table adapted from Langmuir [18].

5.7 Summary

In this chapter, we have introduced three methodologies including TR-SANS, TR-DSC, and FCS for investigating lipid transfer. These techniques can also be extended to the molecules other than lipids. Two model systems, ULV and bicelles, are used to demonstrate the feasibility of accessing lipid transfer rate constants. We have also shown the kinetics models describing the molecular transfer. The goal of this chapter is to provide researchers the appropriate methods to study monomeric molecular transfer.

References

[1] Balla, T. Phosphoinositides: tiny lipids with giant impact on cell regulation. Physiol. Rev. 2013, 93(3), 1019–1137.
[2] Ivanov, A. I. Pharmacological inhibition of endocytic pathways: is it specific enough to be useful? Methods Mol. Biol. 2008, 440, 15–33.
[3] Gruenberg, J.; Maxfield, F. R. Membrane transport in the endocytic pathway. Curr. Opin. Cell Biol. 1995, 7(4), 552–563.
[4] Luchette, P. A.; Vetman, T. N.; Prosser, R. S.; Hancock, R. E.; Nieh, M. P.; Glinka, C. J.; Krueger, S.; Katsaras, J. Morphology of fast-tumbling bicelles: a small angle neutron scattering and NMR study. Biochim. Biophys. Acta. 2001, 1513(2), 83–94.
[5] John Katsaras, T. A. H., Jeremy Pencer, Mu-Ping Nieh. "Bicellar" Lipid mixtures as used in biochemical and biophysical studies. Naturwissenschaften 2005, 92, 355–366.
[6] Dong, H.; Lund, R.; Xu, T. Micelle stabilization via entropic repulsion: balance of force directionality and geometric packing of subunit. Biomacromolecules 2015, 16(3), 743–747.
[7] Kastantin, M.; Ananthanarayanan, B.; Karmali, P.; Ruoslahti, E.; Tirrell, M. Effect of the lipid chain melting transition on the stability of DSPE-PEG(2000) micelles. Langmuir 2009, 25(13), 7279–7286.
[8] Dong, H.; Shu, J. Y.; Dube, N.; Ma, Y.; Tirrell, M. V.; Downing, K. H.; Xu, T. 3-Helix micelles stabilized by polymer springs. J Am Chem Soc 2012, 134(28), 11807–11814.
[9] Sumeet Jain, F. S. B. Consequences of Nonergodicity in aqueous binary PEO-PB micellar dispersions. Macromolecules 2004, 37(4), 1511–1523.
[10] Lu, J., Choi, S., Bates, F. S., Lodge, T. P.. Molecular exchange in diblock copolymer micelles: Bimodal distribution in core-block molecular weights. ACS Macro. Lett. 2012, 1(8), 982–985.
[11] Nichols, J. W.; Pagano, R. E. Kinetics of soluble lipid monomer diffusion between vesicles. Biochemistry 1981, 20(10), 2783–2789.
[12] Hrafnsdottir, S.; Nichols, J. W.; Menon, A. K. Transbilayer movement of fluorescent phospholipids in Bacillus megaterium membrane vesicles. Biochemistry 1997, 36(16), 4969–4978.
[13] J. Wylie Nicholas, R. E. P. Use of Resonance energy transfer to study the kinetics of amphiphile transfer between vesicles. Biochemistry 1982, 21, 1720–1726.
[14] Kornberg, R. D.; McConnell, H. M. Inside-outside transitions of phospholipids in vesicle membranes. Biochemistry 1971, 10(7), 1111–1120.
[15] McNamee, M. G.; McConnell, H. M. Transmembrane potentials and phospholipid flip-flop in excitable membrane vesicles. Biochemistry 1973, 12(16), 2951–2958.

[16] Brown, R. E. Spontaneous lipid transfer between organized lipid assemblies. Biochim. Biophys. Acta. 1992, 1113 (3–4), 375–389.
[17] Wimley, W. C.; Thompson, T. E. Exchange and flip-flop of dimyristoylphosphatidylcholine in liquid-crystalline, gel, and two-component, two-phase large unilamellar vesicles. Biochemistry 1990, 29(5), 1296–1303.
[18] Xia, Y.; Li, M.; Charubin, K.; Liu, Y.; Heberle, F. A.; Katsaras, J.; Jing, B.; Zhu, Y.; Nieh, M. P. Effects of nanoparticle morphology and acyl chain length on spontaneous lipid transfer rates. Langmuir 2015, 31(47), 12920–12928.
[19] Wimley, W. C.; Thompson, T. E. Transbilayer and interbilayer phospholipid exchange in dimyristoylphosphatidylcholine/dimyristoylphosphatidylethanolamine large unilamellar vesicles. Biochemistry 1991, 30(6), 1702–1709.
[20] Yang, P. W.; Lin, T. L.; Hu, Y.; Jeng, U. S. A time-resolved study on the interaction of oppositely charged bicelles – implications on the charged lipid exchange kinetics. Soft Matter 2015, 11(11), 2237–2242.
[21] Liu, Y.; Li, M.; Yang, Y.; Xia, Y.; Nieh, M. P. The effects of temperature, salinity, concentration and PEGylated lipid on the spontaneous nanostructures of bicellar mixtures. Biochim Biophys Acta 2014, 1838(7), 1871–1880.
[22] Zinn, T., Willner, L., Lund, R., Pipich, V., Richter, D.. Equilibrium exchange kinetics in n-alkyl-PEO polymer micelles: single exponential relaxation and chain length dependence. Soft Matter 2012, 8, 623–626.
[23] Nakano, M.; Fukuda, M.; Kudo, T.; Endo, H.; Handa, T. Determination of interbilayer and transbilayer lipid transfers by time-resolved small-angle neutron scattering. Phys. Rev. Lett. 2007, 98 (23), 238101.
[24] Xia, Y.; Charubin, K.; Marquardt, D.; Heberle, F. A.; Katsaras, J.; Tian, J.; Cheng, X.; Liu, Y.; Nieh, M. P. Morphology-induced defects enhance lipid transfer rates. Langmuir 2016, 32(38), 9757–9764.
[25] Liu, Y.; Xia, Y.; Rad, A. T.; Aresh, W.; Nieh, M. P. Stable discoidal bicelles: a platform of lipid nanocarriers for cellular delivery. Methods Mol. Biol. 2017, 1522, 273–282.
[26] Nakano, M.; Fukuda, M.; Kudo, T.; Miyazaki, M.; Wada, Y.; Matsuzaki, N.; Endo, H.; Handa, T. Static and dynamic properties of phospholipid bilayer nanodiscs. J. Am. Chem. Soc. 2009, 131 (23), 8308–8312.
[27] Higgins, J. S.; Benoît, H. Polymers and neutron scattering; Clarendon Press; Oxford University Press: Oxford New York, 1994. p xix, 436p.
[28] Garg, S.; Porcar, L.; Woodka, A. C.; Butler, P. D.; Perez-Salas, U. Noninvasive neutron scattering measurements reveal slower cholesterol transport in model lipid membranes. Biophys. J. 2011, 101(2), 370–377.
[29] Gerelli, Y.; Porcar, L.; Lombardi, L.; Fragneto, G. Lipid exchange and flip-flop in solid supported bilayers. Langmuir 2013, 29(41), 12762–12769.
[30] Tilcock, C. P.; Fisher, D. The interaction of phospholipid membranes with poly(ethylene glycol). Vesicle aggregation and lipid exchange. Biochim. Biophys. Acta. 1982, 688(2), 645–652.
[31] Bayerl, T. M., Schimidt, C. F., Sackmann, E. Kinetics of symmetric and asymmetric phospholipid transfer between small sonicated vesicles studied by high-sensitivity differential scanning calorimetry, NMR, electron microscopy, and dynamic light scattering. Biochemistry 1988, 27, 6078.
[32] Yu, P. H.; Nguyen, T. V. Deuterium isotope effect in the transamination of p-tyrosine by rat liver tyrosine transaminase. Life Sci. 1985, 37(14), 1287–91.
[33] Marsh, D. Handbook of lipid bilayers; 2nd ed.; CRC Press, Taylor & Francis Group: Boca Raton, FL., 2013. p. xxvii, 1,145p.
[34] Shaw, A. W.; McLean, M. A.; Sligar, S. G. Phospholipid phase transitions in homogeneous nanometer scale bilayer discs. FEBS Lett. 2004, 556 (1–3), 260–264.

[35] Marsh, D.; Watts, A.; Knowles, P. F. Cooperativity of the phase transition in single- and multibilayer lipid vesicles. Biochim. Biophys. Acta. 1977, 465(3), 500–514.
[36] Oleg Krichevsky, G. B. Fluorescence correlation spectroscopy: the technique and its applications. Rep. Prog. Phys. 2002, 65, 251–297.
[37] Guido Bose, P. S., Tilman Lamparter. The mobility of phytochrome within protonemal tip cells of the moss ceradodon purpureus, monitored by fluorescence correlation spectroscopy. Biophys. J. 2004, 87, 2013–2021.
[38] Hegener, O., Prenner, L., Runkel, F., Baader, S.L., Kappler, J., Haberlein, H. Dynamics of beta2-adrenergic receptor-ligand complexes on living cells. Biochemistry 2004, 43(20), 6190–6199.
[39] Bacia, K.; Scherfeld, D.; Kahya, N.; Schwille, P. Fluorescence correlation spectroscopy relates rafts in model and native membranes. Biophys. J. 2004, 87(2), 1034–1043.
[40] Kahya, N.; Scherfeld, D.; Bacia, K.; Schwille, P. Lipid domain formation and dynamics in giant unilamellar vesicles explored by fluorescence correlation spectroscopy. J. Struct. Biol. 2004, 147(1), 77–89.
[41] Kahya, N.; Scherfeld, D.; Bacia, K.; Poolman, B.; Schwille, P. Probing lipid mobility of raft-exhibiting model membranes by fluorescence correlation spectroscopy. J. Biol. Chem. 2003, 278(30), 28109–28115.
[42] Armendariz, K. P.; Huckabay, H. A.; Livanec, P. W.; Dunn, R. C. Single molecule probes of membrane structure: orientation of BODIPY probes in DPPC as a function of probe structure. Analyst 2012, 137(6), 1402–1408.
[43] Chen, Y.; Lagerholm, B. C.; Yang, B.; Jacobson, K. Methods to measure the lateral diffusion of membrane lipids and proteins. Methods 2006, 39(2), 147–153.

Eugene Cheung, Darian Cash

6 Fundamentals of Nuclear Magnetic Resonance spectroscopy (NMR) and its applications

Abstract: Nuclear magnetic resonance (NMR) spectroscopy is an eminent technique for the characterization of structure and dynamics of molecules in solution and in the solid state. With applications in a majority of the physical and biological sciences, ranging from chemistry and materials science to mineralogy and molecular biology, the frontiers of NMR spectroscopic analysis seem limited only by the creativity of its users. NMR spectroscopy provides exquisite detail about the nuclear environment of any molecular system with little, if any, perturbation of the system during analysis. In the context of analyzing bilayers, biomimetics, or biological molecules, variants of ^{1}H, ^{13}C, ^{15}N, and ^{31}P NMR experiments are extending the reach of NMR spectroscopy for tackling key physical and chemical aspects of biologic drugs and drug-delivery systems. Pseudo-2D NMR experiments such as pulsed gradient spin echo NMR spectroscopy are exceptionally powerful in studying translational motion directly in biological fluids and under simulated physiological conditions. The NOESY experiment gives correlation between two nuclei that are within several angstroms from each other and is unique in its ability to elucidate membrane-bound entities. New NMR pulse sequences are invented and technological breakthroughs are constantly adopted as the ever expanding arsenal of NMR experiments continues to inspire generations of scientists. For characterization of complex structures such as biomembranes and nanoparticles, NMR spectroscopy is perhaps the only analytical technique that can elucidate structure and dynamics in situ and without dilution.

Keywords: NMR, NOESY, Micelles

6.1 Introduction

Nuclear magnetic resonance (NMR) spectroscopy, based upon the absorption of radio-frequency electromagnetic radiation, is a powerful technique for elucidating chemical structure. Since its discovery by Felix Bloch and Edward Mills Purcell in 1946, for which they were awarded the Nobel Prize in Physics in 1952, NMR has become a pre-eminent characterization technique to obtain the structure and dynamics of molecules in solution and in the solid state [1–3]. With applications in a majority of the physical and biological sciences, ranging from chemistry and materials science to mineralogy

---

Eugene Cheung, Darian Cash, Moderna Therapeutics Inc.

https://doi.org/10.1515/9783110544657-006

and molecular biology, the frontiers of NMR spectroscopic analysis seem limited only by the creativity of its users (selected references [4–12]).

Like other branches of spectroscopy, NMR spectroscopy relies on the interaction between matter and electromagnetic radiation to investigate the nature of atoms and molecules. Typical spectroscopic techniques measure radiation intensity, either absorption or emission, as a function of wavelength. NMR spectroscopy is no different – it is a technique that observes the atomic nuclear absorption of electromagnetic radiation in the radiofrequency range. The positions, intensities, and peak splitting of resonance peaks in the resulting NMR spectra contain quantitative information about atomic structure. The capability of discerning structure from radiofrequency lies in the fundamental phenomenon that the molecular environment of nuclei affects the absorption of radiation by each nucleus when exposed to an external magnetic field, thereby allowing each nucleus, and hence each molecule, to generate a specific and characteristic spectral pattern.

Advances in NMR theory and hardware design have allowed significant growth in answering questions posed by materials science, membranes and biology, mineralogy, and nanoparticles [4–12]. Three considerations are apparent when evaluating NMR spectroscopy: (a) it is comparably hampered by low sensitivity, (b) the study of large molecule systems is difficult because of overlap of resonances, and (c) most users only tap into the very basics of its capability, primarily for structure determination of (small) organic compounds. Stemming from the latter reason, *automated* commercial NMR spectrometers have become workhorses in many laboratories, especially in chemistry laboratories, providing the practicing scientist fairly routine access to one of the most versatile chemical tools developed in the twentieth century.

With the improvement of NMR spectrometer control software using graphical interfaces, the operation of the spectrometer can become as straightforward as requiring only the input of a series of keystrokes or clicks through the spectrometer interface. Little or no understanding of the meaning for each "click" is necessary. Essential steps to NMR data acquisition such as matching, tuning, and shimming, which were once somewhat tedious or laborious manual rituals, are now performed in minutes by users who do not know what these procedures actually do. The state of automation has led some laboratories to rely on standard operating procedures (SOPs) that outline step by step which menus to choose and which buttons to click in the NMR spectrometer software interface. In the extreme and yet very common case, the entire data collection process is automated. The user has only to insert an NMR tube into a sample holder, place the sample holder onto a carousel belt, and click Start in the computer software. For complex self-assembled molecular or biological systems, successful practice of NMR spectroscopic analysis necessitates more than just SOPs and automation.

NMR spectroscopy reference texts generally explain in depth the mathematics upon which NMR is based, and extensively introduce the concepts of the rotating frame, nuclear spin, relaxation, nuclear shielding, Fourier transform, and pulse programs [13–16]. For the purposes of this chapter, the emphasis is on the practicing user who has specific questions to solve by NMR spectroscopy, but who owing to other responsibilities may not have the time or the desire to become fully versed in the fundamentals of NMR spectroscopy. For this reason, the mathematical descriptions and diagrams typically seen in most NMR references are not depicted, except for those concepts which an end user may on occasion view in modern NMR software [17]. The chief purpose is to familiarize the reader with NMR experiments which are beyond those typically used in a routine setting, but which could have significant impact in answering key scientific questions.

Despite the promise of NMR spectroscopy being able to answer many scientific questions, not all materials can be analyzed using this technique. Materials heavily laden with paramagnetic metals show peak broadening. Solid samples require the use of solid-state NMR spectroscopy (SSNMR). Many excellent texts are available for SSNMR, and the reader is encouraged to study those [18, 19]. The fundamental basics of liquid- and solid-state NMR are the same – they differ mainly in the instrumental and technical aspects of data acquisition. Despite the recognized importance of using SSNMR to answer many chemical and materials questions, this technique, in the form of an open walk-up departmental instrument, remains less common to the typical chemist, biologist, or chemical engineer. As a result, the examples in this chapter will focus on liquid NMR spectroscopy, which is the technique primarily available at first glance for practicing scientists.

6.2 Theoretical aspects of NMR spectroscopy

6.2.1 Basic concepts

University undergraduate chemistry courses will normally cover some introduction to NMR spectroscopy and so the expectation here is that current or prospective users of NMR equipment will have a basic understanding of NMR spectroscopy. In its most basic form, NMR spectroscopy is essentially unchanged in concept from the first origins of spectroscopy, which itself is a convolution of the Latin word *spectrum*, meaning "image", and the Greek suffix *-scopy*, meaning "viewing", and which was used by early physicists to describe the splitting of visible light by a prism into a corresponding rainbow of wavelengths [20]. We now know that this dispersion of white light into multiple colors is in fact a form of energy splitting. NMR spectroscopy is based upon this same principle, but applied toward the energy splitting of nuclei exposed to a strong external magnetic field.

Quantum chemistry teaches that a nucleus may rotate around an axis (the so-called "spin"). Only those atomic nuclei of specific elemental isotopes have the characteristic nuclear spin that can be witnessed by NMR radio pulses. Nuclear spin, specified by quantum number (I), is a fundamental property of nature and can range from integral spins (e.g., $I = 1, 2, 3$, etc.) to fractional spins (e.g., $I = 1/2, 3/2, 5/2$, etc.), to no spins (e.g., $I = 0$). NMR spectroscopy can only be performed on elemental isotopes whose spin nuclear value is nonzero (Table 6.1) [21]. Such a spinning nucleus will have magnetic properties, and those nuclei which have spin quantum number, $I = 1/2$, are the most accessible for NMR spectroscopic detection. These nuclei, conveniently, are also the nuclei most commonly incorporated into organic or biological molecules and complexes.

Table 6.1: NMR Properties of spin ½ nuclei found commonly in chemical and biological systems.

Nucleus	Gyromagnetic ratio (radian $T^{-1}s^{-1}$)	Isotopic Abundance %	Relative Sensitivity	Absorption Frequency at 4.7 T (MHz)
^1H	2.68×10^8	99.98	1.00	200
^{13}C	6.73×10^7	1.11	0.016	50.2
^{19}F	2.52×10^8	100	0.83	188
^{31}P	1.08×10^8	100	0.066	81.0
^{15}N	-2.71×10^7	0.37	0.001	20.3

The magnetic moment, μ, of each spinning nucleus is proportional to the angular momentum p and the gyromagnetic ratio γ that is specific to each nucleus. In the absence of an external magnetic field, the nuclei of one isotope will all have the same energy. When exposed to an external magnetic field B_0, nuclei can align along one direction of the magnetic field (denoted α) or in the opposite direction (denoted β). The energies of these two populations are no longer the same, nor are the numbers of nuclei in each population. One population will have higher energy (E_2) than the other (E_1) (Figure 6.1).

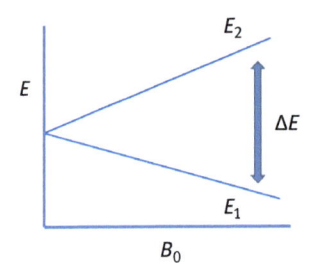

Figure 6.1: The difference in energy (ΔE) between two spin states, the lower energy E_1 and the higher energy E_2, as a function of exposure to an external magnetic field denoted B_0.

The difference in energy between the two populations can be described by equalities which link the magnetic moment μ with the gyromagnetic ratio γ, angular momentum p, and quantum number I in order to derive the differences in energy between states E_2 and E_1 to the external magnetic field B_0:

$$p = \frac{h}{2\pi} I$$

$$\mu = \gamma p$$

$$\Delta E = \frac{\gamma h}{2\pi} B_0$$

where h is Planck's constant

It can be seen from these equations that the difference in energy among nuclei depends solely on the gyromagnetic ratio and the external magnetic field. The gyromagnetic ratio is fixed for any given isotope, with larger gyromagnetic ratios resulting in larger magnetic moments, and greater energy differences in an external magnetic field. Smaller values of γ need a stronger external magnetic field to enlarge the energy gap ΔE between spin states. It can be seen from Table 6.1 that the proton spin ½ nucleus has a gyromagnetic ratio that is four times larger than the ^{13}C nucleus. When the natural abundances of the isotopes are considered, the sensitivity of the ^{1}H nucleus over the ^{13}C nucleus is even more apparent. The difference in energy results in a ^{1}H NMR experiment requiring much less data acquisition time compared to the identical experiment where ^{13}C is observed.

Due to their atomic abundance and detection sensitivity, protons are the most commonly studied nuclei using NMR spectroscopy (^{1}H NMR spectroscopy) for organic and biological molecules and complexes. The high sensitivity of ^{1}H NMR spectroscopy for protons and the large number of protons present in organic molecules make the proton nucleus relatively easy to detect, and it is thus a convenient coincidence that the most prevalent nucleus in organic and biological chemistry is also the one that is easiest for detection. In addition to the ^{1}H proton, a number of other nuclei can be studied by modern NMR probes, with ^{13}C and ^{15}N being the next most common nuclei of interest because of their frequent presence in organic molecules. While sensitivity for these two nuclei is low, isotopic enrichment is not necessary unless measurement times are unacceptably long. The natural abundance of ^{13}C is only 1.1% that of ^{12}C, and that of ^{15}N is even less, at 0.4%. However, one advantage of studying nonproton nuclei is their lower resonance energies, which results in a larger frequency spread for each signal, reducing the likelihood of peak overlap.

6.2.2 Free induction decay

Introductory courses and texts to NMR spectroscopy tend to begin with descriptions of continuous wave NMR spectroscopic data collection in which the external magnetic field is kept constant as the spectrometer scans through the frequency domain. As will be discussed in further detail later in this chapter, it is unlikely that students and scientists now will use such equipment. Almost exclusively (pulse) Fourier transform NMR spectroscopy (FT-NMR) is expected to be available in the modern setting. In a pulse experiment, all the nuclei are excited simultaneously, and their relaxation is monitored overtime. The intensity versus time spectrum is called the free induction decay (FID) spectrum (Figure 6.2). The FID spectrum contains all the information in an NMR experiment. Advances in computing have enabled software to be written that, with just a few mouse clicks, invokes a Fourier transform of the FID spectrum to convert the intensity versus time spectrum into the commonly recognized NMR intensity versus frequency spectrum (Figure 6.3).

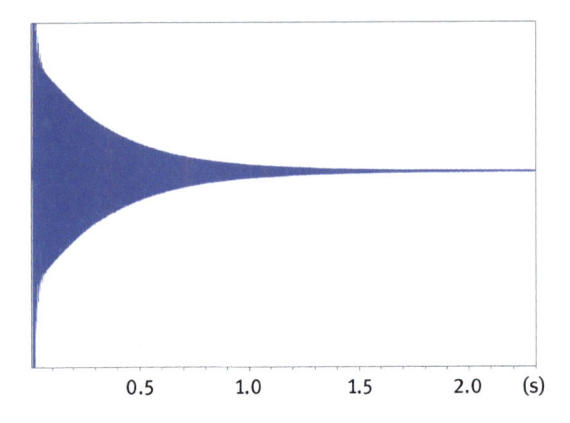

Figure 6.2: An example of an intensity versus time spectrum showing the free induction decay (FID) of the NMR signal.

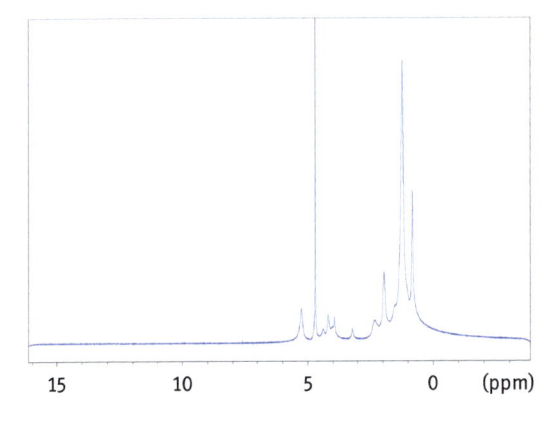

Figure 6.3: The intensity versus frequency spectrum after Fourier transform of the FID.

6.2.3 The chemical shift

In NMR spectroscopy, the difference in energy between populations ΔE is described in terms of frequency v. Instead of using Joules (J) or Tesla (T) to describe the relationship between ΔE and B_0, the differences in energy are expressed by chemical shift δ (Hz and ppm). So typical are these descriptions, commercial magnets are frequently described to by their proton resonance frequency in MHz (or even GHz) rather than the magnetic field they can exhibit in T. A departmental NMR-automated workhorse spectrometer may have a frequency of 400 or 500 MHz, whereas the spectrometer strength for biological NMR may rise to the 800 MHz and GHz range.

The connection between the absolute frequency Hz of the resonance of each nucleus and its chemical shift δ is expressed through the relationship of the individual resonance with the frequency of a reference nucleus:

$$v_0 = Chemical\ shift(\delta)$$
$$= \left(\frac{frequency\ of\ nucleus\ i\ (Hz) - frequency\ of\ reference\ nucleus(Hz)}{frequency\ of\ NMR\ spectrometer(MHz)}\right) \times 10^6$$

The resonance nucleus is trimethysilane (TMS) which is set to 0 ppm in an intensity versus chemical shift spectrum. Resonances downfield of TMS on spectrometers of different magnetic field strength would be different, but their chemical shift would be the same. How is this possible? The chemical shift of a nucleus such as a proton is determined by the adjacent through-bond coupling to other nuclei, and to a limited extent, through spatial coupling. Software has existed for many years that can predict the location of the ^1H and ^{13}C lines, with the ^{13}C predictions being fairly accurate [22].

Chloroform has one proton ($CHCl_3$) which has a resonance frequency of 2912 Hz downfield from TMS at 400 MHz, resulting in a chemical shift δ of 7.28 ppm. On an 800 MHz spectrometer, chloroform will have a resonance frequency of 5824 Hz downfield from TMS, and the chemical shift δ will be the same as it is on a 400 MHz spectrometer, that is, 7.28 ppm. The advantage of the higher field magnet spectrometer is the spectral separation (in Hz) between signals (5824 versus 2912 Hz).

One unique advantage of NMR spectroscopy over other techniques is the direct relationship existing between the number of identical nuclei in an environment and the peak intensity arising from these identical nuclei. As long as peak overlap is not occurring, the area of a peak is representative of the number of protons (for ^1H NMR spectroscopy) that comprises this peak. By knowing the ratios of protons under each peak, the chemical structure of a molecule can be reconstructed from its NMR spectrum. With the use of an internal standard for which the peak area of a proton is known, the number of protons in a sample of unknown concentration can be calculated, thereby allowing NMR to be used quantitatively in an analytical setting [23–27].

6.2.4 The Nuclear Overhauser Effect

Because nuclei behave as magnets, nuclei which are near to one another will interact in a fashion similar to physical magnets brought into close proximity. When a nucleus of one type is irradiated with a specific radiofrequency, nuclei nearby will also experience changes to the population of their energy states, resulting in an observable difference to the intensity of their resonances. This phenomenon in which the Boltzmann distributions of nuclei are altered through space by dipole–dipole interactions of nearby nuclei is known as the Nuclear Overhauser Effect (NOE). The NOE is dependent on the distance between nuclei (falls off at $1/r^6$) and on the mobility of the nuclei (torsional rotation of chemical bonds or Brownian motion for entire molecules), and not on chemical bond connectivity. Intramolecular as well as intermolecular cross-relaxation of nuclei is possible. Traditional NOE experiments are performed in organic chemistry research where the stereochemistry of chiral molecules can be ascertained by selective irradiation of nuclei and subsequent observation of the NOE. The connectivity of the molecule is then inferred from this through-space understanding. The structure of much larger molecular systems such as proteins and nucleic acid structures are also solved with the aid of the NOE. Nuclear Overhauser Spectroscopy (NOESY) that is used to study large complexes such as biomembranes and drug delivery systems are recognized as some of the most structurally revealing NMR spectroscopic experiments [28, 29]. Exploration of the NOE for large complexes is expanded in Section 6.4.4.

6.3 Practical aspects of NMR spectroscopy

6.3.1 Specimen preparation

Before embarking on describing the practical operation of the NMR spectrometer, it is essential to define some basic terms of the samples being studied by NMR spectroscopy. For the purpose of this chapter, the concept of sample and specimen is defined in the context of an analytical laboratory. A *primary* sample is taken from the bulk material. If the primary sample is not entirely analyzed, then by default an *aliquot* must be removed from the primary sample for analysis. The entire aliquot can be analyzed or a portion of the aliquot. The analyzed material is often referred to as the *specimen* [30]. For liquid NMR spectroscopy, the sample being studied in the spectrometer is the specimen that has been dissolved in solvent in an NMR tube. For solid-state NMR spectroscopy, the specimen is a solid, packed into a ceramic rotor. These specimens are manually inserted into the probe of the spectrometer's magnet. Although the quality and construction of NMR tubes and rotors are important for reducing magnetic field inhomogeneity during data acquisition, in practice disposable NMR tubes for liquid NMR spectroscopy are becoming more and more commonly used. When NMR tubes are to be cleaned and reused instead of disposed, it is important to not use oxidative cleaning agents that contain paramagnetic ions which may persist after rinsing, and to not overheat and distort the tubes during drying in the oven.

In solution NMR spectroscopy, the compound of interest is dissolved in a solvent that does not have resonances which would overlap in the spectral region of the compound. A number of deuterated solvents are available. Solvents which are water sensitive with exchangeable protons are also available commercially in single use ampoules. The presence of water in the sample or deuterated solvent often confounds spectral interpretation. The location of the water peaks found in common NMR solvents should be memorized to reduce the likelihood that they are mistakenly identified as peaks belonging to the analyte (Table 6.2) [31]. When selecting the solvent for the NMR experiment, choosing one which dissolves the compound and is relatively easy to remove for sample recovery, is preferred. For this reason, high boiling solvents such as DMSO-d6 and D_2O should be used only after careful consideration. D_2O solutions can be frozen and lyophilized so that the sample is recovered, albeit as a fluffy powder that may be difficult to isolate. The amount of solution added to an NMR tube depends on the type of tube and the probe size. An under-filled tube or a tube containing undissolved solids will affect the magnetic field homogeneity. For this reason, a device called a depth gauge tool exists to assist the user in properly adjusting the sample volume in the probe (see Section 6.3.2.3).

Table 6.2: Water location in common NMR solvents.

Deuterated NMR solvent	Formula	Residual water peak (ppm)
d-Chloroform	$CDCl_3$	1.56
d-DMSO	$(CD_3)_2S=O$	2.81–2.84
d-Acetone	$(CD_3)_2C=O$	3.30–3.33
d-Methanol	CD_3OD	4.87
Deuterium oxide	D_2O	4.79

6.3.2 Instrumentation

With industry consolidation of commercial NMR sources, the selection of NMR equipment, which had never been plentiful, has been reduced. With fewer options available, users and their institutions will have to adapt to only a few dominant players. Fortunately, NMR instrumentation is based around common equipment components, of which the core is the electromagnet itself, which supplies the magnetic field. Electronic components are needed to generate the magnetic pulse and to detect the signal from the sample. The sample itself is loaded into an NMR probe that locates the sample into a preferred position inside the electromagnetic. Smooth operation of all these interconnected parts is facilitated by computers, without which modern NMR spectroscopy would not be possible. The NMR technique for most end-users is now a software endeavor.

As mentioned previously in Section 6.2.2, most introductions to NMR spectroscopy will start with descriptions of continuous wave (CW) NMR spectrometers. The CW spectrometer keeps the magnetic field constant as it scans through frequency. However, these systems have largely been superseded by newer instrumentation that no longer scan through frequency but instead use radiofrequency pulses to simultaneously excite all nuclei. It is unlikely for scientists to encounter CW spectrometers, which are slow and lacking in sensitivity compared to modern instruments, except in educational settings. Far more commonly purchased by institutions are the modern (pulse) Fourier transform NMR spectrometers (FT-NMR). Fourier-transform NMR spectrometers use a pulse of radiofrequency (RF) radiation to flip all the nuclei in the external magnetic field into the higher energy state at the same time. This excitation to the higher energy alignment is possible according to the Heisenberg uncertainty principle, which stipulates that the frequency width of an RF pulse in the range of 1–10 µs is sufficiently broad to excite all nuclei simultaneously. The nuclei will re-emit RF radiation at their respective resonance frequencies as they decay back to the lower energy state. The decay creates a RF interference pattern as a function of time, defined as the free-induction decay (FID). Because each pulse and subsequent FID contains the energy decay information of all the nuclei in all their respective environments, each interference pattern contains the complete information from which a spectrum can be reconstructed. To alleviate the low sensitivity of NMR spectroscopy that was alluded to earlier, pulsing of the sample can be repeated many times, and the FIDs can be summed and averaged to improve the signal-to-noise (S/N) ratio of the collective FID. Nuclei such as ^{13}C and ^{15}N will normally require more scans and longer acquisition times than ^{1}H for this reason.

6.3.2.1 The magnet

Commercial systems have increasingly higher field strengths (>1 GHz for protons) and lower shielding requirements. The higher field superconducting magnets increase the difference in nuclear energy states, and consequently resulting in better resolution (in Hz or ppm). The downside is that higher field superconducting magnets cost significantly more than the lower field magnets and may require some extensive modifications to existing building infrastructure than their lower field counterparts. Nevertheless, shielding technologies continue to improve, and the five Gauss lines (safety perimeter) have steadily moved closer and closer to the magnet itself. The magnet is typically housed in a dewar-within-dewar layout. The magnet lies in a nonmagnetic metallic chamber that is bathed in liquid helium (4.2 K). The outer dewar is filled with liquid nitrogen which insulates the liquid helium.

The magnetic field generated from the NMR spectrometer is not constant, and field drifting is expected to some degree. The changing magnetic field can be referenced to an internal standard in the specimen being analyzed. For solution NMR spectroscopy, the choice is to use deuterated solvent and electronically lock the magnetic field to the deuterium resonances in the standard. The locking process is in itself an NMR experiment in which the lock transmitter is matched to the observed frequency of the deuterium nuclei found in the contents of the NMR tube (deuterium is usually found in the NMR solvent). When the lock transmitter is matched to the deuterium signal, the difference between the frequency generated by the lock transmitter and the observed frequency from the deuterated solvent is zero. Even if the deuterium signal is successfully locked at the beginning of an NMR experiment, the lock can be lost during data acquisition. This may occur when the magnetic field inhomogeneity is large, such as during the study of suspensions.

6.3.2.2 The console

The electronic components are packaged into a unit commonly referred to as "the console." The console will contain the pulse transmitters with frequency synthesizers that can be tuned to the resonances of multiple nuclei, and the power amplifiers. On the receiver side, preamplifiers and broadband receivers amplify and convert the analog signals into digital signals. A computer, nowadays mostly Windows- or Linux-based PCs, oversees the entire data acquisition operation.

6.3.2.3 The NMR probe

Perhaps, the second most described part of an NMR spectrometer after the magnet is the NMR probe. A typical probe consists of an inner coil for one nucleus and an outer coil for other nuclei. For example, HX probe would have one coil for the ^{1}H and the other coil for ^{13}C. The closer the coil to the sample is, the more sensitive the detection is. For an HX probe, where the weaker ^{13}C nucleus is X, the inner coil would be for X. Probe sizes can vary, and so NMR tubes (and rotors for SSNMR) come in various sizes. The liquid height of a standard NMR tube (inner diameter 4 mm) is approximately 3–4 cm when filled with 0.75 mL from an ampoule of deuterated solvent. The larger the probe is, the larger the amount of sample that can be studied and the stronger the signal are. Therefore, large probes are useful when plenty of sample are available, or when the concentration of the species intended for detection is low. When sample availability is limited, low volume NMR tubes can be used to analyze samples of only a few µL, although there is of course a reduction to detection sensitivity. New developments in cryo-probe technology reduce noise in detection, and allow for high sensitivity detection even for small sample sizes, increasing S/N by up to four times.

When purchasing and installing an NMR spectrometer, a depth gauge tool is provided to assist in finding the optimal level to push the NMR tube into the NMR tube holder (the "spinner turbine"). The depth is adjusted during experimental setup to deliver the sample to the most uniform region of the magnetic field. Knowledge of the probe dimensions is essential in setting the depth gauge because very often the depth gauge settings are taken for granted as being correct or invariable. In fact, not only will an improperly set gauge not place the NMR tube to deliver the sample to desired location within the probe it may even result in damage to the probe when the NMR tube depth is set too high so that the tube over extends into sensitive electronic components in the probe. Most end-users of an automated NMR spectrometer system will assume that the tube spinner is set to the proper height, which is false assumption if the magnet has been serviced without readjustment of the depth gauge.

Once the NMR tube (or rotor for solid state NMR spectroscopy) is inserted into the magnet, the further ability to optimize the signal-to-noise relies on electronic components such as the transmitter coils for supplying the resonance frequencies and the receiver coils. Optimization of the transmitter frequency in liquid probes in modern spectrometers takes place almost entirely in the "tune and match" routines of the software. Solid state NMR spectrometers still need, to some degree, the user to manually adjust the tune and match keys on the probe with the aid of an oscilloscope. The "tuning" procedure finds and positions the transmitter coil to the radiofrequency of the desired nucleus. Accurate tuning is necessary for increasing probe sensitivity, especially for nuclei of low natural abundance. The "matching" procedure sets the total resistance of the electronic components to maximize the radiofrequency going from the transmitter to the receiver. The tune and match routines are iterative because adjusting the tune will affect the match and vice versa. Whether using an oscilloscope or the spectrometer's software, the two routines are distinguished by a V-shaped curve, where the horizontal displacement from the center position is the tuning, and the vertical distance from the center position is the match. For two-dimensional NMR experiments, or proton decoupled experiments, the inner and outer coils must be tuned and matched separately. For liquid NMR spectroscopy, the tuning and matching is software-driven, and in automated spectrometers, the entire process is virtually invisible to the end-user. The user only knows there is a problem when the collected spectrum looks poor. Understanding of acquisition parameters is suggested in order to obtain quality data (see Table 6.3 for common parameters) [32].

Table 6.3: Typical NMR data acquisition parameters (1D proton).

Parameter	Description	Typical Values
TD	Time domain size	65536
DS	Dummy scans	4
NS	Number of scans	32
TD0	Number of averages in 1D	1
SW	Sweep width (ppm)	20 ppm
SWH	Sweep width (hertz)	6024 Hz
AQ	Acquisition time	4.53 sec
FIDRES {Hz}	Fid resolution	0.22 Hz
FW {Hz}	Filter width	125000 Hz

Table 6.3 (continued)

Parameter	Description	Typical Values
RG	Receiver gain	128
DW	Dwell time	83 µsec
DWOV	Oversampling dwell time	0.025 µsec
DECIM	Decimation rate of digital filter	3320
DE	Pre scan delay	6.5 µsec
P	Pulse (list of all pulses used, P0, P1, P2, etc)	P1 = 14.5 µsec
D	Delay (list of all delays used, D0, D1, D2, etc)	D1 = 1.0 sec
IN	Delay increment (list of all increments used, IN0, IN1, IN2, etc)	IN1 = 0 sec
HDDUTY %	Homodecoupling duty cycle	20%
PLW	Power level W(list of all power levels in watts, PLW1, PLW2, etc)	PLW1 = 6.084 watts
PLdB	Power level dB (list of all power levels in decibels, PLW1, PLW2, etc)	PLdB1 = −7.84 dB
PLSTEP	Step width for PL switching	0.1
GRADIENT	Gradient parameters(GPX, GPY, GPZ, GPnam)	0
AMP %	Amplitude of pulse (list of all amplitudes used, AMP0, AMP1, AMP2, etc)	AMP1 = 100%

6.3.2.4 Magnetic field and solvent frequency locking

Magnetic field inhomogeneity must be corrected in order to obtain the highest possible spectral resolution. Compensation to the inhomogeneity of the external magnetic field can be made by flowing current into shim coils that are part of the NMR probe. The shim coil will then apply a magnetic field to counteract magnetic field inhomogeneity in a number of user selectable directions – modern software offer a myriad of directions from the core Cartesian *xyz* directions to various higher order combinations. The amount of adjustment that is needed to make in any of these directions depends on the difference between the current settings of the shim coil gradient with the perceived magnetic field inhomogeneity. Like the tune and match routine, the shimming process is iterative. As the shim gradient improves field homogeneity, the lock signal will increase. The lock power may have to be dampened, and then the gradient coil shimmed again. The settings of the shim coil gradient are normally stored in computer memory, and can be recalled or deleted, depending on circumstance. When samples are very poorly prepared and contain solid particles, it is possible that no amount of shimming can compensate for the magnetic field inhomogeneity. One consequence of very poorly shimmed gradient coils is the appearance of spinning sidebands. This is not to say that suspensions cannot be run in liquid NMR mode, but that precautions should be taken to optimize the shims and that there should be no expectation for exceptionally good-looking data coming from such samples.

6.3.3 Data acquisition

With the availability of many adjustable data acquisition parameters in NMR software, the user is often staring at a confusing menu of dials, dialogue boxes, and variables. Copying the experimental template from one experiment to another is often a simple way to start, but knowing what the software is doing can be the difference between collecting useful data or meaningless data. This section offers some insight into the more salient variables which have a direct impact on the acquisition of good data. The practicalities of data processing and data interpretation, which are unique to each sample, are beyond the scope of this chapter.

A successful NMR experiment requires balancing the amount of time available for data collection against the resolution of the data. One of the first acquisition parameters presented in NMR control software is the number of dummy scans (DS). These are scans in which a FID is not recorded but all other acquisition parameters are utilized. Dummy scans are incorporated into the data acquisition routine mainly to ensure that the nuclei in the unknown sample return to full relaxation during the course of each measurement. Though they may seem as a "waste of time" to the inexperienced user, the DSs which return the nuclei to the fully relaxed state are essential for many types of NMR experiments.

Automated NMR spectrometers will often have a preset spectral resolution and a preset spectral width (in ppm). Because the x-axis is spectral resolution in frequency (v), two signals, which are separated by x_2-x_1 in Hz (Δv), require that data acquisition be at least as long as the reciprocal of Δv. Because frequency is measured in units of Hz, $1/\Delta v$ is the amount of needed time in seconds for data acquisition. Although acquiring for longer time improves resolution, there is no need to acquire for periods of time longer than necessary. While it may be prudent to acquire as wide a spectral width as possible, in practice, the user often knows what functional groups are present and can reduce the spectral width. A preset spectral width for ^{1}H-NMR spectra of ca. 12 ppm should cover most scenarios. Spectral width presets which should be sufficient for most compounds are 240 ppm (center at 100 ppm), 236 ppm (center at 100 ppm), and 603 ppm (center at 0 ppm) for ^{13}C, ^{19}F, ^{31}P, respectively. The spectral width is controlled by the number of sampling points (commonly taught as 8K, 16K, 32K, 64K scans). The larger the number of points, the larger the required computer memory to store the points, but modern computers no longer see memory as a limitation. The ppm can be related back to the time domain for sampling the FID by invoking Nyquist sampling, which specifies that identification of a sinusoidal frequency requires sampling twice within one cycle (twice in a period). A frequency of N requires a rate of 2N sampling. The reciprocal of the sampling rate is the dwell time, or the amount of time between samplings. The center of an NMR spectrum is determined by setting the transmitter off-set parameter.

The time required for data acquisition is important for several reasons. From a purely practical point of view, the availability of the spectrometer often dictates the time for a measurement. Departmental workhorse spectrometers suffer particularly from time constraints. The data acquisition time for a NMR experiment should not be set for too long a duration to complete the measurement – too long an acquisition time may result in the contents of the NMR tube chemically degrading over the course of the experiment. Fortunately, the acquisition time can be estimated by adjusting the sweep width and the number of collected data points. A working formula is

$$\text{Time}_{(acquisition)} = (NP)/(2*SW)$$

where NP = number of points, and SW = sweep width in Hz.

For historic reasons, NMR data storage and Fourier transform calculations of the data have used base 2, so the number of data points is usually referred to as 2^k data points. These data points can be treated as bytes of information, where 1024 bytes is 1 kilobyte, and the number of data points can be divided by 1024. For example, when k = 14, 16384 points are recorded (16384/1024 = 16K scans) and when k = 17, 131072 data points (131072/1024 = 128K scans) are recorded.

Once the number of points and sweep width for a scan have been set, the total number of scans (scans are also called *transients*) can be selected based upon the time available for data acquisition and the desire for sufficient signal-to-noise ratio. The signal amplitude increases by N whiles the noise ratio increases by √N.

$$S/N = n/n^{1/2} = n^{1/2} \text{ (for one scan)}$$

Therefore, the difference in the S/N ratio between experiments where NS = 8 and NS = 1024 is √(1024/8) = 11, even though there are 128x more scans for the latter experiment. Quadrupling the number of scans only brings about a doubling in S/N. For weak nuclei such as ^{15}N, doubling the S/N for an 8-h experiment would require 32 more hours of spectrometer time, which is not always possible in a facility with shared instrumentation. In such situations, the user should consider reducing the data collection time by analyzing isotopically enriched custom-made materials.

The S/N and time for an experiment also depend on the sensitivity of the electronic receiver of the spectrometer. The receiver gain (RG) is chosen to maximize detection sensitivity without overloading the electronics. Autogain functions are now built into most spectrometer software so that the user does not have to guess-and-test. Other parameters which influence data collection times are the delay (and pre-scan delay) parameters, which are mainly used to control the selected pulse program.

S/N can be influenced by digital manipulation of the way data are stored. Several possible digitizer modes are available for storage of the analog data. The primary mode used is oversampling followed by digital filtering. Many more points are collected during data collection that what is set by the user. The points are then reduced by digital filtration to the preset number of scans. Oversampling followed by filtration increases the S/N without increasing the storage needs for each FID.

To summarize some of these learnings, to observe differences in two NMR signals with frequencies v_1 and v_2, data acquisition must be carried out for at least $1/\Delta v$ (v is in Hz; correspondingly, $1/\Delta v$ is in seconds). The longer the data acquisition, the bigger the value of Δv and the finer the degree of separation of the NMR signals in Hz. For example, keeping all data collection parameters the same except for acquisition time, lengthening the duration from 4 to 16 s (4 s = $1/\Delta v$ where Δv = 0.25 Hz, and 16 s = $1/\Delta v$, where Δv is now = 0.0625 Hz) will allow signal separation to improve from 0.25 Hz to 0.0625 Hz. The peak detection in chemical shift ppm is also improved equivalently:

$$ppm = \frac{SW}{B}$$

where SW is in Hz and B is in MHz.

6.4 NMR spectroscopic experiments for large complexes

6.4.1 NMR Spectroscopy of bulk biomimetic nanoparticles

Traditional small molecule drugs rely on oral solid dosage forms, such as capsules and tablets, or liquid formulations, often suspensions, as the work horses of drug delivery. For biologics, the route of administration is by injection, with liquid formulations being the primary mode of drug delivery. Biological NMR spectroscopy has been widely used [33–36]. Recent advances in the development of completely new classes of drugs, such as siRNA or mRNA molecules, have altered the drug delivery landscape for therapeutics and vaccines [37–39]. To successfully deliver these non-protein macromolecules, carrier systems now include liposomes and complex lipid nanoparticles which must be biocompatible as complex products. The effectiveness of all these formulations depends on the drug encapsulation efficiency, which measures the amount of drug entrapped within the delivery particle: the encapsulation efficiency (often expressed as %EE) is defined as the ratio of the encapsulated drug to the total drug (encapsulated + free). The difficulty of making these measurements stems from the lack of techniques which can distinguish the encapsulated drug from unencapsulated drug without needing to physically remove the unencapsulated drug. Experimental techniques, such as NMR spectroscopy, which can quantitatively measure the true drug encapsulation of a complex formulation have significant advantages to determining true encapsulation.

Liposomes are a type of delivery vehicle which is modeled after natural phospholipid membranes. Liposomes can be described as generally spherical structures that utilize phospholipid layers to enclose a large solvent cavity which can hold drugs of interest [40–44]. Various classifications of liposomes are designated, from unilamellar vesicles (UMV) to multilamellar vesicles (MLV), with sizes ranging from nm to µm size [45]. Because very small vesicles have high surface curvature with a tendency toward differences between the outer and inner membrane layers [45, 46], large vesicles, with their relatively lower curvatures, have proven to be more similar to model biological membranes [40–46]. The disadvantage of using the larger liposomes is that their molecular tumbling is slow, and analysis by solution NMR spectroscopy tends to result in broad peaks [40–46]. A number of examples are available in the literature for the study of complex systems by high-resolution Magic Angle Spinning solid state NMR spectroscopy (HR-MAS) [47, 48]. The reader is encouraged to explore the application of SSNMR spectroscopy for the study of samples which display slow molecular tumbling, but which can be packed at high enough concentrations into a solid state rotor, and which can survive possible heating effects at high spinning speeds.

Larger aqueous soluble complexes can be studied using solution NMR spectroscopy. Any physical changes to the constituent components of these complexes, such as the formation of micelles, may result in different magnetic environments for the monomers, and manifest as changes to chemical shift. Hence, the transition from monomer to micelle may be monitored to obtain a critical micelle concentration (CMC) value. Those vehicular complexes which encapsulate metal-containing drugs or are large enough to entrap metallo-proteins will experience even larger changes to the chemical shifts owing to the presence of the metal. The relaxation of the protons of the vehicle can also exhibit changes to the T1 relaxation times. With careful analysis of T1 values and chemical shifts, the orientation of the entrapped species can be inferred [49].

6.4.2 Chemical shift and paramagnetic reagents

Chemical shift reagents have fallen out of fashion as a tool used in NMR spectroscopy. The stronger magnetic fields of new spectrometers have improved peak resolution; chemical shift reagents, which were once the only means of separating peaks, have lost much of their raison d'être [50]. Nevertheless, there is still a place for chemical shift reagents in studying large complexes. Chemical shift reagents function by changing the radiofrequency absorption of nuclei which they affect, causing the molecule of interest to have a spectrum that has peaks distributed over a larger frequency range. Many chemical shift reagents are able to change the absorption frequency of near neighbors because of lone pair electron interactions with nearby nuclei. In addition to interacting with the molecule of interest without chemically altering it, the chemical shift reagent should also have few or no resonances in the desired spectral range (Table 6.4), and be soluble and stable in the environment for its planned use. Many of the commercially available chemical shift reagents utilize the chirality of camphor and the lone pairs of a rare metal. Commercial chiral shift kits are also available.

Table 6.4: Selected commercially available shift reagents.

Chemical shift reagent
Europium tris[3-(heptafluoropropylhydroxymethylene)-(+)-camphorate]
Europium tris[3-(trifluoromethylhydroxymethylene)-(–)-camphorate]
Praseodymium(III) tris[3-(heptafluoropropylhydroxymethylene)-d-camphorate]

Even if the use of chemical shift reagents has dropped over the years, their unique signatures in altering the absorption frequencies of nearby nuclei enable them to answer specific questions. For example, the configuration of surfactant molecules in SDS micelles was deduced by comparing the different relaxation rates of the SDS molecule when it encounters paramagnetic ions that are introduced into the aqueous environment [51]. By changing the resonance properties of the species of interest, paramagnetic metals and chemical shift reagents are capable of distinguishing whether a molecular component is on the outside or inside of a large complex. The encapsulation efficiency of drug delivery systems can be monitored by selecting the appropriate shift reagent. One further advantage is that some chemical shift reagents are environmentally sensitive, so altering the pH or temperature can induce changes in the NMR spectrum.

The protons of drug molecules within the carrier system and free floating unencapsulated in the formulation are often so similar in absorption frequency that their chemical shifts are indistinguishable even with extremely powerful magnets. Chemical shift reagents have been found to be particularly useful in separating the protons of amino acid drugs in an internal liposomal environment versus the external environment. Without the chemical shift reagent thulium(III)-1,4,7,10-tetraazacyclododecane-1,4,7,10-tetra(methylene phosphonic acid sodium salt), the proton signals of encapsulated and free homocarnosine were almost indistinguishable, but with the addition of 0.3 mM of the chemical shift reagent, the proton resonances of the imidazole were very clearly separated by 0.05 ppm or more [52]. Moreover, ^{1}H NMR spectroscopy is inherently quantitative, so integration of the resolved peaks gives a direct ratio of the encapsulation efficiency. For species which exchange on the NMR timescale, the appearance of lanthanide-induced chemical shifts overtime is a powerful method for monitoring dynamics [53].

6.4.3 Porosity and surface structure using 1D and 2D diffusion NMR spectroscopy

Diffusion Ordered SpectroscopY (DOSY) NMR using pulsed gradient spin echo (PGSE) is exceptionally powerful in studying translational motion directly in biological fluids and under simulated physiological conditions. There are three general steps in such a pulse field gradient experiment: (1) scatter the NMR signal, (2) allow time for molecular diffusion, and (3) refocus the NMR signal (Figure 6.4). A molecule that moves quickly will be farther away from its original position than a molecule that diffuses slowly. The molecule which has moved farther away from its original position will not be refocused as well in a gradient field, resulting in a decrease in signal intensity. Slower moving molecules will refocus better because they are nearer to their original position, resulting in a higher intensity (Figure 6.5).

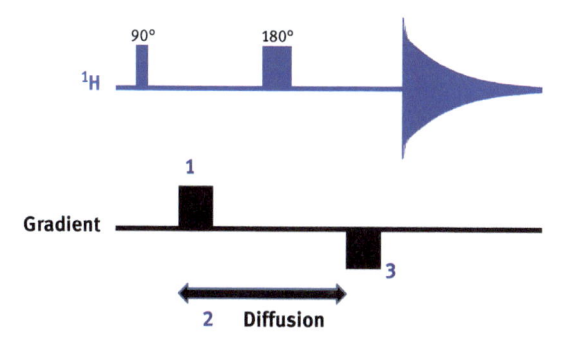

Figure 6.4: The three steps to monitor movement of molecules in a pulse field gradient experiment: (1) Scatter the signal, (2) Allow time for diffusion, (3) Refocus the signal.

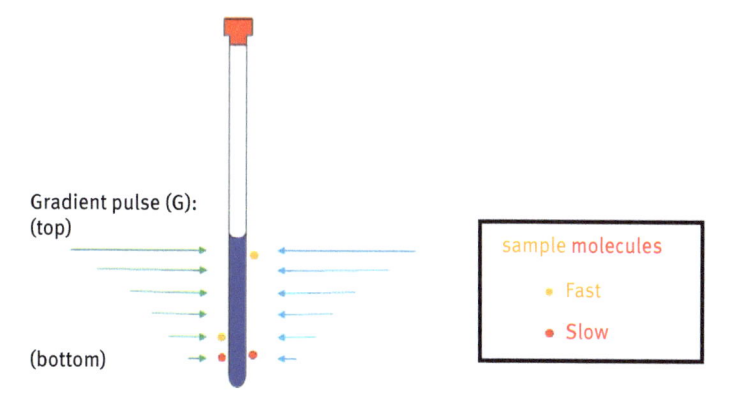

Figure 6.5: Pulse field gradient in an NMR sample tube. If molecule has moved quickly from its original position, it will not be refocused well, resulting in a decrease in signal intensity. Slower moving molecules will refocus well, resulting in a higher intensity.

The DOSY experiment allows for determination of the diffusion coefficient for different nuclei. This is achieved through the application of magnetic field gradients as described. A series of 1D experiments are collected with varying gradient strength. For each peak, the decreasing signal intensity can be plotted at each gradient step to yield a curve (Figure 6.6), which is fit to decay algorithms to extract the diffusion coefficient. Suggested starting parameters for data collection are found in Table 6.5 for a DOSY experiment [32]. The data from such a diffusion experiment are plotted as a pseudo-2D diagram, "pseudo" because the Y-axis is not an NMR measurement but contains the determined diffusion coefficient values. The data can be plotted on a linear scale, although they are more commonly plotted on a logarithmic scale because diffusion rates can vary by orders of magnitude.

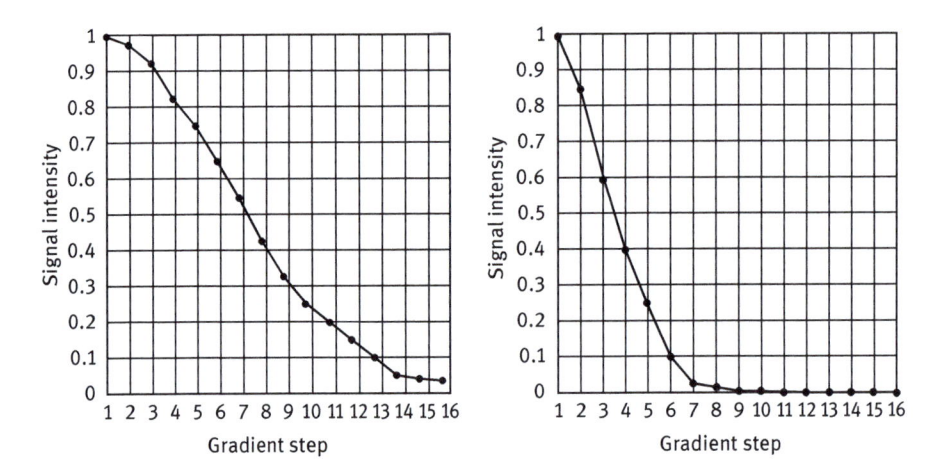

Figure 6.6: Depiction of molecular translation during an NMR diffusion experiment (slow diffusion on left, fast diffusion on right).

The diffusion rate of each peak (individual nuclei) can be read from the pseudo-2D diagram. Typically, when nuclei are on the same molecule, they have the same diffusion rate. The diffusion coefficients from the pulse field experiment can be used (crudely) in the Stokes–Einstein equation to solve for particle radius r:

$$r = \frac{k_B T}{6\pi \eta D}$$

where k_B is the Boltzmann's constant, T = temperature (K), η = dynamic viscosity of the solvent at temperature T, and D is the diffusion coefficient from the NMR experiment (m^2/s).

Table 6.5: Typical NMR data acquisition parameters (DOSY).

Parameter	Description	Typical Values (F2 dimension)	Typical Values (F2 dimension)
TD	Time domain size	5120	16
DS	Dummy scans	4	
NS	Number of scans	32	
TD0	Number of averages in 1D	1	
SW	Sweep width (ppm)	5.63 ppm	10 ppm
SWH	Sweep width (hertz)	1694.9 Hz	3012 Hz
AQ	Acquisition time	1.51 sec	0.0026 sec
FIDRES {Hz}	Fid resolution	0.662 Hz	376.5 Hz
FW {Hz}	Filter width	125000 Hz	
RG	Receiver gain	11.3	
DW	Dwell time	295 µsec	
DWOV	Oversampling dwell time	0.025 µsec	
DECIM	Decimation rate of digital filter	11800	
DE	Pre scan delay	6.5 µsec	
P	Pulse (list of all pulses used, P0, P1, P2, etc)	P1 = 14.5 µsec	
D	Delay (list of all delays used, D0, D1, D2, etc)	D1 = 1.0 sec	
IN	Delay increment(list of all increments used, IN0, IN1, IN2, etc)	IN1 = 0 sec	
HDDUTY %	Homodecoupling duty cycle	20%	
PLW	Power level W (list of all power levels in watts, PLW1, PLW2, etc)	PLW1 = 6.084 watts	
PLdB	Power level dB (list of all power levels in decibels, PLW1, PLW2, etc)	PLdB1 = −7.84 dB	
PLSTEP	Step width for PL switching	0.1	
GRADIENT	Gradient parameters(GPX, GPY, GPZ, GPnam)	GPZ6 = 100%, GPZ7 = −17.13%, GPZ8 = −13.17% GPnam = SMSQ10.100	
AMP %	Amplitude of pulse (list of all amplitudes used, AMP0, AMP1, AMP2, etc)	AMP1 = 100%	

A faster diffusion rate D corresponds to a smaller size r. In terms of molecular context, a molecule with a higher molecular weight will also be larger in size (diameter) and is expected to have a slower diffusion rate (Figure 6.7).

PFG-NMR has been used to measure the diffusion of solvents from porous polymer nanoparticles or of shedding components from biomimetic nanoparticles. The diffusion decay curve may also reveal complex internal morphology

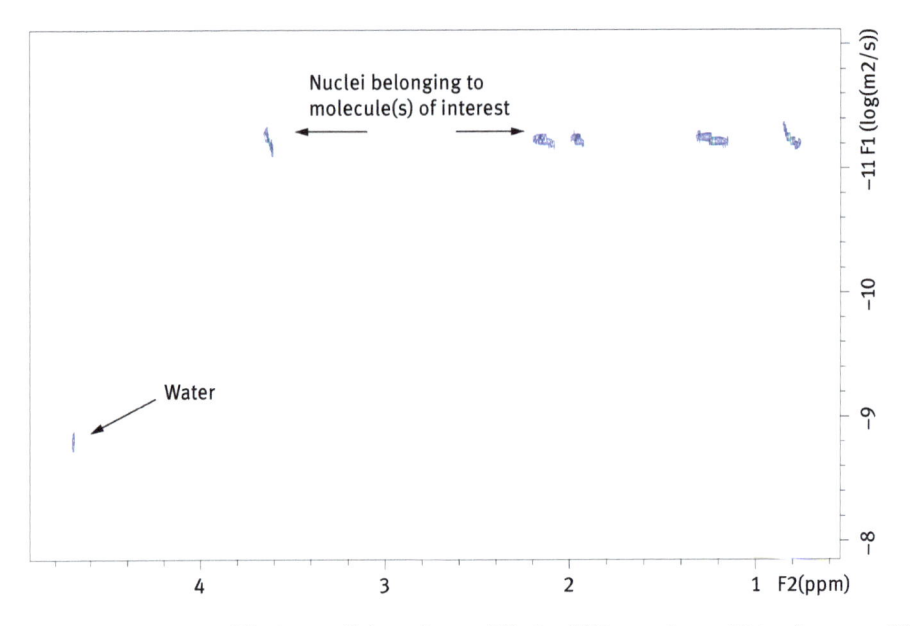

Figure 6.7: Capturing diffusion coefficients from a diffusion NMR experiment. Water, because of its small size, diffuses fast and therefore has a smaller diffusion coefficient that is measurable by a diffusion ordered (DOSY) NMR experiment.

of particles, and lead to conclusions on restricted diffusivity even when nanoparticles appear to homogenously porous. More simply, one can imagine that a drug in a complex nanoparticle vehicle will have a higher diffusion rate when the drug is free, and a slower diffusion rate when it is associated with the nanoparticle. The reasoning for this assumption is that the diffusion coefficient of a species is related to its molecular weight (and indirectly physical size); therefore, a nanoparticle-bound or encapsulated drug is larger and slower moving than the free drug. As stated previously, chemical shift reagents are useful in monitoring internal versus external species, but because PFG NMR diffusion methods are direct and do not alter the chemical system, they have been used extensively for studying molecular translation, which can be particularly useful for the study of colloids [54, 55].

The diffusion of small molecules or solvent in large complexes may have significant effect on the observed activity of such complexes. The rate of component exchange between the intravehicular space and the external bulk environment in liposomal formulations for drug delivery is one parameter that can be obtained using PFG NMR spectroscopy [56]. One example of a PFG study that distinguishes internalized molecules versus external (or bulk) species is the analysis of sodium ions in vesicular formulations. At least two diffusion coefficients are derived (9.04×10^{-10} m^2/s and 2.65×10^{-12} m^2/s, respectively), which are highly suggestive of the presence of two sodium ion populations, one that is free in solution and the other confined or associated with the vesicles [57].

In heterogeneous systems such as suspensions, internal magnetic field gradients may arise because of differences in magnetic susceptibility among the different types of materials being studied. The errors introduced into the diffusion coefficients from the coupling between the applied and internal magnetic field gradients can be reduced or eliminated by introducing bipolar gradients in the pulse sequence, and a comparison of the various methods to compensate for such errors are discussed elsewhere [58–61].

Instead of devising characterization methods for components that diffuse away from biomimetic nanoparticle complexes such as protected liposomes, the study of components which do not shed is equally important. The bioperformance of nanoparticles such as liposomes for drug delivery can be significantly improved by coating these nanoparticles with polymers such as polyethylene glycol (PEG) and PEG-like polymer derivatives [62]. Because PEG polymers are large, they have the effect of physically stabilizing the nanoparticles through interparticle steric repulsions, prolonging the in-use lifetime of the formulation. The PEG layer that surrounds the nanoparticles also appears to reduce immune system recognition. Consequently, uptake by the mononuclear phagocyte system (MPS) is lessened, and the circulation time of the nanoparticles in blood is extended [63–66]. While the exact mechanism of how PEG polymers on the surface of nanoparticles sustains the circulation still requires study, the improved pharmacokinetic properties using PEG polymers in colloidal systems is known to depend on the molecular weight of the PEG and the surface density of the PEG polymers on the nanoparticle [67–70]. Quantifying the amount of PEG polymer on the carrier is an important question that can be answered using proton NMR spectroscopy because the large number of nearly equivalent protons and the simplicity of the PEG proton spectrum make it easily recognizable in a ^{1}H NMR spectrum. Separation of these peaks from other components in the nanoparticle system is possible with simple 2D homonuclear correlation NMR experiments (e.g., COSY, TOCSY), which are available through automation [32]. For PEG-derivatives which are chemically grafted to the surface of a liposomal nanoparticle, the quantification of the PEG-derivative can be been demonstrated reproducibly by purification of the liposomes to remove unreacted starting materials (i.e. ungrafted components), and then studying the NMR spectra of the purified liposomes. Once the amount of ungrafted PEG polymer in the removed fraction is quantified by a similar fashion, the ratio between the grafted versus ungrafted PEG molecules allows the estimation of surface PEG on the nanoparticles. The goal is to ascertain whether successful chemical grafting can be linked to the shielding properties of PEG polymers for improving the bioperformance of novel drug carriers.

6.4.4 Near neighbors and structure determination using NOE

Successful performance of biomembranes and other complex biomolecular systems often require conformational, surface, or structural changes in response to the dynamic environment composed of proteins and biological fluids. NMR spectroscopy is able to discern a range of time scales, so fast bond librations (rates in picoseconds to nanoseconds), slow rigid body rotational diffusion (rates ranging from a few to tens of nanoseconds), and even slower translational motions can be captured. Solution NMR spectroscopy can be used to study surface protons in complex systems where the bulk protons are mobility-restricted and therefore not motionally averaged. The fast rotational motion of the organic functional groups on the surface of a large slowly tumbling complex system means that the reduction in line-broadening of the surface protons allows for these nuclei to be detectable. The mobile surface protons are observed by solution NMR spectroscopy as sharp peaks, and the immobile protons, if observed at all, are broad or poorly defined peaks. Solution NMR techniques have been used to selectively probe the surface structure and composition of functionalized nanoparticles in colloidal solutions [71]. These NMR experiments which monitor changes on the NMR timescale rely on relaxation of nuclei that is induced from the local field fluctuations occurring by molecular motion. For heteronuclear HX spin systems, spin-lattice T1 relaxation (longitudinal magnetization in the z-direction returning to equilibrium) and spin-spin T2 relaxation (transverse magnetization in the xy-plane returning to equilibrium) are the standard relaxation phenomena. For liquid samples, measuring the T2 spin–spin/transverse relaxation time, which represents the loss of coherence as individual nuclear spins experience different resonance frequencies from the local magnetic field, and the steady-state NOE, which represents the changes in the magnetization of one spin species when a second spin is saturated, are very useful experiments [72–75].

The NOESY experiment gives correlation between two nuclei that are within approximately 6 Å distance to each other. As this is a through-space correlation, whether the nuclei are bonded does not matter. Small and large molecules can be studied using NOESY. For hard-to-crystallize biopolymers such as RNA or membrane proteins, NOESY experiments are essential. Elucidation of NOESY data for RNA begins with the assignment of all essential protons in a sequential fashion, termed the "NOESY walk." For each nucleotide, the H1' is close to its own H8 (purine) or H6 (pyrimidine) protons, yielding a NOESY cross-peak. Using the RNA sequence A1-U2-C3-G4-U5-G6 as an example, the A1-H1' and A1-H8 have an intranucleotide correlation, giving one peak in the NOESY spectrum. In the 5' to 3' direction, the H1' of a nucleotide is also close to the next H8/H6 resulting in an internucleotide peak (A1-H1', U2-H6). This pattern continues throughout the RNA strand, alternating between intra- and internucleotide cross-peaks, allowing for the generation of sequential connectivity (Figure 6.8). In addition to the secondary structure connections, the NOESY spectra also contain cross-peaks arising from the tertiary structure of the

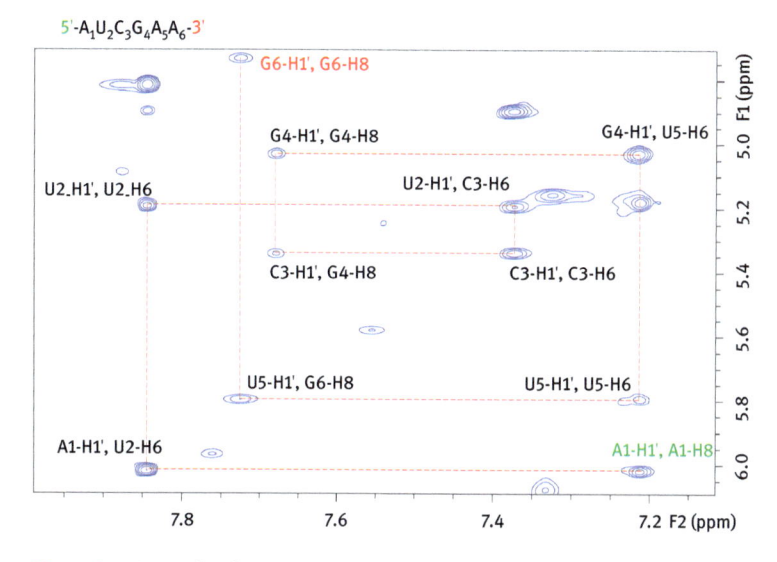

Figure 6.8: Example of a "NOESY walk" to elucidate higher order structure of biomolecules.

molecule. The intensities of the peaks are used to gauge the distance between the two correlated protons. These distance values are used as input data for modeling software to generate the overall RNA structure.

NOEs can probe intramolecular interactions and intermolecular interactions between neighboring molecules in large complexes such as liposomes and biomembranes. To reduce spin diffusion effects, mixing times must be optimized – overly long mixing times will cause strong spin diffusion cross peaks, and these must be recognized and disregarded when necessary. For large systems such as micelles or rigid nanoparticle systems, even short mixing times may be sufficient to cause significant spin diffusion. Best practice for such experiments would be to systematically increase contact times and observe the spectra for the presence of an identifiable residual cross-peak which is associated with spin diffusion. The disadvantage of using very short mixing times is that the data will only show the strongest cross-peaks which correspond to short distances (< 4 Å). These distances will not paint a complete picture of the molecular neighborhood, although may be sufficient to assist in building a crude structural model.

The system described and depicted in the DOSY experiment in Figure 6.7 in Section 6.4.3 serves as a useful illustration of NOESY in determining the formation of a model biomembrane. The DOSY experiment has indicated that there are two species detectable, water and another species. In fact, this sample consists of three components, two species of phospholipids, DOPS and DOPE, and water. The sample was prepared by mixing and agitating these phospholipids together in chloroform and then drying the mixture to a film. The film was redispersed in water to achieve a

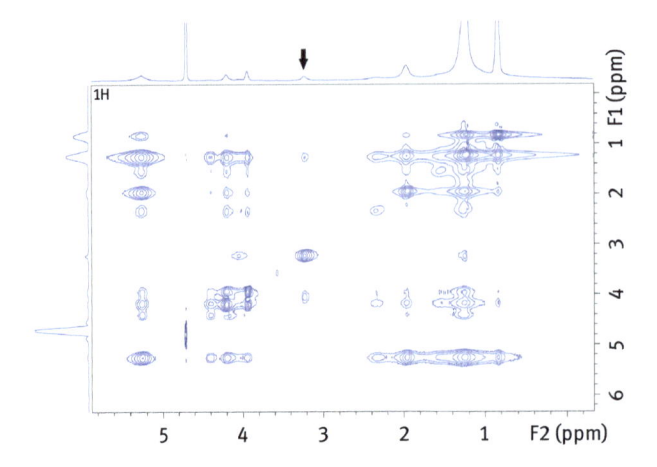

Figure 6.9: The NOESY spectrum of a two component liposome, with the peak at 3.2 ppm (arrow) being the only peak belonging to one component. Because this peak has more than one cross-peak, it is in close proximity to the second component.

clear solution, which is the state of the material used to measure the DOSY (Figure 6.7) and NOESY (Figure 6.9) spectra. From observing the individual ^{1}H NMR spectra of the two phospholipids, the signal at 3.2 ppm in the NOESY spectrum is known to belong to one component, DOPE. DOPE has no other signals in the mixture, suggestive that the DOPE is in a constrained environment where significant chemical shift anisotropy results in no other signal in solution NMR spectroscopy. The signal at 3.2 ppm has a strong cross-peak with itself along the diagonal of the 2D plot and has also two other cross-peaks. Therefore, this NOESY experiment reveals that these two components are in close contact. If we take into consideration that the DOSY data show the two components are diffusing at the same rate, then what the NOESY spectrum is revealing is that the DOPE and DOPS are diffusing *together* within Ångstrom level proximity. In fact, the procedure used to prepare the sample is typical for making liposomes. These two pieces of NMR data are persuasive evidence that the phospholipids are complexed together in close proximity, consistent with a liposome model.

6.4.4.1 Transfer NOE

Large molecules have short correlation times; they build up NOE rapidly and also have extensive spin diffusion. Small molecules build up NOE slowly and have relatively low spin diffusion because of their rigidity. When a small molecule is bound to a large molecule, and where free small molecules are also present in bulk, a NOESY experiment should show NOEs within the bound small molecule accumulating quickly, whereas the NOEs within the free small molecules accumulating slower. Exchange of the small molecule between the bound and free-state will show peaks in the spectrum with sharp peaks at the chemical shifts of the free small molecule, yet have NOEs representative of the bound state (small molecule signal that has been attenuated from the spin diffusion) [76–78].

6.4.4.2 Saturation transfer difference NOE

For large molecules which have fast spin diffusion, saturation of a single large molecule resonance will soon lead to complete or near complete saturation of all the large molecule resonances, regardless of chemical shift. When the NOE is transferred to a bound small molecule, and a difference spectrum is obtained by subtracting the bound state spectrum from the free-state spectrum, the resulting spectrum should illustrate signals of only the small molecules that dissociate from the large molecule faster than the T1 relaxation time of the free small molecule. Extending this concept of saturation transfers further, it may even be possible to locate which part of the small molecule is most proximal to the large molecule because the closest part of the small molecule will have benefited the most from the NOE transfer and therefore have the largest NOEs [79].

6.4.5 Solvent exchange and penetration of small molecules into biomembranes and nanoparticles

Water molecules in the bulk act as small molecules and have quick translational diffusion, which results in short correlation times. Water molecules which are hydrogen-bonded to large molecules or which are sequestered in large molecules may have less translational diffusion, and therefore have longer correlation times. The NOEs from the large molecule to bulk water molecules have short correlation times, whereas the NOEs of bound water molecules have longer correlation times, and are more "large molecule like" [80–84].

Concepts from magnetic resonance imaging (MRI) and solvent relaxation can be applied to the study of water in large complexes which contain solvent. Proton relaxometry can be understood as a rapid exchange between free solvent molecules (water) that have long T2 relaxation times and bound solvent molecules which have shorter T2 relaxation times [85–87]. An averaged relaxation rate between the free and bound states lends itself to the interpretation that the shorter the overall solvent T2 relaxation time, the greater the contribution from bound solvent molecules. The bound state can be any environment that constrains the water molecule. Water molecules within pores of nanoparticles or membranes would show a different relaxation time compared to free water in the bulk, and would also be amenable to solvent relaxation analysis by NMR spectroscopy [88]. Studies of silica nanoparticles which have been titrated with polymer indicate that the water T2 relaxation value decreases with increasing concentration of polymer [89]. This would mean that water is increasingly bound to the surface of the silica at higher polymer concentration, or that the residence time of the water molecule constrained on the surface is longer during the NMR experiment. A structural understanding of the adsorbed surface water layer can be developed from the relaxation data.

6.4.6 Proteins and other components of biological membranes

One of the most difficult areas of study by any methodology is the structure and dynamics of membrane components, in particular membrane proteins, which include transmembrane proteins, transporters, ion channels, and so forth [90–94]. Membrane proteins constitute a significant portion of all proteins translated by mRNA from the organism's genome, and yet of the ca. 120,000 proteins solved structures of proteins, no more than a few percent are membrane proteins [93]. The disproportionate successes are indications of the inherent difficulty in solving questions of membrane structure. Membrane components are stabilized by neighbors and prefer the hydrophobic environment – there is low propensity for these components to crystallize, and even if they do, they may not adopt their nature conformation. The interactions of each membrane protein with its surrounding lipid bilayer is likely vital to the protein folding and stability, and therefore the traditional method of protein characterization by X-ray crystallography which involves removing the protein from the membrane and then crystallizing the protein will analyze a protein that has been severely perturbed from its natural folding. NMR spectroscopy has exceptional potential to tackle membrane structure and behavior because it has no need to crystallize the components of interest and may therefore study molecules in environments similar, if not identical to the native environment [94]. Though it must still overcome various technical challenges such as spectral crowding, solid-state NMR spectroscopy seems to be even more suited for elucidating membrane structure because it can study the protein with very little processing [95]. An abundance of solution NMR methods are also available to interrogate these protein–membrane complexes, though there are molecular tumbling rate limitations (correlation times <100 ns). Regardless of which NMR approach to take, several steps are required to prepare the protein for analysis.

Analyzing components of actual biomembranes by NMR requires isolation of the component using a detergent to disrupt the cellular membrane, followed by isolation and purification. The screening of the most compatible detergents, and their concentrations, is usually performed [93, 96, 97]. The advantage of the detergent-solubilized state is that the protein complex can tumble more rapidly, although solution NMR methods might still be inapplicable. A number of phospholipid detergents have been used to solubilize and capture proteins, among which are dihexanoyl phosphatidylcholine, diheptanoyl phosphatidylcholine, and DMPC [93]. Bicelles, which are ellipsoid in shape that depends on mixture between long and short phosphoplipids, can be tuned through composition to be flatter, and therefore more bilayer-like akin to a true biological membrane [94]. The solubilizing step of the protein into a lipid bilayer (e.g., common surface-active aggregates like micelle, bicelles, etc.) brings the protein into a native-like environment. The downside of this constrained environment is that the mobility of the protein is restricted, possibly to the point at which molecular tumbling is very slow.

To reduce NMR data acquisition time, the component of interest should be isotopically enriched. In this scenario, when the component of interest is a protein, the process would encompass: protein biosynthesis in the presence of labeled ^{15}N or ^{13}C amino acids in E. coli, isolation and purification of the labeled protein, and then solubilization of the protein into a micelle to mimic the native cellular membrane [93, 96–98]. It must be remembered that milligrams of protein are normally required for NMR spectroscopic analysis, so some form of overexpression is needed to produce the protein in sufficient amounts. With sufficient sensitivity, protein–protein interactions can be analyzed [99–102].

As mentioned previously, a specific limitation to solution NMR methods for the study of protein in micelle complexes is the slow tumbling of the protein in its constrained environment. As a result, some types of proteins may be easier to analyze than others. For example, globular proteins, which can be considered to have less anisotropy, are more amenable to structural determination. For transmembrane proteins which can embed and extend beyond the outer surface of the micelle (or other bilayer model membrane), chemical shift or paramagnetic agents which cannot penetrate into the bilayer can be used to distinguish which amino acids are outside of the bilayer [97]. The amino acids inside and outside of the bilayer will be shifted or broadened by the addition of these reagents (e. g., Mn^{2+}). Another strategy to improve spectral properties is to achieve narrower peaks by increased deuteration; the deuterons attached to ^{13}C nuclei will alter the relaxation rate of ^{13}C. However, the increased number of deuterons means that there are fewer opportunities to exploit ^{1}H-N or ^{1}H-C NOEs. When phosphatidyl headgroup-containing lipidic micelles or bicelles are used to capture proteins for NMR studies, ^{31}P NMR can be used to study the phospholipid interactions with the protein [103]. The actual NMR methods employed to study structure and dynamics of these materials are based upon the homonuclear and heteronuclear NOE methods discussed in Section 6.4.4, reiterating the importance of appreciating the versatility of the NOE [72–80].

6.5 Outlook

NMR spectroscopy has truly been a transformational technique that modern chemists cannot live without. It provides exquisite detail about the nuclear environment of any molecular system with little if any perturbation of the system during analysis. For characterization of complex structures such as biomembranes and biomimetic nanoparticulate systems which may not be stable when taken out of the formulation, NMR spectroscopy is perhaps the only analytical technique that can elucidate structure and dynamics *in situ*. The utility of NMR spectroscopy can be broadened by training researchers in the routine use of NMR, and not just for the purpose of chemical elucidation needed for the daily activity in an organic chemistry laboratory. One ramification of practicing scientists indoctrinated with the belief that NMR spectroscopy is either accessible (routine use) or very difficult (requiring the aid of the spectroscopist) is that the understanding of the capability of the technique beyond routine usage is murky. As a result, only small selections of NMR techniques are known and widely adopted by the majority of users. The conceptual obstruction exists is sometimes manifested as a technical barrier, but in reality this is an awareness barrier. The perception that the more elaborate NMR experiments are beyond the capability of nonspecialized scientists can be easily overcome. The first steps are in learning more about NMR spectroscopy and in establishing fruitful NMR collaborations with spectroscopists who are enthusiastic in tackling problems of complexity and relevance.

References

[1] Bloch, F. Nuclear induction. Phys. Rev. 1946, 70, 460–474.
[2] Bloch, F., Hansen, W.W., & Packard, M. The nuclear induction experiment. Phys. Rev. 1946, 70, 474–485.
[3] Purcell, E.M., Torrey, H.C., & Pound, R.V. Resonance absorption by nuclear magnetic moments in a solid. Phys. Rev. 1946, 69, 37–38.
[4] Dong, R.Y. Nuclear magnetic resonance of liquid crystals. New York, NY, USA, Springer, 1997.
[5] Fujisawa, S., Kadoma, Y., & Komoda, Y. ^{1}H and ^{13}C NMR studies of the interaction of eugenol, phenol, and triethyleneglycol dimethacrylate with phospholipid liposomes as a model system for odontoblast membranes. J. Dent. Res. 1988, 67, 1438–1441.
[6] Chapman, D., Kamat, V.B., De Gier, J., & Penkett, S.A. Nuclear magnetic resonance studies of erythrocyte membranes. J. Mol. Biol. 1968, 31, 101–114.
[7] Ashbrook, S.E., & Dawson, D.M., NMR spectroscopy of minerals and allied materials. In: V. Ramesh, ed. Nuclear Magnetic Resonance Vol 45, Cambridge, UK, Royal Society of Chemistry, 2016, 1–52.
[8] Terban MW, Cheung EY, Krolikowski P, Billinge SLJ. Recrystallization, Phase Composition, and local structure of amorphous lactose from the total scattering pair distribution function. Cryst. Growth Des. 2016, 16, 210–220.

[9]   Marbella, L.E., & Millstone, J.E. NMR techniques for noble metal nanoparticles. Chem. Mater. 2015, 27, 2721–2739.
[10] Cash, D.D., Cohen-Zontag, O., Kim N.K., Shefer K., Brown Y., Ulyanov N.B., Tzfati Y., & Feigon J. Pyrimidine motif triple helix in the kluyveromyces lactis telomerase RNA pseudoknot is essential for function in vivo. Proc. Nat. Acad. Sci. 2013, 110, 10970–10975.
[11] Andrew, E.R., & Szczesniak, E. A historical account of NMR in the solid state. Progr. Nucl. Magn. Reson. Spectrosc. 1995, 28, 11–36.
[12] Emsley, J.W., & Feeney, J. Milestones in the first fifty years of NMR. Progr. Nucl. Magn. Reson. Spectrosc. 1995, 28, 1–9.
[13] Günther, H. NMR Spectroscopy: Basic Principles, Concepts and Applications in Chemistry 3rd Edition. Weinheim, Germany, Wiley-VCH, 2013.
[14] Richards, S.A. Essential practical NMR for organic chemistry. New York, USA, Wiley, 2010.
[15] Berger, S., & Braun, S. 200 and More NMR Experiments: A Practical Course. New York, USA, Wiley, 2004.
[16] Mirau, P.K. A Practical Guide to Understanding the NMR of Polymers. New York, USA, Wiley, 2004.
[17] For a light-hearted text on NMR spectroscopy, see Doucleff M, Hatcher-Skeers, M, Crane NJ. Pocket guide to biomolecular NMR. New York, USA, Springer-Verlag, 2011.
[18] Apperley, D.C., Harris, R.K., & Hodgkinson, P. Solid state NMR: Basic principles and practice. New York, USA, Momentum Press, 2012.
[19] Duer, M.J. Introduction to solid-state NMR spectroscopy. Oxford, UK, Blackwell Publishing, 2004.
[20] Newton, I. Opticks: Or a treatise of the reflections, refractions, inflections and colours of light. 4th Edition London 1730. Dover Publications, 2012.
[21] Skoog, D.A., & Leary, J.J. Principles of instrumental analysis 4th ed. New York, USA, Saunders College Publishing, 1992, 312.
[22] Several web-based tools and commercial software are available for ^{1}H and ^{13}C prediction. Chemical drawing programs have built-in, but optional, NMR prediction tools.
[23] Agrahari, V., Meng, J., Purohit, S.S., Oyler, N.A., & Youan, B.C. qNMR real-time analysis of tenofovir release kinetics using quantitative phosphorus (^{31}P) nuclear magnetic resonance spectroscopy. J. Pharm. Sci. 2017, 106, 3005–3015.
[24] Wawer, I., & Diehl, B. NMR spectroscopy in pharmaceutical analysis. New York, USA, Elsevier, 2008.
[25] Holzgrabe, U., Deubner, R., Schollmayer, C., & Waibel, B. Quantitative NMR spectroscopy - applications in drug analysis. J. Pharm. Biomed. Anal. 2005, 38, 806–812.
[26] Griffiths, L., & Irving, A.M. Assay by nuclear magnetic resonance spectroscopy: Quantification limits. Analyst. 1998, 123, 1061–1068.
[27] Saito, T., Nakaie, S., Kinoshita, M., Ihara, T., Kinugasa, S., Nomura, A., & Maeda, T. Practical guide to accurate quantitative solution state NMR analysis. Metrologia. 2004, 41, 213–218.
[28] Neuhaus, D., & Williamson, M.P. The Nuclear Overhauser Effect in structural and conformational analysis. Weinheim, Germany, Wiley-VCH, 2000.
[29] Iwahara, J., Wojciak, J.M., & Clubb, R.T. Improved NMR spectra of a protein–DNA complex through rational mutagenesis and the application of a sensitivity optimized isotope-filtered NOESY experiment. J. Biomol. NMR. 2001, 19, 231–241.
[30] Buhrke, V.E., Jenkins, R., & Smith, D.K. A practical guide for the preparation of specimens for X-ray fluorescence and X-ray diffraction analysis. New York, Wiley-VCH, 1998.
[31] Gottlieb, H.E., Kotlyar, V., & Nudelman, A. NMR chemical shifts of common laboratory solvents as trace impurities. J. Org. Chem. 1997, 62, 7512–7515.
[32] Topspin software version 3.5 from Bruker Biospin, Billerica, USA.

[33] Marion, D. An introduction to biological NMR spectroscopy. Mol. Cell. Proteomics. 2013, 12, 3006–3025.
[34] Evans, J.N.S. Biomolecular NMR Spectroscopy. Oxford, UK, Oxford University Press, 1995.
[35] Cohen, J.S., Jaroszewski, J.W., Kaplan, O., Ruiz-Cabello, J., & Collier, S.W. A history of biological applications of NMR spectroscopy. Progr. Nucl. Magn. Reson. Spectrosc. 1995, 28, 53–85
[36] Dwek, R.A. Nuclear Magnetic Resonance in Biochemistry: Applications to Enzyme Systems. Oxford, UK, Oxford University Press, 1973.
[37] Tabernero, J., Shapiro, G.I., LoRusso, P.M. et al. First-in-humans trial of an RNA interference therapeutic targeting VEGF and KSP in cancer patients with liver involvement. Cancer Discovery. 2013, 3, 406–417.
[38] Akinc, A., Goldberg, M., Qin, J. et al. Development of lipidoid-siRNA formulations for systemic delivery to the liver. Mol Ther. 2009, 17, 872–879.
[39] Lu, K., Miyazaki, Y., & Summers, M.F. Isotope labelling strategies for NMR studies of RNA. J. Biomol. NMR. 2010, 46, 113–125.
[40] Sharma, G., Anabousi, S., Ehrhardt, C., & Ravi Kumar, M.N.V. Liposomes as targeted drug delivery systems in the treatment of breast cancer. J. Drug Targeting. 2006, 14, 301–310.
[41] Drummond, D.C., Meyer, O., Hong, K., Kirpotin, D.B., & Papahadjopoulos, D. Optimizing liposomes for delivery of chemotherapeutic agents to solid tumors. Pharmacol. Rev. 1999, 51, 691–744.
[42] Malam, Y., Loizidou, M., & Seifalian, A.M. Liposomes and nanoparticles: Nanosized vehicles for drug delivery in cancer. Trends in Pharmacol. Sci. 2009, 30, 592–599.
[43] Egbaria, K., & Weiner, N. Liposomes as a topical drug delivery system. Adv. Drug Del. Rev. 1990, 5, 287–300.
[44] Allen, T.M., & Cullis, P.R. Liposomal drug delivery systems: From concept to clinical applications. Adv. Drug Del. Rev. 2013, 65, 36–48.
[45] Cruciani, O., Mannina, L., Sobolev, A.P., Cametti, C., & Segre, A. An improved NMR study of liposomes using 1-palmitoyl-2-oleoyl-sn-glycero-3-phospatidylcholine as model. Molecules. 2006, 11, 334–344.
[46] MacLachlan, I. Liposomal formulations for nucleic acid delivery. In: S.T. Crooke, ed. Antisense Drug Technology: Principles, Strategies, and Applications. Boca Raton, CRC Press, 2001, 237–270.
[47] Sinnige, T., Weingarth, M., Renault, M., Baker, L., Tommassen, J., & Baldus, M. Solid-state NMR studies of full-length bamA in lipid bilayers suggest limited overall POTRA mobility. J. Molecular. Bio. 2014, 426, 2009–2021.
[48] Murray, D.T., Griffin, J., & Cross, T.A. Detergent optimized membrane protein reconstitution in liposomes for solid state NMR. Biochemistry. 2014, 53, 2454–2463.
[49] Mazumdar, S. ^{1}H and ^{13}C Studies on the structure of micelles encapsulating hemes in aqueous sodium dodecyl sulfate solutions. J. Phys. Chem. 1990, 94, 5947–5953.
[50] Parker, D. NMR determination of enantiomeric purity. Chem. Rev. 1991, 91, 1441–1457.
[51] Chevalier, Y., & Chachaty, C. NMR investigation of the micellar properties of monoalkylphosphates. Colloid Polym. Sci. 1984, 262, 489–496.
[52] Zhang, X.M., Patel, A.B., de Graaf, R.A., & Behar, K.L. Determination of liposomal encapsulation efficiency using proton NMR spectroscopy. Chem. Phys. Lipids. 2004, 127, 113–120.
[53] Sanders, J.K.M., & Williams, D.H. Shift reagents in NMR spectroscopy. Nature. 1972, 240, 385–390.
[54] Price, W.S. Pulsed-field gradient nuclear magnetic resonance as a tool for studying translational diffusion: Part 1. Basic theory. Concepts Magn. Reson. A. 1998, 9, 299–336.

[55] Price, W.S. Pulsed-field gradient nuclear magnetic resonance as a tool for studying translational diffusion: Part II. Experimental aspects. Concepts Magn. Reson. A. 1998, 10, 197–237.
[56] Gao, W.G., Langer, R., & Farokhzad, O.C. Poly(ethylene glycol) with observable shedding. Angew. Chem. Int. Ed. 2010, 49, 6567–6571.
[57] Momot, K.I., & Kuchel, P.W. Pulsed field gradient nuclear magnetic resonance as a tool for studying drug delivery systems. Concepts Magnetic Res. A., 2003, 19, 51–64.
[58] Karlicek, R.F., & Lowe, I.J. A modified pulsed gradient technique for measuring diffusion in the presence of large background gradients. J. Magn. Reson. 1980, 37, 74–91.
[59] Sørland, G.H., Aksnes, D., & Gjerdåker, L. A pulsed field gradient spin-echo method for diffusion measurements in the presence of internal gradients. J. Magn. Reson. 1999, 137, 397–401.
[60] Seland, J.G., Sørland, G.H., Zick, K., & Hafskjold, B. Diffusion measurements at long observation times in the presence of spatially variable internal magnetic field gradients. J. Magn. Reson. 2000, 146, 14–19.
[61] Seland, J.G., Ottaviani, M., & Hafskjold, B. A PFG-NMR study of restricted diffusion in heterogenous polymer particles. J. Colloid. Interface. Sci. 2000, 239, 168–177.
[62] Romberg, B., Kettenes-van Den Bosch, J.J., de Vringer, T., Storm, G., & We, H. ^{1}H NMR spectroscopy as a tool for determining the composition of poly(hydroxyethyl-L-asparagine)-coated liposomes. Bioconjugate Chem. 2006, 17, 860–864.
[63] Woodle, M.C., & Lasic, D.D. Sterically stabilized liposomes. Biochim. Biophys. Acta. 1992, 1113, 171–199.
[64] Torchilin, V.P., & Trubetskoy, V.S. Which polymers can make nanoparticulate drug carriers long-circulating?. Adv. Drug. Del. Rev. 1995, 16, 141–155.
[65] Storm, G., Belliot, S.O., Daemen, T., & Dd, L. Surface modification of nanoparticles to oppose uptake by the mononuclear phagocyte system. Adv. Drug. Del. Rev. 1995, 17, 31–48.
[66] Metselaar, J.M., Bruin, P., de Boer, L.W., de Vringer, T., Snel, C., Oussoren, C., Wauben, M.H., Crommelin, D.J., Storm, G., & Hennink, W.E. A novel family of L-amino acid-based biodegradable polymer-lipid conjugates for the development of long-circulating liposomes with effective drug-targeting capacity. Bioconjugate Chem. 2003, 14, 1156–1164.
[67] Vernooij, E.A., Gentry, C.A., Herron, J.N., Crommelin, D.J., & Kettenes-van Den Bosch, J.J. ^{1}H NMR quantification of poly(ethylene glycol)-phosphatidylethanolamine in phospholipid mixtures. Pharm. Res. 1999, 16, 1658–1661.
[68] Torchilin, V.P., Omelyanenko, V.G., Papisov, M.I., Bogdanov, A.A., Trubetskoy, V.S., Herron, J.N., & Gentry, C.A. Poly(ethylene glycol) on the liposome surface: On the mechanism of polymer-coated liposome longevity. Biochim. Biophys. Acta. 1994, 1195, 11–20.
[69] Romberg, B., Hennink, W.E., & Storm, G. Sheddable coatings for long-circulating nanoparticles. Pharm. Res. 2007, 25, 55–71.
[70] Ambegia, E., Ansell, S., Cullis, P., Heyes, J., Palmer, L., et al. Stabilized plasmid-lipid particles containing PEG-diacylglycerols exhibit extended circulation lifetimes and tumour selective gene expression. Biochim. Biophys. Acta. 2005, 1669, 155–163.
[71] Mui, B.L., Tam, Y.K., Jayaraman, M., Ansell, S.M., & Du, Y. Influence of polyethylene glycol lipid desorption rates on pharmacokinetics and pharmacodynamics of siRNA lipid nanoparticles. Mol. Ther. Nucleic Acids. 2013, 2, e139.
[72] Williamson, M.P. Applications of the NOE in molecular biology. Ann. Rep. NMR Spectrosc. 2009, 65, 77–109.
[73] Claridge, T.D.W. Chapter 9 – Correlations Through Space: The Nuclear Overhauser Effect. In: T.D.W Claridge, ed. High-Resolution NMR Techniques in Organic Chemistry. Oxford, Elsevier Science. 2009, 315–380.

[74] Vögeli, B. The nuclear overhauser effect from a quantitative perspective. Progr. Nucl. Magn. Reson. Spectrosc. 2014, 78, 1–46.
[75] Bonnaccio, S., Capitani, D., Segre, A., Walde, P., & Luisi, P. Liposomes from phosphatidyl nucleosides: An NMR investigation. Langmuir. 1997, 13, 1952–1956.
[76] Ni, F. Recent developments in transferred NOE methods. Prog. Nucl. Magn. Reson. Spectrosc. 1994, 26, 517–606.
[77] Meyer, B., Weimar, T., & Peters, T. Screening mixtures for biological activity by NMR. Eur. J. Biochem. 1997, 246, 705–709.
[78] Post, C.B. Exchange-transferred NOE spectroscopy and bound ligand structure determination. Curr. Opin. Struct Biol. 2003, 13, 581–588.
[79] Viegas, A., Manso, J., Nobrega, F.L., & Cabrita, E.J. Saturation-transfer difference (STD) NMR: A simple and fast method for ligand screening and characterization of protein binding. J. Chem. Educ. 2011, 88, 990–994.
[80] Melacini, G., Kaptein, R., & Boelens, R. Editing of chemical exchange-relayed NOEs in NMR experiments for the observation of protein–water interactions. J. Magn. Reson. 1999, 136, 214–218.
[81] Halle, B.J. Cross-relaxation between macromolecular and solvent spins: The role of long-range dipole couplings. Chem. Phys. 2003, 119, 12372–12385.
[82] Halle, B. Protein hydration dynamics in solution: A critical survey. Philos. Trans. Royal Soc. Lond. B. 2004, 359, 1207–1223.
[83] Catoire, L.J., Zoonens, M., van Heijenoort, C., Giusti, F., & Guittet, E. Inter- and intramolecular contacts in a membrane protein/surfactant complex observed by heteronuclear dipole-to-dipole cross-relaxation. J. Magn Reson. 2009, 197, 91–95.
[84] Davis, J.H., & Komljenović, I. Nuclear Overhauser effect as a probe of molecular structure, dynamics and order of axially reorienting molecules in membranes. Biochimica et Biophysica Acta (BBA) – Biomembranes. 2016, 1858, 295–303.
[85] Ward, K.M., & Balaban, R.S. Determination of pH using water protons and chemical exchange dependent saturation transfer (CEST). Magn. Reson. Med. 2000, 4, 799–802.
[86] St Pierre, T.G., Clark, P.R., & Chua-Anusorn, W. Single spin-echo proton transverse relaxometry of iron-loaded liver. NMR Biomed. 2004, 17, 446–458.
[87] Vander Elst, L., Piérart, C., Fossheim, S., Raux, J.C., Roch, A., & Muller, R.N. Enumeration of liposomes by multinuclear NMR and photon correlation spectroscopy. Supramol. Chem. 2002, 14, 411–417.
[88] Hansen, E.W., Fonnum, G., & Weng, E. Pore morphology of porous polymer particles probed by NMR relaxometry and NMR cryoporometry. J. Phys. Chem. B. 2005, 109, 24295–24303.
[89] Mears, S.J., Cosgrove, T., Thompson, L., & Howell, I. NMR surfactant solvent relaxation NMR measurements on polymer, particle, surfactant systems. Langmuir. 1998, 14, 997–1001.
[90] Opella, S.J., & Marassi, F.M. Structure determination of membrane proteins by NMR spectroscopy. Chem. Rev. 2004, 104, 3587–3606.
[91] Hong, M., Zhang, Y., & Hu, F. Membrane protein structure and dynamics from NMR spectroscopy. Ann. Rev. Phys. Chem. 2012, 63, 1–24.
[92] Nietlispach, D., & Gautier, A. Solution NMR studies of polytopic α–helical membrane proteins. Curr. Opin. Struct. Biol. 2011, 21, 497–508.
[93] Liang, B., & Tamm, L.K. NMR as a tool to investigate the structure, dynamics and function of membrane proteins. Nat. Struct. Mol. Biol. 2016, 23, 468–474.
[94] Catoire, L.J., Warnet, X.L., & Warschawski, D.R. Micelles, bicelles, amphipols, nanodiscs, liposomes, or intact cells: The hitchhiker's guide to the study of membrane proteins by NMR. In: Building Model Membranes with Lipids and Proteins: Dangers and Challenges. J. N. Sturgis, ed. New York, USA, Springer, 2014, 315–345.

[95] Gong, X., Franzin, C.M., Thai, K., Yu, J., & Marassi, F.M. Nuclear magnetic resonance structural studies of membrane proteins in micelles and bilayers. Methods Mol Biol. 2007, 400, 515–529.

[96] Kielec, J.M., Valentine, K.G., & Wand, A.J. A method for solution NMR structural studies of large integral membrane proteins: Reverse micelle encapsulation. Biochim. Biophysica. Acta. 2010, 1798, 150–160.

[97] Franzin, C.M., Gong, X., Thai, K., Yu, J., & Marassi, F.M. NMR of membrane proteins in micelles and bilayers: The FXYD family proteins. Methods. 2007, 41, 398–408.

[98] Skinner, A.L., & Laurence, J.S. High-field solution NMR spectroscopy as a tool for assessing protein interactions with small molecule ligands. J. Pharm. Sci. 2008, 97, 4670–4695.

[99] Betz, M., Saxena, K., & Schwalbe, H. Biomolecular NMR: A chaperone to drug discovery. Curr. Opin. Chem. Biol. 2006, 10, 219–225.

[100] O'Connell, M.R., Gamsjaeger, R., & Mackay, J.P. The structural analysis of protein-protein interactions by NMR spectroscopy. Proteomics. 2009, 9, 5224–5232.

[101] Markwick, P.R.L., Malliavin, T., & Nilges, M. Structural biology by NMR: Structure, dynamics, and interactions. PLoS Comput. Biol. 2008, 4, e1000168.

[102] Mulder, F.A.A., Mittermaier, A., Hon, B., Dahlquist, F.W., & Kay, L.E. Studying excited states of proteins by NMR spectroscopy. Nat. Struct. Biol. 2001, 8, 932–935.

[103] Filippov, A., Khakimov, A., Afonin, S., & Antzutkin, O.N. Interaction of prostatic acid phosphatase fragments with a lipid bilayer as studied by NMR spectroscopy. Mendeleev Commun. 2013, 23, 313–315.

[104] Oxenoid, K., & Chou, J.J. The present and future of solution NMR in investigating the structure and dynamics of channels and transporters. Curr. Opin. Struct. Biol. 2013, 23, 547–554.

Michael F. Brown
7 Collective dynamics in lipid membranes

The continuum is that which is divisible into indivisibles that are infinitely divisible
(Aristotle, Physics)

Abstract: In solid-state NMR of biomolecules, the average structure and dynamics are addressed by combining experimental results with theory. Relaxation rates exhibit a functional dependence on order parameters because of molecular motions and/or collective excitations of liquid-crystalline membranes. Mixtures of phospholipids with cholesterol or nonionic surfactants allow the experimental correspondence of ^{2}H NMR observables to be quantitatively tested. For cholesterol-stiffened bilayers, the spin-lattice relaxation rate profile is reduced together with an increased order profile. Bilayer softening due to nonionic surfactants gives an opposite relaxation enhancement accompanied by reduced order parameters. In both cases a square-law functional dependence (Fermi's Golden Rule) explains the relaxation and order profiles in terms of mean-square amplitudes of the lipid fluctuations. Model-free analysis reveals an $\omega^{-1/2}$ frequency law for three-dimensional (3D) fluctuations of the membrane, whereas for two-dimensional (2D) elastic sheets an ω^{-1} dependence is expected. Collective segmental or molecular modes emerge on the mesoscale of the bilayer thickness and smaller that are formulated with continuum elastic theory. Furthermore, the bilayer core resembles a hydrocarbon fluid with a viscosity of only a few centipoises (cP). Magnetic resonance spectroscopy thus reveals properties of the membrane lipids described by a hierarchical energy landscape that affects their polymorphism, phase behavior, and lipid-protein interactions.

Key words: cholesterol, director fluctuations, elastic modulus, molecular dynamics, solid-state NMR, spin-label EPR, order parameter, lipid rafts, relaxation

7.1 Introduction

Nuclear magnetic resonance (NMR) spectroscopy offers one of the premiere biophysical tools for addressing the structural and dynamical properties of biomolecules, including proteins, lipids, and nucleic acids. The spectral lineshapes give us knowledge of the average molecular structure in analogy to X-ray or neutron diffraction, while the nuclear spin relaxation times manifest the fluctuations around the mean conformation.

Michael F. Brown, Department of Chemistry and Biochemistry, University of Arizona, Tucson, USA; Department of Physics, University of Arizona, Tucson, USA.

Together, the solid-state NMR spectral lineshapes and relaxation rates address the problem of disentangling the mean-square amplitudes and rates of the motions versus the forces and potentials that govern the molecular properties. For membrane lipids, the dynamics can include local segmental motions, reorientations of the entangled molecules, and harmonic or elastic excitations of the bilayer [1, 2]. An important question is whether the fluctuations are mainly local, or whether the dynamics are inherently collective and entail a hierarchical energy landscape [1, 3–5]. Here we furnish a perspective of how information of biochemical or biomedical significance is captured in relation to molecular mechanisms and functions of membranous soft matter [6].

In applying magnetic resonance spectroscopy (spin-label electron paramagnetic resonance, EPR, and NMR) to biomolecules the study of lipid membranes is particularly illuminating [5, 7–16]. Proteins and lipids exist within a fluid bilayer matrix that shares many features in common with the liquid-crystalline state [1, 17–19]. A comprehensive view considers both the ordering and dynamics of the molecules [1, 20, 21]. Investigations of phospholipids, cholesterol, peptides, and membrane proteins address their average structures and molecular dynamics within the membrane [21–26] – multiple scales of time and space are involved. Such multi-scale approaches to solid-state NMR spectral lineshape measurements and relaxation times can be uniquely instructive [27]. The convergence of advances in NMR technology with biophysical knowledge emphasizes some of the most challenging and timely questions in contemporary membrane science [6, 28–31]. Combining solid-state NMR spectroscopy with relaxation measurements allows one to investigate the collective dynamics of membrane liquid crystals, in which a unified view of the relaxation frequency (magnetic field) and order parameter dependence suggests the emergence of bilayer elasticity at an atomistic level. Nuclear spin relaxation studies [32] together with spin-label EPR results [3, 5] and complementary scattering approaches [33–37] thus inform the emergent biophysical properties of cellular membranes.

7.2 A brief history of membrane biophysics

Some readers may think the subject of lipid bilayer dynamics – much less biomembranes – is far too complicated to be understood at any semi-rigorous level of physics or chemistry. Perhaps it is best left to the cell biologists. How can such a messy and disorderly biological system with countless degrees of freedom, and with flexible chain molecules that undergo all sorts of internal isomerizations, as well as rotational and lateral diffusion, be possibly understood at any level of rigor, for example, as in other areas of physics or chemistry [38]? We would submit that the answer is: yes. Below we justify this assertion, which rests on a confluence of molecular spectroscopy with comparative thinking about the properties of molecular liquids, solids, and liquid crystals.

7.2.1 X-ray scattering and magnetic resonance spectroscopy

Lipid bilayers are central to structural biology because they are the stuff that cellular membranes are made of. Various proteins exist in contact with the lipid bilayer, beginning from their birth on the endoplasmic reticulum, and ending up with the new membrane material. Because lipid bilayers encapsulate the features of soft matter [6], understanding their dynamics is vital to biophysics (see Figure 7.1). Levine and Wilkins [39] recognized this aspect very early on using X-ray diffraction, and pointed out the liquid-like interior of the lipid bilayer [39]. Yet X-ray scattering [37, 40–42] is handicapped by its utility mainly as a structural tool. By contrast, magnetic resonance spectroscopy (spin-label EPR and NMR) provides site-specific knowledge of both the structure and dynamics [2, 32]. In a series of early papers, Chapman et al. [43, 44] used proton NMR methods (primitive by today's standards) to investigate the gel and liquid crystal states of membrane lipids, thus providing further evidence for a fluid-like bilayer. Current magnetic resonance applications follow these leads, and seek to explain how membrane structure, dynamics, and function are related.

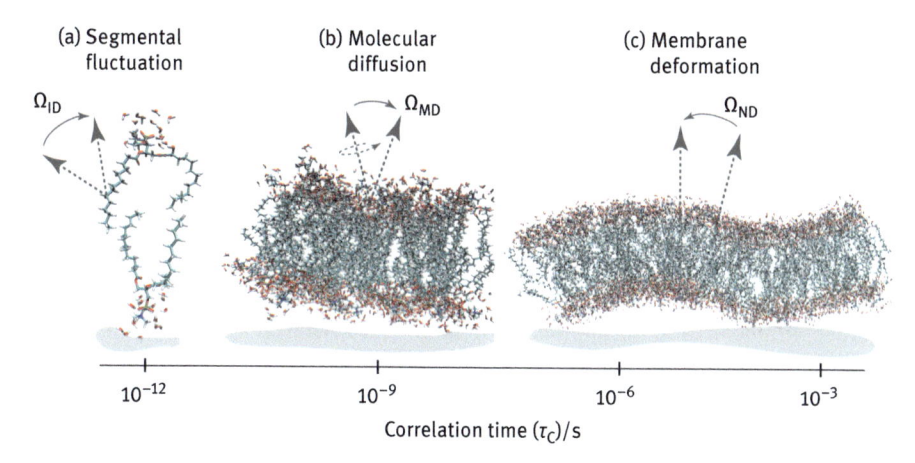

Figure 7.1: Magnetic resonance spectroscopy (NMR and spin-label EPR) explores the hierarchical energy landscapes and fluctuations of lipid membranes. The dynamics include segmental motions, molecular diffusion, and viscoelastic deformation. Fluctuations involve the geometry of the interactions via Euler angles (Ω) for transformations between various coordinate frames, together with the mean-square amplitudes and correlation times (τ_C) of the motions. Coordinate frames are designated as follows: I, intermediate (segmental) frame; M, molecular interaction frame; N, local director frame; and D, bilayer director frame. The multi-scale dynamics of membranous soft matter involve a broad range of time and length scales. Figure adapted with permission from Ref Leftin and Brown [32].

However, perhaps the singular development that established the fluid-like nature of phospholipids was spin-label EPR spectroscopy, as originally introduced by Wayne Hubbell and Harden McConnell at Stanford [7]. In a series of innovations, they produced not only the first nitroxide spin probes of phospholipids, but also worked out the basic resonance theory as put forth in the classic text of Carrington and McLachlan [45]. Closely related work by Joachim Seelig for membrane liquid crystals [46] was likewise seminal in establishing the flexibility gradient of the lipid chains of membrane bilayers. And in more or less concurrent proton NMR studies, Sunney Chan and his students at Caltech discovered site-specific differences in relaxation times of membrane lipids [47, 48]. The natural-abundance ^{13}C NMR studies of Yehudi Levine et al. further established the mobility gradient in terms of lipid dynamics [49], where the relaxation times were found to vary moving away from the glycerol backbone toward either the polar head group or the acyl chain termini [49]. Additional insights by Melvin Klein et al. at Berkeley [50] and Ulrich Häberlen and Hans Spiess at Heidelberg [51, 52], together with pioneering work involving liquid crystals by Pier Luigi Nordio and his group at Padova [53] and by Jack Freed at Cornell [20], set the stage for the ideas brought forth in this chapter.

7.2.2 Order parameters and relaxation rates of phospholipid liquid crystals

Nuclear spin relaxation is notable among the various biophysical methods because both the amplitudes and rates of the motions (spectral densities) come naturally into play. That is to say, the mean-square amplitudes and correlation times at a site-directed level are involved. For membrane lipids, to establish whether the previously mentioned flexibility gradient is observed – or alternatively a mobility (fluidity) gradient – requires detailed experimental investigations. Even so, the above-mentioned ^{1}H or ^{13}C NMR relaxation studies do not readily separate the contributions from the motional amplitude (order parameter) and the motional correlation times. One breakthrough came with the implementation of solid-state ^{2}H NMR spectroscopy by Seelig et al., who established detailed order parameter profiles for both phospholipids [54] and membrane liquid crystals (soap-like bilayers) [46, 55]. However, the correspondence of the flexibility gradient with the mobility gradient had to await a different type of experiment. This involved the combined solid-state ^{2}H NMR order parameter and spin-lattice relaxation time measurements of Brown, Seelig, and Häberlen [56]. These authors first showed that both the order parameters and relaxation rates had gradients along the chains that paralleled one another. Besides the order profile (the flexibility gradient), there is also a relaxation profile (mobility gradient) of the lipids (Figure 7.2). The question is then whether the relaxation profile is because of the motional amplitudes, or the rates of the motions, or both [57]. That is to say: What is the functional correspondence of the two profiles, namely, the flexibility gradient and the mobility gradient?

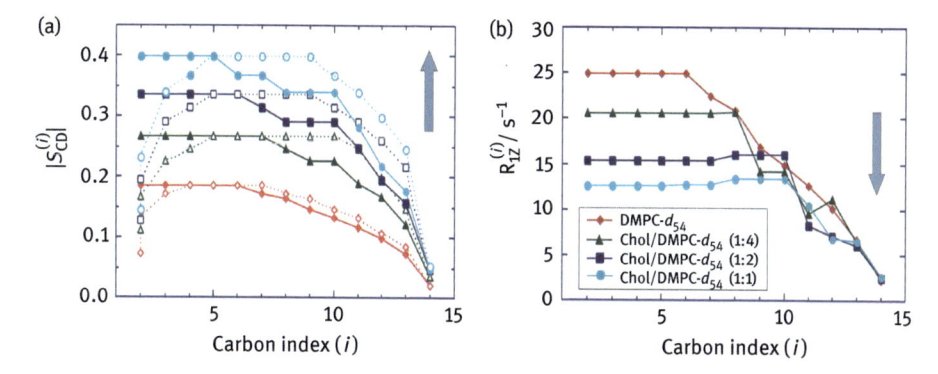

Figure 7.2: Experimental observables from solid-state ^{2}H NMR spectroscopy reveal both the structure and dynamics of membrane liquid crystals. Results are summarized for binary mixtures of DMPC-d_{54}/cholesterol in the liquid-ordered (lo) phase. (a) Orientational order parameters $|S_{CD}^{(i)}|$ are plotted versus acyl chain segment position (i) for various mole fractions (X_C) of cholesterol (Chol). Note that greater X_C yields an increase in order parameters due to loss of configurational degrees of freedom. For the initial part of the chains a plateau is seen followed by a decrease within the bilayer core. (b) Plots of corresponding spin-lattice $R_{1Z}^{(i)}$ relaxation rates against carbon index (i) for mixtures of DMPC-d_{54}/cholesterol in the lo phase. With greater X_C an opposite decrease in the relaxation rates is evident. The ^{2}H NMR measurements were conducted for unoriented multilamellar dispersions at $T = 44$ °C and Larmor frequency $\nu_0 = 76.8$ MHz (11.7 T). Solid-state ^{2}H NMR reveals changes in both the flexibility gradient and the mobility gradient for the membrane lipids at an atomistic level. Data are from Ref Martinez et al. [60].

Here we offer a new twist on the question of the mobility gradient of membrane liquid crystals: NMR relaxation studies detect the emergence of collective thermal excitations that underlie the bulk elasticity due to chain conformations over a broad time range (10^{-11} to 10^{-6} s). The connection of the flexibility gradient with the mobility gradient is due to collective interactions of the lipids. Order fluctuations represent elastic (harmonic) excitations of the bilayer; for example, those associated with the local chain tilt and other types of slowly relaxing or long-lived local structures (on the timescale of $\approx 10^{-8}$ s). In consequence, the bilayer flexibility gradient depends on the timescale of the lipid deformations. Slowly relaxing conformations occur due to long-lived collective lipid interactions involving tilt of portions of the chains [37], together with other rotational isomeric configurations. Through recognizing the frequency–amplitude correspondence, one is then able to establish a connection to bulk membrane properties [58–61].

7.3 Flexibility and mobility of lipid bilayers as seen by magnetic resonance spectroscopy

At this point, we can briefly summarize the results of previous magnetic resonance studies (NMR and spin-label EPR) of membrane lipids as follows:

1. Spin-label EPR spectroscopy shows there is an increase in motional averaging of the hyperfine tensor moving toward the bilayer center, which Hubbell and McConnell [7] called the flexibility gradient. McConnell et al. [7, 8] first interpreted the flexibility gradient for phospholipids and Seelig et al. for smectic liquid crystals (soap-like bilayers) [46] in terms of orientational order parameters. The spin-label order parameters correspond to the unpaired electron of the paramagnetic oxazolidine ring (π-orbital z-axis is normal to the H–C–H plane of a polymethylene chain) [8, 55, 62, 63].
2. Analogously, proton and ^{13}C NMR studies of lipids reveal a gradient of the spin-lattice relaxation times moving away from the glycerol backbone toward either the polar head groups or the nonpolar bilayer core, as shown by Chan et al. [47, 48, 64] and Levine et al. [49]. Still, the separation of the motional amplitudes versus the motional rates is less clear-cut than for spin-label EPR spectroscopy [8].
3. Solid-state ^{2}H NMR spectroscopy as introduced by Seelig et al. determines orientational order parameters and correspondingly the motional amplitudes as in spin-label EPR spectroscopy [54, 55, 65]. However, rather than an exponential decrease along the chains [7, 46], an approximate plateau is found [54]. Coupled rotational isomerizations of the acyl groups occur [55], together with statistical chain terminations moving away from the lipid aqueous interface (end effects), as pointed out by Dill and Flory [66].
4. The mobility gradient from solid-state ^{2}H NMR relaxation reveals an equivalent plateau as seen for the flexibility gradient. As first investigated by Brown et al. [56], both the order parameters and the relaxation rates show a functional correspondence, which differs from spin-label EPR results. Because the relaxation depends on both the mean-square amplitudes and the correlation times, the question is whether the relaxation gradient mirrors the order profiles (flexibility gradient) [7, 54], or whether there is a corresponding mobility gradient along the chains [56].
5. One can reduce the many-body problem of a lipid bilayer to considering the order parameters and relaxation times of the individual segments of the flexible molecules at a site-resolved level. According to this view, collective properties of the membrane lipids are significant for interpreting the results. Each of the segments is treated analogously to a nematic liquid crystal within the mean field of the bilayer, as opposed to considering the cumulative internal chain dynamics.

7.3.1 The flexibility gradient of a lipid bilayer

The understanding of lipid membranes clearly benefits from knowledge of both the flexibility gradient and the mobility gradient, for example, as encapsulated by molecular dynamics simulations [67–70]. To explain the flexibility gradient, one must consider the packing of the ensemble of phospholipids, together with the chain travel (flux) away from the aqueous interface [66, 71–73]. Because of their amphiphilic character, the lipid molecules are anchored to the aqueous interface by their polar ends. The lipids are moreover subject to a balance of hard repulsive forces acting at short range that govern the average structure, together with soft van der Waals attractions at longer range, that fix the density at approximately equal to that of liquid hydrocarbon. Either a lattice model [66] or a mean-torque model [73] for the distribution of acyl chain segments is informative. The chain flux away from the aqueous interface depends on the area per molecule, which is only weakly affected by the acyl length, yet rather strongly by the polar head groups as shown by Brown et al. [73]. *Gauche* isomers shorten the individual chain length projections along the bilayer normal [55], giving a broad distribution of end-to-end lengths. Beyond a certain point (the so-called plateau), the chains with more *gauche* isomers terminate, so that the number extending deeper into the bilayer core is correspondingly reduced. The surrounding chains are more disordered to keep the density nearly the same as liquid hydrocarbon [58, 66, 71–75]. Hence the length of the order parameter plateau increases with the acyl chain length [73], and the decrease in chain order near the bilayer midplane is approximately the same for various chain-length lipids [73], as also noted by Hubbell and McConnell [7]. Packing of the chains thus accounts for the disorder (flexibility) gradient for the lipid ensemble [66, 71, 73].

Additionally, besides the much-discussed flexibility gradient, we must also consider the mobility gradient of the membrane lipids [49, 56]. Clearly the relaxation rates require further consideration, because they depend on both the motional amplitude (i.e., mean-square amplitude) and the rates of the motions [57]. One cannot a priori expect any relation between the mean-square amplitude and the rate, except perhaps for simple harmonic or anharmonic motions. Yet such a relation has indeed been discovered in the case of lipid bilayers [1]. Most surprising, the order fluctuations within the bilayer core manifest the continuum elastic properties of the bulk material (e.g., due to a distribution of harmonic excitations). Classical liquid crystal physics is uncovered, inasmuch as the results show the collective behavior of the ordered chains. For the case of phospholipids, the long acyl groups (e.g., polymethylene) are tethered to the aqueous interface. Lack of water penetration means both the configurational properties [7, 62, 66] and the dynamical properties [49, 76] of the lipids come into play, as described by the force field of the bilayer [68].

7.3.2 The mobility gradient of a lipid bilayer in the liquid-crystalline state

With the above-mentioned discussion in mind, the reader may ask: How is the mobility gradient of a lipid bilayer in the liquid-crystalline state connected to the flexibility gradient? In fact the author first reported the correspondence at a conference in 1979 at

Stanford University (in a session chaired by McConnell) [1]. A simple square-law dependence of the relaxation rates and order parameters was uncovered for the first time. It immediately suggested how the NMR relaxation is governed by the order fluctuations. The characteristic relation between the mean-square amplitude and the relaxation clearly points to modulating the local ordering of the bilayer lipids [1, 77]. The question then becomes: Are the order fluctuations because of non-collective molecular rotations, or to quasi-elastic (harmonic) excitations of the lipids due to their collective interactions? Indeed, the dependence of the relaxation on frequency (magnetic field strength) and the order parameters argues that collective motions account for the spin-lattice relaxation. The slow dynamics are formulated in terms of a local director axis and correspond to so-called order-director fluctuations (ODF). Of course, there is no a priori reason why this has to be so – we all know that NMR is a site-specific molecular technique [57]. So why are molecularly specific details not included?

It turns out that the concept of a director frame [71] is key to considering the bilayer relaxation due to the time-dependent lipid motions. For membrane liquid crystals, it is an axis of rotational symmetry that is perpendicular to the plane of the bilayer, that is, it is the lamellar normal [54]. Molecular motions are expressed by rotations either around (longitudinal) or perpendicular (transverse) to the principal molecular axis within the director frame. The director axis is a collective property of the lipid assembly rather than the single molecules. It implies that the ensemble-averaged motions are rotationally symmetric around a preferred axis or direction in space. Rigid body motions of the individual molecules around their long axes would be one example, but that is neither assumed nor required. Rather, the director manifests the collective rotations of the molecules and/or their flexible segments. By introducing a local director axis, one can readily account for the multi-scale dynamics of the chains within a continuum approximation, as first introduced for lipids by this author [1] (see Figure 7.1). This immediately suggests a connection between the relatively long-lived structures seen by spin-label EPR (e.g., due to rotational isomerism or local chain tilt) and the collective order fluctuations put forth to explain the spin-lattice relaxation in lipid bilayers [58]. For NMR spin-lattice relaxation, slow motions may involve transitions (timescale of $\approx 10^{-8}$ s or longer) corresponding to those detected with spin-label EPR (also with lifetimes in the range of $\approx 10^{-8}$ s) – the timescale is similar for both methods. Still, the proposal of local director fluctuations spanning a broad range of scales in space and time has remained under discussion or debate even today.

Perhaps the reader might also be inclined to ask: Did the notion of local director fluctuations emerge fully formed, just like Athena from the head of Zeus [78]? The answer is: Of course not – it was a gradual evolution, brought about by considering the various possible alternative formulations, for example, segmental versus molecular motions and so forth, beginning with the site-specific viewpoint of NMR spectroscopy. It was enforced by the fact that the spin-lattice relaxation times differ for every magnetic field strength or nucleus studied [58]. At the time, some researchers thought it impossible to disentangle the types of motions that govern the relaxation, for example, given the debate about the spin-label flexibility gradient. How can one possibly sort out the mobility gradient measured with NMR spin-lattice relaxation?

7.4 The role of time and space in membrane biophysics

To continue further, how do we go about separating the influences of the motional mean-square amplitudes from the rates of the motions for liquid-crystalline lipid systems? Sometimes a lack of foreordained knowledge can work to one's advantage. The conventional wisdom is that the dynamics problem for membrane lipids is intractable in analytical closed form. Yet according to a Chinese fortune cookie, one should: "Avoid unchallenging occupations – they waste your talents." Indeed, the test is to disentangle the various motional contributions, and here the benefit of NMR spectroscopy is the ability to determine site-specific information about both order and dynamics [57].

7.4.1 Separation of dynamical and spatial variables

Now with regard to the nuclear spin relaxation, an analysis in terms of mean-square amplitudes and reduced correlation functions (or spectral densities) entails separating the coupling Hamiltonian into a time-averaged (secular) part, and a time-dependent (nonsecular) part. The average or secular part of the coupling Hamiltonian $\langle \hat{H} \rangle$ commutes with the main Zeeman Hamiltonian, and it causes the well-known energy level shifts that affect the spectral lineshape [2, 79, 80]. It corresponds to the effective Hamiltonian of McConnell and coworkers [7, 81]. The fluctuating part of the Hamiltonian $\hat{H}'(t) = \hat{H}(t) - \langle \hat{H} \rangle$ amounts to a time-dependent perturbation, brought about by the nonsecular terms that induce the nuclear spin transitions [45]. For membrane lipids, the nonsecular terms cause transitions among the energy levels, and affect the spin-lattice relaxation rate [8]. Let us first consider the average Hamiltonian: what left an early impression on this author was how both McConnell and Seelig were able to adapt the concept of an order tensor from the liquid crystal literature [82, 83] to lipid bilayers [7, 46]. For spin-labeled phospholipids [7] or soap-like bilayers (smectic liquid crystals) [46], the oxazolidine moiety is treated analogously to a nematic liquid crystal, albeit connected by covalent bonds to the rest of the chain, that is, within the force field of the membrane lipid bilayer. Then from the measured order parameter profile or flexibility gradient, one can infer the average behavior of the ensemble of lipid chains [7, 46, 84], independently of models for the cumulative dynamics.

7.4.2 Nuclear spin relaxation as a probe of lipid membrane dynamics

Continuing along these lines, provided there is a director axis, then local director excursions can also occur, which manifest the collective motions of the aligned molecules [20, 85]. Because there is an average director axis for the individual lipid segments, fluctuations of the instantaneous director can emerge from the local structure. Introducing a director frame informs the presence of collective interactions of the lipid molecules, and gives us a route to the energy landscape and the hierarchy of motions for membrane lipid bilayers. In this way, we are led to the proposal that the spin-lattice relaxation is governed by the local director fluctuations, that is, due to slowly relaxing local structures [1]. By separating the coupling Hamiltonian into a secular (effective or averaged) part and a time-dependent perturbative part, the transition probability and hence the nuclear spin relaxation rates depend on the mean-square interaction strength. The molecular fluctuations close to the resonance frequency ω_0 govern the rate of the relaxation – that is, the transition probability depends on the squared matrix elements of the coupling Hamiltonian. Matching the spectral density of the fluctuations to the coherent nuclear resonance frequency ω_0 (and twice ω_0) causes transitions among the nuclear spin energy levels, and hence the spin-lattice relaxation. For multi-scale (or composite) motions, one can then iteratively separate the coupling Hamiltonian into various parts with characteristic timescales. What is secular for the faster timescale motion becomes nonsecular for the next slower motion in the hierarchy, and so forth [1]. Of course, cross-correlations can also be considered, but what do we gain by making a complicated problem even more complicated? Rather, our aim is simplification, leaving the details for subsequent refinement.

For the sake of illustration, let us assume a broad separation of the dynamics into faster local motions and slower motions (see Figure 7.1). The local motions (with timescales $\approx$ 5–20 ps) entail rotational isomerization [86] of the lipids, which can be affected by collective interactions of the assembly due to the energy landscape [3, 56]. As an example, it is known that lateral diffusive jumps of the lipids can occur on a longer timescale ($\approx 10^{-7}$ s) [87] that can alter the rotational isomeric state or orientations of molecules. Provided the local order parameters are proportional to the observed values, the slower dynamics would affect all positions approximately equally. The secular (averaged) part of the Hamiltonian for the slow motions then scales with the local order parameter. Because the spin-lattice transition probability depends on the squared Hamiltonian matrix element, that immediately leads us to a square-law functional dependence of the relaxation profile (i.e., the mobility gradient) on the observed order parameter profile (i.e., the flexibility gradient) [58]. Notably the functional dependence of the relaxation rates on the squared order parameters along the chains [1] is completely model free. It can be further interpreted in closed mathematical form, as shown for the case of proteins in solution [88], and concurrently and more generally for aligned systems [1].

7.4.3 Non-collective or collective order fluctuations?

In hindsight, the above treatment is just a simple consequence of Fermi's golden rule – the relaxation depends on the interaction strength (hence the coupling Hamiltonian) squared. We shall return to this point below. Given that the local isomerizations set up an ensemble of structures formulated by a local director (e.g., which might include transient local tilting of portions of the chains [58, 62, 84, 89–91]), the spin-lattice transition probability depends on the squared order parameter. It is assumed the collective director fluctuations are approximately the same within the bilayer. A posteriori the experimental observation of a simple square-law functionality of the mobility gradient on the flexibility gradient can thus be simply rationalized. Most arresting, a direct connection to lipid material properties exists, thus hinting to the physical significance of the findings. Still there is no a priori reason why this has to be so – and some would say (and did) that the problem is intractable at any level of physicochemical rigor. Why not just average the results over the whole bilayer, and calculate an average (micro)viscosity and be done with it? But that is clearly untenable as an oversimplification [58].

It is probably safe to state that many scientists would agree that the relaxation rates of lipid bilayers do not predominantly involve local segmental motions. There may be a mobility (fluidity) gradient analogous to the flexibility gradient of McConnell, but that is not the major feature [92]. In effect, the discussion revolves around the cutoff for the distance or timescales for the local director fluctuations, including whether a single type of relaxation dispersion covers the whole frequency range. The cutoff for the ODF can be on the order of the bilayer thickness, in which case the dispersion at higher frequencies can be due to whole-molecule motions. Alternatively, the various segments can be considered analogously to a nematic liquid crystal, subject to an orienting potential with a cutoff at the segmental size, yielding the emergence of collective thermal fluctuations of the local director. A break in dimensionality can occur from 3D order fluctuations to 2D order fluctuations of lower amplitude, for example, connected with undulations of the membrane lipid bilayer at low (kHz) frequencies. (For now, we put aside whether the small-angle approximation for the ODF is applicable [93].) Assuming the bilayer dynamics involve both local motions (e.g., *trans-gauche* isomerizations) as well as order fluctuations, the issue boils down to whether the slow dynamics can be formulated by a non-collective model, for example, rigid-body molecular rotations in a potential of mean force, or whether a broad distribution of quasi-elastic excitations is present. Indeed, the author extensively discussed both alternatives [1], and it was argued that collective lipid dynamics are the most plausible interpretation.

Basically the reasoning goes as follows: effectively, there is a continuous frequency dispersion of the relaxation, yet the temperature variation by contrast implies the fast motional (white noise) limit. How can such apparently contradictory observations be reconciled, for example, by a model of rotational diffusion? Then again, evidence for distinct rotational modes in lipid bilayers (as shown by a minimum in the spin-lattice relaxation times) has never been obtained for phospholipids in the liquid-crystalline, that is, liquid-disordered (ld) state. (Such a relaxation minimum is observed due to matching of the spectral density of the rotational modes to the nuclear resonance frequency.) It is only for the phosphodiester moiety that such a minimum is seen [94], or for bilayers containing cholesterol [95, 96]. Even in the case of inertial averaging of the lipids over the internal (fast) degrees of freedom, whole-molecule rotations of the lipids are not detected in the liquid-crystalline state. Apparently a reduction in the degrees of freedom is needed for rotational modes to be observed. How could such discrete whole molecule motions exist, and not give rise to a relaxation minimum at some combination of frequency and temperature?

7.5 The energy landscape of a membrane lipid bilayer

Outwardly the more interesting (in our view) alternative is a hierarchical energy landscape for the dynamics of lipid membranes. The various tiers correspond to the fast and slow motional modes [1, 60, 97, 98] that occur within the basins of attraction. Here, we are much inspired by work involving protein dynamics by Hans Frauenfelder and coworkers [99–101]. To all appearances, the hierarchy shown in Figure 7.1 can be viewed as a low-dimensional representation of the energy landscape for the flexible membrane lipids in the liquid-crystalline (i.e., ld) state. The individual C–H bonds can fluctuate due to vibrational motions, *trans-gauche* isomerizations, and restricted segmental reorientations (ps time range). Fluctuations in local ordering of the lipids can also occur due to molecular motions (e.g., involving inertial averaging, ns range), or collective lipid dynamics over a broad span of timescales (ns–μs or ms range, and beyond). Structural and dynamic features are explored using various biophysical techniques, including small-angle X-ray and neutron scattering (SAXS, SANS), molecular dynamics (MD) simulations, and magnetic resonance (spin-label EPR and NMR) spectroscopy.

7.5.1 Hierarchical models for lipid membrane dynamics

This thinking leads us to the idea of a separation of relatively fast (local) and slow (collective) fluctuations, that in a first approximation are uncoupled (statistically independent) [10]. What are the classes of models that can be considered? Broadly speaking, the fast motions set up the local ordering that is further modulated by slower collective disturbances. Yet as mentioned above, for membrane lipids, the slower order fluctuations have been the topic of intense debate. Some authors [1] claim that a distribution of collective excitations occurs in analogy with nematic and smectic liquid crystals. The relatively slow motions are formulated as order-director fluctuations, spanning the segmental dimensions on up to the macroscopic bilayer. Still other authors consider inertial averaging over the entire lipid molecule, that is, a non-collective molecular model is assumed, whereby the average molecule undergoes rotational diffusion within the mean field of all the other lipids of the bilayer. For the higher frequency spin-lattice relaxation (MHz) regime, many workers in the field accept an interpretation in terms of rotational modes of the lipids [102, 103]. As first proposed by the author for lipid bilayers [1], the possibility of ODF is mainly considered for the lower frequency (kHz) range, where the order fluctuations are relatively small in amplitude [104].

Coming back to Figure 7.1, such motions affect the coupling interactions in NMR spectroscopy, which correspond to a static or time-dependent perturbation of the Zeeman Hamiltonian. The coupling interactions are formulated as second-rank tensors, and are represented in either a Cartesian or spherical (irreducible) basis [2, 80]. By introducing the closure property of the group of rotations [105], consideration of the multi-scale motions then becomes transparent, simplifying the overall treatment of the problem [80, 106]. Besides the local segmental motions of the lipids, molecular motions can occur with respect to a local director axis $n(t)$ that itself can undergo time-dependent reorientation (Figure 7.1). For membrane lipids, the residual quadrupolar couplings (RQCs) or residual dipolar couplings (RDCs) directly give the order parameters for flexible molecules as model-free experimental observables. The nuclear spin relaxation rates depend on the types of the lipid or protein motions, as well as their rates and mean-square amplitudes (related to the order parameters) [1, 21, 23, 24, 27, 57, 88, 107, 108]. Both equilibrium structural properties and dynamical properties are thereby accessible [57]. For liquid-crystalline membranes, the motions encompass local isomerizations of the flexible lipids, rotations of the entire molecules, and collective excitations of the whole bilayer, for example, involving a distribution of quasi-elastic relaxation modes, as in other liquid crystals. Collective director fluctuations occur with respect to the average director n_0 (the bilayer normal), and are due to the distribution of either slowly relaxing local structures (ODF), or non-collective molecular motions [1].

In the case of phospholipid membranes, samples are typically aligned bilayers, multilamellar dispersions, or unilamellar vesicles. Lipid bilayers are ordered systems with a large number of internal degrees of freedom, and they typically reorient slowly on the NMR timescale. What is more, in solid-state NMR spectroscopy the symmetry axis (bilayer normal) can be aligned with respect to the magnetic field. Because the motions are restricted, the transformation between the principal axis system of the coupling interaction (dipolar, quadrupolar, and chemical shift) and the bilayer frame is only partially averaged, for example, by segmental isomerizations or molecular rotations. Larger scale motions involve collective disturbances of the bilayer. Assuming statistically independent fluctuations (e.g., with separate timescales), a hierarchical two-step relaxation model can be introduced. Such a two-step separation of the relaxation is implicit in considering the multi-scale motion [109]. First, relatively fast motions of lesser amplitude modulate the static coupling interaction about the average value. And second, slower motions of greater amplitude further modulate the average or residual interaction over the faster motions, and furnish a second relaxation contribution due to the order fluctuations. In the limit of isotropic or unrestricted motion, a generalized model-free (GMF) approach [27] gives a two-step model, as put forward by Håkan Wennerström et al. for surfactant micelles [110], or the model-free approach of Giovanni Lipari and Attila Szabo for globular proteins in solution [88, 111].

Experimentally, the order parameters and relaxation rates describe the coupling interactions – they are the observables of the solid-state ^{2}H NMR experiments. The GMF reduction of the ^{2}H and ^{13}C NMR lineshape and relaxation measurements involves the motional mean-square amplitudes (order parameters) and reduced spectral densities (correlation times). Direct dipolar couplings and quadrupolar interactions are measured, together with the nuclear spin relaxation rates. The segmental order parameters give information about the average bilayer structural properties, including the area per lipid and volumetric hydrocarbon thickness, as evaluated using a mean-torque model [73]. Fluctuations about the mean value, for example, area or thickness fluctuations, or director fluctuations, yield the multiscale bilayer dynamics. What is more, by combining the RQCs or RDCs with the associated spectral densities, a unified picture of the protein and lipid dynamics can be developed. The NMR relaxation rates give us knowledge of the types, rates, and amplitudes of the collective or molecular motions within the bilayer. Measuring both the experimental ^{13}C–^{1}H and ^{2}H NMR segmental order parameters (S_{CH} or S_{CD}) and relaxation times (T_{1Z}, T_{1Q}, T_2) allows findings to be obtained for membranes that cannot be acquired with other biophysical techniques. In this context, a comprehensive database of experimental NMR results has been developed for membrane phospholipids [32].

7.5.2 Example of bilayers containing cholesterol

To illustrate these ideas further, it is instructive to consider the interactions of phospholipids with cholesterol [61, 70, 112–117], a molecule in which McConnell had a longstanding interest [14, 16, 118–120]. In solid-state NMR, the motional averaging gives a distribution of RQCs, from which the order profiles can be derived (Figure 7.2). The static electric field gradient tensor is essentially the same for all the deuterated positions [2]. Yet the *residual* tensor is different: it varies strikingly as a function of acyl chain position, giving the RQCs that naturally draw our attention. A further aspect is that the solid-state ^{2}H NMR spectra clearly exhibit the axial symmetric signature of the liquid-crystalline (ld) phase of lipid bilayers. As famously shown by Myer Bloom and his coworkers [121], for acyl chain perdeuterated phospholipids, the powder-type spectra of random multilamellar dispersions can be numerically inverted (or de-Paked) to reveal the RQCs most directly [122]. The larger splittings correspond to the groups closer to the aqueous interface, with a progressive diminution along the chains, toward the hydrocarbon core of the bilayer [7, 8, 46]. From the de-Paked ^{2}H NMR spectra, order parameter profiles $S_{CD}^{(i)}$ (where $i \equiv$ chain index) are obtained for the case of random multilamellar dispersions [2].

Some readers will know already that the residual couplings (RQCs and RDCs) are simply related to the average bilayer structure, for example, in terms of a mean-torque potential model [73]. Insights are obtained about the structure and dynamics of the membrane lipids in terms of the mean-square amplitudes and rates of the motions. Returning back to Figure 7.2a, a progressive increase in the order parameters is seen with a greater mole fraction (X_C) of cholesterol. Interaction with the rigid sterol frame yields a pronounced reduction of the degrees of freedom of the flexible phospholipids. The effect of cholesterol is to increase the order parameters proportionately for the various acyl segment positions. The well-known condensing effect (decrease in area per molecule) is due to an increase in the acyl chain projections along the bilayer normal (director) [73]. By contrast, for the polar head group segments there is no such increase in the order parameters, as first shown by Brown and Seelig [112]. Because of the umbrella effect, the rigid sterol moiety is situated deeper in the bilayer – it acts as a spacer insofar as the polar head groups are concerned [112].

Analogous measurements of the nuclear spin relaxation rates are possible by perturbing the spin system with a suitable radiofrequency pulse sequence, and then following the return back to equilibrium [60]. Both the $S_{CD}^{(i)}$ order parameters and $R_{1Z}^{(i)}$ relaxation rates for multilamellar lipid dispersions decrease along the acyl chains as a function of segmental index (i) in the liquid-crystalline (ld) state, that is, on going from the aqueous interfacial region toward the bilayer hydrocarbon core (Figure 7.2). The approximate plateau for both experimental observables as a function of segmental position is because of the effects of tethering the acyl chains to the aqueous interface [55, 56]. The order profiles for DMPC-d_{54} show an increased disorder toward the bilayer center, accompanied by a reduction in the acyl segmental relaxation rates. Most arresting, as the molar ratio of cholesterol in the bilayer increases, there is an opposite effect on the NMR observables: in Figure 7.2a the order parameters increase, while in Figure 7.2b there is a reduction in the spin-lattice relaxation rates. How can we explain these apparently contradictory experimental results?

7.5.3 Space and time revisited

Let us next return to the question: What are the types of motions that govern the properties of phospholipid membranes, including their rates, amplitudes, and energetics? And how can we explain the opposite influences of cholesterol on the acyl segmental order profiles compared to the relaxation rate profiles? In discussing the lipid conformations in liquid-crystalline phases, the various local structures can be expected to have different lifetimes. It is then paramount to consider the spectroscopic timescale used to investigate the distribution. Undoubtedly, in the case of NMR and spin-label EPR, the couplings involve direct and indirect dipolar (hyperfine) interactions, quadrupolar interactions, and chemical shielding (Zeeman) interactions. For comparing spin-label EPR studies to NMR studies of lipid bilayers, the magnitudes of the various couplings need to be kept in mind. The reason is that structural transitions can occur with rates greater than or approximately equal to the anisotropy of the coupling (e.g., dipolar, quadrupolar, and chemical shift), leading to averaging of the interactions. McConnell pointed out this aspect very clearly when he stated:

> The angular brackets... denote a time average over a characteristic time T. For nitroxide spin labels the time T is determined by the reciprocal of the anisotropy of the hyperfine interaction and Zeeman interactions, in frequency units. This places T in the range 10^{-7}–10^{-9} sec [8].

In the example of spin label EPR, the reciprocal anisotropy of the hyperfine coupling is approximately $1/(T_\| - T_\perp)$ and falls in the range of $\approx 10^{-8}$ s [123]. This timescale is comparable to the motions detected by NMR spin-lattice (Zeeman) relaxation, which also detects motions in the range of $1/\nu_0 \approx 10^{-8}$ s [1]. Even so, the anisotropy of the static quadrupolar coupling in solid-state ^{2}H NMR spectroscopy is smaller, and is $1/[(\nu_Q^\pm)_\| - (\nu_Q^\pm)_\perp] \approx 5 \times 10^{-6}$ s for the two individual spectral branches [80].

We can then ask: How are the NMR results connected to those of spin-label EPR spectroscopy? Certainly, it is thought provoking that spin-label EPR and ^{2}H NMR spin-lattice relaxation both detect motions on a similar timescale, while solid-state ^{2}H NMR spectra include additional motions at lower frequencies. Here the introduction of a local director can aid in further explaining the spectroscopic observables. Spin-lattice (T_{1Z}) measurements detect fluctuations near $1/\nu_0 \approx 10^{-8}$ s as already mentioned earlier, which is near the limit of the motional averaging in conventional EPR spectroscopy [8]. Provided the relaxation entails ODF, conformations with lifetimes of $\approx 10^{-7}$–10^{-9} s that are long-lived for spin-label EPR spectroscopy might contribute to the nuclear spin-lattice relaxation, yet be further averaged by the more than 100-fold greater timescale in solid-state ^{2}H NMR spectroscopy. Local perturbations in the vicinity of the spin label may also be involved [54, 124]. Here we simply note that for lipid bilayers the concept of a local director axis [1] implies it can vary in both space and time. One should also recall that the director axis is a property of the phase, rather than an individual molecule (see above discussion). It depends on the assembly of flexible molecules within the bilayer, for example, the local director can vary over small distances, approaching the segmental or molecular size [58]. The presence of a local director frame (e.g., due to slowly relaxing local structures of the lipids) might explain why transient local chain tilting can occur in the vicinity of a spin label, which might not be detected by solid-state ^{2}H NMR because of the longer timescale [1].

7.5.4 Dimensionality of the order-director fluctuations

Moreover, the dimensionality of the order fluctuations can depend on the spectroscopic timescale involved. The next question is whether a 3D membrane deformation model or a 2D flexible surface model is appropriate. As an example, the spin-lattice (T_{1Z}, Zeeman) measurements are sensitive to the spectral density of the motions near $\approx 10^{-8}$ s, and can detect 3D ODF (quasi-nematic) at the level of the local structures of the hydrocarbon chains (e.g., as seen by spin-label EPR). For 3D ODF the elastic fluctuations might not be so strongly constrained by the polar interface with water, in contrast to 2D fluctuations due to the membrane surface. One can also consider that there are high-frequency ODF [1] but their dimensionality is 2D, as for a smectic liquid crystal, rather than 3D as for a nematic liquid crystal. Alternatively there may be no high-frequency ODF at all – rather, a non-collective model describes the order fluctuations in the MHz range, due to whole-molecule rotational modes of flexible lipids.

Still, a bilayer has a finite thickness as investigated experimentally by X-ray and neutron scattering methods [73, 125]. Collective motions over larger distances can then evolve from 3D into 2D ODF (smectic-like) that are governed by the bending rigidity, as in the case of lower frequency transverse relaxation (R_2) measurements (kHz range). Starting at high frequencies, an $\omega^{-1/2}$ law can apply (MHz range) due to quasi-nematic ODF at short distances ($\lambda < t$), as first discussed by the author [1], where the local director might not extend beyond the bilayer thickness (t). As the nuclear resonance frequency (magnetic field strength) decreases over longer distances on the order of the bilayer thickness and beyond ($\lambda > t$), the dynamic behavior can increasingly pick up smectic-like character, due to the boundary condition of the aqueous interface. The fluctuations can then involve smectic-like ODF at lower frequencies, where an ω^{-1} law occurs (kHz range) due to bending of the membrane interface with water [104, 126]. The idea of collective disturbances of the bilayer has been subsequently adopted in work involving transverse nuclear spin relaxation [127–130]. It is assumed that the elastic modes occur at long wavelengths and low frequency (kHz), because the molecular motions yield a cutoff for order fluctuations in the MHz range. The low-frequency motions amount to 2D undulations of a flexible surface of relatively small amplitude, as opposed to 3D ODF of potentially larger amplitude. Such collective motions can be important for processes involving membrane fusion, interbilayer forces, and lipid–protein interactions.

At this stage we conclude that the introduction of a local director can account for slowly relaxing local structures (e.g., transient local tilt of portions of the chains) [1, 58, 131]. The idea of a broad distribution of collective bilayer tilt disturbances suggests there may be two regimes of the relaxation dispersion due to harmonic or elastic excitations. Depending on the cutoff length, either nematic-like or smectic-like fluctuations can occur due to the hydration levels and the types of polar head groups. In the high frequency MHz regime, primarily 3D collective fluctuations of the local structures govern the spin-lattice relaxation, which overlaps the spin-label EPR timescale [32, 132]. Contributions from collectively tilted local structures, with lifetimes intermediate between the ^{2}H NMR ($\approx 10^{-5}$ s) and spin label EPR ($\approx 10^{-7}$–10^{-9} s) timescales, might average the ^{2}H NMR quadrupolar coupling, but not the ^{14}N hyperfine interactions in spin-label EPR spectroscopy [62, 84]. At longer distances, the bilayer aqueous interface can yield a transition from a frequency dispersion governed by 3D ODF to 2D ODF as the limit. In the lower frequency kHz range, a 2D collective model may then be applicable corresponding to transverse NMR relaxation time studies [104, 128, 130].

7.6 The paradigm shift to order-director fluctuations

The preceding discussion builds on the idea of a broad distribution of harmonic or quasi-harmonic (elastic) excitations in lipid bilayers, as originally put forth by the author [1]. As an approximation, we can neglect both the short and long wavelength cutoffs for the distribution, which are taken as zero and infinity, respectively. It is also possible that breaks in the frequency dispersion law (power spectral density) can occur. For example, there can be a transition from the regime of 3D ODF (membrane deformation model) with an $\omega^{-1/2}$ dispersion law to the 2D ODF regime (flexible surface model) having an ω^{-1} power law [126, 132]. In effect, the long wavelength cutoff for the 3D regime becomes the short wavelength cutoff for the 2D fluctuations (e.g., on the order of the bilayer thickness). Moreover, the elastic constants and amplitudes of the fluctuations can be different. According to this view, the 3D regime has an $\omega^{-1/2}$ dependence in the MHz range that overlaps the spin-label EPR timescale. In the lower frequency kHz range, the properties of the aqueous interface give an increasingly smectic-like behavior, with an ω^{-1} dispersion law [126, 132]. The 2D collective motions include flexible surface undulations that contribute mainly to the transverse (T_2) relaxation in NMR, for example, they can show up as inhomogeneous line broadening [8] in spin-label EPR measurements.

The idea of a local director axis that extends over a portion of the lipid or the bilayer thickness can be helpful to understanding the multi-scale physics of the bilayer fluctuations involving slowly relaxing local structures. For a 3D collective model (i.e., membrane deformation model), the relaxation dispersion scales as $\omega^{-1/2}$ [1, 58, 132, 133], whereas for 2D collective motions (i.e., flexible surface model), the relaxation goes as ω^{-1} [104, 126, 130, 132]. One can then ask: What is the contribution to the relaxation rates from ODF assuming plausible values for the various physical constants? For a 3D collective model, the properties of the hydrocarbon core are considered. The theoretical viscoelastic coefficient assuming an $\omega^{-1/2}$ frequency (magnetic field) dispersion law is $D = (3/5)(k_B T/\pi S_{\text{slow}}^2)(\eta/2K^3)^{1/2}$, while the experimental value for an $\omega^{-1/2}$ power-law dispersion is $D \approx 2 \times 10^{-5}$ s$^{1/2}$ [132]. Substituting $S_{\text{slow}} \approx 0.6$ for the slow order parameter [1, 132], a viscosity of $\eta \approx 1$ cP [58, 134], and an elastic constant of $K \approx 0.2 \times 10^{-11}$ N [93] gives a value of $D = 1.7 \times 10^{-5}$ s$^{1/2}$ in agreement with 3D collective motions. (No distinction is made between twist, splay, and bend modes, and moreover the compression modulus B may come into play [41].) It follows that the physical significance of a 3D collective ODF model is supported by its consistency with the relevant material constants.

On the other hand for a 2D collective model (i.e., flexible surface model), properties of the bilayer interface with water may become an important factor. Fluctuations of a deformable surface are involved (despite that the bilayers have a finite thickness) [73, 125]. The theoretical value of the elastic coefficient (there is no viscosity due to lack of coupling to a third dimension) is $D' = (3/5)(k_B T/S_{slow}^2)/2K_C$, while the experimental value assuming an ω^{-1} power-law dispersion is $D' \approx 0.5$ [132]. Inserting experimental values or estimates of $S_{slow} \approx 0.6$ [132] and a curvature elastic modulus (bending rigidity) of $K_C = 0.6 \times 10^{-19}$ J [135] yields $D' = 0.06$, which is about an order of magnitude smaller than the experimental value. Apparently an ω^{-1} power-law dispersion in terms of undulations does not fit the experimental frequency (magnetic field) dependence of the spin-lattice relaxation in the high-frequency MHz regime [132, 136]. Then again, it may be appropriate for lower frequency (kHz range) motions, as studied by transverse Carr–Purcell–Meiboon–Gill (CPMG) relaxation dispersion experiments [104, 128, 137].

As a matter of fact, the idea of collective motions as governing the relaxation has been surprisingly controversial. Although the formulation of ODF for lipid bilayers as first proposed for high-frequency motions [1] has been debated, the identical concept has been reintroduced subsequently when it comes to lower frequency motions. In keeping with this view, the high frequency dispersion (MHz range) includes non-collective molecular rotations within the force field of the membrane lipid bilayer. Collective membrane motions contribute to the relaxation, but only at low frequencies (kHz range), where their dimensionality is 2D and they are of small amplitude. Evidently, the debate centers around the amplitude and timescales of the collective disturbances, and their connection to magnetic resonance observables (solid-state NMR or spin-label EPR spectroscopy). A further aspect is that a simple physical significance of the theoretical approach should emerge in combination with experimental measurements [58].

7.6.1 Fermi's golden rule

Consistent with the above-mentioned thinking, NMR spectroscopy provides us with essential experimental information about the structural dynamics of biomolecules that is largely unobtainable with other biophysical methods. For flexible membrane lipids, a distribution of RQCs or RDCs is evident that manifests the bilayer structural quantities, such as the area per lipid at the aqueous interface and the volumetric bilayer thickness [73]. Most significantly, the NMR relaxation times tell us about the types of motions that yield the motional averaging within the membrane bilayer, including their amplitudes and rates. The question is then whether the NMR relaxation reports on the local segmental lipid motions, or rather the collective or molecular fluctuations of the lipids. Put in the language of NMR spectroscopy: Does the relaxation entail modulation of the static coupling, or the residual couplings pre-averaged by faster segmental motions? Moreover: What is the connection to bilayer material properties and/or biological functions?

It is here that the influences of bilayer additives such as cholesterol [138] or nonionic surfactants [139] can further inform the analysis of bilayer fluctuations, as seen by magnetic resonance spectroscopy. Undeniably, cholesterol is one of the most important chemical compounds found in association with the lipid hydrocarbon chains in a biological setting [61, 138, 140, 141]. Very few biomolecules have been scrutinized to the same extent as cholesterol, for example, as in the studies conducted by McConnell et al. [14, 142]. The sterol intercalates between the lipids, and causes a dramatic stiffening of the bilayer in conjunction with the well-known condensing effect. Coming back to Figure 7.2 as a specific example, a clear influence of cholesterol is evident for both the order profiles and the relaxation profiles. Most conspicuous, there are opposite effects of cholesterol on the order parameters (S_{CD}) and spin-lattice (R_{1Z}) relaxation rates, as already noted. For cholesterol-stiffened bilayers, large absolute order parameters and low relaxation rates are measured. But as the mole fraction of cholesterol X_C in the lipid mixtures increases, the R_{1Z} rates become smaller – exactly opposite to the effect of cholesterol on the order profiles in Figure 7.2a. How can we explain these apparently contradictory findings?

Indeed, Fermi's golden rule gives a simple explanation of the correspondence of the relaxation rates and the order parameters, whereby the spectral transition probabilities depend on the squared interaction strength. Referring now to Figure 7.3, we see that theory predicts – and experiments confirm – a remarkably simple square-law functional dependence for the relaxation due to slow order fluctuations. For if we transform the same data in Figure 7.2 into a square-law plot, the previously confusing results fall neatly into line with the stiffening effect of cholesterol on the bending rigidity of the membrane [135]. The bilayer lipids modulate the residual couplings left over from the local segmental isomerizations, for example, ODF, as first proposed by the author [1]. Most arresting, the experimental results in Figure 7.3 are totally model free. We can conclude from the square-law signature [32, 77] that the relatively slow order fluctuations govern the NMR relaxation of lipid bilayers [1, 132]. Local segmental motions modulate the same static coupling tensor for all the segment positions, while the slow order fluctuations involve the residual couplings pre-averaged by the faster local motions. Modulation of the site-specific couplings remaining from the local segmental motions is what produces the NMR relaxation for the membrane lipids – what we call order fluctuations [1].

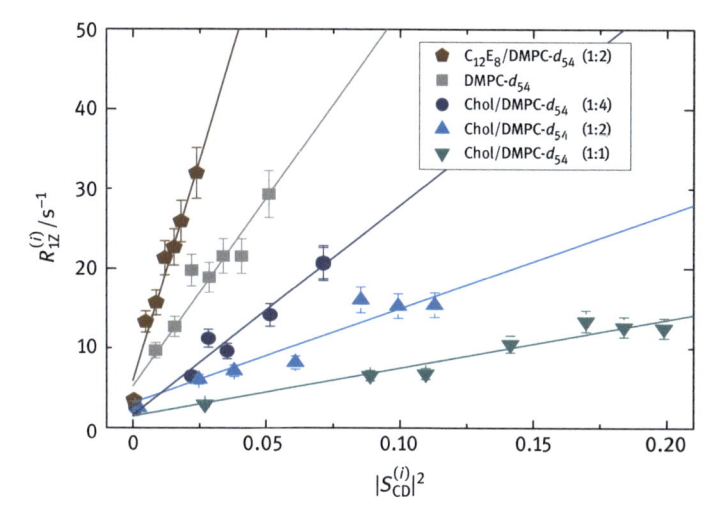

Figure 7.3: Functional dependence of solid-state ^{2}H NMR relaxation rates and order parameters reveals collective fluctuations in lipid membranes. Experimental data are the same as in Figure 7.2. The ^{2}H $R_{1Z}^{(i)}$ relaxation rates are plotted versus the corresponding $|S_{CD}^{(i)}|^2$ values for acyl segments of DMPC-d_{54} with different molar ratios of cholesterol or detergent. The spin-lattice relaxation rates $R_{1Z}^{(i)}$ (mobility gradient) depend on the squared order parameters $S_{CD}^{(i)}$ (flexibility gradient) for the individual acyl segments (index i). Changes in square-law slope correspond to the macroscopic bilayer elasticity. Lipid membranes are stiffened by cholesterol (Chol) and yield a progressive reduction in slope in the lo phase. By contrast, softening by the nonionic surfactant $C_{12}E_8$ gives an opposite increase in square-law slope. The ^{2}H NMR measurements were conducted at $T = 42$ or $44\,°C$ and 76.8 MHz (11.7 T). Data are replotted from Refs Martinez et al. [60] and Otten et al. [149].

7.6.2 Quasi-elastic bilayer deformation

What is the physical significance of the model-free square-law functional dependence? To continue, Figure 7.3 illustrates how the square-law slopes observed for DMPC-d_{54} bilayers depend on the presence of either cholesterol or nonionic surfactants. One indication is a progressive diminution in the square-law slopes for the bilayer liquid-ordered (lo) phase with increasing mole fraction of the rigid cholesterol molecules, matching its effects on the bulk bilayer elasticity (Figure 7.3). Lanosterol is known to stiffen the lipid bilayer less than cholesterol [143, 144], which also agrees with the solid-state ^{2}H NMR analysis [145]. A plausible inference involves the elastic properties of bilayers containing lanosterol versus cholesterol [146, 147]. Then again, an opposite softening is produced by nonionic surfactants such as $C_{12}E_8$, which are chemical compounds important for pharmaceutical and cosmetic formulations [139, 148]. Here also a clear parallel exists with the NMR square-law slope for the combined order parameter and relaxation observables (Figure 7.3) [98, 145, 149]. Low absolute order parameters are observed for surfactant-softened bilayers, along with enhanced relaxation rates, giving a large square-law slope (Figure 7.3). The increase is due to the greater number of degrees of freedom arising from the chain configurational disorder. Opposite influences of detergents and sterols underlie detergent-resistant, raft-like lipid microdomains in biomembranes [150]. (Here we recall that like dissolves like as in our introductory chemistry courses.)

From the previously mentioned results, it follows that the model-free, square-law functional dependence of the relaxation (R_{1Z}) rates and the order parameters (S_{CD}) can be interpreted by quasi-elastic deformations of the lipid bilayer. Such harmonic excitations occur on the mesoscopic length scale of the bilayer thickness, and even less [1]. The combined order parameters (S_{CD}) and relaxation rates (R_{1Z}) inform the stiffening or softening of the membrane that emerges from the local atomistic level interactions. Changes in membrane properties are explained using a theoretically predicted, square-law proportionality between the RQCs and relaxation rates, as first proposed [1]. The combined ^{2}H NMR relaxation and order parameter data manifest the order fluctuations due to the collective interactions of the entangled membrane lipids. Even more striking is the correspondence of the atomistic NMR observables with bulk material properties, for example, as investigated using micropipette deformation [135], or shape fluctuations of giant unilamellar vesicles [151, 152]. The opposite effects of cholesterol and detergents – while maintaining a square-law dependence – are an arresting discovery. It supports our original contention that the NMR observables manifest quasi-elastic bilayer deformation on the mesoscopic and even atomistic level [1]. Hydration-mediated collective dynamics of phospholipid membranes are similarly revealed by ^{2}H NMR transverse relaxation rate (R_2) measurements [153]. Essentially the same intermolecular forces and potentials govern the atomistic-level fluctuations, as well as deformation of the membranes by an external disturbance (fluctuation–dissipation theorem).

7.6.3 Relaxation dispersion and unification of experimental relaxation laws

Up until now we have tended to focus on the square-law signature of relatively slow order fluctuations of membranous soft matter. Still, in membrane biophysics the dependencies of the relaxation rates on frequency (magnetic field strength) and ordering (mean-square amplitude) are likewise pivotal to interpreting the spectroscopic observables. We next put forth a unified view of the combined frequency (magnetic field) and order dependence of the relaxation. Clearly, the dynamics of membranes are the same when studied using either ^{2}H or ^{13}C spin probes. Yet the power spectral densities of the fluctuations for the two nuclei near the resonance (Larmor) frequency vary on account of the different coupling mechanisms. The frequency dispersive behavior of the relaxation depends on the nucleus studied. In consequence, by analyzing the relaxation data for the two nuclei, we can cover a larger effective frequency interval than for a single nucleus alone. Fits of the ^{2}H and ^{13}C relaxation rates to a composite membrane model yield the contributions of fast local segmental motions (R_{1Z}^{fast}) and slower motions (R_{1Z}^{slow}) to the frequency dispersion. Cross-correlations are neglected as a first approximation, giving $R_{1Z} = R_{1Z}^{fast} + R_{1Z}^{slow}$ for the observed relaxation rate [1, 106]. The order parameter S_{fast} describes local segmental motions, and the S_{slow} order parameter includes the slower motions (either molecular or collective). Here $S_\lambda = S_{fast} S_{slow} = \langle P_2(\cos\beta_{PD})\rangle$ where $S_\lambda \equiv S_{CD}$ or S_{CH} and the rotational correlation time τ_C is on the order of $\approx 10^3$ bond vibrational periods. The local dynamics do not appreciably affect the frequency dispersion in the short correlation time limit. However, slower motions such as non-collective molecular rotations and collective bilayer excitations contribute to the frequency dispersion of the R_{1Z}^{slow} relaxation rate.

Additionally, the range of the frequency dispersion can be expanded through consideration of the multinuclear spin relaxation rates. The overall relaxation rates arising because of molecular motions R_{1Z}^{mol} and/or collective motions R_{1Z}^{col} can be investigated in terms of a unified ^{13}C and ^{2}H frequency scaling law [32, 154] (see Figure 7.4). The scaled rates (denoted here by $\tilde{R}_{1Z}$) show the unification of the quadrupolar and dipolar relaxation for the acyl group $(CH_2)_n$ resonance of the DMPC bilayer in the ld phase. It is arresting that the ^{2}H and ^{13}C spin-lattice relaxation rate dispersions measured for the same lipid segments collapse to a single plot as shown in Figure 7.4a. A scaled or reduced relaxation rate $\tilde{R}_{1Z}$ is obtained, such that a

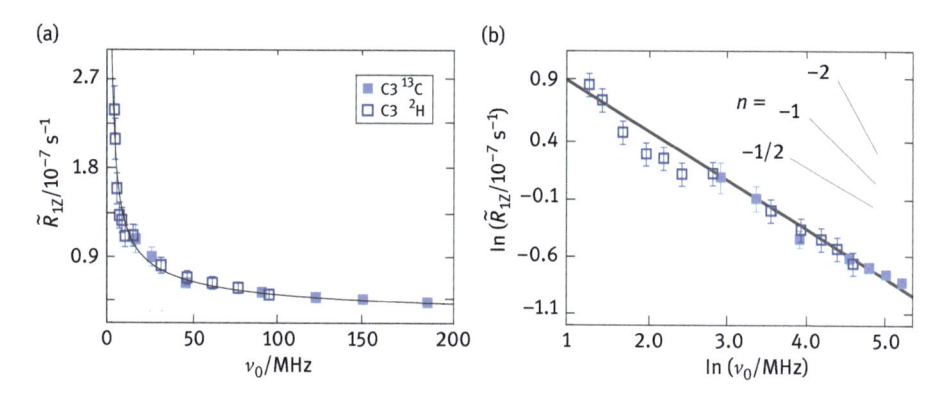

Figure 7.4: Multinuclear spin relaxation rates for liquid-crystalline bilayers are unified by a simple frequency power-law. Scaled ^{2}H NMR and ^{13}C NMR spin-lattice relaxation rates $\tilde{R}_{1Z}$ are compared ($\nu_0 \equiv \nu_C, \nu_D$) and collapse to a single curve. (a) Relaxation rate dispersions for natural abundance DMPC and isotopically enriched 1, 2[3′, 3′-^{2}H] DMPC are shown at $T = 30$ °C. The power-law dispersions for the C3 position in ^{13}C NMR and ^{2}H NMR are fit by a single power-law function (———) with $n = -1/2$ as the exponent for 3D collective order fluctuations. (b) Double-logarithmic plots of scaled relaxation rates fit to various power-law frequency scalings are shown. Power-law exponents are shown for $n = -2, -1$, and $-1/2$ due to non-collective molecular rotations, 2D collective order fluctuations (flexible surface model), and collective 3D order fluctuations (membrane deformation model). A 3D collective model describes the frequency (magnetic field) dispersion of the relaxation. Figure adapted from Ref Leftin and Brown [32].

simple power-law trend is followed for the collective dynamics, as shown in Figure 7.4b. Notably the combined dispersions obey a three-dimensional power-law ($\omega^{-1/2}$; $d = 3$) spanning nearly the full MHz regime, from $\omega_D/2\pi = 2$ MHz to $(\omega_C + \omega_H)/2\pi = 939$ MHz for the carbon acyl chain segments [132]. Evidently it follows that a simple 3D collective model (membrane deformation model) can unify the combined ^{2}H and ^{13}C spin-lattice (R_{1Z}) relaxation rates, consistent with quasi-elastic excitations as the origin of the relaxation in the liquid-crystalline state. Although an ω^{-1} frequency dispersion has also been proposed [96, 126, 130], it does not seem to be supported for the acyl chain segments. We can conclude at this point that the measurable relaxation rates are described by a simple relaxation law of the form $R_{1Z}^{(i)} = A\tau_C^{(i)} + B|S_{CD}^{(i)}|^2 \omega^{-1/2}$ where A, B are constants, which encapsulates the dependence on the order parameter and the resonance frequency [1].

Nevertheless with regard to the phospholipid head groups, a different experimental relaxation law has been observed for lipid bilayers in the case of transverse relaxation time measurements. A 2D power-law scaling (ω^{-1}; $d = 2$) for the ^{31}P NMR phospholipid head group dynamics has been identified by transverse relaxation (T_2) measurements in the kHz regime by Gerd Kothe and his coworkers [130, 137], while for the acyl chains such a relaxation law is not supported by the experimental data as shown in Figure 7.4b. Interestingly, the difference in power-law scaling may reflect the greater sensitivity of the head group ^{31}P nucleus to the surface or smectic-like undulations. By contrast, measurements using ^{2}H or ^{13}C NMR as a probe of the acyl chain dynamics may be indicative of the 3D interior of the bilayer hydrocarbon core. Despite the changed power-law exponent, the analysis suggests that the nonexponential relaxation dispersion arises from low-frequency motions due to highly damped ODF, as first proposed [1]. An alternative involves distinct rotational modes of the head group segments [94, 102], as discussed subsequently.

7.7 Collective or non-collective lipid motions revisited

In the case of phospholipids the acyl chains are highly entangled, and additionally the polar head groups have dipolar and hydrogen-bonding interactions with the neighboring molecules. What are the types of molecular motions that occur in fluid, liquid-crystalline bilayers? Let us now return to the question of whether we can identify discrete modes for the rotational dynamics of phospholipids in the liquid-crystalline state. The most arresting distinction is whether the dynamics correspond to a non-collective molecular model, for example, as in the case of molecular rotations within the potential of mean force of the lipid bilayer. Here the moments of inertia are averaged over the internal degrees of freedom of the molecule. Alternatively, as described above, we can consider the quasi-elastic (harmonic) excitations of the bilayer lipids that are broadly distributed in space and time. The dynamics can be inherently collective due to slowly relaxing structures of the flexible lipids. Both views have been discussed [1] yet a clear distinction remains under debate. How can we formulate the various lipid motions in terms of their types, amplitudes, and rates as characteristic of the molecular dynamics within the possible nanostructures?

7.7.1 Rotational modes and inertial averaging

Possibly the simplest extension of well-established concepts for simple liquids or complex fluids is to approximate the molecular motions as those of a rigid body, with discrete rotational modes [105]. The motions of the flexible phospholipid molecules might then be understood by inertial averaging over the internal degrees of freedom. One can assume a single averaged inertial tensor over the various length and timescales considered, that is, corresponding to the molecular size and/or the bilayer thickness. Non-collective molecular lipid rotations and wobble can occur around the principal axes of the motion-averaged inertial tensor. By considering the averaged moments of inertia, the molecular rotations around the principal axes of the diffusion tensor would entail the mean-torque potential due to all the other lipids of the membrane [155]. But what are the internal coordinates for the internal averaging – segments, chains, entire molecules, or collections of molecules? The flip side is that this approximation bypasses the molecular flexibility. As noted earlier, both spin-label EPR [7] and solid-state ^{2}H NMR spectroscopy [54, 55] give a profile (flexibility gradient) of the order parameters versus the depth within the bilayer. Entanglements of the membrane phospholipid molecules can occur over the various length scales. Even if we can make a timescale separation, it is still open to question whether the integration should extend over all the internal degrees of freedom of the molecules for a given time interval. Alternatively, one can consider a distribution of length and timescales, which for a continuum approximation essentially brings us back to a simple picture in terms of ODF [1, 97].

7.7.2 What is the connection to experimental NMR relaxation data?

For reasons of brevity, the question of molecular rotational modes in lipid bilayers in the liquid-crystalline state will not be addressed here in any detail. For a non-collective molecular model [1, 156, 157], a key prediction is that the modes predict a minimum in the spin-lattice relaxation times (T_{1Z}) versus temperature [1, 158], as shown experimentally for membrane proteins [1, 27, 158]. The relaxation is most efficient when the power spectrum of the stochastic molecular fluctuations (spectral density) is optimally matched to the (coherent) nuclear resonance frequency, which occurs when the inverse correlation time $1/\tau_C \approx \omega_0 = 2\pi\nu_0$ where ν_0 is the Larmor frequency. But if an interpretation in terms of discrete rotational modes applies, why has a minimum in the spin-lattice relaxation times [158] never been observed at any temperature or frequency for lipid bilayers in the liquid-crystalline state? Evidently a relaxation minimum is found mainly for lipid systems with a large reduction in the configurational degrees of freedom – as in the solid-ordered (so) (gel) state, or when cholesterol is present in the lo state [95, 96] – but not for the liquid-disordered (ld) state.

Arriving at this juncture, we can say the following: consideration of a mean-field potential picture is of course possible. But what do we gain by analyzing the non-collective motions (rotational modes) for individual molecules within the mean field due to all the other molecules of the bilayer? In effect, we interpret our assumption rather than testing its validity. To our knowledge, for lipids in the liquid-crystalline state, no experimental proof for rotational modes of the entire lipid molecules or even portions of the flexible lipid molecules has been obtained, except for the phosphodiester moiety [94], for example, in contrast to microwave spectroscopy of small molecules in the gas phase [159]. Perhaps such a relaxation (T_{1Z}) minimum is not observed because distinct rotational modes of the lipid molecules – corresponding to a unique inertial tensor that describes the (non-collective) molecular motions – are not present. Alternatively a broad distribution of relaxation modes can exist, consistent with quasi-elastic (harmonic) excitations, spanning the bilayer dimensions down to the segmental size, as we have proposed [1]. By analogy with liquid crystals, we can treat the latter as ODF.

7.8 The emergence of membrane elasticity

The above-mentioned findings suggest that for bilayers in the liquid-crystalline state, we can put aside our molecular view for a continuum picture, with a distribution of harmonic or quasi-elastic excitations of the flexible lipids. Because the lipid motions are inherently collective, it is open to discussion whether a picture applies in terms of stochastic modes due to non-collective molecular rotations. The alternative is a composite membrane deformation model, whereby the molecules are entangled, with collective interactions due to their tethering to the aqueous interface. Collective fluctuations in the local ordering of the lipids give a broad continuum of elastic, wave-like disturbances. These order fluctuations explain the frequency dependence of the nuclear spin relaxation rates (R_{1Z} and R_2) by a distribution of correlation times for membrane liquid crystals. Seemingly at short distances, a nematic-like 3D model can approximate the ODF of fluid membranes. For distances much greater than the bilayer thickness, however, a transition to 2D smectic-like order fluctuations of an elastic sheet can occur. A remarkably simple square-law functional dependence of the experimental relaxation rates and order parameter profiles of the acyl groups follows in either the 3D or the 2D limits (Figure 7.3). The underlying explanation is the emergence of bilayer elasticity over the short distances [58]. Invoking the equipartition theorem, with first-order (exponential) (over)damping of the modes, we readily get to the predicted frequency dependence of the relaxation rates in a continuum approximation. In this way, we find an $\omega^{-1/2}$ frequency dependence for 3D nematic-like fluctuations of the hydrocarbon core; for a 2D elastic sheet, an ω^{-1} law is obtained, and lastly for elastic fluctuations of a 1D string an $\omega^{-3/2}$ dependence is present [132]. Collective modes emerge on the mesoscale of the bilayer thickness and less, and are connected with the bulk membrane elasticity [151, 160–162] by a distribution of slowly relaxing structures, that we formulate as ODF [60, 97]. Hence, we arrive at a new view of the lipid bilayer as a membrane liquid crystal [58] – albeit, one endowed with physicochemical properties of high biological importance.

7.8.1 Physical significance of the order-director fluctuations

At any rate, the most telling aspect entails the following question: What is the physical significance of the combined NMR order parameter and relaxation approach in terms of the multi-scale lipid dynamics? And how are the order fluctuations related to the corresponding lipid membrane properties? To this end, Figure 7.5 gives us a summary of previous data for the fatty acyl chains of multilamellar dispersions and small vesicles of 1,2-dipalmitoyl-sn-glycero-3-phosphocholine (DPPC) in the liquid-crystalline state [58]. In Figure 7.5a, the square-law functional dependence of the ^{2}H R_{1Z} relaxation rates and ^{2}H NMR order parameters along the acyl chains is depicted for DPPC multilamellar dispersions. Remarkably, in accord with the equipartition theorem from statistical mechanics, the harmonic excitations extend down to a length scale approaching

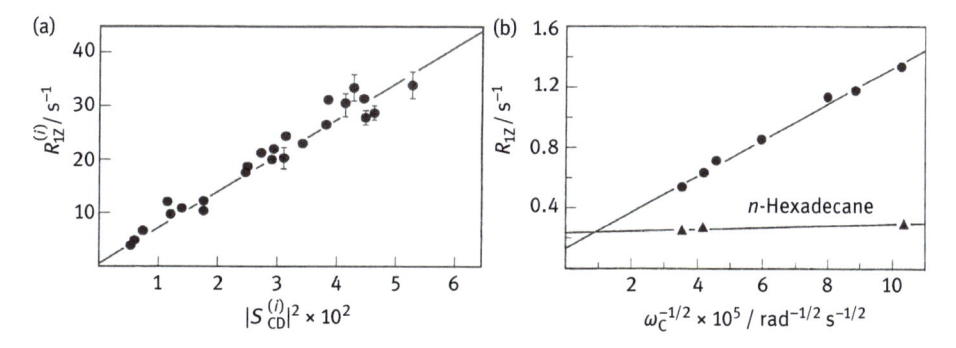

Figure 7.5: Experimental results from NMR spectroscopy indicate the physical significance of bilayer fluctuations. (a) Plot of ^{2}H spin-lattice relaxation rates $R_{1Z}^{(i)}$ versus squared order parameters] $|S_{CD}^{(i)}|^2$ for acyl segments of multilamellar dispersions. (b) Plot of ^{13}C spin-lattice relaxation rates R_{1Z} for (CH$_2$)$_n$ resonance against $\omega_C^{-1/2}$ for small unilamellar vesicles. Data are summarized for DPPC in the liquid-crystalline (also known as liquid-disordered, ld) state at $T = 50$ °C. Theory indicates that both dependencies arise from relatively slow order fluctuations due to collective bilayer excitations. ODF are due to collective lipid motions and correspond to emergent viscoelastic properties of the lipid bilayer. Figure adapted with permission from Ref Brown et al. [58].

the segmental or atomistic dimensions [60, 97]. In addition, the magnetic field dependence (frequency dispersion) of the NMR relaxation rates corresponds to various simplified power laws that manifest the types of fluctuations. As a further example, the $\omega_C^{-1/2}$ frequency-dispersion found in ^{13}C R_{1Z} relaxation studies of small DPPC vesicles is shown in Figure 7.5b. Both dependencies are indicative of the fluid, ld phase of membrane lipid bilayers. To all appearances, for harmonic excitations, the integration over the entire bilayer (neglecting the high- and low-frequency cutoffs) explains the simple $\omega^{-1/2}$ frequency dispersion law [1]. By contrast, Figure 7.5b shows the ^{13}C R_{1Z} rates of liquid paraffins such as n-Hexadecane are approximately independent of frequency – that is to say, the magnetic field strength – over the entire range is considered.

7.8.2 The microviscosity of a lipid bilayer

Equally important, according to Figure 7.5b there is clearly an enhancement in the relaxation of lipid bilayers versus simple hydrocarbon fluids. The local ordering of the lipids, for example, due to segmental isomerizations, is modulated by additional fluctuations of larger orientational amplitude. Apparently, the contributions from the order fluctuations make the lipid relaxation in the NMR frequency range (kHz–MHz) more effective than for hydrocarbon fluids. In simple physical terms, the enhancement depends on both the frequency, that is, the magnetic field strength, as well as the ordering or mean-square amplitudes of the fluctuations. Slower motions not found in simple hydrocarbon fluids are tied to the collective properties of the assembly. The spectral density due to the internal rapid chain fluctuations extends to relatively high frequencies (correlation times $\tau_C \approx$ 5–20 ps). Extrapolating the relaxation rates to zero ordering or infinite frequency gives us the local contribution to the relaxation rates, which matches liquid hydrocarbon taken as a reference. Consequently, the local microviscosity of the bilayer hydrocarbon core – where a bulk viscosity cannot be measured – amounts to a fluidity of only a few centipoises (cP), as originally proposed [1], and supported by the highly influential molecular dynamics simulations of Richard Pastor et al. [134].

It follows that collective bilayer excitations emerge over mesoscopic length scales between the molecular and bilayer dimensions, and are important for the lipid organization and protein interactions. A schematic cartoon (Figure 7.6) illustrates the collective excitations that may explain the frequency- and order-dependent relaxation enhancement of membrane bilayers (as drawn by the author) [58, 131]. By analogy to nematic liquid crystals [163], the collective dynamics are approximated by twist, splay, and bend deformations in the high frequency (or free membrane) limit, or alternatively by smectic undulations involving splay in the low frequency (or strongly coupled) limit [130]. Extension of such a continuum approach to distances approaching the acyl segmental dimensions approximates the many-body problem encapsulated by molecular dynamics simulations.

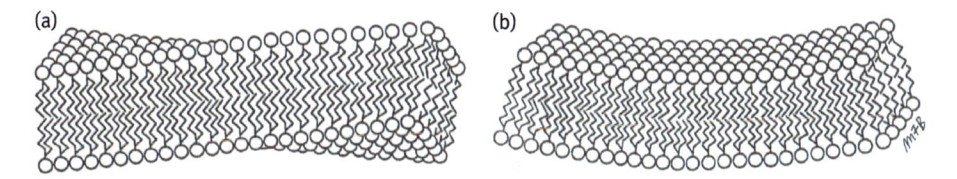

Figure 7.6: Examples of collective bilayer excitations showing the emergence of membrane elasticity: (a) twist and (b) splay. For clarity long wavelength modes are depicted. Figure adapted with permission from Ref Brown et al. [58].

7.9 Return to the future

Looking back, what made McConnell such an exemplary scientist and an inspiration to his many admiring colleagues entailed the interplay of experimental approaches with fundamental principles. By focusing on self-taught concepts and useful applications of physical chemistry, he emphasized simplifying generalizations. Following his approach, NMR spectroscopy allows us to explore the multi-scale structure and dynamics of biomembranes, giving complementary new insights to other biophysical methods. Solid-state NMR not only yields access to the lipid components of membranes but also the proteins [26, 158, 164–167]; and the same is true for spin-label EPR spectroscopy [22, 25, 168–170]. Increasingly, we can expect relaxation times to be combined with molecular spectroscopy [171–173], bearing down on the essential questions of membrane biophysics. Together with theoretical approaches, new experimental innovations bring with them a heightened awareness of how biomolecular structure, dynamics, and function are interrelated. Following the path set forth by McConnell and his colleagues, magnetic resonance approaches will continue to inspire and inform our appreciation of lipid biophysics both now and in the times to come.

Acknowledgments

This contribution is dedicated to the memory of Professor Harden M. McConnell. The author has also benefitted from discussions with Wayne Hubbell, Richard Pastor, and Attila Szabo. The U.S. National Institutes of Health and the U.S. National Science Foundation support research from the laboratory of the author.

References

[1] Brown, M.F. Theory of spin-lattice relaxation in lipid bilayers and biological membranes. ^{2}H and ^{14}N quadrupolar relaxation. J. Chem. Phys. 1982, 77, 1576–1599.
[2] Molugu, T.R., Lee, S., & Brown, M.F. Concepts and methods of solid-state NMR spectroscopy applied to biomembranes. Chem. Rev. 2017, 117, 12087–12132.

[3] Ferrarini, A., Nordio, P.L., Moro, G.J., Crepeau, R.H., & Freed, J.H. A theoretical model of phospholipid dynamics in membranes. J. Chem. Phys. 1989, 91, 5707–5721.
[4] Rheinstädter, M.C., Ollinger, C., Fragneto, G., Demmel, F., & Salditt, T. Collective dynamics of lipid membranes studied by inelastic neutron scattering. Phys. Rev. Lett. 2004, 93, 108107-1–108107-4.
[5] Fresch, B., Frezzato, D., Moro, G.J., Kothe, G., & Freed, J.H. Collective fluctuations in ordered fluids investigated by two-dimensional electron-electron double resonance spectroscopy. J. Phys. Chem. B. 2006, 110, 24238–24254.
[6] Brown, M.F. Soft matter in lipid-protein interactions. Annu. Rev. Biophys. 2017, 46, 379–410.
[7] Hubbell, W.L., & McConnell, H.M. Molecular motion in spin-labeled phospholipids and membranes. J. Am. Chem. Soc. 1971, 93, 314–326.
[8] McConnell, H.M. Molecular motion in biological membranes. In: L. J. Berliner, ed. Spin Labeling. Theory and Applications. New York, Academic Press, 1976, 525–560.
[9] Freed, J.H. Theory of slow tumbling ESR spectra for nitroxides. In: L. J. Berliner, ed. Spin Labeling. Theory and Applications. New York, Academic Press, 1976, 53–132.
[10] Brown, M.F., & Chan, S.I. Bilayer membranes: deuterium & carbon-13 NMR. In: D. M. Grant & R. K. Harris, eds. Encyclopedia of Nuclear Magnetic Resonance. New York, John Wiley & Sons, 1996, 871–885.
[11] Ge, M., & Freed, J.H. Hydration, structure, and molecular interactions in the headgroup region of dioleoylphosphatidylcholine bilayers: an electron spin resonance study. Biophys. J. 2003, 85, 4023–4040.
[12] Chiang, Y.-W., Shimoyama, Y., Feigenson, G.W., & Freed, J.H. Dynamic molecular structure of DPPC-DLPC-cholesterol ternary lipid system by spin-label electron spin resonance. Biophys. J. 2004, 87, 2483–2496.
[13] Smith, A.K., & Freed, J.H. Determination of tie-line fields for coexisting lipid phases: an ESR study. J. Phys. Chem. B. 2009, 113, 3957–3971.
[14] McConnell, H. Nuclear relaxation and critical fluctuations in membranes containing cholesterol. J. Chem. Phys. 2009, 130, 165103-1–165103-7.
[15] Smith, A.K., & Freed, J.H. Dynamics and ordering of lipid spin-labels along the coexistence curve of two membrane phases: an ESR study. Chem. Phys. Lipids. 2012, 165, 348–361.
[16] McConnell, H.M. Research studies of Harden M. McConnell. 2014. (Accessed February 5, 2019, at http://www.hardenmcconnell.org/.)
[17] Rommel, E., Noack, F., Meier, P., & Kothe, G. Proton spin relaxation dispersion studies of phospholipid membranes. J. Phys. Chem. 1988, 92, 2981–2987.
[18] Brown, M.F., Salmon, A., Henriksson, U., & Söderman, O. Frequency dependent ^{2}H N.M.R. relaxation rates of small unilamellar phospholipid vesicles. Mol. Phys. 1990, 69, 379–383.
[19] Halle, B. ^{2}H NMR relaxation in phospholipid bilayers. Towards a consistent molecular interpretation. J. Phys. Chem. 1991, 95, 6724–6733.
[20] Freed, J.H. Stochastic-molecular theory of spin-relaxation for liquid crystals. J. Chem. Phys. 1977, 66, 4183–4199.
[21] Meirovitch, E., Shapiro, Y.E., Polimeno, A., & Freed, J.H. Structural dynamics of biomacromolecules by NMR: the slowly relaxing local structure approach. Prog. Nucl. Magn. Reson. Spectr. 2010, 56, 360–405.
[22] Tombolato, F., Ferrarini, A., & Freed, J.H. Dynamics of the nitroxide side chain in spin-labeled proteins. J. Phys. Chem. B. 2006, 110, 26248–26259.
[23] Meirovitch, E., Shapiro, Y.E., Polimeno, A., & Freed, J.H. An improved picture of methyl dynamics in proteins from slowly relaxing local structure analysis of ^{2}H spin relaxation. J. Phys. Chem. B. 2007, 111, 12865–12875.

[24] Halle, B. The physical basis of model-free analysis of NMR relaxation data from proteins and complex fluids. J. Chem. Phys. 2009, 131, 224507-1–224507-22.
[25] Lerch, M.T., López, C.J., Yang, Z., Kreitman, M.J., Horwitz, J., & Hubbell, W.L. Structure-relaxation mechanism for the response of T4 lysozyme cavity mutants to hydrostatic pressure. Proc. Natl. Acad. Sci. U.S.A. 2015, 112, E2437–E2446.
[26] Meirovitch, E., Liang, Z., & Freed, J.H. Protein dynamics in the solid state from ^{2}H NMR line shape analysis: a consistent perspective. J. Phys. Chem. B. 2015, 119, 2857–2868.
[27] Xu, X., Struts, A.V., & Brown, M.F. Generalized model-free analysis of nuclear spin relaxation experiments. eMagRes. 2014, 3, 275–286.
[28] Engelman, D.M. Membranes are more mosaic than fluid. Nature. 2005, 438, 578–580.
[29] Brown, M.F. Curvature forces in membrane lipid-protein interactions. Biochemistry. 2012, 51, 9782–9795.
[30] Goñi, F.M. The basic structure and dynamics of cell membranes: an update of the Singer–Nicolson model. Biochim. Biophys. Acta. 2014, 1838, 1467–1476.
[31] Nicolson, G.L. The fluid—mosaic model of membrane structure: still relevant to understanding the structure, function and dynamics of biological membranes after more than 40 years. Biochim. Biophys. Acta. 2014, 1838, 1451–1466.
[32] Leftin, A., & Brown, M.F. An NMR data base for simulations of membrane dynamics. Biochim. Biophys. Acta. 2011, 1808, 818–839.
[33] Pabst, G., Rappolt, M., Amenitsch, H., & Laggner, P. Structural information from multilamellar liposomes at full hydration: full q-range fitting with high quality x-ray data. Phys. Rev. E. 2000, 62, 4000–4009.
[34] Kučerka, N., Nagle, J.F., Sachs, J.N., et al. Lipid bilayer structure determined by the simultaneous analysis of neutron and X-Ray scattering data. Biophys. J. 2008, 95 2356–2367.
[35] Pabst, G., Kučerka, N., Nieh, M.-P., Rheinstädter, M.C., & Katsaras, J. Applications of neutron and X-ray scattering to the study of biologically relevant model membranes. Chem. Phys. Lipids. 2010, 163, 460–479.
[36] Heberle, F.A., Marquardt, D., Doktorova, M., et al. Subnanometer structure of an asymmetric model membrane: interleaflet coupling influences domain properties. Langmuir. 2016, 32, 5195–5200.
[37] Nagle, J.F. Experimentally determined tilt and bending moduli of single-component lipid bilayers. Chem. Phys. Lipids. 2017, 205, 18–24.
[38] Sanov, A. Laboratory-frame photoelectron angular distributions in anion photodetachment: insight into electronic structure and intermolecular interactions. Annu. Rev. Phys. Chem. 2014, 65, 341–363.
[39] Levine, Y.K., & Wilkins, M.H.F. Structure of oriented lipid bilayers. Nature New Biol. 1971, 230, 69–72.
[40] Petrache, H.I., & Brown, M.F. X-ray scattering and solid-state deuterium nuclear magnetic resonance probes of structural fluctuations in lipid membranes. In: A. M. Dopico, ed. Methods in Molecular Biology. Totowa, Humana, 2007, 341–353.
[41] Nagle, J.F., & Tristram-Nagle, S. Structure of lipid bilayers. Biochim. Biophys. Acta. 2000, 1469, 159–195.
[42] Nagle, J.F. Introductory lecture: basic quantities in model biomembranes. Faraday Discuss. 2013, 161, 11–29.
[43] Salsbury, N.J., & Chapman, D. Physical studies of phospholipids. VIII. Nuclear magnetic resonance studies of diacyl-L-phosphatidylcholines (lecithins). Biochim. Biophys. Acta. 1968, 163, 314–324.
[44] Penkett, S.A., Flook, A.G., & Chapman, D. Physical studies of phospholipids. IX. Nuclear resonance studies of lipid-water systems. Chem. Phys. Lipids. 1968, 2, 273–290.
[45] Carrington, A., & McLachlan, A.D. Introduction to Magnetic Resonance. Harper & Row, New York, 1967.

[46] Seelig, J. Spin label studies of oriented smectic liquid crystals (a model system for bilayer membranes). J. Am. Chem. Soc. 1970, 92, 3881–3887.
[47] Chan, S.I., Feigenson, G.W., & Seiter, C.H.A. Nuclear relaxation studies of lecithin bilayers. Nature. 1971, 231, 110–112.
[48] Feigenson, G.W., & Chan, S.I. Nuclear magnetic relaxation behavior of lecithin multilayers. J. Am. Chem. Soc. 1974, 96, 1312–1319.
[49] Levine, Y.K., Birdsall, N.J.M., Lee, A.G., & Metcalfe, J.C. ^{13}C Nuclear magnetic resonance relaxation measurements of synthetic lecithins and the effect of spin-labeled lipids. Biochemistry. 1972, 11, 1416–1421.
[50] Horwitz, A.F., Horsley, W.J., & Klein, M.P. Magnetic resonance studies on membrane and model membrane systems: proton magnetic relaxation rates in sonicated lecithin dispersions. Proc. Natl. Acad. Sci. U.S.A. 1972, 69, 590–593.
[51] Häberlen, U. High Resolution NMR in Solids. Selective Averaging. Academic Press, New York, 1976.
[52] Spiess, H.W. Rotation of molecules and nuclear spin relaxation. In: P. Diehl, E. Fluck, & R. Kosfeld, eds. NMR Basic Principles and Progress. Heidelberg, Springer-Verlag, 1978, 55–214.
[53] Nordio, P.L., & Segre, U. Rotational dynamics. In: G. R. Luckhurst & G. W. Gray, eds. The Molecular Physics of Liquid Crystals. New York, Academic Press, 1979, 411–426.
[54] Seelig, A., & Seelig, J. The dynamic structure of fatty acyl chains in a phospholipid bilayer measured by deuterium magnetic resonance. Biochemistry. 1974, 13, 4839–4845.
[55] Seelig, J., & Niederberger, W. Two pictures of a lipid bilayer. A comparison between deuterium label and spin-label experiments. Biochemistry. 1974, 13, 1585–1588.
[56] Brown, M.F., Seelig, J., & Häberlen, U. Structural dynamics in phospholipid bilayers from deuterium spin-lattice relaxation time measurements. J. Chem. Phys. 1979, 70, 5045–5053.
[57] Brown, M.F. Deuterium relaxation and molecular dynamics in lipid bilayers. J. Magn. Reson. 1979, 35, 203–215.
[58] Brown, M.F., Ribeiro, A.A., & Williams, G.D. New view of lipid bilayer dynamics from ^{2}H and ^{13}C NMR relaxation time measurements. Proc. Natl. Acad. Sci. U.S.A. 1983, 80, 4325–4329.
[59] Brown, M.F., Thurmond, R.L., Dodd, S.W., Otten, D., & Beyer, K. Composite membrane deformation on the mesoscopic length scale. Phys. Rev. E. 2001, 64, 010901-1–010901-4.
[60] Martinez, G.V., Dykstra, E.M., Lope-Piedrafita, S., Job, C., & Brown, M.F. NMR elastometry of fluid membranes in the mesoscopic regime. Phys. Rev. E. 2002, 66, 050902-1–050902-4.
[61] Molugu, T.R., & Brown, M.F. Cholesterol-induced suppression of membrane elastic fluctuations at the atomistic level. Chem. Phys. Lipids. 2016, 199, 39–51.
[62] McFarland, B.G., & McConnell, H.M. Bent fatty acid chains in lecithin bilayers. Proc. Natl. Acad. Sci. U.S.A. 1971, 68, 1274–1278.
[63] Griffith, O.H., & Jost, P.C. Lipid spin labels in biological membranes. In: L. J. Berliner, ed. Spin Labeling. Theory and Applications. New York, Academic Press, 1976, 453–523.
[64] Kroon, P.A., Kainosho, M., & Chan, S.I. Proton magnetic resonance studies of lipid bilayer membranes. Experimental determination of inter- and intramolecular nuclear relaxation rates in sonicated phosphatidylcholine bilayer vesicles. Biochim. Biophys. Acta. 1976, 433, 282–293.
[65] Seelig, J. Deuterium magnetic resonance: theory and application to lipid membranes. Q. Rev. Biophys. 1977, 10, 353–418.
[66] Dill, K.A., & Flory, P.J. Interphases of chain molecules: monolayers and lipid bilayer membranes. Proc. Natl. Acad. Sci. U.S.A. 1980, 77, 3115–3119.
[67] Högberg, C.-J., & Lyubartsev, A.P. A molecular dynamics investigation of the influence of hydration and temperature on structural and dynamical properties of a dimyristoylphosphatidylcholine bilayer. J. Phys. Chem. B. 2006, 110, 14326–14336.
[68] Klauda, J.B., Venable, R.M., Freites, J.A., et al. Update of the CHARMM all-atom additive force field for lipids: validation on six lipid types. J. Phys. Chem. B. 2010, 114, 7830–7843.

[69] Sodt, A.J., & Pastor, R.W. Bending free energy from simulation: correspondence of planar and inverse hexagonal lipid phases. Biophys. J. 2013, 104, 2202–2211.
[70] Sodt, A.J., Sandar, M.L., Gawrisch, K., Pastor, R.W., & Lyman, E. The molecular structure of the liquid-ordered phase of lipid bilayers. J. Am. Chem. Soc. 2014, 136, 725–732.
[71] de Gennes, P.G., & Prost, J. The Physics of Liquid Crystals, 2nd Ed. Oxford University Press, Oxford, 1993.
[72] Salmon, A., Dodd, S.W., Williams, G.D., Beach, J.M., & Brown, M.F. Configurational statistics of acyl chains in polyunsaturated lipid bilayers from ^{2}H NMR. J. Am. Chem. Soc. 1987, 109, 2600–2609.
[73] Petrache, H.I., Dodd, S.W., & Brown, M.F. Area per lipid and acyl length distributions in fluid phosphatidylcholines determined by ^{2}H NMR spectroscopy. Biophys. J. 2000, 79, 3172–3192.
[74] Dill, K.A., & Flory, P.J. Molecular organization in micelles and vesicles. Proc. Natl. Acad. Sci. U.S.A. 1981, 78, 676–680.
[75] Dill, K.A., Naghizadeh, J., & Marqusee, J.A. Chain molecules at high densities at interfaces. Annu. Rev. Phys. Chem. 1988, 39, 425–461.
[76] Levine, Y.K., Partington, P., & Roberts, G.C.K. Calculation of dipolar nuclear magnetic relaxation times in molecules with multiple internal rotations. I. Isotropic overall motion of the molecule. Mol. Phys. 1973, 25, 497–514.
[77] Williams, G.D., Beach, J.M., Dodd, S.W., & Brown, M.F. Dependence of deuterium spin-lattice relaxation rates of multilamellar phospholipid dispersions on orientational order. J. Am. Chem. Soc. 1985, 107, 6868–6873.
[78] Hornblower, S., & Spawforth, A. The Oxford Classical Dictionary, 3rd Ed. Oxford University Press, Oxford, 2003.
[79] Davis, J.H. The description of membrane lipid conformation, order and dynamics by ^{2}H-NMR. Biochim. Biophys. Acta. 1983, 737, 117–171.
[80] Brown, M.F. Membrane structure and dynamics studied with NMR spectroscopy. In: K. M. Merz Jr. & B. Roux, eds. Biological Membranes. A Molecular Perspective from Computation and Experiment. Basel, Birkhäuser, 1996, 175–252.
[81] Gaffney, B.J., & McConnell, H.M. The paramagnetic resonance spectra of spin labels in phospholipid membranes. J. Magn. Reson. 1974, 16, 1–28.
[82] Maier, W., & Saupe, A. Eine einfache molekular-statistische theorie der nematischen kristallinflussigen phase. Teil II. Z. Naturforsch. Teil A. 1960, 15a, 287–292.
[83] Zannoni, C. Quantitative description of orientational order: rigid molecules. In: J. W. Emsley, ed. Nuclear Magnetic Resonance of Liquid Crystals. Dordrecht, D. Reidel Publishing Company, 1985, 1–34.
[84] McConnell, H.M., & McFarland, B.G. The flexibility gradient in biological membranes. Ann. N.Y. Acad. Sci. 1972, 195, 207–217.
[85] Ukleja, P., Pirs, J., & Doane, J.W. Theory for spin-lattice relaxation in nematic liquid crystals. Phys. Rev. A. 1976, 14, 414–423.
[86] Flory, P.J. Statistical Mechanics of Chain Molecules. Interscience, New York, 1969.
[87] Devaux, P., & McConnell, H.M. Lateral diffusion in spin-labeled phosphatidylcholine multilayers. J. Am. Chem. Soc. 1972, 94, 4475–4481.
[88] Lipari, G., & Szabo, A. Model-free approach to the interpretation of nuclear magnetic resonance relaxation in macromolecules. I. Theory and range of validity. J. Am. Chem. Soc. 1982, 104, 4546–4559.
[89] Hamm, M., & Kozlov, M.M. Elastic energy of tilt and bending of fluid membranes. Eur. Phys. J. E. 2000, 3, 323–335.
[90] Jablin, M.S., Akabori, K., & Nagle, J.F. Experimental support for tilt-dependent theory of biomembrane mechanics. Phys. Rev. Lett. 2014, 113, 248102-1–248102-5.

[91] Terzi, M.M., & Deserno, M. Novel tilt-curvature coupling in lipid membranes. J. Chem. Phys. 2017, 147, 084702-1–084702-14.
[92] Brown, M.F. Theory of spin-lattice relaxation in lipid bilayers and biological membranes. Dipolar relaxation. J. Chem. Phys. 1984, 80, 2808–2831.
[93] Vold, R.L., Vold, R.R., & Warner, M. Higher-order director fluctuations. J. Chem. Soc. Faraday Trans. 2 1988, 84, 997–1013.
[94] Milburn, M.P., & Jeffrey, K.R. Dynamics of the phosphate group in phospholipid bilayers. A ^{31}P nuclear relaxation time study. Biophys. J. 1987, 52, 791–799.
[95] Bonmatin, J.-M., Smith, I.C.P., Jarrell, H.C., & Siminovitch, D.J. Use of a comprehensive approach to molecular dynamics in ordered lipid systems: cholesterol reorientation in ordered lipid bilayers. A ^{2}H NMR relaxation case study. J. Am. Chem. Soc. 1990, 112, 1697–1704.
[96] Weisz, K., Gröbner, G., Mayer, C., Stohrer, J., & Kothe, G. Deuteron nuclear magnetic resonance study of the dynamic organization of phospholipid/cholesterol bilayer membranes: molecular properties and viscoelastic behavior. Biochemistry. 1992, 31, 1100–1112.
[97] Brown, M.F., Thurmond, R.L., Dodd, S.W., Otten, D., & Beyer, K. Elastic deformation of membrane bilayers probed by deuterium NMR relaxation. J. Am. Chem. Soc. 2002, 124, 8471–8484.
[98] Orädd, G., Shahedi, V., & Lindblom, G. Effect of sterol structure on the bending rigidity of lipid membranes: a ^{2}H NMR transverse relaxation study. Biochim. Biophys. Acta. 2009, 1788, 1762–1771.
[99] Fenimore, P.W., Frauenfelder, H., McMahon, B.H., & Young, R.D. Bulk-solvent and hydration-shell fluctuations, similar to α- and β-fluctuations in glasses, control protein motions and functions. Proc. Natl. Acad. Sci. U.S.A. 2004, 101, 14408–14413.
[100] Frauenfelder, H., Chen, G., Berendzen, J., et al. A unified model of protein dynamics. Proc. Natl. Acad. Sci. U.S.A. 2009, 106, 5129–5134.
[101] Frauenfelder, H., Young, R.D., & Fenimore, P.W. A wave-mechanical model of incoherent quasielastic scattering in complex systems. Proc. Natl. Acad. Sci. U.S.A. 2014, 111, 12764–12768.
[102] Klauda, J.B., Roberts, M.F., Redfield, A.G., Brooks, B.R., & Pastor, R.W. Rotation of lipids in membranes: molecular dynamics simulation, ^{31}P spin-lattice relaxation, and rigid-body dynamics. Biophys. J. 2008, 94, 3074–3083.
[103] Sivanandam, V.N., Cai, J., Redfield, A.G., & Roberts, M.F. Phosphatidylcholine "wobble" in vesicles assessed by high-resolution ^{13}C field cycling NMR spectroscopy. J. Am. Chem. Soc. 2009, 131, 3420–3421.
[104] Stohrer, J., Gröbner, G., Reimer, D., Weisz, K., Mayer, C., & Kothe, G. Collective lipid motions in bilayer membranes studied by transverse deuteron spin relaxation. J. Chem. Phys. 1991, 95, 672–678.
[105] Rose, M.E. Elementary Theory of Angular Momentum. John Wiley & Sons, New York, 1957.
[106] Leftin, A., Xu, X., & Brown, M.F. Phospholipid bilayer membranes: deuterium and carbon-13 NMR spectroscopy. eMagRes. 2014, 3, 199–214.
[107] Meirovitch, E., Polimeno, A., & Freed, J.H.. Comment on "The physical basis of model-free analysis of NMR relaxation data from proteins and complex fluids" [J. Chem. Phys. 131, 224507 (2009)]. J. Chem. Phys. 2010, 132, 207101-1–207101-2.
[108] Shapiro, Y.E., & Meirovitch, E. The time correlation function perspective of NMR relaxation in proteins. J. Chem. Phys. 2013, 139, 084107-1–084107-13.
[109] Davis, J.H., Jeffrey, K.R., & Bloom, M. Spin-lattice relaxation as a function of chain position in perdeuterated potassium palmitate. J. Magn. Reson. 1978, 29, 191–199.
[110] Wennerström, H., Lindman, B., Söderman, O., Drakenberg, T., & Rosenholm, J.B. ^{13}C magnetic relaxation in micellar solutions. Influences of aggregate motion on T_1. J. Am. Chem. Soc. 1979, 101, 6860–6864.

[111] Palmer III, A.G. NMR characterization of the dynamics of biomacromolecules. Chem. Rev. 2004, 104, 3623–3640.
[112] Brown, M.F., & Seelig, J. Influence of cholesterol on the polar region of phosphatidylcholine and phosphatidylethanolamine bilayers. Biochemistry. 1978, 17, 381–384.
[113] Vist, M.R., & Davis, J.H. Phase equilibria of cholesterol/dipalmitoylphosphatidylcholine mixtures: ^{2}H nuclear magnetic resonance and differential scanning calorimetry. Biochemistry. 1990, 29, 451–464.
[114] Veatch, S.L., Soubias, O., Keller, S.L., & Gawrisch, K. Critical fluctuations in domain-forming lipid mixtures. Proc. Natl. Acad. Sci. U.S.A. 2007, 104, 17650–17655.
[115] Goñi, F., Alonso, A., Bagatolli, L.A., et al. Phase diagrams of lipid mixtures relevant to the study of membrane rafts. Biochim. Biophys. Acta. 2008, 1781, 665–684.
[116] Parisio, G., Sperotto, M.M., & Ferrarini, A. Flip-flop of steroids in phospholipid bilayers: effects of the chemical structure on transbilayer diffusion. J. Am. Chem. Soc. 2012, 134, 2198–12208.
[117] Toppozini, L., Meinhardt, S., Armstrong, C.L., et al. Structure of cholesterol in lipid rafts. Phys. Rev. Lett. 2014, 113, 228101-1–228101-5.
[118] Keller, S.L., Pitcher III, W.H., Huestis, W.H., & McConnell, H.M. Red blood cell lipids form immiscible liquids. Phys. Rev. Lett. 1998, 81, 5019–5022.
[119] Radhakrishnan, A., Anderson, T.G., & McConnell, H.M. Condensed complexes, rafts, and the chemical activity of cholesterol in membranes. Proc. Natl. Acad. Sci. U.S.A. 2000, 97, 12422–12427.
[120] McConnell, H. Communication: critical dynamics and nuclear relaxation in lipid bilayers. J. Chem. Phys. 2011, 134, 011102-1–011102-4.
[121] Sternin, E., Bloom, M., & MacKay, A.L. De-Pake-ing of NMR spectra. J. Magn. Reson. 1983, 55, 274–282.
[122] McCabe, M.A., & Wassall, S.R. Fast-Fourier-transform dePakeing. J. Magn. Reson. B. 1995, 106, 80–82.
[123] Gaffney, B.J. Practical considerations for the calculation of order parameters for fatty acid or phospholipid spin labels in membranes. In: L. J. Berliner, ed. Spin Labeling. Theory and Applications. New York, Academic Press, 1976, 567–571.
[124] Taylor, M.G., & Smith, I.C.P. The conformations of nitroxide-labelled fatty acid probes of membrane structure as studied by ^{2}H-NMR. Biochim. Biophys. Acta. 1983, 733, 256–263.
[125] Kinnun, J.J., Mallikarjunaiah, K.J., Petrache, H.I., & Brown, M.F. Elastic deformation and area per lipid of membranes: atomistic view from solid-state deuterium NMR spectroscopy. Biochim. Biophys. Acta. 2015, 1848, 246–259.
[126] Marqusee, J.A., Warner, M., & Dill, K.A. Frequency dependence of NMR spin lattice relaxation in bilayer membranes. J. Chem. Phys. 1984, 81, 6404–6405.
[127] Watnick, P.I., Dea, P., Nayeem, A., & Chan, S.I. Cooperative lengths and elastic constants in lipid bilayers: the chlorophyll *a* /dimyristoyllecithin system. J. Chem. Phys. 1987, 86, 5789–5800.
[128] Watnick, P.I., Dea, P., & Chan, S.I. Characterization of the transverse relaxation rates in lipid bilayers. Proc. Natl. Acad. Sci. U.S.A. 1990, 87, 2082–2086.
[129] Althoff, G., Heaton, N.J., Gröbner, G., Prosser, R.S., & Kothe, G. NMR relaxation study of collective motions and viscoelastic properties in biomembranes. Coll. Surf. A. 1996, 115, 31–37.
[130] Althoff, G., Frezzato, D., Vilfan, M., et al. Transverse nuclear spin relaxation studies of viscoelastic properties of membrane vesicles. I. Theory. J. Phys. Chem. B. 2002, 106, 5506–5516.
[131] Adamson, A.W. Physical Chemistry of Surfaces, 5th Ed. New York, John Wiley & Sons, 1990, 185.
[132] Nevzorov, A.A., & Brown, M.F. Dynamics of lipid bilayers from comparative analysis of ^{2}H and ^{13}C nuclear magnetic resonance relaxation data as a function of frequency and temperature. J. Chem. Phys. 1997, 107, 10288–10310.

[133] Pace, R.J., & Chan, S.I. Molecular motions in lipid bilayers. II. Magnetic resonance of multilamellar and vesicle systems. J. Chem. Phys. 1982, 76, 4228–4240.
[134] Venable, R.M., Zhang, Y., Hardy, B.J., & Pastor, R.W. Molecular dynamics simulations of a lipid bilayer and of hexadecane: an investigation of membrane fluidity. Science. 1993, 262, 223–226.
[135] Rawicz, W., Olbrich, K.C., McIntosh, T., Needham, D., & Evans, E. Effect of chain length and unsaturation on elasticity of lipid bilayers. Biophys. J. 2000, 79, 328–339.
[136] Brown, M.F., Ellena, J.F., Trindle, C., & Williams, G.D. Frequency dependence of spin-lattice relaxation times of lipid bilayers. J. Chem. Phys. 1986, 84, 465–470.
[137] Althoff, G., Stauch, O., Vilfan, M., et al. Transverse nuclear spin relaxation studies of viscoelastic properties of membrane vesicles. II. Experimental results. J. Phys. Chem. B. 2002, 106, 5517–5526.
[138] van Meer, G., Voelker, D.R., & Feigenson, G.W. Membrane lipids: where they are and how they behave. Nat. Rev. Mol. Cell Biol. 2008, 9, 112–124.
[139] Olsson, U., & Wennerström, H. Globular and bicontinuous phases of nonionic surfactant films. Adv. Colloid Interface Sci. 1994, 49, 113–146.
[140] Heberle, F.A., Doktorova, M., Goh, S.L., Standaert, R.F., Katsaras, J., & Feigenson, G.W. Hybrid and nonhybrid lipids exert common effects on membrane raft size and morphology. J. Am. Chem. Soc. 2013, 135, 14932–14935.
[141] Ackerman, D.G., & Feigenson, G.W. Multiscale modeling of four-component lipid mixtures: domain composition, size, alignment, and properties of the phase interface. J. Phys. Chem. B. 2015, 119, 4240–4250.
[142] McConnell, H.M., & Radhakrishnan, A. Condensed complexes of cholesterol and phospholipids. Biochim Biophys Acta. 2003, 1610, 159–173.
[143] Endress, E., Heller, H., Brown, M.F., & Bayerl, T.M. Differences of the molecular motions of cholesterol and lanosterol in oriented phospholipid bilayers: a quasi-elastic neutron scattering, MD simulation, and deuterium NMR study. Biophys. J. 2001, 80, 528a.
[144] Endress, E., Heller, H., Casalta, H., Brown, M.F., & Bayerl, T. Anisotropic motion and molecular dynamics of cholesterol, lanosterol, and ergosterol in lecithin bilayers studied by quasielastic neutron scattering. Biochemistry. 2002, 41, 13078–13086.
[145] Martinez, G.V., Dykstra, E.M., Lope-Piedrafita, S., & Brown, M.F. Lanosterol and cholesterol-induced variations in bilayer elasticity probed by ^{2}H NMR relaxation. Langmuir. 2004, 20, 1043–1046.
[146] Bloom, M., Evans, E., & Mouritsen, O.G. Physical properties of the fluid lipid-bilayer component of cell membranes: a perspective. Q. Rev. Biophys. 1991, 24, 293–397.
[147] Miao, L., Nielsen, M., Thewalt, J., et al. From lanosterol to cholesterol: structural evolution and differential effects on lipid bilayers. Biophys. J. 2002, 82, 1429–1444.
[148] Gentile, L., Behrens, M.A., Balog, S., Mortensen, K., Ranieri, G.A., & Olsson, U. Dynamic phase diagram of a nonionic surfactant lamellar phase. J. Phys. Chem. B. 2014, 118, 3622–3629.
[149] Otten, D., Brown, M.F., & Beyer, K. Softening of membrane bilayers by detergents elucidated by deuterium NMR spectroscopy. J. Phys. Chem. B. 2000, 104, 12119–12129.
[150] Simons, K., & Gerl, M.J. Revitalizing membrane rafts: new tools and insights. Nat. Rev. Mol. Cell Biol. 2010, 11, 688–699.
[151] Méléard, P., Gerbaud, C., Pott, T., et al. Bending elasticities of model membranes: influences of temperature and sterol content. Biophys. J. 1997, 72, 2616–2629.
[152] Arriaga, L.R., López-Montero, I., Monroy, F., Orts-Gil, G., & Hellweg, T. Stiffening effect of cholesterol on disordered lipid phases: a combined neutron spin echo + dynamic light scattering analysis of the bending elasticity of large unilamellar vesicles. Biophys. J. 2009, 96, 3629–3637.
[153] Molugu, T.R., Lee, S.K., Job, C., & Brown, M.F. Hydration-modulated collective dynamics of membrane lipids are revealed by solid-state ^{2}H NMR relaxation. Biophys. J. 2015, 108, 77a.

[154] Brown, M.F. Unified picture for spin-lattice relaxation of lipid bilayers and biomembranes. J. Chem. Phys. 1984, 80, 2832–2836.
[155] Pastor, R.W., Venable, R.M., & Feller, S.E. Lipid bilayers, NMR relaxation, and computer simulations. Acc. Chem. Res. 2002, 35, 438–446.
[156] Pastor, R.W., Venable, R.M., & Karplus, M. Brownian dynamics simulation of a lipid chain in a membrane bilayer. J. Chem. Phys. 1988, 89, 1112–1127.
[157] Klauda, J.B., Eldho, N.V., Gawrisch, K., Brooks, B.R., & Pastor, R.W. Collective and noncollective models of NMR relaxation in lipid vesicles and multilayers. J. Phys. Chem. B. 2008, 112, 5924–5929.
[158] Struts, A.V., Salgado, G.F.J., & Brown, M.F. Solid-state ^{2}H NMR relaxation illuminates functional dynamics of retinal cofactor in membrane activation of rhodopsin. Proc. Natl. Acad. Sci. U.S.A. 2011, 108, 8263–8268.
[159] Ziurys, L.M. Millimeter and submillimeter wave spectroscopy and astrophysical applications. In: M. Quack & F. Merkt, eds. Handbook of High-Resolution Spectroscopy. Chichester, UK, John Wiley & Sons, 2011, 939–963.
[160] Fernandez-Puente, L., Bivas, I., Mitov, M.D., & Méléard, P. Temperature and chain length effects on bending elasticity of phosphatidylcholine bilayers. Europhys. Lett. 1994, 28, 181–186.
[161] Baumgart, T., Capraro, B.R., Zhu, C., & Das, S.L. Thermodynamics and mechanics of membrane curvature generation and sensing by proteins and lipids. Annu. Rev. Phys. Chem. 2011, 62, 483–506.
[162] Schneider, M.D., Jenkins, J.T., & Webb, W.W. Thermal fluctuations of large quasi-spherical phospholipid vesicles. J. Phys. (Paris). 1984, 45, 1457–1472.
[163] Ferrarini, A. The theory of elastic constants. Liquid Crystals. 2010, 37, 811–823.
[164] McDermott, A. Structure and dynamics of membrane proteins by magic angle spinning solid-state NMR. Ann. Rev. Biophys. 2009, 38, 385–403.
[165] Bhate, M.P., Wylie, B.J., Tian, L., & McDermott, A.E. Conformational dynamics in the selectivity filter of KcsA in response to potassium ion concentration. J. Mol. Biol. 2010, 401, 155–166.
[166] Park, S.H., Das, B.B., Casagrande, F., et al. Structure of the chemokine receptor CXCR1 in phospholipid bilayers. Nature. 2012, 491, 779–783.
[167] Lalli, D., Idso, M.N., Andreas, L.B., et al. Proton-based structural analysis of a heptahelical transmembrane protein in lipid bilayers. J. Am. Chem. Soc. 2017, 139, 13006–13012.
[168] Altenbach, C., Kusnetzow, A.K., Ernst, O.P., Hofmann, K.P., & Hubbell, W.L. High-resolution distance mapping in rhodopsin reveals the pattern of helix movement due to activation. Proc. Natl. Acad. Sci. U.S.A. 2008, 105, 7439–7444.
[169] Manglik, A., Kim, T.H., Masureel, M., et al. Structural insights into the dynamic process of β_2-adrenergic receptor signaling. Cell. 2015, 161, 1101–1111.
[170] Georgieva, E.R., Borbat, P.P., Norman, H.D., & Freed, J.H. Mechanism of influenza A M2 transmembrane domain assembly in lipid membranes. Sci. Rep. 2015, 5, 11757; doi: 11710.11038/srep11757.
[171] Mahalingam, M., Martínez-Mayorga, K., Brown, M.F., & Vogel, R. Two protonation switches control rhodopsin activation in membranes. Proc. Natl. Acad. Sci. U.S.A. 2008, 105, 17795–17800.
[172] Struts, A.V., Salgado, G.F.J., Martínez-Mayorga, K., & Brown, M.F. Retinal dynamics underlie its switch from inverse agonist to agonist during rhodopsin activation. Nat. Struct. Mol. Biol. 2011, 18, 392–394.
[173] Struts, A.V., & Brown, M.F. Structural dynamics of retinal in rhodopsin activation viewed by solid-state ^{2}H NMR spectroscopy. In: F. Separovic & A. Naito, eds. Advances in Biological Solid-State NMR: Proteins and Membrane-Active Peptides. Cambridge, The Royal Society of Chemistry, 2014, 320–352.

Laura N. Poloni, Amy Won, Christopher M. Yip

8 Mapping protein– and peptide–membrane interactions by atomic force microscopy: strategies and opportunities

Abstract: The interactions of proteins and peptides with lipid bilayer membranes are critical for a wide breadth of biological phenomena and processes, from protein clustering and subsequent membrane reordering and restructuring to membrane-facilitated protein aggregation and folding. Atomic force microscopy (AFM) offers a unique platform to investigate the complexities and diversities of protein– and peptide–membrane interactions because it can capture the spatio-temporal dynamics of these processes under physiologically relevant conditions and high spatial resolution. The relative ease with which the AFM can be integrated with other imaging modalities and spectroscopies yields a powerful and flexible platform to investigate these complex processes. In this chapter, we present a detailed protocol for the preparation of support lipid bilayers (SLBs) and subsequent imaging by in situ AFM, along with a discussion of experimental design for the investigation of protein- and peptide-membrane interactions and the potential of combinatorial platforms for multi-modal image acquisition and interpretation.

Keywords: lipid bilayer, atomic force microscopy, vesicle fusion, peptide–membrane interactions, protein–membrane interactions

8.1 Introduction

Membrane proteins govern processes that are essential for cell physiology [1]. It has been estimated that 20–30% of the genomes encode transmembrane proteins [2] and membrane proteins are the targets of over 50% of all modern medicinal drugs [3]. Membranes also mediate processes that impart antimicrobial peptides (AMPs) with their specificity and activity [4]. Lipid bilayers – once though thought of as simply an adaptable matrix for proteins comprising the membrane – have been found to influence local structure, dynamics, and even the activity of membrane proteins [5]. Thus,

---

Laura N. Poloni, Department of Chemical Engineering and Applied Chemistry, University of Toronto; Donnelly Centre for Cellular and Biomolecular Research, University of Toronto
Amy Won, Christopher M. Yip, Department of Chemical Engineering and Applied Chemistry, University of Toronto; Donnelly Centre for Cellular and Biomolecular Research, University of Toronto; Department of Biochemistry, University of Toronto, Canada

https://doi.org/10.1515/9783110544657-008

the ability to study protein– and peptide–membrane interactions in physiologically relevant environments is critical to understanding their function.

(Figure 8.1) Protein– and peptide–membrane interactions can be studied via model lipid membranes, typically supported lipid bilayers (SLBs) (Figure 8.1). SLBs are typically formed by vesicle fusion or Langmuir–Blodgett/Langmuir–Schaefer methods. Lipid composition can be varied to form membrane mimics of various organisms. Mammalian membranes are typically composed of neutral phospholipids such as phosphatidylcholine and sphingomyelin, whereas bacterial cell membranes are composed of negatively charged lipids such as phosphatidylglycerol (PG), cardiolipin, and the zwitterionic phosphatidylethanolamine (PE) [6]. Overall, lipids are

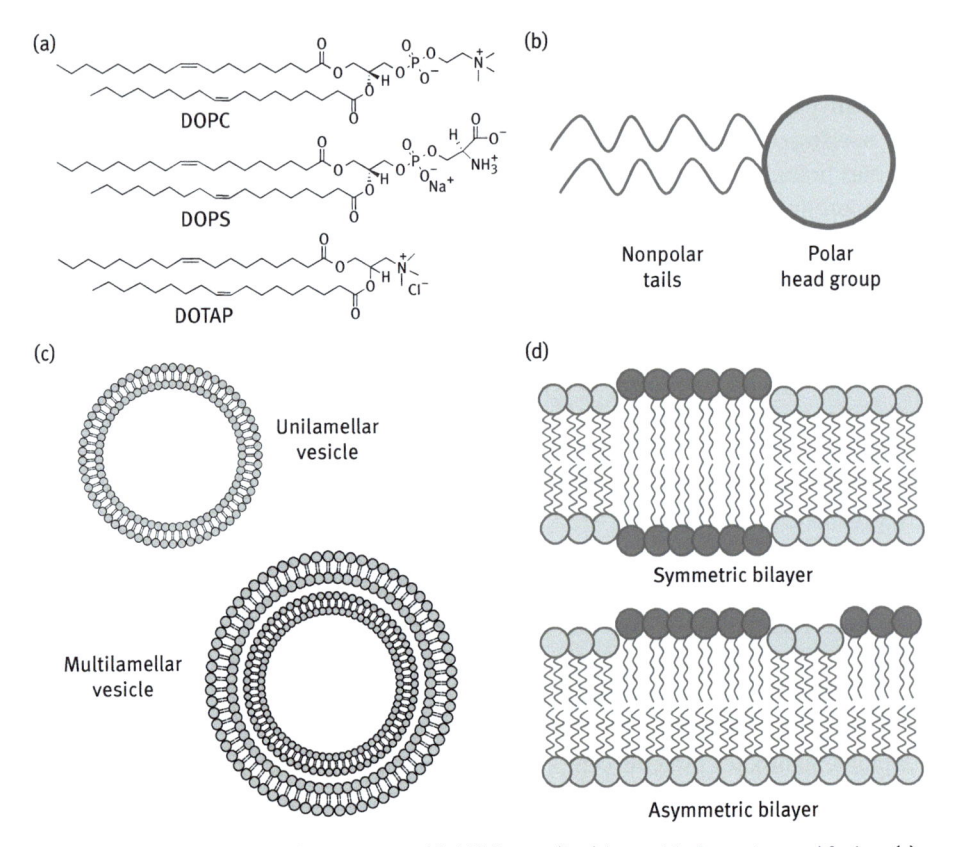

Figure 8.1: Amphiphilic lipids form supported lipid bilayers (SLB) by vesicle formation and fusion. (a) Chemical structures of lipids commonly used in SLBs: 1,2-dioleoyl-sn-glycero-3-phosphocholine (DOPC), 1,2-dioleoyl-sn-glycro-3-phospho-L-serine (sodium salt) (DOPS), and 1,2-dioleoyl-3-trimethylammonium-propane (chloride salt) (DOTAP). (b) Common representation of a lipid molecule depicting the polar "head" and nonpolar "tails." (c) Organization of lipids into vesicles upon addition to an aqueous solution can result in unilamellar (i.e., encapsulated by a single lipid bilayer) or multilamellar (i.e., encapsulated by multiple lipid bilayers) vesicles. (d) Symmetric and asymmetric bilayers.

characterized by a hydrophilic (i.e., polar) headgroup and long hydrocarbon tails. In aqueous environments, lipids organize into bilayer structures composed of two lipid monolayers (leaflets) in which the lipids are oriented such that their polar headgroups interact with the aqueous environment and the hydrocarbon chains form the core of the membrane. Bilayer leaflets can differ in both lipid composition and physical properties, resulting in bilayer asymmetry. Differences in lipid composition as well as symmetry/asymmetry can lead to different phase segregation and mechanical properties that affect the interactions between membranes and proteins, highlighting the importance of investigating proteins in their native membrane environment.

The flexibility of atomic force microscopy (AFM) as a platform for spatial and temporal mapping of lipid bilayers in physiological conditions is unrivaled. The various AFM modes enable direct observation of lipid organization in bilayers and their mechanical properties in fluid environments [7]. The ability to easily modify the imaging environment during the course of an experiment allows for the examination of the effects of exogenous agents such as proteins and peptides [8–11]. Of the many techniques used to characterize SLBs and their interaction with exogenous agents [12,13], AFM is unique in its versatility of configurations that facilitates its use in combination with other imaging/spectroscopic techniques [14,15].

In this protocol, we outline the preparation of model SLBs and subsequent imaging by tapping-mode AFM in fluid. Preparation of model SLBs by vesicle fusion and Langmuir–Blodgett/Langmuir–Schaefer methods are described, along with useful considerations regarding substrate properties (surface charge, chemical composition, and roughness) and medium (buffer composition, pH, and ionic strength) [16]. We also describe the use of a flow-through cell to introduce protein and peptide solutions to the imaging environment to study their effects on membrane organization and mechanical properties. Opportunities for combinatorial experiments with the AFM platform are discussed.

8.2 Materials

8.2.1 Materials for buffer solutions

HEPES (4-(2-hydroxyethyl)-1-piperazineethanesulfonic acid)
NaCl
Deionized water
Glass beaker
NaOH
$CaCl_2$

8.2.2 Materials for substrate preparation

Muscovite mica (see Note 8.4.3)
Scotch tape or razor blade
Double-sided tape or optical adhesive (see Note 8.4.3)
Magnetic metal specimen disc (Ted Pella) or glass-bottomed dish (WillCo) (see Note 8.4.3)

8.2.3 Materials for bilayer preparation via fusion of small unilamellar vesicles (SUVs) on a solid support

Lipids (see Note 8.4.3)
Chloroform or 3:1 v/v chloroform:methanol
Rotary evaporator
Nitrogen stream
Vacuum oven
Ultrasonicator
Incubator

8.2.4 Materials for bilayer preparation by Langmuir–Blodgett or Langmuir–Schaefer techniques

Langmuir trough equipped with surface pressure sensor (Wilhelmy plate or round rod) and dipping mechanism for Langmuir–Blodgett (vertical) or Langmuir–Schaefer (horizontal)
Lipids (see Note 8.4.1)
Chloroform or 3:1 v/v chloroform:methanol (see Note 8.4.6)
Deionized water
Micropipette
Aspirator

8.2.5 Materials for in-situ AFM imaging

Plasma cleaner or UV light source
AFM probes (see Note 8.4.4)
AFM equipped with a cell or cantilever mount compatible with imaging in fluid
Disposable syringes
For flow-through experiments: (see Note 8.4.8)
Tygon tubing

Disposable syringes
Syringe pump(s)
Luers

8.3 Methods

8.3.1 Preparation of buffer solutions

(See Note 8.4.2)
HEPES buffered saline solution (10 mM HEPES, 150 mM NaCl, pH 7.45)
   Add 200 mL deionized water into a glass beaker.
   Weigh and transfer 0.595 g HEPES and 2.19 g NaCl into the glass beaker.
   Mix well and adjust the pH to 7.45 with NaOH.
   Note: Make up to 250 mL with deionized water and store at 4 °C up to a week.

$CaCl_2$ solution (1 M)
   Add 25 mL deionized water to a glass beaker.
   Weigh 3.675 g $CaCl_2$ and transfer it into the glass beaker.
   Store at 4 °C.

8.3.2 Preparation of substrate and AFM components

1. Prepare a freshly cleaved mica substrate using a razor blade or by removing several layers of the material using adhesive tape. There are several alternative substrates for bilayer formation (see Note 8.4.3).
2. Secure the mica substrate onto a magnetic metal specimen disc or glass-bottomed dish. Depending on the configuration of the AFM being used, mica is either secured to a metal puck with a piece of double-sided tape or optical adhesive, or on a glass-bottomed dish with optical glue. (See Note 8.4.3).
3. Clean the AFM fluid cell by plasma cleaning for a few minutes or UV irradiation for 30 min to remove all possible organic contaminants.
4. If performing a flow-through experiment, assemble the necessary components for the closed fluid cell (i.e., tygon tubing, luers, disposable syringes and syringe pump) or an open fluid cell (i.e., tygon tubing, luers, disposable syringes and *dual* syringe pump). (See Note 8.4.8)

8.3.3 Preparation of lipid bilayers

Two protocols for the preparation of SLBs are described below. Symmetric SLBs can be prepared by the fusion of vesicles (Section 8.3.3.1), whereas successful preparation of asymmetric SLBs typically requires Langmuir–Blodgett or Langmuir–Schaefer techniques (Section 8.3.3.2).

8.3.3.1 Fusion of small unilamellar vesicles on a solid support

1. Weigh or measure the required amounts of lipids to achieve the desired composition of the bilayer (See Note 8.4.1) and transfer to a test tube.
2. Add chloroform to the test tube until all lipids are dissolved. It is sometimes necessary to use 3:1 v/v chloroform/methanol to ensure complete dissolution of some lipid mixtures.
3. Remove the chloroform (or 3:1 v/v chloroform/methanol) by evaporation using a rotary evaporator or by subjecting the solution to a stream of nitrogen. Place the test tube in the vacuum oven for 30 min to ensure complete removal of the organic solvent(s). After removal of the test tube from the vacuum oven, a thin, dry film will be observed on the inner surface of the test tube.
4. Add buffer to the test tube containing the lipid film to achieve the desired concentration. Incubate the solution at a temperature above the lipid transition temperature for 15 min.
5. Subject the solution to ultra-sonication at a temperature above the lipid transition temperature until the solution becomes transparent (see Note 8.4.5). For mixtures containing anionic lipids, add $CaCl_2$ (to a final concentration of 1 mM) to aid in bilayer formation (see Note 8.4.2).
6. Add the lipid vesicle solution to either the liquid cell of the AFM or the WillCo dish. Incubate at a temperature above room temperature for 20 min.
7. Wash the sample surface to remove any unfused vesicles by repeated flushing of lipid-free buffer through the AFM fluid cell or by buffer exchange through the open WillCo dishes (see Note 8.4.5). Exposure to air during the washing process should be avoided, as SLBs are only stable in solution.

8.3.3.2 Deposition of bilayer by Langmuir–Blodgett or Langmuir–Schaefer techniques

1. Thoroughly clean the Langmuir trough and barriers using chloroform or 3:1 v/v chloroform:methanol to remove any organic contaminants. Rinse the trough three times with deionized water.

2. Fill the trough with water and run the aspirator along the barriers to remove any contaminants on the water surface that are picked up by the barrier.
3. Ensure the Wilhelmy plate is clean. Dip the Wilhelmy plate in water and hang it from the balance hook. Ensure that two-thirds of the plate is submerged in the water.
4. Allow the water to stabilize until the pressure no longer decreases, usually a period of 20 min.
5. Tare the surface pressure sensor to zero. Ensure that the water surface is free of contaminants by running a blank isotherm. If the blank isotherm is flat (i.e., zero surface pressure), then the water surface is free of contamination. If there are fluctuations in the blank isotherm, repeat steps 1–5.
6. Thoroughly clean the mica or glass substrate and treat the substrate to ensure that the surface is hydrophilic (see Note 8.4.2).
7. Load the substrate onto the dipper mechanism. Carefully lower the substrate into the water, ensuring that the area intended to be covered with the SLB is submerged.
8. Weigh or measure the required amounts of lipids to achieve the desired composition of the bilayer (See Note 8.4.1) and transfer to a test tube.
9. Add chloroform to the test tube until all lipids are dissolved. It is sometimes necessary to use 3:1 v/v chloroform/methanol to ensure complete dissolution of some lipid mixtures. The required concentration will be dependent on the geometry of the Langmuir trough and the amount of solution applied to the Langmuir trough.
10. Add the lipid solution on the subphase surface using a clean micropipette and depositing the solution dropwise, holding the micropipette tip end a few millimeters from the water surface.
11. Wait for 10–15 min to allow for solvent evaporation and surface stabilization.
12. Perform a surface pressure-area isotherm. Stop the movement of the barriers when the critical surface pressure is reached (see Note 8.4.6).
13. Transfer the Langmuir film (Figure 8.2a) to the substrate by moving the substrate up and out of the subphase through the lipid film using the dipper mechanism provided with the Langmuir trough (Figure 8.2b).
14. Repeat step 12 to create a second Langmuir film on the surface of the subphase.
15. Lower the substrate to form the second leaflet of the SLB either using the Langmuir–Blodgett method (vertically lowering the substrate through the film into the subphase, Figure 8.2c) or the Langmuir–Schaefer method (horizontally lowering the substrate to contact the film, Figure 8.2d).
16. Raise the substrate and transfer to the AFM, taking care not to expose the SLB to air.
17. Clean the trough.

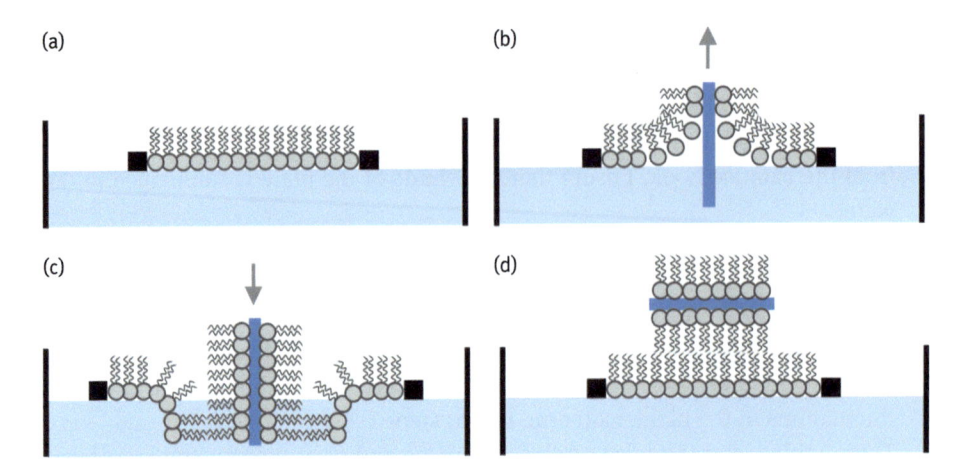

Figure 8.2: Supported lipid bilayer (SLB) formation by Langmuir–Blodgett and Langmuir–Schaefer techniques. (a) Formation of a Langmuir lipid film on the surface of the subphase (typically water) water in a Langmuir trough. (b) Formation of a lipid bilayer on a substrate by transfer of a Langmuir lipid film. (c) Formation of an SLB using the Langmuir–Blodgett technique in which the second leaflet of the bilayer is formed by pushing the substrate with lipid monolayer coating either side vertically through a Langmuir lipid film on the subphase of a Langmuir trough. (d) Formation of an SLB using the Langmuir–Schaefer technique in which the second leaflet of the bilayer is formed by contact of the substrate with lipid monolayer coating either side horizontally on top of a Langmuir lipid film on the subphase of a Langmuir trough.

8.3.4 Imaging the lipid bilayer by AFM

1. Load a cantilever onto the fluid-compatible cantilever holder [see Note 8.4.4]. Align the laser onto the AFM tip.
2. If performing the experiment in an open fluid cell, it may be useful to place a small drop of the buffer solution on the AFM tip to avoid any issues with surface tension effects and bubble formation when the tip is placed over the sample.
3. If in tapping or intermittent-contact mode, determine the optimal resonant frequency for imaging. It is important to recognize that there may be more than one resonant peak and that optimum imaging may not occur at the peak resonant frequency. Evaluating imaging performance at different resonant frequencies and amplitudes is important for optimal resolution.
4. Ensure proper fusion of the bilayer by scanning an area of interest (see Note 8.4.7).
5. Ensure the presence of a bilayer (or multilayer structure) by performing a force curve measurement or creating of a hole in the bilayer to measure the height difference between the substrate and the bilayer. The symmetry of the bilayer should also be verified if critical for the application under investigation (see Note 8.4.7).

6. Retract the tip. Introduce the peptide solution (at least 200 μL to ensure that the solution is exchanged over the entire bilayer) to the SLB and continue to image (see Note 8.4.4). Depending on the configuration of the AFM fluid cell, it may be possible to set up a flow-through system using appropriately configured fluid handling (syringe pump and tubing). For a closed fluid cell, a single syringe pump is required to push the solution through the cell, whereas for an open Petri-dish type configuration, two syringe pumps are necessary to fill and evacuate the cell to maintain a constant liquid volume and avoid spills (see Note 8.4.8).

8.4 Notes

8.4.1 Selection of lipids

The choice of lipid composition will primarily depend on the system being studied [17]. Most conventional lipids can be used to form bilayers or vesicle patches. Preparation of SLBs by fusion of SUVs may require a diluent lipid or modification of buffers used for preparation of the SLB or for imaging to ensure successful fusion. Variation of lipid composition affects the topology and spatial organization of lipid rafts in SLBs, and must be carefully selected for the system of interest, especially when investigating the interactions of peptides and proteins with SLBs, as these are affected by membrane organization [18].

8.4.2 Selection of buffer

The selection of buffer used for preparation of lipid solutions is dependent on the lipid type, charge, and nature of imaging substrate (i.e., mica and glass). It may be necessary to modify the salt concentrations and pH of the buffer to optimize vesicle fusion, bilayer formation and stability, and image quality. [19] This is particularly important for preparation of SLBs by Langmuir–Blodgett and Langmuir–Schaefer techniques, for which substrates must be hydrophilic. To aid in the fusion of negatively charged lipids on a mica substrate, it is useful to wash the mica with a divalent cation-containing buffer (i.e., $CaCl_2$ and $MgCl_2$) prior to introduction of liposomes for fusion and bilayer formation or include a divalent cation in the imaging buffer [20a-b].

8.4.3 Selection of substrate

It should be noted that there are two substrates that need to be considered when preparing lipid bilayers: (1) the *bilayer substrate* and the (2) *sample substrate*

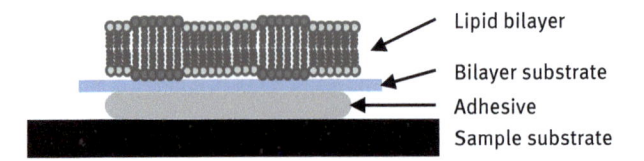

Figure 8.3: Selection of *bilayer substrate* can impact the properties of the bilayer, while the selection of *sample substrate* can be determined by the AFM configuration or correlated imaging/spectroscopic modality.

(Figure 8.3). The bilayer substrate is the substrate on which the lipid bilayer is directly formed. The sample substrate depends on the AFM configuration, such as when the sample must be secured to a magnetic steel disc to be held in place during AFM imaging.

Typical bilayer substrates include mica, silicon oxide (i.e., glass), gold-coated surfaces, and highly ordered pyrolytic graphite (HOPG). The substrate on which the lipid bilayer is formed can impact the bilayer structure [21]. Mica is often the preferred substrate for AFM imaging because of the ease of creating atomically flat surfaces by cleavage using a razor blade or removal of layers using adhesive tape. To aid in the fusion of negatively charged lipids to mica, it is useful to wash the mica with a divalent cation-containing buffer (i.e., $CaCl_2$ and $MgCl_2$) prior to introduction of liposomes for fusion and bilayer formation or include a divalent cation in the imaging buffer [20a-b]. Silicon oxide (i.e., glass) is another commonly used hydrophilic substrate but often requires rigorous cleaning and is rougher than mica. Gold-coated surfaces and HOPG are commonly used as hydrophobic substrates when prepared appropriately to achieve atomically flat surfaces (i.e., template stripped gold on mica or cleaved HOPG). Gold-coated surfaces can be used to form tethered lipid bilayers (tBLMs) by fusion of liposomes on self-assembled monolayers (SAMs) or thiolipids, [22] and the presence of a gold surface presents opportunities for performing correlated surface spectroscopy such as surface plasmon resonance.

The sample substrate can be governed by the AFM configurations (Figure 8.4). Sample substrates are typically magnetic steel discs when the sample substrate is magnetically held in place on the AFM. In this case, the bilayer substrate is adhered to the magnetic steel disc using double-sided tape or optical adhesive. Some AFM sample stages have vacuum holes that can require large sample substrates. Typically, these stages are also magnetic, so the sample substrate can vary from magnetic steel discs to anything that is sufficiently large to cover the vacuum holes such that the sample does not move during imaging. In AFM configurations in which the sample sits on top of an inverted microscope or when the AFM configuration has been modified to be used in combination with another imaging or spectroscopic technique that requires access to the sample from below, the sample substrate should be optically transparent (i.e., glass, for example, a glass-bottomed WillCo dish). If the lipid bilayer is to be prepared on a substrate that is not glass (i.e., mica), the bilayer substrate should be adhered to the

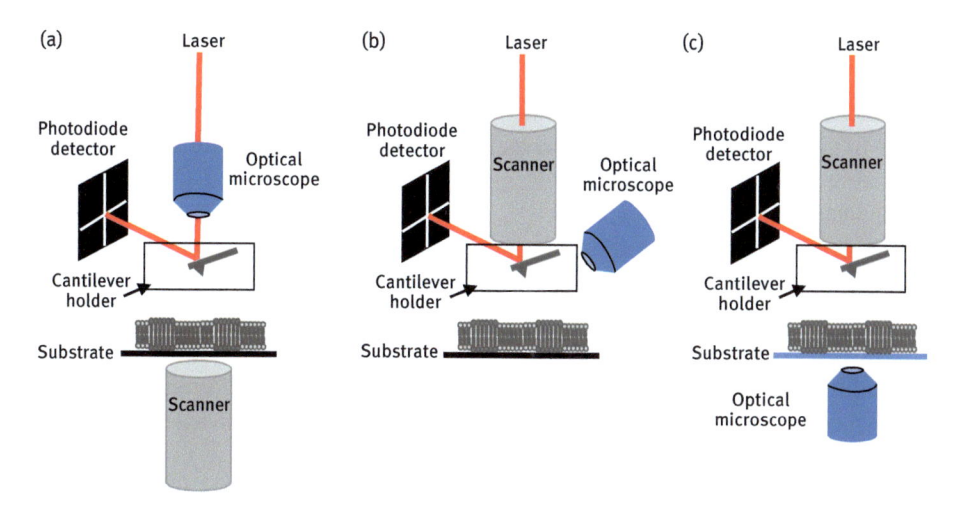

Figure 8.4: Configuration of the AFM platform being used can impact the types of experiments that can be performed. (a) Sample sits on top of the piezoelectric scanner and the sample substrate is typically a magnetic steel disc and is coupled to the scanner magnetically. (b) Sample sits on top of a magnetic stage that is sometimes also equipped with vacuum holes to hold the sample in place. The sample substrate can be a magnetic steel disc or anything sufficiently large to cover the vacuum holes. (c) Sample sits on top of an inverted microscope and requires a transparent sample substrate for optical imaging.

glass sample substrate using optical adhesive with good optical clarity (rather than use double-sided tape), taking care to use as little adhesive as possible to secure the sample. A commonly used UV-curable optical adhesive with good optical clarity between 320 and 3,000 nm is Norland Optical Adhesive 63 (Norland Products Incorporated, NJ, USA).

8.4.4 Selection of AFM probe

It is important to consider the mechanical properties of the AFM tip when imaging in fluid. V-shaped probes are generally preferable to use for liquid imaging as lateral bending is minimized. The tip must be sufficiently stiff to image in fluid but sufficiently soft to detect the presence of the bilayer. If a cantilever with too high a spring constant is chosen, it may not be possible to image the bilayer without causing damage to the bilayer itself. It should be noted that the mechanical coupling between the fluid and the tip will result in an effectively higher tip spring constant than what is reported in air. Repeated high-force scanning of an area can result in tip damage as well as possible accumulation of lipid on the tip, resulting in decreased spatial resolution. It is difficult to completely prevent proteins, peptides, or lipids from binding to the AFM tip when imaging in solution. Chemical modification of the AFM tip may be useful to minimize adsorption of species in solution [23].

8.4.5 Considerations for vesicle formation and fusion

Unilamellar – rather than multilamellar – vesicles are required for the formation of a single bilayer (Figure 8.1). While many methods exist for vesicle formation [24], simply sonicating the vesicle solution until it is clear typically ensures that all vesicles are unilamellar. Lipid degradation, especially upon storage, can make it difficult to create unilamellar vesicles via sonication. It is highly recommended that fresh lipid be used if at all possible. Checking for lipid degradation via TLC can be useful. Filtration of the lipid suspension through a polycarbonate membrane with a defined pore size, referred to as extrusion, can be a useful strategy for creating unilamellar vesicles for some lipids.

Unfused vesicles sitting on the bilayers can be removed by vigorous washing. If unfused vesicles persist, a lower incubation concentration and increased incubation temperature should be used.

If the vesicle solution is added to the substrate via the AFM liquid cell for the incubation period, it is common to observe the changes in the photodiode deflection during heating because of thermal effects on the cantilever.

8.4.6 Considerations for Langmuir film formation

The solvent in which the lipid mixture is dissolved (i.e., the carrying solvent) must be immiscible with water in order to facilitate formation of Langmuir film in the Langmuir trough and to limit loss of material to the water subphase. Additionally, the carrying solvent should be a very volatile solvent as it is removed from the surface by evaporation.

The concentration of lipid solution applied to the Langmuir trough to generate a film and the amount of the solution to be applied is related to the initial surface area of the trough at maximum opening of the balance. Concentration of the lipid solution in the organic solvent is typically 0.1–2 mg/mL [25]. Dilute lipid solutions can require the application of large volumes of the solution, which can result in long wait times for evaporation of the carrying solvent or can lead to increased residual solvent in the SLB. Lipid solutions above 2 mg/mL can be difficult to apply to the subphase with an increased potential for loss of material by sinking through the subphase during application.

The critical surface pressure indicating the formation of a Langmuir film is dependent on the lipid composition, but typically corresponds to a sudden drop or prolonged flat portion of the isotherm and is usually ~30 mN/m [25].

8.4.7 Verifying bilayer formation and symmetry by AFM

Two methods are typically used to confirm the presence of a single-component SLB or well-formed multicomponent bilayer that appears to be uniform. One method involves performing a simple force curve measurement and confirming a breakthrough event in the approach curve and subsequent adhesion in the retract curve. The other method involves repeatedly scanning a small area (i.e., $1 \times 1\,\mu m$) with high force at a high scan rate (~10 Hz) to create a hole in the bilayer for subsequent measurement of the bilayer height. The typical thickness of SLBs depends on the lipid composition and typically ranges from 5–7 nm.

Transmembrane asymmetry is typical of cell membranes, whereas domains generally observed by AFM on model SLBs display transmembrane symmetry. One method to obtain compositional asymmetry is to vary the temperature [26]. Height measurements can be used to infer the presence of a symmetric or asymmetric bilayer.

Combinatorial configurations in which a fluorescence imaging modality is added to the AFM platform are particularly useful for assessing bilayer quality. In such an experiment, an appropriate membrane dye (i.e., DiI) can be added to the lipid solution and used as an indicator of bilayer uniformity and coverage using fluorescence microscopy.

8.4.8 Considerations for in situ AFM

AFM imaging in fluid is routine on most instruments, but care must be taken to prevent damage to electronic components. Fluid imaging must be conducted with a cantilever designed for imaging in fluid. Most AFM configurations can support either an open or closed fluid cell.

If a fluid cell is not available for the AFM model, a fluid-compatible cantilever can be used and imaging can be conducted in a droplet that was carefully added between the tip and sample. It should be noted that the concentration of the fluid can be variable if experiments are conducted in this manner because of evaporation of the solvent (i.e., water).

If a closed fluid cell is available, a syringe pump can be used to inject fluid and to exchange the fluid throughout the course of the experiment. Some AFM configurations only offer an open fluid cell. In this case, a dual syringe pump must be used to drain the fluid cell during fluid exchange to prevent leaking of fluid out of the fluid cell.

8.4.9 Considerations for combinatorial experiments

The ability to incorporate additional imaging/spectroscopic techniques into an AFM platform can largely depend on the configuration of the AFM (Figure 8.4). Additionally, different models of AFMs have different configurations that affect the method of sample preparation that is best for imaging in fluid. A configuration in which the bottom of the sample can be imaged from below is ideal for combinatorial techniques. Many different fluorescence techniques can be combined with AFM imaging, including bright field, phase contrast, differential interference contrast (DIC), modulation contrast, epifluorescence, near-field scanning optical microscopy

(NSOM) confocal laser scanning microscopy (CLSM), total internal reflection fluorescence (TIRF), "F-techniques" (FCS, FRET, FLIM, FRAP), and super-resolution techniques (PALM, STED, STORM). Spectroscopic techniques including infrared (IR) spectroscopy, Raman spectroscopy, and surface plasmon resonance can also be used in combination with AFM [15].

Combined AFM-fluorescence microscopy or spectroscopy typically requires the incorporation of an artificial chromophore such as BODIPY, anthracene, or pyrene into one of the fatty acid side chains of selected lipids to generate lipid analogues for the system of study [27]. It should be noted that when using membrane dyes in correlated AFM-fluorescence microscopy experiments, it is important to choose fluorescent analogues with similar acyl chain lengths as the lipids in the bilayer to avoid unexpected effects due to hydrophobic mismatch or other structural effects. It is also important to consider the chemical structure of the fluorophore and the effects it may have on partitioning in the bilayer. For example, it has been shown that many fluorescent lipid probes, including derivatives of known raft-associated lipids, preferentially partitioned into topographical features consistent with nonraft domains, suggesting that the covalent attachment of a small fluorophore to a lipid molecule can abolish its ability to associate with rafts [28].

Combined AFM-fluorescence techniques can suffer from background fluorescence from the bulk solution and backscattering of the excitation source from the AFM cantilever. These issues are eliminated through the combination of AFM with TIRF microscopy, where fluorescence is limited to structures within ~100 nm of the surface [29].

The ability to assess molecular orientation and conformation for molecules that insert into the hydrophobic cores of SLBs is also of interest. Polarized fluorescence techniques can be used to measure the order parameter of molecules, a time- and ensemble-averaged quantity that enables the assessment of the orientational order of a molecule. This family of techniques uses the dichroic absorption and/or the anisotropic emission of polarized light by fluorescent lipids or proteins to spatially detect and infer their orientation, organization, conformation, and rotational mobility. This has been demonstrated using polarized TIRF microscopy with AFM, [30] a technique through which the measurement of the order parameter of a fluorescent lipid analogue has been used to investigate the interaction of a model AMP with bacterial membrane-mimetic SLBs, revealing rapid reorganization of the membrane into domains upon interaction with the peptide [31,32].

Modification of AFM probes can facilitate new combinatorial characterization methods.[33] By coupling excitation lasers to optical fiber AFM cantilevers, AFM imaging can be performed in combination with NSOM, in which the evanescent field at the apex of the AFM tip excites lipid and peptide/protein analogs in the SLB and excitation light is collected from below the sample with an objective lens [34].

Vibrational spectroscopies have also been combined with AFM imaging and offer the possibility of chemical identification without the use of fluorescent lipid and peptide/protein analogs that might perturb the native phase behavior of SLBs. Relatively straightforward implementation of combined AFM and IR spectroscopy is in combination with an attenuated total reflectance (ATR)-IR platform, in which the ATR crystal serves as both the sample support and the spectral sensing element such that the IR spectra are collected over a large sample area. ATR-Fourier Transfer (FT) IR-AFM enables tracking of topographical changes while simultaneously following protein-membrane bonding interactions and secondary structure changes through IR spectroscopy, such as thermal transitions in lipid bilayers and direct evidence of lipid-induced conformational changes in adsorbed proteins [35]. Super-resolved FT-IR spectroscopy has been demonstrated by confining the IR sampling to the region immediately under the scanning AFM tip by either using the tip as the source or detector [36]. The use of the AFM tip as an IR source has been demonstrated in techniques such as apertureless near-field scanning infrared microscopy (ANSIM), based on scattering of IR radiation by metal-coated AFM tip as it scans over a surface, or IR-SNOM, which uses an IR-compatible optical fiber tip. The AFM can also be used as a detector through photothermal-induced resonance (PTIR), in which absorption of infrared light by the sample results in a thermal–mechanical expansion of the sample, which in turn induces a transient excitation of the AFM tip, and FFT analysis of this response is used to recreate the absorption spectrum [37].

AFM can also be combined with Raman spectroscopy despite the lack of sensitivity of conventional nonresonant Raman spectroscopy for imaging of SLBs because of weak Raman scattering. Surface-enhanced Raman spectroscopy (SERS) has been used in combination with AFM by preparation of SLBs on two-dimensional arrays of metallic nanostructures formed by thin-film evaporation over hexagonally closed-packed polystyrene spheres [38]. Tip-enhanced Raman spectroscopy (TERS), in which the metallic or metallized AFM tip is used to confine and enhance an electromagnetic field in close proximity to the sample surface, and has been used to image segregated lipid domains in SLBs with full spectral information at each pixel [39].

References

[1] Cournia, Z., Allen, T.W., Andricioaei, I., Antonny, B., Baum, D., Brannigan, G., Buchete, N.-V., Deckman, J.T., Delemotte, L., del Val, C., Friedman, R., Gkeka, P., Hege, H.-C., Hénin, J., Kasimova, M. A., Kolocouris, A., Klein, M. L., Khalid, S., Lemieux, M. J., Lindow, N., Roy, M., Selent, J., Tarek, M., Tofoleanu, F., Vanni, S., Urban, S., Wales, D. J., Smith, J. C., & Bondar, A.-N. Membrane protein structure, function, and dynamics: a perspective from experiments and theory. *The Journal of Membrane Biology.* **2015**, *248*(4), 611–640.

[2] Wallin, E., & Heijne, G. V. Genome-wide analysis of integral membrane proteins from eubacterial, archaean, and eukaryotic organisms. *Protein. Science.* **1998**, *7*(4), 1029–1038.

[3] Overington, J. P., Al-Lazikani, B., & Hopkins, A. L. How many drug targets are there?. *Nature Reviews Drug Discovery.* **2006**, *5*, 993.

[4] Hancock, R. E. W., & Scott, M. G. The role of antimicrobial peptides in animal defenses. *Proceedings of the National Academy of Sciences.* **2000**, *97*(16), 8856-8861.
[5] Phillips, R., Ursell, T., Wiggins, P., & Sens, P., Emerging roles for lipids in shaping membrane-protein function. *Nature.* **2009**, *459*, 379.
[6] Guilhelmelli, F., Vilela, N., Albuquerque, P., Derengowski, L.d. S., Silva-Pereira, I., & Kyaw, C. M. Antibiotic development challenges: the various mechanisms of action of antimicrobial peptides and of bacterial resistance. *Frontiers in Microbiology.* **2013**, *4*, 353.
[7] Picas, L., Milhiet, P.-E., & Hernández-Borrell, J. Atomic force microscopy: A versatile tool to probe the physical and chemical properties of supported membranes at the nanoscale. *Chemistry and Physics of Lipids.* **2012**, *165*(8), 845–860.
[8] Morandat, S., Azouzi, S., Beauvais, E., Mastouri, A., & El Kirat, K. Atomic force microscopy of model lipid membranes. *Analytical and Bioanalytical Chemistry.* **2013**, *405*(5), 1445–1461.
[9] Whited, A. M., & Park, P. S. H. Atomic force microscopy: a multifaceted tool to study membrane proteins and their interactions with ligands. *Biochimica et Biophysica Acta(BBA) – Biomembranes.* **2014**, *1838* (1, Part A), 56–68.
[10] de Planque, M. R. R., & Killian*, J. A. Protein–lipid interactions studied with designed transmembrane peptides: role of hydrophobic matching and interfacial anchoring (Review). *Molecular Membrane Biology.* **2003**, *20*(4), 271–284.
[11] Brogden, K. A. Antimicrobial peptides: pore formers or metabolic inhibitors in bacteria? *Nature Reviews Microbiology.* **2005**, *3*, 238.
[12] Gözen, I., & Jesorka, A. Instrumental methods to characterize molecular phospholipid films on solid supports. *Analytical Chemistry.* **2012**, *84*(2), 822–838.
[13] Simons, K., & Toomre, D. Lipid rafts and signal transduction. *Nature Reviews Molecular Cell Biology* **2000**, *1*, 31.
[14] Alessandrini, A., & Facci, P. Unraveling lipid/protein interaction in model lipid bilayers by Atomic Force Microscopy. *Journal of Molecular Recognition.* **2011**, *24*(3), 387–396.
[15] Moreno Flores, S., & Toca-Herrera, J. L. The new future of scanning probe microscopy: Combining atomic force microscopy with other surface-sensitive techniques, optical microscopy and fluorescence techniques. *Nanoscale.* **2009**, *1*(1), 40–49.
[16] Richter, R. P., Bérat, R., & Brisson, A. R. Formation of solid-supported lipid bilayers: An integrated view. *Langmuir.* **2006**, *22*(8), 3497–3505.
[17] van Meer, G., Voelker, D. R., & Feigenson, G. W. Membrane lipids: where they are and how they behave. *Nature Reviews Molecular Cell Biology.* **2008**, *9*, 112.
[18] de Almeida, R. F. M., Fedorov, A., & Prieto, M. "Sphingomyelin/phosphatidylcholine/cholesterol:" boundaries and composition of lipid rafts. *Biophysical Journal.* **2003**, *85*(4), 2406–2416.
[19] Suresh, S., & Edwardson, J. M. Phase separation in lipid bilayers triggered by low pH. *Biochemical and Biophysical Research Communications.* **2010**, *399*(4), 571–574.
[20] (a) Richter, R. P., & Brisson, A. R. Following the formation of supported lipid bilayers on mica: A study combining AFM, QCM-D, and ellipsometry. *Biophysical Journal.* **2005**, *88*(5), 3422–3433
[20] (b) Richter, R., Mukhopadhyay, A., & Brisson, A. Pathways of lipid vesicle deposition on solid surfaces: A combined QCM-D and AFM study. *Biophysical Journal.* **2003**, *85*(5), 3035–3047.
[21] Seeger, H. M., Cerbo, A. D., Alessandrini, A., & Facci, P. Supported lipid bilayers on mica and silicon oxide: Comparison of the main phase transition behavior. *The Journal of Physical Chemistry B.* **2010**, *114*(27), 8926–8933.
[22] Naumann, R., Schiller, S. M., Giess, F., Grohe, B., Hartman, K. B., Kärcher, I., Köper, I., Lübben, J., Vasilev, K., & Knoll, W. Tethered lipid bilayers on ultraflat gold surfaces. *Langmuir.* **2003**, *19*(13), 5435–5443.

[23] Barattin, R., & Voyer, N. Chemical modifications of atomic force microscopy tips. In *Atomic Force Microscopy in Biomedical Research: Methods and Protocols*, Braga, P. C.; Ricci, D., Eds. Humana Press: Totowa, NJ, 2011, pp 457–483.

[24] F Szoka, J., & Papahadjopoulos, D. Comparative properties and methods of preparation of lipid vesicles (Liposomes). *Annual Review of Biophysics and Bioengineering*. **1980**, *9*(1), 467–508.

[25] Cruz, A., & Pérez-Gil, J. Langmuir films to determine lateral surface pressure on lipid segregation. In *Methods in Molecular Biology, vol. 400: Methods in Membrane Lipids*, Dopico, A. M., Ed. Humana Press Inc.: Totowa, New Jersey, 2007, pp 439–457.

[26] Lin, W.-C., Blanchette, C. D., Ratto, T. V., & Longo, M. L. Lipid asymmetry in DLPC/DSPC-supported lipid bilayers: A combined AFM and fluorescence microscopy study. *Biophysical Journal*. **2006**, *90*(1), 228–237.

[27] Kuerschner, L., Ejsing, C. S., Ekroos, K., Shevchenko, A., Anderson, K. I., Thiele, C. Polyene-lipids: A new tool to image lipids. *Nature Methods*. **2004**, *2*, 39.

[28] Shaw, J. E., Epand, R. F., Epand, R. M., Li, Z., Bittman, R., Yip, C. M. Correlated fluorescence-atomic force microscopy of membrane domains: Structure of fluorescence probes determines lipid localization. *Biophysical Journal*. **2006**, *90*(6), 2170–2178.

[29] Shaw, J. E., Slade, A., Yip, C. M. Simultaneous in situ total internal reflectance fluorescence/atomic force microscopy studies of DPPC/dPOPC microdomains in supported planar lipid bilayers. *Journal of the American Chemical Society*. **2003**, *125*(39), 11838–11839.

[30] Oreopoulos, J., & Yip, C. M. Probing membrane order and topography in supported lipid bilayers by combined polarized total internal reflection fluorescence-atbomic force microscopy. *Biophysical Journal*. **2009**, *96*(5), 1970–1984.

[31] Oreopoulos, J., Epand, R. F., Epand, R. M., & Yip, C. M. Peptide-induced domain formation in supported lipid bilayers: Direct evidence by combined atomic force and polarized total internal reflection fluorescence microscopy. *Biophysical Journal*. **2010**, *98*(5), 815–823.

[32] Oreopoulos, J., & Yip, C. M. Combinatorial microscopy for the study of protein–membrane interactions in supported lipid bilayers: Order parameter measurements by combined polarized TIRFM/AFM. *Journal of Structural Biology*. **2009**, *168*(1), 21–36.

[33] Lieberman, K., Lewis, A., Fish, G., Shalom, S., Jovin, T. M., Schaper, A., & Cohen, S. R. Multifunctional, micropipette based force cantilevers for scanned probe microscopy. *Applied Physics Letters*. **1994**, *65*(5), 648–650.

[34] Hollars, C. W., & Dunn, R. C. Submicron fluorescence, topography, and compliance measurements of phase-separated lipid monolayers using tapping-mode near-field scanning optical microscopy. *The Journal of Physical Chemistry B*. **1997**, *101*(33), 6313–6317.

[35] Verity, J. E., Chhabra, N., Sinnathamby, K., & Yip, C. M. Tracking molecular interactions in membranes by simultaneous ATR-FTIR-AFM. *Biophysical Journal*. **2009**, *97*(4), 1225–1231.

[36] Li, J. J., & Yip, C. M. Super-resolved FT-IR spectroscopy: Strategies, challenges, and opportunities for membrane biophysics. *Biochimica et Biophysica Acta (BBA) – Biomembranes*. **2013**, *1828*(10), 2272–2282.

[37] Dazzi, A., Prazeres, R., Glotin, F., & Ortega, J. M. Local infrared microspectroscopy with subwavelength spatial resolution with an atomic force microscope tip used as a photothermal sensor. *Optics Letter*. **2005**, *30*(18), 2388–2390.

[38] Sweetenham, C. S., Larraona-Puy, M., Notingher, I. Simultaneous surface-enhanced raman spectroscopy (SERS) and atomic force microscopy (AFM) for label-free physicochemical analysis of lipid bilayers. *Applied Spectroscopy*. **2011**, *65*(12), 1387–1392.

[39] Opilik, L., Bauer, T., Schmid, T., Stadler, J., & Zenobi, R. Nanoscale chemical imaging of segregated lipid domains using tip-enhanced Raman spectroscopy. *Physical Chemistry Chemical Physics*. **2011**, *13*(21), 9978–9981.

Brittney L. Gorman, Ashley N. Yeager, Corryn E. Chini, Mary L. Kraft

9 Imaging the distributions of lipids and proteins in the plasma membrane with high-resolution secondary ion mass spectrometry

Abstract: Characterizing the distributions of distinct lipid and protein species within biological membranes is an important step toward elucidating the roles of these molecules in membrane function. Ideally, the lipids and proteins of interest could be imaged under native conditions with high spatial resolution and without labels that might alter their distributions and functions. While no single technique meets all these criteria, a variety of complementary imaging modalities, each with its own unique benefits and limitations, have yielded new insight into biological membrane structure. In this chapter, we review one of these techniques, high-resolution secondary ion mass spectrometry (SIMS) performed on a commercial Cameca NanoSIMS 50(L) instrument. High-resolution SIMS reveals the elemental and isotopic distributions at the surface of a sample with a lateral resolution better than 100 nm and a depth resolution as good as a few nanometers. First, we explain the fundamental principles of this technique, emphasizing the aspects that affect the application of SIMS to imaging the distributions of distinct lipid and protein species in the plasma membranes of intact mammalian cells. Then we describe the sample preparation, imaging conditions, and data analysis strategies that have been developed for cell membrane analysis. The data analysis strategies include statistical approaches for quantitatively defining membrane domains enriched with a component of interest, and quantifying co-localization between different membrane species. We also discuss the potential artifacts that may arise when imaging component distributions in the membranes of mammalian cells with high-resolution SIMS, control experiments for detecting these artifacts, and strategies for avoiding them. Examples drawn from our reported results are provided to illustrate key concepts in this chapter and demonstrate how this technique may be used to improve understanding of plasma membrane organization. Finally, we discuss

---

Brittney L. Gorman, Center for Biophysics and Computational Biology, University of Illinois at Urbana-Champaign, Urbana, IL
Ashley N. Yeager, Department of Chemical and Biomolecular Engineering, University of Illinois at Urbana-Champaign, Urbana, IL
Corryn E. Chini, Department of Chemistry, University of Illinois at Urbana-Champaign, Urbana, IL
Mary L. Kraft, Center for Biophysics and Computational Biology, University of Illinois at Urbana-Champaign, Urbana, IL; Department of Chemical and Biomolecular Engineering, University of Illinois at Urbana-Champaign, Urbana, IL; Department of Chemistry, University of Illinois at Urbana-Champaign, Urbana, IL

https://doi.org/10.1515/9783110544657-009

future directions for using high-resolution SIMS to elucidate membrane structure-function relationships.

Keywords: SIMS, sphingolipids, cholesterol, lipid domain, raft, metabolic labeling, stable isotope incorporation, cell membrane, fixation, artifacts

9.1 Introduction

Cellular membranes are highly organized structures. Multitudes of different lipid species establish the membrane's bilayer structure, and numerous protein components are embedded in this bilayer or associated with its periphery. Mammalian cell membranes also contain cholesterol, which modulates their biophysical properties. In addition to the bilayer structure, the plasma membranes of mammalian cells are organized into domains with distinct protein and lipid compositions [1–5]. The distributions of specific proteins within plasma membranes have been visualized by labeling with genetically encoded fluorescent proteins or functionalized affinity tags, and imaging with advanced fluorescence microscopy techniques or immunoelectron microscopy [4, 6–14]. Many lipid species are also expected to be non-randomly distributed within the plasma membrane [5, 15–17], but their distributions typically cannot be visualized with genetic approaches or functionalized affinity tags. The distributions of a few lipid species have been assessed by incorporating fluorophore-tagged lipids into cells and imaging them with fluorescence microscopy [18–21]. Some drawbacks of this approach are only a small fraction of the lipid species of interest will bear the fluorophore that enables direct visualization, and the fluorophore may alter the intracellular trafficking, subcellular distribution, and molecular interactions of the lipids they label [22–25].

Secondary ion mass spectrometry (SIMS) is a complementary approach to fluorescence-based lipid detection, enabling cholesterol and distinct lipid species to be directly imaged in the plasma membrane without using fluorophores or other bulky labels. SIMS combines the direct component detection and identification afforded by mass spectrometry with a scanning primary ion beam that provides location specificity. Depending on the instrumentation, elemental or isotopic composition may be mapped with as high as 50 nm-lateral resolution, whereas molecular composition may be visualized on somewhat longer length scales [26–29]. Because the majority of the secondary ions detected with SIMS are ejected from the outermost ~5 nm of the sample [30], SIMS has a shallow sampling depth that enables restricting the analysis to the plasma membrane. However, SIMS is performed under ultrahigh vacuum (UHV), so it cannot be applied to living cells.

SIMS instruments with various primary ion sources, ion optics, and mass analyzers have been used to image component distributions in cellular membranes with differing levels of specificity, sensitivity, and lateral resolution [27, 29, 31, 32]. Regardless of the instrumentation, SIMS analysis is performed as follows. A transfer rod is used to load the sample into the SIMS analysis chamber, which is under UHV. Various ion sources (*e.g.*, a duoplasmatron, liquid metal ion gun, gas cluster, or polyatomic thermal effusive source) may be used to produce the primary ion beam [31, 33, 34]. This beam is focused onto the sample, producing a spot with a diameter that varies from 30 nm to a few microns, depending on the primary ion source and ion-focusing optics [30, 35]. When the primary ions impact the sample, they fragment the molecules on its surface into charged and neutral species. The resulting charged fragments, called secondary ions, range from monoatomic and diatomic species to nearly intact molecular ions, and their mass-to-charge ratios (m/z) are usually between 1 and 1500, depending on the sample and instrument [27, 29, 31]. The secondary ions ejected from the sample are extracted and directed to a mass spectrometer that uses flight time or magnetic deflection to separate them according to their m/z [27–29]. As the primary ion beam is scanned across the sample, spectral data indicative of sample composition are recorded at each pixel in the raster scan. The intensities of the secondary ion species of interest are used to create a map that reveals the two-dimensional (2D) distributions of the molecules on the surface of the sample.

This chapter focuses on the application of high-resolution SIMS performed using a Cameca NanoSIMS 50(L) instrument to imaging the distributions of specific components in the plasma membranes of mammalian cells. An image of a NanoSIMS 50 instrument is shown in Figure 9.1. Like other SIMS modalities, the

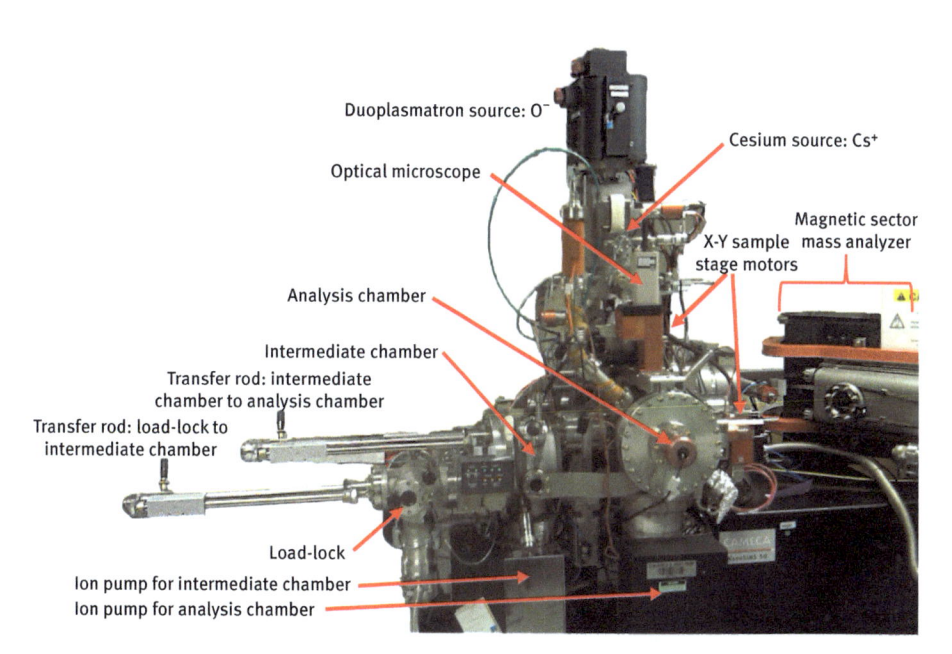

Figure 9.1: Photograph of a NanoSIMS 50 instrument, labeled to show key components. Only a portion of the magnetic sector mass analyzer is visible in this image.

NanoSIMS can achieve a shallow sampling depth that enables imaging component distributions in the plasma membrane while minimizing the detection of intracellular components. NanoSIMS instruments routinely achieve the small primary ion beam diameters (30–100 nm) and high secondary ion signal intensities that are required to image with a high (<100 nm) lateral resolution [35]. These features are particularly well suited for detecting and imaging submicron-sized domains enriched with one or more distinct lipid, protein, or sterol species of interest within the plasma membrane.

The NanoSIMS instrument uses a beam of reactive primary ions, namely positively charged cesium ions (Cs^+) or negatively charged oxygen ions (O^-), to eject monoatomic and diatomic secondary ions from the surface of the sample [36]. The implantation of these reactive primary ions into the sample increases the probability of generating secondary ions with an opposite polarity [36]. Specifically, Cs^+ primary ions promote the formation of negatively charged secondary ions, whereas O^- primary ions promote the formation of positively charged secondary ions [37]. Although the monoatomic and diatomic secondary ions produced by the reactive Cs^+ and O^- primary ions lack molecular specificity, their high yields help to ensure that the working lateral resolution of the resulting SIMS image is not limited by low signal intensity [35].

The primary ion beam and the resulting secondary ions are focused with coaxial optics [35]. This configuration allows the primary ion beam to have a normal incidence to the surface, which improves focusing and produces a smaller diameter spot with a higher current than that obtained with the conventional SIMS configuration [35]. The coaxial configuration also improves secondary ion collection efficiency by reducing the distance between the sample and ion focusing optics [35]. Noteworthy, in order to use colinear optics to direct the primary and secondary ions toward and away from the sample, respectively, the primary and secondary ions must have opposite polarities [35].

The secondary ions extracted from the sample are accelerated into a drift tube with a magnetic field that deflects the ions along a radius of curvature that depends on the ion's m/z. Depending on the NanoSIMS model, the magnetic sector mass analyzer has either five or seven individual ion counting detectors for the simultaneous collection of ions with five or seven different preselected m/z, respectively [37]. The secondary ions are collected with a mass resolving power ($M/\Delta M$) that is high enough to separate multiple interfering isobaric species with the same nominal mass, such as $^{12}C^{15}N^-$ (26.9996 amu) and $^{13}C^{14}N^-$ (27.0059 amu) [27]. Maps that show the monoatomic or diatomic secondary ion counts detected at each pixel are created, revealing the elemental and isotopic composition on the surface of the sample. Because the biomolecules within native plasma membranes share a common elemental and isotopic composition, the components of interest must be labeled with nonnative elements or rare stable isotopes so they produce characteristic secondary ions. Strategies to accomplish this are described in Section 9.2.1.

Typically, high-resolution SIMS analysis with a NanoSIMS instrument is performed by rastering the primary ion beam across the same location multiple times [38]. Because a thin layer of material is sputtered from the sample each time the primary ion beam is scanned across its surface, each image produced in a subsequent pass shows the secondary ion distribution successively deeper within the sample. The data from several sequential raster scans are typically added together to create a single image that has higher signal-to-noise than any individual image plane [38]. The resulting compiled image shows the component distribution within the thin layer of material that was sputtered from the sample during the analysis. The thickness of this layer is the sputter depth. If a very large number of raster scans are acquired at the same location, the individual image planes can be stacked to form a three-dimensional (3D) representation of the component distribution at different depths within the sample [39].

In this chapter, we describe the application of high-resolution SIMS performed with a NanoSIMS to imaging component distribution in the plasma membranes of mammalian cells with a lateral resolution <100 nm. Mammalian cells must be carefully prepared so the analytes of interest can be detected with high-resolution SIMS, and their distributions in the cell are not perturbed. Sample preparation strategies that address these requirements are presented in Section 9.2. In Section 9.3, we describe how the design of the NanoSIMS instrument affects the acquisition of SIMS data from intact cells, including the sensitivity and spatial resolution of the measurement. The construction of high-resolution SIMS images that reveal component distribution in the plasma membrane and the quantitative interpretation of these images are covered in Section 9.4. We have devoted a significant portion of Section 9.4 to statistical methods for unambiguously testing hypotheses concerning plasma membrane domains enriched with specific components. Throughout these sections, we discuss potential artifacts that may be encountered, experiments for detecting them, and strategies for avoiding them. We also present examples that are drawn from our published work because we can describe this work in detail. We conclude the chapter with a brief discussion of future directions for high-resolution SIMS analysis of cell membranes.

9.2 Sample preparation

To acquire meaningful data from a sample using a NanoSIMS instrument, strict sample requirements must be met. First, because the NanoSIMS instrument detects monoatomic and diatomic secondary ions, the molecules of interest must contain distinctive elements or isotopes so they will produce characteristic secondary ions that can be attributed to the parent molecule. Second, samples cannot contain volatile substances that would outgas upon exposure to the UHV environment in the NanoSIMS instrument's analysis chamber. Finally, the sample should be conductive to increase the likelihood that it will produce the high secondary ion signal intensities that are required to detect the molecules of interest with high sensitivity and lateral resolution. Although biological membranes in their native state do not satisfy these requirements, the sample preparation strategies described below can be employed to enable imaging component distribution within biological membranes using a NanoSIMS instrument.

9.2.1 Labeling the components of interest

As mentioned above, the monoatomic and diatomic secondary ions that are detected with a NanoSIMS instrument reveal the elemental and isotopic composition at the surface of the sample. Proteins and lipids generally contain the most abundant isotopes of the same elements, namely, carbon, hydrogen, oxygen, nitrogen, phosphorous (most lipids and phosphorylated proteins), and sulfur (most proteins and sulfatide lipids). Therefore, to image a specific lipid or protein species with a NanoSIMS instrument, a distinct stable isotope or nonnative element must be incorporated into the species of interest so that it will fragment into elementally or isotopically distinct secondary ions that can be used for identification. Here, we briefly review established strategies for labeling lipids and proteins with tags that permit imaging their distributions with a NanoSIMS.

For visualizing specific lipid species with a NanoSIMS, the replacement of one or more atoms in the lipid with a rare stable isotope of the same element is preferable to substitution with a nonnative element. This is because elemental substitution (e.g., replacement of a hydrogen atom with fluorine) may alter the molecular interactions that determine the lipid's mixing with other membrane components or its intracellular trafficking. In contrast, a rare stable isotope-labeled lipid has the same chemical composition and structure as the native lipid molecule. Consequently, substituting the naturally abundant isotope of carbon, nitrogen, or oxygen with its rare stable isotope (*i.e.*, ^{12}C with ^{13}C, ^{14}N with ^{15}N, or ^{16}O with ^{17}O or ^{18}O, respectively) does not alter the lipid's interactions with other biomolecules, distribution in a membrane, or intracellular trafficking. However, care should be taken when deuterium (^{2}H) is substituted for hydrogen (^{1}H). The polarizability of carbon-hydrogen versus carbon-deuterium bonds differ [40], although high levels of deuterium substitution into the lipids' fatty acid tails may only negligibly affect lipid mixing behavior in phase-separated model membranes [41–44]. Nonetheless, deuterium substitution affects hydrogen bond stability [45], and it can markedly perturb the kinetics of reactions that involve breaking a covalent bond to hydrogen in the rate-limiting step [46, 47]. This perturbation of reaction kinetics arises because the vibrational energy of a bond depends on the reduced mass of the connected atoms, and deuterium is twice as heavy as hydrogen. Therefore, deuterium substitution into sites involved in hydrogen bonding or elimination reactions could alter native cell function.

A high fraction of the lipid molecules of interest must contain the rare stable isotope so that the isotopically distinctive secondary ions produced during NanoSIMS analysis reveal the global distribution of the lipid of interest in the plasma membrane. This is easy to accomplish when studying model lipid membranes because rare stable isotope-labeled lipids can be purchased from commercial sources or synthesized in-house and used for model membrane formation. For studies of native cell membranes, rare isotope-labeled lipids and cholesterol can be incorporated into cellular membranes using lipid carriers, such as cyclodextrins, to exchange the naturally abundant lipid molecules in the cell membrane with the isotope-labeled analog [48–51]. However, some exogenous lipids have a tendency to adhere to the cell surface without inserting into the plasma membrane [52]. This material may be removed by transferring the cells to label-free medium for a sufficient time to allow them to internalize any lipid aggregates attached to their surfaces.

As an alternative to lipid carriers, metabolic labeling can be used to biosynthetically incorporate rare stable isotopes into specific cellular lipid species. This entails culturing the cells in the presence of rare isotope-labeled lipid precursors, which the cells take up, metabolize into an isotope-labeled lipid, and traffic in the same manner as the endogenous lipid molecule. To ensure that the rare isotope distribution imaged by the NanoSIMS instrument reveals the distribution of the lipid species of interest, the metabolic precursor that is used must be large and complex enough to direct its incorporation into the targeted lipid species, but not other membrane components. Methods to metabolically incorporate rare isotopes and other labels into specific lipid classes, including gangliosides [53–56], sphingolipids [38, 53, 57–59], phosphatidylethanolamines [60, 61], phosphatidylcholines [61], and lipids that contain specific fatty acids [61, 62], are well established. Because mammalian cells have receptors for cholesterol uptake, rare isotope-labeled cholesterol can be naturally incorporated into mammalian cells by complexing it with a serum carrier and adding it to the culture medium [63, 64]. Mammalian cells will internalize the isotope-labeled cholesterol and transport it to the appropriate subcellular location via the natural cholesterol transport pathway [63, 64]. This method has been used to replace 50–80% of the unlabeled endogenous cholesterol in mammalian cells from different lines with rare isotope-labeled cholesterol [64, 65]. Isotope incorporation into lipids is assessed by extracting the lipids from cells labeled identically to those that will be imaged using high-resolution SIMS, and measuring the fraction of the lipid species of interest that contains the distinctive isotope with liquid chromatography-mass spectrometry (LC-MS) [38, 66–68]. An analogous method is used to assess isotope incorporation into cellular cholesterol, but gas chromatography-mass spectrometry (GC-MS) is used instead of LC-MS [64].

Unlike lipids, proteins of interest cannot be selectively labeled with metabolic precursors that contain distinct isotopes or elements because all proteins contain the same amino acids. Instead, antibodies functionalized with nonnative elements or rare isotope tags can be used to selectively label the protein of interest so that it can be detected using a NanoSIMS instrument [65, 69–71]. Commercially available colloidal gold-labeled antibodies produce ^{197}Au$^-$ secondary ions that enable detection of the proteins they label using high-resolution SIMS [70, 71]. However, the mass dispersion that the NanoSIMS 50 instrument attains at the high magnetic field strength required to detect ^{197}Au$^-$ ions does not permit the simultaneous detection of secondary ions that are separated by one mass unit (i.e., $^{12}C^{14}N^-$ and $^{12}C^{15}N^-$, or $^{12}C^1H^-$ and $^{13}C^1H^-$). This is problematic because proteins must be imaged in parallel with lipids in order to assess hypotheses that predict the colocalization of specific proteins with distinct lipids, and pairs of secondary ions that differ by one mass unit are used to detect lipids that contain the rare stable isotopes of nitrogen, carbon, or hydrogen. To circumvent this obstacle, immunolabels consisting of a secondary antibody covalently conjugated to a colloidal gold nanoparticle functionalized with fluorinated alkanethiols have been developed [71]. The $^{19}F^-$ secondary ions produced by these immunolabels enable detecting the proteins they label in parallel with the adjacent secondary ion pairs that have been used to detect isotope-labeled lipids with a NanoSIMS instrument [65, 71]. However, the size of the colloidal gold nanoparticles may hinder the labeling of intracellular target proteins. An alternative strategy uses genetic approaches to insert a reactive unnatural amino acid into the protein of interest, which can be conjugated to an elementally or isotopically labeled tag for detection using a NanoSIMS instrument [72, 73]. Correlated imaging approaches that involve using fluorescence or electron microscopy to characterize sample structure, and high-resolution SIMS to identify the component abundance within these structures have also been developed [74–78].

9.2.2 Fixation

For compatibility with the UHV environment within the NanoSIMS analysis chamber, water and other volatile components must be removed from the sample prior to analysis. When water is drawn out of an unfixed cell or tissue during dehydration, it pulls dispersed molecules along with it, altering the molecular distributions in the sample, and sample ultrastructure [79]. To preserve the native molecular distributions in the sample on the 50–100 nm-length scale that is resolved with a NanoSIMS instrument, the molecules in the sample must be immobilized before the sample is dehydrated. This can be accomplished with freeze-drying, chemical fixation, or high-pressure freezing with chemical fixation, as discussed below.

Freeze-drying has primarily been used to prepare supported lipid membranes and lipid vesicles for high-resolution SIMS analysis [80–86]. This method entails rapidly freezing the sample to immobilize its molecular constituents, and subliming off the ice. Sample freezing must be very rapid to stop molecular motion and promote the formation of amorphous ice [87]. In contrast, slow cooling allows the water to expand into a lattice arrangement as it freezes, forming crystalline ice [87]. The water expansion that occurs during crystalline ice formation would rupture lipid membranes and other cellular structures. Rapid freezing, which is also called flash freezing, is accomplished by plunging the sample into liquid ethane (−188 °C) or liquid propane (−185 °C) that is cooled with liquid nitrogen. Importantly, though liquid nitrogen is also very cold (−195 °C), it should not be used for flash freezing because its heat capacity is not sufficient to rapidly freeze water. Once frozen, the sample is transferred to a precooled chamber, and a vacuum is applied to sublime the vitreous ice. After the ice has sublimed, the sample must be kept under vacuum as it is warmed to room temperature because water would condense on its surface if it were exposed to atmospheric conditions while cold. Once at room temperature, the sample can be removed from the vacuum chamber and stored under ambient conditions until analysis. Prior reports have demonstrated that this dehydration process stops the lateral diffusion of lipids within the lipid membrane, but lipid diffusion resumes when the sample is rehydrated [83].

Flash freezing is not effective for thicker samples, such as cells and tissues, because crystalline ice begins to form a few microns from the surface of the sample [87]. Fortunately, methods that use chemical fixatives to prepare cells for electron microscopy, which is also performed under UHV, have been under development, testing, and optimization since the first electron micrographs were acquired from cultured cells in 1945 [88]. In this approach, reactive compounds are used to form covalent bonds between the molecules in the sample. These covalent bonds "fix" the molecule in place so it cannot migrate within the sample during or after dehydration. In addition to mechanically stabilizing biomolecular assemblies within the cell, fixation also prevents decay by inactivating intrinsic proteolytic enzymes and chemically altering the cellular components so they cannot be consumed by microorganisms. Noteworthy, the fixation-induced covalent bonds do not affect the generation of monoatomic and diatomic secondary ions during high-resolution SIMS analysis.

Multiple chemical fixatives are required to immobilize all the biomolecules within the sample. The first step in chemical fixation is to treat the biological sample with aldehydes, such as formaldehyde and glutaraldehyde, that crosslink the cellular proteins [89]. Formaldehyde is typically used to prepare samples for immunolabeling or fluorescence microscopy because it does not alter protein antigenicity or produce a strong background fluorescence signal [90]. However, formaldehyde fixation may not prevent the redistribution of cell surface receptors or other membrane components [90]. In contrast, glutaraldehyde fixation is irreversible and preserves fine structures better than formaldehyde [89–91]. For this reason, glutaraldehyde is used to prepare biological samples for electron microscopy and other analyses performed under UHV, including high-resolution SIMS [65, 89], even when the sample is already fixed with formaldehyde to permit immunolabeling [90]. Glutaraldehyde crosslinks proteins by reacting with their primary amine, amide, and thiol functionalities [89, 92–95], and may also crosslink lipids with headgroups that contain a primary amine [96]. We have demonstrated that glutaraldehyde fixation does not induce artifactual lipid clustering or allow lipid redistribution on the length scales resolved by total internal reflection fluorescence microscopy (TIRFM) [97]. Specifically, TIRFM imaging of a fibroblast cell showed the distributions of metabolically incorporated fluorescent lipids in the plasma membrane of the living cell were identical to those observed in the same cell after glutaraldehyde fixation (Figure 9.2) [97].

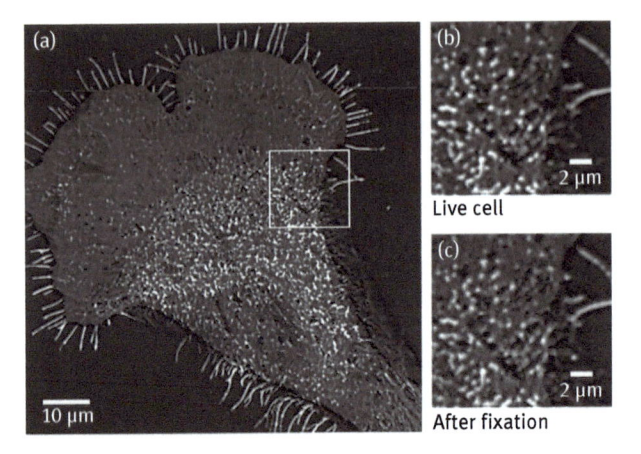

Figure 9.2: (a) Background subtracted TIRFM image shows the bottom of a living fibroblast cell that was metabolically labeled so its plasma membrane contained a fluorescent lipid (BODIPY-sphingolipid). Domains enriched in the fluorescent lipid are visible in the plasma membrane of the living cell. The outlined region is enlarged in (b), which shows the fluorescent lipid domains in the living cell. (c) TIRFM image of the same region on the cell following glutaraldehyde fixation shows the fluorescent lipid domains observed after fixation are identical to those at that location when the cell was alive. All images were background subtracted to improve signal-to-noise. Figure is reproduced with permission from Ref Frisz et al. [97].

A prior spin-labeling study of nerve bundles showed that glutaraldehyde fixation does not eliminate all lipid motion within the plasma membrane [98]. Therefore, glutaraldehyde-fixed samples are subsequently fixed with osmium tetroxide to crosslink the lipids [99, 100]. Osmium tetroxide reacts with the double bonds in unsaturated lipids [101–105], forming crosslinks that stop lipid motion, as demonstrated in multi-lamellar membranes [98, 106] and nerve tissue [98]. X-ray diffraction measurements performed on lipid extracts before and after fixation, dehydration, and embedding demonstrated that osmium tetroxide fixation preserves the lamellar structure in lipid membranes, and prevents lipid reorganization during dehydration [107].

In the past few decades, a preservation technique for cells and tissues called high-pressure freezing has become more popular. This technique uses freezing to rapidly immobilize molecules in the sample, followed by chemical fixation of the frozen sample to lock the molecules in place. As mentioned above, the water in a biological sample must be rapidly frozen so that it does not have time to expand and form crystalline ice, which would rupture membranes and other cellular structures [87]. The rate of sample cooling decreases with increasing distance from the cryogenic liquid, so vitreous ice only forms to a depth of ~10 µm from the sample's surface when freezing is performed at atmospheric pressure [87]. High pressure hinders the water expansion that occurs when crystalline ice forms, allowing samples up to hundreds of microns in thickness to be vitrified [87]. Commercially available high pressure freezing devices have been optimized to freeze biological samples at high pressure (roughly 2000 atm) without crushing them. Once the sample is frozen, a freeze substitution step is performed to dehydrate and chemically fix it. Fixatives, namely glutaraldehyde and osmium tetroxide, dissolved in acetone are introduced to the sample at very low temperatures (−90– −78 °C), and the sample is slowly warmed [108]. Glutaraldehyde and osmium tetroxide fixation are likely to begin when the sample reaches approximately −50 °C and −30 °C, respectively, which is too cold for the lipid and protein molecules to move within the sample [109]. This fixation immobilizes the molecules so they are unable to move during the subsequent dehydration process.

9.2.3 Strategies to increase sample conductivity

Under ideal conditions, the charge introduced to the sample by the primary ion beam during high-resolution SIMS analysis escapes by passage to ground and through the ejection of charged particles from the sample. Charging occurs when the number of charged primary ions impinging on the sample exceeds those that escape it, and it occurs more frequently on lipid membranes, cells, and other insulating samples. This charging reduces the number of secondary ions that may be collected from the sample, which is detrimental to the working lateral resolution of the analysis. To reduce charging, cells should be supported by conductive or semi-conductive substrates, such as metals, silicon, or indium tin oxide-coated glass, because they provide a conductive ground for charge dissipation. In our experience, cell samples must also be coated with a 2–3 nm-thick metal layer to reduce sample charging. This thickness is a balance between thicker layers that ensure complete coverage and promote charge dissipation, and thinner layers that allow the primary ions to pass through the metal layer and interrogate the sample. Prior studies demonstrate that metal layers produced with the sputter coaters that are used to prepare samples for scanning electron microscopy do not alter lipid distribution in the plasma membrane [97].

9.3 High-resolution SIMS performed with a Cameca NanoSIMS 50(L) instrument

To image specific components in the plasma membrane with high-resolution SIMS, secondary ions related to the desired components' concentrations in the membrane must be collected with sufficient intensities and lateral resolution to visualize the features of interest. The design of the Cameca NanoSIMS 50(L) instrument has been described in several reviews [27, 29, 36, 37, 110], and a protocol for its application to imaging component distributions in plasma membranes can be found in a prior publication [38]. Here, we focus on the aspects of the NanoSIMS instrument that affect imaging the relative abundances of isotope-labeled components in the plasma membranes of intact cells. We first describe the basic instrumental specifications and constraints that influence the combinations of different secondary ion species that can be simultaneously collected, and the lateral resolution of the resulting images. Then, we discuss the challenges that are specific to semi-quantitatively imaging component distributions in the plasma membranes of mammalian cells with a NanoSIMS 50(L). Focus is placed on the potential artifacts that may be encountered, and the strategies to identify and avoid them.

9.3.1 Specifications and constraints of secondary ion generation and collection

As mentioned, the primary and secondary ions must have opposite polarities for compatibility with the colinear optics that are used to focus the primary and secondary ions. Therefore, whether imaging with the NanoSIMS instrument should be performed using Cs^+ or O^- primary ions is dictated by the polarities of the desired secondary ions. The polarities of the monoatomic and diatomic secondary ions that are produced by a parent molecule depend on whether the molecule is composed of electronegative or electropositive elements. Electronegative elements have a high electron affinity, which promotes the formation of negatively charged secondary ions. Conversely, electropositive elements have a high ionization potential that favors the formation of positively charged secondary ions. Lipids and proteins primarily contain electronegative elements, such as carbon, hydrogen, oxygen, nitrogen, phosphorous, and sulfur, so they predominantly fragment into negatively charged monoatomic and diatomic secondary ions, such as H^-, C^-, CH^-, C_2^-, P^-, O^-, S^-, and CN^- [27]. Consequently, the Cs^+ primary ion beam should be employed to image lipid and protein components in plasma membranes.

The NanoSIMS instrument is equipped with individual ion counting detectors for the collection of secondary ions, and a secondary electron detector for collecting secondary electrons in parallel with negatively charged secondary ions. Importantly, secondary electron images cannot be acquired when the imaging is performed with O⁻ primary ions. During instrument setup, each of the five or seven ion-counting detectors in the NanoSIMS 50 or 50L, respectively, is positioned to collect secondary ions with a preselected m/z. The combinations of different m/z that can be simultaneously collected are constrained in two ways. First, the mass dispersion, which is the ratio of the highest mass to the lowest mass (M_{max}/M_{min}) that can be detected in parallel, is limited by the size of the magnet in the instrument's mass spectrometer. The maximum mass dispersion that can be achieved with a NanoSIMS 50 is 13.2, whereas the NanoSIMS 50L, which has a larger magnet, can achieve a maximum mass dispersion of 21 [37]. Second, due to the size of the ion detectors and the mass dispersion achieved by the instrument, the secondary ions collected by adjacent detectors must have a mass (M) difference of at least M/30 or M/58 for the NanoSIMS 50 or 50L, respectively [37]. This means that secondary ions that differ by one atomic mass unit (amu) cannot be collected in parallel by the NanoSIMS 50 or 50L if their mass is greater than 30 or 58, respectively. One experimental consequence of this limitation is that $^{31}P^-$ and $^{32}S^-$ can be simultaneously collected with a NanoSIMS 50L, but not with a NanoSIMS 50 unless the setup of the individual ion-counting detectors is carefully optimized. Noteworthy, although neither NanoSIMS models can simultaneously collect two secondary ion species that differ in mass by <1 amu on separate ion-counting detectors, both models can selectively collect a single secondary ion species and exclude signals from interfering isobars with the same nominal mass [27, 37]. For example, an ion counting detector can selectively collect $^{12}C^{15}N^-$ (26.9996 amu) ions without interference from $^{13}C^{14}N^-$ (27.0059 amu), but $^{12}C^{15}N^-$ and $^{13}C^{14}N^-$ cannot be simultaneously collected on two separate ion-counting detectors [38].

If the desired secondary ion species cannot be collected in parallel due to the aforementioned constraints, or the number of desired ion species exceeds the number of ion counting detectors, magnetic peak switching may be used to sequentially collect the desired ion species. Magnetic peak switching cycles the instrument between two magnetic fields, which changes the masses of the secondary ions that are directed to each ion-counting detector. However, magnetic peak switching significantly increases the time required to complete the analysis, and the secondary ions of interest are generated, but not collected, when the alternate magnetic field is in effect.

9.3.2 Determinants of lateral resolution

The NanoSIMS instrument uses electrostatic lenses, sectors, and deflection plates to direct the primary ions so that they strike the sample perpendicular to its surface [35, 37, 110]. The diameter of this primary ion beam determines the best possible lateral resolution that can be achieved, and it is modulated through the use of a primary beam limiting aperture that is positioned near the sample. When the smallest (150 μm-diameter) aperture is used, the Cs^+ primary ion beam can be focused to a spot with a diameter that is conservatively estimated to be 50 nm, though beam diameters as small as 30 nm have been reported [36, 111]. The effective diameter of the primary ion beam is experimentally determined by taking a line scan across a small feature in the SIMS image, and measuring the distance over which the signal intensity changes from 16% to 84% [38]. Because random noise in the secondary ion counts may produce artifactually sharp line scans, multiple scans on multiple images should be used to assess the lateral resolution of the SIMS measurement. In addition, the pixels should overlap by roughly 50%, which means the pixel size should be half the diameter of the primary ion beam, to produce continuous line scans with good fidelity [38].

To obtain a pixel size that is half the diameter of the primary ion beam, the width of the analysis region cannot exceed the number of pixels multiplied by half of the primary ion beam diameter. For example, if a 50 nm-diameter primary ion beam is used to acquire 512-by-512 pixel images of a square region on the sample, the height and width of the analysis region cannot exceed 12,800 nm (12.8 μm). To image entire cells that are larger than the maximum analysis region, multiple adjacent images must be acquired and stitched together with imaging processing software. In our experience, allowing each new image to overlap the previous analysis region by 2–3 μm facilitates image alignment.

The number of molecules within the primary ion beam's focal area, and thus, the number of secondary ions that may be collected from the analysis region, decreases as the diameter of the primary ion beam decreases. Consequently, when the beam diameter is reduced, the counts of the component-specific secondary ions may become too low to detect the species of interest within the beam's focal area. When this occurs, the working lateral resolution of the measurement is limited by the signal intensity, and not the diameter of the primary ion beam. Because insulating samples such as cells produce lower secondary ion yields than conducting samples, the signal intensity is often insufficient to resolve the features of interest in membrane samples when the smallest diameter Cs^+ ion beam is employed [39, 112]. This necessitates performing the acquisition with a larger diameter primary ion beam, or using an image processing technique, such as pixel binning or smoothing, to increase the signal-to-noise ratio (S/N) in the image [38]. Pixel binning combines the ion counts detected at adjacent pixels to produce larger pixels with higher ion counts. Pixel smoothing replaces the raw secondary ion counts measured at a given pixel with the average counts of that secondary ion within a n-by-n-pixel region centered around the given pixel [113]. Noteworthy, if the new pixel size in the binned image, or the smoothing width in the smoothed image is greater than the diameter of the primary ion beam, the working lateral resolution of the resulting image is defined by the new pixel size or smoothing width, respectively.

9.3.3 Strategies for minimizing signal intensity variations induced by concentration-independent factors

To assess whether a component of interest is concentrated at specific sites in the plasma membrane, the intensities of the component-specific secondary ions must be related to their local abundance in the membrane. To achieve this goal, the intensity of each component-specific secondary ion must be translated into a signal that is proportional to the component's abundance within the analysis volume. Additionally, the component's abundance within the analysis volume must be proportional to its concentration within the plasma membrane. Here, we describe strategies for acquiring information about the relative concentrations of the components of interest within the plasma membrane using a NanoSIMS 50(L) instrument.

The intensities of the component-specific secondary ions detected at each position on a cell are strongly dependent on the component of interest's local concentration. However, secondary ion signal intensities are further modulated by the chemical composition of the matrix and sample topography [37, 114–117]. The chemical composition within the analysis region affects secondary ion signal intensities by altering the probability that the component of interest will produce ionized species. This effect, which is called a chemical matrix effect, may either enhance or suppress secondary ion yields [114–116]. Sample topography affects secondary ion signal intensity by changing the depth at which secondary ions are generated within the sample, which influences the likelihood that they will be ejected into the vacuum [117]. The primary ions in the analysis beam strike the sample and penetrate into its surface, which produces collision cascades that fragment molecules into neutral and ionized species [37]. Although the primary ions may implant more than 10 nm below the sample surface, only the secondary ions generated <5 nm from the surface have sufficient energy to be ejected from the sample [118]. Topographical features that increase the primary ion's angle of incidence produce collision cascades that are closer to the sample surface, which increases the sputtering yield, and consequently, the number of secondary ions that may be detected [117]. Topographical features may also impede the collection of the secondary ions that are ejected from the sample by blocking the path that the secondary ions must take to reach the detector [117]. When this occurs, a "shadow" of low secondary ion signal intensities will be visibly adjacent to the feature [117].

The extent that chemical matrix effects and sample topography affect ionization and sputtering may vary substantially between secondary ion species composed of different elements [37]. In contrast, the changes in ion yields induced by the matrix and sample topography are very similar for species composed of the same elements but different isotopes [119]. Ratioing the counts of the component-specific rare isotope-containing secondary ion to those of the corresponding naturally abundant species detected at each pixel "cancels out" the effects of the concentration-independent factors on signal intensity [97]. Consequently, a signal that is proportional to the local abundance of the rare isotope-labeled component is produced by dividing the counts of the rare isotope-containing secondary ion detected at each pixel by those of the naturally abundant species [97, 119]. In order to perform this normalization, both the component-specific secondary ion species that contains the rare stable isotope and the naturally abundant isotopologue of this ion must be collected in parallel during high-resolution SIMS analysis. Additionally, the same rare and abundant secondary ion pairs should be collected from unlabeled cell samples to show that the isotope-ratioed images are featureless, and the isotopic composition on the cell is uniform and equal to the terrestrial natural abundance ratio. This confirms that ratioing to the naturally abundant isotopologue produces a signal that is not affected by topography or chemical matrix effects [97]. Noteworthy, an equivalent normalization process does not exist for component-specific secondary ion signals that contain a nonnative element instead of a rare stable isotope. Although ratioing to a ubiquitous ion signal, such as $^{12}C^-$, has been used to normalize signals collected from model lipid membranes [80–83], whether this approach is applicable to actual cell membranes has not been assessed. Variations in the intensities of secondary ions that contain elemental labels are more likely to signify changes in the elementally labeled component's local abundance when they do not correlate with topographical features or the intensities of other secondary ion species.

The magnitude of the isotope ratio signal, which is proportional to the local abundance of the rare isotope-labeled component, is influenced by both the labeled molecule's local concentration in the membrane, and the amount of membrane sampled at each pixel. The amount of membrane sampled at each pixel affects the isotope ratio because rare isotope-labeled secondary ions are only produced by a labeled component that resides in the membrane, whereas the abundant secondary ion species used for ratioing will be produced throughout the cell. Therefore, damage to the plasma membrane, and imaging with a sputter depth that exceeds the thickness of the plasma membrane, may cause the amount of membrane sampled at each pixel, and thus the isotope ratio, to vary significantly [97, 112]. For instance, the isotope ratio signal would be relatively low at sites where the membrane was torn, or relatively high at sites where sample preparation induced artifactual elevations in lipid density [97]. When the sputter depth exceeds the plasma membrane's thickness, the fraction of the total secondary ions produced by membrane components, and thus the isotope ratio signal, would vary due to the cell's 3D structure (Figure 9.3). At most sites on the cell, secondary ions would be collected from both the plasma membrane and the underlying cytosol. However, if intracellular vesicles or membrane-bound organelles are directly adjacent to the plasma membrane, a larger fraction of the total secondary ions collected at that site would be produced by membrane components. This detection of intracellular membranes would produce an elevated isotope ratio, as shown in Figure 9.3 [97]. Similarly, the fraction of the total secondary ions collected from the plasma

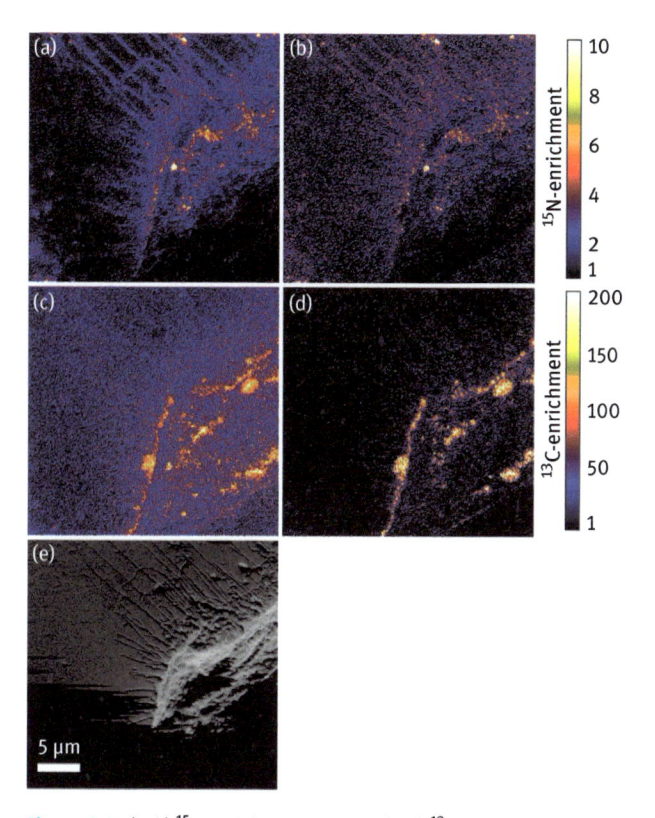

Figure 9.3: (a, b) ^{15}N-enrichment images, (c, d) ^{13}C-enrichment images, and (e) secondary electron image acquired from a mouse cell metabolically labeled with ^{15}N-sphingolipid precursors and uniformly ^{13}C-labeled fatty acids using a NanoSIMS 50 instrument. The ^{15}N- and ^{13}C-enrichment images were created by dividing the rare isotope ratio (^{12}C^{15}N$^-$/^{12}C^{14}N$^-$ and ^{13}C^{1}H$^-$/^{12}C^{1}H$^-$, respectively) by their standard terrestrial abundances, which produces an isotope enrichment factor that equals the rare isotope's enrichment compared to an unlabeled sample. The color bar to the right of the images applies to both images in the row. Images (a) and (c) were compiled from the data acquired in ten image planes, which corresponds to an estimated sputter depth of approximately 13 nm. Because this sputter depth is approximately twice the thickness of the plasma membrane, the regions with high ^{15}N-enrichment in (a) may signify the detection of intracellular membranes and not an elevated concentration of ^{15}N-sphingolipids. Comparison to (c) reveals that the ^{13}C-enrichment was elevated at many of the same sites as the elevated ^{15}N-enrichment, which indicates intracellular membranes were detected at these sites. Images (b) and (d) were created using data collected in image planes 8 – 10, which corresponds to an approximately 2.5 nm-thick layer that spanned from approximately 10.5 nm to 13 nm below the surface of the sample. Regions of elevated ^{15}N-enrichment in (b) and ^{13}C-enrichment in (d) signify the detection of ^{15}N-sphingolipids and ^{13}C-lipids, respectively, within intracellular membranes. Comparison of (a) and (c) to (b) and (d) reveal that the isotope ratio signals produced by ^{15}N-sphingolipids and ^{13}C-lipids within intracellular membranes contributed to the elevations in ^{15}N-enrichment and ^{13}C-enrichment in (a) and (c), respectively. A more shallow sputtering depth should have been employed to better ensure that the isotope-enriched sites detected in the images acquired at the surface of the cell correspond to local elevations in the isotope-labeled component's local concentration in the plasma membrane. Figure is adapted with permission from research originally published in Ref Frisz et al. [97].

membrane, and thus the isotope ratio, would be elevated at the cell's periphery (Figure 9.3) and at sites with tiny 3D membrane structures, such as microvilli, ruffles, or microscale folds. The detection of excess membrane might lead to the erroneous conclusion that the isotope-labeled component's concentration in the plasma membrane is elevated at these sites, when in actuality, all lipid species are more abundant at these locations.

To help ensure that changes in the abundance of the isotope-labeled lipid species are due to differences in its local concentration in the plasma membrane, and not the amount of cellular membrane sampled, the sputter depth should be less than the thickness of the plasma membrane, which is roughly 7.5 nm [120]. Sputter depth can be estimated using the sputtering rate measured on comparable biological samples, the raster area, and the primary ion beam current, as previously reported [38]. Whether the amount of cellular membranes sampled at each pixel varies can be experimentally assessed. In these experiments, uniformly ^{13}C-labeled fatty acids are metabolically incorporated into all lipid species within the cells, and the abundances of the ^{13}C-lipids at various locations on the cell surface are visualized according to the normalized ^{13}C-labeled secondary ion signal (*i.e.*, $^{13}C^1H^-/^{12}C^1H^-$) [38, 97]. A continuous and relatively uniform ^{13}C-enrichment on the cell surface would indicate the plasma membrane was intact, and the amount of cellular membrane sampled at each pixel was constant. In contrast, ^{13}C-deficient regions on the cell would signify tears in the plasma membrane, as outlined with dashed lines in Figure 9.4. Alternatively, ^{13}C-enriched hotspots would suggest an excess of lipid (Figure 9.4), which could be caused by artifactual lipid clustering (Figure 9.4), or the detection of intracellular membranes adjacent to the plasma membrane due to a high sputter depth. When the sputtering depth exceeds the membrane thickness, it may be reduced by discarding some of the image planes.

9.4 Data analysis

The conversion of data acquired with a NanoSIMS 50(L) into isotope enrichment images that reveal isotope-labeled component distributions is performed with custom software, such as the OpenMIMS plugin for ImageJ [121] or the L'IMAGE program (L.R. Nittler, Carnegie Institution of Washington, Washington, DC) that runs on the PV-Wave platform (Rogue Wave Software, Boulder, CO). The capabilities of these software packages can be found in Nuñez *et al* [26]. Here, we emphasize the quantitative statistical analyses that are applied to high-resolution SIMS data acquired using a NanoSIMS instrument in order to unambiguously test hypotheses concerning the existence of plasma membrane domains enriched with components of interest. We first briefly describe how rare isotope enrichment images are constructed. Then, we present statistical methods for definitively identifying local elevations in rare isotope abundance that signify component-enriched

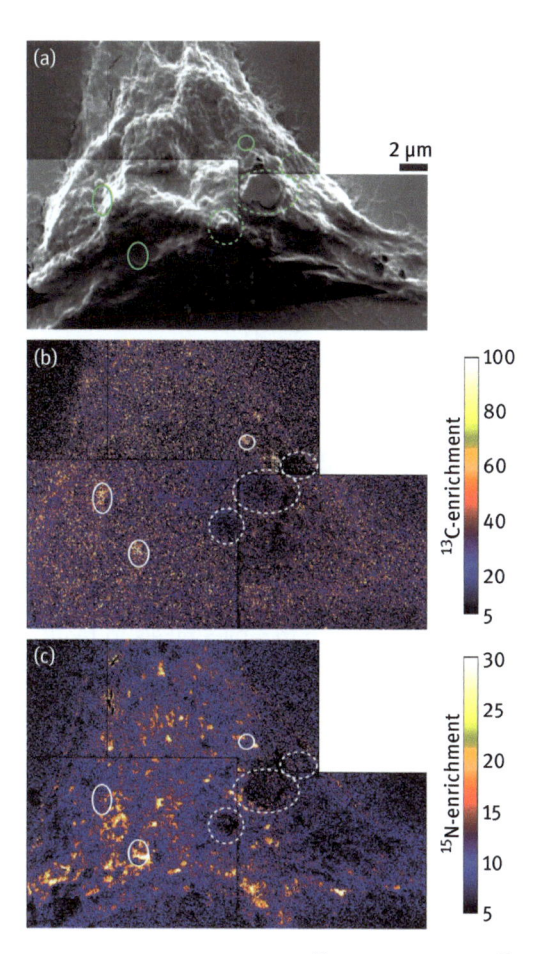

Figure 9.4: Secondary electron, ^{15}N-enrichment, and ^{13}C-enrichment images were acquired in parallel from 15 µm-by-15 µm areas on a mouse fibroblast cell that was metabolically labeled so >60% of its lipids contained uniformly ^{13}C-labeled fatty acids, and ~80% of its sphingolipids contained ^{15}N. The individual images were collected with a 70 nm-diameter beam, smoothed to a lateral resolution of 87 nm, and stitched together to show a larger region of the sample. (a) The montage of secondary electron images shows cell morphology. The montages of (b) ^{13}C-enrichment and (c) ^{15}N-enrichment images show the ^{13}C-lipid and ^{15}N-sphingolipid abundances, respectively, on the cell. The regions encircled by dashed lines have low ^{13}C-enrichments, and thus, low ^{13}C-lipid abundances that indicate the membrane was torn at these sites. The solid lines encircle areas with high ^{13}C-enrichments that signify the detection of excess membrane. Variations in the ^{15}N-enrichment within the outlined regions may reflect differences in the local abundance of all lipid species, and not variations in the local concentration of ^{15}N-sphingolipids in the plasma membrane. Figure is adapted with permission from Ref Frisz et al. [97].

plasma membrane domains. Statistical hypothesis tests for elevated concentrations of a second isotope-labeled species within the component-enriched membrane domains are also described.

9.4.1 Creation of rare isotope abundance images

Each NanoSIMS acquisition file contains multiple raster planes that were acquired at the same sample location. These individual image planes must be spatially aligned relative to one another to correct for any drift in the position of the sample or primary ion beam that may have occurred during the acquisition. Because the pixel size is half the diameter of the primary ion beam, random fluctuations in the secondary ion counts detected at adjacent pixels may produce false heterogeneity in the resulting images. To reduce the effects of these random variations in signal intensity, a boxcar smoothing algorithm is applied to the secondary ion images. As mentioned this boxcar smoothing algorithm replaces the secondary ion counts at a given pixel with the average of the counts of that ion measured within a n-by-n-pixel region centered around the given pixel [113]. We typically use a 3-by-3 pixel smoothing width for 512-by-512-pixel high-resolution SIMS images acquired from 15-μm-by-15-μm regions using a primary ion beam with a diameter < 100 nm [38, 64, 65, 71, 97]. A higher smoothing width may be required if the secondary ion signal intensity, and thus, the S/N, is low. For example, the low secondary ion signals we obtained when imaging Madin-Darby canine kidney (MDCK) cells necessitated increasing the smoothing width to 5-by-5 pixels to reduce random variations in signal intensity [39, 112].

When alignment and smoothing is complete, the secondary ion signals are converted into a ratio that is proportional to the rare isotope-labeled component's local abundance. As mentioned above, this is done by ratioing the counts of the rare isotope-containing secondary ions detected at each pixel to the counts of the naturally abundant isotopologue of that secondary ion. Then, the resulting ratio is encoded in a color scale and mapped to its corresponding pixel. To facilitate interpreting whether the measured isotope ratios are elevated compared to an unlabeled sample, we divide the isotope ratios by the rare isotope's standard natural abundance ratio. This division yields an isotope enrichment factor that equals the number of times the rare isotope's abundance is greater than that in an unlabeled sample. The standard natural abundance ratios for several rare isotopes are listed in Table 9.1. Noteworthy, if multiple images of adjacent regions on the

Table 9.1: Standard natural abundance ratios for rare isotopes commonly used to label membrane components. Note that ^{34}S abundance has a range of 3.98%–4.73% in terrestrial samples (from Böhlke et al. [122]).

Rare Isotope	% Abundance	Ratioed Secondary Ions
Nitrogen-15	0.367	$^{12}C^{15}N^-/^{12}C^{14}N^-$
Carbon-13	1.1237	$^{13}C^{1}H^-/^{12}C^{1}H^-$
Oxygen-18	0.2005	$^{18}O^-/^{16}O^-$
Deuterium	0.0156	$^{12}C\,^{2}H^-/^{12}C^{1}H^-$
Sulfur-34	4.5045*	$^{34}S^-/^{32}S^-$

sample need to be stitched together to form a mosaic, the range encoded in the color scale must be the same for all of the images.

As an example, Figure 9.5 shows mosaics of secondary electron images, ^{15}N-enrichment images, and ^{13}C-enrichment images acquired from a metabolically

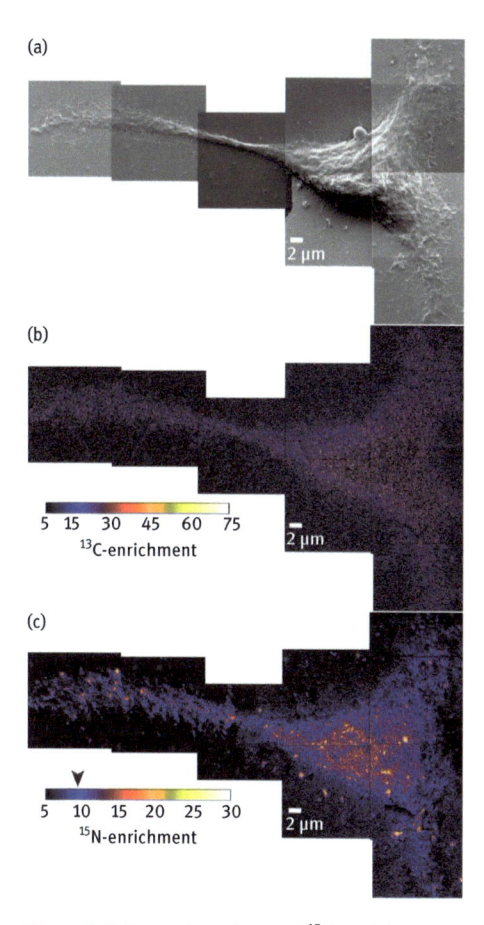

Figure 9.5: Secondary electron, ^{15}N-enrichment, and ^{13}C-enrichment images were collected in parallel at 15 µm-by-15 µm regions on a metabolically labeled mouse fibroblast cell that had uniformly ^{13}C-labeled fatty acids in >60% of its lipids, and the ^{15}N isotope in ~80% of its sphingolipids. (a) The individual secondary electron images were stitched together to show the morphology of the entire cell. Montages of the (b) ^{13}C-enrichment and (c) ^{15}N-enrichment images show the abundances of ^{13}C-lipids and ^{15}N-sphingolipids on the surface of the cell, respectively. The black arrowhead on the color bar in (c) indicates the threshold for a statistically significant local elevation in ^{15}N-enrichment that cannot be caused by noise. The uniform ^{13}C-enrichment, and thus, ^{13}C-lipid distribution on the cell mean the statistically significant local elevations in ^{15}N-enrichment on the cell are plasma membrane domains that are enriched with ^{15}N-sphingolipids. The individual isotope enrichment images were acquired with 87 nm-lateral resolution, but are presented here with lower resolution due to space limitations. Figure is adapted with permission from Ref Frisz et al. [97].

labeled mouse fibroblast cell using a NanoSIMS instrument [97]. Each mosaic consists of multiple 512 pixel-by-512 pixel images that were acquired at partially overlapping 15 μm-by-15 μm regions on the sample using a 70 nm-diameter Cs$^+$ primary ion beam. The cell had been metabolically labeled so that approximately 80% of its sphingolipids contained the rare ^{15}N isotope, and about 60% of the fatty acids in its lipids were uniformly ^{13}C-labeled. The component-specific secondary ions that contained the rare isotopes and the naturally abundant isotopologues of these ions were collected in parallel using a NanoSIMS 50. After applying a 3 pixel-by-3-pixel boxcar smoothing algorithm to the secondary ion images, ^{15}N-enrichment and ^{13}C-enrichment images that showed the local abundances of ^{15}N-sphingolipids and ^{13}C-lipids, respectively, were constructed. The ^{15}N-enrichment images were made by ratioing the ^{12}C^{15}N$^-$ counts to the ^{12}C^{14}N$^-$ counts detected at each pixel, and dividing by the standard natural abundance ratio for ^{15}N. Similarly, ^{13}C-enrichment images were created by ratioing the ^{13}C^{1}H$^-$ counts to the ^{12}C^{1}H$^-$ counts at each pixel, and dividing by the standard natural abundance ratio for ^{13}C. Mosaics that show the entire cell were produced by tiling the individual images together with the Adobe Illustrator graphic-editing program. The lateral resolution of the unsmoothed secondary electron images was defined by the 70 nm primary ion beam diameter, whereas the working lateral resolution of the isotope enrichment images equaled the 87 nm smoothing width [65].

9.4.2 Detection of statistically significant local elevations in rare isotope abundance

High-resolution SIMS performed using a NanoSIMS instrument may be employed to probe whether a labeled species of interest, such as a lipid [29, 38, 39, 64, 65, 97, 112], protein [65, 123], nutrient, or toxin [124–126] is heterogeneously distributed in the plasma membrane. Uneven component distribution on the surface of a cell may be visible in the rare isotope enrichment images. However, because visual assessment is subjective, quantitative methods are required to demonstrate the significance of any local elevations in rare isotope abundance. We have developed a statistical approach to establish the threshold for the lowest isotope enrichment that denotes a statistically significant elevation in rare isotope abundance. Regions on the cell that exceed this threshold are enriched in the rare isotope-labeled component.

When a homogeneous sample is analyzed using SIMS, the ion intensities measured at each pixel will vary due to random noise. Though the random variation in a secondary ion signal is Poisson, the variation in a large number (10^3–10^5) of SIMS measurements acquired from a homogenous sample with a uniform isotopic composition can be approximated by a Gaussian distribution [127]. Thus, if a rare isotope-labeled species was uniformly distributed in the plasma membrane and the signal variation was due to random noise, the isotope enrichments measured at a sufficiently large number of different locations on the cell would fit a Gaussian distribution. In contrast, if the abundance of the isotope-labeled component of interest were elevated at distinct regions in the plasma membrane, the frequency distribution of the isotope enrichments would instead match a mixture of two normal distributions. One of the two distributions in this mixture would correspond to domains where the isotope-labeled component was enriched, and the other to non-domain regions in the plasma membrane that contained relatively lower levels of the isotope-labeled component.

The first step in assessing which distribution better approximates the range of component-specific isotope enrichments within the cell's plasma membrane is to tabulate the isotope enrichments measured at thousands of tiny regions on the cell. NanoSIMS image processing software is used to divide the region of each image that corresponds to the surface of the cell into 3-by-3-pixel regions of interest (ROIs). Then, the isotope enrichment factor for each ROI is exported and tabulated. This process is repeated for every image acquired from the same cell. When the isotope enrichment factors for all the ROIs on the same cell have been tabulated, they are imported into the MATLAB Statistics Toolbox.

Next, the frequency distribution of the experimental enrichment factors measured on the cell are fit with the best-matching probability density function (PDF) for a unimodal normal distribution, and the best-matching PDF for a mixture of two unimodal normal distributions. The maximum likelihood estimation (MLE) function in the MATLAB Statistics Toolbox is first used to estimate the mean (μ) and standard deviation (σ) for the unimodal normal distribution model. Initial values of these parameters must be provided to start the fitting process, so the 50th percentile of the isotope enrichment is used as the initial value of μ, and the square root of the variance is used for the initial value of σ. The MLE function in the MATLAB Statistics Toolbox adjusts these initial parameters so the unimodal normal distribution best fits the frequency distribution for the isotope enrichments experimentally measured on the cell. Next, the MLE function is used to find the parameters for the mixture of two normal distributions that best fit the experimental isotope enrichment measurements. The mixture of two normal distributions has five parameters: the mean and standard deviation for the non-domain (μ_{ND} and σ_{ND}) and domain (μ_D and σ_D) measurements, and a mixing parameter (m), which is the fraction of the total measurements that contribute to the normal distribution with the lower mean. The initial values used for these parameters are based on the assumption that the plasma membrane contains an equal proportion of non-domain and domain regions ($m = 0.5$). The initial values of μ_{ND} and μ_D are set to the 25th and 75th percentiles of the isotope enrichments, respectively, and the initial values of σ_{ND} and σ_D are set to the square root of each population's variance. The MLE function adjusts the initial values of m, μ_{ND}, μ_D, σ_{ND}, and σ_D so the resulting mixture of two normal distributions best fits the frequency distribution for the isotope enrichments experimentally measured on the cell.

Next, data sets that correspond to the best-fit unimodal normal distribution and the best-fit mixture of two normal distributions are simulated with the MATLAB Statistics Toolbox for comparison to the experimentally measured isotope enrichment factors. The μ and σ for the best-fit unimodal normal distribution is used to simulate a *Unimodal normal distribution* data set that contains the same number of data points as the experimental data set. A *Mixture of two normal distributions* data set that also contains the same number of data points as the experimental data set is created by using μ_{ND} and σ_{ND} to simulate a fraction m of the total data points in the set, and μ_D and σ_D to simulate the remaining data points in the set. Then the cumulative distribution function (CDF) for the experimental data, the *Unimodal normal distribution* data set, and the *Mixture of two normal distributions* data set are plotted in the MATLAB Statistics Toolbox.

Two-sample Kolmogrov–Smirnov tests performed in the MATLAB Statistics Toolbox are used to assess which simulated data set better fits the distribution of the isotope enrichment factors experimentally measured on the cell. A two-sample Kolmogorov–Smirnov test compares the CDFs of two sample sets to determine whether their shapes, and thus, the populations they were extracted from, are similar or different [128]. This test returns three outputs, which are the Kolmogrov–Smirnov test statistic, a h value, and a p value. The Kolmogrov–Smirnov test statistic is the maximum distance between the two CDFs that are being compared. The h value indicates whether the null hypothesis that the two data sets were drawn from the same population is rejected at a 95% confidence level (h = 0, do not reject; h = 1, reject). The p-value ($0 \leq p \leq 1$) quantifies the probability of obtaining a test statistic equal or larger than the observed value when the null hypothesis is true. Of these outputs, only the Kolmogrov–Smirnov test statistic is needed to identify which distribution best approximates the experimental data.

The two-sample Kolmogrov–Smirnov test is performed to compare the CDF of the experimental data to that for each simulated distribution, and the resulting Kolmogrov–Smirnov test statistics are compared. The simulated data set with the smallest Kolmogrov–Smirnov test statistic corresponds to the distribution that better models the experimental data [128]. Specifically, a finding that the best-fit unimodal normal distribution has a smaller test statistic, and thus, better approximates the experimental data, would indicate the rare isotope-labeled component is uniformly distributed within the plasma membrane. Conversely, a finding that the best-fit mixture of two normal distributions has a smaller test statistic, and thus, better fits the experimental data, would indicate sites on the cell are enriched with the rare isotope-labeled component, and other regions have a lower abundance of that component. When the data are better approximated by the best-fit mixture of two normal distributions, the threshold for the lowest isotope enrichment that denotes a statistically significant elevation in the rare isotope's abundance, which corresponds to local enrichment, is defined as $\mu_{ND} + 2\sigma_{ND}$.

Figure 9.6 shows the graphs generated when using these statistical methods to test for domains enriched with ^{15}N-sphingolipids in the plasma membrane of the mouse fibroblast cell shown in Figure 9.5 [97]. Figure 9.6a shows the frequency distributions of the ^{15}N-enrichment factors measured on the cell (vertical grey bars), the PDF for the best-fit unimodal normal distribution (solid black line; μ = 8.9, σ = 3.5), and the PDF for the best-fit mixture of two normal distributions (dashed line; μ_{ND} = 7.8, σ_{ND} = 2.1, μ_D = 12.0, and σ_D = 3.7). The corresponding CDFs are shown in Figure 9.6b. The two-sample Kolmogorov–Smirnov test statistic was 0.071 for the unimodal normal distribution, and 0.014 for the mixture of two normal distributions. Therefore, the ^{15}N-enrichment factors experimentally measured on the cell were best approximated by the mixture of two normal distributions. This finding indicates regions on the surface of the mouse fibroblast cell shown in Figure 9.5 contained statistically significant elevations in ^{15}N-sphingolipid abundance.

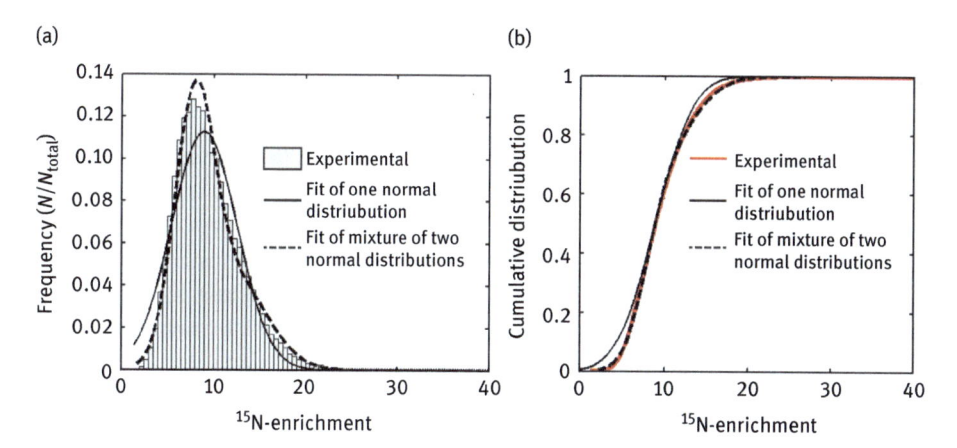

Figure 9.6: (a) The frequency distribution of ^{15}N-enrichment factors measured at 59,301 ROIs on the cell shown in Figure 9.5 was fit with a unimodal normal distribution ($\mu \pm \sigma$ = 8.9 ± 3.5) or a mixture of two normal distributions ($\mu_{ND} \pm \sigma_{ND}$ = 7.8 ± 2.1 and $\mu_D \pm \sigma_D$ = 12.0 ± 3.7). (b) Plot of the empirical and model cumulative distributions shows the data are better fit by the mixture of two normal distributions, which confirms that the plasma membrane of this cell contained sphingolipid-enriched domains.

9.4.3 Test for excess membrane at sites with an elevated rare isotope-labeled component abundance

If statistically significant local elevations in rare isotope abundance are detected on the cell, control experiments should be performed to test whether they signify an increase in the rare isotope-labeled component's concentration in the membrane or an excess of all membrane components. To determine whether more plasma membrane was sampled at the sites where the component of interest is more abundant, both the rare isotope-labeled component of interest, and all cellular lipids in the plasma membrane must be visualized in parallel. This is accomplished by metabolically incorporating uniformly ^{13}C-labeled fatty acids into all cellular lipid species, and a different rare isotope into the component of interest. The rare isotope-containing secondary ion and the corresponding naturally abundant species characteristic of the ^{13}C-lipids (i.e., ^{13}C^{1}H$^-$ and ^{12}C^{1}H$^-$, respectively) and the other rare isotope-labeled component of interest are simultaneously collected from the sample. Then, statistical approaches are used to assess whether the ^{13}C-enrichment, and thus, the abundance of all cellular lipids, within the domains enriched with the rare isotope-labeled component of interest differ from those in the non-domain regions on the cell.

The ^{13}C-enrichments within the component-enriched domains and equally sized non-domain regions must be tabulated so that they can be quantitatively compared. We use a particle definition algorithm in the L'IMAGE program to objectively divide the areas on the cell where the rare isotope ratio exceeds the threshold that signifies a local elevation in isotope-labeled component abundance into individual domains. This algorithm identifies the center of each domain as a pixel where the rare isotope ratio is a local maximum that exceeds the threshold for a statistically significant elevation in the rare isotope's abundance [113]. The domain perimeters are found by expanding out from the center pixel until either another domain is encountered, or the isotope ratio drops below the Gaussian diameter, which is approximately 13.5% ($1/e^2$) of the ratio at the domain center [113]. The number of pixels in each domain and the ^{13}C-enrichment measured within it are extracted from the image, and tabulated (*Domain* data set). For comparison, the ^{13}C-enrichments are also extracted from the same number of non-domain regions on the cell that have the same size distribution as the component-enriched domains, and tabulated (*Non-domain* data set). This process is repeated for every image acquired from the same cell.

A two-sample Kolmogorov–Smirnov test performed in the MATLAB Statistics Toolbox is also used to determine whether the ^{13}C-enrichments measured in the *Domain* and *Non-domain* data sets are statistically different. Unlike the aforementioned use of this test to assess whether the rare isotope-labeled component of interest is uniformly distributed on the cell surface, the h and p values are used to compare the ^{13}C-enrichments in the *Domain* and *Non-domain* data sets. Specifically, an output of $h = 0$ would mean the ^{13}C-enrichment, and thus, the amount of ^{13}C-lipid detected within the component-enriched domains did not differ significantly from the non-domain regions. Such a finding would indicate that the statistically significant local elevations in the abundance of the rare isotope-labeled component detected on the cell signify an increase in the labeled component's local concentration in the membrane, and not the detection of excess membrane. Conversely, an output of $h = 1$ would mean the ^{13}C-enrichment was elevated at the sites where the rare isotope was more abundant. This outcome would indicate the local elevations in rare isotope abundance reflected an excess amount of all membrane lipids, and not an elevated concentration of the rare isotope-labeled component at these sites. As explained in Section 9.3.3, the elevated amounts of membrane lipids detected at these sites may signify artifactual elevations in lipid density induced during sample preparation, or the detection of intracellular membranes, which occurs when the sputtering depth exceeds the thickness of the plasma membrane.

9.4.4 Test for enrichment of a second species within compositionally distinct plasma membrane domains

Numerous hypotheses postulate that specific protein or lipid species are selectively recruited to compositionally distinct domains in the plasma membrane, resulting in a higher concentration of the recruited species within these membrane domains. For example, sphingolipid-associated proteins are hypothetically recruited to, and thus enriched within, sphingolipid domains in the plasma membrane [65, 123]. In another hypothesis, favorable cholesterol-sphingolipid interactions are postulated to drive cholesterol to accumulate within sphingolipid domains in the plasma membrane [16]. A modified version of the statistical approach to test for elevated amounts of all membrane lipids at the rare isotope-enriched sites enables testing for an elevated abundance of a second labeled species in the domains enriched with the first rare isotope-labeled component. Such analyses are performed as described in Section 9.4.3, but the isotope enrichment specific to the second labeled species of interest is used in place of the ^{13}C-enrichment that gauges the ^{13}C-labeled lipid abundance [64, 65]. When the second species of interest is labeled with a distinct nonnative element instead of a different rare isotope, the counts of the secondary ions that contain the distinctive nonnative element are used in place of a rare isotope ratio [65].

As an example, Figure 9.7 shows a secondary electron microscopy image, and mosaics of ^{15}N-enrichment and ^{18}O-enrichment images of a mouse fibroblast cell that was metabolically labeled so ~90% of its sphingolipids contained ^{15}N and ~60% of its cholesterol contained ^{18}O [64]. The montage of ^{15}N-enrichment images shows the ^{15}N-sphingolipids were enriched within small regions on the cell, and statistical analysis indicated ^{15}N-enrichments greater than 14.5 were statistically significant elevations that represent plasma membrane domains enriched with ^{15}N-sphingolipids [64]. In comparison, the montage of ^{18}O-enrichment images suggests the ^{18}O-cholesterol was uniformly distributed within the plasma membrane. A two-sample Kolmogorov–Smirnov test was performed to compare the ^{18}O-enrichments within the ^{15}N-sphingolipid domains that were defined as regions where the ^{15}N-enrichment was greater than 14.5 to the ^{18}O-enrichments within non-domain areas on the cell. This test returned a p-value of 0.80, which means no significant difference in ^{18}O-enrichment was detected between the ^{15}N-sphingolipid domain and

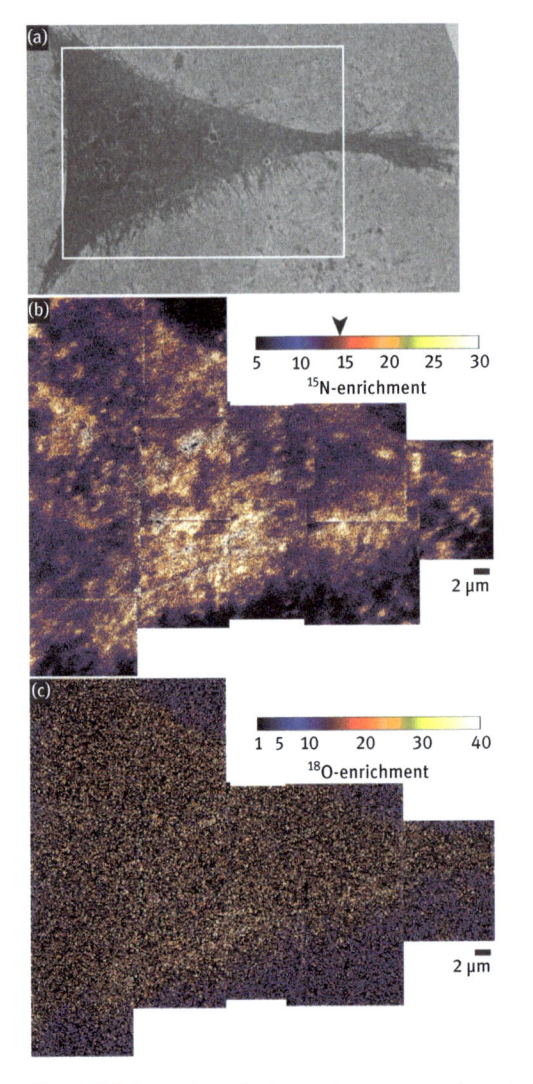

Figure 9.7: Secondary electron microscopy and montages of isotope enrichment images show a metabolically labeled mouse fibroblast that has the rare ^{15}N isotope in ~90% of its sphingolipids, and the rare ^{18}O isotope in ~60% of its cholesterol. (a) The SEM image shows the cell morphology. The approximate area that was subsequently analyzed using a NanoSIMS 50 is outlined in white. (b) Montage of ^{15}N-enrichment images shows regions of the plasma membrane were enriched with ^{15}N-sphingolipids. ^{15}N-enrichments greater than 14.5, indicated by the black arrow on the color scale, were statistically significant elevations that denote ^{15}N-sphingolipid domains. (c) Montage of ^{18}O-enrichment images shows the ^{18}O-cholesterol is relatively evenly distributed in the plasma membrane. A two-sample Kolmogorov-Smirnov test did not find a statistically significant difference in the ^{18}O-enrichments in the ^{15}N-sphingolipid domain and non-domain regions (p = 0.80). Figure is adapted with permission from Ref Frisz et al. [64].

non-domain regions on the surface of the cell [64]. This indicates the ^{18}O-cholesterol was not enriched within the ^{15}N-sphingolipid domains in the plasma membrane of this mouse fibroblast cell.

Hypotheses concerning the coenrichment of two different rare isotope-labeled components of interest may be tested by assessing whether the magnitudes of their component-specific rare isotope ratios are spatially correlated. This involves dividing the regions of the images that correspond to the surface of the cell into 3 pixel-by-3 pixel ROIs, and extracting each of the two component-specific isotope ratios from every ROI. After tabulating the two component-specific isotope ratios for the entire cell surface, the MATLAB Statistics Toolbox is used to calculate the correlation coefficient for the magnitudes of the two component-specific rare isotope ratios detected at every ROI. This correlation coefficient is a measure of the linear dependence between the two variables, and its value ranges from −1 to 1. A correlation coefficient of 1 would signify a strong correlation between the two component-specific isotope ratios, which would strongly suggest a high degree of colocalization between the components that produced them. Conversely, a correlation coefficient of −1 would signify a strong negative correlation between the two component-specific isotope ratios, and thus, the membrane components that produced them. A correlation coefficient of 0 would mean the two component-specific rare isotope ratios were not correlated. This outcome would indicate the distributions of the two components that produced these isotope ratios had no dependency on one another.

9.5 Conclusions and outlook

This chapter has described the high-resolution SIMS analysis of plasma membrane organization. Our coverage began with the basic principles of high-resolution SIMS analysis performed with a Cameca NanoSIMS 50(L) instrument, and proceeded to explain how these basic principles affect imaging component distribution in the plasma membranes of intact mammalian cells. Special attention was paid to the quantitative interpretation of component-specific high-resolution SIMS images constructed from data acquired with a NanoSIMS 50(L). Statistical tools for the detection of component-enriched plasma membrane domains were described, and strategies for detecting and avoiding potential artifacts were presented.

Our application of high-resolution SIMS imaging to cells that contained rare isotope-labeled sphingolipids [38, 39, 64, 65, 97, 112], cholesterol [39, 64, 65, 112], all lipid species [38, 97], and/or affinity labeled proteins [65, 71] has yielded new insights into component distributions within the plasma membranes of mammalian cells. Extension of this approach to visualizing the distributions of additional protein and lipid species will further improve understanding of plasma membrane organization. The distributions of components of interest within intracellular membranes may also be explored by repeatedly imaging the same sample location on a metabolically labeled cell with a NanoSIMS 50(L), producing images at different depths inside the cell. Our preliminary efforts to assess the abundances of cholesterol and sphingolipids within organelle membranes have demonstrated a need for strategies to incorporate distinct elements or isotope labels into specific organelles so that they might be located during NanoSIMS analysis [39]. A simple strategy to incorporate a distinct nonnative element, fluorine, into the endoplasmic reticulum enables detecting this organelle according to the distinctive $^{19}F^-$ ions it produces [129]. The development of additional organelle-specific labels would facilitate future progress in this area.

References

[1] Henis, Y.I., Hancock, J.F., & Prior, I.A. Ras acylation, compartmentalization and signaling nanoclusters. Mol Membr Biol 2009, 26, 80–92.
[2] Laude, A.J., & Prior, I.A. Plasma membrane microdomains: Organization, function and trafficking (Review). Mol Membr Biol 2004, 21, 193–205.
[3] Omerovic, J., & Prior, I.A. Compartmentalized signalling: Ras proteins and signalling nanoclusters. FEBS J 2009, 276, 1817–1825.
[4] Prior, I.A., Muncke, C., Parton, R.G., & Hancock, J.F. Direct visualization of Ras proteins in spatially distinct cell surface microdomains. J Cell Biol 2003, 160, 165–170.
[5] Kraft, M.L. Sphingolipid organization in the plasma membrane and the mechanisms that influence it. Front Cell Dev Biol 2017, 4, 154.
[6] Brown, C.M., Hebert, B., Kolin, D.L., et al. Probing the integrin-actin linkage using high-resolution protein velocity mapping. J Cell Sci 2006, 119, 5204–5214.
[7] Gaietta, G., Deerinck, T.J., Adams, S.R., et al. Multicolor and electron microscopic imaging of connexin trafficking. Science 2002, 296, 503–507.
[8] Wang, L., & Sigworth, F.J. Cryo-EM and single particles. Physiology 2006, 21, 13–18.
[9] Tsien, R.Y. The green fluorescent protein. Annu Rev Biochem 1998, 67, 509–544.
[10] Zhang, J., Leiderman, K., Pfeiffer, J.R., Wilson, B.S., Oliver, J.M., & Steinberg, S.L. Characterizing the topography of membrane receptors and signaling molecules from spatial patterns obtained using nanometer-scale electron-dense probes and electron microscopy. Micron 2006, 37, 14–34.
[11] Kotani, N., Gu, J., Isaji, T., Udaka, K., Taniguchi, N., & Honke, K. Biochemical visualization of cell surface molecular clustering in living cells. Proc Nat Acad Sci USA 2008, 105, 7405–7409.
[12] Maurel, D., Banala, S., Laroche, T., & Johnsson, K.. Photoactivatable and photoconvertible fluorescent probes for protein labeling. ACS Chem Biol 2012, 5, 507–516.
[13] Kenworthy, A.K. Imaging protein-protein interactions using fluorescence resonance energy transfer microscopy. Methods 2001, 24, 289–296.
[14] Hess, S.T., Gould, T.J., Gudheti, M.V., Maas, S.A., Mills, K.D., & Zimmerberg, J. Dynamic clustered distribution of hemagglutinin resolved at 40 nm in living cell membranes discriminates between raft theories Proc Natl Acad Sci USA 2007, 104, 17370–17375.
[15] Kraft, M.L. Plasma membrane organization and function: Moving past lipid rafts. Mol Biol Cell 2013, 24, 2765–2768.

[16] Simons, K. Cell membranes: A subjective perspective. Biochim Biophys Acta 2016, 1858, 2569–2572.
[17] Lu, S.M., & Fairn, G.D. Mesoscale organization of domains in the plasma membrane – beyond the lipid raft. Crit Rev Biochem Mol Biol 2018, 53, 192–207.
[18] Marks, D.L., Bittman, R., & Pagano, R.E. Use of bodipy-labeled sphingolipid and cholesterol analogs to examine membrane microdomains in cells. Histochem Cell Biol 2008, 130, 819–832.
[19] Carquin, M., Conrard, L., Pollet, H., et al. Cholesterol segregates into submicrometric domains at the living erythrocyte membrane: Evidence and regulation. Cell Mol Life Sci 2015, 72, 4633–4651.
[20] D'Auria, L., Fenaux, M., Aleksandrowicz, P., et al. Micrometric segregation of fluorescent membrane lipids: Relevance for endogenous lipids and biogenesis in erythrocytes. J Lipid Res 2013, 54, 1066–1076.
[21] Tyteca, D., D'Auria, L., Smissen, P.V.D., et al. Three unrelated sphingomyelin analogs spontaneously cluster in plasma membrane micrometric domains. Biochim Biophys Acta 2010, 1798, 909–927.
[22] Wüstner, D., & Solanko, K. How cholesterol interacts with proteins and lipids during its intracellular transport. Biochim Biophys Acta 2015, 1848, 1908–1926.
[23] Hughes, L.D., Rawle, R.J., & Boxer, S.G. Choose your label wisely: Water-soluble fluorophores often interact with lipid bilayers. PLoS One 2014, 9, e87649.
[24] Shaw, J.E., Epand, R.F., Epand, R.M., Li, Z., Bittman, R., & Yip, C.M. Correlated fluorescence-atomic force microscopy of membrane domains: Structure of fluorescence probes determines lipid localization. Biophys J 2006, 90, 2170–2178.
[25] Chatelut, M., Leruth, M., Harzer, K., et al. Natural ceramide is unable to escape the lysosome, in contrast to a fluorescent analogue. FEBS Lett 1998, 426, 102–106.
[26] Nunez, J., Renslow, R., Cliff, J.B., & Anderton, C.R. NanoSIMS for biological applications: Current practices and analyses. Biointerphases 2018, 13, 03B301.
[27] Boxer, S.G., Kraft, M.L., & Weber, P.K. Advances in imaging secondary ion mass spectrometry for biological samples. Annu Rev Biophys 2009, 38, 53–74.
[28] Fletcher, J.S., Vickerman, J.C., & Winograd, N. Label free biochemical 2D and 3D imaging using secondary ion mass spectrometry. Curr Opin Chem Biol 2011, 15, 733–740.
[29] Kraft, M.L., & Klitzing, H.A. Imaging lipids with secondary ion mass spectrometry. Biochim Biophys Acta 2014, 1841, 1108–1119.
[30] Fletcher, J.S., & Vickerman, J.C. Secondary ion mass spectrometry: Characterizing complex samples in two and three dimensions. Anal Chem 2013, 85, 610–639.
[31] Winograd, N., & Garrison, B.J. Biological cluster mass spectrometry. Annu Rev Phys Chem 2010, 61, 305–322.
[32] Passarelli, M.K., & Winograd, N. Lipid imaging with time-of-flight secondary ion mass spectrometry (ToF-SIMS). Biochim Biophys Acta Mol Cell Biol Lipids 2011, 1811, 976–990.
[33] Fletcher, J.S. Latest applications of 3D ToF-SIMS bio-imaging. Biointerphases 2015, 10, 018902.
[34] Weibel, D., Wong, S., Lockyer, N., Blenkinsopp, P., Hill, R., & Vickerman, J.C. A C-60 primary ion beam system for time of flight secondary ion mass spectrometry: Its development and secondary ion yield characteristics. Anal Chem 2003, 75, 1754–1764.
[35] Slodzian, G., Daigne, B., Girard, F., Boust, F., & Hillion, F. Scanning secondary ion analytical microscopy with parallel detection. Biol Cell 1992, 74, 43–50.
[36] Williams, P. Biological imaging using secondary ions. J Biol 2006, 5, 18.
[37] Hoppe, P., Cohen, S., & Meibom, A.. NanoSIMS: Technical aspects and applications in cosmochemistry and biological geochemistry. Geostand Geoanal Res 2013, 37, 111–154.

[38] Klitzing, H.A., Weber, P.K., & Kraft, M.L. Secondary Ion Mass Spectrometry Imaging of Biological Membranes at High Spatial Resolution. In: A. A. Sousa & M. J. Kruhlak, eds. Methods in Molecular Biology: Nanoimaging Methods and Protocols. Totowa, New Jersey, Humana Press, 2013, 483–501.
[39] Yeager, A.N., Weber, P.K., & Kraft, M.L. Three-dimensional imaging of cholesterol and sphingolipids within a Madin-Darby canine kidney cell. Biointerphases 2016, 11, 02A309.
[40] Wolfsberg, M., Van Hook, W.A., & Paneth, P. Isotope Effects on Dipole Moments, Polarizability, NMR Shielding, and Molar Volume. Springer, Dordrech, Netherlands, 2009, 389–412.
[41] Tokutake, N., Jing, B., Cao, H., & Regen, S.L. Quantifying the effects of deuterium substitution on phospholipid mixing in bilayer membranes. A nearest-neighbor recognition investigation. J Am Chem Soc 2003, 125, 15764–15766.
[42] Klump, H.H., Gaber, B.P., Peticolas, W.L., & Yager, P. Thermodynamic properties of mixtures of deuterated and undeuterated dipalmitoyl phosphatidylcholines (differential scanning calorimetry/lipid bilayers/membranes). Thermochim Acta 1981, 48, 361–366.
[43] Baldyga, D.D., & Dluhy, R.A. On the use of deuterated phospholipids for infrared spectroscopic studies of monomolecular films: A thermodynamic analysis of single and binary component phospholipid monolayers. Chem Phys Lipids 1998, 96, 81–97.
[44] Bouloussa, O., & Dupeyrat, M. Disturbing effect of perdeuterated fatty acids used as probes in phospholipid monolayers. Biochim Biophys Acta 1987, 896, 239–246.
[45] Calvin, M., Hermans, J., & Scheraga, H.A. Effect of deuterium on the strength of hydrogen bonds. J Am Chem Soc 1959, 81, 5048–5050.
[46] Wiberg, K.B. The deuterium isotope effect. Chem Rev 1955, 55, 713–743.
[47] Westheimer, F.H. The magnitude of the primary kinetic isotope effect for compounds of hydrogen and deuterium. Chem Rev 1961, 61, 265–273.
[48] Tanhuanpää, K., & Somerharju, P. γ-Cyclodextrins greatly enhance translocation of hydrophobic fluorescent phospholipids from vesicles to cells in culture: Importance of molecular hydrophobicity in phospholipid trafficking studies. J Biol Chem 1999, 274, 35359–35366.
[49] Kilsdonk, E.P.C., Yancey, P.G., Stoudt, G.W., et al. Cellular cholesterol efflux mediated by cyclodextrins. J Biol Chem 1995, 270, 17250–17256.
[50] Klein, U., Gimpl, G., & Fahrenholz, F. Alteration of the myometrial plasma membrane cholesterol content with beta-cyclodextrin modulates the binding affinity of the oxytocin receptor. Biochemistry 1995, 34, 13784–13793.
[51] Kainu, V., Hermansson, M., & Somerharju, P. Introduction of phospholipids to cultured cells with cyclodextrin. J Lipid Res 2010, 51, 3533–3541.
[52] Schwarzmann, G., Hoffmann-Bleihauer, P., Schubert, J., Sandhoff, K., & Marsh, D. Incorporation of ganglioside analogs into fibroblast cell membranes. A spin-label study. Biochemistry 1983, 22, 5041–5048.
[53] van Echten, G., Birk, R., Brenner-Weiss, G., Schmidt, R.R., & Sandhoff, K. Modulation of sphingolipid biosynthesis in primary cultured neurons by long chain bases. J Biol Chem 1990, 265, 9333–9339.
[54] Pitto, M., Brunner, J., Ferraretto, A., Ravasi, D., Palestini, P., & Masserini, M. Use of a photoactivable GM1 ganglioside analogue to assess lipid distribution in caveolae bilayer. Glycoconjugate J 2000, 17, 215–222.
[55] Chigorno, V., Tettamanti, G., & Sonnino, S. Metabolic processing of gangliosides by normal and salla human fibroblasts in culture: A study performed by administering radioactive GM3 ganglioside. J Biol Chem 1996, 271, 21738–21744.
[56] Sonnino, S., Chigorno, V., Acquotti, D., Pitto, M., Kirschner, G., & Tettamanti, G. A photoreactive derivative of radiolabeled GM1 ganglioside: Preparation and use to establish the

involvement of specific proteins in GM1 uptake by human fibroblasts in culture. Biochemistry 1989, 28, 77–84.

[57] Prinetti, A., Chigorno, V., Tettamanti, G., & Sonnino, S. Sphingolipid-enriched membrane domains from rat cerebellar granule cells differentiated in culture: A compositional study. J Biol Chem 2000, 275, 11658–11665.

[58] Dolo, V., D'Ascenzo, S., Sorice, M., et al. New approaches to the study of sphingolipid enriched membrane domains: The use of microscopic autoradiography to reveal metabolically tritium labeled sphingolipids in cell cultures. Glycoconjugate J 2000, 17, 261–268.

[59] Chigorno, V., Riva, C., Valsecchi, M., Nicolini, M., Brocca, P., & Sonnino, S. Metabolic processing of gangliosides by human fibroblasts in culture – Formation and recycling of separate pools of sphingosine. Eur J Biochem 1997, 250, 661–669.

[60] Zhu, L., Johnson, C., & Bakovic, M. Stimulation of the human CTP: Phosphoethanolaminecytidylyltransferase gene by early growth response protein 1. J Lipid Res 2008, 49, 2197–2211.

[61] Rouquette-Jazdanian, A.K., Pelassy, C., Breittmayer, J.-P., & Cousin J-L, A.C. Metabolic labelling of membrane microdomains/rafts in Jurkat cells indicates the presence of glycerophospholipids implicated in signal transduction by the CD3 T-cell receptor. Biochem J 2002, 363, 645–655.

[62] Gazi, E., Harvey, T.J., Brown, M.D., Lockyer, N.P., Gardner, P., & Clarke, N.W. A FTIR microspectroscopic study of the uptake and metabolism of isotopically labelled fatty acids by metastatic prostate cancer. Vib Spectrosc 2009, 50, 99–105.

[63] Basu, S.K., Goldstein, J.L., Anderson, G.W., & Brown, M.S. Degradation of cationized low density lipoprotein and regulation of cholesterol metabolism in homozygous familial hypercholesterolemia fibroblasts. Proc Natl Acad Sci USA 1976, 73, 3178–3182.

[64] Frisz, J.F., Klitzing, H.A., Lou, K., et al. Sphingolipid domains in the plasma membranes of fibroblasts are not enriched with cholesterol. J Biol Chem 2013, 288, 16855–16861.

[65] Wilson Robert, L., Frisz Jessica, F., Klitzing Haley, A., Zimmerberg, J., Weber Peter, K., & Kraft Mary, L. Hemagglutinin clusters in the plasma membrane are not enriched with cholesterol and sphingolipids. Biophys J 2015, 108, 1652–1659.

[66] Bleijerveld, O.B., Houweling, M., Thomas, M.J., & Cui, Z. Metabolipidomics: Profiling metabolism of glycerophospholipid species by stable isotopic precursors and tandem mass spectrometry. Anal Biochem 2006, 352, 1–14.

[67] Sullards, M.C., Allegood, J.C., Kelly, S., et al. Structure-Specific, Quantitative Methods for Analysis of Sphingolipids by Liquid Chromatography–Tandem Mass Spectrometry: "Inside-Out" Sphingolipidomics. Methods Enzymol 2007, 432, 83–115.

[68] Gagné, S., Crane, S., Huang, Z., Li, C.S., Bateman, K.P., & Lévesque, J.-F. Rapid measurement of deuterium-labeled long-chain fatty acids in plasma by HPLC-ESI-MS. J Lipid Res 2007, 48, 252–259.

[69] Angelo, M., Bendall, S.C., Finck, R., et al. Multiplexed ion beam imaging of human breast tumors. Nat Med 2014, 20, 436–442.

[70] Thiery-Lavenant, G., Guillermier, C., Wang, M., & Lechene, C. Detection of immunolabels with multi-isotope imaging mass spectrometry. Surf Interface Anal 2014, 46, 147–149.

[71] Wilson, R.L., Frisz, J.F., Hanafin, W.P., et al. Fluorinated colloidal gold immunolabels for imaging select proteins in parallel with lipids using high-resolution secondary ion mass spectrometry. Bioconj Chem 2012, 23, 450–460.

[72] Vreja, I.C., Kabatas, S., Saka, S.K., et al. Secondary ion mass spectrometry of genetically encoded targets. Angew Chem Int Ed 2015, 54, 5784–5788.

[73] Kabatas, S., Vreja, I.C., Saka, S.K., et al. A contamination-insensitive probe for imaging specific biomolecules by secondary ion mass spectrometry. Chem Commun 2015, 51, 13221–13224.

[74] Jiang, H., Kilburn, M.R., Decelle, J., & Musat, N. NanoSIMS chemical imaging combined with correlative microscopy for biological sample analysis. Curr Opin Biotechnol 2016, 41, 130–135.
[75] Richter, K.N., Rizzoli, S.O., Jähne, S., Vogts, A., & Lovric, J. Review of combined isotopic and optical nanoscopy. Neurophotonics 2017, 4, 020901.
[76] Goulbourne Chris, N., Gin, P., Tatar, A., et al. The GPIHBP1–LPL complex is responsible for the margination of triglyceride-rich lipoproteins in capillaries. Cell Metab 2014, 19, 849–860.
[77] Jiang, H., Favaro, E., Goulbourne, C.N., et al. Stable isotope imaging of biological samples with high resolution secondary ion mass spectrometry and complementary techniques. Methods 2014, 68, 317–324.
[78] Saka, S.K., Vogts, A., Kröhnert, K., Hillion, F., Rizzoli, S.O., & Wessels, J.T. Correlated optical and isotopic nanoscopy. Nat Commun 2014, 5, 3664.
[79] Deegan, R.D., Bakajin, O., Dupont, T.F., Huber, G., Nagel, S.R., & Witten, T.A. Capillary flow as the cause of ring stains from dried liquid drops. Nature 1997, 389, 827–820.
[80] Anderton, C.R., Weber, P.K., Lou, K., Hutcheon, I.D., Kraft, M.L., & Correlated, A.F.M. and NanoSIMS imaging to probe cholesterol-induced changes in phase behavior and non-ideal mixing in ternary lipid membranes. Biochim Biophys Acta 2011, 1808, 307–315.
[81] Galli Marxer, C., Kraft, M.L., Weber, P.K., Hutcheon, I.D., & Boxer, S.G. Supported membrane composition analysis by secondary ion mass spectrometry with high lateral resolution. Biophys J 2005, 88, 2965–2975.
[82] Kraft, M.L., Fishel, S.F., Galli Marxer, C., Weber, P.K., Hutcheon, I.D., & Boxer, S.G. Quantitative analysis of supported membrane composition using the NanoSIMS. Appl Surf Sci 2006, 252, 6950–6956.
[83] Kraft, M.L., Weber, P.K., Longo, M.L., Hutcheon, I.D., & Boxer, S.G. Phase separation of lipid membranes analyzed with high-resolution secondary ion mass spectrometry. Science 2006, 313, 1948–1951.
[84] Lozano MM, Hovis JS, Moss FR, Boxer SG. Dynamic Reorganization and Correlation among lipid raft components. J Am Chem Soc 2016, 138, 9996–10001.
[85] Lozano, M.M., Liu, Z., Sunnick, E., Janshoff, A., Kumar, K., & Boxer, S.G. Colocalization of the ganglioside GM1 and cholesterol detected by secondary ion mass spectrometry. J Am Chem Soc 2013, 135, 5620–5630.
[86] Rakowska, P.D., Jiang, H., Ray, S., et al. Nanoscale imaging reveals laterally expanding antimicrobial pores in lipid bilayers. Proc Natl Acad Sci USA 2013, 110, 8918–8923.
[87] Moor, H.. Theory and Practice of High Pressure Freezing. In: S. Ra & K. Zierold, eds. Cryotechniques in Biological Electron Microscopy. Berlin, Heidelberg, Springer Berlin Heidelberg, 1987, 175–191.
[88] Porter, K.R., Claude, A., & Fullam, E.F. A study of tissue culture cells by electron microscopy. J Exp Med 1945, 81, 233–246.
[89] Hayat, M.A. Glutaraldehyde: Role in electron microscopy. Micron Microsc Acta 1986, 17, 115–135.
[90] Tanaka, K.A.K., Suzuki, K.G.N., Shirai, Y.M., et al. Membrane molecules mobile even after chemical fixation. Nat Methods 2010, 7, 865–866.
[91] Mayor, S., Rothberg, K., & Maxfield, F. Sequestration of GPI-anchored proteins in caveolae triggered by cross-linking. Science 1994, 264, 1948–1951.
[92] Bowes, J.H., & Cater, C.W. Crosslinking of collagen. J Appl Chem 1965, 15, 296–304.
[93] Hopwood, D. Theoretical and practical aspects of glutaraldehyde fixation. Histochem J 1972, 4, 267–303.
[94] Hopwood, D., Callen, C.R., & McCabe, M. The reactions between glutaraldehyde and various proteins. An investigation of their kinetics. Histochem J 1970, 2, 137–150.

[95] Migneault, I., Dartiguenave, C., Bertrand, M.J., & Waldron, K.C. Glutaraldehyde: Behavior in aqueous solution, reaction with proteins, and application to enzyme crosslinking. BioTechniques 2004, 37, 790–802.

[96] Roozemond, R.C. The effect of fixation with formaldehyde and glutaraldehyde on the composition of phospholipids extractable from rat hypothalamus. J Histochem Cytochem 1969, 17, 482–486.

[97] Frisz, J.F., Lou, K., Klitzing, H.A., et al. Direct chemical evidence for sphingolipid domains in the plasma membranes of fibroblasts. Proc Natl Acad Sci USA 2013, 110, E613–E622.

[98] Jost, P., Brooks, U.J., & Griffith, O.H. Fluidity of phospholipid bilayers and membranes after exposure to osmium tetroxide and glutaraldehyde. J Mol Biol 1973, 76, 313–318.

[99] Hayat, M.A. Principles and Techniques of Electron Microscopy: Biological Applications. Cambridge, UK, Cambridge University Press; 2000.

[100] Mitchell, C.D. Preservation of the lipids of the human erythrocyte stroma during fixation and dehydration for electron microscopy. J Cell Biol 1969, 40, 869–872.

[101] Bahr, G.F. Osmium tetroxide and ruthenium tetroxide and their reactions with biologically important substances: Electron stains III. Exp Cell Res 1954, 7, 457–479.

[102] Adams, C.W.M. Osmium tetroxide and the Marchi method: Reactions with polar and non-polar lipids, protein and polysaccharide. J Histochem Cytochem 1960, 8, 262–267.

[103] Khan, A.A., Riemersma, J.C., & Booij, H.L. The reactions of osmium tetroxide with lipids and other compounds. J Histochem Cytochem 1961, 9, 560–563.

[104] Korn, E.D., & Chromatographic, A. and Spectrophotometric study of the products of the reaction of osmium tetroxide with unsaturated lipids. J Cell Biol 1967, 34, 627–638.

[105] Hayes, T.L., Lindgren, F.T., & Gofman, J.W. A Quantitative Determination of the Osmium Tetroxide-Lipoprotein Interaction. J Cell Biol 1963, 19, 251–255.

[106] Jost, P.C., & Griffith, O.H. The molecular reorganization of lipid bilayers by osmium tetroxide. Arch Biochem Biophys 1973, 159, 70–81.

[107] Stoeckenius, W., Schulman, J.H., & Prince, L.M. The structure of myelin figures and microemulsions as observed with the electron microscope. Kolloid Z 1960, 169, 170–180.

[108] Mühlfeld, C.. High-Pressure Freezing, Chemical Fixation and Freeze-Substitution for Immuno-electron Microscopy. In: Hewitson TD, Darby IA, eds. Histology Protocols. Totowa, NJ, Humana Press, 2010, 87–101.

[109] McDonald, K.L., & Auer, M. High-pressure freezing, cellular tomography, and structural cell biology. BioTechniques 2006, 41, 137–143.

[110] Kilburn, M.R., & Chapter, W.D. 1 Nanoscale Secondary Ion Mass Spectrometry (NanoSIMS) as an Analytical Tool in the Geosciences. In: K. Grice, ed. Principles and Practice of Analytical Techniques in Geosciences. Cambridge, UK, The Royal Society of Chemistry, 2015, 1–34.

[111] Lechene, C., Hillion, F., McMahon, G., et al. High-resolution quantitative imaging of mammalian and bacterial cells using stable isotope mass spectrometry. J Biol 2006, 5, 20.

[112] Yeager, A.N., Weber, P.K., & Kraft, M.L. Cholesterol is enriched in the sphingolipid patches on the substrate near nonpolarized MDCK cells, but not in the sphingolipid domains in their plasma membranes. Biochim Biophys Acta 2018, 1860, 2004–2011.

[113] Nittler, L.R. Quantitative isotopic ratio ion imaging and its application to studies of preserved stardust in meteorites. Saint Louis: Washington University; 1996.

[114] Yu, M.L. Matrix effects in the work-function dependence of negative-secondary-ion emission. Phys Rev B 1982, 26, 4731–4734.

[115] Yu, M.L. Chemical enhancement effects in SIMS analysis. Nucl Instrum Methods Phys Res, Sect B 1986, 15, 151–158.

[116] Altelaar, A.F.M., van Minnen, J., Heeren, R.M.A., & Piersma, S.R. The influence of the cholesterol microenvironment in tissue sections on molecular ionization efficiencies and distributions in ToF-SIMS. Appl Surf Sci 2006, 252, 6702–6705.

[117] Rangarajan, S., & Tyler, B.J. Topography in secondary ion mass spectroscopy images. J Vac Sci Technol A 2006, 24, 1730–1736.

[118] Muramoto, S., Brison, J., & Castner, D.G. Exploring the surface sensitivity of TOF-secondary ion mass spectrometry by measuring the implantation and sampling depths of bin and C60 ions in organic films. Anal Chem 2012, 84, 365–372.

[119] McMahon, G., Glassner, B.J., & Lechene, C.P. Quantitative imaging of cells with multi-isotope imaging mass spectrometry (MIMS)--Nanoautography with stable isotope tracers. Appl Surf Sci 2006, 252, 6895–6906.

[120] Edidin, M. Lipids on the frontier: A century of cell-membrane bilayers. Nat Rev Mol Cell Biol 2003, 4, 414–418.

[121] Gormanns, P., Reckow, S., Poczatek, J.C., Turck, C.W., & Lechene, C. Segmentation of multi-isotope imaging mass spectrometry data for semi-automatic detection of regions of interest. PLoS One 2012, 7, e30576.

[122] Böhlke, J.K., JRd, L., Bièvre, P.D., et al. Isotopic Compositions of the Elements, 2001. J Phys Chem Ref Data 2005, 34, 57–67.

[123] He, C., Hu, X., Jung, R.S., et al. High-resolution imaging and quantification of plasma membrane cholesterol by NanoSIMS. Proc Natl Acad Sci USA 2017, 114, 2000–2005.

[124] Misson, J., Henner, P., Morello, M., et al. Use of phosphate to avoid uranium toxicity in Arabidopsis thaliana leads to alterations of morphological and physiological responses regulated by phosphate availability. Environ Exp Bot 2009, 67, 353–362.

[125] Clode, P.L., Kilburn, M.R., Jones, D.L., et al. In situ mapping of nutrient uptake in the rhizosphere using nanoscale secondary ion mass spectrometry. Plant Physiol 2009, 151, 1751–1757.

[126] Smart, K.E., Smith, J.A.C., Kilburn, M.R., Martin, B.G.H., Hawes, C., & Grovenor, C.R.M. High-resolution elemental localization in vacuolate plant cells by nanoscale secondary ion mass spectrometry. Plant J 2010, 63, 870–879.

[127] Fitzsimons, I.C.W., Harte, B., & Clark, R.M. SIMS stable isotope measurement: Counting statistics and analytical precision. Mineral Mag 2000, 64, 59–83.

[128] Pratt, J.W., & Gibbons, J.D. Kolmogorov-Smirnov Two-Sample Tests. Concepts of Nonparametric Theory. New York, NY, Springer New York, 1981, 318–344.

[129] Chini, C.E., Fisher, G.L., Johnson, B., Tamkun, M.M., & Kraft, M.L. Observation of endoplasmic reticulum tubules via TOF-SIMS tandem mass spectrometry imaging of transfected cells. Biointerphases 2018, 13, 03B409.

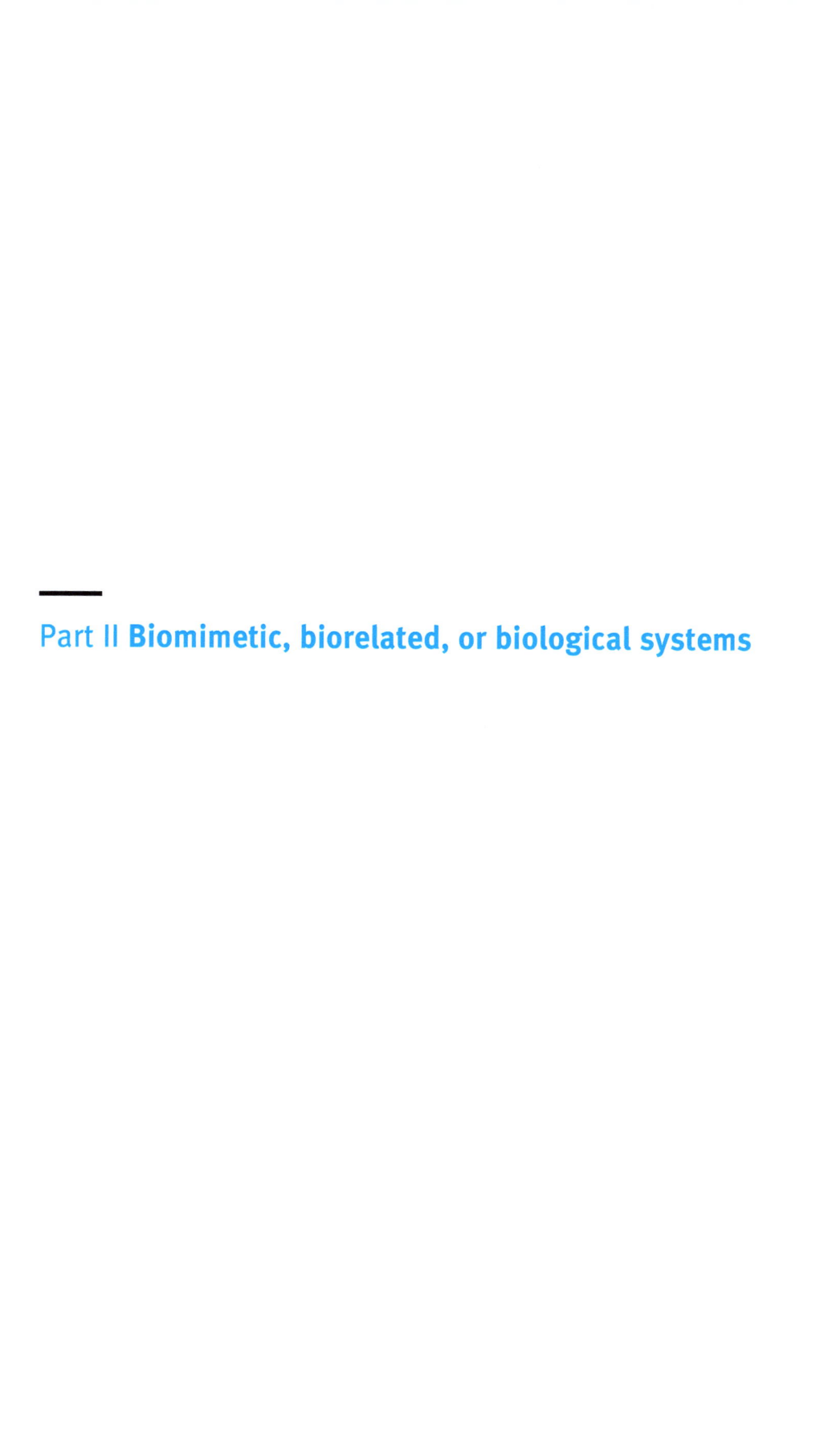

Part II Biomimetic, biorelated, or biological systems

James H. Davis, Miranda L. Schmidt
10 Cholesterol in model membranes

Abstract: Because cholesterol is an essential component of numerous biological membranes, its effect on model membranes has been an active area of research for several decades. The impact of cholesterol on the molecular order, dynamics and phase equilibria in model membranes has been studied by many experimental techniques including nuclear magnetic resonance (NMR). NMR provides quantitative information on the organization and behavior of the molecules making up the membrane and is sensitive to changes in the order and dynamics of these systems. This chapter will provide a broad review of the contributions of NMR experiments to the understanding of the behavior of cholesterol and phospholipids in model membrane systems of varying complexity.

Keywords: cholesterol, model membranes, nuclear magnetic resonance, molecular order, axially symmetric reorientation, critical behavior

10.1 Introduction to cholesterol in membranes

Cholesterol has an enormous impact on the physical properties and biological function of membranes. It is found in high concentrations in many eukaryotic cell membranes. Its' most obvious effects on the physical properties of the membrane include an increase in the lipid hydrocarbon chain order which results in an increase in the thickness of the bilayer, an enhancement of the permeability barrier and a stiffening of the bilayer. This is accomplished with little change in the lateral molecular mobility within the membrane [1, 2]. Cholesterol preferentially interacts with sphingomyelins and long-chain saturated phospholipids which lead to interesting phase equilibria in mixed lipid bilayers containing cholesterol [1, 3–5]. In particular, it is found that at high cholesterol concentrations, there are two distinct "fluid" phases [1, 6, 7], the liquid disordered (ℓ_d), and the liquid ordered (ℓ_o) phases which can coexist at physiological temperatures. Figure 10.1 is a cartoon description of a membrane containing a lipid "raft" (ℓ_o) domain and an ℓ_d region. This discovery led to the membrane "raft" hypothesis [8], which suggests that membranes with high cholesterol concentrations, such as mammalian plasma membranes, contain regions which are rich in cholesterol, and therefore thicker, where certain types of proteins prefer to reside [9]. These include glycosylphosphatidylinositol (GPI)-anchored proteins and acylated proteins. The interaction of these proteins, and others such as G-protein-coupled receptors (GCPRs) with cholesterol, is suggested to play an

---

James H. Davis, Miranda L. Schmidt, Department of Physics, University of Guelph, Ontario, Canada

Raft "phase"

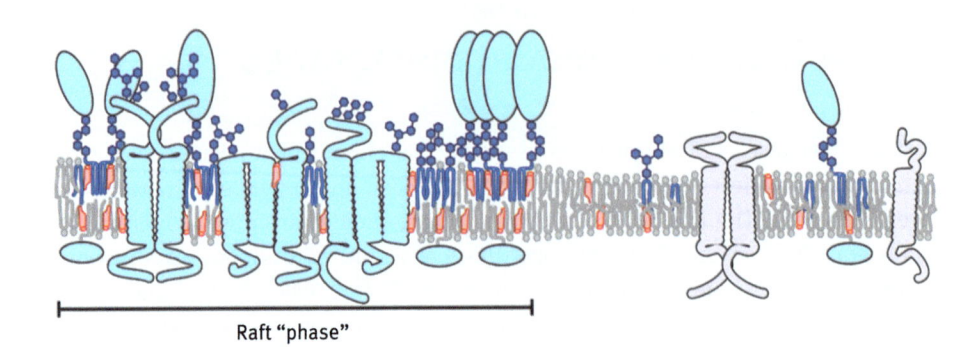

Figure 10.1: The molecular building blocks of the cell membrane. Reproduced from Lingwood and Simons [16], with permission.

important role in signal transduction [10–14]. Trans-bilayer asymmetry, with very different concentrations of cholesterol in the two leaflets, adds a potentially interesting twist to cholesterol's role in signaling [15]. Perhaps most importantly, cholesterol may be implicated in a broad spectrum of human diseases [2, 12]. Even after 200 years, the study of the role of cholesterol in biological membrane function remains an active and fascinating area of research.

Nuclear magnetic resonance (NMR) provides a means of studying molecular structure, orientational order, dynamics, and phase equilibria in both model and biological membrane systems [17–21]. ^{2}H NMR has been particularly useful for measuring both the effects of cholesterol on membrane physical properties [1, 4, 22–24] and the orientational order and dynamics of cholesterol within the membrane [25]. Since deuterium is a spin one nucleus, it possesses an electric quadrupolar moment. Thus, in situations with low symmetry, the NMR spectrum of deuterium is dominated by the electric quadrupolar Hamiltonian [26]. The quadrupolar splittings observed are sensitive to the orientation of the deuterium labeled molecule relative to the applied magnetic field. For deuterium in C-^{2}H bonds in molecules undergoing rapid axially symmetric reorientation (which occurs about the local bilayer normal in fluid membranes), the splitting depends on the orientation of the C-D bond relative to the axis of symmetry for the motion. The magnitudes of the quadrupolar splittings that one measures then provide a description of the orientational order of the system. One of the main effects of the incorporation of cholesterol into a membrane is the dramatic increase in lipid hydrocarbon chain orientational order [22, 23, 27]. Motional averaging, or reorientation, reduces the quadrupolar splittings so that deuterium NMR is also a powerful method for studying molecular dynamics. The changes in mobility which occur at many types of phase transitions can often be readily measured by ^{2}H NMR so it is a powerful tool for studying phase equilibria as well. Finally, nuclear magnetic relaxation studies provide dynamical information on the molecular motions present and ^{2}H relaxation, which is dominated by fluctuations in the quadrupolar Hamiltonian, is extremely sensitive to molecular reorientation.

NMR of other magnetic nuclei, particularly ^{1}H, ^{13}C, and ^{31}P, which have appreciable natural abundance, can also be used to study the molecular environments found in membrane systems [19, 21, 28]. Information complementary to that obtained from ^{2}H NMR is obtained from anisotropic chemical shifts (^{13}C and ^{31}P), heteronuclear chemical shift correlation experiments and both static and dynamic dipolar couplings (^{1}H and ^{13}C) [29, 30].

10.1.1 The structure of cholesterol

The steroid fused ring structure forms the core of the cholesterol molecule. There is a polar hydroxyl group at the C3 position, methyl groups at the C10 and C13 positions and a short, branched hydrocarbon side chain at C17. There is one double bond between carbons C5 and C6. The amphiphilic character of the molecule makes it well suited to the membrane environment with the hydroxyl group oriented toward the lipid water interface and with the hydrocarbon chain toward the center of the bilayer. One surface of the rigid ring moiety of cholesterol is relatively smooth while the two methyl groups protrude from the opposite face. This has been suggested to be a geometric explanation of the strong "ordering" effect of cholesterol on the neighboring saturated lipid chains. Figure 10.2 shows the molecular structure of cholesterol and its orientation relative to the bilayer normal (the vertical line represents the local bilayer normal and the long axis of cholesterol is shown making an angle ϕ relative to the bilayer normal).

The molecular structure of cholesterol has been determined by X-ray and neutron diffraction [31–34] and parameterized by molecular dynamics calculations [35]. Its location and orientation within model lipid bilayers has been studied by low-angle neutron scattering and ^{2}H NMR [36–42]. The nature of the orientational order and the molecular reorientation of cholesterol within a lipid bilayer has also been extensively studied [6, 25, 37, 39–41, 43–45]. The ^{1}H and ^{13}C NMR isotropic chemical shifts of cholesterol in lipid bilayers have been assigned using magic angle spinning (MAS) NMR techniques [41, 46]. This makes it possible to extract detailed dynamical and orientational information from experiments such as ^{13}C–^{1}H separated local field spectroscopy [29] and the nuclear Overhauser effect [30].

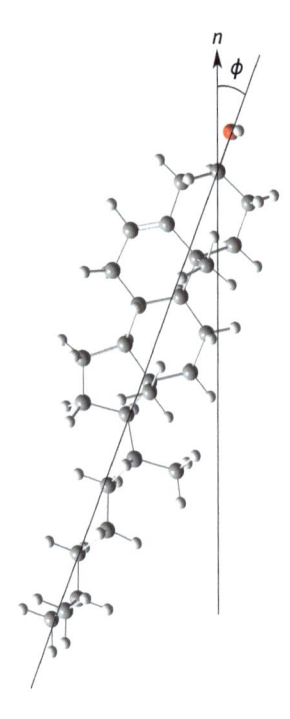

Figure 10.2: The structure of cholesterol and its orientation relative to membrane bilayer normal.

10.2 NMR of phospholipids in the presence of cholesterol

^{2}H NMR was first used to demonstrate the influence of cholesterol on the membrane lipids' hydrocarbon chain mobility by Oldfield and Chapman [23, 47]. The large increase in quadrupolar splittings observed when cholesterol was present corroborated the observations made using electron paramagnetic resonance (EPR) [22]. The introduction of the quadrupolar echo technique [48] made it possible to obtain "high fidelity" ^{2}H NMR line shapes and improved our ability to make quantitative measurements of chain order [1, 26, 49–52]. The two NMR spectra shown in Figure 10.3 have powder pattern line shapes which are typical of a fluid lipid bilayer. The upper spectrum is for a sample of 1,2-dipalmitoyl-d_{31} - sn-glycero-3-phosphocholine (DPPC-d_{62})/cholesterol at a 3:1 mole ratio while the lower spectrum is for DPPC-d_{62}, both spectra were taken at a temperature of 317.5 K and both samples are multilamellar dispersions in an excess of 50 mM phosphate buffer at pH 7.0. The much larger quadrupolar splittings of the spectrum of the sample containing cholesterol are due to the ordering effect of cholestrol on saturated lipid chains.

As we shall see below, there are a number of interesting changes in molecular order, molecular motion and phase behavior occurring in the temperature range

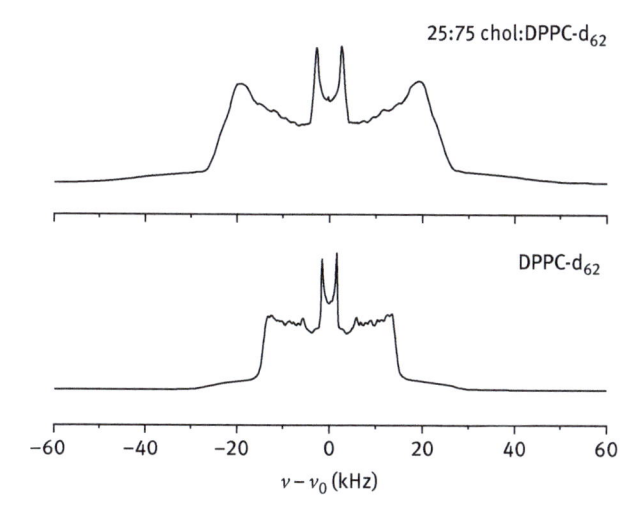

Figure 10.3: The ordering effect of cholesterol as seen in ^{2}H NMR spectra. Top spectrum is of DPPC-d_{62}/cholesterol (75:25 molar ratio) and bottom spectrum is of DPPC-d_{62} with no cholesterol, both are at 317.5 K.

above the chain melting transition of DPPC-d_{62} ($T_m \approx 37.75$ °C (or 310.9 K) [1]). These changes are reflected in a dramatic increase in the hydrocarbon chain order as cholesterol concentration is increased. The deuterium quadrupolar splitting is proportional to the C-^{2}H (or C–D) bond order parameter S_{CD} defined as $S_{CD} = <(3\cos^2\theta - 1)/2>$ where θ is the angle between the C–D bond and the external magnetic field. In this equation, the averages are over the molecular reorientations relative to the direction of the magnetic field which are fast relative to the inverse of the observed quadrupolar splitting. Then,

$$\Delta v_Q = \frac{3e^2qQ}{2\hbar} \cdot S_{CD}$$

In the absence of any motional averaging, the angle θ is fixed and the quadrupolar splitting observed in a powder spectrum will vary from 252 kHz, for C-^{2}H bonds which are parallel to the magnetic field, to 126 kHz for bonds which are perpendicular to the field. For bonds oriented at the magic angle, $\theta_m = 54.7°$, the quadrupolar splitting is zero. Since all orientations occur in a random powder, the NMR spectrum takes the form of the Pake doublet which has maximum intensity at the perpendicular orientation since more molecules will have their C-^{2}H bonds aligned perpendicular to the field than for any other angle.

In the fluid lipid bilayer, the local normal to the bilayer is an axis of symmetry for the motion. This axially symmetric motion projects the quadrupolar interaction onto the axis so that the average angle, β, between the C-^{2}H bond and the bilayer normal determines the value of the quadrupolar splitting

$$\delta\nu_Q = <(3\cos^2\beta - 1)/2> \cdot [(3\cos^2\gamma - 1)/2]\frac{3e^2qQ}{2\hbar}$$

where γ is the angle between the bilayer normal and the magnetic field. Now the maximum quadrupolar splitting for C–^{2}H bonds oriented perpendicular to the local bilayer normal is 63 kHz, which corresponds to the 90° peaks in the powder pattern spectrum, while the 0° shoulders have a maximum splitting of 126 kHz. Internal chain motions, such as *gauche/trans* isomerization about the lipid chain C–C bonds, further averages the quadrupolar splittings so that there is a variation in splitting with chain position. This hydrocarbon chain flexibility gradient leads to very small quadrupolar splittings for positions near the methyl terminus. These vary from about 2–3 kHz for the methyl group which also undergoes rapid reorientation about the terminal C-C bond, to as much as 20–25 kHz for positions near the lipid/water interface [49, 53–55]. The interaction of the lipid chains with the rigid ring moiety of cholesterol reduces this isomerization substantially so that splittings near the headgroup can be as large as 50–55 kHz for cholesterol concentrations above about 25 mol%.

The use of chain perdeuterated lipids, such as DPPC-d_{62}, results in a superposition of Pake doublets as shown in the spectra of Figure 10.3, one for each labeled position on the chain. With so many overlapping powder patterns in one spectrum, it is often difficult to separately assign each pair of peaks in the powder pattern. To simplify the analysis of such spectra, we introduced the use of spectral moments to describe the distribution of quadrupolar splittings [26, 49]. The nth moment is defined by

$$M_n = \frac{\int_{-\infty}^{\infty} \omega^n f(\omega) d\omega}{\int_{-\infty}^{\infty} f(\omega) d\omega}$$

where the integrals are over all angular frequencies ω. In practice, the integrals are replaced by sums over the discretized spectrum, $f(\omega)$, and the center of the spectrum is taken to be the ^{2}H Larmor frequency ω_0.

The most useful of these is the first moment, M_1, which gives the average quadrupolar splitting. Figure 10.4 shows that there is a roughly linear increase in the average quadrupolar splitting as cholesterol is added at temperatures above the chain melting transition of pure DPPC-d_{62}. Figure 10.5 shows the temperature dependence of the first moment for pure DPPC-d_{62}, data shown as blue squares, and for a 3:1 mole ratio mixture of DPPC-d_{62}/cholesterol, data shown as red circles. The chain melting transition at 37.5°C results in an abrupt increase in M_1 as the temperature is lowered through the transition for the pure DPPC-d_{62} sample while for the sample containing cholesterol, the change is much more gradual. Furthermore, below the transition temperature, the lipid chain order in the presence of cholesterol is reduced compared to that of the pure lipid.

Clearly, the deuterium NMR spectrum is very sensitive to the degree of motional averaging occurring at the labeled position. This motion is heavily influenced by the

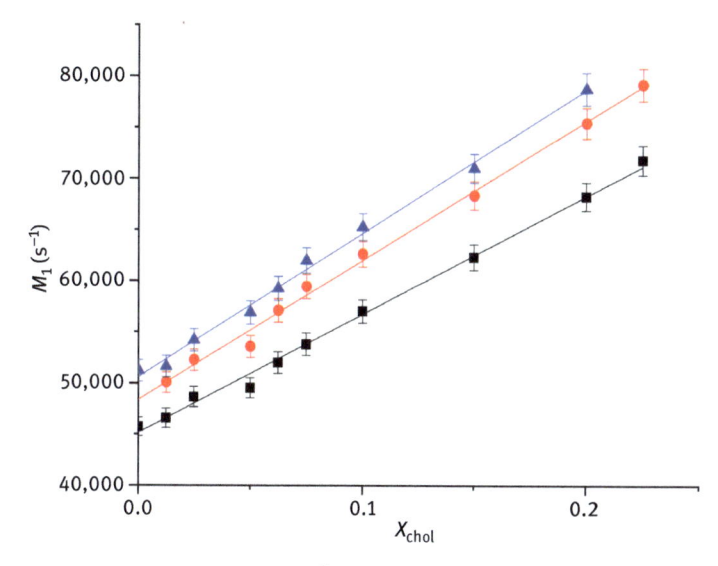

Figure 10.4: First moment of the ^{2}H NMR spectra as a function of cholesterol concentration in chain perdeuterated DPPC-d$_{62}$/cholesterol mixtures. Blue triangles are at a temperature of 315 K, red circles are at 317 K, and black squares are at 323 K. (Data are from Vist and Davis [1].)

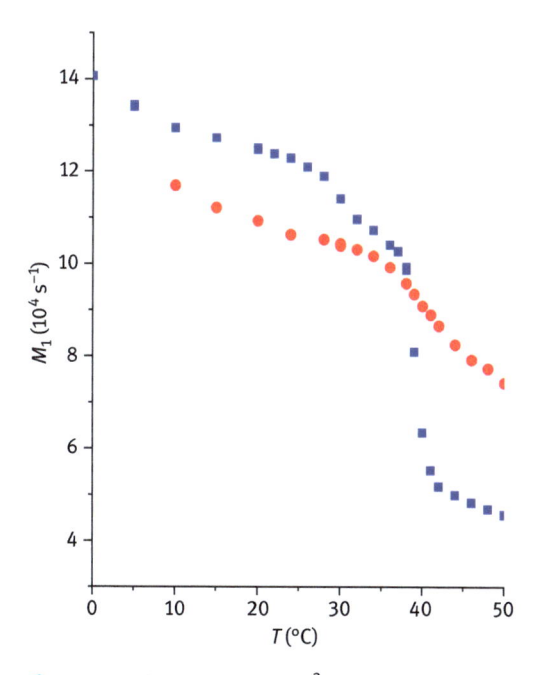

Figure 10.5: First moment of the ^{2}H NMR spectra of DPPC-d$_{62}$ and DPPC-d$_{62}$/cholesterol (at 75:25 molar ratio) as a function of temperature. The fluid to gel phase, or chain melting transition of DPPC-d$_{62}$ is clearly evident.

nature and symmetry of the phase in which the molecules are found. The effect of the chain melting transition on the first moment, shown in Figure 10.5, is easily seen in the spectra. For mixtures of DPPC-d$_{62}$ or DMPC-d$_{54}$ and cholesterol, we observe spectra which are characteristic of the three different phases that we find. Figure 10.6 shows the types of spectra we observe. In part (a) of the figure, we show the spectrum of the liquid-disordered (ℓ_d) phase, in part (b) we show the spectrum of the liquid-ordered (ℓ_o) phase and in (c) we show the spectrum typical of the gel phase (sometimes referred to as the solid ordered phase).

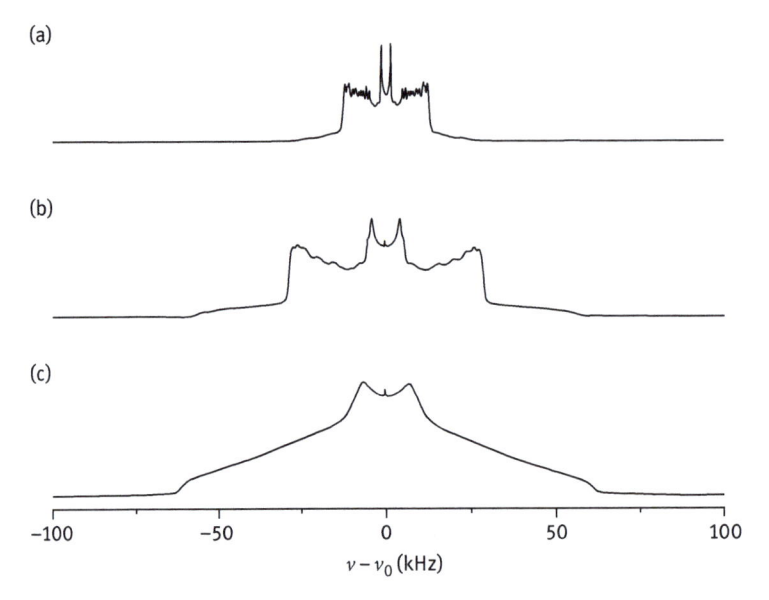

Figure 10.6: ^{2}H NMR spectra of the three phases commonly observed in fully hydrated phospholipid/cholesterol mixtures: (a) DMPC-d$_{54}$ in the fluid or liquid-disordered (ℓ_d) phase at T = 324.5 K, (b) DMPC-d$_{54}$/cholesterol (75:25) in the liquid-ordered (ℓ_o) phase at T = 287.2 K, and (c) DMPC-d$_{54}$ in the gel phase at T = 271 K.

At the magic angle, θ_m = 54.7°, the function $3\cos^2\theta_m - 1 = 0$ so the quadrupolar splitting vanishes. Just as rapid axial reorientation about the bilayer normal projects, the quadrupolar interaction onto the symmetry axis for the motion, rapid rotation of the sample about an axis inclined at the magic angle relative to the external magnetic field will project the residual quadrupolar interaction onto the rotation axis. In this way, it would be possible to obtain high resolution ^{2}H NMR spectra by MAS. These spectra would be dominated by the isotropic parts of the chemical shift and the scalar part of the indirect dipolar interaction (or J-coupling). Unfortunately, the range of isotropic chemical shifts for deuterium is only about 10 ppm (as for its lighter isotope hydrogen) so that at the typical magnetic fields used (11.6 Tesla for a 500 MHz spectrometer, making the Larmor frequency for deuterium 76.7 MHz) the range of isotropic shifts is only ≈ 800 Hz, which is comparable to the ^{2}H line width. On the other hand, hydrogen has a Larmor frequency of 500 MHz in the same magnet so it has a chemical shift range of ≈ 5 kHz and hydrogen line widths are of the order of 10–30 Hz. While this is a much more usable frequency range, there is an added difficulty. The ^{1}H NMR spectrum is dominated by ^{1}H–^{1}H homonuclear dipolar couplings. Although these couplings are not as strong as the quadrupolar couplings, there are far more of them since ^{1}H is such an abundant nucleus in biology. In static systems, MAS does not work particularly well in removing the homonuclear dipolar couplings even though they also depend on orientation as $3\cos^2\theta - 1$ [56]. In fluid phase bilayers, however, ^{1}H MAS can work quite well provided that the axially symmetric molecular reorientations are rapid enough [57, 58]. Figure 10.7 shows a ^{1}H MAS NMR spectrum of a 70:30 mole ratio mixture of DMPC-d_{54} and cholesterol. The deuteration of the phospholipid hydrocarbon chains reduces the intensity of the DMPC hydrogen peaks in the aliphatic region (from 0 to 3 ppm), whereas the lipid glycerol and headgroup resonances as well as those from cholesterol are readily detected. There is still a lot of spectral overlap, especially for the cholesterol hydrogens so the one-dimensional ^{1}H spectra alone are of limited utility.

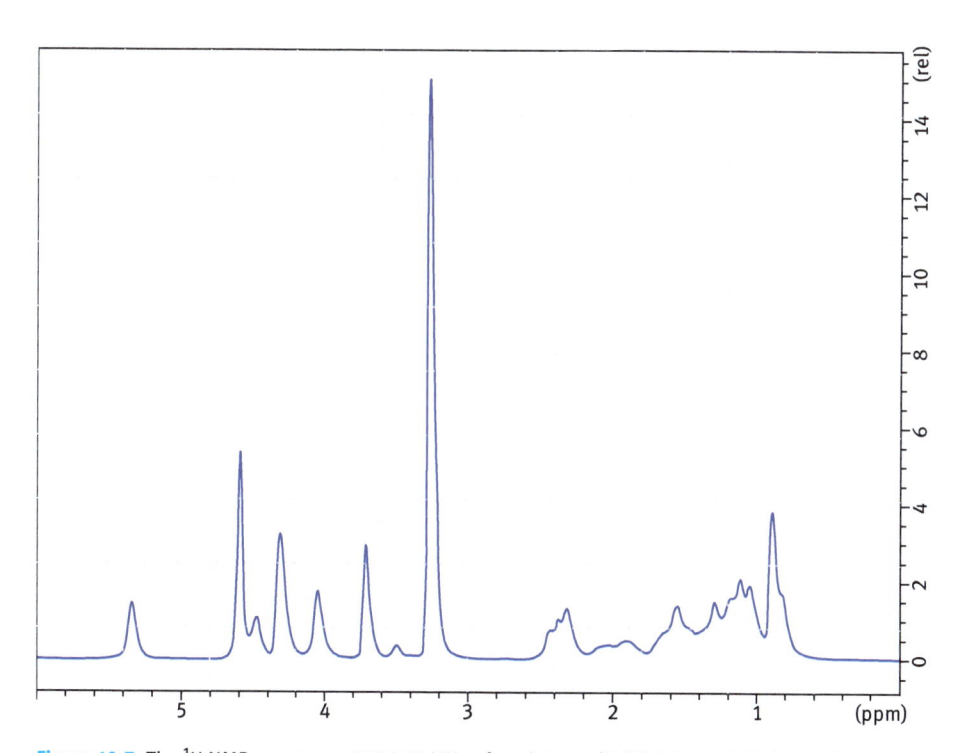

Figure 10.7: The ^{1}H NMR spectrum at 500.13 MHz of a mixture of DMPC-d_{54} and cholesterol at a 70:30 molar ratio. Temperature is 317.5 K at a MAS spinning rate of $v = 15$ kHz.

Although the natural abundance of the magnetic isotope of carbon, ^{13}C, is only 1%, it is still a very useful nucleus for NMR investigations of biological systems. The static ^{13}C NMR spectrum is dominated by the anisotropic ^{13}C chemical shift and the heteronuclear ^{1}H–^{13}C dipolar couplings. Under MAS, both of these interactions are effectively reduced and high resolution ^{13}C spectra can be obtained. The isotropic chemical shifts remaining after MAS have a range of as much as 200 ppm, which at a Larmor frequency of 125.7 MHz (at 11.6 T), corresponds to roughly 25 kHz. With ^{1}H decoupling to remove the heteronuclear ^{1}H–^{13}C J-couplings line widths of 10 Hz are readily obtained so that resolution of the large number of ^{13}C resonances is actually quite good. Figure 10.8 shows a ^{13}C MAS NMR spectrum of a 70:30 mole ratio mixture of DMPC-d_{54} and cholesterol at 317.5 K. This spectrum was obtained using cross-polarization (CP) transfer from ^{1}H to ^{13}C to enhance the signal of those carbons which are strongly interacting with neighboring protons (via the heteronuclear dipolar coupling) [59]. Because the phospholipid hydrocarbon chains are perdeuterated, there is very little signal arising from the DMPC-d_{54} chains. For this reason, most of the signals in the chemical shift range from 10 to 45 ppm are due to cholesterol. The lipid headgroup is undeuterated, however, so the glycerol backbone and the choline moiety are readily detected in the region from 50 to 70 ppm. The signals in

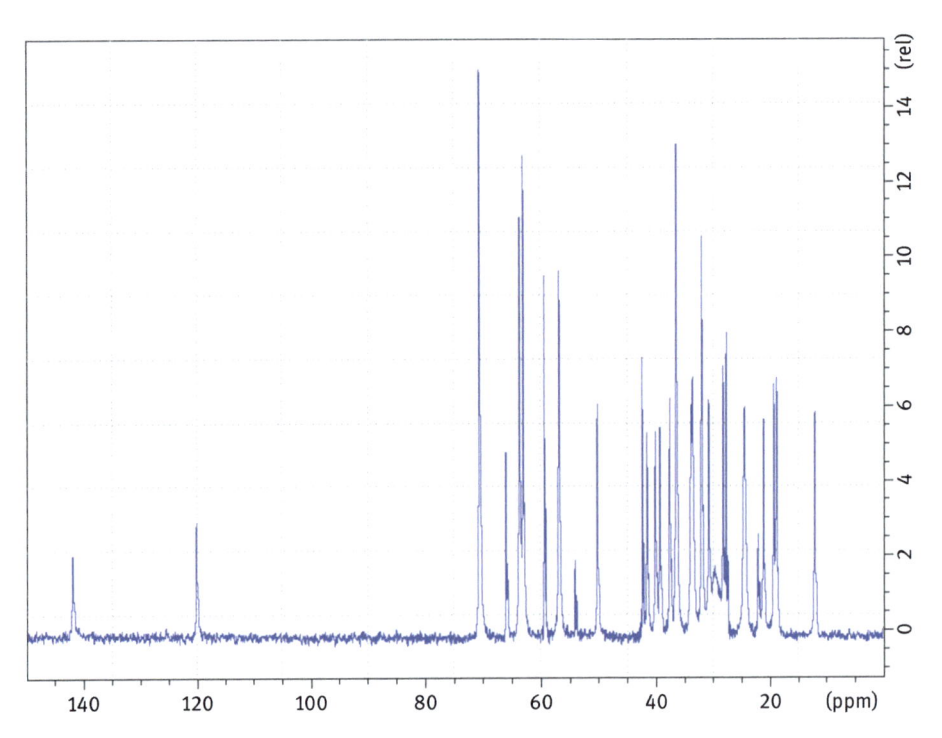

Figure 10.8: The ^{13}C NMR spectrum at 125.77 MHz of a mixture of DMPC-d_{54} and cholesterol at a 70:30 molar ratio. Temperature is 317.5 K at a MAS spinning rate of $v = 15$ kHz.

the aromatic region, near 120 and 140 ppm, are from cholesterol's C5 and C6 carbons, which are connected by a double bond.

To improve resolution even further, we can combine ^{1}H and ^{13}C into a two-dimensional heteronuclear correlation (HETCOR) experiment as illustrated by Figure 10.9. Nearly all of the peaks in this DMPC-d$_{54}$/cholesterol mixture are resolved (with the exception of the deuterated lipid chains) allowing the complete assignment of the cholesterol signals [46]. While the isotropic chemical shifts obtained in such spectra in themselves do not contain much information about molecular orientation or molecular dynamics, as we shall see below, such spectra make further, more informative, experiments possible [30].

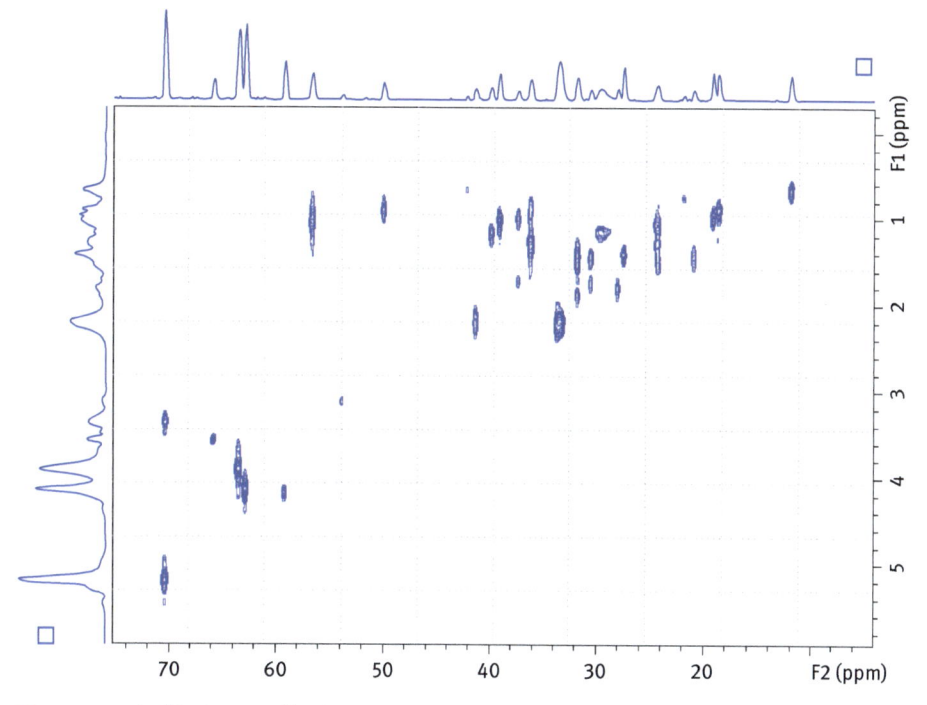

Figure 10.9: The ^{13}C detected ^{13}C–^{1}H heteronuclear correlation spectrum of a mixture of DMPC-d$_{54}$ and cholesterol at a 70:30 molar ratio. Temperature is 313 K at a MAS spinning rate of $v = 15$ kHz.

10.2.1 Comparison of mechanically or magnetically oriented samples with powders

Lipid samples used for static solid-state NMR experiments can be divided into two categories: powder and oriented. The so-called powder samples are not dry powders, but rather hydrated mixtures of lipids in water or buffer in which all orientations of the lipids are present (e.g., as in multilamellar dispersions). Oriented lipid samples are useful because the static NMR spectra for oriented samples are much simpler to interpret, with the signal coming from a single orientation, rather than all orientations as in the corresponding powder patterns. Oriented samples can be prepared by either mechanically or magnetically aligning the sample such that most of the lipids have their bilayer normal oriented in a single direction. Mechanically aligned samples can be made by depositing the lipids onto glass slides, stacking them and hydrating the lipids in a humidity chamber. These types of samples are useful because the alignment is generally very reliable and the orientation of the bilayer normal with respect to the external magnetic field can be controlled by the experimenter. However, much of the sample volume is taken up by the supporting material (e.g., glass) so the signal per volume of sample is greatly reduced.

Magnetically aligned samples have their own set of advantages and challenges. These samples do not need a supporting substrate, so the signal per volume ratio is larger than for mechanically aligned samples. Bicelles are hydrated mixtures of long-chain and short-chain phospholipids that spontaneously align when placed in an external magnetic field due to the anisotropic magnetic susceptibility of the molecules [60–62]. Commonly used bicelles, such as those made with DMPC (di(14:0)PC) with 1,2-dicaproyl-sn-glycero-3-phosphocholine (DCPC or di(6:0)PC, also sometimes referred to as DHPC (1,2-dihexanoyl-sn-glycero-3-phosphocholine)) or DPPC (di(16:0)PC) with either DHPC or 1,2-diheptanoyl-sn-glycero-3-phosphocholine, align such that the bilayer normal is perpendicular to the external magnetic field. It is possible to magnetically align samples with the bilayer normal parallel to the magnetic field by adding paramagnetic lanthanide ions [63], or using a long-chain biphenyl phospholipid [64–66] when making the bicelles. Many different bicelle mixtures have been made using one or more species of long-chain lipids along with either short-chain lipids or detergents [64, 67–75]. The morphology of these orientable systems depends on several factors, most notably the molar ratio "q" (mol long-chain lipid/mol short-chain lipid), overall composition, buffer/water content, and temperature [71, 76–78].

Figure 10.10 shows potential morphologies of bicelle mixtures which depend on the ratio of the long- to short-chain lipids: q. Since there is relatively little of the short-chain lipid at high q values, $q \geq 7.5$, multilamellar vesicles are formed as shown in Figure 10.10 (a). At low q values, $q \leq 2.5$, fast tumbling, isotropic bicelles such as those depicted in Figure 10.10 (d) are formed. Magnetically aligned bilayers occur for moderate values of q from ~2.5 to ~7.5. These bicelles can exhibit different morphologies including disk-shaped bicelles, chiral nematic ribbons, and perforated lamellae. The long-chain lipids make up the flat part of the disk-shaped bicelles, while the short-chain lipids fill in the edges where higher curvature is required. Similarly, the holes of the perforated lamellae are lined with the short-chain lipids as shown in Figure 10.10 (b) and (c) [71, 73].

Cholesterol can also be incorporated into bicelle mixtures of various types [74, 79–83]. When up to ~15 mol% cholesterol (with respect to the long chain lipid

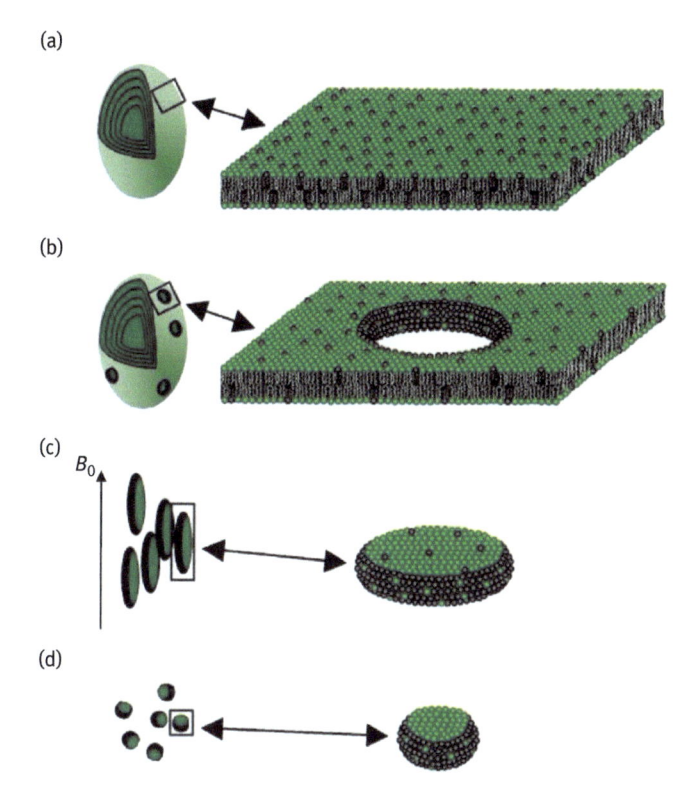

Figure 10.10: Mixtures of long- and short-chain lipids can form a variety of structures, some of which spontaneously orient in a large magnetic field. The shorter chain phosphatidylcholines, shown in grey, tend to form highly curved surfaces while the long-chain lipids, shown in green, prefer a planar structure. For a large ratio, q, of long- to short-chain lipids aggregates from perforated bilayers, at medium values of q (near 3) disk-shaped objects which orient magnetically can form if there is sufficient water. For small values of q, the disks are small and tend to reorient in the magnetic field. Reprinted from Warschawski et al. [71], with permission from Elsevier.

DMPC) is added to DMPC/DCPC bicelles, phase behavior similar to that of DMPC/cholesterol powder mixtures is observed. At high temperatures, the first moment increases as a function of cholesterol concentration and as the temperature is lowered, the first moment increases due to the changes in the molecular order of the system. Figure 10.11 shows the ^{2}H NMR spectra of pure DMPC-d_{54}/DCPC bicelles (a) and DMPC-d_{54}/cholesterol/DCPC bicelles (15 mol% cholesterol relative to DMPC) (b) both with $q = 3.2$ at 317.5 K. Notice that the quadrupolar splittings are increased by the presence of cholesterol in the mixture in (b). In both cases shown, the sample is well aligned and there is a single phase. When the cholesterol concentration was increased to 20 mol% relative to DMPC, we observed the onset of the liquid-ordered phase. At low temperatures, there was a conversion to gel and/or isotropic phases for all cholesterol concentrations. The phase behavior of more

(a)

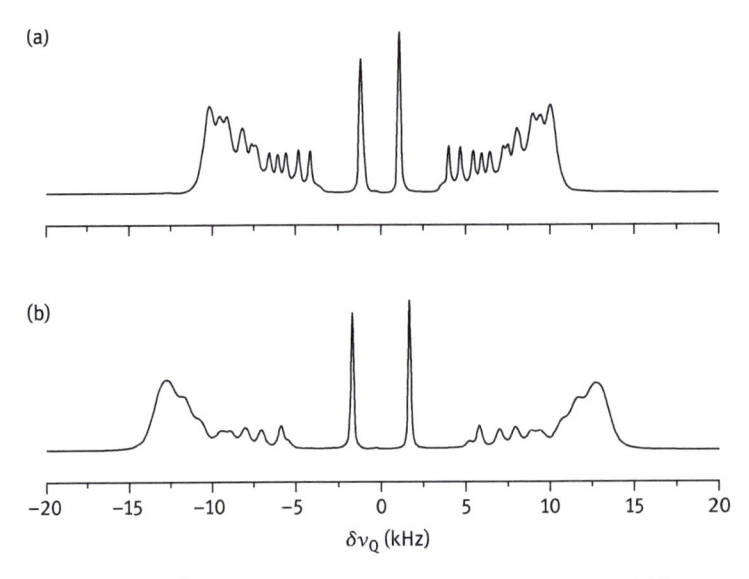

(b)

Figure 10.11: The ^{2}H NMR spectra of (a) DMPC-d_{54}/DCPC bicelles and (b) DMPC-$_{54}$/cholesterol/DCPC bicelles (15 mol% chol with respect to amount of DMPC-d_{54}). The ratio of DMPC-d_{54} to DCPC is $q = 3.2$ in both cases. The spectra were taken at a temperature of 317.5 K.

complex mixtures which include both saturated and unsaturated long-chain lipids and cholesterol has been investigated as well, and an example will be discussed below [74, 79–83].

10.3 Orientation, order, and dynamics of cholesterol in membranes

NMR has provided a lot of information concerning the effect of cholesterol on the behavior of lipids in fluid-phase bilayers. It can also give a detailed description of the behavior of cholesterol itself in a membrane environment. The ^{2}H NMR spectrum of [2,2,3,4,4,6-d_6]-cholesterol in 1,2-dioleoyl- sn-glycero-3-phosphocholine (DOPC) at 273 K is shown in Figure 10.12 [42]. This spectrum can be interpreted assuming that cholesterol undergoes a rapid diffusive motion about its long axis (defined using the principal values of the moment of inertia tensor calculated from the atomic structure) and a slower reorientation of this diffusion axis relative to the bilayer normal. The labeled positions are all in the rigid first ring of cholesterol so that the ratios of the splittings observed can be used to determine the orientation of the diffusion axis as well as the "whole molecule" order parameter S_{mol} [6, 25, 30, 37–39, 41, 42]. A typical value for S_{mol} is 0.85 but values from 0.7 to 0.95 are observed depending on temperature and lipid

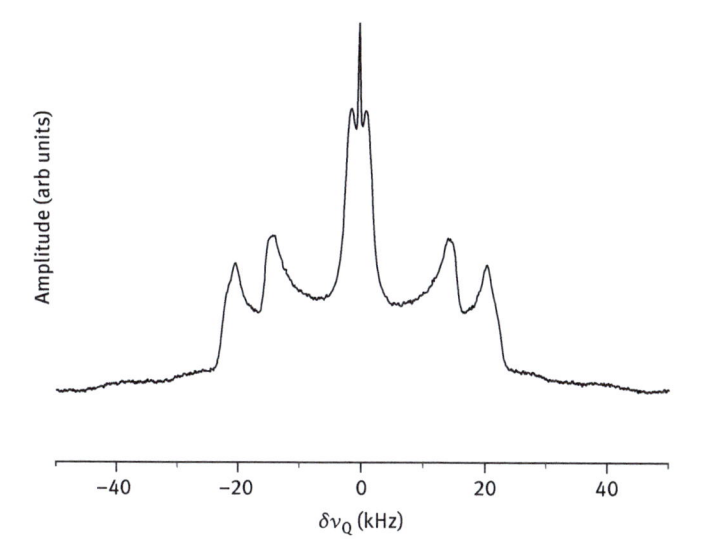

Figure 10.12: The ^{2}H NMR spectrum of [2,2,3,4,4,6-d$_6$]-cholesterol in DOPC (20:80) at 271 K.

composition. From these measurements, we conclude that cholesterol is aligned roughly along the bilayer normal but the value of S$_{mol}$ indicates that it undergoes significant reorientation relative to that direction. To complete the picture, neutron diffraction [36] and paramagnetic ion-induced relaxation [40] show that the cholesterol hydroxyl group is located near the lipid/water interface.

Measurement of the ^{2}H spin lattice relaxation time, T$_1$, of deuterium labeled cholesterol in membranes and liquid crystals has found T$_1$ minima in the temperature range from 20 to 30 °C [6, 38, 44]. This allows us to determine the correlation time for axial reorientation to be roughly $\tau_\parallel \approx 3-4$ ns. The correlation time for the slower reorientation of the long axis relative to the bilayer normal (often described as a wobbling motion) is somewhat longer at about $\tau_\perp \approx 30$ ns [30, 43, 45]. It is this slower motion which leads to the reduction in S$_{mol}$.

The nuclear Overhauser effect (NOE) [84] can be used to study the orientational order and dynamics of cholesterol or other rapidly axially reorienting molecules in a membrane. Since we already have an accurate description of cholesterol's behavior in a bilayer, we can use it as a test case for evaluating the potential for using NOE spectroscopy (NOESY) experiments to obtain this information. Figure 10.13 shows a two-dimensional HETCOR-NOESY spectrum of a sample of [3,4-^{13}C]-cholesterol/DMPC-d$_{54}$/DCPC bicelles with a cholesterol to DMPC-d$_{54}$ molar ratio of 15:85, and a DMPC-d$_{54}$ to DCPC ratio $q = 3.2$ [30]. The two strong parent peaks are for the C$_3$ carbon and its directly bonded hydrogen at [70, 3.5] ppm (the [^{13}C, ^{1}H] chemical shift coordinates) and for the C$_4$ carbon and its two directly bonded hydrogens at [41, 2.2] ppm. In this experiment, the hydrogens which are directly bonded to the labeled carbons, that is, on carbons 3 and 4 of

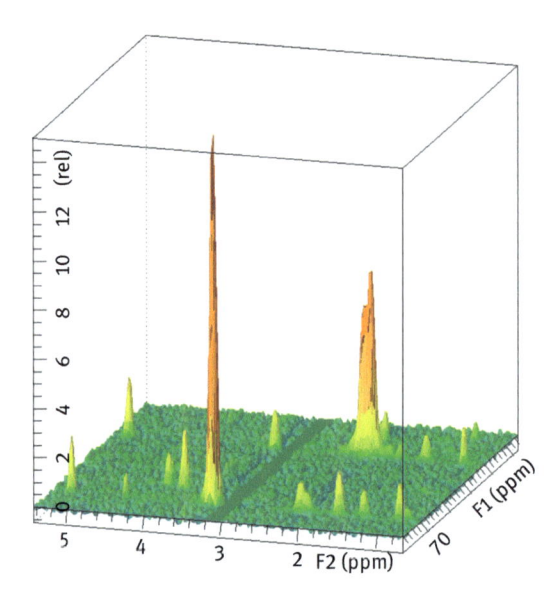

Figure 10.13: ^{1}H detected HETCOR-NOESY spectrum of [3, 4]-2-^{13}C labeled cholesterol in DMPC-d$_{54}$. The spectrum was taken with a NOESY mixing time τ_m = 20 ms at a MAS spinning frequency of v_R = 15 kHz and a temperature of 317.5 K.

cholesterol, are allowed to exchange energy for a time τ_m via mutual spin flips mediated by the ^{1}H–^{1}H dipole–dipole interaction. In the absence of this energy exchange, the only two peaks in the spectrum (aside from a few natural abundance phospholipid peaks) would be those of the parent peaks mentioned above. As the mixing time is increased, a number of daughter peaks grow at ^{1}H chemical shifts corresponding to other positions on the cholesterol molecule. Since the source of this magnetic polarization was the hydrogen bonded to one of the parent carbons, the daughter peaks share the same ^{13}C chemical shift. By fitting the intensities of the parent and daughter peaks as a function of mixing time to the predictions of a simple two-motion model for cholesterol (involving rapid axial reorientation and a slower "wobble" of the reorientation axis relative to the bilayer normal), we can obtain values for the correlation times for the motion and the molecular order parameter S_{mol} which are in complete agreement with those detailed above [30].

The spectrum in Figure 10.13 was obtained using ^{13}C-labeled cholesterol. The spectra in Figures 10.9 and 10.14 were obtained using natural abundance ^{13}C NMR on a sample of unlabeled cholesterol in DMPC-d$_{54}$ (molar ratio of 15:85). In this case, although the signal intensity is much lower, there are a large number of cross-peaks whose intensities can be measured as a function of the NOESY mixing time. Analysis of the complete set of data can be performed to provide the complete set of motional and orientational parameters described above. The same approach can be applied to other types of molecules undergoing similar types of motion within a membrane.

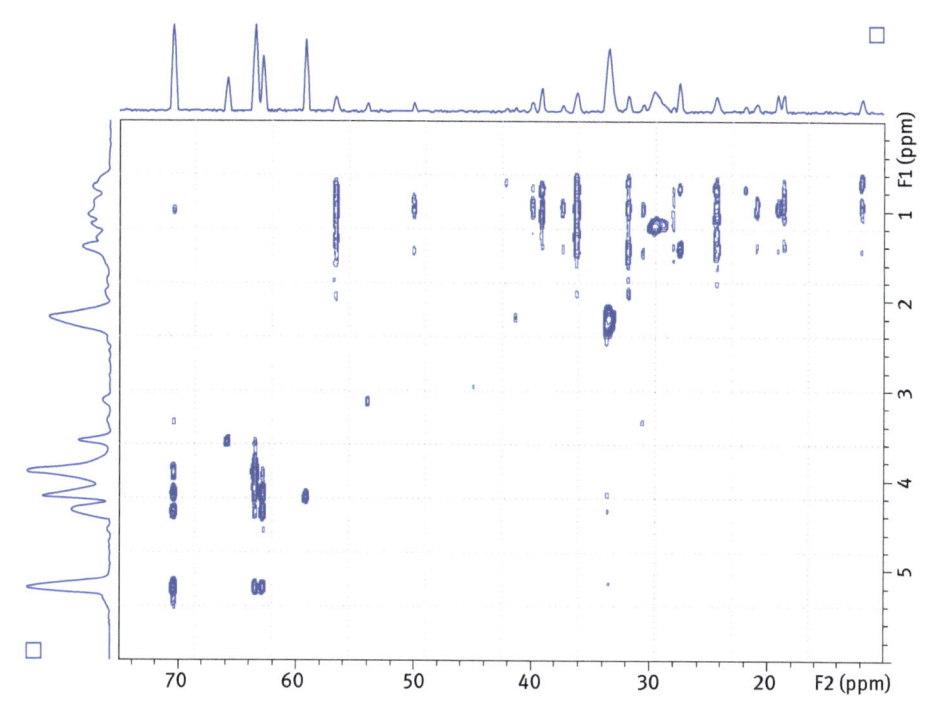

Figure 10.14: ^{13}C detected NOESY-HETCOR spectrum of natural abundance cholesterol in DMPC-d$_{54}$. The spectrum was taken with a NOESY mixing time $\tau_m = 75$ ms at a MAS spinning frequency of $\nu_R = 15$ kHz and a temperature of 317.5 K.

10.3.1 Comparison of cholesterol in saturated and unsaturated lipids

Cholesterol is found in high concentrations in a large number of mammalian cell membranes so it is not surprising that it can be incorporated into a wide variety of different model membrane systems. While cholesterol seems to interact preferentially with longer chain saturated lipids [7] such as DPPC [1, 4] and sphingomyelin [85], it is readily incorporated into bilayers of unsaturated lipids [42, 86]. Figure 10.15 shows ^{2}H NMR spectra of [2,2,3,4,4,6-^{2}H]-cholesterol in oriented bilayers of DOPC (di-C18:1 PC), 1,2-dilinoleoyl- sn-glycero-3-phosphocholine (di-C18:2 PC or DLoPC), 1,2-diarachydonyl- sn-glycero-3-phosphocholine (di-C20:4 PC or DAPC), and 1,2-didocosahexaenoyl- sn-glycero-3-phosphocholine (di-C22:6 PC or DDPC). All spectra are consistent with the same orientation of the cholesterol-ring structure relative to the bilayer normal. The only difference observed is a gradual reduction of the whole molecule order parameter S_{mol} as the level of lipid unsaturation is increased [42].

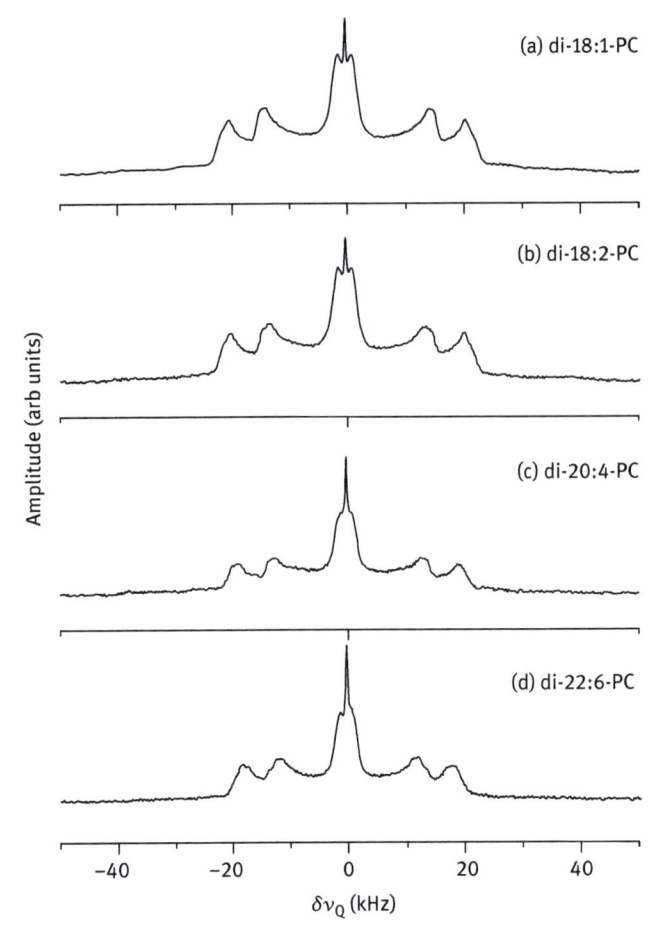

Figure 10.15: The ^{2}H NMR spectra of 20 mol% cholesterol-d_6 in (a) di-18:1-PC (di-oleoyl PC), (b) di-18:2-PC (di-linoleoyl PC), (c) di-20:4-PC (di-arachidonoyl PC) and (d) di-22:6-PC (di-docosahexaenoyl PC). The spectra were taken at a temperature of 271 K.

10.4 Phase behavior in two and three component mixtures

10.4.1 DPPC/cholesterol and DMPC/cholesterol

When cholesterol is added to a single phospholipid, it extends the region of fluidity of the lipid by disrupting the tight packing of the lipid chains within the gel phase. In Figure 10.16, we see several ^{2}H NMR spectra as a function of temperature near the gel-fluid phase transition for DPPC-d_{62} on its own (a) and with 1.25 mol% cholesterol

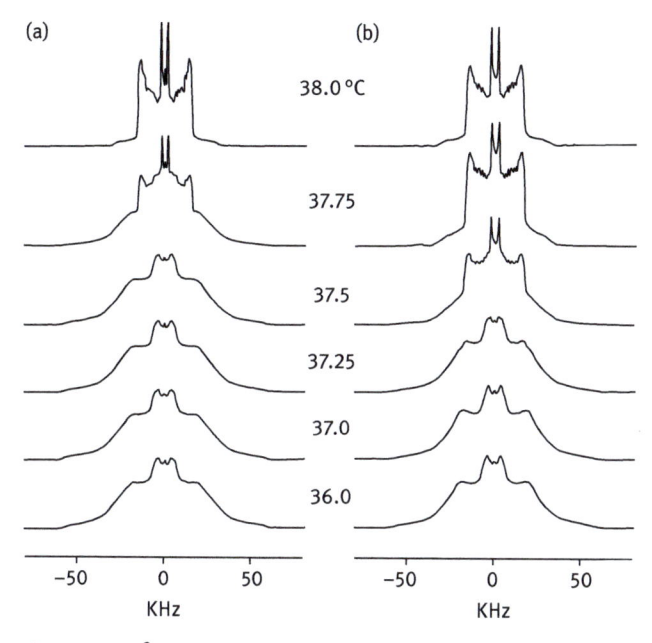

Figure 10.16: ^{2}H NMR spectra of (a) DPPC-d$_{62}$ and (b) 1.25 mol% cholesterol in DPPC-d$_{62}$ for a series of temperatures near the chain melting (fluid to gel) phase transition of DPPC-d$_{62}$. Reprinted from Vist and Davis [1]. Copyright 1990. American Chemical Society.

added (b). Notice that 1.25 mol% cholesterol already results in a measurable difference in the gel-fluid phase transition temperature. The presence of cholesterol not only affects the gel to fluid-phase transition, but also the nature of the fluid-phase behavior. Because cholesterol interacts with the chains of the lipids, it reduces the population of *gauche* conformers (increasing the hydrocarbon chain order of the fluid phase) and can result in the formation of the "liquid ordered" phase. Figure 10.17 shows the phase diagram of fully hydrated binary DPPC-d$_{62}$/cholesterol mixtures [1]. The boundaries of the two-phase gel-liquid ordered (referred to as the β phase in the figure) coexistence region below T_m for DPPC for cholesterol concentration of ~7.5 % and higher were established using spectral subtractions to determine the end-point spectra [1, 87]. The narrow two-phase region for lower cholesterol concentrations was determined by careful inspection of the spectra. Note that the three phase line, as well as the boundaries of the two fluid-phase coexistence region (above T_m) were determined both from differential scanning calorimetry and inspection of the ^{2}H NMR spectra. The existence of this three-phase line near 310 K is key to understanding the phase behavior as it requires three distinct phases to be in equilibrium at that temperature. Other binary lipid/cholesterol systems such as DSPC/cholesterol [88], palmitoyl sphingomyelin (PSM)/cholesterol [85], and DMPC/cholesterol [89] have also been investigated using NMR

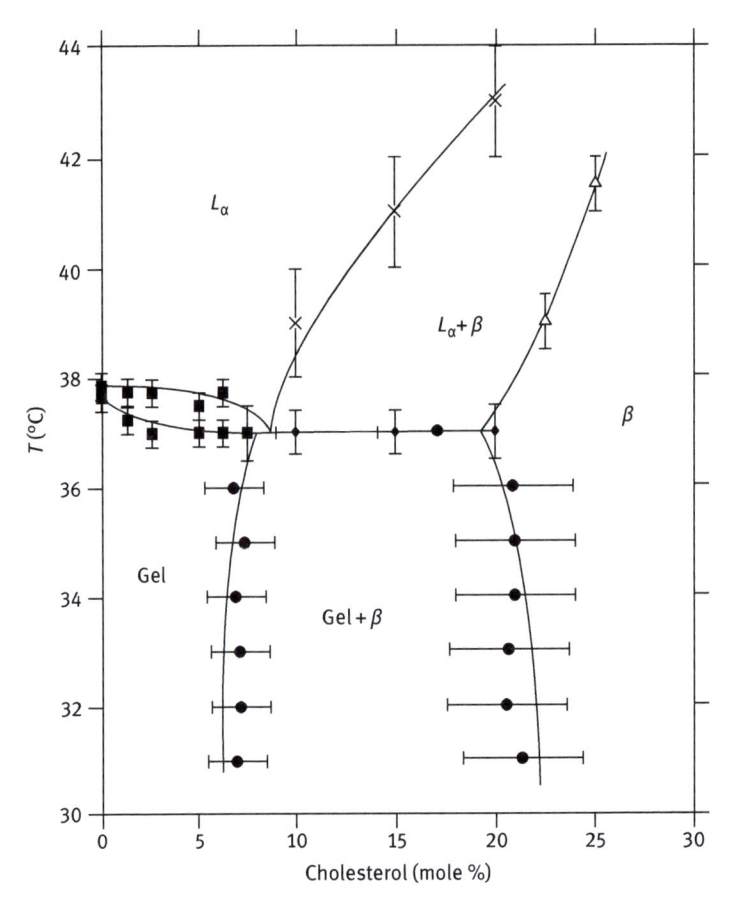

Figure 10.17: A partial temperature/composition phase diagram for mixtures of DPPC-d_{62}/cholesterol in excess water. The L_α phase is frequently referred to as the "fluid" phase or the "liquid-disordered" phase. The β phase is now commonly referred to as the "liquid-ordered" phase. The "gel" phase is a more ordered phase characterized by slow molecular motion. Reprinted from Vist and Davis [1]. Copyright 1990. American Chemical Society.

or other biophysical techniques [24] and show similar phase behavior to the DPPC/cholesterol system.

The effect of cholesterol on bicelles composed of DMPC and the short-chain lipid DCPC has also been investigated [74, 80, 81]. Stable, well-aligned bicelles are formed when cholesterol is added up to ~15 mol% cholesterol (with respect to the long-chain lipid DMPC). The phase behavior of these systems is very similar to that of standard multilamellar dispersions of DMPC/cholesterol. At high temperatures, the addition of cholesterol increases the phospholipid hydrocarbon chain order and, as a result, the first moment of the ^{2}H NMR spectrum. As the temperature is lowered, the first moment increases due to the changes in the orientational averaging, which occur as the temperature is decreased. Indeed, the phase behavior and changes in the first moment of the ^{2}H spectra are very similar to the changes shown in Figure 10.4 for powder DPPC/cholesterol mixtures. When the cholesterol concentration is increased above about 20 mol% relative to DMPC, the onset of the liquid-ordered phase is observed [74].

10.4.2 DOPC/DPPC/cholesterol and DPoPC/DMPC/cholesterol

Ternary mixtures which exhibit liquid ordered/liquid disordered two fluid-phase coexistence can be formed by combining a low melting temperature lipid (such as an unsaturated long-chain phospholipid), a high melting temperature lipid (such as a saturated long-chain phospholipid) and cholesterol. DOPC/DPPC/cholesterol is a well-studied [4, 90] ternary mixture in which two fluid phase coexistence has been observed over a broad range of temperatures and compositions. Figure 10.18 shows a set of nine ^{2}H NMR spectra for various compositions within the two phase ℓ_d-ℓ_o region at 22°C [4]. In two-phase regions, the spectra are actually composed of a superposition of subspectra for each of the two phases. Different sample compositions result in different proportions of the subspectra from the two phases at a given temperature. As the fraction of saturated lipid increases, so does the fraction of ℓ_o phase component in the spectrum. Similarly, for a 1:1 DOPC/DPPC-d_{62} ratio, as cholesterol is added the amount of ℓ_o phase increases.

If we look at the spectra from DOPC/DPPC-d_{62}/cholesterol mixtures, such as those shown in Figure 10.19, more carefully we can see the features arising from each of the two phases. In Figure 10.19, the two inner-dotted vertical lines point out the inequivalence of the very sharp methyl peaks for the two phases, while the two outer-dotted lines indicate the plateau edges of each phase (with the smaller δv_Q corresponding to the ℓ_d phase). The spectra for different compositions contain different proportions of the two subspectra depending on the composition. Since the spectral features in all three spectra shown in Figure 10.19 line up with one another, these three compositions lie on the same tie-line meaning that it is possible to perform pairwise spectral subtractions with them in order to determine the boundaries of the two-phase region in the phase diagram [4]. A phase diagram for the DOPC/DPPC-d_{62}/cholesterol ternary mixture is shown in Figure 10.20. The process for determining this phase diagram is described in detail in Davis et al. [4]. There are several different single-phase, two-phase, and three-phase regions depicted in this diagram. One of the key features of the phase behavior of this ternary system is the line of critical points which is shown in magenta in Figure 10.20. This critical line lies on the boundary between the two fluid-phase coexistence region at lower temperatures and the single-phase region at higher temperatures. At a critical point (any point on this line of critical compositions), the liquid-disordered phase and liquid-

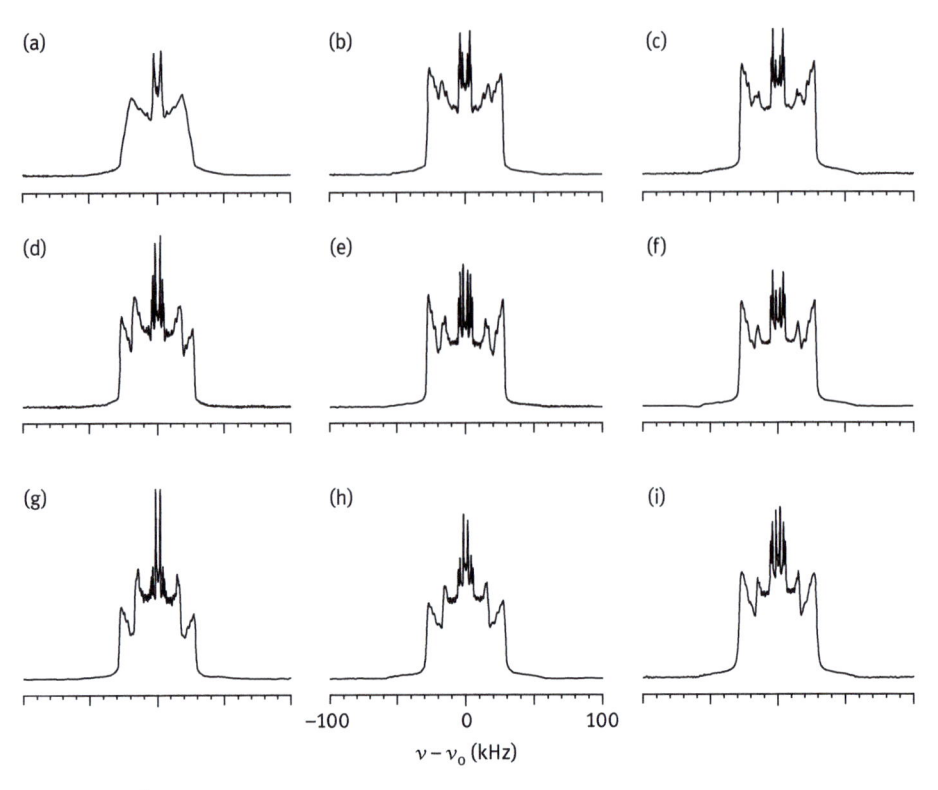

Figure 10.18: ^{2}H NMR spectra of mixtures of DOPC/DPPC-d_{62}/cholesterol in two-phase region at 22° C. The molar proportions are (a) 42:28:30, (b) 35:35:30, (c) 28:42:30, (d) 45:30:25, (e) 37.5:37.5:25, (f) 30:45:25, (g) 48:32:20, (h) 40:40:20, and (i) 32:48:20. Reprinted from Davis, et al. [4], with permission from the Biophysical Society.

ordered phase become indistinguishable. The critical behavior of ternary lipid/cholesterol systems is discussed in more detail below.

As discussed earlier, DMPC is the preferred lipid to work with for making bicelles. The ternary mixture made up of the long-chain unsaturated lipid DPoPC, saturated DMPC and cholesterol is analogous to the longer chain mixture DOPC/DPPC/cholesterol. Indeed, samples composed of DPoPC/DMPC/cholesterol exhibit two fluid phase coexistence as expected. Just as in the DOPC/DPPC/cholesterol mixtures, if the concentration of the saturated lipid in the DPoPC/DMPC/cholesterol mixtures is increased the two-phase region occurs at a higher temperature. Figure 10.21 shows ^{2}H NMR spectra for the three different sample types, multilamellar dispersions, bicelles, and bilayers oriented on glass slides at different temperatures showing both single ℓ_d phase, and ℓ_d-ℓ_o coexistence [74, 91]. Other ternary cholesterol containing systems, including 1-palmitoyl 2-oleoyl- sn-glycero-3-phosphocholine (POPC)/sphingomyelin (SM)/cholesterol [92–95], POPC/DSPC/cholesterol [96], 1-

(a)

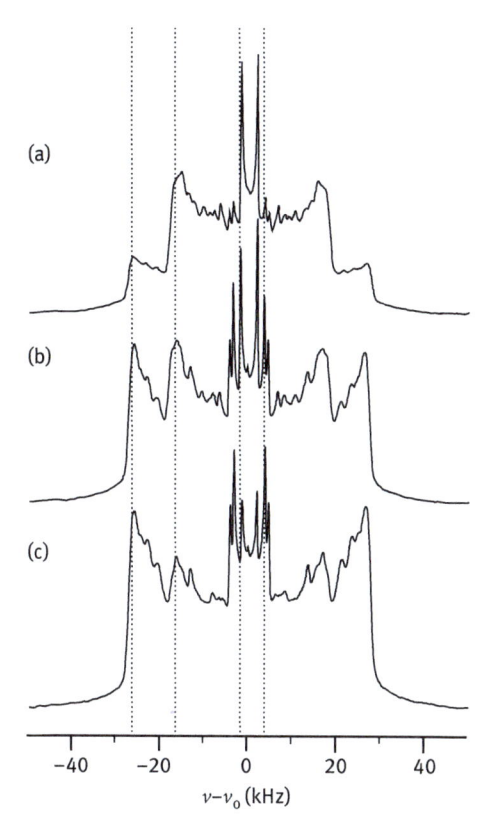

(b)

(c)

−40   −20   0   20   40
$\nu-\nu_0$ (kHz)

Figure 10.19: ^{2}H NMR spectra of DOPC/DPPC-d$_{62}$/cholesterol mixtures at 26°C. The molar proportions are (a) 48:32:20, (b) 37.5:37.5:25, and (c) 28:42:30. The vertical dotted lines indicate either the 90° edges of the methyl group patterns or the "plateau" region of the spectra of either the ℓ_d or ℓ_o phase. Reprinted from Davis et al. [4], with permission from the Biophysical Society.

steroyl 2-oleoyl-it sn-glycero-3-phosphocholine (SOPC)/1,2-disteroyl- sn-glycero-3-phosphocholine (DSPC)/cholesterol [97], 1-palmitoyl-d$_{31}$,2-eicosapentaenoyl-PC (PEPC-d$_{31}$)/SM/cholesterol [98], and 1-palmitoyl-d$_{31}$,2-docosahexaenoyl-PC (PDPC-d$_{31}$)/SM/cholesterol [98], have been investigated.

10.5 Critical behavior in membranes

In the ternary DOPC/DPPC/cholesterol phase diagram described in the previous section, the two coexisting fluid phases, ℓ_d and ℓ_o, differ in composition. The liquid-ordered phase domains are rich in the saturated lipid and cholesterol while the liquid-disordered phase domains are rich in the unsaturated lipid. There is a

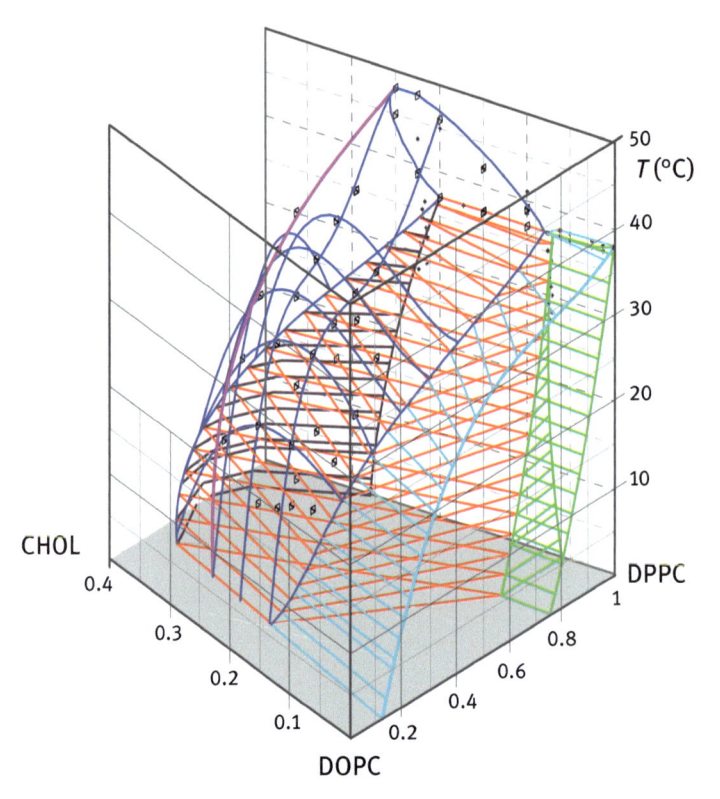

Figure 10.20: A partial temperature/composition phase diagram of the ternary mixture DOPC/ DPPC-d_{62}/cholesterol. Reprinted from Davis et al. [4], with permission from the Biophysical Society.

surface in the phase diagram which separates the ℓ_d-ℓ_o two-phase region below from the one-phase region above it. As the sample temperature is increased, a point is reached where either the ℓ_o phase domains undergo the transition into the ℓ_d phase, and the sample is entirely in the ℓ_d phase, or the ℓ_d phase domains undergo the transition into the ℓ_o phase and the sample is entirely in the ℓ_o phase. Which of these depend upon the sample composition. However, there exists a line of critical points lying on the surface bounding the two-phase region. For sample compositions along this line, as the temperature approaches the surface from below (i.e., at temperatures and compositions directly below the line of critical points), the difference between the ℓ_o and ℓ_d phases becomes insignificant and at and above the critical line, the phases are indistinguishable. For temperatures and compositions near the line of critical points, we observe fluctuations in the composition of the sample both by NMR [4, 90, 99] and by fluorescence microscopy [99, 100]. Figure 10.22 illustrates this effect using confocal fluorescence microscopy (CFM) of giant unilamellar vesicles (GUVs) made of a mixture of 32:48:20 molar proportions of DOPC/DPPC/cholesterol [100]. The pattern of light and dark areas fluctuates with time during the

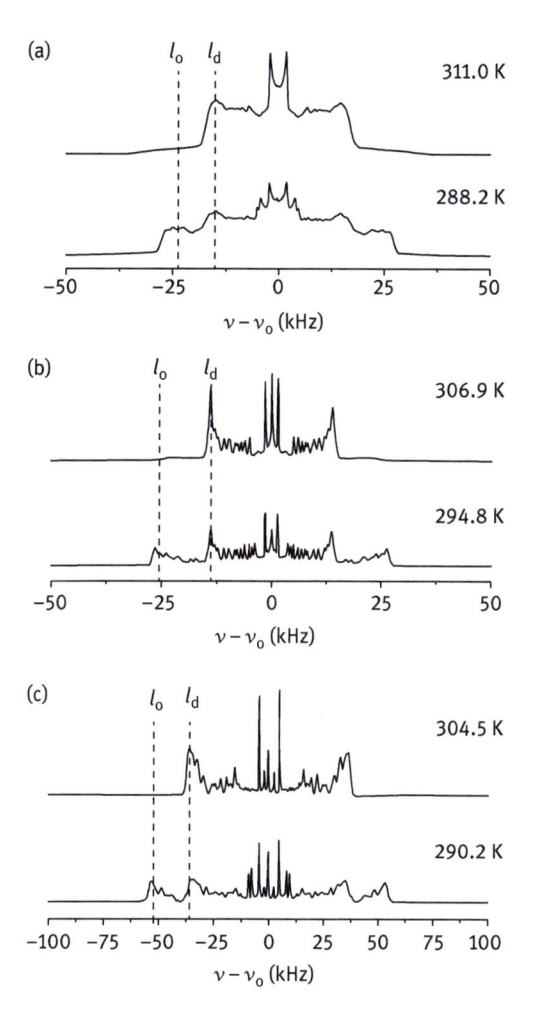

Figure 10.21: Two-phase coexistence in DPoPC/DMPC-d_{54}/cholesterol (32:48:20) (a) multilamellar dispersion, (b) bicelles with DCPC at $q=3.5$, and (c) oriented on glass plates with normal to plates parallel to external magnetic field.

measurements, and the pattern is different from one vesicle to the next [100]. One of the difficulties in using CFM to study these fluctuations is that the images are taken by scanning across the sample and therefore require a certain amount of time to acquire an image (of the order of a second for the images shown). During this scanning process, the pattern is continually changing. Below the critical temperature, there is a much more stable pattern of light and dark domains making quantitative analysis by CFM relatively straightforward and a direct comparison with the analysis of ^{2}H NMR spectra is possible [100].

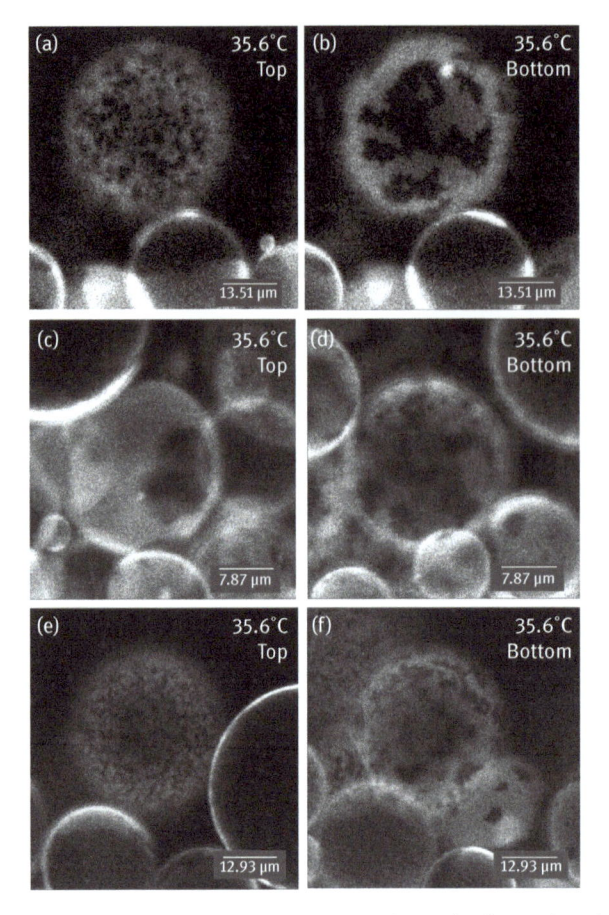

Figure 10.22: Fluorescence microscopy shows the fluctuations in local composition in a sample of (32:48:20) DOPC/DPPC/cholesterol at a temperature above but close to the critical temperature at different times. Reprinted from Juhasz et al. [100], with permission from Elsevier.

NMR has a characteristic timescale as well. There are situations where we can tune that timescale to match the characteristic timescale of the dynamic processes occurring within the sample. The ^{2}H spectra in Figures 10.18 and 10.19 were taken under "static" conditions and the experimental timescale is related to the characteristic features present in the spectra. For example, the vertical lines in Figure 10.19 define the 90° edges in the powder pattern spectra. The frequency difference between the positions of these edges defines an experimental time scale, in this case, the difference is about 5–10 kHz, so the timescale is roughly 100–200 μs. If the DPPC-d_{62} molecules sample both domains during this timescale, then we would see only an average spectrum with the 90° edge in between the two vertical lines in the figure.

Since the molecules do not exchange rapidly between the two types of domain, we see two separate but overlapping spectra.

Figure 10.23(a) shows a similar static spectrum having two distinct but overlapping components corresponding to ℓ_o and ℓ_d domains. In part (b) of the figure, we show a ^{2}H MAS NMR spectrum of the same sample at approximately the same temperature. Under MAS conditions, the NMR spectrum breaks up into a center band (located at the center of the static powder pattern) and a number of spinning side bands separated by the spinning frequency (3 kHz in this example). In Figure 10.24, we focus on the spinning side bands found in the region of the spectrum indicated by the horizontal bracket in Figure 10.23. At temperatures well above the critical point, the side bands are very sharp but as we lower the sample temperature, and approach the critical temperature, the width of the side bands increases dramatically as shown in the figure. Once the sample is below the critical point, the side bands are very sharp again. The width of the side bands depends on the intensity of dynamical fluctuations occurring at frequencies close to the MAS spinning rate [101, 102]. Thus, by changing the spinning rate, we can follow the dynamical changes which occur as the critical point is approached from above.

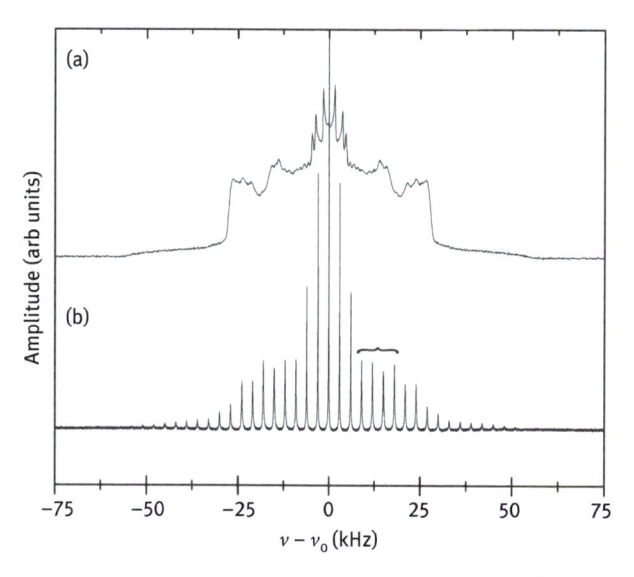

Figure 10.23: Comparison of (a) the static (nonspinning) ^{2}H NMR spectrum of DOPC/DPPC-d_{62}/cholesterol (30:45:25) at 300 K with (b) the MAS NMR spectrum of the same sample spinning at v_R = 3 kHz at 299.7 K. The horizontal bracket in (b) shows the region where we have analyzed the spinning side band line widths as a function of spinning rate. Reprinted from Davis et al. [102], with the permission of AIP Publishing.

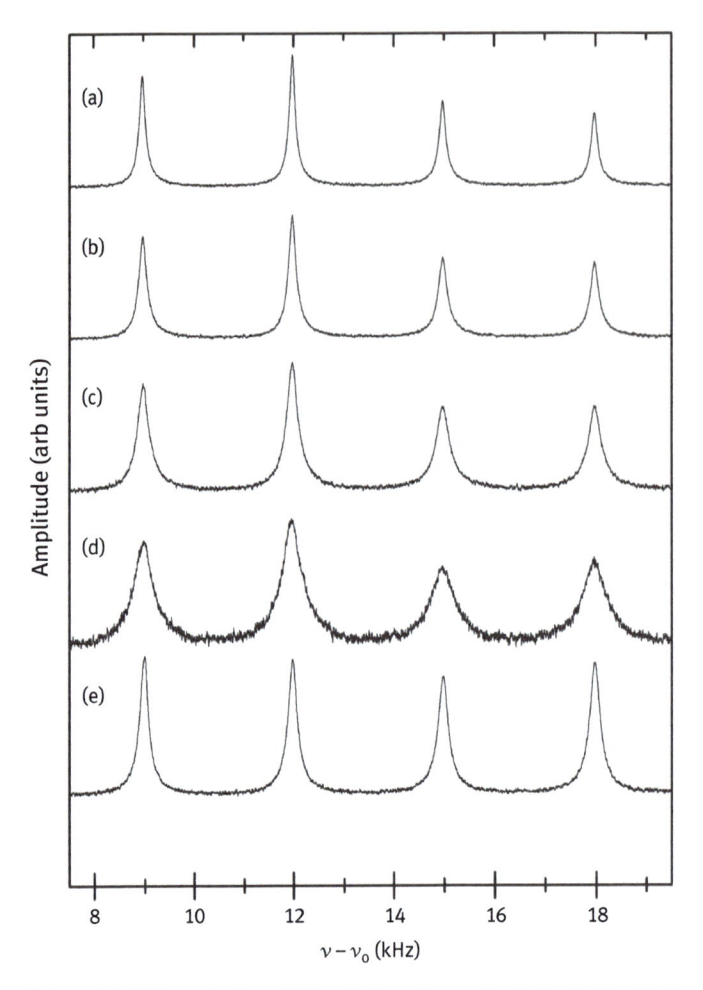

Figure 10.24: The spinning sidebands from the region indicated by the horizontal bracket in Figure 10.23 as a function of temperature while spinning at $v_R = 3$ kHz. (a) T = 12.3 K, (b) 310 K, (c) 308.5 K, (d) 304.8 K, and (e) 299.7 K. Reprinted from Davis et al. [102], with the permission of AIP Publishing.

The theory of critical phenomena predicts that many properties of the system obey simple power law dependences as the critical temperature is approached [103–106]. For example, the correlation length increases according to the power law $\xi = \xi_0 (T/T_c - 1)^{-\nu}$ where ξ_0 is a length which is of the order of the molecular dimensions, T is the absolute temperature, T_c is the critical temperature, and ν is a critical exponent whose predicted value depends on the universality class to which the system belongs [104]. The fluctuations which we observe at temperatures above but close to the critical point are fluctuations in local composition (which serves as the order parameter for the phase transition). The correlation length can be thought of as a measure of the size of regions of uniform composition (order parameter). A timescale for the fluctuations can be defined as the length of time required for molecular self-diffusion either to create or to eliminate the compositional differences represented by those fluctuations. As the correlation length increases (as the critical point is approached), this timescale increases. As discussed above, we expect that the maximum MAS sideband line width for a given spinning rate will occur at the temperature where this characteristic time matches the inverse of the spinning rate. As we lower the spinning rate, the temperature at which this maximum occurs will decrease and approach the critical temperature. Figure 10.25 shows line width data as a function of temperature for four different MAS spinning rates. The solid curves are the results of a simultaneous fit to the expression for the line width [101] to these four data sets. The best–fit yields a value for the critical exponent for the correlation length $v = 0.628 \pm 0.01$ and a critical temperature $T_c = 294.7°K$.

Below the critical point, the order parameter, η, which is defined by the difference in composition of the coexisting phases, is nonzero. The area per lipid, bilayer thickness, and hydrocarbon chain orientational order are closely interrelated and all are sensitive to cholesterol concentration. If we assume that the order parameter η is proportional to the cholesterol concentration then the static ^{2}H NMR spectra allow us to measure directly the temperature dependence of η since, as shown in Figure 10.4 above, the average ^{2}H quadrupolar splitting varies linearly with cholesterol concentration. Figure 10.26 shows a series of ^{2}H NMR spectra of chain methyl labeled DPPC-d_6 in a 37.5:37.5:25 molar proportion mixture of DOPC/DPPC-d_6/cholesterol at temperatures near the critical point. Above T_c, there is a single Pake doublet characteristic of the fluid phase. At temperatures well below the critical point, there are three overlapping powder patterns. One is for the remaining ℓ_d phase (the component with the smallest quadrupolar splitting) and two for the inequivalent chains characteristic of DPPC in the ℓ_o phase. The variations with temperature of the quadrupolar splittings of all components are shown in Figure 10.27. The difference between the average methyl group splitting observed in the ℓ_o phase and the splitting of the ℓ_d phase component is shown in Figure 10.28. It is this difference which we assume is proportional to the critical order parameter η. A fit to the expected power law dependence for the order parameter $\eta \propto [(T_c - T)/T_c]^\beta$, shown by the solid line in the figure, gives a value for the critical exponent $\beta = 0.338 \pm 0.009$ and a critical temperature of $T_c = 303.78 \pm 0.07$ K. Notice that the critical temperature is higher in this case than that determined from the analysis of the data in Figure 10.25 partly due to the lower degree of chain deuteration (the chain melting transition of DPPC-d_6 is close to 314 K while that of DPPC-d_{62} is near 310 K). The sample compositions are also slightly different.

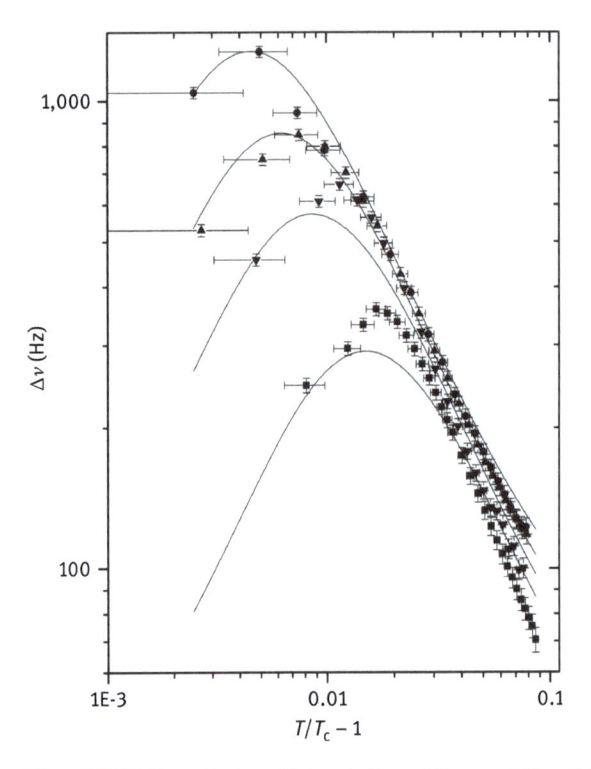

Figure 10.25: The spinning sidebands line widths vs $(T/T_c - 1)$ for four different MAS spinning frequencies: The filled circles are for $v_R = 2.0$ kHz, the upward point triangles are the data at 3.0kHz, the downward pointing triangles are at 4.5kHz and the filled squares are the data at a spinning rate of 9.0kHz. The solid lines are from a simultaneous fit to all the data. The fit gives the correlation length critical exponent $v_c \approx 0.628$ and a critical temperature of $T_c \approx 294.7$ K for the DOPC/DPPC-d_{62}/cholesterol sample with composition 37:37:26. Reprinted from Davis et al. [102], with the permission of AIP Publishing.

Data from fluorescence microscopy experiments on GUVs, which are large unilamellar vesicles, have been interpreted in terms of the 2D Ising model [5, 90, 105, 106]. The results we have shown here, from ^{2}H NMR of multilamellar dispersions, fit much more closely to the 3D Ising model, which predicts a critical exponent for the correlation length $v_{3D} = 0.630$ and for the order parameter $\beta_{3D} = 0.325$. The 2D Ising model predicts values of $v_{2D} = 1$ and $\beta_{2D} = 0.125$. It certainly seems plausible that a unilamellar vesicle will behave more like a 2D system than a multilamellar dispersion. There are a number of experimental challenges involved in obtaining reliable values for critical exponents. In addition to precise temperature control, one of the most important, in the present case of ternary lipid/cholesterol mixtures, is obtaining a sample composition which will actually pass through the line of critical points as the temperature is varied. Having a sample

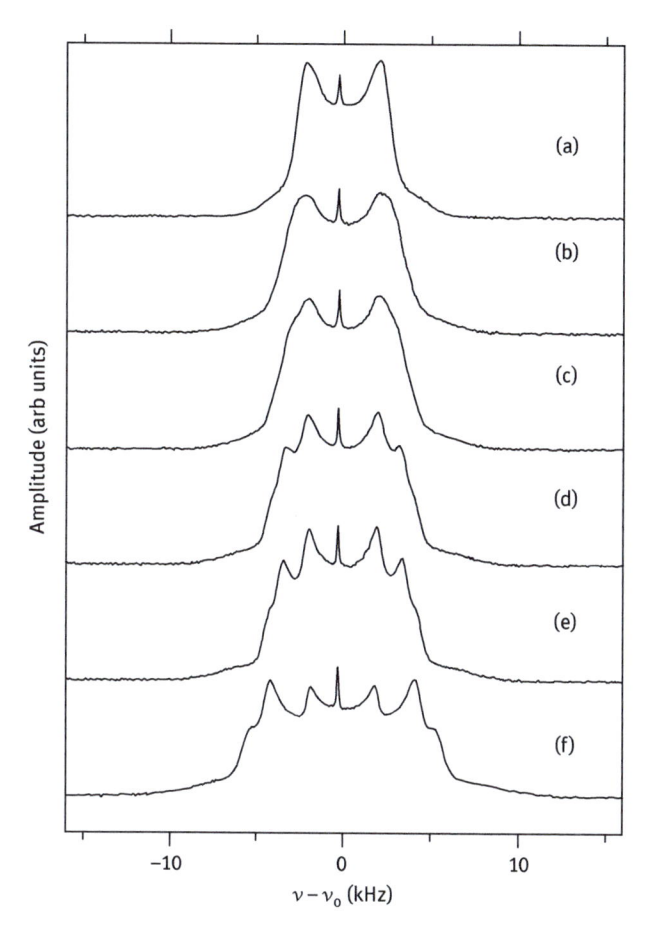

Figure 10.26: Static ^{2}H NMR spectra of chain methyl labeled DPPC-d_6 in a 37.5:37.5:25 molar proportion mixture of DOPC/DPPC-d_6/cholesterol at different temperatures near the critical point. The temperature is (a) 310.4 K, (b) 305.56 K, (c) 303.86 K, (d) 303.38 K, (e) 302.32 K, and (f) 300.85 K. Reprinted from Davis and Schmidt [18], with permission from the Biophysical Society.

composition which is even slightly off can be expected to lead to a significant deviation in the values of the exponents measured.

10.6 Conclusion

NMR studies of mixtures of phospholipids, cholesterol, and water have provided a description of the molecular orientation, order and dynamics of the components of these relatively simple model membranes. Much has also been learned about the influence of cholesterol on the phase behavior and the characteristics of two and

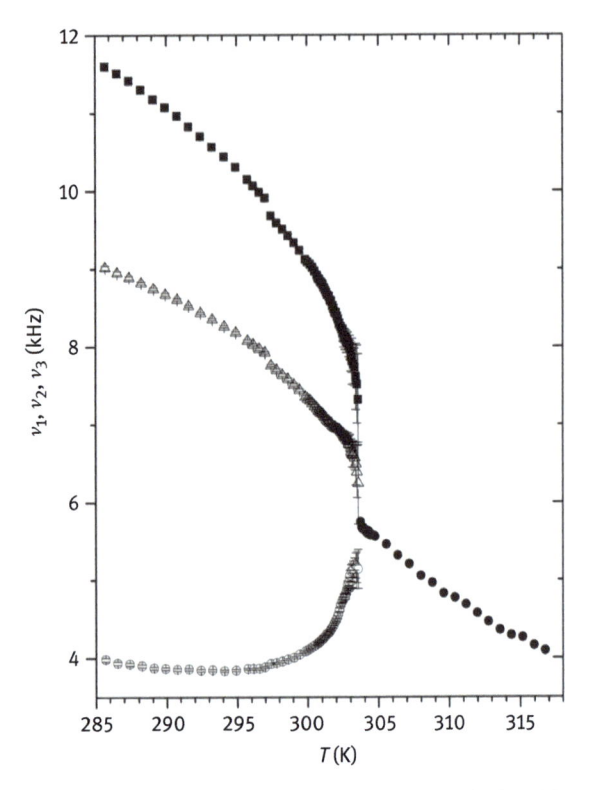

Figure 10.27: Quadrupolar splittings of the DPPC-d_6 ^{2}H NMR spectra of a 37.5:37.5:25 molar proportion mixture of DOPC/DPPC-d_6/cholesterol as a function of temperature. Solid circles are for the ℓ_d component above the critical temperature while the open circles are for the ℓ_d component below T_c. The open triangles and squares are for the ℓ_o component below T_c. Reprinted from Davis and Schmidt [18], with permission from the Biophysical Society.

three component mixtures. Cholesterol greatly increases phospholipid chain order, thickens and stiffens the lipid bilayer but without seriously impeding lateral mobility. At high enough concentrations, cholesterol results in the formation of a distinct liquid-ordered phase which can coexist with either the liquid-disordered or fluid phase of phospholipids above their chain melting transition, or with the much more motionally restricted gel phases typically found below the phospholipid chain melting transition. The position, orientation, and dynamics of cholesterol in phospholipid bilayers are well understood. Further studies using nuclear magnetic relaxation and techniques such as the NOESY-HETCOR experiments on rapidly axially reorienting molecules can refine our knowledge of the roles that cholesterol and other membrane-associated components play. The factors which determine whether and under what conditions a more complex mixture of lipids and cholesterol will form macroscopic domains, rafts, nanodomains or a microemulsion are still unclear and provide a fascinating opportunity for further investigation. Finally, lipid/cholesterol mixtures are rather complex

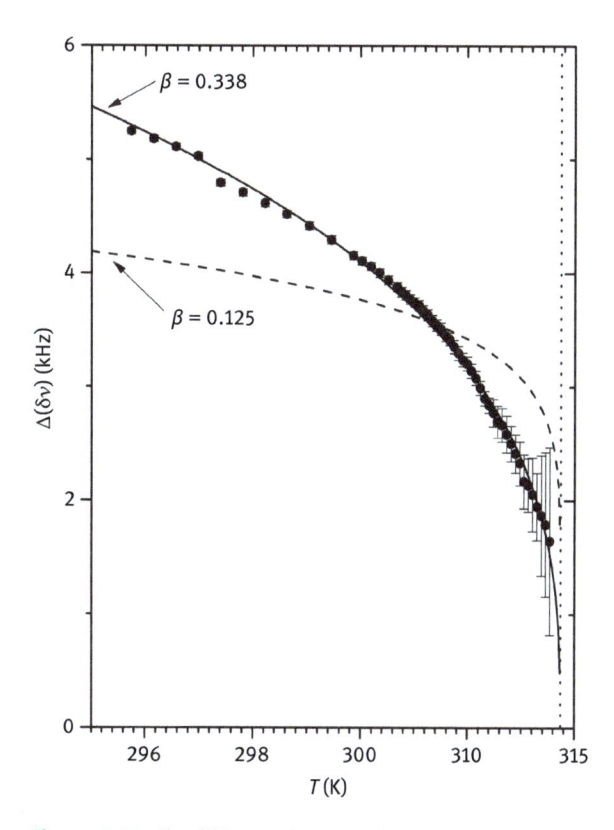

Figure 10.28: The difference between the average methyl group quadrupolar splittings in the ℓ_o phase and the quadrupolar splitting in the ℓ_d phase below the critical temperature for the 37.5:37.5:25 molar proportion sample of DOPC/DPPC-d_6/cholesterol sample. The best-fit, shown by the solid line, gives a critical exponent for the order parameter $\beta = 0.338 \pm 0.009$ and a critical temperature $T_c = 303.784 \pm 0.07$ K. The vertical dotted line indicates the critical temperature. Reprinted from Davis and Schmidt [18], with permission from the Biophysical Society.

physical systems which display critical behavior. Additional careful experimentation and analysis are necessary before we can claim to understand the physics of these systems or its relevance to biological function in membranes.

Acknowledgment

This work was supported by grants from the Natural Sciences and Engineering Research Council of Canada, the Canada Foundation for Innovation and the Ontario Ministry of Research and Innovation. We would like to thank the staff of the University of Guelph NMR Centre for their help with the instrumentation and analysis.

References

[1] Vist, M.R., & Davis, J.H. Phase equilibria of cholesterol/ dipalmitoylphosphatidylcholine mixtures: H nuclear magnetic resonance and differential scanning calorimetry, Biochemistry. 29 (1990) 451464. doi:10.1021/bi00454a021.

[2] Maxfield, F.R., & Tabas, I. Role of cholesterol and lipid organization in disease, Nature. 438 (2005) 612621. doi:10.1038/nature04399.

[3] Veatch, S.L., Gawrisch, K., & Keller, S.L. Closed-loop miscibility gap and quantitative tie-lines in ternary membranes containing diphytanoyl PC, Biophys J. 90 (2006) 44284436. doi:10.1529/biophysj.105.080283.

[4] Davis, J.H., Clair, J.J., & Juhasz, J. Phase equilibria in DOPC/DPPC-d_{62}/cholesterol mixtures, J. Biophys. 96 (2) (2009) 521539. doi:10.1016/j.bpj.2008.09.042.

[5] Veatch, S.L., Polozov, I.V., Gawrisch, K., & Keller, S.L. Liquid domains in vesicles investigated by NMR and fluorescence microscopy, J. Biophys. 86 (5) (2004) 29102922. doi:10.1016/S0006-3495(04)74342-8.

[6] Davis, J.H. NMR studies of cholesterol orientational order and dynamics, and the phase equilibria of cholesterol/phospholipid mixtures, in: B. Maraviglia (Ed.), Physics of NMR Spectroscopy in Biology and Medicine, Soc. Itialiana di Fisica, Bologna, 1988, pp. 302312.

[7] Ipsen, J.H., Karlstrom, G., Mouritsen, O.G., Wennerstrom, H., & Zuckermann, M.J. Phase equilibria in the phosphatidylcholine-cholesterol system, Biophys. Biochim. Acta. 905 (1987) 162172. doi:10.1016/0005-2736(87)90020-4.

[8] Simons, K., & Ikonen, E. Functional rafts in cell membranes, Nature. 387 (1997) 569572. doi:10.1038/42408.

[9] Andersen, O.S., & Koeppe, I.R.E. Bilayer thickness and membrane protein function: An energetic perspective, Annu. Rev. Biophys. Biomol. Struct. 36 (2007) 107130. doi:10.1146/annurev.biophys.36.040306.132643.

[10] Oates, J., & Watts, A. Uncovering the intimate relationship between lipids, cholesterol and GPCR activation, Curr. Opin. Struct. Biol. 21 (2011) 802807. doi:10.1016/j.sbi.2011.09.007.

[11] Sezgin, E., Levental, I., Mayor, S., & Eggeling, C. The mystery of membrane organization: composition, regulation and roles of lipid rafts, Nat. Rev. Mol. Cell Biol. 18 (2017) 361–374. doi:10.1038/nrm.2017.16.

[12] Kumar, G.A., & Chattopadhyay, A. Cholesterol: An evergreen molecule in biology, Biomed. Spectrosc. Imaging. 5 (2016) 855–866. doi:10.3233/BSI-160159.

[13] Paila, Y.D., & Chattopadhyay, A. Membrane cholesterol in the function and organization of G-protein coupled receptors, Subcell. Biochem. 51 (2010) 439–466. doi:10.1007/978-90-481-8622-8_16.

[14] Simons, K., & Toomre, D. Lipid rafts and signal transduction, Nat. Rev. Mol. Cell Biol. 1 (2000) 32–41. doi:10.1038/35036052.

[15] Liu, S.-L., Shent, R., Jung, J.H., Want, L., O'Connor, M.J., Song, S., Bikkavilli, R.K., Winn, R.A., Lee, D., Baek, K., Ueda, K., Levitan, I., Kim, K.-P., & Cho, W. Orthogonal lipid sensors identify transbilayer asymmetry of plasma membrane cholesterol, Nat. Chem. Biol. 13 (2016) 268–274. doi:10.1038/NCHEMBIO.2268.

[16] Lingwood, D., & Simons, K. Lipid rafts as a membrane-organizing principle, Science. 327 (2010) 46–50. doi:10.1126/science.1174621.

[17] Weingarth, M., & Baldus, M. Introduction to biological solid-state NMR, in: F. Separovic & A. Naito (Eds.), Advances in biological solid-state NMR: Proteins and membrane-active peptides, The Royal Society of Chemistry, Cambridge, UK, 2014, pp. 1–17.

[18] Davis, J.H., & Schmidt, M.L. Critical behaviour in DOPC/DPPC/cholesterol mixtures: Static ^{2}H NMR line shapes near the critical point, J. Biophys.. 106 (2014) 1970–1978. doi:10.1016/j.bpj.2014.03.037.
[19] Davis, J.H., & Auger, M. Static and magic angle spinning NMR of membrane peptides and proteins, Prog. NMR Spect. 35 (1999) 1–84. doi:10.1016/50079-6565(99)00009-6.
[20] Molugu, T.R., Xu, X., Leftin, A., Lope-Piedrafita, S., Martinex, G.V., Petrach, H.I., & Brown, M.F. Solid-state deuterium NMR spectroscopy of Membranes, in: G. A. Webb (Ed.), Modern magnetic resonance, Springer Int. Publ. AG, 2017, pp. 1–23. doi:10.1007/978-3-319-28275-6_89-1.
[21] Pfeil, M.-P., & Watts, A. Magnetic resonance studies of antimicrobial peptides in membranes, Amino Acids, Pept. Proteins. 42 (2018) 146–189. doi:10.1039/9781788010627-00146.
[22] Oldfield, E., Chapman, D., & Derbyshire, W. Deuteron resonance: A novel approach to the study of hydrocarbon chain mobility in membrane systems, F. E. B. S. Lett. 16 (1971) 102–104. doi:10.1016/0014-5793(71)80343-5.
[23] Oldfield, E., Chapman, D., & Derbyshire, W. Lipid mobility in *acholeplasma* membranes using deuteron magnetic resonance, Chem. Phys. Lipids. 9 (1972) 69–81. doi:10.1016/0009-3084(72)90034-5.
[24] Marsh, D. Handbook of lipid bilayers, Second Edition, CRC Press, Taylor & Francis Group, Boca Raton, FL, 2013.
[25] Davis, J.H. The molecular dynamics, orientational order, and thermodynamic phase equilibria of cholesterol/phosphatidylcholine mixtures: ^{2}H nuclear magnetic resonance, in: L. Finegold (Ed.), Cholesterol in membrane models, CRC Press, Boca Raton, 1993, Ch. 4, pp. 67–135.
[26] Davis, J.H. The description of membrane lipid conformation, order and dynamics by ^{2}H-NMR, Biochim. Biophys. Acta. 737 (1983) 117–171. doi:10.1016/0304-4157(83)90015-1.
[27] Davis, J.H., Bloom, M., Butler, K.W., & Smith, I.C.P. The temperature dependence of molecular order and the influence of cholesterol in acholeplasma laidlawii membranes, Biochim. Biophys. Acta. 597 (1980) 166–176. doi:10.10116/0005-2736(80)90221-7.
[28] Gopinath, T., Mote, K.R., & Veglia, G. Sensitivity and resolution enhancement of oriented solid-state NMR: Application to membrane proteins, Prog. NMR Spect. 75 (2013) 50–68. doi:10.1016/j.pnmrs.2013.07.004.
[29] Ferreira, T.M., Coreta-Gomes, F., Ollila, O.H.S., Morena, M.J., Vaz, W.L., & Topgaard, D. Cholesterol and POPC segmental order parameters in lipid membranes: solid state ^{1}H-^{13}C NMR and MD simulation studies, Phys. Chem. Chem. Phys. 15 (2013) 1976–1989. doi:10.1039/c2cp42738a.
[30] Davis, J.H., & Komljenovic, I. Nuclear Overhauser effect as a probe of molecular structure, dynamics and order of axially reorienting molecules in membranes, Biochim. Biophys. Acta. 1858 (2016) 395–303. doi:10.1016/j.bbamem.2015.11.016.
[31] Craven, B.M. Crystal structure of cholesterol monohydrate, Nature. 260 (1976) 727–729. doi:10.1038/260727a0.
[32] Weber, H.P., Craven, B.M., Sawzik, P., & McMullan, R.K. Crystal structure and thermal vibrations of cholesteryl acetate from neutron diffraction at 123 and 20 K, Acta Cryst. B47 (1991) 116–127. doi:10.1107/S0108276819009739.
[33] Bocskei, Z., Szendi, Z., & Sweet, F. A side-chain substituted cholesterol analog, Acta Cryst. C52 (1996) 1302–1304. doi:10.1107/S0108270195016295.
[34] Hsu, L.-Y., Kampf, J.W., & Nordman, C.E. Structure and pseudosymmetry of cholesterol at 310 K, Acta Cryst. B58 (2002) 260–264. doi:10.1107/S0108768101018729.
[35] Jambeck, J.P.M., & Lyubartsev, A.P. Another piece of the membrane puzzle. extending lipids further, J. Chem. Theory Comput. 9 (2013) 774–784. doi:10.1021/ct300777p.

[36] Leonard, A., Escrive, C., Laguerre, M., Pebay-Peyroula, E., Neri, W., Pott, T., Katsaras, J., & Dufourc, E.J. Location of cholesterol in DMPC membranes. A comparative study by neutron diffraction and molecular mechanics simulation, Langmuir. 17 (2001) 2019–2030. doi:10.1021/la001382p.

[37] Taylor, M.G., Akiyama, T., & Smith, I.C.P. The molecular dynamics of cholesterol in bilayer membranes: a deuterium NMR study, Chem. Phys. Lipids. 29 (1981) 327–339. doi:10.1016/0009-3084(81)90066-9.

[38] Taylor, M.G., Akiyama, T., Saito, H., & Smith, I.C.P. Direct observation of the properties of cholesterol in membranes by deuterium NMR, Chem. Phys. Lipids. 31 (1982) 359–379. doi:10.1016/0009-3084(82)90072-X.

[39] Dufourc, E.J., Parish, E.J., Chitrakorn, S., & Smith, I.C.P. Structural and dynamical details of cholesterol-lipid interaction as viewed by deuterium NMR, Biochemistry. 23 (1984) 6062–6071. doi:10.1021/bi00320a025.

[40] Villalain, J. Location of cholesterol in model membranes by magic-angle-sample-spinning N. M. R. Eur J. Biochem. 241 (1996) 586–593. doi:10.1111/j.1432-1033.1996.00586.x.

[41] Marsan, M.P., Muller, I., Ramos, C., Rodriguez, F., Dufourc, E.J., Czaplicki, J., & Milon, A. Cholesterol orientation and dynamics in dimyristoylphosphatidylcholine bilayers: a solid state deuterium NMR analysis, J. Biophys. 76 (1999) 351–359. doi:10.1016/S0006-3495(99)77202-4.

[42] Braithwaite, I.M. Investigating the molecular order and orientation of cholesterol in mixtures of polyunsaturated phospholipids. MSc Thesis. 2011. University of Guelph, Guelph, ON, Canada N1G 2W1.

[43] Dufourc, E.J., Parish, E.J., Chitrakorn, S., & Smith, I.C.P. A detailed analysis of the motions of cholesterol in biological membranes by ^{2}H NMR relaxation, Chem. Phys. Lipids. 41 (1986) 123–135. doi:10.1016/0009-3084(86)90004-6.

[44] Bonmatin, J.M., Smith, I.C.P., Jarrell, H.C., & Siminovitch, D.J. Use of a comprehensive approach to molecular dynamics in ordered lipid systems: cholesterol reorientation in oriented lipid bilayers. A deuterium NMR relaxation case study, J. Am. Chem. Soc. 112 (1990) 1697–1704. doi:10.1021/ja00161a007.

[45] Weisz, K., Grobner, G., Mayer, C., Stohrer, J., & Kothe, G. Deuteron nuclear magnetic resonance study of the dynamic organization of phospholipid/cholesterol bilayer membranes: Molecular properties and viscoelastic behavior, Biochemistry. 31 (1992) 1100–1112. doi:10.1021/bi00119a019.

[46] Soubias, O., Jolibois, F., Reat, V., & Milon, A. Understanding sterol-membrane interactions Part II: Complete ^{1}H and ^{13}C assignments by solid-state NMR spectroscopy and determination of the hydrogen-bonding partners of cholesterol in a lipid bilayer, Chem. Eur. J. 10 (2004) 6005–6014. doi:10.1002/chem.200400246.

[47] Oldfield, E., & Chapman, D. Effects of cholesterol and cholesterol derivatives on hydrocarbon chain mobility in lipids, biochem. Biophys. Res. Commun. 43 (1971) 610–616. doi:10.1016/0006-291X(71)90658-9.

[48] Davis, J.H., Jeffrey, K.R., Bloom, M., Valic, M.I., & Higgs, T.P. Quadrupolar echo deuteron magnetic resonance spectroscopy in ordered hydrocarbon chains, Chem. Phys. Lett. 42 (2) (1976) 390–394. doi:10.1016/0009-2614(76)80392-2.

[49] Davis, J.H. Deuterium magnetic resonance study of the gel and liquid crystalline phases of dipalmitoyl phosphatidylcholine, Biophys. J. 27 (1979) 339–358. doi:10.1016/S0006-3495(70)85222-4.

[50] Davis, J.H., Maraviglia, B., Bloom, M., & Godin, D.V. Bilayer rigidity of the erythrocyte membrane: ^{2}H NMR of a perdeuterated palmitic acid probe, Biochim. Biophys. Acta. 550 (1979) 362–366. doi:10.1016/0005-2736(79)90222-0.

[51] Davis, J.H., Bloom, M., Butler, K.W., & Smith, I.C.P. The temperature dependence of molecular order and the influence of cholesterol in *Acholeplasma laidlawii* membranes, Biochim. Biophys. Acta. 597 (1980) 477–491. doi:10.1016/0005-2736(80)90221-7.

[52] Maraviglia, B., Davis, J.H., Bloom, M., Westerman, J., & Wirtz, K.W.A. Human erythrocytes are fluid down to −5° C, Biochim. Biophys. Acta. 686 (1982) 137–140. doi:10.1016/0005-2736(82)90160-2.

[53] Seelig, A., & Seelig, J. The dynamic structure of fatty acyl chains in a phos-pholipid bilayer measured by deuterium magnetic resonance, Biochemistry. 13 (1974) 4839–4845. doi:10.1021/bi00720a024.

[54] Seelig, A., & Seelig, J. Effect of a single *cis* double bond on the structure of a phospholipid bilayer, Biochemistry. 16 (1977) 45–50. doi:10.1021/bi00620a008.

[55] Seelig, J., & Waespe-Sarcevic, N. Molecular order in *cis* and *trans* unsaturated phospholipid bilayers, Biochemistry. 17 (1978) 3310–3315. doi:10.1021/bi00609a021.

[56] Maricq, M.M., & Waugh, J.S. NMR in rotating solids, J. Chem. Phys. 70 (1979) 3300–3316. doi:10.1063/1.437915.

[57] Oldfield, E., Bowers, J.L., & Forbes, J. High resolution proton and carbon-13 NMR of membranes: Why sonicate?, Biochemistry. 26 (1987) 6919–6923. doi:10.1021/bi00396a009.

[58] Davis, J.H., Auger, M., & Hodges, R.S. High resolution ^{1}H nuclear magnetic resonance of a transmembrane peptide, Biophys. J. 69 (1995) 1917–1932. doi:10.1016/S0006-3495(95)80062-7.

[59] Pines, A., Gibby, M.G., & Waugh, J.S. Proton enhanced nuclear induction spectroscopy: A method for high resolution NMR of dilute spins in solids, J. Chem. Phys. 56 (1972) 1776–1777. doi:10.1063/1.1677439.

[60] Sanders, C.R., & Schwonek, J.P. Characterization of magnetically orientable bilayers in mixtures of dihexanoylphosphatidylcholine and dimyristoylphosphatidylcholine by solid-state NMR, Biochemistry. 31 (37) (1992) 8898–8905. doi:10.1021/bi00152a029.

[61] Vold, R.R., & Prosser, R.S. Magnetically oriented phospholipid bilayered micelles for structural studies of polypeptides. Does the ideal bicelle exist?, J. Magn. Reson., Series B. 113 (1996) 267–271. doi:10.1006/jmrb.1996.0187.

[62] Sanders, C.R., & Prosser, R.S. Bicelles: a model membrane system for all seasons?, Structure. 6 (10) (1998) 1227–1234. doi:10.1016/S0969-2126(98)00123-3.

[63] Prosser, R.S., Hwang, J.S., & Vold, R.R. Magnetically aligned phospholipid bilayers with positive ordering: A new model membrane system, Biophys. J. 74 (5) (1998) 2405–2418. doi: http://dx.doi.org/10.1016/S0006-3495(98)77949-4.

[64] Loudet, C., Manet, S., Gineste, S., Oda, R., Achard, M.-F., & Dufourc, E. Biphenyl bicelle disks align perpendicular to magnetic fields on large temperature scales: a study combining synthesis, solid-state NMR, TEM, and SAXS, Biophys. J. 92 (2007) 3949–3959. doi:10.1529/biophysj.106.097758.

[65] Diller, A., Loudet, C., Aussenac, F., Raffard, G., Fournier, S., Laguerre, M., Grelard, A., Opella, S., Marassi, F., & Dufourc, E. Bicelles: a natural molecular goniometer for structural, dynamical and topological studies of molecules in membranes, Biochimie. 91 (2009) 744 751. doi:10.1016/j.biochi.2009.02.003.

[66] Liebi, M., Kohlbrecher, J., Ishikawa, T., Fischer, P., Walde, P., & Windhab, E. Cholesterol increases the magnetic aligning of bicellar disks from an aqueous mixture of DMPC and DMPE DTPA with complexed thulium ions, Langmuir. 28 (2012) 10905–10915. doi:10.1021/la3019327.

[67] Raffard, G., Steinbruckner, S., Arnold, A., Davis, J.H., & Dufourc, E.J. Temperature-composition diagram of dimyristoylphosphatidylcholine-dicaproylphosphatidylcholine "bicelles" self-orienting in the magnetic field. a solid state 2H and 31P NMR study, Langmuir. 16 (20) (2000) 7655–7662. doi:10.1021/la000564g.

[68] Marcotte, I., & Auger, M. Bicelles as model membranes for solid-and solution-state NMR studies of membrane peptides and proteins, Concepts Magn. Reson. Part A. 24 (1) (2005) 17–37. doi:10.1002/cmr.a.20025.

[69] Triba, M.N., Warschawski, D.E., & Devaux, P.F. Reinvestigation by phosphorus NMR of lipid distribution in bicelles, Biophys. J. 88 (2005) 1887–1901. doi:10.1529/biophysj.104.055061.

[70] Prosser, R.S., Evanics, F., Kitevski, J.L., & Al-Abdul-Wahid, M.S. Current applications of bicelles in NMR studies of membrane-associated amphiphiles and proteins, Biochemistry. 45 (2006) 8453–8464. doi:10.1021/bi060615u.

[71] Warschawski, D.E., Arnold, A.A., Beaugrand, M., Gravel, A., Chartrand, E., & Marcotte, I. Choosing membrane mimetics for NMR structural studies of transmembrane proteins, Biochem. Et. Biophys. Acta. 1801 (2011) 1957–1974. doi:10.1016/j.bbamem.2011.03.016.

[72] MacEachern, L., Sylvester, A., Flynn, A., Rahmani, A., & Morrow, M. Dependence of bicellar system phase behavior and dynamics on anionic lipid concentration, Langmuir. 29 (2013) 3688–3699. doi:10.1021/la305136q.

[73] Dürr, U.H.N., Soong, R., & Ramamoorthy, A. When detergent meets bilayer: Birth and coming of age of lipid bicelles, Prog. NMR Spect. 69 (2013) 1–22. doi:10.1016/j.pnmrs.2013.01.001.

[74] Schmidt, M.L., & Davis, J.H. Liquid disordered-liquid ordered phase coexistence in bicelles containing unsaturated lipids and cholesterol, Biochim. Biophys. Acta. 1858 (2016) 619–626. doi:10.1017/j.bbamem.2015.12.016.

[75] Beaugrand, M., Arnold, A., Juneau, A., Gambaro, A., Warschawski, D., Williamson, P., & Marcotte, I. Magnetically oriented bicelles with monoalk- ylphosphocholines: versatile membrane mimetics for nuclear magnetic resonance applications, Langmuir. 32 (2016) 13244–13251. doi:10.1021/acs.langmuir.6b03099.

[76] Katsaras, J., Harroun, T.A., Pencer, J., & Nieh, M.-P. "Bicellar" lipid mixtures as used in biochemical and biophysical studies, Naturwissenschaften. 92 (2005) 355–366. doi:10.1007/s00114-005-0641-1.

[77] Triba, M.N., Devaux, P.F., & Warschawski, D.E. Effects of lipid chain length and unsaturation on bicelles stability. a phosphorus NMR study, Biophys. J. 91 (4) (2006) 1357–1367. doi:10.1529/biophysj.106.085118.

[78] Macdonald, P.M., Saleem, Q., Lai, A., & Morales, H.H. NMR methods for measuring lateral diffusion in membranes, Chem. Phys. Lipids. 166 (2013) 31–44. doi:10.1016/j.chemphyslip.2012.12.004.

[79] Tiburu, E.K., Dave, P.C., & Lorigan, G.A. Solid-state ^{2}H NMR studies of the effects of cholesterol on the acyl chain dynamics of magnetically aligned phospholipid bilayers, Magn. Reson. Chem. 42 (2) (2004) 132–138. doi:10.1002/mrc.1324.

[80] Minto, R., Adhikari,, P., & Lorigan, G.A. ^{2}H solid-state NMR spectroscopic investigation of biomimetic bicelles containing cholesterol and polyunsaturated phosphatidylcholine, Chem. Phys. Lipids. 132 (2004) 55 64. doi:10.1016/j.chemphyslip.2004.09.005.

[81] Lu, J.X., Caporini, M.A., & Lorigan, G.A. The effects of cholesterol on magnetically aligned phospholipid bilayers: a solid-state NMR and EPR spectroscopy study, J. Magn. Reson. 168 (1) (2004) 18–30. doi:10.1016/j.jmr.2004.01.013.

[82] Cho, H.S., Dominick, J.L., & Spence, M.M. Lipid domains in bicelles containing unsaturated lipids and cholesterol, J. Phys. Chem. B 114 (28) (2010) 9238 9245. doi:10.1021/jp100276u.

[83] Schmidt, M.L. NMR studies of liquid disordered and liquid ordered phase coexistence in model membranes, Ph.D. thesis, University of Guelph, Guelph, ON, Canada (2016).

[84] Neuhaus, D., & Williamson, M. The nuclear overhauser effect in structural and C. analysis, V. C. H. Publishers, & N. Y. New York, USA, 1989.

[85] Keyvanloo, A., Shaghaghi, M., Zuckermann, J.J., & Thewalt, J.L. The phase behavior and organization of sphingomyelin/cholesterol membranes: A deuterium NMR study, Biophys. J. 114 114 (2018) 1344–1356. doi:10.1016/j.bpj.2018.01.024.

[86] Kinnun, J.J., Bittman, R., Shaikh, S.F., & Wassall, S.R. DHA modifies the size and composition of raftlike domains: A solid-state ^{2}H NMR study, Biophys. J. 114 (2018) 380–391. doi:10.1016/j.bpj.2017.11.023.

[87] Huschilt, J.C., Hodges, R.S., & Davis, J.H. Phase equilibria in an amphiphilic peptide-phospholipid model membrane by deuterium nuclear magnetic resonance difference spectroscopy, Biochemistry. 24 (1985) 1377–1386. doi:10.1021/bi00327a015.

[88] Huang, T.H., Lee, C.W.B., Das Gupta, S.K., Blume, A., & Griffin, R.G. A carbon-13 and deuterium nuclear magnetic resonance study of phosphatidylcholine/cholesterol interactions: Characterization of liquid-gel phases, Biochemistry. 32 (48) (1993) 13277–13287. doi:10.1021/bi00211a041.

[89] Almeida, P.F.F., Vaz, W.L.C., & Thompson, T.E. Lateral diffusion in the liquid phases of dimyristoylphosphatidylcholine/cholesterol lipid bilayers: a free volume analysis, Biochemistry. 31 (29) (1992) 6739–6747.

[90] Veatch, S.L., Soubias, O., Keller, S.L., & Gawrisch, K. Critical fluctuations in domain-forming lipid mixtures, Proc. Natl. Acad. Sci. (USA). 104 (2007) 17650–17655. doi:10.1073/pnas.0703513104.

[91] Schmidt, M.L., & Davis, J.H. Liquid disordered - liquid ordered phase coexistence in lipid/cholestrol mixtures: A deuterium 2D NMR exchange study, Langmuir. 33 (2017) 1881–1890. doi:10.1021/acs.langmuir.6b02834.

[92] Bartels, T., Lankalapalli, R.S., Bittman, R., Beyer, K., & Brown, M.F. Raftlike mixtures of sphingomyelin and cholesterol investigated by solid-state ^{2}H NMR spectroscopy, J. Am. Chem. Soc. 130 (44) (2008) 14521–14532. doi:10.1021/ja801789t.

[93] de Almeida, R.F.M., Fedorov, A., & Prieto, M. Sphingomyelin/phosphatidylcholine/cholesterol phase diagram: boundaries and composition of lipid rafts, Biophys. J. 85 (4) (2003) 2406–2416. doi:10.101l6/50006-3495(03)74664-5.

[94] Veatch, S.L., & Keller, S.L. Miscibility phase diagrams of giant vesicles containing sphingomyelin, Phys. Rev. Lett. 94 (2005) 148101.

[95] Ionova, I.V., Livshits, V.A., & Marsh, D. Phase diagram of ternary cholesterol/palmitoylsphingomnyelin/palmitoyloleoyl-phosphatidylcholine mixtures: spin-label EPR study of lipid-raft formation, Biophys. J. 102 (2012) 1856–1865. doi:10.101l6/j.bpj.2012.03.043.

[96] Zhao, J., Wu, J., Shao, H., Kong, F., Jain, N., Hunt, G., & Feigenson, G. Phase studies of model biomembranes: macroscopic coexistence of Lalpha + Lbeta, with light-induced coexistence of Lalpha + Lo phases, Biochim. Biophys. Acta. 1768 (2007) 2777–2786. doi:10.1016/j.bbamem.2007.07.009.

[97] Heberle, F.A., Wu, J., Goh, S.L., Petruzielo, R.S., & Feigenson, G.W. Comparison of three ternary lipid bilayer mixtures: FRET and ESR reveal nanodomains, Biophys. J. 99 (2010) 3309–3318. doi:10.1016/j.bpj.2010.09.064.

[98] Williams, J.A., Batten, S.E., Harris, M., Rockett, B.D., Shaikh, S.R., Stillwell, W., & Wassell, S.R. Docosahexaenoic and eicosapentaenoic acids segregate differently between raft and nonraft domains, Biophys. J. 103 (2012) 228–237. doi:10.1016.j.bpj.2012.06.016.

[99] Veatch, S.L., Polozov, I.V., Gawrisch, K., & Keller, S.L., Liquid domains in vesicles investigated by NMR and fluorescence microscopy, J. Biophys.. 86 (2004) 2910–2922. doi:10.1016/S0006-3495(04)74342-8.

[100] Juhasz, J., Sharom, F.J., & Davis, J.H. Quantitative characterization of coexisting phases in DOPC/DPPC/cholesterol mixtures: Comparing confocal fluorescence microscopy and

deuterium nuclear magnetic resonance, Biochim. Biophys. Acta. 1788 (2009) 2541–2552. doi:1016/j.bbamem.2009.10.006.

[101] Suwelack, D., Rothwell, W.P., & Waugh, J.S. Slow molecular motion detected in the NMR spectra of rotating solids, J. Chem. Phys. 73 (1980) 2559–2569. doi:10.1063/1.440491.

[102] Davis, J.H., Ziani, L., & Schmidt, M.L., Critical fluctuations in DOPC/DPPC-d_{62}/cholesterol mixtures: ^{2}H magnetic resonance and relaxation, J. Chem. Phys. 139 (2013) 045104. doi:10.1063/1.4816366.

[103] Cardy, J. Scaling and renormalization in statistical physics, Cambridge University Press, Cambridge, UK, 1996.

[104] Goldenfeld, N. Lectures on phase transitions and the renormalization group, Westview Press, Boulder, CO, USA, 1992.

[105] Honerkamp-Smith, A.R., Veatch, S.L., & Keller, S.L. An introduction to critical points for biophysicists; observations of compositional heterogeneity in lipid membranes, Biochim. Biophys. Acta. 1788 (2009) 53–63. doi:10.1016/j.bbamem.2008.09.010.

[106] Honerkamp-Smith, A.R., Cicuta, P., Collins, M.D., Veatch, S.L., Den Nijs, M., Schick, M., & Keller, S.L. Line tensions, correlation lengths, and critical exponents in lipid membranes near critical points, Biophys. J. 95 (2008) 236–246. doi:10.1529/biophysj.107.128421.

José Carlos Bozelli Junior, Richard M. Epand
11 Study of mitochondrial membrane structure and dynamics on the molecular mechanism of mitochondrial membrane processes

Abstract: Mitochondria play a crucial role in ATP generation and cellular energy homeostasis as well as being recognized for their other vital functions in various biosynthetic and cell signaling pathways. It is now accepted that mitochondria were acquired in an endosymbiotic event about 2 billion years ago when an alpha-proteobacterium was engulfed by a eukaryotic progenitor. Like their bacterial ancestor, mitochondria are composed of two separate and functionally distinct outer and inner membranes (MOM and MIM, respectively) that encapsulate the intermembrane space and matrix compartments, respectively. The presence of two distinct bilayers makes the mitochondrial membrane structure complex. The major phospholipid composition of the smooth MOM of most mammalian cells is phosphatidylcholine (PC), phosphatidylethanolamine (PE), and phosphatidylinositol (PI). On the other hand, the highly folded MIM is enriched in nonbilayer forming lipids such as cardiolipin (CL) and PE. In addition, the physical interaction of MOM with other organelles (ER and peroxisomes) and MIM allows a highly dynamic trafficking of molecules among these locations, adding an additional intricacy to the system. Here we review the structure of mitochondrial membranes as well as the dynamics of its lipid constituents and how these are involved in the molecular mechanisms of mitochondrial membrane processes in pathophysiological conditions. Furthermore, a discussion is made on how the lipid-phase state, driven by membrane curvature or other properties related to the packing of lipids, can lead to substrate acyl chain specificity of a MIM enzyme, tafazzin. Finally, we present some of the extensive information one can get about the membrane structure and dynamics using the spectroscopic technique, nuclear magnetic resonance (NMR), focusing on the study of mitochondrial membranes.

Keywords: Transmembrane phospholipid asymmetry, mitochondria, tafazzin, cardiolipin, cristae, Barth Syndrome, NMR, lipid phase polymorphism

---

José Carlos Bozelli Junior, Richard M. Epand, Department of Biochemistry and Biomedical Sciences, McMaster University, Ontario, Canada

https://doi.org/10.1515/9783110544657-011

11.1 Mitochondria

It is currently and widely acknowledged that mitochondria were acquired when an alpha-proteobacterium was engulfed by a eukaryotic progenitor around 2 billion years ago [1]. The two primitive organisms lived in symbiosis. The eukaryotic ancestor was endowed with the ability to generate higher amounts of energy, whereas the bacterial one gained a layer of protection in addition to a continuous flux of substrate. The evolution of the primitive alpha-proteobacterium led to what we currently know as the mitochondrion. Mitochondria, like their bacterial ancestor, are organelles composed of two separate and functionally distinct outer and inner membranes (mitochondrial outer membrane [MOM] and mitochondrial inner membrane [MIM], respectively) that encapsulate the intermembrane space and matrix compartments, respectively (Figure 11.1) [2].

In a cellular context, the energy needed for anabolism is stored in the form of high-energy molecules. ATP (Adenosine triphosphate) is the major form of storing energy within a cell. In 1961, Mitchell proposed the chemiosmotic theory in which the oxidation of nutrients via metabolic pathways was coupled to the generation of an electrochemical gradient across a membrane, which in turn used the energy of this gradient to drive ATP synthesis [3]. This process, known as oxidative phosphorylation, was for long thought to be the only function carried out by mitochondria in aerobic eukaryotic cells. Indeed, mitochondria through oxidative phosphorylation are responsible for the generation of more than 80% of the energy demands within a cell in normal conditions [4, 5].

Although we cannot disregard the importance of mitochondria as "the powerhouse of the cell," we now understand that this organelle presents a highly dynamic network of

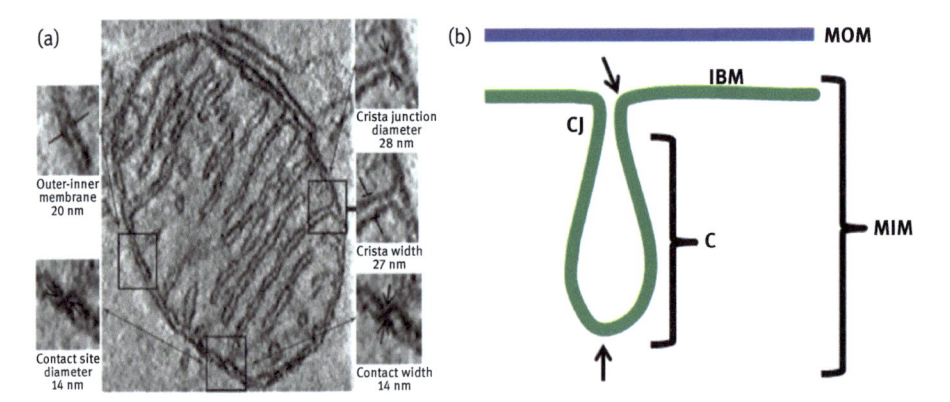

Figure 11.1: Mitochondrion membrane architecture. (a) A section of a 3D tomogram of chicken cerebellum mitochondrion. Examples of its structural features enlarged in the left and right panels with its average dimensions. Figure was taken from Ref [2]. and used with permission of Elsevier. (b) A cartoon highlighting some aspects of the organization of mitochondrial membranes. In blue, mitochondrion outer membrane (MOM). In green, mitochondrion inner membrane (MIM) with its structural features: inner boundary membrane (IBM), cristae junctions (CJ), and cristae (C). Arrows point to regions where membrane presents high curvature.

interactions and, therefore, is involved in several other crucial cellular functions. This includes contributing to Ca^{2+} homeostasis and signaling, important metabolic pathways (amino acids, lipids, and iron), generation and control of reactive oxygen species (ROS), mitophagy (a mitochondrial quality control used to keep health cells), aging, and cell death (necrosis and apoptosis) [5]. In this context, it is reasonable to think that mitochondria dysfunction can lead to pathological conditions. Indeed this is the case, mitochondria has been reported to play a role in age-related disorders, cardiac pathologies, diabetes, neurodegenerative disorders, and various forms of cancer [5–7]. Mitochondrial membrane lipids have an important role in all health and disease states.

From the late nineteenth century until 1970's, there were significant advancements in the understanding of membrane structure and dynamics. In their seminal work, Singer and Nicolson proposed a model of the membrane architecture, the fluid mosaic model, in which the membrane is depicted as "a matrix made up of a mostly fluid bilayer of phospholipids with mobile globular integral membrane proteins and glycoproteins that are intercalated into the fluid lipid bilayer" [8, 9]. Although this model remained for four decades of active research on membranes, it describes lipids playing a rather passive role. We now understand that lipids (and, therefore, membranes) do not only play a structural role but rather actively participate in signaling processes. The role of lipids in signaling can occur because of (i) an increase/decrease in the concentration of a particular lipid species, (ii) the presence/formation of lipid domains in the plane of the membrane, (iii) change in transversal lipid asymmetry, and (iv) membrane curvature modulation [10–13]. A remarkable example of how membranes through their structure and dynamics actively participate in biological processes is the case of mitochondrial membranes. In this chapter, we will review the structure of mitochondrial membranes as well as the dynamics of their lipid constituents in health and disease states. First, we will review mitochondrial membrane architecture, their phospholipid composition and asymmetry, and how their lipids move between mitochondrial membranes as well as between mitochondria and other organelle membranes. In section 11.3, we will give emphasis on how membrane properties can drive substrate acyl chain specificity of a mitochondrial enzyme, tafazzin. We conclude this chapter by describing how one can use nuclear magnetic resonance (NMR) to retrieve structural and dynamic properties of biological and model membranes with a focus on mitochondrial membranes (see section 11.4).

11.2 Mitochondrial membranes

11.2.1 Architecture

Mitochondria are double membrane-enclosed organelles (Figure 11.1). The presence of two distinct bilayers makes the mitochondrial membrane structure complex. The MOM is the frontier between the organelle and its environment and, therefore, controls the entry and exit of metabolites, ions, lipids, and mitochondrial proteins. The integrity of the MOM is vital for the cell. In the mitochondrial pathway of apoptosis, formation of pores in the MOM by the Bcl-2 family of proteins releases proapoptotic proteins (like, cytochrome c, a hemeprotein that is the final protein of the electron transport chain), which, in turn, leads to signaling cascades committing the cell to die [14]. MOM protein/phospholipid ratio is lower than that of the MIM (Table 11.1) [15]. The major protein component of the MOM is VDAC, a voltage-dependent anion channel, which allows the passage of small anionic molecules. Although mitochondria have their own DNA, only 1% of mitochondrial proteins are encoded by mitochondrial DNA [16]. The remainder of its proteins are encoded by the nuclear genome and must, therefore, be imported into the mitochondria. Two major protein complexes (sorting and assembly machinery [SAM], and translocase of the outer mitochondrial membrane [TOM]) present in MOM are the major entry of nuclear-encoded proteins into mitochondria [17]. The lipid-rich MOM surface is smooth; however, its shape changes. This is a consequence between the balance of fusion and fission events mitochondria undergo [6]. Hyperfusion leads to an interconnected mitochondrial network presenting a rod-shaped morphology, whereas hyperfission results in smaller and spherical mitochondria [6]. Impairment of mitochondrial fusion and fission machinery has been associated with several human diseases [18]. Membrane fusion and fission events involve changes in membrane curvature [19].

The MIM is richer in protein than the MOM (Table 11.1). The majority of the proteins embedded into MIM takes part in the oxidative phosphorylation process. In this process, a series of redox reactions catalyzed by protein complexes in the MIM lead to protons being pumped from the mitochondrial matrix to the intermembrane space. The result is the build-up of an electrochemical gradient, the free energy of which is used to drive ATP synthesis through ATP synthase. Several pathophysiological conditions have been reported to either cause or result in a more "leaky" MIM, which results in the breakdown of the electrochemical gradient [5, 7, 20, 21]. MIM is also a second barrier for nuclear-encoded proteins and metabolites. After transposing MOM, mitochondrial proteins can either interact with the TIM 22 (translocase of the inner membrane of mitochondria, built

Table 11.1: Compositional analysis of mitochondrial membrane components. Values for the protein/lipid, saturated/unsaturated acyl chains, and cholesterol/phospholipid ratios of acyl chains, and cholesterol/phospholipid ratio are presented for mitochondrial outer membrane (MOM) and mitochondrial inner membrane (MIM) as well as the contact sites derived from those membranes, MOM (CS) and MIM (CS). The table was built using the values reported in Ref [15].

	MOM	MOM (CS)	MIM (CS)	MIM
Protein/lipid (w/w)	1.7	2.1	2.2	2.9
Saturated/unsaturated	1.33	0.97	0.80	0.78
Cholesterol/phopholipid (mol/mol)	0.092	0.099	0.004	0.025

around TIM 22 subunit) or TIM23 (a complex formed from the channel TIM23) complexes for integration into MIM or having access to the mitochondrial matrix, respectively; while metabolites such as ADP and ATP pass through MIM using the ADP/ATP carrier [17, 22].

One of the most striking characteristics of MIM is its morphology. This mitochondrial membrane presents a larger surface area than MOM (in the liver cell, it can constitute up to one-third of total cell membrane [23]) and contains structural features referred as cristae (Figure 11.1). For a long time, cristae have been thought of as random MIM wide infoldings. However, more recently the use of electron tomography allowed a better description of these structural features, which are now acknowledged to be internal compartments formed by MIM invaginations [2, 24]. The connection of the cristae with the boundary region of MIM (the MIM region that parallels MOM) is made through cristae junctions, which are narrow (20–50 nm) neck-like membrane segments (Figure 11.1). The membrane region in the cristae and cristae junctions presents high curvature (Figure 11.1). The membrane curvature in these regions is likely formed by a fine-tuned interplay between protein and lipid components. It has been shown that mutations in yeast ATP synthase hampering its dimerization presents altered cristae morphology [25]. More recently, a combined cryo-electron microscopy and X-ray crystallography study showed that ATP synthase dimer structure induces a strong membrane curvature [26]. This protein is an integral membrane protein and when dimerized the angle between the two monomers has been shown to be ~ 100°, which, in turn, imposes a high curvature on the membrane [26]. Mutations in proteins involved in MIM fusion and fission processes have also resulted in morphological changes in cristae [27, 28]. In addition, the presence of a large protein complex, named MINOS/MICOS/MitOS, has been implicated in the stability of crista junctions [22]. A thermodynamic model has been proposed simply based on the theory of lipid spontaneous curvature [29]. In this model, the formation of crista junctions has been described as a consequence of a higher rate of MIM biogenesis in comparison to the one presented by MOM. In studies with liposomes containing the mitochondrial lipid cardiolipin (CL) and in the absence of proteins, it has been shown that a pH gradient (as occurs in MIM) can lead to the formation of cristae-like structures, which was later ascribed to a pH-induced modulation of lipid packing in these liposomes [30, 31]. Finally, molecular dynamic simulations of lipid bilayers showed that an asymmetric distribution of CL, and also likely PE, (as occurs in biological MIM, see section 11.2.2) induces membrane curvature, which can also take part in the process of cristae and cristae junction formation [32].

The amazing complex architecture of mitochondrial membranes is strengthened by the presence of physical interactions between membranes. The points of physical interaction between two membranes are called contact sites. Mitochondrial contact sites are believed to play a role in protein, lipid, and ion channeling as well as in mitochondria fusion and fission events and apoptosis [33–35]. As the outermost membrane, MOM is the main player in the formation of mitochondrial contact sites. MOM forms contact sites with MIM as well as membranes of other cellular organelles, that is, endoplasmic reticulum (ER) and peroxisomes. The first evidence of MOM–MIM contact sites came from transmission electron microscopy studies of chemically fixed and dehydrated mitochondrial specimens [36]. Although originally thought to be a direct physical contact between MOM and MIM, studies with frozen-hydrated mitochondria showed the presence of 10–15 nm particles that were bridging MOM and MIM at contact sites [37–40]. At least two protein complexes are known to be responsible for bridging MOM and MIM. The oligomeric form of two mitochondrial kinases, the nucleoside diphosphate kinase D (NDPK-D/Nm23-H4) and mitochondrial creatine kinases (MtCK), were reported to be enriched in MOM–MIM contact sites as well as presenting the ability to bridge two membranes pointing to their role in the formation of MOM–MIM contact sites [41, 42]. The MINOS/MICOS/MitOS protein complex has also been proposed to bridge MOM and MIM as a result of their interaction with MOM proteins (TOM and SAM complexes, VDAC, and proteins of the fusion machinery) [22]. The use of electron microscopy has also allowed the identification of MOM–ER contact sites, which by far are the most well-characterized membrane contact sites between mitochondria and other organelles [43, 44]. In yeast the tether between MOM and ER membrane is mediated by a multi-protein complex named ERMES, ER-mitochondria encounter structure, which is composed of MOM and ER proteins as well as cross-linking and regulatory proteins [44]. In mammals the identity of the proteins bridging those membranes are less well described; however, it is believed to occur either through direct interaction between MOM and ER proteins or indirectly through the use of a cross-linker protein, which bridges MOM and ER proteins [44, 45]. A physical interaction between MOM and peroxisomes has also been reported [46]. In yeast, peroxisomes interact with ERMES via its Pex 11 protein in glucose growth cellular conditions, whereas in mammals the interaction between MOM and peroxisomes has been reported to be a consequence between the physical interaction of the peroxisome protein acyl coenzyme A-binding domain 2 (ACBD2) and the TOM complex of MOM [47–49].

11.2.2 Mitochondrial membrane lipid composition

Mitochondrial lipids have been reported to play a role in several health and disease states [5, 50]. In different tissues and organs as well as in pathological conditions the specific mitochondrial phospholipid composition varies [5, 51]. However, the major phospholipids found in mitochondrial membranes are phosphatidylcholine (PC) and phosphatidylethanolamine (PE) (Figure 11.2) [52]. In some organs like heart and brain the pool of PC and PE contains a significant fraction of plasmalogens, a phospholipid with a vinyl ether, and ester bonds in the glycerol sn-1 and sn-2 positions, respectively [52]. In addition to PC and PE, mitochondrial membranes have a high content of CL, a phospholipid mainly found in mitochondrial membranes and considered as the mitochondrial signature phospholipid (Figure 11.2). Mitochondria are capable of full or partial syntheses of some of their lipids (full: CL and phophatidylglycerol [PG]; partial: PE, phosphatidic acid [PA], and CDP-diacylglycerol), while the remaining lipids (PC,

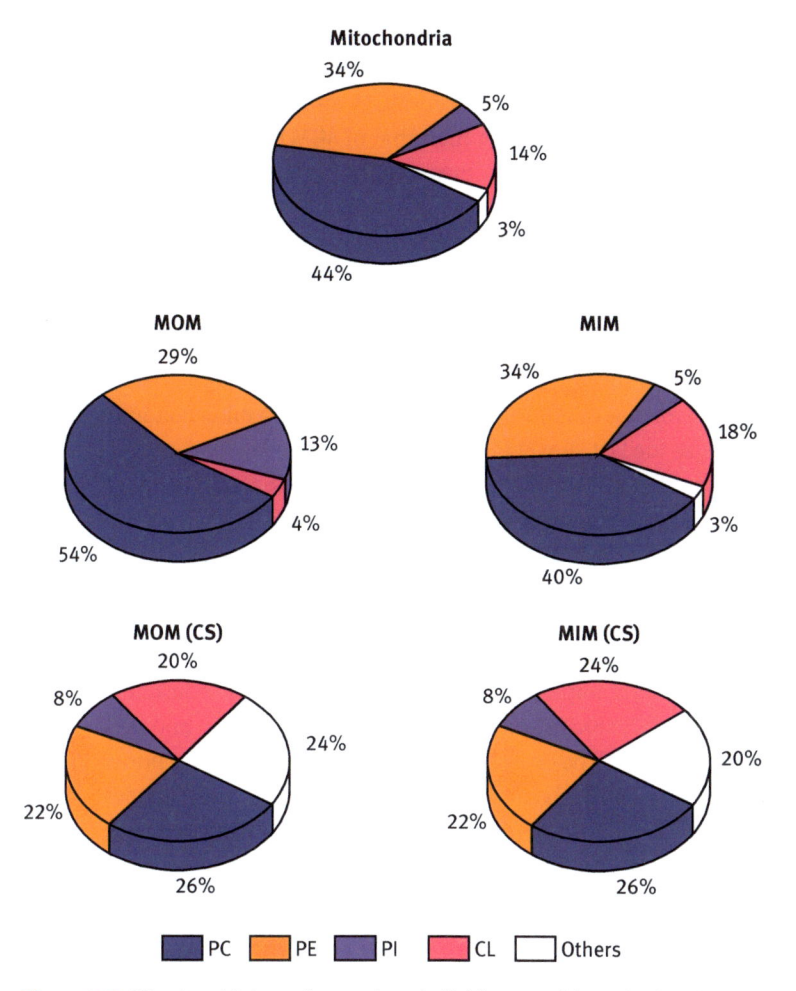

Figure 11.2: Mitochondrial membrane phospholipid composition. The data is presented as percentage of total phospholipid for whole mitochondria, mitochondria outer membrane (MOM), mitochondria inner membrane (MIM), and MOM–MIM contact sites – MOM (CS) and MIM (CS). Pie charts were built using the values reported for mouse liver mitochondria in Ref [52].

phosphatidylserine [PS], phosphatidylinositol [PI], sphingolipids, and cholesterol) need to be imported, mainly from ER. One of the functional importances of contact sites between mitochondria and ER or peroxisomes is to facilitate the movement of lipids or lipid intermediates.

The lipid composition also varies between MOM and MIM. While PC and PE are the major phospholipids in both the membranes, the majority of PI within mitochondria is found at MOM and that of CL is found at MIM (Figure 11.2). The fatty acid composition between MOM and MIM also varies. In mitochondria from mouse liver, MOM lipids are enriched with saturated acyl chains, while MIM is enriched with unsaturated acyl chains (Table 11.1) [15]. Although the amount of cholesterol in mitochondrial membranes is the smallest when compared to other cellular membranes, its distribution is not evenly between MOM and MIM [52]. The amount of cholesterol at MOM is significantly higher than that of MIM (Table 11.1). It is interesting to note that fluorescent anisotropic measurements of a probe intercalated in isolated MOM and MIM showed a higher degree of probe motion within MIM when compared to MOM [15]. Although measurements in these biological membranes also should take into account the presence of intrinsic proteins in both systems, the differences in acyl chain composition and cholesterol content discussed previously would be enough to explain the higher fluidity of MIM in comparison to MOM.

The best characterized roles of a mitochondrial lipid on mitochondrial functional properties are those involving CL. In humans, mutations in the mitochondrial enzyme tafazzin, which catalyze the last step in CL synthesis (the remodeling of its acyl chains), leads to a disease condition, named Barth syndrome (see Section 11.3.4) [53]. The increase in CL and its oxidation/hydrolysis products in MOM have been reported to play a role in the signaling of mitophagy and apoptosis processes, respectively [32, 54]. CL physically interacts with protein complexes involved in the oxidative phosphorylation process of electron transport and proton translocation [55, 56]. The protein complexes of the electron transport chain in the MIM associate into high-order organizations, named supercomplexes, in order to have a more efficient oxidative phosphorylation process [57, 58]. The physical interaction between CL and these supercomplexes has been reported to be critical for their stability as well as for preventing CL degradation [59]. CL and PE are both lipids with negative curvature tendencies. In mitochondria, CL and PE are proposed to play a role in the formation of high curved membrane regions, such as contact sites and cristae/cristae junctions [34, 60]. They have also been reported to play a role in mitochondrial membrane remodeling during fusion and fission events [61].

11.2.3 Nonhomogeneous distribution of mitochondrial membrane components

The study of lateral heterogeneities in the plane of the membrane, so-called membrane domains, has been related to several signaling processes and is an area of intensive research in the field of membranes since their proposal [11, 62]. Although most of the studies focus on the cell plasma membrane, the importance of membrane domains has shed light on other pathophysiological processes including the killing of Gram-negative bacteria by antimicrobial agents [63]. Along this line, in mitochondrial membranes in addition to differences between the components present in MOM and MIM, these membranes also have been shown to contain lateral heterogeneities, which were described to play a role in pathophysiological conditions. In mouse liver mitochondria, the phospholipid composition has been reported to differ between whole MOM and that of MOM regions forming contact sites with MIM, highlighting the presence of lipid lateral heterogeneities within this membrane [15]. The most remarkable difference was the enrichment of MOM contact sites with CL, a lipid mainly found in MIM (Figure 11.2). In model membranes mimicking MOM contact sites, it has been shown that the proapoptotic protein tBid can induce alterations in membrane curvature, suggesting a role of MOM domains as sites of membrane remodeling [64]. It has also been suggested that the lipid composition of this domain can stabilize the local arrangement of membrane junctions [52]. The enrichment of CL in this region has also been shown to act as an activating platform for the apoptotic signal through the activation of caspase 8 [65]. Membrane contact sites between mitochondria and other organelles also likely lead to heterogeneities in the plane of the membrane. It has been shown that MOM–ER contact sites determine the point where mitochondrial fission machinery will assemble during mitochondrial fission [66].

The intricate morphology of MIM facilitates even further the existence of lateral heterogeneities in the plane of the membrane. MIM proteins have been shown to partition differently between the boundary region of MIM and cristae. For instance, the boundary region of MIM contains preferentially TIM complexes, the MINOS/MitOS/MICOS complex, and proteins involved in MIM fusion, while the cristae are enriched with protein complexes involved in the oxidative phosphorylation, ATP synthase, and ADP/ATP carrier [22]. The presence of highly curved membrane regions in cristae/cristae junctions also have been proposed to lead to an enrichment of non-lamellar lipids (CL and PE) on those regions, which would result in the formation of domains [30, 60]. Finally, studies had shown the tight association of CL with protein complexes and supercomplexes of oxidative phosphorylation, leading to membrane domains [55, 56, 59].

Biological membranes present transversal distribution of its lipid and protein components. The phospholipid transversal asymmetry is known since 1972, but it is still not well understood the role of lipid asymmetry to membrane protein function or membrane morphology [67]. Both MOM and MIM present phospholipid asymmetry (Figure 11.3). In rat liver mitochondria the MOM has all its CL and 67% of its PE on the outer leaflet (the cytoplasmic side), while in the MIM 80% and 60% of these lipids, respectively, are on the inner leaflet (the matrix side). One of the most well-known effects of loss of asymmetry is the exposure of PS on the plasma membrane exofacial leaflet [68]. It has been shown that PS exposure signals the cell for apoptosis and blood coagulation events by driving the recruitment of proteins involved on those processes. In mitochondria, the loss of lipid asymmetry has also been reported to work as a signal to mark the mitochondria to undergo certain pathophysiological events. The most well-understood break of lipid asymmetry triggering a cellular response in mitochondria is that reported for CL. As discussed previously the major content of CL is found in the inner leaflet of MIM. Some cellular responses have been described to be triggered by CL exposure in the outer leaflet of MOM. The specific nature of the cellular response triggered depends on the exact molecular species of CL exposed. For instance, exposure of mature CL triggers mitophagy by acting as platforms to recruit proteins involved in this process [54]. On the other hand,

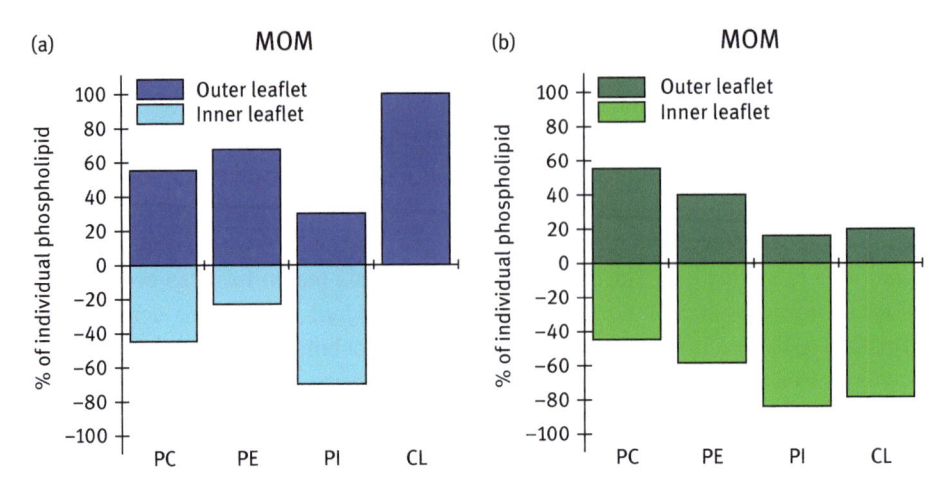

Figure 11.3: Mitochondrial membrane transverse phospholipid asymmetry of (a) MOM and (b) MIM. The data is presented as percentage of individual phospholipid present on that particular membrane. Positive and negative values are arbitrary and represent the percentage on outer and inner leaflets, respectively. Graphs were built using the values reported for mouse liver mitochondria in Ref [52].

if the molecular species of CL exposed contains oxidized acyl chains, the process triggered is apoptosis instead [54].

The physical properties of membranes (permeability, shape, curvature, stability, membrane potential, and surface charge) have also been predicted to be affected by phospholipid asymmetry [69]. The high degree of transversal asymmetry presented by non-bilayer lipids (CL and PE) in the MIM (Figure 11.3) has been hypothesized to contribute to the formation of cristae/cristae junctions. Recently, molecular dynamic simulations showed the induction of membrane curvature by bilayers where CL was asymmetrically distributed [32]. However, there still lacks experimental evidence for this finding. It is still technically challenging to evaluate the effect of phospholipid asymmetry on the physical properties of biological membranes. This is due to their complex nature and the presence of many components. One possibility to address those questions is the use of model membranes bearing phospholipid asymmetry. For long time, even preparing model membranes bearing phospholipid asymmetry was challenging, but recently special methods have been described on how to prepare them, the most used currently is the methy-β-cyclodextrin exchange method [70, 71]. Recently, the preparation of an asymmetric model membrane mimicking MOM has been described for the first time (Bozelli et al., manuscript in preparation). The role of phospholipid asymmetry in these model membranes can be ascribed by a direct comparison with a symmetric version of the model membrane. It was possible to show that the phospholipid asymmetry led to a change in the surface potential of the model membrane, as predicted, and, by consequence, affected the formation of pores by the apoptotic protein Bax.

11.2.4 Dynamics of mitochondrial membrane phospholipids

The lipids of mitochondrial membranes are highly dynamic, involving movement of lipids between mitochondria and different organelles as well as among mitochondrial membranes. This lipid dynamics is tightly associated with their roles in mitochondrial pathophysiological conditions. For instance, lipids and lipid intermediates synthesized in other organelles, such as the ER, must translocate from those membranes to mitochondrial ones, while the role of CL in signaling transduction involves its movements from the inner leaflet of MIM to the outer leaflet of MOM. Lipids present many kinds of motions in a broad range of timescales within a membrane; however, they are usually confined to the membrane due to their low water solubility. However, there are mechanisms for the trafficking of lipids from one membrane to another.

Vesicle trafficking is one of the most well-described mechanisms of lipid transfer between membranes; however, in mitochondrial membranes this does not seem to be the major mechanism. One way to facilitate lipid movement between mitochondrial membranes is to increase the water solubility of the lipids. This can be done through mechanisms (enzymatically or nonenzymatically) that render the lipid molecule more hydrophilic. It has been reported that the formation of either hydroperoxide derivatives of CL or monolysocardiolipin (MLCL, a product formed by the deacylation of CL and, therefore, presenting three acyl chains) facilitate the translocation of these CL molecular species from the inner leaflet of MIM to the outer leaflet of MOM triggering the apoptotic process [72, 73]. The observed facilitated movement of MLCL from MIM to MOM in comparison to CL is in agreement with recent findings showing that the former lipid is extracted relatively more easily than the later from bilayers [74]. Nevertheless, the majority of the studies suggest that lipid transfer involving mitochondrial membranes occurs mainly through a protein-assisted mechanism either indirectly or directly (Figure 11.4) [45].

Indirectly, proteins can facilitate the movement of lipids usually by two mechanisms. The first one is by bridging two membranes together (Figure 11.4). In this way proteins or protein complexes bring the two membranes in close apposition favoring a direct flip of lipids between those membranes, especially if there is alteration of membrane curvature. This mechanism is favored by the presence of membrane positive curvature. In addition to bridging two membranes, these proteins or protein complexes can have intrinsic activities of lipid extraction and lipid transport. This has been shown to be the case for the oligomeric kinases, NPDK-D/NM23-H4 and MtCk, as well as the MINOS/MICOS/MitOS complex bridging MOM–MIM and the ERMES complex bridging MOM–ER [75–77]. The second mechanism involves the formation of non-bilayer phases, which will connect two adjacent membranes. The presence of such structures allows the lipid trafficking by simple diffusion between leaflets of the two membranes. It has been reported that the formation of these structures occurs as intermediates of membrane remodeling during fission/fusion events [19]. Finally, proteins can directly facilitate lipid trafficking by specifically binding to lipid

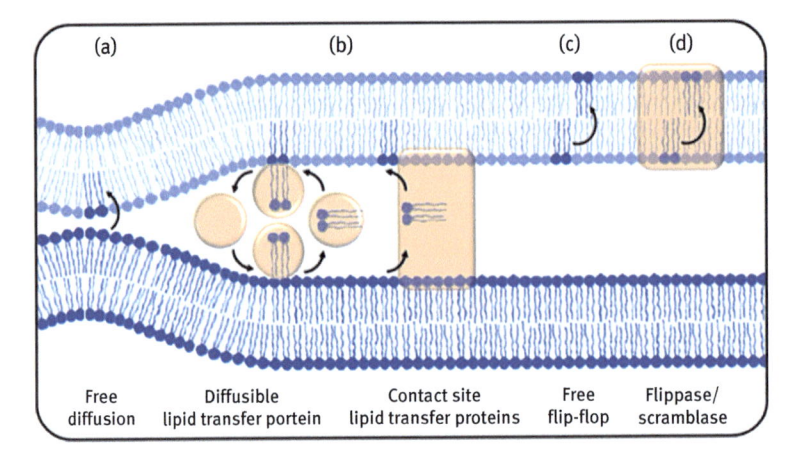

Figure 11.4: Proposed mechanisms of lipid trafficking involving mitochondrial membranes. (a) Spontaneous exchange of lipids in regions of close apposition of membranes. (b) Lipid transfer mediated by diffusible proteins or by proteins/protein complexes tethering two membranes. (c) Spontaneous lipid movement between the two leaflets of the membrane. (d) Protein mediated lipid movement between the two leaflets of the membrane. Lipids are presented in blue, while proteins are presented in pink. Figure was taken from Ref [45]. and used with the permission of Elsevier.

molecules within their hydrophobic cavities and drive their movement from one membrane to another. This has been proposed to be the mechanism of lipid transfer between MOM–MIM by the Ups/PRELI protein family and between MOM–ER by the apoptotic protein Bid [78, 79].

Among the many kinds of motions lipids present within a membrane, lipid flip-flop (lipid movement between leaflets) is the slowest (half-life of minutes to days) (Figure 11.4). However, the rate of flip-flop can be greatly enhanced by catalytically supported mechanisms. There are two main mechanisms to induce flip-flop. The first one is through the insertion of proteins in the membrane, which will disrupt the lipid bilayer and, therefore, facilitate lipid flip-flop. This has been shown for a number of pore-forming compounds and in mitochondria it has been shown to occur by the insertion of the apoptotic protein Bax in model membranes of MOM [80]. The other one is through the action of specific enzymes that catalyze lipid movement from one leaflet to another. There are three types of enzymes that transfer lipids between membrane leaflets, these are as follows: flippases, floppases, and scramblases [81]. The first two types actively translocate lipids requiring an energy-dependent process with the energy derived from ATP hydrolysis. The last one, scramblases, carries out the passive movement of lipids down a concentration gradient. A mitochondrial scramblase is the phospholipid scramblase 3, which is activated by phosphorylation and calcium and is hypothesized to be found in MIM where it would enhance the flip-flop rate of CL from inner to the outer leaflet of MIM during apoptosis or mitophagy processes [45, 82].

11.3 Tafazzin

11.3.1 Function

Tafazzin is an enzyme that is important to mitochondrial structure and function. This enzyme catalyzes the transfer of acyl chains between a phospholipid and a lysophospholipid. One of the reasons for the interest in this enzyme is that its mutation in humans results in a genetic disease called Barth Syndrome. In vivo, tafazzin is known to be responsible for the acyl chain remodeling of cardiolipin.

11.3.2 Cellular localization

Tafazzin is located exclusively in the mitochondria. Tafazzin is an interfacial membrane protein that localizes to both the outer and inner membranes, lining the intermembrane space. This unique topology results from a novel import pathway in which tafazzin crosses the outer membrane via the TOM complex and then uses the regions of the Tim9p–Tim10p complex in the intermembrane space to insert into the mitochondrial outer membrane [83]. Yeast tafazzin is then transported to membranes of an intermediate density to reach a location in the inner membrane. Critical targeting information for tafazzin resides in the membrane anchor and flanking sequences, which are often mutated in Barth syndrome patients [83]. Cardiolipin is synthesized on the matrix-facing leaflet of the inner membrane. Thus, the remodeling of cardiolipin by tafazzin requires that cardiolipin translocates from the matrix side of the inner mitochondrial membrane to the intermembrane side where it is remodeled and returned to the matrix side of the inner membrane, where it is normally found [84].

11.3.3 Acyl chain specificity

In the mammalian heart the predominant molecular species of cardiolipin is tetralinoleoyl-cardiolipin. The fraction of cardiolipin that is the tetralinoleoyl species is markedly decreased in the heart of tissues taken from tafazzin-defective animals, including human patients with Barth Syndrome. However, several observations provide evidence that the specificity of tafazzin is not solely responsible for the high enrichment of tetralinoleoyl-cardiolipin in the heart. In the same normal animal there is much less or no enrichment of cardiolipin with the tetralinoleoyl species in other organs, although tafazzin is expressed in these organs. Additionally, if human tafazzin is expressed in *Drosophila*, the cardiolipin synthesized resembles that of the fly and not that of the human. Furthermore, in vitro assays of tafazzin activity generally exhibit little or no specificity for either the phospholipid or the lysophospholipid in acyl exchange reactions [85, 86]. Finally, with certain lipid mixtures containing cardiolipin, it has been shown that the addition of Ca^{2+} to convert the lipid to the hexagonal phase, increased the specificity of the acyl exchange for linoleoyl chains, indicating that the substrate acyl chain specificity was not inherent to the enzyme, but rather to the membrane to which the enzyme binds [86]. It was pointed out that the reaction catalyzed by tafazzin has a phospholipid and a lysolipid as both substrate and product; therefore, tafazzin catalyzes a reaction that has an equilibrium constant not far from one. As a result the products of one tafazzin-catalyzed reaction can become the substrate for a subsequent tafazzin-catalyzed acyl exchange and that the series of reactions would continue until it reached an equilibrium state. Hence, the reactions were governed not only by the rate of the reaction but also by the position of equilibrium [85]. This view was challenged by Abe et al [87, 88]. who showed that in assays of tafazzin-catalyzed acyl exchange of phospholipids and lysolipids present in liposomes, rather than micelles, there was specificity for transfer of a linoleoyl chain from the *sn-2* position of PC to a free hydroxyl group in monolysocardiolipin. It was then shown, however, that the tafazzin-catalyzed reaction in liposomes required the presence of detergent [89] and that in agreement with previous studies [86], the rate of tafazzin-catalyzed reactions was slow on stable bilayer membranes. The controversy is not yet completely resolved. Even at reduced detergent concentration, there is still acyl chain specificity with liposomes [88]. Two interpretations of these findings have been presented. One is that the enzyme loses activity in the complete absence of detergent and therefore no activity is found when the enzyme is assayed in the complete absence of detergent. The other interpretation is that the kinetics is modified by the presence of even a minor contamination with detergent. The detergent concentration is generally not zero and it is not known what detergent to tafazzin molar ratio is required to alter the properties of the enzyme. In addition, for a reversible reaction with an equilibrium constant not very far from one, the rate of reaction must also depend on the position of equilibrium. It appears that the position of equilibrium must be sensitive to the presence of certain membrane additives and/or minor changes in the structure of the substrates and products.

11.3.4 Barth syndrome

11.3.4.1 Species of cardiolipin

The only known genetic defect in individuals with Barth syndrome is the mutation of tafazzin [53]. It is found that the molecular species of cardiolipin are altered in several different tissues of individuals with Barth syndrome [90]. This finding is also supported by lipidomic analysis of a mouse model of Barth syndrome in which the enzyme tafazzin had been knocked down [91]. In particular, cardiolipin in the heart becomes more heterogeneous.

11.3.4.2 Loss of cardiolipin

In addition to the change in the acyl chain composition of cardiolipin with the loss of tafazzin function, there is also a loss in the content of cardiolipin [90, 91]. This loss of cardiolipin cannot be explained solely by the loss of the enzyme, tafazzin, that catalyzes the remodeling of cardiolipin. Tafazzin only changes the acyl chains on lipids and it does not synthesize or degrade lipids. However, one of the steps in cardiolipin remodeling is the acylation of monolysocardiolipin catalyzed by tafazzin. In the absence of a functional tafazzin, there are other enzymes that can acylate monolysocardiolipin. However, the monolysocardiolipin can also be further degraded. This degradation is made more likely by the fact that monolysocardiolipin does not interact strongly with membrane proteins [59].

11.3.4.3 Accumulation of monolysocardiolipin

Monolysocardiolipin is not normally detected in tissue samples, but it is present in blood or tissue from individuals with Barth syndrome [92]. Hence the rate of formation of monolysocardiolipin is greater than its rate of degradation. The results also suggest that tafazzin-catalyzed transacylation is the major pathway for the conversion of monolysocardiolipin to cardiolipin and other enzymes cannot completely substitute for tafazzin. There are also changes in the molecular species of PC and PE in Barth syndrome [90] as well as plasmenylcholine (Kimura et al., manuscript submitted for publication [102]). It is not yet clear how the loss of these other lipids is related to tafazzin, but if they can serve as acyl donors to convert monolysocardiolipin to cardiolipin, then their depletion in Barth syndrome can contribute to the reduced reacylation of monolysocardiolipin.

11.4 NMR

NMR is a very versatile technique and there are many variations of this method. In this review, we will emphasize applications to membranes, particularly to mitochondrial membranes. A major limitation of using NMR to study membrane systems is that membranes are very large multimolecular aggregates that generally tumble slowly in suspension, giving rise to broad lines with overlapping signals. There are two general methods to circumvent this problem. Both are often referred to as solid-state NMR, but they correspond to different approaches and experimental methods. One of these methods is to use oriented samples [93]. Thus, although the molecule in the membrane may tumble slowly when they are suspended in liquid solvent, by orienting the specimen on a solid support, one limits its orientation with respect to the external magnetic field due to the unique orientation of the underlying solid substrate. This method can be particularly useful for determining the orientation of helical segments with respect to the bilayer in which they are embedded. The other solid-state NMR method is called magic angle spinning NMR (MAS/NMR). In this method the sample is introduced into a capsule that is inserted into a rotor that is spun at high speeds in the cavity of the NMR magnet. The principle of MAS/NMR is that the spinning sample is oriented at the magic angle of 54.74°, so that $3\cos^2\theta_m - 1 = 0$, resulting in the elimination of broadening due to dipolar interactions. The method of MAS/NMR is rapidly developing and has many applications to membrane proteins and lipid mixtures [94].

In addition to oriented samples and to MAS/NMR, there is also the determination of powder patterns from suspensions of lipids. The shape of these powder patterns are interpreted assuming that the overall tumbling time of the vesicle is long with respect to the spin-lattice relaxation time, so that the vesicles do not have time to reorient with respect to the external magnetic field within the lifetime of the spin state and they therefore act like solids and for this reason these spectra are also referred to as solid-state NMR. Two isotopes are primarily used to obtain these NMR powder patterns, ^{2}H and ^{31}P. ^{2}H is an isotope with a spin quantum number $I = 1$. As a result the energy of its spin states are split by interaction with the electric field dipole (nuclear–quadrupole interactions) and the strength of this interaction is dependent on the average angle between the C–^{2}H bond and the electric field gradient. Hence it can be used to measure motional order. In comparison, ^{31}P, like ^{1}H or ^{13}C, has an $I = ½$ nucleus $I = ½$ nuclei. Because of the nonsymmetrical distribution of electrons around the ^{31}P, this isotope has relatively large chemical shift anisotropy. This chemical shift anisotropy results in a broadening of the spectra that can be narrowed by molecular motion.

A common application of ^{2}H NMR is to obtain the order parameter profile for the acyl chain protons. It is easier to prepare a fully deuterated fatty acid that can be synthetically incorporated into a phospholipid, than to chemically deuterate a specific position of the fatty acid. As a result the ^{2}H–NMR spectra that is measured is a superimposition of the spectrum of the ^{2}H–NMR corresponding to each position of the acyl chain. There are mathematical procedures, referred to as "de-Paking" to resolve each contribution [95]. Larger splitting corresponds to a more restricted motion. Introduction of double bonds results in greater disordering as well as a shortening of the acyl chain length.

^{31}P-NMR of lipids is frequently used to define lipid phases [96]. In particular, lamellar phases exhibit powder patterns that are distinct from hexagonal phases. One assumes that the lipid aggregates are sufficiently large not to allow motional averaging of the chemical shift anisotropy by tumbling of the whole particle; this is almost always the case with multilamelar vesicles or with hexagonal phase aggregates. With such samples there are only two types of motion of individual molecules that are fast enough to cause averaging of the chemical shift anisotropy. These motions are rapid spinning around the long axis of the phospholipid (this is also assumed in analyzing the shape of ^{2}H–NMR) and lateral diffusion along the surface of the membrane. The first type of motion is common to all lipid phases and can only partially average the chemical shift anisotropy. The lateral diffusion provides evidence about the morphology of the lipid aggregate. In the case of a flat bilayer, that is essentially the arrangement in large multilamellar vesicles, lateral diffusion does not change the orientation of the phospholipid with respect to the external magnetic field within the spin-lattice relaxation time of ^{31}P. A structure with curvature is required. In the hexagonal phase, diffusion along the length of the cylinder also does not change the orientation of the molecule. However, diffusion along the circumference of the cylinder can occur within the lifetime of the spin state and average out the orientation, but only with respect to that dimension. Hence, the powder pattern of the ^{31}P–NMR of a hexagonal phase is narrower than that for a bilayer phase. In contrast, a bicontinuous cubic phase contains curvatures of opposite sign at saddle points. A phospholipid molecule can thus access all orientations with respect to the external magnetic field within the spin-lattice relaxation time of ^{31}P and thus present a sharp, isotropic signal. The limitation of using this powder pattern as evidence for a cubic phase is that it is not specific. Any structure that allows the complete averaging of the chemical shift anisotropy would give rise to a similar ^{31}P–NMR powder pattern. In addition to cubic phases, such structures would include micelles, small membrane fragments, or small vesicles as well as disordered regions of a bilayer or membrane fusion intermediates.

In addition to measuring ^{31}P–NMR powder patterns with a static, non-spinning sample, it is also possible to use MAS/NMR to take advantage of the fact that a series of spinning side bands in MAS/NMR will trace out the shape of the NMR powder pattern (Figure 11.5). This can be qualitatively seen in Figure 11.5b where the major spinning side bands from CL become gradually smaller in going from right to left, while for the case of CL + Ca^{2+}, the opposite trend is observed. The same asymmetric patterns are observed for the ^{31}P–NMR static powder patterns of these two samples, corresponding to bilayer versus hexagonal phase, respectively. An advantage of using spinning side bands is that these are sharp peaks, rather than the broad envelopes of static powder patterns. It is thus more accurate to define the upper and lower limits of the powder pattern using spinning side bands. In addition, one can vary the spinning speed to change the spinning sideband pattern, thus allowing more accurate chemical shift anisotropy measurements by combining data from several spectra (Figure 11.5). Furthermore, the method becomes more sensitive since it requires less material to measure the sharp spinning side-band peaks. Lastly, in more complicated cases in which there are several phases, the relative contribution of each phase can be calculated more accurately from spinning sideband data. This behavior is shown in the samples of CL/LPC without Ca^{2+}. There can also be signals from isotropic motion, as is indicated by the arrows for CL/LPC (Figure 11.5a). Such signals do not fit into a series of spinning side bands, but rather occur as individual peaks at the position of their isotropic resonance. These small peaks in a complex spinning sideband pattern with other phases are sometimes difficult to identify. In contrast, such a small isotropic peak will standout over a broad powder pattern in a static ^{31}P–NMR.

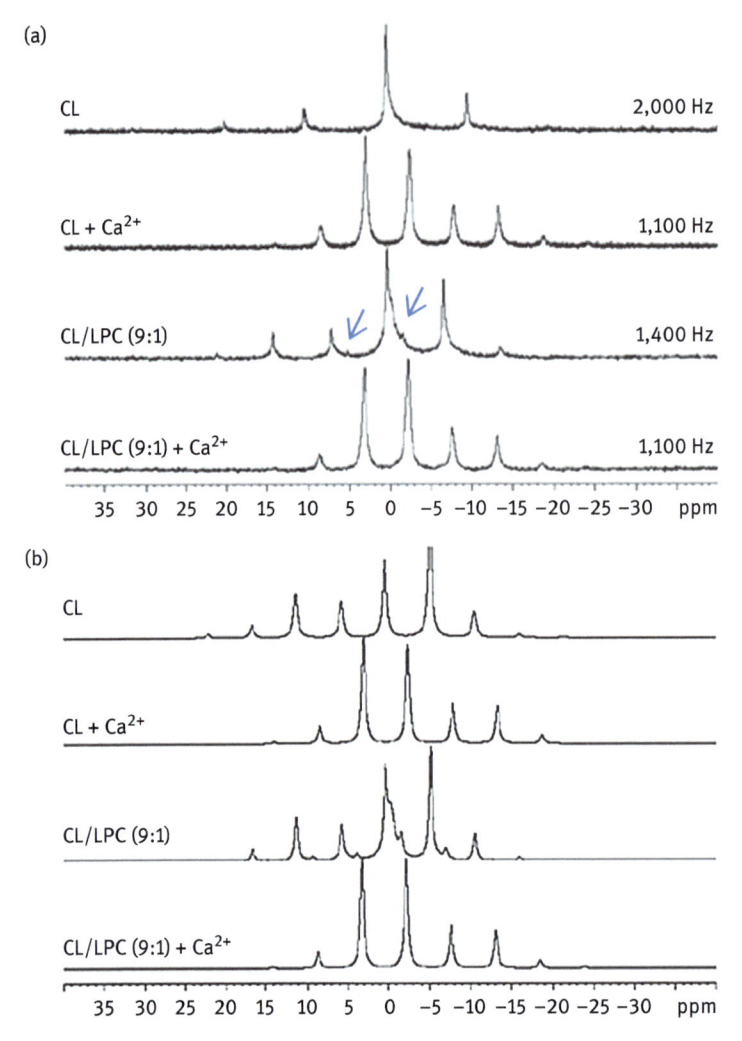

Figure 11.5: MAS ^{31}P NMR spectra of CL/LPC mixtures. Spectra of bovine heart cardiolipin (CL) and a mixture of CL with 14:0-LPC (9:1) were recorded in the absence and presence of 20 mM CaCl$_2$. (a) The panel shows stacked spectra with the spinning rate of each spectrum given at the right-hand side. Ca^{2+} induces a spinning sideband pattern indicative of the presence of a hexagonal phase in both CL and CL/LPC. CL/LPC contains more than one spectral component (arrows) in the absence but not in the presence of Ca^{2+}. (b) The panel shows simulated spectra generated from the data of panel A. All simulations were performed for a uniform spinning rate (1,100Hz). Simulations clearly show the heterogeneous nature of the CL/LPC sample in the absence of Ca^{2+}. Figure was taken from Supplementary Materials of Ref [86].

We will now apply some NMR methods to contribute to our understanding of the organization and properties of mitochondria both through studies of model systems as well as measurements using isolated, purified mitochondria.

11.5 Sidedness

One of the developing themes in the use of model membranes to study the properties of mitochondria is to prepare these membranes with transbilayer asymmetry of phospholipids, as is known to occur in mitochondrial membranes. After preparing such asymmetric membranes, it is important to characterize the degree of transbilayer asymmetry that has been achieved. One also has to measure what the spontaneous flip-flop rates of the lipids are. In principle, this can be done by NMR with the application of paramagnetic ions that will eliminate signals from the exposed leaflet of the liposome or by a similar strategy one can use chemical shift reagents that alter the resonance position of the lipids in the exposed leaflet of the liposome. A recent example using a chemical shift reagent and monitoring the strong signal from the nine equivalent quaternary ammonium methyl protons in PC has been presented [97].

11.5.1 Polymorphism

11.5.1.1 Lysolipids/phospholipids

Spinning sideband patterns in the MAS/NMR of mixtures of phospholipids and lysolipids (Figure 11.5) were used to identify lipid phases [86]. Several parameters can be calculated from these spectra and in several cases the spinning sideband pattern can be considered as the sum of two or in some cases three different phases. From the series of peaks, we can determine the chemical shift anisotropy of each component. This can identify which lipid species were associated with a particular series of spinning sidebands from known isotropic chemical shifts of the lipids used. In the mixtures the components were not purely segregated and as a result the isotropic chemical shifts were modified as a result of lipid mixing. The relative amount of each lipid domain was determined from the relative peak area. In addition, for each component the chemical shift anisotropy can be calculated. It was negative for lipid in the hexagonal phase, positive for lipid in the lamellar phase and zero for lipid giving isotropic motion. In addition, the calculated linewidth of a spinning sideband peak gave information about the molecular motion of that species, with lipids having more rapid motion giving rise to narrower lines. The characteristics of the spinning sideband pattern can be compared with the extent of the tafazzin-catalyzed acyl exchange [86].

11.5.1.2 Model membranes

LUVs are generally not ideal for measuring ^{31}P NMR powder patterns because the tumbling time of the entire vesicle can narrow the chemical shift anisotropy. Generally, MLV or multilamellar vesicles are used for this purpose. They are also simple to prepare by just hydrating and vortexing a lipid film and facilitating the preparation of samples at high lipid concentration. The powder patterns corresponding to lamellar and hexagonal phases are well documented. However, sharp narrow isotropic peaks, although easy to detect, are not specific for a particular lipid structure but can arise from lipid in the cubic phase, micellar lipid, or small lipid particles, including vesicles, that can average the chemical shift anisotropy by motion of the whole particle.

11.5.1.3 Mitochondria

The mitochondrial inner membrane forms crista that has high curvature, typical of non-bilayer membranes. Furthermore, this membrane has a high content of cardiolipin and PE, both lipids having a high propensity to form inverted phases. Whether such non-lamellar phases exist in intact mitochondria is still uncertain. Initial studies with ^{31}P NMR suggested that isolated inner mitochondrial membranes had significant isotropic signals, indicating the presence of non-lamellar phases [98]. However, subsequent studies using intact mitochondria suggested that >95% of the lipid was in a lamellar phase [99]. It still left opened the possibility that there was a minor fraction of non-bilayer lipid in mitochondria, but concluded that most, if not all of the lipid was in a bilayer arrangement. This latter work suggested that the greater amount of isotropic signal in isolated inner mitochondrial membranes was a result of fragmentation of this membrane during isolation, possibly producing small membrane particles that can undergo rotation of the whole particle within the lifetime of the excited state [99]. However, one of the problems intrinsic to studying isolated mitochondria is that they contain water-soluble organic phosphates and inorganic phosphates that give rise to sharp signals, which can be confused with signals coming from phospholipids in an isotropic phase [100]. In order to eliminate these water-soluble materials, the mitochondria have to be osmotically shocked. A recent publication has shown significant isotropic peaks in such shocked mitochondria that increase when the temperature was increased from 8° to 37 °C or by lowering the pH to 3 [101]. It was further suggested that the appearance of these non-bilayer structures stimulated the activity of ATP synthase [101]. Future testing of these findings will be important to validate these results.

References

[1] Margulis, L. Archaeal-eubacterial mergers in the origin of eukarya: phylogenetic classification of life. Proc Natl Acad Sci U S A 1996;93(3):1071–1076.
[2] Frey, T.G., & Mannella, C.A. – The internal structure of mitochondria. Trends Biochem Sci [Internet] 2000;4:1–6. Available from: papers2://publication/uuid/07A16125-F493-43E7-8CEE-EF344B83E701

[3] Mitchell, P. Coupling of phosphorylation to electron and hydrogen transfer by a chemi-osmotic type of mechanism. Nature 1961;191(4784):144–148.
[4] Papa, S. Mitochondrial oxidative phosphorylation changes in the life span. Molecular aspects and physiopathological implications. Biochim Biophys Acta 1996;1276(2):87–105.
[5] Monteiro, J.P., Oliveira, P.J., & Jurado, A.S. Mitochondrial membrane lipid remodeling in pathophysiology: A new target for diet and therapeutic interventions. Prog Lipid Res 2013;52 (4):513–528.
[6] Wai, T., & Langer, T. Mitochondrial dynamics and metabolic regulation. Trends Endocrinol Metab [Internet] 2016;27(2):105–117. Available from: http://dx.doi.org/10.1016/j.tem.2015.12.001
[7] Sebastián, D., Palacín, M., & Zorzano, A. Mitochondrial dynamics: Coupling mitochondrial fitness with healthy aging. Trends Mol Med [Internet] 2017;23(3):201–215. Available from: http://dx.doi.org/10.1016/j.molmed.2017.01.003
[8] Singer, S.J., & Nicolson, G.L. The fluid mosaic model of the structure of cell membranes. Science [Internet] 1972;175(4023):720–731. Available from: http://www.sciencemag.org/cgi/doi/10.1126/science.175.4023.720
[9] Nicolson, G.L. The fluid – mosaic model of membrane structure: Still relevant to understanding the structure, function and dynamics of biological membranes after more than 40 years. Biochim Biophys Acta – Biomembr [Internet] 2014;1838(6):1451–1466. Available from: http://dx.doi.org/10.1016/j.bbamem.2013.10.019
[10] McLaughlin, S., & Murray, D. Plasma membrane phosphoinositide organization by protein electrostatics. Nature 2005;438(7068):605–611.
[11] Lingwood, D., & Simons, K. Lipid rafts as a membrane-. Science 2010;327:46–50.
[12] van Meer, G. Dynamic transbilayer lipid asymmetry. Cold Spring Harb Perspect Biol 2011;3(5):1–11.
[13] McMahon, H.T., & Gallop, J.L. Membrane curvature and mechanisms of dynamic cell membrane remodelling. Nature 2005;438(7068):590–596.
[14] Tait, S., & Green, D. Intrinsic and extrinsic pathways of apoptosis. Nat Rev Mol Cell Biol 2010;11:621–3.
[15] Ardail, D., Privat, J.P., Egret-Charlier, M., Levrat, C., Lerme, F., & Louisot, P. Mitochondrial contact sites. lipid composition and dynamics. J Biol Chem [Internet] 1990;265(31):18797–18802. Available from: http://www.ncbi.nlm.nih.gov/pubmed/2172233
[16] Anderson, S., Bankier, A.T., Barrell, B.G., et al. Sequence and organization of the human mitochondrial genome. Nature [Internet] 1981;290(5806):457–465. Available from: http://www.ncbi.nlm.nih.gov/pubmed/7219534%5Cnhttp://www.nature.com.libproxy.ucl.ac.uk/nature/journal/v290/n5806/pdf/290457a0.pdf
[17] Dolezal, P., Likic, V., Tachezy, J., & Lithgow, T. Evolution of the molecular machines for protein import into mitochondria. Science [Internet] 2006;313(5785):314–318. Available from: http://www.ncbi.nlm.nih.gov/pubmed/16857931
[18] Archer, S.L., & Dynamics, M. — Mitochondrial fission and fusion in human diseases. N Engl J Med [Internet] 2013;369(23):2236–2251. Available from: http://www.nejm.org/doi/10.1056/NEJMra1215233
[19] Kozlov, M.M., McMahon, H.T., & Chernomordik, L.V. Protein-driven membrane stresses in fusion and fission. Trends Biochem Sci 2010;35(12):699–706.
[20] Tait, S.W.G., & Green, D.R. Mitochondria and cell death: outer membrane permeabilization and beyond. Nat Rev Mol Cell Biol [Internet] 2010;11(9):621–632. Available from: http://www.nature.com/doifinder/10.1038/nrm2952
[21] Schlattner, U., Tokarska-Schlattner, M., Epand, R.M., Boissan, M., Lacombe, M.-L., & Kagan, V.E. NME4/nucleoside diphosphate kinase D in cardiolipin signaling and mitophagy. Lab Investig

2018;98(2):228–232. Available from: http://www.nature.com/doifinder/10.1038/labinvest.2017.113

[22] van der Laan, M., Bohnert, M., Wiedemann, N., & Pfanner, N. Role of MINOS in mitochondrial membrane architecture and biogenesis. Trends Cell Biol [Internet] 2012;22(4):185–192. Available from: http://dx.doi.org/10.1016/j.tcb.2012.01.004

[23] Alberts, B., Johnson, A., Lewis, J., Raff, M., Roberts, K., & Walter, P. Molecular Biology of the Cell. New York: Garland Science, 2002.

[24] Mannella, C.A. Structure and dynamics of the mitochondrial inner membrane cristae. Biochim Biophys Acta – Mol Cell Res 2006;1763(5–6):542–548.

[25] Strauss, M., Hofhaus, G., Schröder, R.R., & Kühlbrandt, W. Dimer ribbons of ATP synthase shape the inner mitochondrial membrane. EMBO J 2008;27(7):1154–1160.

[26] Hahn, A., Parey, K., Bublitz, M., et al. Structure of a complete ATP synthase dimer reveals the molecular basis of inner mitochondrial membrane morphology. Mol Cell 2016;63(3):445–456.

[27] Cipolat, S., Om, D.B., Dal Zilio, B., & Scorrano, L. OPA1 requires mitofusin 1 to promote mitochondrial fusion. Proc Natl Acad Sci [Internet] 2004;101(45):15927–15932. Available from: http://www.pnas.org/cgi/doi/10.1073/pnas.0407043101

[28] Messerschmitt, M., Jakobs, S., Vogel, F., et al. The inner membrane protein Mdm33 controls mitochondrial morphology in yeast. J Cell Biol 2003;160(4):553–564.

[29] Renken, C., Siragusa, G., Perkins, G., et al. A thermodynamic model describing the nature of the crista junction: A structural motif in the mitochondrion. In: Journal of Structural Biology 2002; 138(1–2):137–144.

[30] Khalifat, N., Puff, N., Bonneau, S., Fournier, J.B., & Angelova, M.I. Membrane deformation under local pH gradient: Mimicking mitochondrial cristae dynamics. Biophys J 2008;95(10):4924–4933.

[31] Khalifat, N., Fournier, J.B., Angelova, M.I., & Puff, N. Lipid packing variations induced by pH in cardiolipin-containing bilayers: The driving force for the cristae-like shape instability. Biochim Biophys Acta – Biomembr 2011;1808(11):2724–2733.

[32] Kagan, V.E., Tyurina, Y.Y., Tyurin, V.A., et al. Cardiolipin signaling mechanisms: Collapse of asymmetry and oxidation. Antioxid Redox Signal [Internet] 2015;22(18):1667–1680. Available from: http://online.liebertpub.com/doi/10.1089/ars.2014.6219

[33] Mannella, C.A. The relevance of mitochondrial membrane topology to mitochondrial function. Biochim Biophys Acta – Mol Basis Dis 2006;1762(2):140–147.

[34] Murley, A., & Nunnari, J. The emerging network of mitochondria-organelle contacts. Mol Cell 2016;61(5):648–653.

[35] Brdiczka, D.G., Zorov, D.B., & Sheu, S.S. Mitochondrial contact sites: Their role in energy metabolism and apoptosis. Biochim Biophys Acta – Mol Basis Dis 2006;1762(2):148–163.

[36] Hackenbrock, C.R. Chemical and physical fixation of isolated mitochondria in low-energy and high-energy states. Proc Natl Acad Sci [Internet] 1968;61(2):598–605. Available from: http://www.pnas.org/cgi/doi/10.1073/pnas.61.2.598

[37] Mannella, C.A., Pfeiffer, D.R., Bradshaw, P.C., et al. Topology of the mitochondrial inner membrane: Dynamics and bioenergetic implications. IUBMB Life 2002;52(3–5):93–100.

[38] Nicastro, D., Frangakis, A.S., Typke, D., & Baumeister, W. Cryo-electron tomography of neurospora mitochondria. J Struct Biol 2000;129(1):48–56.

[39] Pfanner, N., Wiedemann, N., & Meisinger, C. Double membrane fusion japan bats a triple. Science 2004;305:1723–4.

[40] Senda, T., & Yoshinaga-Hirabayashi, T. Revealed by quick-freeze deep-etch. Anatomical Record 1998;251(3):339–345

[41] Adams, V., Bosch, W., Schlegel, J., Wallimann, T., & Brdiczka, D. Further characterization of contact sites from mitochondria of different tissues: topology of peripheral kinases. BBA – Biomembr 1989;981(2):213–225.

[42] Schlattner, U., Tokarska-Schlattner, M., Ramirez, S., et al. Mitochondrial kinases and their molecular interaction with cardiolipin. Biochim. Biophys. Acta - Biomembr 2009;1788(10):2032–2047.

[43] Copeland, D.E. An association between mitochondria and the endoplasmic reticulum in cells of the pseudobranch gland of a teleost. J Cell Biol [Internet] 1959;5(3):393–396. Available from: http://www.jcb.org/cgi/doi/10.1083/jcb.5.3.393

[44] Elbaz-Alon, Y. Mitochondria–organelle contact sites: the plot thickens. Biochem Soc Trans [Internet] 2017;45(2):477–488. Available from: http://biochemsoctrans.org/lookup/doi/10.1042/BST20160130

[45] Schlattner, U., Tokarska-Schlattner, M., & Rousseau, D., et al. Mitochondrial cardiolipin/phospholipid trafficking: The role of membrane contact site complexes and lipid transfer proteins. Chem Phys Lipids [Internet] 2014;179:32–41. Available from: http://dx.doi.org/10.1016/j.chemphyslip.2013.12.008

[46] Hicks, L., & Fahimi, H.D. Peroxisomes (microbodies) in the myocardium of rodents and primates - A comparative ultrastructural cytochemical study. Cell Tissue Res 1977;175(4):467–481.

[47] Cohen, Y., Klug, Y.A., Dimitrov, L., et al. Peroxisomes are juxtaposed to strategic sites on mitochondria. Mol BioSyst [Internet] 2014;10(7):1742–1748. Available from: http://xlink.rsc.org/?DOI=C4MB00001C

[48] Mattiazzi U., Brložnik, M., Kaferle, P., et al. Genome-wide localization study of yeast pex11 identifies peroxisome-mitochondria interactions through the ERMES complex. J Mol Biol 2015;427(11):2072–2087.

[49] Fan, J., Li, X., Issop, L., Culty, M., & Papadopoulos, V. ACBD2/ECI2-Mediated peroxisome-mitochondria interactions in leydig. Cell Steroid Biosynthesis. Mol Endocrinol [Internet] 2016;30(7):763–782. Available from: https://academic.oup.com/mend/article-lookup/doi/10.1210/me.2016-1008

[50] Osman, C., Voelker, D.R., & Langer, T. Making heads or tails of phospholipids in mitochondria. J Cell Biol 2011;192(1):7–16.

[51] Wahle, K. (1983). Fatty acid modification and membrane lipids. Proceedings of the Nutrition Society, 42(2), 273–287. doi:10.1079/PNS19830032

[52] Horvath, S.E., & Daum, G. Lipids of mitochondria. Prog Lipid Res [Internet] 2013;52(4):590–614. Available from: http://dx.doi.org/10.1016/j.plipres.2013.07.002

[53] Bione, S., D'Adamo, P., Maestrini, E., Gedeon, A.K., Bolhuis, P.A., & Toniolo, D. A novel X-linked gene, G4.5. is responsible for barth syndrome. Nat Genet 1996;12(4):385–389.

[54] Maguire, J.J., Tyurina, Y.Y., Mohammadyani, D., et al. Known unknowns of cardiolipin signaling: The best is yet to come. Biochim Biophys Acta - Mol Cell Biol Lipids 2017;1862(1):8–24.

[55] Arnarez, C., Mazat, J., Elezgaray, J., Marrink, S., & Periole, X. Evidence for Cardiolipin Binding Sites on the Membrane-Exposed Surface of the Cytochrome bc 1. J Am Chem Soc 2013;135(8):3112–3120.

[56] Pöyry, S., Cramariuc, O., Postila, P.A., et al. Atomistic simulations indicate cardiolipin to have an integral role in the structure of the cytochrome bc1 complex. Biochim Biophys Acta - Bioenerg [Internet] 2013;1827(6):769–778. Available from: http://dx.doi.org/10.1016/j.bbabio.2013.03.005

[57] Gu, J., Wu, M., R, G., et al. The architecture of the mammalian respirasome. Nature [Internet] 2016;537(7622):639–643. Available from: http://dx.doi.org/10.1038/nature19359

[58] Letts, J.A., Fiedorczuk, K., & Sazanov, L.A. The architecture of respiratory supercomplexes. Nature [Internet] 2016;537(7622):644–648. Available from: http://dx.doi.org/10.1038/nature19774

[59] Xu, Y., Phoon, C.K.L., Berno, B., et al. Loss of protein association causes cardiolipin degradation in barth syndrome. Nat Chem Biol 2016;12(8):641–647.

[60] Phan, M.D., & Shin, K. Effects of cardiolipin on membrane morphology: A langmuir monolayer study. Biophys J 2015;108(8):1977–1986.
[61] Ha, E.E.-J., & Frohman, M.A. Regulation of mitochondrial morphology by lipids. Biofactors [Internet] 2014;40(4):419–424. Available from: http://www.pubmedcentral.nih.gov/article render.fcgi?artid=4146713&tool=pmcentrez&rendertype=abstract.
[62] Simons, K., & Ikonen, E. Functional rafts in cell membranes. Nature 1997;387(6633):569–572.
[63] Epand, R.M., & Epand, R.F. Lipid domains in bacterial membranes and the action of antimicrobial agents. Biochim. Biophys. Acta - Biomembr 2009;1788(1):289–294.
[64] Epand, R.F., Martinou, J.C., Fornallaz-Mulhauser, M., Hughes, D.W., & Epand, R.M. The apoptotic protein tBid promotes leakage by altering membrane curvature. J Biol Chem 2002;277(36):32632–32639.
[65] Sorice, M., Manganelli, V., Matarrese, P., et al. Cardiolipin-enriched raft-like microdomains are essential activating platforms for apoptotic signals on mitochondria. FEBS Lett 2009;583 (15):2447–2450.
[66] Marchi, S., Patergnani, S., & Pinton, P. The endoplasmic reticulum-mitochondria connection: One touch, multiple functions. Biochim Biophys Acta - Bioenerg [Internet] 2014;1837(4):461–469. Available from: http://dx.doi.org/10.1016/j.bbabio.2013.10.015
[67] Bretscher, M.S. Asymmetrical lipid bilayer structure for biological membranes. Nat New Biol 1972;236(61):11–12.
[68] Bevers, E.M., & Williamson, P.L. Getting to the outer leaflet: Physiology of phosphatidylserine exposure at the plasma membrane. Physiol Rev [Internet] 2016;96(2):605–645. Available from: http://physrev.physiology.org/lookup/doi/10.1152/physrev.00020.2015
[69] Devaux, P.F. Static and dynamic lipid asymmetry in cell membranes. Biochemistry 1991;30 (5):1163–1173.
[70] Cheng, H.T., & Megha, L.E. Preparation and properties of asymmetric vesicles that mimic cell membranes. effect upon lipid raft formation and transmembrane helix orientation. J Biol Chem 2009;284(10):6079–6092.
[71] Heberle, F.A., Marquardt, D., Doktorova, M., et al. Subnanometer structure of an asymmetric model membrane: Interleaflet coupling influences domain properties. Langmuir 2016;32 (20):5195–5200.
[72] Degli Esposti, M., Cristea, I.M., Gaskell, S.J., Nakao, Y., & Dive, C. Proapoptotic bid binds to monolysocardiolipin, a new molecular connection between mitochondrial membranes and cell death. Cell Death Differ 2003;10(12):1300–1309.
[73] Korytowski, W., Basova, L.V., Pilat, A., Kernstock, R.M., & Girotti, A.W. Permeabilization of the mitochondrial outer membrane by bax/truncated Bid (tBid) proteins as sensitized by cardiolipin hydroperoxide translocation: Mechanistic implications for the intrinsic pathway of oxidative apoptosis. J Biol Chem 2011;286(30):26334–26343.
[74] Bozelli, J.C., Hou, Y.H., & Epand, R.M. Thermodynamics of methyl-β-cyclodextrin-induced lipid vesicle solubilization: Effect of lipid headgroup and backbone. Langmuir 2017;33(48):13882–13891.
[75] Schlattner, U., Tokarska-Schlattner, M., Ramirez, S., et al. Dual function of mitochondrial Nm23-H4 protein in phosphotransfer and intermembrane lipid transfer: A cardiolipin-dependent switch. J Biol Chem 2013;288(1):111–121.
[76] Epand, R.F., Schlattner, U., Wallimann, T., Lacombe, M.L., & Epand, R.M. Novel lipid transfer property of two mitochondrial proteins that bridge the inner and outer membranes. Biophys J [Internet] 2007;92(1):126–137. Available from: http://dx.doi.org/10.1529/biophysj.106.092353
[77] Kawano, S., Tamura, Y., Kojima, R., et al. Structure – function insights into direct lipid transfer between membranes by Mmm1 – Mdm12 of ERMES. Journal of Cell Biology 2018;217(3):959–974.

[78] Tatsuta, T., Scharwey, M., & Langer, T. Mitochondrial lipid trafficking. Trends Cell Biol 2014;24(1):44–52.
[79] Esposti, M.D., Erler, J.T., Hickman, J.A., & Dive, C. Bid, a widely expressed proapoptotic protein of the Bcl-2 family, displays lipid transfer activity. Mol Cell Biol [Internet] 2001;21(21):7268–7276. Available from: http://mcb.asm.org/cgi/doi/10.1128/MCB.21.21.7268-7276.2001
[80] Epand, R.F., Martinou, J.C., Montessuit, S., & Epand, R.M. Transbilayer lipid diffusion promoted by bax: Implications for apoptosis. Biochemistry 2003;42(49):14576–14582.
[81] Devaux, P.F., Herrmann, A., Ohlwein, N., & Kozlov, M.M. How lipid flippases can modulate membrane structure. Biochim Biophys Acta - Biomembr 2008;1778(7–8):1591–1600.
[82] He, Y., Liu, J., Grossman, D., et al. Phosphorylation of mitochondrial phospholipid scramblase 3 by protein kinase C-δ induces its activation and facilitates mitochondrial targeting of tBid. J Cell Biochem 2007;101(5):1210–1221.
[83] Herndon, J.D., Claypool, S.M., & Koehler, M. The taz1p transacylase is imported and sorted into the outer mitochondrial membrane via a membrane anchor domain. Eukaryotic Cell 2013;12(12):1600–1608.
[84] Baile, M.G., Whited, K., & Claypool, S.M. Deacylation on the matrix side of the mitochondrial inner membrane regulates cardiolipin remodeling. Mol Biol Cell 2013;24(12):2008–2020.
[85] Malhotra, A., Xu, Y., Ren, M., & Schlame, M. Formation of molecular species of mitochondrial cardiolipin. 1. A novel transacylation mechanism to shuttle fatty acids between sn-1 and sn-2 positions of multiple phospholipid species. Biochim Biophys Acta - Mol Cell Biol Lipids [Internet] 2009;1791(4):314–320. Available from: http://dx.doi.org/10.1016/j.bbalip.2009.01.004
[86] Schlame, M., Acehan, D., Berno, B., et al. The physical state of lipid substrates provides transacylation specificity for tafazzin. Nat Chem Biol [Internet] 2012;8(10):862–869. Available from: http://dx.doi.org/10.1038/nchembio.1064
[87] Abe, M., Hasegawa, Y., Oku, M., et al. Mechanism for remodeling of the acyl chain composition of cardiolipin catalyzed by saccharomyces cerevisiae tafazzin. J Biol Chem 2016;291(30):15491–15502.
[88] Abe, M., Sawada, Y., Uno, S., et al. Role of acyl chain composition of phosphatidylcholine in tafazzin-mediated remodeling of cardiolipin in liposomes. Biochemistry 2017;56(47):6268–6280.
[89] Schlame, M., Xu, Y., & Ren, M. The basis for acyl specificity in the tafazzin reaction. J Biol Chem 2017;292(13):5499–5506.
[90] Schlame, M., Kelley, R.I., Feigenbaum, A., et al. Phospholipid abnormalities in children with barth syndrome. J Am Coll Cardiol [Internet] 2003;42(11):1994–1999. Available from: http://dx.doi.org/10.1016/j.jacc.2003.06.015
[91] Kiebish, M.A., Yang, K., Liu, X., et al. Dysfunctional cardiac mitochondrial bioenergetic, lipidomic, and signaling in a murine model of barth syndrome. J Lipid Res [Internet] 2013;54(5):1312–1325. Available from: http://www.jlr.org/lookup/doi/10.1194/jlr.M034728
[92] Houtkooper, R.H., Rodenburg, R.J., Thiels, C., et al. Cardiolipin and monolysocardiolipin analysis in fibroblasts, lymphocytes, and tissues using high-performance liquid chromatography-mass spectrometry as a diagnostic test for barth syndrome. Anal Biochem [Internet] 2009;387(2):230–237. Available from: http://dx.doi.org/10.1016/j.ab.2009.01.032
[93] Cross, T.A., Ekanayake, V., Paulino, J., & Wright, A. Solid state NMR: The essential technology for helical membrane protein structural characterization. J Magn Reson [Internet] 2014;239:100–109. Available from: http://dx.doi.org/10.1016/j.jmr.2013.12.006
[94] Wylie, B.J., Do, H.Q., Borcik, C.G., & Hardy, E.P. Advances in solid-state NMR of membrane proteins. Mol Phys [Internet] 2016;114(24):3598–3609. Available from: https://doi.org/10.1080/00268976.2016.1252470

[95] Holte, L.L., Peter, S.A., Sinnwell, T.M., & Gawrisch, K. 2H nuclear magnetic resonance order parameter profiles suggest a change of molecular shape for phosphatidylcholines containing a polyunsaturated acyl chain. Biophys J [Internet] 1995;68(6):2396–2403. Available from: http://dx.doi.org/10.1016/S0006-3495(95)80422-4

[96] Epand, R.M., D'Souza, K., Berno, B., & Schlame, M. Membrane curvature modulation of protein activity determined by NMR. Biochim Biophys Acta - Biomembr [Internet] 2015;1848(1):220–228. Available from: http://dx.doi.org/10.1016/j.bbamem.2014.05.004

[97] Marquardt, D., Heberle, F.A., Miti, T., et al. 1H NMR shows slow phospholipid flip-flop in gel and fluid bilayers. Langmuir 2017;33(15):3731–3741.

[98] Cullis, P.R., de Kruijff, B., Hope, M.J., Nayar, R., Rietveld, A., & Verkleij, A.J. Structural properties of phospholipids in the rat liver inner mitochondrial membrane. A 31P-NMR study. Biochim Biophys Acta - Biomembr 1980;600(3):625–635.

[99] de Kruijff, B., Reitveld, A., & Cullis, P.R. 31P-NMR studies on membrane phospholipids in microsomes, rat liver slices and intact perfused rat liver. Biochim Biophys Acta [Internet] 1980;600(2):343–357. Available from: http://www.ncbi.nlm.nih.gov/pubmed/7407118

[100] Ogawa, S., Rottenberg, H., Brown, T.R., Shulman, R.G., Castillo, C.L., & Glynn, P. High-resolution 31P nuclear magnetic resonance study of rat liver mitochondria. Proc Natl Acad Sci USA 1978;75 (4)(4):1796–1800.

[101] Gasanov, S.E., Kim, A.A., Yaguzhinsky, L.S., & Dagda, R.K. Non-bilayer structures in mitochondrial membranes regulate ATP synthase activity. Biochim Biophys Acta - Biomembr [Internet] 2018;1860(2):586–599. Available from: https://doi.org/10.1016/j.bbamem.2017.11.014

[102] Kimura, T., Kimura, A.K., Ren. M., Berno. B., Xu. Y., Schlame. M., Epand. R.M. Substantial Decrease in Plasmalogen in the Heart Associated with Tafazzin Deficiency. Biochemistry 2018; 57:2162–2175

Mitchell DiPasquale, Michael H.L. Nguyen, Thad A. Harroun, Drew Marquardt

12 Monitoring oxygen-sensitive membranes and vitamin E as an antioxidant

Abstract: The investigation of biomembrane mimics already possesses significant technical hurdles without considering the bilayer's susceptibility to reactive oxygen species. Phospholipids that contain polyunsaturated fatty acids (PUFAs) are highly susceptible to oxidative damage due to reactive oxygen species. PUFA's vulnerability to oxidation creates unique biological and technical problems, both structurally and in terms of bioavailability. Structurally, PUFA oxidation products can grossly alter the physical properties of a bilayer, which can ultimately lead to the malfunction of integral proteins. In terms of bioavailability, PUFAs can only be obtained through dietary consumption; therefore, there must be a mechanism for PUFA preservation, such as the inclusion of antioxidants into the bilayer. We highlight the need for a physical understanding of the oxidation/antioxidant relation through a case study of vitamin E. Vitamin E correlates strongly with its physical location in a model lipid bilayer which has been overlooked due to the problem of the physical distance between the vitamin's reducing hydrogen and lipid acyl chain radicals. Our case study demonstrates the need for combined data from complimentary techniques (neutron diffraction, NMR, and UV spectroscopy) to probe the physiochemical nature of vitamin E. Ultimately, measurements of PUFA and antioxidant oxidation kinetics, and products, should be interpreted by taking into consideration the physical properties of the membrane in which the PUFA and antioxidant reside.

Keywords: vitamin E, model membranes, oxygen uptake

12.1 Introduction

12.1.1 Polyunsaturated fatty acids

At the level of whole organisms, a long history of biochemical investigations has shown how many complex interrelated metabolic pathways are involved in the

---

Mitchell DiPasquale, Michael H.L. Nguyen, Drew Marquardt, Department of Chemistry and Biochemistry, University of Windsor, 401 Sunset Ave. Windsor, Ontario, Canada N9B 3P4
Thad A. Harroun, Department of Physics, Brock University, 1812 Sir Isaac Brock Way, St. Catherines, ON L2S 3A1

https://doi.org/10.1515/9783110544657-012

biosynthesis and degradation of lipids in the cells of all living things. In these in vitro and in vivo experiments, oxidative stress via reactive oxidative species plays an important role in cellular health through lipid degradation, altering the lipid composition and consequently protein function. Several disease states can be traced to the regulation of lipid composition, including the presence and quantity of mono- and polyunsaturated acyl chains of the lipids. The presence of double bonds makes these lipids particularly susceptible to oxidation degradation, which alters the physical properties of the membrane. Thus, numerous biochemical methods have been developed to monitor the oxidation state of membranes in vitro, as well as chemical methods to attain a deeper understanding of the molecular mechanism of lipid oxidation at the quantum mechanical level.

The field of membrane biophysics uses a combination of physical and computational methods to study biological membrane structure and its relation to cellular functions. A molecular understanding of the interaction of proteins with lipid bilayers requires a fundamental experimental understanding of the structure of the membrane bilayer itself. Typically, simpler model and mimetic systems are used to tightly control lipid composition under regulated conditions; several examples are given in other chapters of this book. Since the dominant composition of acyl chains in many mammalian cells are the saturated 16:0 and monounsaturated 18:1, such lipids have formed the vast majority of basic biophysical experiments. Their general biological relevancy and general stability against oxidation, means that their molecular interactions are useful and stable for the duration of most conceivable experiments.

However, in the paradigm of how lipid structure affects protein function, the uniqueness of a polyunsaturated fatty acid (PUFA) adds an additional dimension to the field of membrane biophysics. Primarily, the increased conformational flexibility of the allylic carbons of PUFA lipids dramatically alters the fluidity and structure of the bilayer. This, in turn, can have profound effects on protein function.

Interestingly, the fragility of PUFA lipids to oxidation changes the understanding of membrane structure into a dynamic question. The kinetics of lipid oxidation in controlled environments reveals how membrane structure changes in time and that, in turn, influences the rates of cellular functions such as energy generation in mitochondria, a process which generates many reactive oxidants. Thus, biophysical techniques have been developed around lipid oxidation that can monitor the unique changes to model PUFA membranes, with the goal to reveal new structure–function relationships apart from their saturated counterparts.

Numerous soluble and membrane-bound small molecules have been demonstrated to afford antioxidant protection to membrane lipids. These include the vitamin E family of tocopherols and tocotrienols as a prime example of a lipid soluble antioxidant, and one that has found wide spread use in commercial products for its antioxidant potency. Incorporation of vitamin E into a lipid study slows or halts lipid oxidation, preserving PUFA-containing membrane structure and function. Hence, an assessment of the oxidative state of both the membrane and antioxidant is important.

Here, we briefly describe a number of biophysical methods that have been recently employed to reveal the structural and oxidative state of membranes in the presence of the α-tocopherol member of the vitamin E family. These have provided insights into the physical significance of tocopherol in membranes and moreover to its roles and mechanisms in the biological system.

12.1.2 Oxidative stress

The healthy function of most cells actually requires the presence of oxidants, created through various natural enzymatic and nonenzymatic pathways [1, 2]. Naturally occurring oxidants found in cells are almost exclusively oxygen bearing, and the free radical forms have reactivity orders of magnitude greater than nonradical forms [3, 4]. By far, the two most prevalent free radicals in vivo are the superoxide anion (O_2^-) and nitric oxide (NO).

Enzymatic processes that produce free radicals nearly always occur in cytosolic and peripheral membrane proteins [5,6]. For example, NADPH oxidase is a peripheral membrane protein that is activated through an immune response that produces a large percentage of the reactive oxygen species (ROS) (O_2^- and H_2O_2) found in the cell. The exceptions to cytosolic free radical and oxidant production are found in the membrane of the mitochondria, where cellular respiration occurs [3–6], as well as in the PUFA metabolization enzyme lipoxygenase, which commonly associates with the nuclear envelope [2, 7]. Valko et al. note that up to 3% of the electrons in the electron transport chain of cellular respiration "leak" out of the chain to create ROS superoxide radical anions [4]. Cadenas and Sies agree and indicate that the majority of superoxide formation occurs within the mitochondria of a cell [8].

Lipoxygenase is activated in the presence of inflammation; the hydroperoxyl lipid radical product is a mediator of the inflammatory response [7]. Free PUFAs are required as a substrate; however, an impediment emerges as they do not exist in the cell as free fatty acids. In this case, liberation of PUFAs by phospholipase A2, for example, must occur prior to the activation of lipoxygenase. Several researchers believe the radical intermediate formed by lipoxygenase remains within the enzyme complex, thereby posing minimal threat to the surrounding membrane [9].

Nonenzymatic sources of free radicals are metal ions in the intercellular and intracellular fluid, and from ionizing radiation. Fenton chemistry can occur with a variety of transition metals, most commonly Fe(II), which is found relatively abundant in vivo [4,6], but can also include copper, cobalt, and chromium [10].

For the plasma membrane to be threatened by ROS formed by cytosolic enzymes or from the mitochondrial machinery, virtually all the ROS must approach the cell membrane from the cytoplasm. Likewise, Fenton reactants are present in aqueous solution (in the cytosol) rather than the hydrophobic media (the center of a lipid bilayer).

The majority of naturally occurring carbon-based free radicals are formed from the alkyl and allylic carbons from lipid acyl chains. Allylic carbon radicals quickly isomerize, conjugating the double bonds in the PUFA. Alkyl carbon radicals react quickly with O_2 to form peroxyl radicals. Then, the radicals can quickly terminate with a hydrogen from another PUFA or with another PUFA-based radical. This carbon radical formation is very uncommon under standard physiological conditions, with the exception of the presence of ionizing radiation. However, it is not until O_2 is present that the chain reaction propagation occurs more rapidly causing uncontrollable damage to the PUFA bilayer. The generally accepted mechanism of PUFA oxidation includes initiation and propagation, as illustrated in Figure 12.1 [11].

12.1.3 Vitamin E

Several of these conditions of oxidative stress may be present in many in vitro cellular or even biophysical experiments. Thus, caution and care must be taken to monitor and prevent oxidation. A number of molecules and methods are available that can manage specimen oxidation through termination of the lipid radicals and prevention of autoxidation [13].

There are several methods of radical termination for PUFA; however, one of particular interests is the antioxidant termination mechanism. A number of known hydrophilic and lipophilic antioxidants exist in vivo including vitamin C, vitamin A, ubiquinol, β-carotene, lutein, lycopene, selenium, and vitamin E [14]. Discovered in 1922 by Evans and Bishop, it was observed that without vitamin E as a dietary component, rats could no longer reproduce [15]; however, its role in maintaining general health is fervently debated. Vitamin E is really two families of molecules known as tocopherols and tocotrienols, each with four members α, β, γ, and δ which refer to substitutions on the chromanol ring (Figure 12.2).

What is known about them is that these compounds can act as fat-soluble antioxidants and therefore are commonly used as preservatives for such things as cosmetics and foods. Although all eight members of vitamin E share many similarities, and are all consumed in our diets, interestingly, α-tocopherol is the only

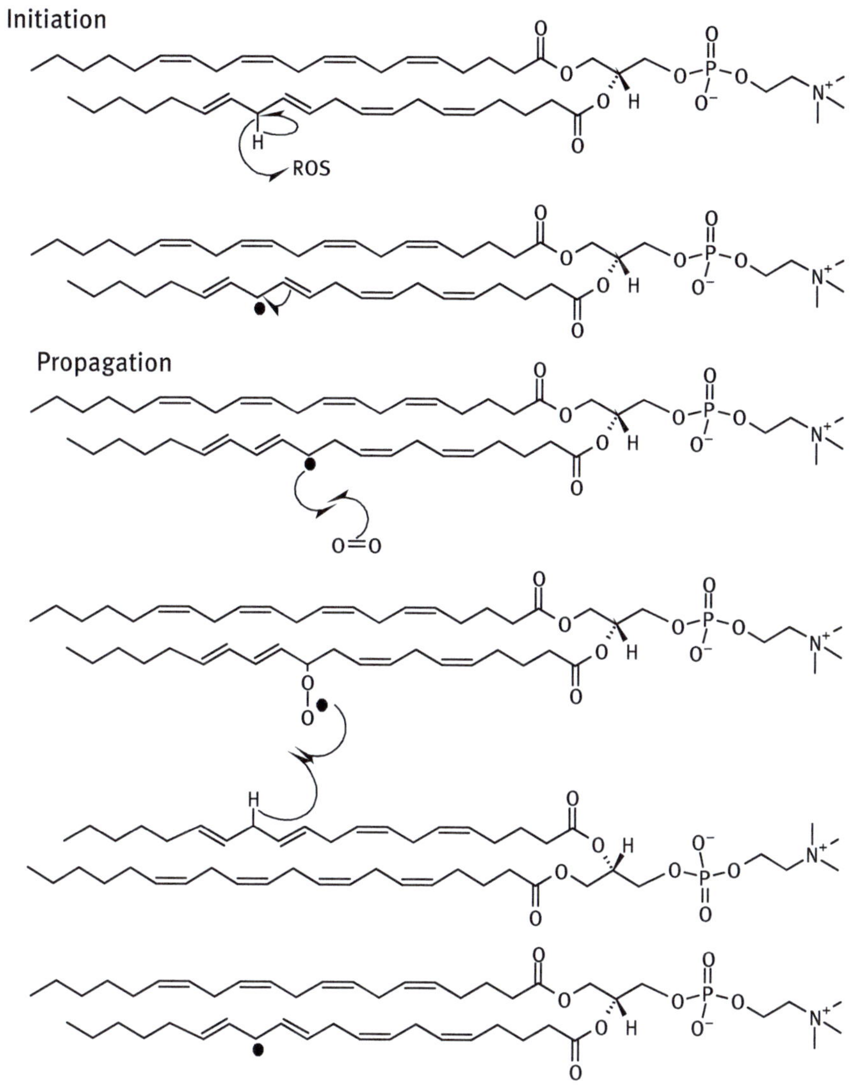

Figure 12.1: Generally accepted mechanism of PUFA oxidation. Oxidation is initiated through abstraction of a proton by a reactive oxygen species. The process propagates through the uptake of molecular oxygen. Image reproduced from Marquard [12].

component taken up by the human body for use. A growing body of research, in vivo, in vitro, and in silico, is seeking to determine what purpose α-tocopherol serves in the human body. Here, a number of experimental strategies are outlined, which are commonly employed to investigate the chemistry, biochemistry, and biology of vitamin E.

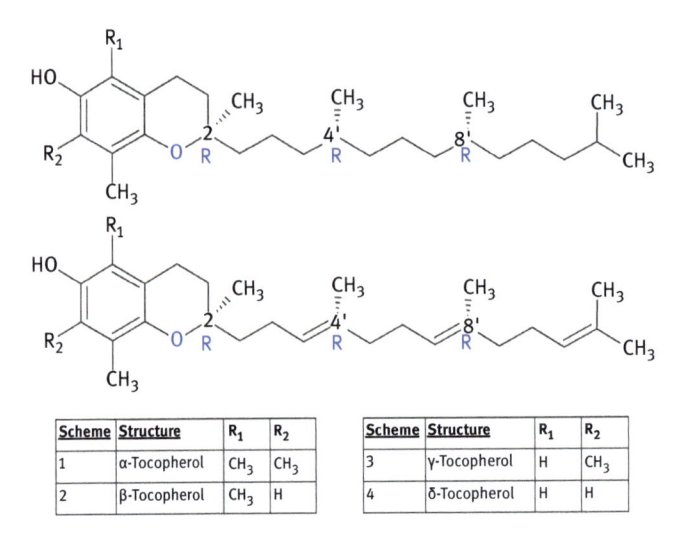

Scheme	Structure	R₁	R₂
1	α-Tocopherol	CH₃	CH₃
2	β-Tocopherol	CH₃	H
3	γ-Tocopherol	H	CH₃
4	δ-Tocopherol	H	H

Figure 12.2: Chemical structures of the vitamin E family. Absolute stereochemistry is noted as blue "R". Figure reproduced from Marquardt et al. [16].

12.2 Structural methods

12.2.1 Neutron lamellar diffraction

The high thermal disorder of fluid bilayers prevents atomic-resolution three-dimensional crystallographic images of lipids in the membrane. However, multilamellar bilayers obtained from phospholipids by dispersion in water or aligned to surfaces are highly periodic along the bilayer normal. This one-dimensional liquid crystallinity allows measurement of the distribution of matter along the bilayer normal to be determined from neutron and X-ray diffraction.

Neutron scattering studies of phospholipid membranes, or biologically relevant samples in general, are advantageous since these molecules are rich in hydrogen. The abundance of hydrogen in these systems can be used as intrinsic labels, where hydrogen (^1H) atoms can be replaced (labeled) by deuterium (^2H). The substitution of ^1H for ^2H, at selective locations, can provide enough contrast in scattering length density to yield the location and distribution of the ^2H label [17–25].

In the study of oriented membrane structure and label location, the scattering intensity is detected when the Bragg condition is satisfied. The relationship between the scattering angle (θ) and the unit cell length (d) is $h\lambda = 2d\sin(\theta)$, where λ is the wavelength of the incident neutron, h is an integer (Bragg order), and d includes the thickness of both leaflets of the lipid bilayer and one full water layer. The coherent scattering of neutrons provides information from spatial correlation of nuclei, which is contained in the amplitude of the scattered neutron wave called the scattering amplitude ($F(q)$). $F(q)$ is the sum of the coherent scattering length (b^{coh}) of all the atoms in the sample (eq. 12.1) and is proportional to the measured intensity of scattered neutrons (eq. 12.2).

$$F(q) = \sum_{i}^{atoms} b_i^{coh} e^{iq \cdot r_i} \tag{12.1}$$

$$\sqrt{I(q)} \propto |F(\mathbf{q})| \tag{12.2}$$

Experimental considerations must be taken into account when determining the experimental $F(q)$. When generating the real-space distribution of scattering lengths (ρ, scattering length density), one must take the Fourier transform of the $F(\mathbf{q})$. Experimentally, discrete values of $F(q)$ are observed; thus, the scattering length density is determined by a Fourier cosine series using the experimental $F(q) = F_h^{exp}$.

When reconstructing the $\rho(r)$, some instrumental and sample considerations and corrections must be applied to F_h^{exp} values to account for the instrument geometry and size of the neutron beam, and neutron absorption of the sample as outlined in many works investigating membrane structure.

$$\rho(r) = \frac{2}{d} \sum_{h=1}^{h} (F_h^{exp}) \cos\left(\frac{2\pi h z}{d}\right) \tag{12.3}$$

Neutron diffraction has proven vital in the determination of α-tocopherol's location and direct evidence showing that the location of α-tocopherol (Figure 12.3) in a membrane is consistent with an antioxidant function. Marquardt et al. have put forward a possible antioxidant mechanism for vitamin E which demonstrates that its physical location in the bilayer directly correlates with its antioxidant activity [22, 23]. The Marquardt–Harroun mechanism postulates that ROS and lipid radicals are terminated at the membrane's hydrophobic–hydrophilic interface. This radical termination occurs through a synergistic relationship between vitamin E's location and the unique property of PUFA acyl chains to "snorkel" to the membrane–water interface from the hydrocarbon core [26, 27]. Furthermore, this work suggests that the investigation of vitamin E should be reexamined taking into consideration the properties of the membrane in which the vitamin resides [23].

12.2.2 Fluorescence quenching

The location and distribution of vitamin E have also been monitored using fluorescence quenching [28, 29]. The quenching of a fluorescent probe by a guest molecule is

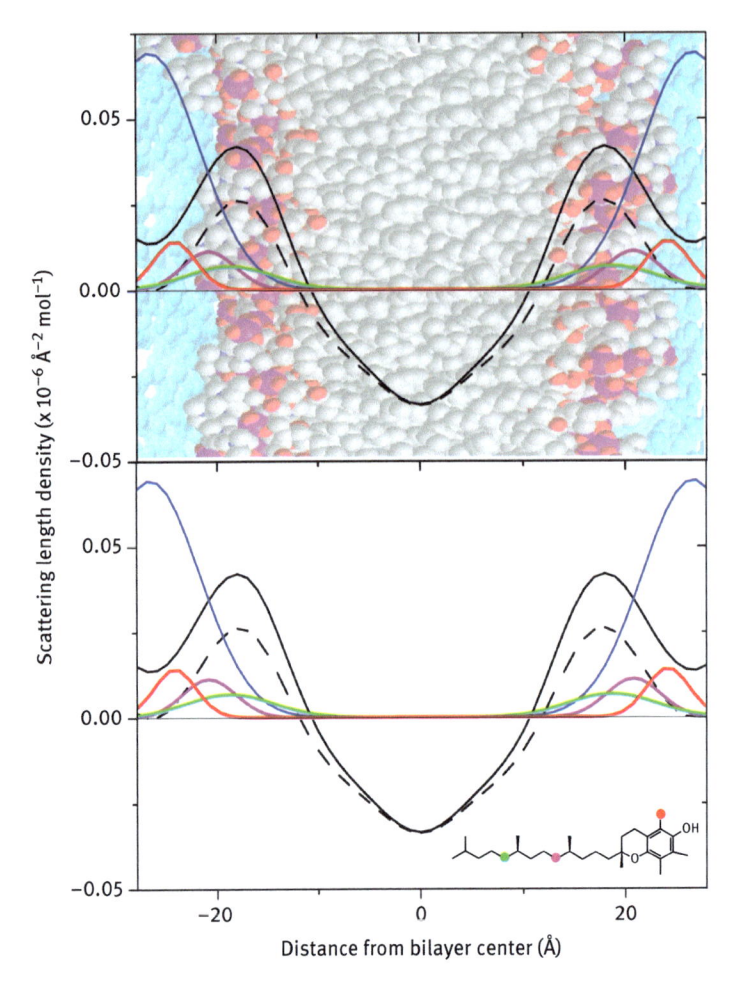

Figure 12.3: The dashed black line is the NSLD profile for 16:0–18:1PC containing 10 mol% protiated α-tocopherol hydrated with 8% H_2O. The solid black line is an example of an NSLD prole containing labeled α-tocopherol; in this case, it is 16:0–18:1PC containing 10 mol% α-tocopherol-C9′d2 hydrated with 8% H_2O. The 2H distributions for the three labels examined in 16:0–18:1PC are shown by the colored lines: α-tocopherol-C5d3 (red), α-tocopherol-C5′d2 (magenta), and α-tocopherol-C9′d2 (green). The top and bottom panel are the same plot [24].

also known as the parallax method and is one of the most straightforward fluorescent techniques when studying a molecule's presence in a given environment [30–34].

Fluorescence quenching occurs when an excited fluorophore has contact with an atom or molecule that allows the excited fluorophore to relax to its ground-state in a nonradiative manner. Fluorescence quenching processes typically come in two flavors: (1) collisional quenching and (2) static quenching. Collisional, or dynamic, quenching occurs when the excited fluorescence molecule and the quencher come in contact with each other without forming a stable molecular complex. Common quenchers relevant to vitamin E include O_2, acrylamide, and n-doxyl-PC. Static quenching occurs when the fluorophore forms a stable complex with a molecule and the relaxation from an excited state to the grounds-state is nonradiative [35]. Conveniently, both cases can be treated in very similar fashions, using the Stern–Volmer relationship.

The Stern–Volmer relationship (eq. 12.4) relates the fluorescence intensities observed in the absence and presence of quencher, I_0 and I, respectively, to the concentration of quencher $[Q]$ and the Stern–Volmer quenching constant, K_{SV}. The plot of I_0/I versus $[Q]$, known as a Stern–Volmer plot, yields a straight line with a slope equal to K_{SV}. For purely dynamic quenching, $I_0/I = \tau_0/\tau$, where τ_0 and τ are the excited state lifetime in the absence and presence of quencher, respectively. Thus, dynamic quenching can be used to determine molecular distributions as well as diffusion. Diffusional information is determined by expressing K_{SV} as $k_q \tau_0$, where k_q is proportional to the sum of the diffusion coefficients for fluorophore and quencher.

$$\frac{I_0}{I} = 1 + K_{SV}[Q] \qquad (12.4)$$

Static quenching differs in the quenching constant. Since quenching occurs when a stable complex is formed between the fluorophore and quencher, K_{SV} becomes K_a, the association constant of the complex. Figure 12.4 illustrates the Stern–Volmer plots for both intensity and relaxation times for both types of quenching.

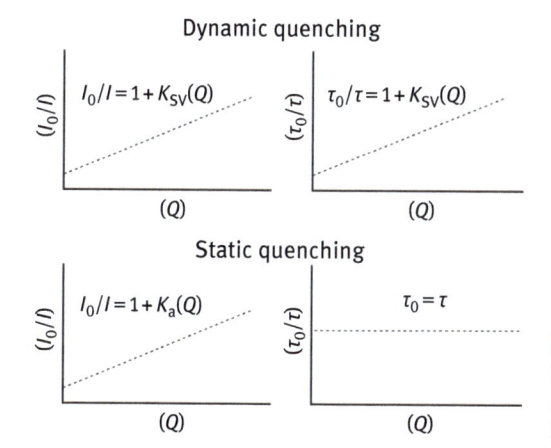

Figure 12.4: Idealized Stern–Volmer plots for dynamic and static fluorescence quenching. Indicated on the plots are the linear relationships to extract K_{SV} and K_a.

Much effort has been invested in attempts to determine the location of vitamin E in membranes using quenching techniques. Typically n-doxyl or anthroyloxyl stearic acid would be used, exploiting the intrinsic fluorescence of vitamin E [14, 36, 37]. The resulting quenching efficiencies suggest vitamin E resides high in the phospholipid bilayer. Most recently, the work of Ausili et al. [28] attempts to elucidate the location of α-tocopherol using both water soluble acrylamide and n-doxyl-PC labeled at different locations along the fatty acid chain [28, 29]. Although these works are able to recover, qualitatively, the probe-free measurements carried out by Marquardt et al., one must proceed with caution. The quenching agents do have the potential to disrupt the local environment of the phospholipid membrane; however, if the quencher is present at low concentration, there are minimal artifacts [38].

12.2.3 Electron paramagnetic resonance

The diffusion and transport rate of O_2 in both in vitro and in vivo membranes is a vital parameter when investigating the mechanisms of membrane oxidation and the role of fat-soluble antioxidants. Electron paramagnetic resonance (EPR) offers a straightforward and relatively simple strategy for monitoring O_2 diffusion and transport within and across membranes.

EPR can monitor the collisional frequency between oxygen and a spin labeled molecule. In the study of membranes, nitroxide-labeled phospholipids serve as an effective probe and lipids can be readily purchased with nitroxide labels at various positions (Figure 12.5). This strategy is sensitive to the spin-lattice relaxation times T_1 of the nitroxide group and it changes in the presence of O_2. The oxygen transport parameter (W, eq. 12.5) can be expressed in terms of the T_1 in the presence and absence of air, T_1 (air) and T_1 (N_2), respectively [39, 40].

$$W = T_1^{-1}(\text{air}) - T_1^{-1}(N_2) \propto D_0 [O_2] \quad (12.5)$$

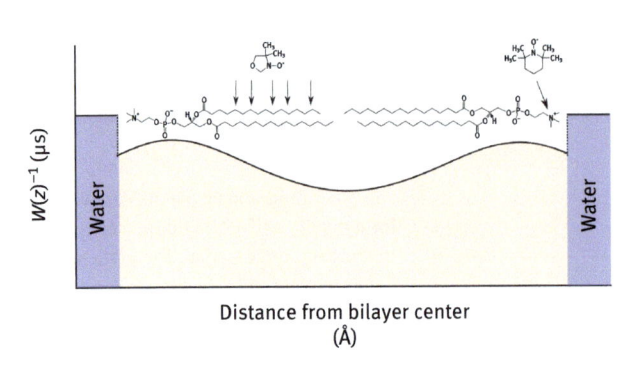

Figure 12.5: Idealized O_2 permeability plot. Lipid structures are shown with arrows indicating where commercially available nitroxide labels are located along the phospholipid.

Perhaps a more physically useful parameter would be the membrane permeability coefficient, P_M, which is a physical characteristic of the membrane as a whole. P_M is essentially the flux of O_2 across the membrane.

$$P_M = \frac{1}{8\pi p r_0 C_w \text{(air)}} \left(\int_{-D_B/2}^{D_B/2} \frac{dz}{W(z)} \right) \quad (12.6)$$

Here, r_0 and p are the interaction distance between O_2 and the nitroxide label (4.5 Å) and p is the probability where an observable interaction occurs [41].

Recent studies have shown that low levels of lipid oxidation increase the passive permeability of lipid bilayers [42]. In fact, with as little as 2.5% oxidized lipid, the passive permeability increased an order of magnitude, from 1.5×10^{-6} to 1.5×10^{-5} cm s^{-1} [42]. Such a result highlights the importance of mitigating bilayer oxidation.

Similarly, the EPR strategy can be used for nitric oxide (NO˙) and it has been shown that NO˙ has a greater permeability than O_2 through a lipid bilayer [43].

12.3 Oxidation and oxygen consumption

12.3.1 UV/Vis spectroscopy

Oxidative stress causes great harm to biological membranes by initiating a series of reactions that form propagative radical products leading to bilayer breakdown. To better understand these reactions and how the physical structure of a membrane affects them, a more straightforward way to monitor the progress of oxidation would be advantageous. Due to the fact that oxidation reactions create and destroy carbon–carbon double bonds, ultraviolet (UV) spectroscopy is an ideal experimental technique to study these reactions.

In a transmission style ultraviolet and visible (UV/vis) spectrophotometer, the absorbance (A) is measured via the percent of incident light that is transmitted (T) through the sample. The absorbance is related to the percent transmittance (%T) as

$$A = -\log\left(\frac{\%T}{100\%}\right) \quad (12.7)$$

Absorbance can also be related to the concentration of solute in the solution (assuming that the solute is UV/vis active), which is known as the Beer–Lambert law:

$$A = \varepsilon C l \quad (12.8)$$

where ε is the extinction coefficient, C is the concentration of the solute, and l is the path length of the sample. The principle of relating absorbance to concentration can

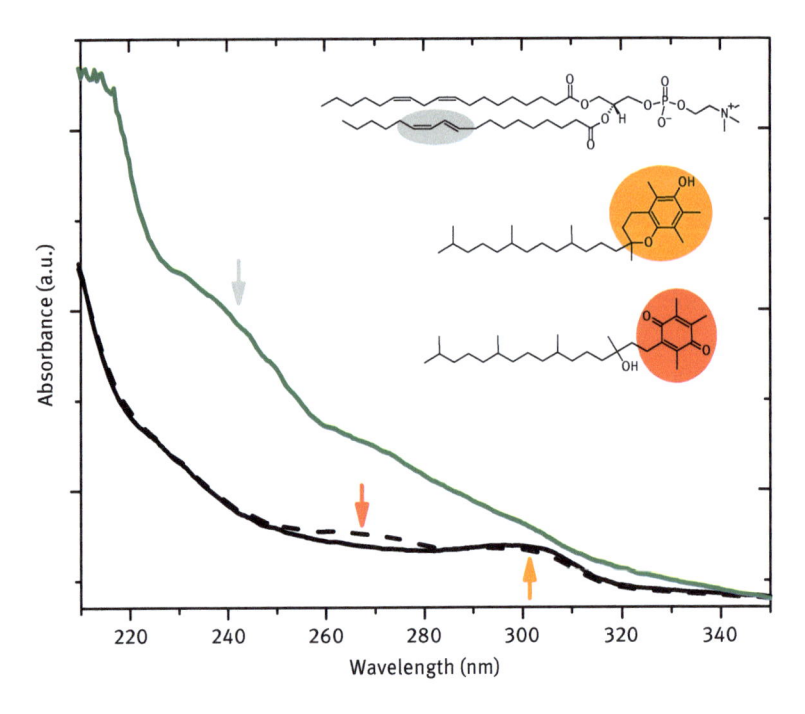

Figure 12.6: UV absorption of 16:0–16:0PC + α-tocopherol for oxidation at time $t = 0$ (solid black line) and at $t = 8$ h (dashed black line) using the Fenton reagent. Oxidized 20:4–20:4PC bilayers are shown a solid green line. The functionality responsible for the different absorbances is highlighted on the chemical structures with the color-coded arrows identifying where the absorbance is on the graph [22].

be used to monitor qualitative increases of oxidation products or decreases in α-tocopherol concentration [22].

Throughout the years, UV/vis spectroscopy has been a popular technique for measuring the level of lipid peroxidation due to its relative technical ease and low costs, despite its potential for inaccurate readings. Lipid peroxide formation is often implied but not required for absorption at the relevant wavelengths. Despite this, this technique is able to do so, albeit indirectly, by measuring the levels of oxidized lipid byproducts: lipid conjugated dienes, lipid hydroperoxides, thiobarbituric acid reactive species, and many more.

Past work on simple phospholipid systems studied a variety of phospholipid bilayers possessing varying degrees of oxidation susceptibility. Oxidation can be initiated from both the water using Fenton chemistry ($Fe^{2+} + H_2O_2 \rightarrow Fe^{2+} + HO\cdot$) and from within the hydrophobic region using azo-compounds and radical initiators such as AMVN [2,2′-azobis-(2,4-dimethylvaleronitrile)]. These types of peroxidized systems have since served as a basis for oxidation-related biological studies.

Scientists have applied these techniques to study biomolecules and their relation to oxidative stress, antioxidation capabilities, and disease. Work done by Fabre et al.

[44] aimed to determine the peroxidation inhibiting potential of lipocarbazole, a lipophilic molecule extracted from the Tsukamurella pseudospumae Acta 1857 bacteria, by comparing it to a known natural antioxidant, vitamin E. The oxidation of PUFAs was monitored by UV/vis absorption at a wavelength of 233 nm. Specifically, this wavelength is the absorbance of the isomerization of the PUFA to the conjugated diene form [45]. It was seen that conjugated diene formation was greater when vitamin E was present as compared to lipocarbazole. In other words, lipocarbazole showed greater antioxidation capabilities than vitamin E, demonstrating its potential to be a lipid peroxidation inhibitor to prevent oxidative stress-induced diseases (e.g., atherosclerosis, Alzheimer's disease, and cirrhosis).

12.3.2 Ozawa–Flynn–Wall method

Studying lipid autoxidation can often be quite a slow measurement and it is difficult to monitor if double bond conjugation is not present. In a series of articles, a differential scanning calorimetry (DSC) strategy was developed that the foodstuff industry has adopted as a work-around to this problem. This strategy employs the steady-state approximation and the Ozawa–Flynn–Wall method [46, 47] to determine the overall Arrhenius rate constant (k) and the activation energy (E_a) from eq. (12.9) using experimental DSC data.[1]

The rate constant, k, of oxidation can be calculated using the Arrhenius equation and extrapolated oxidation onset temperatures from the plots outlined by the Ozawa–Flynn–Wall method. In a nutshell, the method takes a series of nonisothermal DSC scans of the same system at different rates of heating $\beta = \partial T/\partial t$, as shown in Figure 12.7. The inverse of the oxidation onset temperatures is plotted against the logarithm of the heating rate ($\log \beta$). This plot gives a straight line with a slope of A and the y-intercept is B.

The rate of conversion of the bis-allylic hydrogen [R–H] is related to the heating rate by the differential equation

$$\frac{\partial [R-H]}{\partial t} = \beta \frac{\partial [R-H]}{\partial T} = -k[R-H] \quad (12.9)$$

where

$$k = Ze^{(-E_a/RT)} \quad (12.10)$$

Ozawa, Flynn, and Wall wrote eqs. (12.9) and (12.10) in an integral form, followed by a replacement of the integrand by an approximation function, which leads to the following:

---

[1] The Ozawa-Flynn-Wall method assumes $E_a/RT \geq 20$.

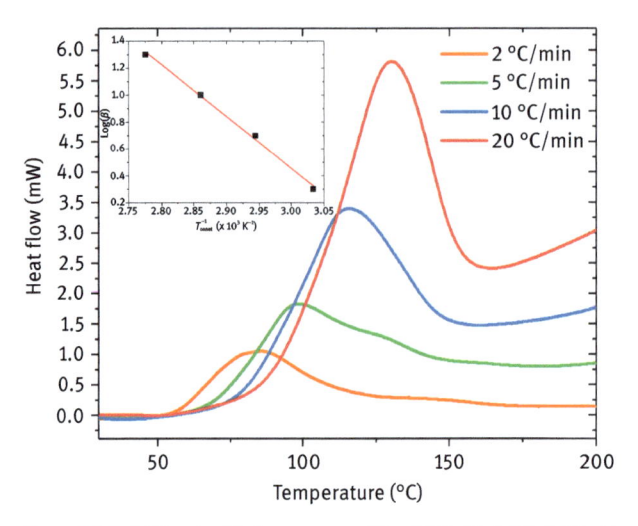

Figure 12.7: DSC autoxidation curves of the PUFA 20:4/20:4 PC at heating rates of 2, 5, 10, and 20 °C min^{-1}. Inset Ozawa–Flynn–Wall plot for 20:4/20:4 PC to obtain the kinetic parameters using the interpolated onset temperatures. Adapted from reference [51].

$$\log \beta = \frac{A}{T} + B \qquad (12.11)$$

The parameters for the Arrhenius equation are taken from the slope and the y-intercept of the plot as outlined in eqs. (12.12) and (12.13).

$$E_a = -2.19RA \qquad (12.12)$$

$$Z = \frac{e^{B+2.315}R}{E_a} \qquad (12.13)$$

The value of k can be plotted to find its value for any given temperature. This method also allows for the study of an antioxidant's effect on the activation energy and the rate constant for lipid autoxidation.

This method has been used for studying autoxidation stability of saturated and unsaturated fatty acids and their esters in the absence of solvent [48–50] and PUFAs in the presence and absence of α-tocopherol [51].

12.3.3 Oxygen uptake

Oxygen uptake studies offer the possibility of monitoring time-dependent autoxidation through the consumption of molecular oxygen. Oxygen consumption rates allow for the extraction of various kinetic parameters that can be attributed to molecular antioxidation pathways and can be used to quantify the effectiveness of a peroxidation inhibitor (antioxidant). This method tracks peroxide forming events, as opposed to the fluctuation of carbon–carbon double bonds observed by UV/vis spectroscopic techniques. Depicted in Figure 12.8, the experimental design is flexible, permitting comparisons between different antioxidants and different peroxidation pathways. Variation of experimental conditions can also suggest patterns such as kinetic solvent effects.

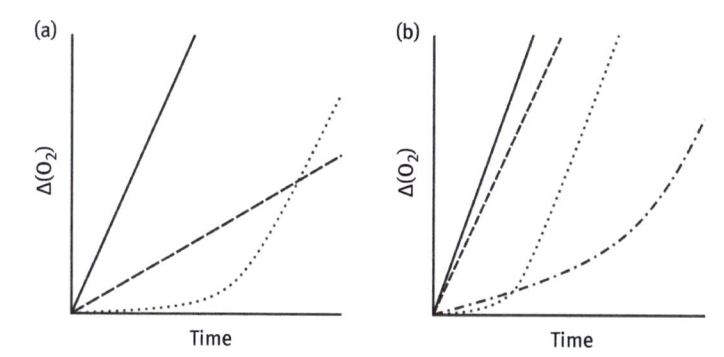

Figure 12.8: Oxygen uptake plots for systems subjected to lipid peroxidation. (A) Oxygen uptake rates for autoxidation in an unprotected system (solid line), as compared to a system doped with a chain-breaking antioxidant (dotted line) or an interfering antioxidant (dashed line). (B) Effectiveness of a chain-breaking antioxidant against alkylperoxyl (ROO˙) radical propagation in uninhibited (solid line) or doped systems (dotted line) compared to hydroperoxyl (HOO˙) radical propagation in uninhibited (dashed line) or doped (dot-dashed line).

The slope of the tangent to the oxygen uptake plot can be quantified as the rate of oxygen consumption, and thus, the rate of peroxidation, R. The induction period, τ_{ind}, where the antioxidant actively inhibits peroxidation, can be determined from the integral:

$$\tau_{ind} = \int_0^\infty \left\{ 1 - \left(\frac{R}{R_{ox}} \right)^2 \right\} dt \qquad (12.14)$$

where R is the rate of oxygen consumption and R_{ox} corresponds to the maximum rate of oxygen consumption, after the induction period.

The determined length of the induction period can be related to the rate of initiation R_i through the method proposed by Hammond and coworkers [52].

$$R_i = \frac{n[\text{inh}]_0}{\tau_{ind}} \qquad (12.15)$$

where $[\text{inh}]_0$ is the initial concentration of the autoxidation inhibitor (antioxidant) and n is the stoichiometric factor of inhibitor, represented physically as the number of peroxyl radicals scavenged per molecule of antioxidant. Further, the parameter of radical chain length v can be defined as the ratio of peroxidation rate R to the rate of initiation R_i, as shown in eq. (12.16). This is physically attributed to the number of peroxyl radicals formed per initiating radical.

$$v = \frac{R_{ox}}{R_i} \tag{12.16}$$

Absolute rate constants of inhibition k_{inh} are derived through the integrated expression of oxidation kinetics during the induction period, as described by Barclay et al., eq. (12.17) [53].

$$\Delta[O_2](t) = -\frac{k_p}{k_{inh}}[\text{LH}] \ln\left(1 - \frac{t}{\tau_{ind}}\right) \tag{12.17}$$

k_p is the rate constant of propagation and $[\text{LH}]$ is the concentration of lipid. A plot of eq. (12.17) yields a linear correlation with slope $k_p/k_{inh}[\text{LH}]$, from which the rate constant of inhibition can be extracted.

Basal oxygen consumption in living systems presents an obstacle when measuring autoxidation events in vivo. Maintenance of cellular homeostasis requires variability of respiration, and thus an inconsistency in oxygen uptake emerges. It is for this caveat that studies using oxygen uptake to track membrane peroxidation are most applicable in vitro, exploiting the control offered by model membrane systems.

Work by Jodko-Piorecka et al. [54] measured oxygen uptake in lipid emulsions to assess the antioxidant ability of the abundant catecholamines that are capable of crossing the blood–brain barrier, such as L-DOPA and dopamine. Based on the oxygen consumption rates, it was determined that catecholamines retard peroxidation by reacting with initiating radicals (Figure 12.8a, dashed line), as opposed to behaving as a chain-breaking antioxidant, such as α-tocopherol (Figure 12.8a, dotted line) [54]. Remarkably, peroxidation of lipid membranes protected by both catecholamines and α-tocopherol exhibits a significant increase in τ_{ind}, suggesting a synergistic effect between the two peroxidation inhibition mechanisms. Parallel experiments indicated that these antioxidants perform most effectively at an intermediate pH, with activity breaking down in alkaline solvents as a consequence of ionization.

Recently, oxygen uptake methods were used to compare the effectiveness of α-tocopherol against hydroperoxyl (HOO⁻) and alkylperoxyl-borne (ROO⁻) radicals [55]. Similar to the trend seen in Figure 12.8b, the chain-breaking antioxidant initially inhibits induction from ROO⁻ but is far more effective at scavenging HOO⁻. This trend is rationalized by the potential for HOO⁻ to reduce the α-tocopheroxyl radical, thereby extending τ_{ind}. In addition, Cedrowski and coworkers indicate the dependence of peroxidation rate on the solvents' hydrogen-bonding tendencies. Solvents with a propensity to accept hydrogen bonds are likely able to stabilize HOO$^-$, thereby facilitating the reduction of α-tocopheroxyl radicals and slowing the peroxidation rate [55].

12.4 Computational approaches

12.4.1 Molecular dynamic simulations

Given the complexity of a plasma membrane, where the constituents encompass a wide range of biomolecules (e.g., lipids, proteins, and vitamins), simpler membrane mimetic models are required to study basic structural and functional membrane properties, biomolecular interactions, and other biophysical properties. In the realm of biomembrane research, the intricacy of a plasma membrane can be reduced using molecular dynamic simulations.

Computer simulations of lipid membranes, with and without proteins, continue their exponential growth in size and complexity. Some very recent examples looking at the membrane–water interface, conformational landscapes of membrane proteins, and extending to complex glycolipids include Refs [56–61]. Simulations of the effect of oxidized lipids on bilayer structure, however, make up a minority of this field.

Atomistic molecular dynamic simulations have been previously used to examine the physical presence of vitamin E and its interactions with neighboring particles. Fabre et al. [44] showed that α-tocopherol preferentially locates toward phospholipid headgroups, specifically in the interfacial region. This is an ideal location to stop ROS from penetrating the bilayer, as well as to inhibit lipid peroxidation from continuing within the core. Furthermore, α-tocopherol was seen to be capable of flip-flop, going from one leaflet to the other, giving it a greater propensity to encounter peroxidized lipids. Lastly, the oxidized product, α-tocopheroxyl, showed temporary noncovalent binding with vitamin C, a crucial step in regenerating α-tocopherol for further antioxidant activity.

In a study that shows the versatility of computer simulations, Leng et al. [62] coupled MD simulations with solid-state ^{2}H nuclear magnetic resonance (NMR) and neutron diffraction experiments to illustrate α-tocopherol's possible antioxidation activity in PUFA bilayers. First, through MD simulations which reflected experimental NMR results, the order parameter (S_{CD}, a measure of the degree of anisotropy and, to a larger extent, membrane disorder) of methyl and methylene groups in polyunsaturated phospholipid bilayers was found. It revealed that, though α-tocopherol possessed a slight condensing effect, the polyunsaturated acyl-chain was still able to fluctuate between different conformational states, permitting the chain to ascend toward the lipid–water interface where the antioxidant moiety of α-tocopherol seems to predominately reside.

Leng et al. were also able to construct neutron scattering length density (NSLD) profiles, a technique typically associated with neutron scattering, using MD simulations that consisted of 100 lipid molecules and 2,000 water molecules. Simulated results were seen to be in agreement with experimental neutron diffraction NSLD profiles, where both situated the chromanol group of α-tocopherol near the bilayer surface but differed in its exact location. Computational results placed the chromanol group in the interfacial region (i.e., hydrophobic–hydrophilic interface) of the bilayer, whereas experimental results had it almost adjacent to the aqueous environment. This is an interesting development which further compounds the debate regarding vitamin E localization and demands more research to be conducted.

MD has also revealed insights on oxidized PUFA lipid membranes; however, hybrid QM/MD simulations of membranes undergoing oxidation have not been reported. Rather, pre-oxidized peroxyl and hydroperoxide lipids were incorporated into membranes as set concentrations. Garrec et al. showed that both moieties of dilinoleic acid PC exhibit large displacements, exploring different depths inside the bilayer [63].

Many force field parameters for oxidized lipids are traced to Wong-ekkabut et al. whose study of hydroperoxidized or an aldehydezed palmitoyl–linoleoyl PC showed substantial structural and dynamical effects, including that the oxidized moiety is generally found close to the lipid headgroup region and forms hydrogen bonds mainly with water [64]. Boonnoy et al. extended this simulation to peroxide and aldehyde palmitoyl–decanoyl PC lipids, which at high concentrations cause water pores that may become unstable leading to micellization [65].

All things considered, this difference (and the similarities mentioned above) demonstrates the additional perspective and benefits MD simulations can provide to experimental work. Essentially, computational studies have become an integral part of science, affirming ideas of the past and serving as a platform for future experiments and innovations; in silico work will certainly continue to play a large role in the progress of vitamin E understanding.

12.4.2 Quantum calculations

A variety of theoretical studies have been performed to obtain mechanistic insights into antioxidant reactions. Because α-tocopherol's antioxidant reaction involves a hydrogen transfer step, there have been many attempts to ascertain whether quantum effects play some role in its antioxidant efficiency by computational quantum chemistry methods.

A number of proposed mechanisms have been put forward based on density functional theory (DFT) calculations; one step hydrogen atom transfer, charge transfer with simultaneous proton tunneling, stepwise single electron transfer–proton transfer, or sequential proton loss-electron transfer. In hydrogen atom transfer, the H atom is moved to an acceptor in a single reaction step. The single electron transfer–proton transfer pathway involves electron transfer and α-tocopheroxyl formation followed by deprotonation in the following reaction step. The sequential proton loss electron transfer begins with deprotonation of antioxidant molecule and the electron transfer from the resulting phenoxide anion in the second reaction step [55, 63, 66, 67].

A wide variety of basis functionals have been employed; however, there seems to be no preferred basis or degree to the methods. In nearly all cases, however, the optimized geometry of α-tocopherol was in vacuum or implicit solvent, and never in the presence of lipids, presumably due to the computational difficulties of too many atoms. Indeed, some like Cedrowski et al. used a simplified nonsolvated phenol as a partial model of α-tocopherol to estimate the hydrogen atom transfer activation energy [55].

DFT has also been used to estimate the oil–water partitioning coefficients of simple models of peroxyl radicals and hydroperoxide derivatives of PUFA lipids, $CH_3OO^{\cdot}$, and CH_3OOH [63]. The basis and calculation method did not affect the results when using a polarizable continuum model mimicking either bulk water or an oil phase.

12.5 Concluding remarks

We have discussed a variety of techniques that prove valuable in the study of oxidation, autoxidation, and antioxidation in both in vitro and in vivo systems. Despite over 90 years of research efforts direct toward vitamin E, an understanding of the biological role and underlying molecular mechanisms remains elusive.

References

[1] Niki, E. Oxidant-specific biomarkers of oxidative stress. association with atherosclerosis and implication for antioxidant effects, Free Radical Biology and Medicine. 120 (2018). doi:doi.org/10.1016/j.freeradbiomed.2018.04.001.
[2] Niki, E. Lipid peroxidation: Physiological levels and dual biological effects, Free Radical Biology & Medicine. 47 (2009) 469–484.
[3] Genestra, M. Oxyl radicals, redox-sensitive signalling cascades and antioxidants, Cellullar Signalling. 19 (2007) 1807–1819.
[4] Valko, M., Leibfritz, D., Moncol, J., Cronin, M.T., Mazur, M., & Telser, J. Free radicals and antioxidants in normal physiological functions and human disease, International Journal of Biochemistry and Cell Biology. 39 (2007) 44–84.

[5] Droge, W. Free radicals in the physiological control of cell function, Physiological Reviews. 82 (2002) 47–95.
[6] Pham-Huy, L.A., He, H., & Huy, C.P. Free radicals, antioxidants in disease and health, International, Journal of Biomedical Science. 4 (2008) 89–96.
[7] Seitz, S.P., & Nelson, M.J. The structure and function of lipoxygenase, Current Opinion in Structural Biology. 4 (1994) 878–884.
[8] Cadenas, E., & Sies, H. The lag phase, Free Radical Research. 28 (1998) 601–609.
[9] Spiteller, G. Peroxyl radicals: Inductors of neurodegenerative and other inflammatory diseases. their origin and how they transform cholesterol, phospholipids, plasmalogens, polyunsaturated fatty acids, sugars, and proteins into deleterious products, Free Radical Biology & Medicine. 41 (2006) 362–387.
[10] Jomova, K., & Valko, M. Advances in metal-induced oxidative stress and human disease, Toxicology. 283 (2011) 65–87.
[11] Barclay, L.R.C., Locke, S.J., & Macneil, J.M. The autoxidation of unsaturated lipids in micelles, synergism of inhibitors vitamins C and E, Canadian Journal of Chemistry. 61 (1983) 1288–1290.
[12] Marquardt, D. α-tocopherol's antioxidant role: A biophysical perspective, phdthesis, Brock University, St. Catherines, Ontario, Canada (2014).
[13] Alam, M.N., Bristi, N.J., & Rafiquzzaman, M. Review on in vivo and in vitro methods evaluation of antioxidant activity, Saudi Pharmaceutical Journal. 21 (2) (2013) 143–152.
[14] Fukuzawa, K. Dynamics of lipid peroxidation and autoxidation of alpha-tocopherol, Journal of Nutritional Science and Vitaminology. 54 (2008) 273–285.
[15] Evans, H.M., & Bishop, K.S. On the existence of a hitherto unrecognized dietary factor essential for reproduction, Science. 56 (1458) (1922) 650–651.
[16] Marquardt, D., Van Oosten, B.J., Ghelfi, M., Atkinson, J., & Harroun, T.A. Vitamin E circular dichroism studies: Insights into conformational changes induced by the solvent's polarity, Membranes. 6 (4) (2016) 56. doi:10.3390/membranes6040056.
[17] Kucerka, N., Marquardt, D., Harroun, T.A., Nieh, M.-P., Wassall, S.R., de Jong, D.H., Schafer, L.V., Marrink, S.J., & Katsaras, J. Cholesterol in bilayers with pufa chains: Doping with dmpc or popc results in sterol reorientation and membrane-domain formation, Biochemistry. 49 (2010) 7485–7493.
[18] Harroun, T.A., Katsaras, J., & Wassall, S.R. Cholesterol hydroxyl group is found to reside in the center of a polyunsaturated lipid membrane, Biochemistry. 45 (2006) 1227–1233.
[19] Harroun, T.A., Katsaras, J., & Wassall, S.R. Cholesterol is found to reside in the center of a polyunsaturated lipid membrane, Biochemistry. 47 (27) (2008) 7090–7096.
[20] Kucerka, N., Marquardt, D., Harroun, T.A., Nieh, M.-P., Wassall, S.R., & Katsaras, J. The functional significance of lipid diversity: Orientation of cholesterol in bilayers is determined by lipid species, Journal of the American Chemical Society. 131 (2009) 16358–16359.
[21] Komljenovic, I., Marquardt, D., Harroun, T.A., & Sternin, E. Location of chlorhexidine in dmpc model membranes: a neutron diffraction study, Chemistry and Physics of Lipids. 163 (2010) 480–487.
[22] Marquardt, D., Williams, J.A., Kucerka, N., Atkinson, J., Wassall, S.R., Katsaras, J., & Harroun, T.A. Tocopherol activity correlates with its location in a membrane: A new perspective on the antioxidant vitamin E, Journal of the American Chemical Society. 135 (20) (2013) 7523–7533.
[23] Marquardt, D., Williams, J.A., Kinnun, J.J., Kucerka, N., Atkinson, J., Wassall, S.R., Katsaras, J., & Harroun, T.A. Dimyristoyl phosphatidylcholine: A remarkable exception to α-tocopherol's membrane presence, Journal of the American Chemical Society. 136 (1) (2014) 203–210.
[24] Marquardt, D., Kucerka, N., Katsaras, J., & Harroun, T.A. α-tocopherol's location in membranes is not affected by their composition, Langmuir. 31 (15) (2015) 4464–4472.

[25] Marquardt, D., Frontzek, M.D., Zhao, Y., Chakoumakos, B.C., & Katsaras, J. Neutron diffraction from aligned stacks of lipid bilayers using the WAND instrument, Journal of Applied Crystallography. 51 (2018) 235–241.

[26] Eldho, N.V., Feller, S.E., Tristram-Nagle, S., Polozov, I.V., & Gawrisch, K. Polyunsaturated docosahexaenoic vs docosapentaenoic acid - Differences in lipid matrix properties from the loss of one double bond, Journal of the American Chemical Society. 125 (2003) 6409–6421.

[27] Soubias, O., & Gawrisch, K. Docosahexaenoyl chains isomerize on the subnanosecond time scale, Journal of the American Chemical Society. 129 (2007) 6678–6679.

[28] Ausili, A., de Godos, A.M., Torrecillas, A., Aranda, F.J., Corbalan-Garcia, S., & Gomez-Fernandez, J.C. The vertical location of α-tocopherol in phosphatidylcholine membranes is not altered as a function of the degree of unsaturation of the fatty acyl chains, Phys. Chem. Chem. Phys. 19 (2017) 6731–6742.

[29] Ausili, A., Torrecillas, A., de Godos, A.M., Corbalán-García, S., & Gómez-Fernández, J.C. Phenolic group of α-tocopherol anchors at the lipid-water interface of fully saturated membranes, Langmuir. 34 (10) (2018) 3336–3348.

[30] Kachel, K., Asuncion-Punzalan, E., & London, E. The location of fluorescence probes with charged groups in model membranes, Biochimica et Biophysica Acta-biomembranes. 1374 (1–2) (1998) 63–76.

[31] Boldyrev, I.A., Zhai, X., Momsen, M.M., Brockman, H.L., Brown, R.E., & Molotkovsky, J.G. New BODIPY lipid probes for fluorescence studies of membranes, Journal Of Lipid Research. 48 (7) (2007) 1518–1532.

[32] Kondo, M., Mehiri, M., & Regen, S.L. Viewing membrane-bound molecular umbrellas by parallax analyses, Journal of the American Chemical Society. 130 (41) (2008) 13771–13777.

[33] Shrivastava, S., Haldar, S., Gimpl, G., & Chattopadhyay, A. Orientation and dynamics of a novel fluorescent cholesterol analogue in membranes of varying phase, Journal Of Physical Chemistry B. 113 (13) (2009) 4475–4481.

[34] Marquardt, D., & Harroun, T.A. Locations of small biomolecules in model membranes, in: G. Pabst, N. Kucerka, M.-P. Nieh, & J. Katsaras (Eds.), Liposomes, Lipid Bilayers and Model Membranes: From Basic Research to Application, CRC Press, Boca Raton, London, New York, 2014, Ch. 11, pp. 199–216.

[35] Weber, G. The quenching of fluorescence in liquids by complex formation. determination of the mean life of the complex, Trans. Faraday Soc. 44 (1948) 185–189.

[36] Fukuzawa, K., Ikebata, W., Shibata, A., Kumadaki, I., Sakanaka, T., & Urano, S. Location and dynamics of a-tocopherol in model phospholipid membranes with different charges, Chemistry and Physics of Lipids. 63 (1992) 69–75.

[37] Fukuzawa, K., Ikebata, W., & Sohmi, K. Location, antioxidant and recycling dynamics of alpha-tocopherol in liposome membranes, Journal of Nutritional Science and Vitaminology. 39 (1993) 9–22.

[38] Ackerman, D.G., Heberle, F.A., & Feigenson, G.W. Limited perturbation of a dppc bilayer by fluorescent lipid probes: A molecular dynamics study, The Journal of Physical Chemistry B. 117 (17) (2013) 4844–4852.

[39] Kusumi, A., Subczynski, W.K., & Hyde, J.S. Oxygen transport parameter in membranes as deduced by saturation recovery measurements of spin-lattice relaxation times of spin labels, Proceedings of the National Academy of Sciences. 79 (6) (1982) 1854–1858.

[40] Subczynski, W.K., & Swartz, H.M. Epr oximetry in biological and model samples, in: S. S. Eaton, G. R. Eaton, & L. J. Berliner (Eds.), Biomedical EPR - Part A: Free Radicals, Metals, Medicine and Physiology, Kluwer Academic/Plenum Publishers, New York, 2005, Ch. 10, pp. 229–282.

[41] Subczynski, W., & Hyde, J.S. The diffusion-concentration product of oxygen in lipid bilayers using the spin-label t1 method, Biochimica et Biophysica Acta (BBA) - Biomembranes. 643 (2) (1981) 283–291.
[42] Runas, K.A., & Malmstadt, N. Low levels of lipid oxidation radically increase the passive permeability of lipid bilayers, Soft Matter. 11 (2015) 499–505.
[43] Subczynski, W.K., Lomnicka, M., & Hyde, J.S. Permeability of nitric oxide through lipid bilayer membranes, Free Radical Research. 24 (5) (1996) 343–349.
[44] Fabre, G., Hänchen, A., Calliste, C.A., Berka, K., Banala, S., Otyepka, M., Süssmuth, R.D., & Trouillas, P. Lipocarbazole, an efficient lipid peroxidation inhibitor anchored in the membrane, Bioorganic and Medicinal Chemistry. 23 (15) (2015) 4866–4870.
[45] Kim, R.S., & LaBelia, F.S. Comparison of analytical methods for monitoring autoxidation profiles of authentic lipids, Journal of Lipid Research. 28 (1987) 1110–1117.
[46] Flynn, J., & Wall, L. A quick, direct method for the determination of activation energy from thermogravimetric data, Polymer Letters. 4 (1966) 323–328.
[47] Ozawa, T. Non-isothermal kinetics of diffusion and its application to thermal analysis, Journal of Thermal Analysis. 5 (1973) 563–576.
[48] Litwinienko, G., Daniluk, A., & Kasprzycka-Guttman, T. A differential scanning calorimetry study on the oxidation of c12–c18 saturated fatty acids and their esters, Journal of the American Oil Chemists' Society. 76 (6) (1999) 655–657.
[49] Litwinienko, G., & Kasprzycka-Guttman, T. Study on the autoxidation kinetics of fat components by differential scanning calorimetry. 2. Unsaturated fatty acids and their esters, Industrial and Engineering Chemistry Research. 39 (2000) 13–17.
[50] Litwinienko, G., Daniluk, A., & Kasprzycka-Guttman, T. Study on autoxidation kinetics of fats by differential scanning calorimetry. 1. Saturated C12–C18 fatty acids and their esters, Industrial Engineering Chemistry Research. 39 (2000) 7–12.
[51] Marquardt, D. α-toopherol: a poly-unsaturated lipid antioxidant, Honours thesis, Brock University, St Catharines, Ontario, Canada (2010).
[52] Hammond, G.S., Boozer, C.E., Hamilton, C.E., & Sen, J.N. Air oxidation of hydrocarbons. iii. mechanism of inhibitor action in benzene and chlorobenzene solutions, Journal of the American Chemical Society. 77 (12) (1955) 3238–3244.
[53] Barclay, L.R.C., Edwards, C.D., Mukai, K., Egawa, Y., & Nishi, T. Chain-breaking naphtholic antioxidants: Antioxidant activities of a polyalkylbenzochromanol, a polyalkylbenzochromenol, and 2,3-dihydro-5-hydroxy-2,2,4-trimethylnaphtho[1,2-b]furan compared to an .alpha.-tocopherol model in sodium dodecyl sulfate micelles, The Journal of Organic Chemistry. 60 (9) (1995) 2739–2744.
[54] Jodko-Piorecka, K., & Litwinienko, G. Antioxidant activity of dopamine and l-dopa in lipid micelles and their cooperation with an analogue of a-tocopherol, Free Radical Biology and Medicine. 83 (2015) 1–11.
[55] Cedrowski, J., Litwinienko, G., Baschieri, A., & Amorati, R. Hydroperoxyl radicals (hoo): Vitamin E regeneration and h-bond effects on the hydrogen atom transfer, Chemistry: A European Journal. 22 (46) (2016) 16441–16445.
[56] Venable, R.M., Brown, F.L.H., & Pastor, R.W. Mechanical properties of lipid bilayers from molecular dynamics simulation, Chemistry and Physics of Lipids. 192 (SI) (2015) 60–74.
[57] Shorthouse, D., Hedger, G., Koldso, H., & Sansom, M.S.P. Molecular simulations of glycolipids: Towards mammalian cell membrane models, Biochimie. 120 (SI) (2016) 105–109.
[58] Pasenkiewicz-Gierula, M., Baczynski, K., Markiewicz, M., & Murzyn, K. Computer modelling studies of the bilayer/water interface, Biochimica et Biophysica Acta-Biomembranes. 1858 (10, SI) (2016) 2305–2321.

[59] Marin-Medina, N., Alejandro Ramirez, D., Trier, S., & Leidy, C. Mechanical properties that influence antimicrobial peptide activity in lipid membranes, Applied Microbiology and Biotechnology. 100 (24) (2016) 10251–10263.

[60] Howard, R.J., Carnevale, V., Delemotte, L., Hellmich, U.A., & Rothberg, B.S. Permeating disciplines: Overcoming barriers between molecular simulations and classical structure-function approaches in biological ion transport, Biochimica et Biophysica Acta-Biomembranes. 1860 (4, SI) (2018) 927–942.

[61] Harpole, T.J., & Delemotte, L. Conformational landscapes of membrane proteins delineated by enhanced sampling molecular dynamics simulations, Biochimica et Biophysica Acta-Biomembranes. 1860 (4, SI) (2018) 909–926.

[62] Leng, X., Kinnun, J.J., Cavazos, A.T., Canner, S.W., Shaikh, S.R., Feller, S.E., & Wassall, S.R. All n-3 PUFA are not the same: MD simulations reveal differences in membrane organization for EPA, DHA and DPA, Biochimica et Biophysica Acta - Biomembranes. 1860 (5) (2018) 1125–1134.

[63] Garrec, J., Monari, A., Assfeld, X., Mir, L.M., & Tarek, M. Lipid peroxidation in membranes: The peroxyl radical does not "float"?, The Journal of Physical Chemistry Letters. 5 (10) (2014) 1653–1658.

[64] Wong-Ekkabut, J., Xu, Z., Triampo, W., Tang, I.-M., Tieleman, D.P., & Monticelli, L. Effect of lipid peroxidation on the properties of lipid bilayers: A molecular dynamics study, Biophysical Journal. 93 (2007) 4225–4236.

[65] Boonnoy, P., Jarerattanachat, V., Karttunen, M., & Wong-Ekkabut, J. Bilayer deformation, pores, and micellation induced by oxidized lipids, The Journal of Physical Chemistry. 6 (24) (2015) 4884–4888.

[66] Inagaki, T., & Yamamoto, T. Critical role of deep hydrogen tunneling to accelerate the antioxidant reaction of ubiquinol and vitamin E, The Journal of Physical Chemistry B. 118 (4) (2014) 937–950.

[67] Fabijanić, I., Fabijanić, C., Jakobušić Brala, V., & Pilepić. The dft local reactivity descriptors of α-tocopherol, Journal of Molecular Modeling. 21 (4) (2015) 99.

Jan Steinkühler, Rumiana Dimova

13 Giant vesicles: A biomimetic tool for assessing membrane material properties and interactions

Abstract: Giant unilamellar vesicles (GUVs) have sizes in the range of 10–100 µm, which defines their unique property: they are visible under a light microscope. GUVs provide a handy biomimetic tool for directly displaying the response of the membrane on the cell-size scale. They represent model biomembrane systems for systematic measurements of mechanical and rheological properties of lipid bilayers as a function of membrane composition and phase state, surrounding media, and temperature. Here, we first summarize methods for preparing GUVs and their observation. Then, we introduce different experimental techniques which can yield precise values of membrane material characteristics such as mechanical properties (bending rigidity, stretching elasticity, lysis tension, and spontaneous curvature) and rheology (fluidity and viscosity of the membrane). Design, setup, practical tips, and evaluation of such experiments are discussed. An example on vesicle immobilization facilitating such measurements is also introduced.

Keywords: Giant unilamellar vesicles (GUVs), elastic deformations, mechanical properties, optical microscopy

13.1 Introduction

Compartmentalization is a hallmark of biological systems. Ubiquitous building blocks of cellular compartments are fluid membranes made of lipids and proteins. Biomembranes are basically impermeable to ions, some solutes (such as sugars), and biomolecules (e.g., cytosolic proteins and DNA) and contribute in this way to the compartmentalization and organization of cells. Eukaryotic cells rely on the asymmetry maintained by membranes in their various organelles like the Golgi apparatus and the mitochondria. Evolution of biomembranes led to the occurrence of additional functional roles, for example, the cellular sensory system located in the plasma membrane, which allows cells to sense cues from their environment. Biological functions are usually linked to proteins and in fact about 30% of proteins encoding eukaryotic genome are membrane proteins. In addition to the functional motifs supported by membrane proteins, there is increasing interest in lipid–protein interactions, the signaling pathways of lipids like PIP_2, and

---

Jan Steinkühler, Rumiana Dimova, Max Planck Institute of Colloids and Interfaces, 14424 Potsdam, Germany

https://doi.org/10.1515/9783110544657-013

the functional role of the membrane conformation and structure. Studies of biomembranes are often performed on reconstituted (model) systems. Experimental models allow access to aspects that in practice may not be possible to assess in the original system (e.g., because active processes within a cell will adapt to an experimental perturbation in a complex manner). Employing a model system implicates neglecting or emphasizing certain aspects of the original system, which should lead to greater insight of the role of the individual components. Commonly used model systems for biomembranes are synthetic membranes made of a limited set of components. In this chapter, we discuss the use and study of closed membrane sheets (vesicles), which have diameters in the size range of cells or some organelles (vesicle diameters 1–100 µm). Due to their size, these vesicles are often referred to as *giant* (compared to conventional 100 nm vesicles). And because their membrane consists of only a single closed lipid lamella, they are called giant unilamellar vesicles (GUVs). Lipid and polymer GUVs receive applications in the fundamental study of membranes and proteins, as microreactors for chemical synthesis, and they represent an important constituent for the bottom up assembly of synthetic (proto-)cells [1–6]. From the experimental point of view, an attractive aspect is the compatibility with light microscopy, allowing direct visual access to the morphologies and properties of the nanometer thin lipid bilayer. GUVs can be observed by label-free microscopy techniques like phase contrast (Figure 13.1a,f), differential interference contrast (Figure 13.1b), or by using fluorescent dyes embedded into the membrane using epi- or confocal microscopy (Figure 13.1c–e). GUVs can be also studied using super-resolution techniques (e.g., using stimulated emission depletion microscopy). More complex morphologies and surface patterns are obtained when the GUV components undergo liquid–liquid phase separation (Figure 13.1d) [7]. However in this chapter, we will consider homogenous, fully miscible lipid membranes only. We will focus on the use of GUVs to study mechanical and rheological properties of lipid bilayers and vesicles. The chapter is divided into five sections: In Section 13.2, we briefly discuss and define some of the material parameters measured in the later sections. In Section 13.3, GUV formation is discussed, and in Sections 13.4–13.7, we describe methods to measure response to stretching and bending deformation, the spontaneous curvature, and immobilization of GUVs for measurements of lipid diffusion and rheology.

13.2 Some membrane material properties

The material parameters describing deformations of a lipid bilayer and membrane dynamics (Figure 13.2) that we will consider here are the bending rigidity κ, area compressibility modulus K_A, shear elastic modulus μ, and the viscosity η. To define these material parameters, we implicitly assume that the membrane response is in the linear regime, manipulations take place slowly enough to allow for equilibration,

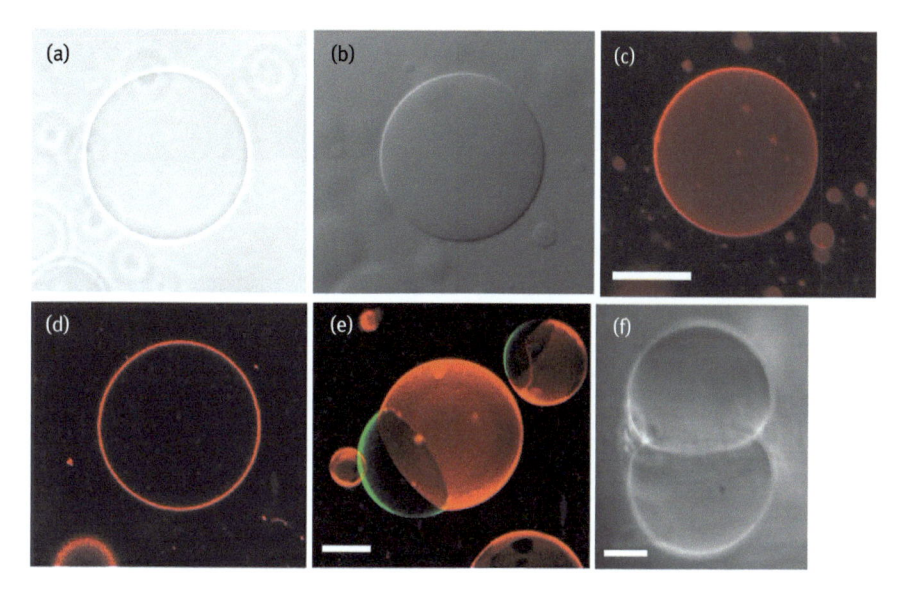

Figure 13.1: Snapshots of giant vesicles observed under different microscopy modes: (a) phase contrast, (b) differential interference contrast, (c, d) confocal microscopy where (c) is a projection averaged image and (d) is an equatorial section image. (a–d) Snapshots of the same vesicle. (e) Confocal 3D projection image of vesicles with immiscible fluid domains visualized with fluorescent dyes, which preferentially partition in one of the lipid phases. (f) Phase contracts side-view image of a vesicle sitting on a glass substrate (the mirror image of the vesicle reflected from the glass is also visible). The vesicle is deformed because of the density difference of the enclosed sucrose and external glucose solutions with osmolarity 200 mOsm/g. (a–d, f) Reproduced from Dimova et al. [1] by permission of IOP Publishing Ltd.

and we neglect any boundary effects due to finite size (later we will see what happens when, e.g., bilayers are stretched above their elastic limit).

Biomembranes are mostly in the liquid crystalline phase, implying that they do not sustain any shear stresses ($\mu = 0$), and here, we do not consider lipid bilayers in the gel phase. When fluid bilayers are manipulated, the molecules are free to diffuse, rearrange, and relax which gives rise to shear and dilatation surface viscosity η_S, η_D, with typical values for η_S of $(1-5) \times 10^{-9}$ N s/m [8, 9] and η_D in the order of 3.5×10^{-7} N s/m [10]. This means that the lipid bilayer is rather viscous; the shear surface viscosity of an equivalently thin water film would be on the order of 10^{-12} N s/m. Even though membranes are very viscous, diffusion within the bilayer is fast because the membrane is thin and dissipation occurs in the bathing solution.

When bilayers experience a mechanical tension Σ, the area per lipid is increased according to the elastic modulus K_A corresponding to Hook's law

$$\frac{A - A_0}{A_0} = \frac{\sigma}{K_A} \qquad (13.1)$$

where A_0 corresponds to the relaxed area of the bilayer corresponding to the limit case, where the bilayer is completely flat (i.e., fluctuations are smoothed out) and any

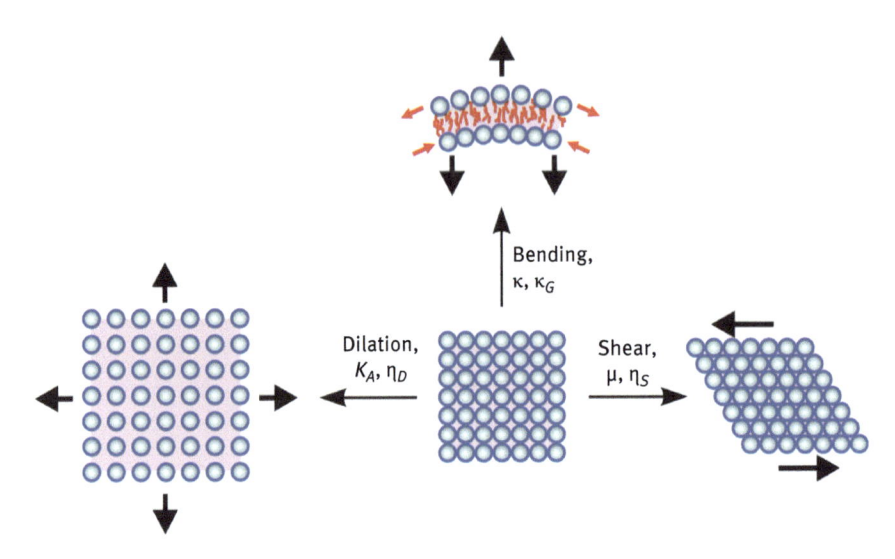

Figure 13.2: Schematic presentation of some modes of bilayer deformations: dilation, bending, shear, and the constants characterizing the membrane response to these deformations. Figure reproduced with permission from Dimova et al. [30].

stretching deformations act directly to increase the area per lipid on the expense of thinning the membrane. We will discuss the effect of thermal fluctuations and effects of other area reservoirs further below.

Above we could have introduced lipid bilayer stretching as analogous to in-pane stretching of an infinitely thin, flat film. For the case of lipid bilayer bending, we need to assume a certain bilayer thickness h. This becomes obvious when we consider that bending deformation must compress one monolayer and dilate the other (Figure 13.2, red arrows). Based on this idea, the relation between bending and stretching moduli is found to be

$$K_A = \frac{\alpha \kappa}{h^2} \tag{13.2}$$

where the proportionally constant α depends on the molecular details of the coupling between the leaflets [11]. For completely bound leaflets, $\alpha = 12$ [12], whereas for freely sliding monolayers, $\alpha = 48$ [13]. A polymer brush model for the bilayer predicts the intermediate value $\alpha = 24$ [14]. For pure dioleoyphosphatidylcholine (DOPC) bilayers, a recent study found $\alpha \approx 18$ [15].

Typical values for the bending rigidity of fluid bilayers are between 10 and 70 $k_B T$, the stiffer one corresponding to membranes containing a certain fraction of cholesterol and exhibiting some tail order (this phase is also referred to as liquid ordered phase). The head-group-to-head-group thickness of lipid membranes h is typically in the range between about 3.5 and 4.5 nm (see, e.g., Ref [16].). The elastic stretching modulus is found to be in the range between 200 and 260 mN/m [14, 17]. We see that the typical lipid bilayer appear rather stiff to stretching deformations (a 1% area increase of a bilayer patch of 1 μm² involves an energetic cost of well above 1 000 $k_B T$) but are very flexible to bending deformations. Because the bending rigidity of some bilayers is close to the thermal energy $k_B T$, Brownian motion in the water is reflected in visible fluctuations in vesicles with excess area (one of the techniques for measuring the bending rigidity, name fluctuation spectroscopy, relies on detecting, and analyzing these fluctuation). The molecular origin of the high-stretching modulus lies in the amphiphilic structure of the lipid molecules, which is also the driving force for the bilayer assembly in the first place. Increasing the area per lipid would expose a fraction of the hydrophobic core to the surrounding water and in this way directly affect the structural integrity of the bilayer. Indeed, typical lipid membranes can only be stretched up to about 5% of their preferred area before they rupture (at lysis tension σ_l of the order of 5–10 mN/m [18]); thus, in general, it is reasonable to assume that bilayers rest around their preferred (relaxed) area per molecule.

When using giant vesicles, we can assume that the lipid membrane is flat on molecular scales. This is because the mean curvature of a vesicle with radius R, that is, $1/R$, is much larger compared to the bilayer thickness h. If we now consider the geometry of a vesicle (in other words, allow for finite size of the system, i.e., that the bilayer is closed), the notion of relaxed areas per lipids reveals a potentially important detail. Imagine a closed vesicle where the lipid bilayer exhibits a (slight) difference between the numbers of lipids in each monolayer, essentially a packing defect that could take place during the vesicle assembly. This defect will lead to stresses between the two monolayers, as the total area of the membrane is fixed by the vesicle geometry but the monolayers cannot adapt their preferred area (which is given by the number of molecules). If we wait long enough, we could expect that these stresses relax by exchange of molecules between the bilayer leaflets. However, as it turns out, membranes made of pure phospholipids exchange their molecules between the bilayer leaflets only rather slowly and a typical half-time for the "flip-flop" motion of a lipid through the bilayer core is on the order of a day. The consequences of a persistent area mismatch between leaflets are well studied [19, 20] and show that vesicles can adopt a number of various shapes (see, e.g., the morphological diagram in Ref [21].). In the rest of this chapter, we will usually assume that the monolayers have obtained their preferred areas. This is a realistic assumption for most biomembrane mimetic systems as they contain cholesterol which undergoes flip-flops in the millisecond timescale; hence, cholesterol will distribute between leaflets relaxing any area mismatches.

As follows from the energy estimates for stretching and bending deformations, we consider the vesicle area as fixed and constant as long as it is not subject to external forces. Additionally, the vesicle volume is usually regulated by osmotic constrains, which we describe in more detail in Section 13.3. Details of the assembly or fusion and/or pinching off small vesicles from the GUV during handling.

In the conditions of relaxed bilayer differences, fluid membrane, fixed vesicle area-to-volume, and fixed topology (i.e. in the absence of fusion or fission of the vesicles), the morphologies of vesicles can be described by the spontaneous-curvature model as introduced by Helfrich in 1973 [19, 22] and is formulated here according to Ref [23].

$$F_{\text{bend}} = 2\kappa \oint dA (M-m)^2 \quad (13.3)$$

where the mean curvature is $M = (c_1 + c_2)/2$, c_1 and c_2 are the principal curvatures, m is the spontaneous curvature (which we will discuss below), and the integral runs over the whole vesicle surface and some constant (fixed) terms are omitted. Minimization of the bending energy F_{bend}, with the constrains of the enclosed vesicle volume and for constant membrane area, defines the vesicle shape. Here we discuss only some simple and instructive cases, for full discussion and derivation of eq (13.3) see for example Refs [19,23,24].

Principal curvatures of a surface are a measure of the surface bending at a given point and are rigorously defined in differential geometry. An important property of the principle curvatures is their invariance toward the particular surface parameterization. This makes them a natural choice for calculations of the bending energy, as it would be unphysical if the energy would depend on the description of the membrane. Intuitive examples are flat surfaces where $c_1 = c_2 = 0$ or spheres of radius R where $c_1 = c_2 = 1/R$.

A convenient quantity to express constrains on area A and volume V is the reduced volume, defined as

$$v = \frac{3\sqrt{4\pi} V}{A^{3/2}} \quad (13.4)$$

The reduced volume reflects the excess area a vesicle would have when compared to a spherical vesicle with the same volume and unperturbed area per lipid (the reduced volume of a sphere is $v_{\text{sphere}} = 1$, while, e.g., a stomatocyte shape has a reduced volume slightly smaller than 0.6).

The two parameters on which the bending energy depends, see eq. (13.3), are then the bending rigidity κ and the spontaneous curvature m. We have already learned about the material parameter κ, but what is the meaning of the spontaneous curvature? We immediately see that a flat bilayer exhibiting spontaneous curvature of $m = 0$ will have the lowest bending energy as the integral in eq. (13.3) yields zero. This is the intuitive case for a planar bilayer consisting of two identical monolayers. Now imagine a lipid bilayer wrapped up to form an (open) cylindrical tube (such structures appear in cells, e.g., as intracellular tunneling tubes [25]). From differential geometry, the principal curvatures of an (open) cylinder are $c_1 = 0$ and $c_2 = 1/R_{\text{cyl}}$. As we see from eq. (13.3), if the membrane happens to exhibit the spontaneous curvature $m = 1/2R_{\text{cyl}}$, the bilayer will be again at an energetic minimum. In contrast, a tube made of a lipid bilayer exhibiting zero spontaneous curvature $m = 0$

will have nonzero bending energy. We see that a mismatch between the actual geometry of the lipid bilayer and the preferred (or spontaneous) curvature will lead to a non-zero bending energy. For closed vesicles such a mismatch may be imposed by constraints on the vesicle shape such as the fixed volume and area of a vesicle. In general, the mechanism leading to spontaneous curvature of a bilayer is an asymmetry between the bilayer leaflets. One source of membrane asymmetry could be the unrelaxed area difference between leaflets discussed above, but in biomembranes, spontaneous curvature generation due to binding proteins and adsorption or depletion of small molecules such as sugars or ions (Figure 13.3b,c,f) might be more common. Only perfectly symmetry between bilayer leaflets and transmembrane solution results in zero spontaneous curvature. This ideal case is seldom true (if ever) for biomembranes, because the biological membrane experiences asymmetry of various origins as exemplified. Experimentally, spontaneous curvatures of a few dozens inverse nanometers to micrometers [23] have been observed, additionally demonstrating that the continuum mechanics approach eq. (13.3) is quite successful in describing membrane morphologies over several length scales. Further reading about mechanisms of spontaneous curvature generation can be found in Refs [23, 26–29].

With this we conclude the short introduction into the bilayer material properties (elastic modulus, bending rigidity, and viscosities) and parameters describing the configuration and state of a vesicle and its surrounding environment (reduced volume, tension, and spontaneous curvature). In the following, we will describe experimental approaches to manipulate giant vesicles and assess these properties.

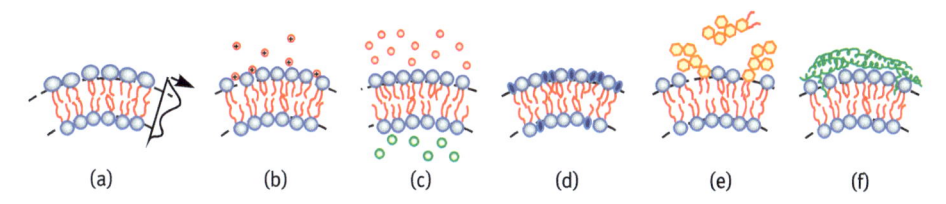

Figure 13.3: Sources of nonzero spontaneous curvature include (a) differences in the effective headgroup size (and respective molecular area) of the lipids in the bilayer, for example, as a result of differences in hydration, pH, or molecular structure of the constituting species; (b) asymmetric ion distribution leading either to condensing or expanding the lipids in one of the bilayer leaflets; (c) asymmetric distribution of nonadsorbing particles or (bio)molecules of different sizes; (d) amphiphilic molecules or lipid species asymmetrically distributed in the membrane; (e) partially water-soluble molecules (such as glycolipids or peripheral proteins) asymmetrically inserting in the membrane; and (f) asymmetrically inserted/anchored proteins with specific geometry. Note that spontaneous curvature has a sign, and when considering a vesicle, the sign convention is such that positive spontaneous curvature is toward the vesicle exterior and negative spontaneous curvature points toward its interior. Figure adapted from section by Dimova in Ref. [30].

Typical energies of biomembrane remodeling

Typically, it is assumed that shape transformations on the size scale of the cell are governed by interaction between the (active, energy consuming) forces of the cytoskeleton, motor proteins, and the cellular membrane. However, the energetic contribution of membrane bending becomes increasingly important on smaller scales, for example, during endocytosis. One contribution to bending of endocytic vesicles (which are typically below 100 nm in diameter) is provided from coating by the clathrin protein, which releases an energy of about 0.08 k_BT/nm^2, contributing to the bending of the initially flat plasma membrane. This energy is comparable to the bending energy of a 100 nm vesicle in diameter ($\kappa = 50\ k_BT$) [31], making membrane bending one of the major energetic costs during endocytosis. Another example is the pulling force of motor proteins which can exert forces of about 20 pN, similar to the forces needed for nanotube formation from a flat membrane sheet [32] (see also Section 13.4). All the estimates above assume zero (plasma) membrane spontaneous curvature, in lieu of a better estimate. As we can see from eq. (13.3) the contribution of spontaneous curvature can be quite significant. It is of current interest to understand how spontaneous curvature contributes to the shaping and remodeling of biomembranes.

13.3 Basics of GUV preparation and handling

To begin with, we remind that GUVs are not in a true thermodynamic equilibrium. The typical equilibrium phases of fluid bilayers in excess of water are lamellar lipid bilayer stacks [33] (exceptions here are some membranes of pure charged lipids [34]). Bilayer stacks can form multilamellar, onion-like vesicles; however, single lamellar vesicles should be unstable and disintegrate over time. Still, GUVs are remarkably stable (for days or even weeks) as disassembly involves high energy barriers, such as exposing hydrophobic edges of the bilayer core or membrane fission. It is instructive to consider GUV formation as an assembly process, requiring some sort of driving force or template. To supply the needed energy, various approaches have been developed, and depending on the application of the GUVs, each method comes with its own strength and weaknesses. Arguably the simplest way to form GUVs is to deposit small volumes (typically 1–100 μL) of phospholipids in an organic solvent (typically chloroform) onto a substrate, either a rough Teflon plate or an electrode (platinum wires and indium tin oxide are common materials) (Figure 13.4a). Often GUVs are prepared with trace amount of lipophilic dyes that integrate into the membrane and provide the fluorescent signal for epi-fluorescence or confocal microscopy. The solvent is evaporated from the deposited solution. Typical evaporation times are 1–2 h in in-house vacuum. Note that, even if chloroform evaporates quickly from the bulk liquid, evaporation of chloroform bound to the lipid molecule and substrate proceeds much more slowly. Optionally, to enhance formation of bilayer stacks, the substrate is hydrated in a saturated water atmosphere (Figure 13.4b). Finally, the substrate is immersed into the hydration solution which will be

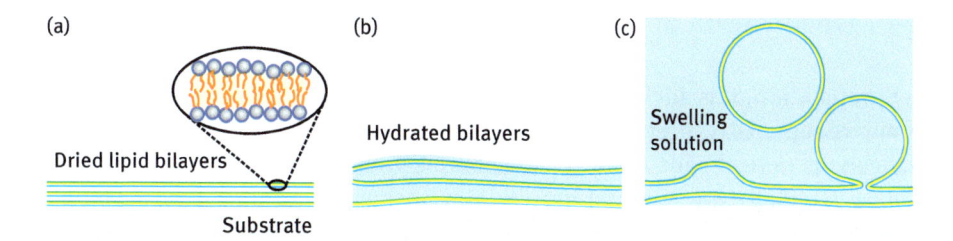

Figure 13.4: Experimental steps during the GUV swelling protocol. (a) A homogenous lipid film is spread on the substrate and dried from any residual solvent. (b) The dried lipid film is then pre-swollen in a water-saturated atmosphere to facilitate the bilayer hydration. (c) The pre-swollen lipid film finally becomes fully hydrated by the addition of the desired swelling solution. During spontaneous swelling, an osmotic pressure difference leads to separation of the bilayer stacks and GUV formation. Alternatively, an AC electric field is applied between two parallel substrates to enhance GUV formation.

encapsulated in the GUVs (Figure 13.4c). Upon hydration of the bilayer stack, water flows transport the solvated osmolyte. The gradient in osmotic pressure leads to separation of the bilayers in the stack and swelling of the GUVs. Often the osmolytes used are sugars such as sucrose. The benefit of depositing the bilayer stacks on a conductive material (electrode) is that an electric field can be applied during GUV swelling, contributing to the faster separation of the lipid bilayers and GUV formation. The so-called electroformation leads to an increased size and quality of the GUVs and is commonly used [35].

Apart from contributing to the GUV swelling process, the osmolytes also provide an osmotic pressure to essentially fix the GUV volume as the membrane is impermeable to sugars. Any change in the outside solution osmolarity will lead to water flows (to which the lipid bilayer is permeable) to equilibrate the osmolyte concentration inside the GUV. Thus, the osmotic conditions can be used to regulate the reduced volume v of the GUVs, similar to hypo- and hypertonic manipulations of red blood cells (RBCs) or other cells. For the observation under an inverted microscope, it is often beneficial to sediment the GUVs on the bottom of the observation chamber. This can be accomplished by dilution of the harvested GUVs in an isosmolar solution of smaller density than that of the encapsulated hydration solution. Similarly, the optical density difference between GUV interior and outside solution can be used to enhance images when using phase contrast microscopy (Figure 13.1a). The observed contrast provides also a good way to check for the GUV bilayer integrity. Stable pores will cause vanishing phase contrast of the GUVs.

For osmotic pressure regulation, sedimentation, and observation under phase contrast microscopy, often encapsulation of sucrose and dilution in glucose are used. More recently developed approaches rely on swelling on polymer cushions [36, 37], or transfer of lipids from an oil-based to an aqueous phase as in Refs [38, 39]; for an overview, see Ref [3]. Our group published a detailed protocol for the gentle hydration and microfluidic handling of GUVs available as a video protocol in Ref [40].

13.3.1 Observation of GUVs and their morphologies

After preparation, it is advisable to check the quality of the obtained GUVs under the microscope. GUV prepared from miscible components should appear with homogenous optical contrast of the membrane (from phase contrast or epi-fluorescence imaging) and minimal defects from inclusions or other lipid material. When studying effects of spontaneous curvature, it is desirable to include a negative control, in other words conditions where GUV exhibits zero spontaneous curvature. A simple way to check for spontaneous curvature is to observe the GUV morphologies upon osmotic deflation ($v<1$). The excess area generated by the osmotic deflation will allow the GUV to adapt a morphology minimizing its free energy. Vesicle populations with close to zero spontaneous curvature should predominately exhibit freely fluctuating nearly spherical GUVs and upon further deflation prolate and oblate shapes [19]. Spontaneous curvature leads to the formation of GUV shapes with buds or nanotubes [19, 28], which can be inward or outward. When preparing GUVs of pure phospholipids, lacking cholesterol, GUV morphologies indicating nonzero spontaneous curvature are occasionally observed. This is because GUV assembly involves some transient nonbilayer structures; during closing and resealing of lamellar membranes, lipids might be unevenly distributed between leaflets, leading to stresses in the bilayer. Even small differences in the number of lipid molecules in a membrane leaflet lead to substantial membrane remodeling [20]. This is one reason for the usually observed richness in morphologies obtained in swelling GUV from lipid that exhibits slow flip-flop rates (e.g., pure phospholipids without cholesterol). Depending on the GUV preparation conditions, additional mechanisms generating membrane asymmetry might be present. For example, desorption of certain lipids [65] or leaflet-dependent sorting of changed lipids.

13.4 Measuring the bending rigidity and spontaneous curvature by fluctuation analysis

The bending rigidity of liquid lipid membranes is on the order of the thermal energy and GUV membranes should be subject to thermal agitation, similar to the diffusive motion of small particles. Indeed, membrane fluctuations can be readily observed under the microscope when the bending rigidity is below about $50\,k_B T$ and GUVs are deflated to exhibit some excess area ($v < 1$) stored in the membrane fluctuations. Experimentally, membrane fluctuations are quantified by acquiring a time sequence of GUV membrane contours. The contour is then presented in polar coordinates (R,φ) and expanded in a Fourier series around the equivalent sphere radius, R_0, of the vesicle [41] (Figure 13.5a):

$$R(\varphi) = R_0 \left[1 + \sum_n a_n \cos(n\varphi) + \sum_n b_n \sin(n\varphi)\right] \quad (13.5)$$

Note that the fluctuations are obtained around the GUV equator and thus only depend on the angle φ. The coefficients a_n and b_n describe deviations from a circle. For (in mean) axisymmetric shapes considered here, the coefficients $<b_n>$ are zero. The remaining coefficients a_n describe the geometry of the membrane, for example, for prolate vesicles, the average value of the coefficient $\langle a_2 \rangle$ describes the elongation of the vesicle, while $\langle a_3 \rangle$ describes the up/down symmetry of the vesicle [42]. As the spontaneous curvature will act on the shape of a vesicle, m is related to the modes of the fluctuation spectra [43]:

$$m \sim \frac{\langle a_3^2 \rangle}{\langle a_2^2 \rangle} \quad (13.6)$$

Using this approach, the change in spontaneous curvature by grafting DNA strands to the GUV membrane was measured [43]. However, note that this approach only allows to quantify the relative change in spontaneous curvature as the proportionality constant is not measured using this approach, except if the mode distributions are mapped onto a reference data from Monte Carlo simulations [44]. Additionally, only small values of spontaneous curvature can be accessed using this approach.

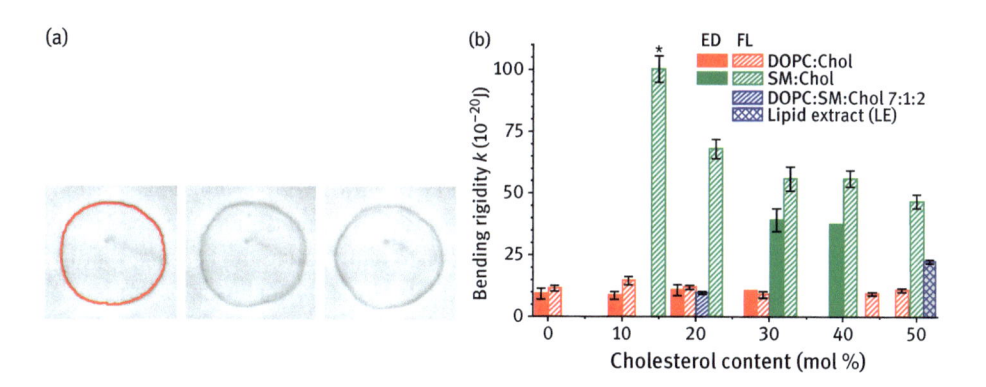

Figure 13.5: (a) Example phase contrast micrographs of a DOPC:DOPG GUV in symmetric sucrose buffer imagined using a fast camera of 200 μs exposure time. The GUV had some excess area to store in membrane fluctuations. In the first image, the detected contour is shown in red as an overlay. (b) GUV bending rigidity values obtained for lipid compositions DOPC:chol SM:chol and DOPC:SM:chol and red blood cell lipid extract (LE). The hatched data were obtained by fluctuations analysis (FL) while the solid filled data were obtained by the electrodeformation method and are not discussed in the main text. Figure reproduced from Gracia et al. [41].

Observation of membrane fluctuations

The detection of the membrane contours is complicated by the finite acquisition time of digital cameras. For example, the mode number 50 of a typical GUV decays with a time constant of 300 μs [41]. Standard cameras used for microscopy have exposure times in the order of milliseconds, which implies averaging out the higher fluctuation modes and interfering with the analysis. Only specialized cameras with μs-exposure times are suitable for direct observation of the GUV fluctuations. Another limit is the sampling of the long, slowly decaying, modes: to obtain good statistics, the total number of contours should be several thousands, that is, the observation time should be in the minute range.

The study of membrane fluctuations turns out to be most useful for bending rigidity measurements. From relation (13.5), the theoretical fluctuation power spectrum for a vesicle can be obtained [41, 45, 46]. The fitting of this expression to the experimentally obtained spectrum allows for quantification of the bending rigidity of individual vesicles. Examples for the application of this method include the quantification of the effect of cholesterol on the bending rigidity of lipid bilayers made of phospholipids of different degree of saturation in their acyl chains. The conventional belief was that above the lipid phase transition temperature, cholesterol orders the acyl chains and leads to a rather strong increase in bending rigidity independently of the exact lipid type [47, 48]. The question might be then why the plasma membrane of cells appears to be rather soft and how cells can tolerate changes in cholesterol level during, for example, nutritional uptake of cholesterol. For instance, the bending rigidity of palmitoyloleoyphosphatidylcholine bilayers increases from about 38 to 86 k_BT upon introducing 30% cholesterol in the membrane, a stiffening effect which might interfere with other functions of biomembranes. Later, it was then demonstrated using X-ray measurements [49, 50] and fluctuation analysis [41] (Figure 13.5b), that the bending rigidity of membranes made of DOPC and cholesterol do not show any significant correlation with the cholesterol content. Additionally, cholesterol mixtures with sphingomyelin (SM) were also explored using fluctuation analysis. The results demonstrated that increasing the amount of cholesterol in SM membranes leads to reduction in the bending rigidity. Thus, the conclusion is that the response of lipid bilayers to the inclusion of cholesterol is not universal but depends on the specific molecular interactions: it turns out that unsaturated lipids can *buffer* the effect of cholesterol on the bending rigidity to some extent. This might be one reason why lipid extracts of RBCs are found to be rather soft with a bending rigidity of about $22\,k_BT$. RBC membranes are both high in cholesterol (50%) and lipids with polyunsaturated bonds. In this way, the lipid composition of the plasma membrane might be contributing to the homeostasis of the cell. The studies exploring lipid specific effects on the interaction with cholesterol also serve as a nice example how lipid bilayers of rather simple composition contribute to the understanding of complex biomembranes by selectively modeling certain aspects of the original system.

The measurement of GUV bending rigidities can be also used to monitor the function of active proteins [51] and physiochemical changes in the membrane composition. It is a common concern that lipids with multiple unsaturated bonds might undergo oxidation during GUV preparation or prolong storage. Because of the quadratic dependence on membrane thicknesses (eq. 13.2), the bending rigidity is a rather sensitive measure of changes in the lipid conformation due to oxidation. In this way, the absence of significant oxidation artifacts during overnight storage of DOPC:DOPG GUVs was demonstrated in Ref [52].

13.5 Micropipette aspiration

The micropipette aspiration technique is used extensively for the measurements of membrane elastic properties [14, 47, 54–56]. The basic measurement setup is shown in Figure 13.6a. The method works by applying a pressure difference into a thin (approximately 1 - 10 µm in diameter) water-filled glass capillary immersed into the GUV sample. By applying negative suction pressure, a membrane segment of a slightly deflated GUV is partly aspirated into the glass capillary to obtain a geometry shown in Figure 13.6b. Now the membrane experiences a tension due to the Laplace pressure acting on the membrane. The membrane tension can be estimated from the pressure difference and the vesicle geometry:

$$\Sigma = \frac{\Delta p R_p}{2\left(1 - \frac{R_p}{R_v}\right)} \tag{13.7}$$

Note that the initial tension of a free (unperturbed) GUV is typically not known, as the exact area-to-volume ratio of a GUV depends on the "history" of this vesicle, that is, e.g. because of budding and fission of small vesicles by hydrodynamic flows during handling. However, when a vesicle shows visible membrane undulations we can safely assume that it exhibits a low initial tension typically below a few µN/m. Thus the applied tension (13.7) will be the main contribution to the total vesicle tension.

Here, R_p is the radius of the micropipette, R_v is the radius of the aspirated GUV, and Δp the applied pressure difference. The micropipette aspiration method can be used to measure the bending rigidity, the elastic stretching modulus, and the membrane lysis tension (at which the vesicle ruptures). In a typical experiment, the GUV area change is measured at varying tensions in the range of 0.001–10 mN/m.

$$\alpha = \frac{A - A_0}{A_0} \approx \frac{\pi R_p}{2}\left(1 - \frac{R_p}{R_v}\right)\Delta L \tag{13.8}$$

Here, A is the area at certain applied tension, A_0 is the GUV area before aspiration, and the formula assumes that the GUV volume is fixed by osmolytes and that experiments are conducted slowly enough to assume equilibrium and fast enough to

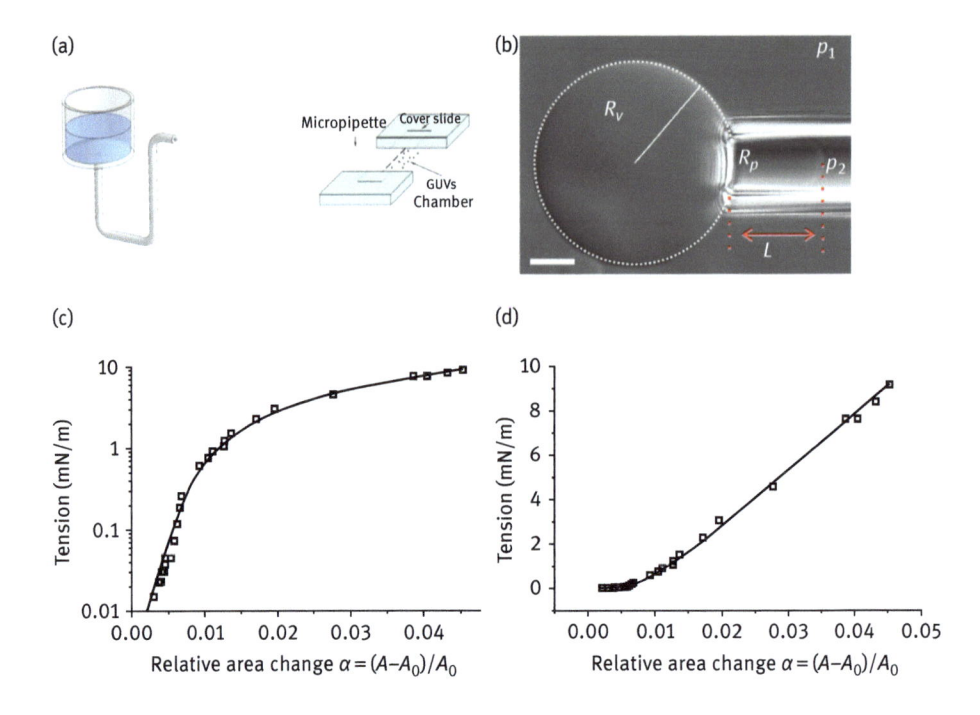

Figure 13.6: Micropipette aspiration of GUVs. (a) Setup of a hydrostatic pressure system. A water reservoir is displaced with micrometer precision in height, to apply a hydrostatic pressure difference into a glass capillary immersed into to a chamber containing the aqueous GUV solution. The chamber is mounted on a microscope stage. Image editing courtesy of Marzie Karimi and adapted from Ref [53]. (b) Phase contrast snapshot of a GUV (radius of the spherical segment R_v) aspirated by an glass capillary (micropipette) of radius R_p and the tongue length L. Scale bar 10 µm. (c, d) Linear and logarithmic plot of area to applied tension (open rectangles) and fit to the model of eq. (13.10) (solid line).

neglect drifts in the experimental setup (e.g., evaporation of water) [57]. We see that by the geometry of the aspirated GUV segment (also referred to as tongue), the estimated area change is linear to the measured tongue length. This enables precise measurements of area changes of below 1% while the area of spherical vesicles can only be estimated with much greater error (due to the quadratic dependence of the area on the measured radius).

GUVs often exhibit small area reservoirs, which could represent small folds or other defects in the membrane [58]. To pull out such membrane defects, a tension of about 2 mN/m is applied (pre-stretching); then, the tension is reduced to the initial (low tension) value. This step should pull out membrane defects before performing the measurements of area change as a function of tension [59].

The bending rigidity is assessed in the low-tension (entropic) regime where the area change is mainly due to the pulling out of thermal membrane undulation. The applied tension has to compete with the entropy stored in the thermal undulations [56, 60].

$$\frac{(A-A_0)}{A_0} \approx \frac{k_B T}{8\pi\kappa} \ln\left(\frac{\Sigma}{\Sigma_0}\right) \tag{13.9}$$

where the definition of the initial area of A_0 at tension Σ_0 provides a valid reference state. The optically resolvable membrane undulations already vanish at low tension of about 10^{-3} mN/m [17]. However, there is still a substantial amount of area stored in undulations of higher modes.

In the high-tension regime (>0.5 mN/m), the membrane area per lipid molecule starts to increase and an elastic response is found according to

$$\frac{(A-A_0)}{A_0} \approx \frac{\Sigma}{K_{app}} \tag{13.10}$$

Even if the stretching and entropic effects appear to be well separated, some membrane undulations persist at high tensions, and the real elastic modulus of the membrane can be found by an iterative procedure to first identify the linear entropic regime in a log(tension) to area change plot, estimate the bending rigidity from eq. (13.9), and then fit eq. (13.11) to the full tension range with fixed bending rigidity (Figure 13.6c,d) [14, 17, 56]:

$$\alpha \approx \frac{k_B T}{8\pi\kappa} \ln\left(\frac{\Sigma}{\Sigma_0}\right) + \frac{\Sigma}{K_A} \tag{13.11}$$

The correction turns out to be important for the understanding of the elastic behavior of lipid bilayers; K_{app} can vary greatly between various lipid types, while the real K_A of lipid bilayers is rather independent of the exact acyl chain configuration of the lipids [14].

At high tensions, the bilayer can undergo rupture or lysis; the loss of some of the GUV volume leads to quick aspiration of the whole GUV into the micropipette. Note that membrane lysis (or pore formation) is a stochastic process due to the fluid nature of the lipid bilayer. In other words, an increased tension leads to a higher likelihood of pore formation and the measured lysis tension depends on the rate of tension change. When the loading rate is low, the system has more time to explore more of its possible configurations at a given tension (and one of the configurations could be the pore forming state) and the measured lysis tension is lower than under a fast-loading rate [56, 61].

Hints for micropipette measurements

Beginners to the micropipette technique often experience some troubles in their first experiments. One common reason is the fragile nature of the glass capillary which can be dealt with only by careful handling and an optimized experimental setup to minimize unnecessary manipulation of the glass capillary. Sometimes the applied water pressure appears to be unstable which is often due to air bubbles in the tubing between the water reservoir and the glass capillary. Further stability can be archived by a rigid connection between pipette micromanipulator and microscope stage. Rigorous micropipette coating with bovine serum albumin or casein solution minimizes sticking or adhesion of the GUV to the glass surface. A complete protocol for micropipette handling can be found in Ref [62].

13.6 Spontaneous curvature measurements using optical tweezers

For the study of highly curved membrane segments and membrane spontaneous curvature, the method of membrane tube pulling using optical tweezers is often employed. Optical tweezers allow for the controlled manipulation of small (e.g., 2 µm in diameter) polystyrene beads in aqueous solutions by focusing a strong laser beam onto a diffraction limited spot. Because the electromagnetic field energy associated with the laser beam is proportional to the dielectric constant, a polystyrene bead in water (relative dielectric constant of about 3, compared to that of water of about 78) will experience a force pulling the bead into the focus point of the laser beam (for more details on the working and applications of optical traps, see, e.g., Ref [63, 64].). The pulling force can be described by a harmonic potential k_{trap} around the center of the trap and the force f_t on the bead follows approximately

$$f_t \approx k_{\text{trap}} \Delta x \qquad (13.12)$$

In this way, optical tracking of the bead displacement from the trap center, Δx, allows for force measurements. When the bead is brought into close contact with the GUV, the lipid bilayer can stick to the bead surface (this can be either accomplished by specific linkers between the membrane and bead, e.g., the biotin–streptavidin couple, or unspecific adhesion). When the bead is displaced away from the vesicle, the lipid bilayer remodels to avoid rupture and forms a cylindrical membrane tube between the spherical GUV and the bead. To study the relation between the force f_t measured on the bead and membrane parameters, eq. (13.3) is extended by the force acting on a membrane tube of length L connected to the GUV.

$$F = 2\kappa \oint dA (M - m)^2 - f_t L + \Sigma_{\text{asp}} A \qquad (13.13)$$

Energy minimization leads to an expression for the (static) pulling force for an outward tube as a function of the spontaneous curvature m, the bending rigidity κ, and the total membrane tension (the sum of the mechanical and the spontaneous tension), which is approximately equal to the aspiration tension Σ_{asp} [23, 65].

$$f_t = 2\pi\sqrt{2\kappa\Sigma_{asp}} - 4\pi\kappa m \tag{13.14}$$

Note that expression (13.14) does not depend on the tube length. As discussed further above, the tension of individual vesicles is not known a priory; thus, it is advantageous to control the GUV tension via the micropipette aspiration technique from Section 13.7. Additionally, the aspiration provides some stability to the tube-pulling process. An aspiration pressure of 10^{-2} mN/m leads to a tube diameter $2R_{tube} = \sqrt{2\kappa/\Sigma_{asp}}$ (assuming $\kappa \approx 10\,k_BT$ and zero spontaneous curvature) of around 90 nm. Thus, the pulled "nanotubes" will be below the optical diffraction limit. Exceptions are tubes pulled at very low tension using hydrodynamic forces [66]. Typical forces exerted on the bead during nanotube pulling are on the order of 10 pN. To extract the unknown parameters from eq. (13.14), the measured force on the pulling bead is plotted as a function of the square root of the applied tension. Now the slope will yield the bending rigidity of the membrane and a (nonzero) intercept can be used to estimate the membrane spontaneous curvature. From these experiments, we also see an intuitive relation for the appearance and stability of nanotubes. At a positive spontaneous curvature m (note that we pull the tube to the outside of the GUV), the extrapolation indicates that at a finite tension, the force measured on the tube will be close to zero. In other words, the nanotube is stabilized spontaneously, and vice versa. This latter can be directly observed when membranes exhibiting high negative spontaneous curvature form inward nanotubes upon osmotic deflation (Figure 13.7b).

13.7 GUV immobilization by electric fields (electro wetting of vesicles) and rheology measurements

Membrane viscosities can be estimated in several ways, including tracking of domains diffusing on the GUV surface [67], falling ball rheometry [9], and the diffusion of probe molecules in the membrane [68]. Here, we will illustrate measurements of the diffusion of fluorescent molecules embedded in the membrane using fluorescent recovery after photobleaching (FRAP). In FRAP measurements, prolonged imaging of the same membrane segment is required. This is complicated by the convective flows often present in GUV experimental chambers. It is therefore desirable to efficiently immobilize GUV during experiments, while at the same time, the immobilization should not disturb the measured properties and characteristics.

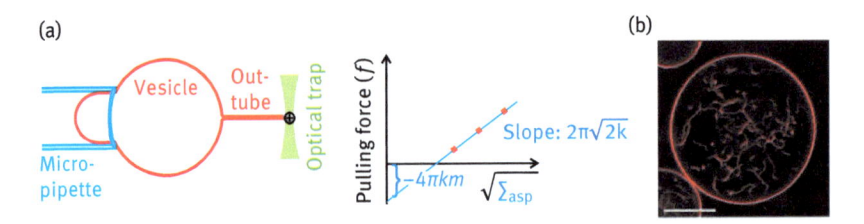

Figure 13.7: Measurements of membrane-bending rigidity and spontaneous curvature by nanotube extraction. (a) Left: Geometry of an aspirated GUV and an optical trap used to pull a nanotube out from the GUV; right: schematic plot of the square root of the applied tension and the measured force. The data slope can be used to deduce the membrane bending rigidity and y-axis intercept yields the spontaneous curvature. (b) GUV exhibiting spontaneous curvature due to polymer absorption to the membrane from the GUV interior and stable appearance of nanotubes (appearing as thick red lines).

GUVs can be immobilized using three-dimensional polymer meshes [69], by specific or unspecific surface coating [70], or using electric fields [52, 71]. In the following, we will illustrate immobilization via reversible adhesion of GUVs induced by means of an externally applied electric potential. The advantage of the approach is that the adhesion strength can be varied easily and gradually for the same vesicle. In this way, the optimal compromise between effective immobilization and minimal perturbation of the measurement can be found for each GUV.

The setup consists of two ITO-coated cover glasses, which are separated by a rubber spacer. An external DC potential is used to charge the bottom electrode positively (Figure 13.8a). At no or low applied voltages, the vesicles are free to displace. The vesicle starts to adhere at a threshold voltage of about 0.8 V and undergoes a shape transition from an unbound to a bound state. Now, the GUV is effectively immobilized. One can see how the thermal membrane undulations in the lower part of the vesicle in close contact with the ITO surface are suppressed (Figure 13.8b,c). With further increase of the external voltage, adhesion is smoothly regulated by the applied voltage and the vesicles spread further over the surface increasing the area of the adhering membrane segment (Figure 13.8d). The shape change induced by adhesion is completely reversible. When a GUV with significant excess area (like the one shown in Figure 13.8b) undergoes adhesion, the vesicle effectively exhibits two membrane segments (the unbound cap and adhered membrane segment). Lipids can redistribute according to their affinity for either of the segments and this behavior was studied in detail in Ref [52]. For GUV immobilization and mechanical measurement, this behavior is undesirable because the membrane composition of the two segments would not be exactly known.

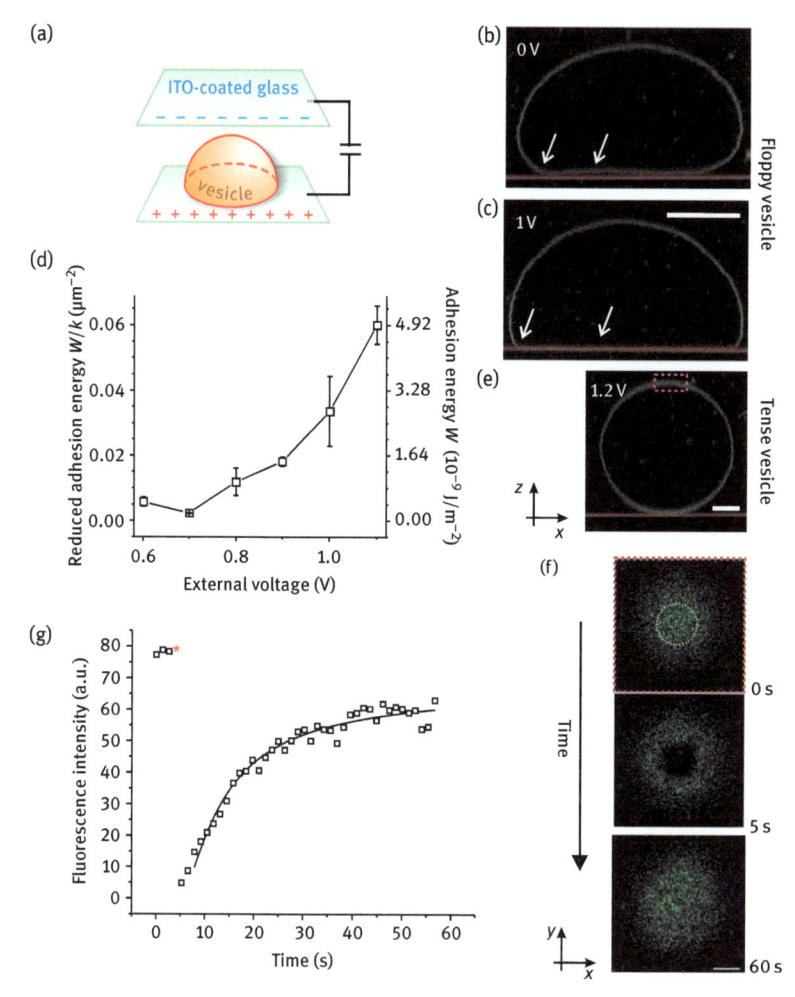

Figure 13.8: Immobilization of GUVs and FRAP measurements. (a) Sketch of the experimental setup to induce electrostatic adhesion of negatively charged GUVs (electrowetting of vesicles) on an ITO-coated glass support. Connection to a DC voltage source is indicated on the right. (b, c) Confocal images of the vertical cross-sections (side view) of a GUV made of DOPC/DOPG 80:20. The membrane fluorescence is shown in green and the reflection from the ITO surface is shown in red. (b) Non-adhering vesicle in the absence of applied voltage. Note the smoothly curved membrane (left arrow). (Right arrow) Undulation of the bound membrane segment, indicating a relatively large separation of this segment from the ITO substrate. (c) Same vesicle adhering to the substrate upon application of 1 V DC field. (Left arrow) Appearance of an effective contact angle; (right arrow) absence of any visible undulation in the vicinity of the surface, which also demonstrates the increased adhesion. The scale bar represents 20 µm. (e) Adhered and immobilized GUV exhibiting high membrane tension at an applied voltage of 1.2 V DC. (g, f) FRAP experiment at the pole of an immobilized GUV, the red star indicates the bleaching pulse applied to the membrane segment shown in the dotted circle. Open rectangles indicate the measured fluorescent intensity and line is a fit to eq. (13.15). Scale bar is 5 µm. Micrographs (b, c) reproduced from Steinkühler et al. [52].

Comparison to electrowetting of aqueous droplets

The vesicle adhesion process described here resembles what is known as electrowetting of aqueous droplets on a solid support (see, e.g., Refs [72, 73]). Indeed, the equilibrium shapes of (strongly) adhering vesicles can be descried by an equation with a formal relation to the Young equation, where the effective contact angle θ of the spherical cap is related to the membrane tension by $W = \Sigma(1 - \cos(\theta))$. The equation balances the energy gain per unit area W supplied by the surface–membrane interaction and the area change constrained by the membrane tension Σ. Both in electrowetting and in the vesicle-substrate system discussed here, W is a function of the applied voltage. However, an important and instructive difference between vesicles and droplets is that droplets can adjust their surface area according to their surface tension, which is a material parameter, while the tension Σ of a GUV is set by the area-to-volume ratio and is thus a property depending on the thermodynamic state of the vesicle (e.g., the number of molecules).

However, when GUVs appear tense (exhibit little stored area in fluctuations and shape), the GUVs basically retain their spherical shape upon voltage application and exhibit only a small contact area with the surface, thus the composition of the membrane segments is changed only very little. Even in this case, the GUV is still efficiently immobilized and does not displace during imaging (Figure 13.8e).

Now, the diffusion coefficient of a labeled molecule embedded in the membrane can be measured using FRAP. This technique exploits the fact that fluorescent molecules only exhibit fluorescence within certain limits: under too strong illumination, dyes become nonfluorescent. This phenomenon is also called photobleaching and is often considered an artifact, interfering with prolonged imaging using low concentration of dyes. During FRAP, photo bleaching is induced on purpose by application of strong laser irradiation to a confined area of the GUV membrane. Subsequently, the bleached membrane area is imagined under nonbleaching conditions. The bleached molecules will now diffuse out of the bleached zone to mix with the reservoir of functional fluorophores that remain on the GUV membrane; similarly, the nonbleached molecules will also diffuse into the bleached zone. Solutions of the diffusion equation are fitted to the experimentally obtained time trace of fluorescence recovery (Figure 13.7f,g). For bleaching of a circular spot of radius r_{frap}, the recovery trajectory is approximated by fitting

$$\frac{I_{\text{free}}(t)}{I_0} = \exp\left(-\frac{\tau_d}{2t}\right)\left(I_0\left(\frac{\tau_d}{2t}\right) + I_1\left(\frac{\tau_d}{2t}\right)\right) \tag{13.15}$$

where I_0 is the fluorescent intensity before bleaching, $I_{0,1}$ are modified Bessel functions of first kind, and τ is the fit parameter which is related to the diffusion coefficient $D = r_{\text{frap}}^2/4\tau_d$. To apply the equation, it is important to match the initial conditions of the derivation. It is assumed that the bleaching is instantaneous, in other words that the bleached species did not diffuse out of the bleached area *during* the bleaching plus. To approximate this idealization, the bleaching pulse should be much shorter than the subsequent recovery time [74]. In this way, negligible diffusion will have taken place during application of the bleaching pulse. It is also assumed that an infinite reservoir of unbleached molecules exists; in other words that the bleached molecules are diluted to zero over the course of fluorescent recovery experiment. This is approximately true for GUVs in which the bleached spot area is much smaller than the total GUV area. Finally, the imaging should be performed under nonbleaching conditions to guarantee full recovery of the initial fluorescent signal. When these conditions are violated, the fit to eq. (13.15) might be still satisfactory but the measured diffusion constant will not reflect the true molecular motion but will be an "apparent" diffusion coefficient characteristic for the measurement setup. These apparent coefficients are hard, if not impossible, to compare between different systems [74]. A good consistency check is to verify that the measured diffusion coefficient is independent of the bleached spot size. Fitting of relation (13.15) to diffusion of a tracer molecule (DiI-C18) in a SM:cholesterol 8:2 GUV yields a diffusion constant of about 0.4 µm²/s (±5%) (Figure 13.8g). Due to the slow tail of the recovery, the complete recovery of the bleached area is not measured but the fit to eq. (13.15) indicates recovery curve up to 90% of the initial fluorescent intensity.

In principle, the measured diffusion constants can be used to estimate the membrane viscosity. In bulk liquids, the Stokes–Einstein relation connects viscosity and diffusion. The situation in membranes is more complex because of the two-dimensional confinement and the boundary effects between membrane and bulk fluid. Saffman and Delbrück (S–D) were the first to find an approximate leading-order solution of the corresponding hydrodynamic problem. An approximate relation between diffusion coefficient and membrane viscosity η_s of a cylindrical (rigid) membrane inclusion of radius R_{in} [75] is

$$D \approx \frac{k_B T}{4\pi \eta_s} \left[\ln\left(\frac{\eta_s}{R_{in} \eta_w}\right) - 5.77 \right] \tag{13.16}$$

Note that also the viscosity η_w of the surrounding medium enters the calculation. An important feature is the logarithmic dependence on the size of the inclusion. If we imagine proteins as rigid cylinders embedded into the fluid lipid bilayer, one would expect only a weak $\ln(1/R_{in})$ dependence between diffusion and protein diameter, which was also experimentally confirmed [68, 76]. The S–D model is only valid for inclusions that allow for a definition of a cylinder with a well-defined radius. It is somewhat unclear if the tracer particle used here (DiI-C18) fulfills this requirement, as it integrates into the membrane rather than appearing as a rigid cylinder [77]. Motivated by good agreement between S–D approximation and a similar tracer to ours [76], we proceeded applying eq. (13.16) to the measured diffusion coefficient. We find a membrane shear viscosity of $\eta_s \approx 16 \times 10^{-9}$ N s/m, which is in agreement with previous estimates for liquid-ordered membrane phases (similar to the one measured here) [67].

References

[1] Dimova, R., Aranda, S., Bezlyepkina, N., Nikolov, V., Riske, K.A., Lipowsky, R. *A practical guide to giant vesicles. Probing the membrane nanoregime via optical microscopy.* Journal of Physics: Condensed Matter, 2006. **18**(28):p. S1151.

[2] Bucher, P., et al. *Giant vesicles as biochemical compartments: the use of microinjection techniques.* Langmuir, 1998. **14**(10):p. 2712–2721.

[3] Walde, P., et al. *Giant vesicles: preparations and applications.* ChemBioChem, 2010. **11**(7): p. 848–865.

[4] Trantidou, T., et al. *Engineering compartmentalized biomimetic micro- and nanocontainers.* ACS Nano, 2017 11(7):p. 6549–656.

[5] Fenz, S.F., & Sengupta, K. *Giant vesicles as cell models.* Integrative Biology, 2012. **4**(9): p. 982–995.

[6] Le Meins, J.F., Sandre, O., & Lecommandoux, S. *Recent trends in the tuning of polymersomes' membrane properties.* The European Physical Journal E, 2011. **34**(2):p. 14.

[7] Veatch, S.L., & Keller, S.L. *Seeing spots: complex phase behavior in simple membranes.* Biochimica et Biophysica Acta - Molecular Cell Research, 2005. **1746**(3):p. 172–185.

[8] Dimova, R., Pouligny, B., & Dietrich, C. *Pretransitional effects in dimyristoylphosphatidylcholine vesicle membranes: optical dynamometry study.* Biophysical Journal, 2000. **79**(1):p. 340–356.

[9] Dimova, R., Dietrich, C., Hadjiisky, A., Danov, K., Pouligny, B. *Falling ball viscosimetry of giant vesicle membranes: finite-size effects.*
The European Physical Journal B-Condensed Matter and Complex Systems, 1999. **12**(4):p. 589–598.

[10] Brochard-Wyart, F., De Gennes, P., & Sandre, O. *Transient pores in stretched vesicles: role of leak-out.* Physica A: Statistical Mechanics and its Applications, 2000. **278**(1-2):p. 32–51.

[11] Bermudez, H., Hammer, D.A., & Discher, D.E. *Effect of bilayer thickness on membrane bending rigidity.* Langmuir, 2004. **20**(3):p. 540–543.

[12] Evans, E.A. *Bending resistance and chemically-induced moments in membrane bilayers.* Biophysical Journal, 1974. **14**(12):p. 923–931.

[13] Goetz, R., Gompper, G., & Lipowsky, R. *Mobility and elasticity of self-assembled membranes.* Physical Review Letters, 1999. **82**(1):p. 221–224.

[14] Rawicz, W., et al. *Effect of chain length and unsaturation on elasticity of lipid bilayers.* Biophysical Journal, 2000. **79**(1):p. 328–339.

[15] Shchelokovskyy, P., Tristram-Nagle, S., & Dimova, R. *Effect of the HIV-1 fusion peptide on the mechanical properties and leaflet coupling of lipid bilayers.* New Journal of Physics, 2011. **13**(2): p. 025004.

[16] Nagle, J.F., & Tristram-Nagle, S. *Structure of lipid bilayers.* Biochimica Et Biophysica Acta, 2000. **1469**(3):p. 159–195.

[17] Evans, E., Rawicz, W., & Smith, B.A. *Concluding remarks back to the future: mechanics and thermodynamics of lipid biomembranes.* Faraday Discussions, 2013. **161**(0):p. 591–611.

[18] Olbrich, K., et al. *Water permeability and mechanical strength of polyunsaturated lipid bilayers.* Biophysical Journal, 2000. **79**(1):p. 321–327.

[19] Seifert, U., & Lipowsky, R. *Morphology of vesicles.* Handbook of Biological Physics, 1995. **1**: p. 403–464.

[20] Mui, B.L., et al. *Influence of transbilayer area asymmetry on the morphology of large unilamellar vesicles.* Biophysical Journal, 1995. **69**(3):p. 930–941.

[21] Dobereiner, H.G. *Properties of giant vesicles.* Current Opinion in Colloid & Interface Science, 2000. **5**(3-4):p. 256–263.

[22] Helfrich, W. *Elastic properties of lipid bilayers: theory and possible experiments.* Zeitschrift für Naturforschung C, 1973. **28**(11-12):p. 693–703.
[23] Lipowsky, R. *Spontaneous tubulation of membranes and vesicles reveals membrane tension generated by spontaneous curvature.* Faraday Discussions, 2013. **161**(0):p. 305–331.
[24] Kamien, R.D. *The geometry of soft materials: a primer.* Reviews of Modern physics, 2002. **74**(4): p. 953.
[25] Sherer, N.M. & Mothes, W., *Cytonemes and tunneling nanotubules in cell–cell communication and viral pathogenesis.* Trends in Cell Biology, 2008. **18**(9):p. 414–420.
[26] Lipowsky, R. "Remodeling of membrane compartments: some consequences of membrane fluidity."Biological chemistry 395, no. 3 (2014): 253-274. DOI: https://doi.org/10.1515/hsz-2013-0244
[27] Różycki, B., & Lipowsky, R. *Spontaneous curvature of bilayer membranes from molecular simulations: Asymmetric lipid densities and asymmetric adsorption.* The Journal of Chemical Physics, 2015. **142**(5):p. 054101.
[28] Liu, Y., et al. *Patterns of flexible nanotubes formed by liquid-ordered and liquid-disordered membranes.* ACS Nano, 2016. **10**(1):p. 463–474.
[29] Stachowiak, J.C., et al. *Membrane bending by protein–protein crowding.* Nature Cell Biology, 2012. **14**(9):p. 944–949.
[30] P. Bassereau, R. Jin, T. Baumgart, M. Deserno, R. Dimova, V.A. Frolov, P.V. Bashkirov, H. Grubmüller, R. Jahn, H.J. Risselada, L. Johannes, M.M. Kozlov, R. Lipowsky, T.J. Pucadyil, W.F. Zeno, J.C. Stachowiak, D. Stamou, A. Breuer, L. Lauritsen, C. Simon, C. Sykes, G.A. Voth, T.R. Weikl, The 2018 biomembrane curvature and remodeling roadmap, J. Phys. D: Appl. Phys., 51 (2018) 343001. DOI: 10.1088/1361-6463/aacb98
[31] Stachowiak, J.C., Brodsky, F.M., & Miller, E.A. *A cost–benefit analysis of the physical mechanisms of membrane curvature.* Nature Cell Biology, 2013. **15**(9):p. 1019.
[32] Zimmerberg, J., & Kozlov, M.M. *How proteins produce cellular membrane curvature.* Nature Reviews Molecular Cell Biology, 2006. **7**(1):p. 9.
[33] Lasic, D.D. *The mechanism of vesicle formation.* Biochemical Journal, 1988. **256**(1):p. 1–11.
[34] Hauser, H., & Gains, N. *Spontaneous vesiculation of phospholipids: a simple and quick method of forming unilamellar vesicles.* Proceedings of the National Academy of Sciences, 1982. **79**(6): p. 1683–1687.
[35] Dimitrov, D., & Angelova, M. *Lipid swelling and liposome formation mediated by electric fields.* Journal of Electroanalytical Chemistry And Interfacial Electrochemistry, 1988. **253**(2): p. 323–336.
[36] Horger, K.S., et al. *Films of agarose enable rapid formation of giant liposomes in solutions of physiologic ionic strength.* Journal of the American Chemical Society, 2009. **131**(5): p. 1810–1819.
[37] Weinberger, A., et al. *Gel-assisted formation of giant unilamellar vesicles.* Biophysical Journal, 2013. **105**(1):p. 154–164.
[38] van Swaay, D., & deMello, A. *Microfluidic methods for forming liposomes.* Lab on a Chip, 2013. **13**(5):p. 752–767.
[39] Stein, H., et al. *Production of isolated giant unilamellar vesicles under high salt concentrations.* Frontiers in Physiology, 2017. **8**:p. 63.
[40] Kubsch, B., Robinson, T., Steinkühler, J., & Dimova, R. (2017). *Phase behavior of charged vesicles under symmetric and asymmetric solution conditions monitored with fluorescence microscopy.* Journal of visualized experiments: JoVE, (128): p. e56034.
[41] Gracia, R.S., Bezlyepkina, N., Knorr, R.L., Lipowsky, R., Dimova, R. *Effect of cholesterol on the rigidity of saturated and unsaturated membranes: fluctuation and electrodeformation analysis of giant vesicles.* Soft Matter, 2010. **6**(7): p. 1472–1482.

[42] Dobereiner, H.-G. *The budding transition of phospholipid vesicles: a quantitative study via phase contrast microscopy.* 1995, (Doctoral dissertation, Theses (Dept. of Physics)/Simon Fraser University). Burnaby, Canada.
[43] Nikolov, V., Lipowsky, R., & Dimova, R. *Behavior of giant vesicles with anchored DNA molecules.* Biophysical Journal, 2007. **92**(12):p. 4356–4368.
[44] Dobereiner, H.G., et al. *Advanced flicker spectroscopy of fluid membranes.* Physical Review Letters, 2003. **91**(4):p. 4.
[45] Milner, S.T., & Safran, S. *Dynamical fluctuations of droplet microemulsions and vesicles.* Physical Review A, 1987. **36**(9):p. 4371.
[46] Henriksen, J.R., & Ipsen, J.H. *Thermal undulations of quasi-spherical vesicles stabilized by gravity.* European Physical Journal E, 2002. **9**(4):p. 365–374.
[47] Henriksen, J., et al. *Universal behavior of membranes with sterols.* Biophysical Journal, 2006. **90**(5):p. 1639–1649.
[48] Duwe, H.P., Kaes, J., & Sackmann, E. *Bending elastic-moduli of lipid bilayers - modulation by solutes.* Journal de Physique, 1990. **51**(10):p. 945–962.
[49] Pan, J., et al. *Cholesterol perturbs lipid bilayers nonuniversally.* Physical Review Letters, 2008. **100**(19):p. 198103.
[50] Pan, J., Tristram-Nagle, S., & Nagle, J.F. *Effect of cholesterol on structural and mechanical properties of membranes depends on lipid chain saturation.* Physical Review E, 2009. **80**(2): p. 021931.
[51] Girard, P., Prost, J., & Bassereau, P. *Passive or active fluctuations in membranes containing proteins.* Physical Review Letters, 2005. **94**(8):p. 088102.
[52] Steinkühler, J., Agudo-Canalejo, J., Lipowsky, R., Dimova, R. *Modulating vesicle adhesion by electric fields.* Biophysical Journal, 2016. **111**(7):p. 1454–1464.
[53] Lee, C.-Y., Herant, M., & Heinrich, V. *Target-specific mechanics of phagocytosis: protrusive neutrophil response to zymosan differs from the uptake of antibody-tagged pathogens.* Journal of Cell Science, 2011. **124**(7):p. 1106–1114.
[54] Needham, D., & Nunn, R.S. *Elastic deformation and failure of lipid bilayer membranes containing cholesterol.* Biophysical Journal, 1990. **58**(4):p. 997–1009.
[55] Henriksen, J.R., & Ipsen, J.H. *Measurement of membrane elasticity by micro-pipette aspiration.* The European Physical Journal E, 2004. **14**(2):p. 149–167.
[56] Evans, E., & Rawicz, W. *Entropy-driven tension and bending elasticity in condensed-fluid membranes.* Physical Review Letters, 1990. **64**(17):p. 2094–2097.
[57] Needham, D., McIntosh, T., & Evans, E. *Thermomechanical and transition properties of dimyristoylphosphatidylcholine/cholesterol bilayers.* Biochemistry, 1988. **27**(13): p. 4668–4673.
[58] Vrhovec, S., et al. *A microfluidic diffusion chamber for reversible environmental changes around flaccid lipid vesicles.* Lab on a Chip, 2011. **11**(24):p. 4200–4206.
[59] Vitkova, V., Genova, J., & Bivas, I. *Permeability and the hidden area of lipid bilayers.* European Biophysics Journal with Biophysics Letters, 2004. **33**(8):p. 706–714.
[60] Helfrich, W., & Servuss, R.-M. *Undulations, steric interaction and cohesion of fluid membranes.* Il Nuovo Cimento D, 1984. **3**(1):p. 137–151.
[61] Evans, E., et al. *Dynamic tension spectroscopy and strength of biomembranes.* Biophysical Journal, 2003. **85**(4):p. 2342–2350.
[62] Longo, M.L., & Ly, H.V. *Micropipet Aspiration for Measuring Elastic Properties of Lipid Bilayers*, in *Methods in Membrane Lipids*, A. M. Dopico & . 2007, Humana Press: Totowa, NJ. p. 421–437.
[63] Neuman, K.C., & Block, S.M. *Optical trapping.* Review of Scientific Instruments, 2004. **75**(9): p. 2787–2809.

[64] Moffitt, J. R., Chemla, Y. R., Smith, S. B., & Bustamante, C. (2008). *Recent advances in optical tweezers*. Annual review of biochemistry, **77**. https://doi.org/10.1146/annurev.biochem.77.043007.090225.

[65] Dasgupta, R., Miettinen, M. S., Fricke, N., Lipowsky, R., & Dimova, R. (2018). The glycolipid GM1 reshapes asymmetric biomembranes and giant vesicles by curvature generation. Proceedings of the National Academy of Sciences, 115(22), 5756–5761. DOI: 10.1073/pnas.1722320115

[66] Dasgupta, R., & Dimova, R. *Inward and outward membrane tubes pulled from giant vesicles*. Journal of Physics D: Applied Physics, 2014. **47**(28):p. 282001.

[67] Cicuta, P., Keller, S.L., & Veatch, S.L. *Diffusion of liquid domains in lipid bilayer membranes*. The Journal of Physical Chemistry B, 2007. **111**(13):p. 3328–3331.

[68] Ramadurai, S., et al. *Lateral diffusion of membrane proteins*. Journal of the American Chemical Society, 2009. **131**(35):p. 12650–12656.

[69] Lira, R.B., Steinkühler, J., Knorr, R.L., Dimova, R., Riske, K.A. *Posing for a picture: vesicle immobilization in agarose gel*. Scientific Reports, 2016. **6**:p. 25254.

[70] Kuhn, P., et al. *A facile protocol for the immobilisation of vesicles, virus particles, bacteria, and yeast cells*. Integrative Biology, 2012. **4**(12):p. 1550–1555.

[71] Korlach, J., et al. *Trapping, deformation, and rotation of giant unilamellar vesicles in octode dielectrophoretic field cages*. BioPhysical Journal, 2005. **89**(1):p. 554–562.

[72] Mugele, F., & Baret, J.C. *Electrowetting: from basics to applications*. Journal of Physics-Condensed Matter, 2005. **17**(28):p. R705-R774.

[73] Mugele, F. *Fundamental challenges in electrowetting: from equilibrium shapes to contact angle saturation and drop dynamics*. Soft Matter, 2009. **5**(18):p. 3377–3384.

[74] Weiss, M. *Challenges and artifacts in quantitative photobleaching experiments*. Traffic, 2004. **5**(9):p. 662–671.

[75] Saffman, P.G., & Delbrück, M. *Brownian motion in biological membranes*. Proceedings of the National Academy of Sciences, 1975. **72**(8):p. 3111–3113.

[76] Weiß, K., et al. *Quantifying the diffusion of membrane proteins and peptides in black lipid membranes with 2-focus fluorescence correlation spectroscopy*. Biophysical Journal, 2013. **105**(2):p. 455–462.

[77] Gullapalli, R.R., Demirel, M.C., & Butler, P.J. *Molecular dynamics simulations of DiI-C 18 (3) in a DPPC lipid bilayer*. Physical Chemistry Chemical Physics, 2008. **10**(24):p. 3548–3560.

Erwin London

14 Formation and properties of asymmetric lipid vesicles prepared using cyclodextrin-catalyzed lipid exchange

Abstract: Lipid bilayers have been important models for biological membranes. The recent development of methods to prepare bilayers with a different lipid composition in each leaflet, that is, asymmetric membranes, has made it possible to prepare model membranes that more closely mimic cellular membranes. Cyclodextrin-catalyzed lipid exchange has proven to be a versatile method for the formation of solvent-free asymmetric small, large, and giant unilamellar vesicles. The principles of asymmetric membrane preparation with cyclodextrins and the properties of the resulting asymmetric vesicles are described. Applications of the method to studies of lipid–protein interaction and to alter the lipid composition and asymmetry of living cells are also briefly discussed.

Keywords: Liquid ordered, liquid disordered, membrane domains, lipid asymmetry, interleaflet coupling, lipid flip-flop

14.1 Introduction: lipid asymmetry, membrane domains, and controversy

Many cellular membranes are composed of an asymmetric lipid bilayer with cytofacial and exofacial lipid monolayers (leaflets) having different compositions both in terms of lipid polar headgroup and acyl chain compositions. (Sometimes, the exofacial leaflet is called the outer leaflet and the cytofacial leaflet the inner leaflet. This is appropriate for the plasma membrane but does not work for the membranes of internal organelles, in which the cytofacial leaflet is on the outside of an organelle.)

The most well-characterized examples of membrane lipid asymmetry are the plasma membrane of eukaryotic cells and the outer membranes of Gram-negative bacteria. The most familiar case is the mammalian erythrocyte, in which the exofacial leaflet is enriched in sphingolipids (including the predominant species, sphingomyelin [SM]), and phosphatidylcholines (PC), while the inner leaflet is enriched in phosphatidylethanolamine (PE) and phosphatidylserine (PS) [1, 2]. It should be noted that there appears to be significant amounts of PC, and a smaller amount of

---

Erwin London, Department of Biochemistry and Cell Biology, Stony Brook University, Stony Brook, NY

https://doi.org/10.1515/9783110544657-014

SM in the inner leaflet, and a significant amount of PE in the outer leaflet. These lipid asymmetry patterns are probably very similar in other mammalian cell types [3]. In addition, phosphorylated versions of phosphatidylinositol (PI) likely reside in the cytosolic leaflet, while almost all glycosphingolipids are in the exofacial leaflet.

Because, as discussed below in detail, most lipids undergo only very slow transverse diffusion ("flip-flop") between leaflets, lipid asymmetry results from a combination of largely protein-driven metabolic processes: lipid synthesis, lipid breakdown, and lipid transport. The transporter proteins involved can be divided into energy-driven flippases (which transport lipids from the exofacial to cytofacial leaflet) and floppases (which transport lipids from the cytofacial to exofacial leaflet) and facilitated diffusion catalyzing scramblases [4].

The major exception to slow flip-flop appears to be cholesterol, which is able to spontaneously flip across lipid bilayers very quickly. This presents an obstacle for measurements of cholesterol asymmetry, and as a result, the degree of asymmetry of cholesterol in membranes, including plasma membranes, remains a controversial issue. Both an early electron microscopy and a recent study using a cholesterol-binding protein suggest most cholesterol is in the outer leaflet [5, 6]. However, studies mainly based on the distribution of the fluorescent sterol probe dehydroergosterol support the majority of cholesterol being in the inner leaflet [7].

The stronger affinity of cholesterol for lipids rich in the outer leaflet, SM and PC, as compared to PE [8], which is inner leaflet rich, would taken by itself predict an equilibrium established by fast flip-flop in which cholesterol is preferentially located in the outer leaflet. However, this is too simplistic a view, and one widely unappreciated factor that could be important for determining cholesterol asymmetry is lipid balance. If, for whatever reason, the net lateral area occupied by phospholipids, sphingolipids, and proteins in the outer leaflet is not equal to that occupied by such lipids and proteins occupied in the inner leaflet, there would exist a net area occupancy deficit in one or the other leaflets. A deficit of this type could exist in the steady state due to unbalanced metabolic processes. Whatever its origin, this type of imbalance would give rise to a free energy difference that could drive the spontaneous flow of membrane cholesterol into the leaflet with a net deficit. Because of this, there is no a priori reason why one necessarily would predict that the higher affinity of cholesterol for lipids SM and PC versus PE must lead to higher cholesterol levels in the sphingolipid-rich outer leaflet.

Relatively few aspects of the functional importance of lipid asymmetry have been explored. Among those that have been are apoptosis, blood clotting and some types of viral uptake by cells, cases in which the breakdown of PS asymmetry, causing PS to appear on the exofacial leaflet, plays a critical role [4, 9–11]. As will be described below, membrane protein conformation, and thus function, can also be influenced by lipid asymmetry. This topic has barely begun to be explored. The functional role of lipid asymmetry has been very little studied for the simple reason that asymmetry is very hard to measure and very hard to control. As will be described, studies are finally starting to tackle this problem.

One area in which advances are just beginning to be made involves the consequences of lipid asymmetry for the physical state of membrane lipids in a bilayer. There are three common physical states commonly formed by membrane phospholipids and sphingolipids [12]. The most familiar state is the liquid disordered (Ld), in the past often called the liquid crystalline state and denoted as the lipid alpha (Lα) state. In this state, lipids are relatively loosely packed, with acyl chains having kinks due to occasional gauche rotamers, and the lipids undergoing relatively fast lateral diffusion (along the plane of the lipid bilayer). Below a (freezing/melting) transition temperature (T_m), membrane lipids spontaneously convert into a more highly ordered and solid-like gel (G or Lβ) state. In this state, van der Waals interactions between neighboring acyl chains result in tight lipid packing. There is very little lateral diffusion. Finally, in the presence of sterols, the liquid ordered (Lo) state can form. In the Lo state, lipids are tightly packed with very few gauche rotamers, but lateral diffusion is fast.

What makes these physical states of biological relevance arises from the fact that the physical state of a bilayer at physiological temperatures is highly dependent on lipid structure. Both polar headgroup and acyl chain structure can influence the temperature at which the gel to Ld melting occurs (or if enough cholesterol is present, the thermal transition from Lo to Ld, see below). One very important factor is acyl chain double bond content (unsaturation). In general, membrane lipids with (unbranched) saturated acyl chains form bilayers with higher melting transition temperatures than those with unsaturated acyl chains, which generally contain *cis* double bonds. Sphingolipids tend to have a saturated acyl chain and a relatively saturated hydrocarbon chain in their sphingoid base (see below). As a result, they form bilayers with very high-T_m values of 37–60 °C, or even higher depending on the specific sphingolipid [13]. In contrast, phospholipids commonly have a mono or polyunsaturated acyl chain attached to carbon 2 of the glycerol backbone. Even though they also have a saturated acyl chain attached glycerol carbon-1, this combination of chains leads to very low-T_m values, generally close to or below 0 °C [14]. In mixtures of high-T_m lipids, low-T_m lipids, and (sufficient) cholesterol, bilayers form with coexisting Ld and Lo regions (domains) [12, 15, 16]. Thus, the basic question is: In biological membranes that have high levels of sphingolipids, unsaturated phospholipids, and cholesterol (such as plasma membranes), under what conditions do coexisting Lo and Ld state lipid domains form, and what are their properties?

It should be noted that when one talks of the tendency of saturated lipids and cholesterol to form the liquid ordered state, it is a simplification that ignores the fact that sphingolipids can occasionally have unsaturated acyl chains and (in mammals) usually have a trans double bound near the polar end of their sphingoid base backbone. Thus, such sphingolipids are not truly saturated lipids, even when they contain a saturated acyl chain. However, this does not negate the fact that such sphingolipids have a high T_m and a much higher tendency to form the

Lo state with cholesterol than ordinary unsaturated phospholipids. The trans double bond in the sphingoid base has only a modest effect on lipid behavior. The reasons for this are that trans double bond stereochemistry does not disturb the bond angles optimal for tight packing as much as do *cis* double bonds, and the fact that there is a plus a double bond located toward an end of a lipid hydrocarbon chain, as is the case in the sphingoid base, disturbs packing much less than one located at the center of the hydrocarbon chain [17]. In addition, the C24:1 acyl chain, which has a *cis* double bond at carbon 15 and is a commonly found fatty acyl chain in SM, still has a moderate ability to pack tightly due to its very long chain length and off-center double bond. This ability to pack tightly is reflected in the observation that di-C24:1 PC has a T_m at about 23 °C, a T_m almost equal to that of the saturated acyl chain lipid di-C14:0 PC [14].

Whether lipid domains are considered true phases or not is related to their size. Domains large enough to visualize by ordinary microscopy are usually considered to be phases. Whether stable nanodomains in the 5–200 nm size range are phases or not is more controversial. What is not controversial is that Ld and Lo domains have similar properties both when they are large and nanodomain sized [15]. The term lipid domains will be used here for both large domains and nanodomains.

In model membrane lipid bilayers, studies of coexisting gel and Ld lipid domains, detected by a wide variety of methods, have been carried out for over 50 years (e.g., Ref. [18]). Lo/Ld coexistence was first considered in binary mixtures of saturated lipids and cholesterol 30 years ago [19, 20]. It is still very poorly understood in binary mixtures. However, Lo/Ld coexistence in ternary mixtures of saturated phospholipids or sphingolipids, unsaturated acyl chain lipids, and cholesterol is much more well understood and has been characterized extensively starting roughly 20 years ago [16, 21–23].

Until very recently, the impact of membrane asymmetry upon membrane domain formation has been very little studied. The reason is that it is relatively difficult to form asymmetric bilayers. However, asymmetry is crucial to understand the principles of domain formation. In plasma membranes, the sphingolipids are very largely restricted to the outer leaflet. The inner leaflet phospholipids do not have a structure that allows them to spontaneously form Lo domains with cholesterol [24]. However, there is a good reason to think that membrane domains would span the entire membrane bilayer in the plasma membrane. An obvious possibility is that lipids in Lo domains in the outer leaflet could be physically coupled to lipids in the inner leaflet, such that they are able to induce the inner leaflet lipids with which they are in contact to form Lo domains. Experiments in cells have suggested that some sort of interleaflet coupling can occur in plasma membranes [25], while other studies have suggested that an asymmetry in the amino acid composition of transmembrane helices in the exofacial and cytofacial leaflets reflects an adaptation to overall tighter lipid packing in the outer leaflet relative to that in the inner leaflet [26].

To fully understand such interactions and their influence on lipid domains, asymmetric biomimetic membranes composed of various lipids are needed. Thus, defining the domain-forming properties of artificial asymmetric membranes has become a crucial part of solving the question of domain formation in cells.

Understanding membrane domain formation in artificial lipid bilayers is all the more important because membrane domains are hard to study in cells. The first widely applicable method to investigate domain formation was based on the insolubility of Lo domains in nonionic detergents such as Triton X-100 [27]. The insoluble membranes are usually called "detergent-resistant membranes". The method is easily used, and thus led to an explosion of studies, but it has important limitations, perhaps foremost of which is probably the need to carry out solubilization of cells at 4 °C, which might induce the formation of Lo domains that would not exist at higher temperature [28]. Although variations of the technique have been proposed for use at higher temperature, they have not come into wide use [29]. It has also been proposed that Triton X-100 might cause Lo domains to form when it interacts with membranes [30], but studies from our lab indicate that what happens is not induction of domain formation, but rather Triton X-100 induced clustering of preexisting nanodomains [31].

The hypothesis that there would be Ld/Lo coexistence in cells was first hypothesized in a collaboration of our group with that of Dr. Deborah Brown, based on detergent solubilization studies of artificial membrane vesicles composed of various lipids and a GPI-anchored protein, plus the observation that Lo domains are insoluble in Triton X-100 while Ld domains are soluble [21]. Studies varying lipid composition showed that in model membrane vesicles, the onset of Lo domain formation detected spectroscopically occurred at the same composition as the onset of insolubility in detergent [16]. The name lipid rafts was coined for sphingolipid-rich, cholesterol-containing membrane domains in cells [32], and lipid rafts were soon equated with cellular Lo domains. Unfortunately, while detergent insolubility is a relatively simple method, detecting membrane domains in cells by other methods is not. This is probably largely a reflection of the small size of domains in cells under most conditions. The lack of facile techniques for studying rafts in cells has proven to be a major experimental barrier. Nevertheless, recent studies with improved methods and less ambiguous biological systems have provided ever stronger support for the presence of lipid rafts in yeast [33–35], mammalian cells [36–39], and even cholesterol-containing bacteria [40, 41].

To summarize, progress in studies of biomembrane structure and function has been hampered both by the difficulty of experimentally assessing and manipulating lipid asymmetry and domain formation in cells and the difficulty in reproducing lipid asymmetry in membrane-mimetic systems. Cyclodextrin (CD)-based lipid exchange techniques that have the ability to help address both of these issues are discussed below.

14.2 Preparation of asymmetric biomimetic membranes: methods based on bringing monolayers of different compositions together or use of phospholipases

Methods have been developed to prepare asymmetric model membranes of various types. Asymmetric planar bilayers have been used to study the effects of lipid asymmetry in a few studies, for which they are particularly valuable when prepared in unsupported or cushioned formats [42–44] that eliminate or minimize interactions with a support layer, which can greatly disturb lipid behavior. Droplet interface methods, in which water droplets are dispersed in oil, can be used to prepare asymmetric bilayers when two aqueous droplets coated with two different lipid monolayers are brought into contact [45, 46]. For some of these procedures, residual organic solvent, especially when it accumulates in the bilayer midplane, is a concern.

The most widely applicable model membranes are lipid vesicles. Various water-in oil-methods have been developed in which an oil solution contains water droplets covered in a monolayer destined to become the inner leaflet of a vesicle, and the water droplets are pushed into aqueous solution while passing through a leaflet of lipid (which forms the outer leaflet of the vesicle) at the oil/aqueous solution interface [47]. Again, residual oil in the vesicles is a concern, although improved methods to minimize this have been reported [48]. So far, water-in-oil vesicles have not been employed in many practical applications [49]. Phospholipase-based methods can also be used to prepare asymmetric vesicles. This method uses digestion or exchange of the polar headgroup alcohol to create headgroup asymmetry. Phospholipase D-catalyzed transformation of outer leaflet PC into PE and PS via headgroup exchange using a mixture of ethanolamine and serine was recently demonstrated [50]. The method has some significant advantages but does not allow independent control of both lipid acyl chain and headgroup composition.

14.3 CD-based lipid exchange to prepare asymmetric vesicles

14.3.1 Methods and limitations

CD-catalyzed lipid exchange, via a method developed by our lab [51], has proven to be a versatile method to prepare asymmetric vesicles and has already seen a variety of applications. Using this approach, asymmetric small unilamellar vesicles (SUV), large unilamellar vesicles (LUV), and giant unilamellar vesicles (GUV) have all been prepared [51–54], as have asymmetric-supported bilayers [55]. A wide variety of lipid compositions have been studied, and methods to prepare vesicles with any desired cholesterol concentration have been developed [56]. Other groups have begun to use and improve this lipid exchange method for various applications [57–59].

CDs are cyclic oligomers generally composed of six (α-CDs), seven (β-CDs), or eight (γ-CDs) individual glucose units connected to each other by a glycosidic bond. The cyclic oligomer forms a barrel-like structure with a relatively hydrophobic interior. This interior readily binds various hydrophobic molecules, including lipids. CDs have been widely used to extract sterols from, or deliver sterols to, membranes [60, 61]. However, they can also bind phospholipids and exchange them between membranes [62–64], presumably because they can bind lipid acyl chains in their hydrophobic interior.

The general principles of asymmetric vesicle preparation with CDs are straightforward. Two populations of lipid vesicles with different lipid compositions are mixed together with an appropriate CD. One population of vesicles is the donor vesicles. It is the population that contains the lipid that will become the outer leaflet of the exchange vesicles. The other population is the acceptor vesicles. It contains the lipid that will be the inner leaflet of the exchange vesicles. The CD binds lipids, but not very strongly, so it generally picks up a lipid molecule from one vesicle and delivers it to another vesicle. If the donor vesicles are in excess, the outer leaflet of the acceptor vesicles will be replaced with donor lipid, resulting in the desired asymmetric vesicle population. Then using size chromatography, or more commonly ultracentrifugation, these exchanged acceptor vesicles are isolated from CD and (partly exchanged) donor vesicles. A complication is that the fact that the rate of exchange is highly dependent upon CD concentration. This might suggest use of as high a CD concentration as possible, but high concentrations of some CDs dissolve lipid vesicles. So, a useful protocol is often to first mix the CD with an excess, saturating donor vesicle population, at which point some, but not all, donor lipid vesicles dissolve. This should minimize any solubilization of acceptor vesicles when the donor lipid–CD mixture is added to acceptor lipid vesicles. Of course, high CD concentration increases solution density, which can complicate centrifugation protocols. Also, if acceptor vesicles are loaded with high sucrose concentrations to aid centrifugation, then dispersal of vesicles in solution with physiological salt levels will result in an osmotic pressure gradient across the membrane. This can be balanced out with external sucrose or avoided by using a protocol in which the donor lipid is pelleted and/or removed by filtration [54].

Control of exchange efficiency is not trivial. The information to predict exchange efficiency is generally not available. This would include both the intrinsic-binding constants of CDs for different lipids and the effect of membrane lipid composition upon the preference of CDs to extract various lipids from vesicles. However, simple experiments comparing the amount of each type of lipid exchanged into acceptor vesicles using donor vesicles containing a mixture of lipids [56], or comparison of the amount of CDs needed to dissolve lipid vesicles with different lipid compositions [65], indicate that the dependence of CD–lipid interaction upon lipid structure is modest, although significant, and it has been possible to efficiently transfer many lipids with CDs. Using donor lipid in excess, or adjusting the ratio of lipids in the donor vesicles, can circumvent differences in exchange efficiency and achieve the desired extent of exchange. In this regard, it should be noted that even when lipid is present in saturating concentration, only a small fraction of the CD in solution seems to have bound lipid. In any case, exchange efficiency and a more rigorous analysis of CD–lipid interaction is a topic that requires more systematic investigation, which some studies have begun to carry out [59, 65].

The ability or lack of ability to exchange cholesterol using CDs deserves additional comment. It is highly desirable to prepare asymmetric vesicles with various cholesterol concentrations. Since β-CDs readily remove or deliver sterols to membranes, they cannot easily be used to exchange phospholipids without altering cholesterol levels. It possible to introduce cholesterol after phospholipid exchange in a second exchange step by using the methyl-β-CDs at a concentration high enough to bind sterol but too low to bind phospholipids [51]. A better way to circumvent this problem is to use α-CDs, which appear to have a cavity too small to accommodate cholesterol [56]. The desired cholesterol concentration can then be incorporated into the acceptor vesicles prior to exchange [56].

It should also be noted that the modifications on CDs affect what exchange conditions are used. Hydroxypropyl CDs can catalyze exchange without solubilization of lipid vesicles [56, 65] but generally require use of higher CD concentrations than for methyl CDs. For some applications making use of CDs that dissolve lipid vesicles (methyl modified CDs), removing excess donor vesicles before adding donor–CD mixtures to acceptor vesicles, that is, carrying out exchange using only the CD–lipid complexes, is desirable to minimize contamination of asymmetric vesicles with donor lipid vesicles [53, 66]. However, this can decrease the efficiency of exchange.

14.3.2 Lipid balance upon lipid exchange and how imbalance might influence membrane properties

One underexplored concern when carrying out exchange is lipid balance. Delivery of more or less than one molecule of lipid for every lipid extracted from a vesicle by CD could have pronounced effects upon membrane structure. Evidence to date strongly suggests that the level of exchange obtained is very close to 1:1. First, gross imbalance in delivery would destroy membrane integrity. This is not observed, at least for the population of asymmetric vesicles recovered after exchange. Second, vesicle size does not seem to change much after lipid exchange, which suggests that total lipid content is similar before and after exchange[51, 56]. Third, lipid composition and asymmetry after exchange are consistent with simple replacement of acceptor vesicle outer leaflet lipids with donor lipids. Donor lipid content in the asymmetric vesicles after exchange is never over 100% replacement of the outer leaflet, as it could be if more than one lipid was being introduced for every lipid removed by exchange. (There is an exception to this when lipid flip-flop is fast, but this is due to exchange of inner leaflet lipid, see below.)

A good question is: Why should exchange be close to 1:1? A likely explanation is that the relative affinities of lipid binding to CD and vesicles are very highly dependent upon the balance of lipid content between the inner and outer leaflets of a membrane. If the surface areas covered by the inner and outer leaflets are in balance initially, so neither leaflet is stretched or compressed, then removal of a lipid molecule from the outer leaflet would be a stress that increases the affinity of the outer leaflet for lipid. This would increase the tendency of a lipid–CD complex to deliver a lipid to, and decrease the tendency of an empty CD to extract an additional lipid from, the outer leaflet. The situation would be the opposite if the stress is due to the delivery of an excess lipid into the outer leaflet. In this case, the balance of affinities would favor extraction of an outer leaflet lipid. As a result, balance between inner and outer leaflet lipid content would tend to be maintained.

This does not mean that exchange is exactly 1:1. One straightforward factor to consider is that different lipids take up different surface areas, especially if they take on different physical states. So, a more precise statement is that after exchange, the surface area of the outer leaflet should be in balance with the inner leaflet. That is to say, after correcting for differences in lipid content due to membrane curvature, the lipids in each leaflet should not be overly compressed or stretched for the physical state in which they exist. In addition, because cholesterol can flip across membranes rapidly, in membranes with cholesterol, any imbalance between inner and outer leaflet lipid content could be balanced by net flow of cholesterol into the leaflet that has a lipid deficit.

In any case, differences in the surface area taken up by lipids in each leaflet could have some interesting consequences. Consider the case in which lipid exchange is used to introduce SM into the outer leaflet of an acceptor vesicle composed of unsaturated lipids and is carried out at high temperature, that is, under conditions all lipids are in the Ld state. What would happen if there were lipid balance after exchange, but then the asymmetric vesicles were cooled sufficiently so that the SM transitioned to form ordered state domains? The outer leaflet SM would take up less lateral surface area, giving rise to a compressive stress on the inner leaflet. Either the inner leaflet would have to compress, or would have to suppress the compression of the outer leaflet, presumably due to repressing formation of the ordered state by the outer leaflet lipids. If cholesterol is present, then an alternative result would be a net cholesterol flow from the inner to outer leaflet to maintain lipid balance.

On the other hand, if the exchange to introduce SM is carried out at low temperature, the outer and inner leaflet imbalance might appear at high temperature, at which point ordered SM domains would tend to melt. In this case, the inner leaflet would resist the melting of the ordered domains, stabilizing them (or if cholesterol is present, a net cholesterol flow toward the inner leaflet might occur).

Unfortunately, detailed studies of these issues may be premature. Non-perturbing methods to measure membrane cholesterol in a leaflet precisely are missing. Although methods to assay cholesterol in a leaflet have been reported, as noted above, they disagree in their conclusions and seem unlikely to be applicable to very precise measures of absolute cholesterol concentrations in the leaflet of an artificial membrane. Given the controversy about cholesterol levels in each leaflet in cells, it is impossible to even know what composition and cholesterol concentration one would like to achieve in an artificial membrane to mimic natural membranes!

14.3.3 Assays of lipid asymmetry applicable to asymmetric vesicles

One of the most difficult tasks is determining the level of asymmetry after lipid exchange. The methods that can be used depend on asymmetric vesicle lipid composition and vesicle size. We initially made use of the observation that asymmetric vesicles with ordered domain forming lipids in the outer leaflet form more thermally stable ordered domains than the corresponding symmetric vesicles with the same overall lipid composition [51, 52]. Thus, a melting temperature higher than that in symmetric vesicles is indicative of asymmetry. The level of asymmetry cannot be quantified in this way, but the time dependence of the decay of asymmetry can, at least roughly, as T_m will decrease as asymmetry decays. We also found that in cases in which the outer leaflet forms ordered domains independently of the inner leaflet, the level of outer leaflet order is a function of the amount of ordered domain-forming lipid in the outer leaflet. The level of order in the outer leaflet (e.g., assayed by fluorescence anisotropy of a probe that located exclusively in the outer leaflet) can be compared to a standard curve of order versus percentage-ordered domain forming lipid using symmetric vesicles composed of the same lipids as in the outer leaflet of the asymmetric vesicle. In many cases, this has worked well, showing good agreement with other assays [51, 67], but cannot be used when the inner leaflet destabilizes asymmetric vesicle outer leaflet-ordered domains.

More direct methods can also be used to assay asymmetry. Nuclear magnetic resonance (NMR) has proven useful in conditions in which inner and outer leaflet lipids give resolved resonances with different chemical shifts. This can done in the case of SUVs, in which case, the choline methyl proton signal from inner and outer leaflet PC or SM can be distinguished [68]. The method can be extended to LUV by adding lanthanide cations to preformed vesicles as a shift reagent to alter the chemical shift of the outer leaflet choline methyl proton signal [54]. Chemical labeling methods can be used for other lipids. For example, reaction with the relatively membrane impermeable reagent trinitrobenezenesulfonate (TNBS) can be used to measure the levels of external PE. However, conditions that give maximal reaction and minimal leakage of TNBS into vesicles must be used. These are not always trivial to identify. For anionic lipids, the ability to interact with cationic molecules externally added to the asymmetric vesicles can be a useful method to measure asymmetry in terms of outer leaflet anionic lipid content. Studies have used calcium cations, cationic peptides, or cationic fluorescent probes for this purpose. In general, the greater the fraction of anionic lipids in the outer leaflet of the asymmetric vesicles, the greater the degree of cation binding [51]. Other studies have more directly measured outer leaflet charge [59]. Standard curves composed of symmetric vesicles with various fractions of anionic lipids are needed for these experiments. Overall, as noted above, the methods used to assay asymmetry have to be tailored to the lipid composition of the asymmetric vesicles. Too often, what works well with one lipid composition fails with another.

14.3.4 Lipid transverse diffusion (flip-flop) and its relationship to lipid asymmetry

Maintenance of lipid asymmetry requires slow flip-flop of membrane lipids. For this reason, what determines the rate of spontaneous lipid flip-flop in a bilayer and how is it affected by lipid composition and asymmetry are important questions. The simplest case to consider is flip-flop in a lipid bilayer composed predominantly of a single lipid species. The rate of flip-flop should be determined by the energy of the lipid in its usual equilibrium conformation relative to that in the transition state for flip-flop, which presumably would be when the polar headgroup was dissolved in the hydrocarbon core of the bilayer. Thus, there are two factors that would determine lipid flip-flop rates. One is lipid chemical structure. The more polar and charged groups on a lipid, the slower it should flip across a membrane. The second is the structure of the bilayer. The more polar the membrane interior, or looser the lipid packing, the easier for a polar headgroup to be accommodated within the bilayer core, that is, the lower the free energy of the transition state. Thus, flip-flop rates should reflect the ability of a lipid to dissolve within the bilayer core, and the physical properties of the bilayer formed by that lipid. When using a nonnatural lipid labeled as a probe of lipid flip in a bilayer composed of a natural lipid, the differences in its flip rate in different lipids will reflect the latter factor, but its absolute flip rate may differ from the natural lipid due to the former factor. Careful probe choice can minimize the difference between the absolute flip rate for the probe lipid and that of the corresponding unlabeled lipids.

It should be noted that the free energy of the transition state might be affected if it also involves co-diffusion of a bound species, such as a counterion interacting with a charged lipid group, or if it involves transient bilayer defects. With regard to the latter, defects are present when the gel and Ld state coexist, and it has been found that lipid flip is accelerated at temperatures close to the gel-to-Ld melting temperature [69]. Headgroup and acyl chain structure can also impact the free energy of the lipid in its normal bilayer conformation, and thus the rate of flip.

Studies of transverse diffusion sometimes have reported conflicting values for flip-flop rates. Nevertheless, there is a clear consensus for lipid vesicles based on many studies. In general, zwitterionic lipid flip-flop has a half-time of days [54, 67, 69–71], but this can decrease to minutes in bilayers composed of lipids with two highly polyunsaturated acyl chains [67, 72]. This is true even if the species flipping is not itself highly polyunsaturated. The fast flip in such bilayers is not surprising given the looser lipid packing and increases in polarity imparted by double bonds [73], but it must be emphasized that lipids with two polyunsaturated acyl chains are not abundant in nature, at least in mammalian cells.

It has also been found that cholesterol decreases flip-flop rates [74], including in the case of flip-flop accelerated by the presence of transmembrane peptides [75]. This presumably reflects the effect of cholesterol on tight lipid packing and is interesting in view of the ability of cholesterol itself to rapidly flip across the bilayer [63].

Results for lipid flip in asymmetric vesicles are consistent with these observations. Membranes composed of PC, PE, PS, or SM with physiological acyl chain combinations form asymmetric membranes having asymmetry that can be stable for days [56, 67, 68]. However, vesicles prepared by lipid exchange using acceptor vesicles containing high amounts of lipids with polyunsaturated acyl chains or very short chain lipids exhibit fast flip-flop and a loss of asymmetry [67]. When SM was exchanged into acceptor vesicles composed of 100% anionic lipids, other than PS, flip-flop was also significant, and asymmetry partly decayed over hours [68]. This loss of asymmetry was not observed when acceptor vesicles had a mixture of anionic lipid and PE prior to exchange. Most anionic lipids may be able to flip quickly since they have fewer charged groups (or in the case of cardiolipin, fewer changed groups per two acyl chains) than zwitterionic lipids (PC, PE, and SM) or than PS. However, this may not be the whole explanation for faster flip in such vesicles. The loss of asymmetry should be rate limited by the flip of SM molecules, which suggests that a change in lipid packing in vesicles with such anionic lipids is also important. This might be due to electrostatic repulsions when anionic lipids are present at high levels in a leaflet.

When preparing asymmetric membranes, it should be kept in mind that donor lipid is initially exchanged into acceptor vesicles composed fully of acceptor lipids. This will determine the initial rate of flip. After exchange is terminated, the lipid composition of the vesicles will have changed, so the rate of flip-flop toward the end of exchange and after exchange may be quite different from that when lipid exchange was initiated.

It is also not yet clear if CDs can alter lipid flip-flop. This possibility has been proposed [54, 76] and could influence asymmetry. Although CDs are hydrophilic, they could potentially bind to membranes to some degree and alter lipid packing during the exchange step in a fashion that influences flip-flop and thus impacts asymmetry.

The presence of coexisting ordered and disordered domains might also influence the stability of asymmetry. As noted above, there could be accelerated flip due to packing defects at the boundary between domains. In addition, the rate-limiting step for destroying asymmetry would be the rate of flip in the domains allowing the fastest flip-flop. For example, consider a membrane composed of a mixture of a lipid with saturated acyl chains, such as SM, and polyunsaturated PC. The flip-flop rate might be modest when the membrane is homogenous, some sort of average for SM, and polyunsaturated PC, but be accelerated when the membrane segregates into ordered domains rich in saturated lipid and disordered domains rich in polyunsaturated lipids. Assuming that lateral movements of lipids between domains are rapid, the fast flip across the polyunsaturated lipid domains would presumably be the rate-limiting step in loss of asymmetry after domains form.

How flip rates would compare in asymmetric membranes and symmetric membranes of the same overall composition is not clear. Suppose one prepares a vesicle in which one leaflet was composed of a lipid that by itself forms bilayers across which flip is slow and the other leaflet is composed of lipids that by themselves form bilayers across which flip is fast. One might predict that flip across the leaflet containing the lipid that slowly flips would be rate limiting, and so overall flip would be slow, with a rate close to that of the more slowly flipping lipid.

If a lipid mixture contains cholesterol (in both leaflets), flip behavior could be further complicated. In a bilayer composed of a more slowly flipping and more quickly flipping lipid, the movement of the more quickly flipping lipid into a leaflet could be compensated for by flipping of cholesterol into the opposite leaflet, creating a transmembrane cholesterol gradient. This might be of some practical use in creating artificial vesicles with cholesterol asymmetry, at least when there is agreement about what the cholesterol gradients are biological membranes!

14.4 Interleaflet coupling studies in asymmetric membranes

A long-standing question is: How are the physical state of the two leaflets in an asymmetric bilayer coupled? This is of interest because if formation of membrane domains in one leaflet induces domain formation in the opposite leaflet, followed by reorganization of proteins interacting with that leaflet, it could result in lipid-mediated signal transduction, a process quite distinct from ordinary transmembrane protein driven signal transduction. Studies in cells suggest that coupling of some sort can occur [25, 77]. Furthermore, studies in asymmetric planar bilayers in which domains are large enough to visualize by light microscopy have shown that interleaflet coupling of physical state can occur and is highly lipid composition dependent [43, 44]. Nevertheless, such studies have just scratched the surface of questions surrounding interleaflet coupling. The details of how acyl chain and headgroup structure and lipid composition influence interleaflet coupling are yet to be explored. In addition, interleaflet coupling needs to be characterized in those lipid compositions that are of perhaps most interest, in which domains are nanoscale in size.

Studies in asymmetric vesicles have already added significantly to our understanding of interleaflet coupling. In small and large asymmetric vesicles, it has been found that outer leaflets composed primarily of high-T_m lipids (e.g., SM) form an ordered state as thermally stable as in symmetric vesicles even when the inner leaflet is composed of unsaturated lipids such as dioleoyl PC (DOPC). This was true both in the absence and presence of cholesterol in the asymmetric vesicles [51, 52]. The average order in the unsaturated lipids in the inner leaflet did not seem to increase much due to the presence of high T_m in the outer leaflet, suggesting no more than weak coupling of leaflet membrane order, at least at room temperature. A somewhat different picture was derived from fluorescence correlation spectroscopy-based lateral diffusion measurements in analogous giant asymmetric vesicles lacking cholesterol. In this case, coupling of inner and outer leaflet diffusion (i.e., reduced diffusion in both leaflets relative to that prior to introduction of SM into the outer leaflet of unsaturated PC vesicles) was observed in some, but not all, cases and was dependent on the acyl chain structure of both the inner and outer leaflet lipids. Surprisingly, this coupling was observed even in lipid compositions in which measurements of average membrane order in the unsaturated lipids forming the inner leaflet did not seem to increase upon introduction of high-T_m lipid into the outer leaflet [66]. Perhaps order and diffusion are not coupled in the same fashion. It is also possible that inner leaflet behavior reflected the fact that membrane order measures the average order in a leaflet (when using a probe restricted to a single leaflet). A change in leaflet structure could occur so that upon domain formation a homogenous leaflet with uniform low order is transformed into one with a combination of domains, some of which have higher order than in the homogenous state, while the remainder has even lower order than in the homogeneous state. It should be pointed out that although a decrease in outer leaflet diffusion was observed in these studies, it was not demonstrated that diffusion was reduced to the same extent it would have been in corresponding symmetric membranes with saturated lipids in both leaflets. In other words, it was possible that the unsaturated lipid in the inner leaflet increased diffusion in the asymmetric vesicles' outer leaflet relative to that in symmetric vesicles. This possibility is supported by neutron scattering studies in asymmetric vesicles [54]. The lipid packing of a dipalmitoyl PC (DPPC)-rich ordered domains appeared to be significantly less than in gel state dipalmitoyl PC (DPPC), suggestive of an effect on DPPC packing arising from coupling to the unsaturated lipid 1-palmitoyl-2-oleoyl PC (POPC)-containing opposite leaflet.

Some of the complexities noted above may reflect the fact that the parameters used to infer asymmetric vesicle physical properties did not directly examine ordered domain formation. Subsequent studies in asymmetric vesicles (that contain cholesterol) have turned to methods that more directly detect the formation of ordered domains. Studies of giant asymmetric vesicles in which egg SM or milk SM was introduced into the outer leaflet of GUV composed of DOPC and cholesterol demonstrated that ordered domains formed, in register with each other, in both the outer leaflet and SM-lacking inner leaflet. The physical properties of the inner and outer leaflet were more tightly coupled using milk SM than using egg SM, as judged by the extent to which fluorescent probe partition coefficients (i.e., the ratio of the concentrations of probe in ordered and disordered domains) did or did not differ between the two leaflets. This could reflect interdigitation between the very long acyl chains found on some milk SM molecules and inner leaflet lipid. The observation that a lipid such as DOPC could ever form ordered domains was surprising. A likely hypothesis is that inner leaflet ordered domain formation was aided by migration of inner leaflet cholesterol into the inner leaflet ordered domains so that they have an unusually high cholesterol content. The observation that inner and outer leaflet ordered domains could have different probe partition properties suggests that Lo states with different levels of order can exist in two leaflets.

To further characterize the influence of interleaflet coupling upon ordered domain formation, additional approaches must be used. These must directly respond to segregation of lipids into different domains and circumvent the inability of ordinary microscopy to detect nanodomains. (It is also important to note that fluorescence microscopy can fail to detect domain if probes that partition nearly equally between Ld and Lo domains are used.) Fluorescence resonance energy transfer (FRET), which can detect nanodomains down to 5 nm size or even smaller, is promising in this regard [31]. In general terms, FRET can be used to detect domains by using two probes with different affinities for different domains. When detecting FRET via quenching of donor fluorescence, a FRET acceptor must have a preferential localization in either Lo or Ld domains. Depending both on whether FRET acceptor preferentially partitions into Ld or Lo domains and on the localization of FRET donor, domain formation can either lead to an overall increase of FRET acceptor density near FRET donor, and so increase FRET, or an overall decrease of FRET acceptor density near FRET donors, and so decrease FRET [78].

FRET has important limitations. The magnitude of the change in FRET under conditions of domain formation is complicated by its sensitivity to acceptor concentration, domain size, and the values of the domain partition coefficients for FRET donor and FRET acceptor, as well as the fraction of the bilayer in the Lo and Ld state [78]. The most unambiguous parameter that can be derived from FRET is the thermal or compositional stability of ordered domain formation, that is, it can detect the compositions in which there are domains and the temperatures above which ordered domain disappear, and lipids in the Ld and Lo domains mix.

Of course, protocols that restrict FRET probes to one leaflet must also be used for studies of domains in one specific leaflet of an asymmetric vesicle. In the case of measuring FRET donor quenching by a FRET acceptor, it is most important that the FRET acceptor be localized to the desired leaflet, to probe domain formation in that leaflet. Specific localization of the FRET donor in the same leaflet is not as important in this case.

14.5 Potential scenarios for the nature of interleaflet coupling

There are many possible variations for interleaflet coupling of lipid physical state in asymmetric membranes in which one leaflet has lipids favorable for formation of ordered domains, in other words, in which one leaflet which contains enough high-T_m phospho- or sphingolipid (mixed with low-T_m phospholipid) to support ordered state formation in the absence of interactions with the inner leaflet, while the other leaflet has only low-T_m phospholipids. We will consider the case in which the vesicles also have sufficient cholesterol for the Lo state to form in place of the gel state. We will also oversimplify the situation to assume that in the presence of cholesterol, the inner and outer leaflet lipids have T_m values above which they transition from the Lo to the Ld state.

If there is no coupling, or more precisely stated, if there is no difference between ordered domain stability in the individual leaflets of an asymmetric vesicle and that in symmetric vesicles with the same outer or same inner leaflet compositions, then the outer and inner leaflets in the asymmetric membranes should melt at with T_m values that are the same as in the corresponding symmetric vesicles with either the outer leaflet or inner leaflet composition. Below the T_m for both leaflets, the entire bilayer would be Lo, above the T_m for both leaflets, the bilayer would be all Ld, and at temperatures between the T_m values for the inner and outer leaflet, the bilayer would have one leaflet (the one with high-T_m lipid) with Lo domains (or all Lo depending on its composition) and one leaflet (the one with low-T_m lipid) in the Ld state.

If there is coupling, then one or both leaflets should show altered physical properties in the asymmetric vesicles relative to the corresponding symmetric ones. There could be strong coupling, in which the physical properties of the two leaflets are identical (or at least very similar) at all temperatures. At one extreme would be strong coupling in which the leaflet with the tendency to form ordered domains dominates, so that ordered domains are present in register (i.e., at the same positions) in both leaflets almost up to the T_m at which the higher T_m leaflet would melt in symmetric vesicles. This type of coupling has been noted in several studies [43, 44, 79, 80]. There could also be strong coupling in which the leaflet that tends to form only Ld state dominates and destroys ordered domain formation in the opposite leaflet, when the temperature is above the T_m in symmetric vesicles with a composition equivalent to that of the leaflet having a lower T_m in the absence of coupling.

This type of coupling has also been reported [43]. The identity of the dominant leaflet could also be highly temperature dependent, with the ordered domain favoring leaflet dominating at lower temperature, and the disordered state favoring leaflet dominating at higher temperatures, so that the T_m in the asymmetric vesicle is between that for the symmetric vesicles composed of only inner or only outer leaflet lipids.

Another variation would be intermediate coupling in which the physical properties of the two leaflets influence each other over some range of temperatures but become uncoupled when temperature is increased or decreased. For example, there could be cases in which the leaflet with high-T_m lipids dominates physical behavior at lower temperatures, forcing the opposite leaflet to form the Lo state even if formed of lipids that would form the Ld state in symmetric vesicles. At higher temperatures, the two leaflets could become uncoupled, so that the leaflet with lower T_m lipids melts while the leaflet with high-T_m lipid still contains Lo domains. Analogously, in some cases, the low-T_m leaflet could dominate at higher temperatures so that the entire bilayer in is the Ld state, while at lower temperatures, the leaflets become uncoupled, so that the leaflet with high-T_m lipids forms the Lo state.

There could also be intermediate levels of coupling in which, due to coupling, one leaflet undergoes a partial change in physical state, for example, from Ld (or Lo) to something intermediate in physical properties relative to Lo and Ld. Another variant of intermediate coupling would be a case in which the physical behavior of lipids was only fully coupled at the bilayer midplane, at which point, the tails of lipids in opposite leaflets come into contact. One could imagine a transverse order gradient across the bilayer when one leaflet is composed of lipids that tend to form Lo domains by themselves and the opposite leaflet would tend to form Ld domains by themselves. In this case, the physical state of the lipids could gradually change from almost fully being in one physical state to almost in the other as one moves across the bilayer. This is somewhat analogous to a bimetallic strip composed of two strips of different metals fused to each other, each strip of which when by itself has a different coefficient of thermal expansion. The coefficient of expansion must match at some intermediate value where the two metals are in contact, while the entire strip bends so the parts of the strips farthest from the midplane have expanded less than at the midplane on the side composed of the metal with the smaller coefficient of expansion and expanded more than at the midplane on the other side, composed of the metal with a larger coefficient of expansion.

It is possible that different levels of coupling might even be present in the same bilayer. This has been reported in planar bilayers [43]. Heterogeneity that might represent different degrees of coupling has been also been observed in asymmetric vesicles [80].

It should also be noted that factors other than lipid composition and temperature might affect coupling. It has recently been reported that leaflet curvature can strongly influence lipid coupling [58].

14.6 Studying lipid–protein interaction in asymmetric vesicles

Asymmetric vesicles also have applications for studies of protein–lipid interactions. Most obvious applications are in reconstituted systems in which asymmetric vesicles with Lo and Ld domains and containing proteins are used to study how protein affinity for ordered domains is affected by lipid composition and asymmetry. A few studies of this type have been carried out using asymmetric planar bilayers [81, 82].

Another aspect of lipid–protein interaction that can be studied in asymmetric vesicles is the effect of lipid asymmetry and the composition of each leaflet on membrane protein conformation and function. An effect of asymmetry on the fraction of a modestly hydrophobic peptide in the transmembrane state was noted in one study [51]. In other studies involving the membrane-inserting toxin protein perfringolysin O, we found that membrane asymmetry can have dramatic effects. The insertion and assembly process for this protein has many steps, including binding to membranes via cholesterol, oligomerization, membrane insertion, and assembly of a transmembrane β-barrel pore [83]. The cholesterol dependence of these processes in asymmetric vesicles that mimic plasma membrane in their lipid asymmetry differed both from that in symmetric vesicles with the same lipid composition, and from that for symmetric membranes mimicking the plasma membrane outer leaflet or plasma membrane inner leaflet. Strikingly, in the asymmetric vesicles, an intermediate conformation involving formation of transmembrane oligomers but lacking a pore was found to predominate at intermediate cholesterol conformations [84]. This conformation could have important implications for how perfringolysin O assembly proceeds in vivo [84]. Such studies illustrate the potential for analysis of protein–lipid interaction in asymmetric membranes. It seems likely that the behavior of many proteins will show marked differences between their conformational behavior in symmetric membranes and asymmetric membranes that mimic natural membranes.

14.7 CD-catalyzed lipid exchange to alter membrane lipid composition and asymmetry in living cells

The studies above describe uses of CD-catalyzed lipid exchange in artificial model membranes. CD-catalyzed exchange can also be applied to cells. The vast majority of these involve use of CDs to change sterol levels or substitute one type of sterol with another [60, 61]. However, phospholipids can also be exchanged into cells [85, 86]. We recently demonstrated that it is possible to achieve efficient exchange of plasma membrane outer leaflet lipids in cultured cells [3]. This allows replacement of the entire plasma membrane outer leaflet with exogenous lipids. The method opens the door to many studies, including studies modulating ordered domain levels in cells, studies of lipid–protein interaction in living cells, and measurements of dynamic changes in membrane asymmetry. In particular, the ability to make changes more rapidly than with metabolic inhibitors and changing a wider range of lipid structural features, by introducing unnatural lipids or altering natural lipid headgroup and acyl chain at the same time, promise novel insights into lipid function. One could also envision subtractive experiments in which exchange is carried out with a natural lipid donor mixture that is missing one specific type of lipid. Although the method is limited to the plasma membrane outer leaflet, it has the potential to lead to some exciting breakthroughs.

Acknowledgment

This work was supported by N.I.H. grant GM 122493.

References

[1] Verkleij, A.J., Zwaal, R.F., Roelofsen, B., Comfurius, P., Kastelijn, D., & van Deenen, L.L. The asymmetric distribution of phospholipids in the human red cell membrane. A combined study using phospholipases and freeze-etch electron microscopy. Biochim Biophys Acta 1973;323:178–193.
[2] Bretscher, M.S. Asymmetrical lipid bilayer structure for biological membranes. Nat New Biol 1972;236:11–12.
[3] Li, G., Kim, J., Huang, Z., St Clair, J.R., Brown, D.A., & London, E. Efficient replacement of plasma membrane outer leaflet phospholipids and sphingolipids in cells with exogenous lipids. Proceedings of the National Academy of Sciences of the United States of America 2016;113:14025–14030.
[4] Hankins, H.M., Baldridge, R.D., Xu, P., & Graham, T.R. Role of flippases, scramblases and transfer proteins in phosphatidylserine subcellular distribution. Traffic 2015;16:35–47.
[5] Liu, S.L., Sheng, R., Jung, J.H., et al. Orthogonal lipid sensors identify transbilayer asymmetry of plasma membrane cholesterol. Nat Chem Biol 2017;13:268–274.
[6] Fisher, K.A. Analysis of membrane halves: cholesterol. Proceedings of the National Academy of Sciences of the United States of America 1976;73:173–177.
[7] Mondal, M., Mesmin, B., Mukherjee, S., & Maxfield, F.R. Sterols are mainly in the cytoplasmic leaflet of the plasma membrane and the endocytic recycling compartment in CHO cells. Mol Biol Cell 2009;20:581–588.
[8] Ohvo-Rekila, H., Ramstedt, B., Leppimaki, P., & Slotte, J.P. Cholesterol interactions with phospholipids in membranes. Prog Lipid Res 2002;41:66–97.
[9] Mercer, J., & Helenius, A. Vaccinia virus uses macropinocytosis and apoptotic mimicry to enter host cells. Science 2008;320:531–535.

[10] Lentz, B.R. Exposure of platelet membrane phosphatidylserine regulates blood coagulation. Prog Lipid Res 2003;42:423–438.
[11] Morizono, K., & Chen, I.S. Role of phosphatidylserine receptors in enveloped virus infection. J Virol 2014;88:4275–4290.
[12] Brown, D.A., & London, E. Structure and origin of ordered lipid domains in biological membranes. J Membr Biol 1998;164:103–114.
[13] Koynova, R., & Caffrey, M. Phases and phase transitions of the sphingolipids. Biochim Biophys Acta 1995;1255:213–236.
[14] Koynova, R., & Caffrey, M. Phases and phase transitions of the phosphatidylcholines. Biochim Biophys Acta 1998;1376:91–145.
[15] Konyakhina, T.M., Wu, J., Mastroianni, JD, Heberle FA, & Feigenson GW. Phase diagram of a 4-component lipid mixture: DSPC/DOPC/POPC/chol. Biochim Biophys Acta 2013;1828:2204–2214.
[16] Ahmed, S.N., Brown, D.A., & London, E. On the origin of sphingolipid/cholesterol-rich detergent-insoluble cell membranes: physiological concentrations of cholesterol and sphingolipid induce formation of a detergent-insoluble, liquid-ordered lipid phase in model membranes. Biochemistry 1997;36:10944–10953.
[17] Barton, P.G., & Gunstone, F.D. Hydrocarbon chain packing and molecular motion in phospholipid bilayers formed from unsaturated lecithins. Synthesis and properties of sixteen positional isomers of 1,2-dioctadecenoyl-sn-glycero-3-phosphorylcholine. J Biol Chem 1975;250:4470–4476.
[18] Thompson, T.E., & Tillack, T.W. Organization of glycosphingolipids in bilayers and plasma membranes of mammalian cells. Ann Rev Biophys Biophys Chem 1985;14:361–386.
[19] Ipsen, J.H., Karlstrom, G., Mouritsen, O.G., Wennerstrom, H., & Zuckermann, M.J. Phase equilibria in the phosphatidylcholine-cholesterol system. Biochim Biophys Acta 1987;905:162–172.
[20] Vist, M.R., & Davis, J.H. Phase equilibria of cholesterol/dipalmitoylphosphatidylcholine mixtures: 2H nuclear magnetic resonance and differential scanning calorimetry. Biochemistry 1990;29:451–464.
[21] Schroeder, R., London, E., & Brown, D. Interactions between saturated acyl chains confer detergent resistance on lipids and glycosylphosphatidylinositol (GPI)-anchored proteins: GPI-anchored proteins in liposomes and cells show similar behavior. Proceedings of the National Academy of Sciences of the United States of America 1994;91:12130–12134.
[22] Silvius, J.R., del Giudice, D., & Lafleur, M. Cholesterol at different bilayer concentrations can promote or antagonize lateral segregation of phospholipids of differing acyl chain length. Biochemistry 1996;35:15198–15208.
[23] Korlach, J., Schwille, P., Webb, W.W., & Feigenson, G.W. Characterization of lipid bilayer phases by confocal microscopy and fluorescence correlation spectroscopy. Proceedings of the National Academy of Sciences of the United States of America 1999;96:8461–8466.
[24] Wang, T.Y., & Silvius, J.R. Cholesterol does not induce segregation of liquid-ordered domains in bilayers modeling the inner leaflet of the plasma membrane. Biophys J 2001;81:2762–2773.
[25] Raghupathy, R., Anilkumar, A.A., Polley, A., et al. Transbilayer lipid interactions mediate nanoclustering of lipid-anchored proteins. Cell 2015;161:581–594.
[26] Sharpe, H.J., Stevens, T.J., & Munro, S. A comprehensive comparison of transmembrane domains reveals organelle-specific properties. Cell 2010;142:158–169.
[27] Brown, D.A., & Rose, J.K. Sorting of GPI-anchored proteins to glycolipid-enriched membrane subdomains during transport to the apical cell surface. Cell 1992;68:533–544.
[28] London, E., & Brown, D.A. Insolubility of lipids in triton X-100: physical origin and relationship to sphingolipid/cholesterol membrane domains (rafts). Biochim Biophys Acta 2000;1508:182–195.

[29] Drevot, P., Langlet, C., Guo, X.J., et al. TCR signal initiation machinery is pre-assembled and activated in a subset of membrane rafts. EMBO J 2002;21:1899–1908.
[30] Heerklotz, H., Szadkowska, H., Anderson, T., & Seelig, J. The sensitivity of lipid domains to small perturbations demonstrated by the effect of Triton. J Mol Biol 2003;329:793–799.
[31] Pathak, P., & London, E. Measurement of lipid nanodomain (raft) formation and size in sphingomyelin/POPC/cholesterol vesicles shows TX-100 and transmembrane helices increase domain size by coalescing preexisting nanodomains but do not induce domain formation. Biophys J 2011;101:2417–2425.
[32] Simons, K., & Ikonen, E. Functional rafts in cell membranes. Nature 1997;387:569–572.
[33] Toulmay, A., & Prinz, W.A. Direct imaging reveals stable, micrometer-scale lipid domains that segregate proteins in live cells. J Cell Biol 2013;202:35–44.
[34] Rayermann, S.P., Rayermann, G.E., Cornell, C.E., Merz, A.J., & Keller, S.L. Hallmarks of reversible separation of living, unperturbed cell membranes into two liquid phases. Biophys J 2017;113:2425–2432.
[35] Tsuji, T., Fujimoto, M., Tatematsu, T., et al. Niemann-Pick type C proteins promote microautophagy by expanding raft-like membrane domains in the yeast vacuole. eLife 2017;6: e25960.
[36] Stone, M.B., Shelby, S.A., Nunez, M.F., Wisser, K., & Veatch, S.L. Protein sorting by lipid phase-like domains supports emergent signaling function in B lymphocyte plasma membranes. eLife 2017;6:e19891.
[37] Kinoshita, M., Suzuki, K.G., Matsumori, N., et al. Raft-based sphingomyelin interactions revealed by new fluorescent sphingomyelin analogs. J Cell Biol 2017;216:1183–1204.
[38] Komura, N., Suzuki, K.G., Ando, H., et al. Raft-based interactions of gangliosides with a GPI-anchored receptor. Nat Chem Biol 2016;12:402–410.
[39] Dinic, J., Riehl, A., Adler, J., & Parmryd, I. The T cell receptor resides in ordered plasma membrane nanodomains that aggregate upon patching of the receptor. Sci Rep 2015;5:10082.
[40] LaRocca, T.J., Crowley, J.T., Cusack, B.J., et al. Cholesterol lipids of Borrelia burgdorferi form lipid rafts and are required for the bactericidal activity of a complement-independent antibody. Cell Host Microbe 2010;8:331–342.
[41] LaRocca, T.J., Pathak, P., Chiantia, S., et al. Proving lipid rafts exist: membrane domains in the prokaryote Borrelia burgdorferi have the same properties as eukaryotic lipid rafts. PLoS Pathogens 2013;9:e1003353.
[42] Garg, S., Ruhe, J., Ludtke, K., Jordan, R., & Naumann, C.A. Domain registration in raft-mimicking lipid mixtures studied using polymer-tethered lipid bilayers. Biophys J 2007;92:1263–1270.
[43] Collins, M.D., & Keller, S.L. Tuning lipid mixtures to induce or suppress domain formation across leaflets of unsupported asymmetric bilayers. Proceedings of the National Academy of Sciences of the United States of America 2008;105:124–128.
[44] Wan, C., Kiessling, V., & Tamm, L.K. Coupling of cholesterol-rich lipid phases in asymmetric bilayers. Biochemistry 2008;47:2190–2198.
[45] Bayley, H., Cronin, B., Heron, A., et al. Droplet interface bilayers. Mol BioSyst 2008;4:1191–1208.
[46] Hwang, W.L., Chen, M., Cronin, B., Holden, M.A., & Bayley, H. Asymmetric droplet interface bilayers. J Am Chem Soc 2008;130:5878–5879.
[47] Pautot, S., Frisken, B.J., & Weitz, D.A. Engineering asymmetric vesicles. Proceedings of the National Academy of Sciences of the United States of America 2003;100:10718–10721.
[48] Kamiya, K., Kawano, R., Osaki, T., Akiyoshi, K., & Takeuchi, S. Cell-sized asymmetric lipid vesicles facilitate the investigation of asymmetric membranes. Nat Chem 2016;8:881–889.
[49] Elani, Y., Purushothaman, S., & Booth, P.J., et al. Measurements of the effect of membrane asymmetry on the mechanical properties of lipid bilayers. Chem Commun (Camb) 2015;51:6976–6979.

[50] Takaoka, R., Kurosaki, H., Nakao, H., Ikeda, K., & Nakano, M. Formation of asymmetric vesicles via phospholipase D-mediated transphosphatidylation. Biochim Biophys Acta 2018;1860:245–249.

[51] Cheng, H.T., Megha, & London E. Preparation and properties of asymmetric vesicles that mimic cell membranes: effect upon lipid raft formation and transmembrane helix orientation. J Biol Chem 2009;284:6079–6092.

[52] Cheng, H.T., & London, E. Preparation and properties of asymmetric large unilamellar vesicles: interleaflet coupling in asymmetric vesicles is dependent on temperature but not curvature. Biophys J 2011;100:2671–2678.

[53] Chiantia, S., Schwille, P., Klymchenko, A.S., & London, E. Asymmetric GUVs prepared by MbetaCD-mediated lipid exchange: an FCS study. Biophys J 2011;100:L1–L3.

[54] Heberle, F.A., Marquardt, D., Doktorova, M., et al. Sub-nanometer structure of an asymmetric model membrane: interleaflet coupling influences domain properties. Langmuir : ACS J Surf Colloids 2016; 32: 5195–5200.

[55] Visco, I., Chiantia, S., & Schwille, P. Asymmetric supported lipid bilayer formation via methyl-beta-cyclodextrin mediated lipid exchange: influence of asymmetry on lipid dynamics and phase behavior. Langmuir : ACS J Surf Colloids 2014;30:7475–7484.

[56] Lin, Q., & London, E. Preparation of artificial plasma membrane mimicking vesicles with lipid asymmetry. PloS One 2014;9:e87903.

[57] Vitrac, H., MacLean, D.M., Jayaraman, V., Bogdanov, M., & Dowhan, W. Dynamic membrane protein topological switching upon changes in phospholipid environment. Proceedings of the National Academy of Sciences of the United States of America 2015;112:13874–13879.

[58] Eicher, B., Marquardt, D., Heberle, F.A., et al. Intrinsic curvature-mediated transbilayer coupling in asymmetric lipid vesicles. Biophys J 2018;114:146–157.

[59] Markones, M., Drechsler, C., Kaiser, M., Kalie, L., Heerklotz, H., & Fiedler, S. Engineering asymmetric lipid vesicles: accurate and convenient control of the outer leaflet lipid composition. Langmuir : ACS J Surf Colloids 2018;34:1999–2005.

[60] Zidovetzki, R., & Levitan, I. Use of cyclodextrins to manipulate plasma membrane cholesterol content: evidence, misconceptions and control strategies. Biochim Biophys Acta 2007;1768:1311–1324.

[61] Kim, J., & London, E. Using sterol substitution to probe the role of membrane domains in membrane functions. Lipids 2015;50:721–734.

[62] Anderson, T.G., Tan, A., Ganz, P., & Seelig, J. Calorimetric measurement of phospholipid interaction with methyl-beta-cyclodextrin. Biochemistry 2004;43:2251–2261.

[63] Leventis, R., & Silvius, J.R. Use of cyclodextrins to monitor transbilayer movement and differential lipid affinities of cholesterol. Biophys J 2001;81:2257–2267.

[64] Niu, S.L., & Litman, B.J. Determination of membrane cholesterol partition coefficient using a lipid vesicle-cyclodextrin binary system: effect of phospholipid acyl chain unsaturation and headgroup composition. Biophys J 2002;83:3408–3415.

[65] Huang, Z., &London, E. Effect of cyclodextrin and membrane lipid structure upon cyclodextrin-lipid interaction. Langmuir : ACS J Surf Colloids 2013;29:14631–14638.

[66] Chiantia, S., & London, E. Acyl chain length and saturation modulate interleaflet coupling in asymmetric bilayers: effects on dynamics and structural order. Biophys J 2012;103:2311–2319.

[67] Son, M., & London, E. The dependence of lipid asymmetry upon phosphatidylcholine acyl chain structure. J Lipid Res 2013;54:223–231.

[68] Son, M., & London, E. The dependence of lipid asymmetry upon polar headgroup structure. J Lipid Res 2013;54:3385–3393.

[69] Marquardt, D., Heberle, F.A., Miti, T., et al. (1)H NMR Shows slow phospholipid flip-flop in gel and fluid bilayers. Langmuir : ACS J Surf Colloids 2017;33:3731–3741.

[70] Rothman, J.E., & Dawidowicz, E.A. Asymmetric exchange of vesicle phospholipids catalyzed by the phosphatidylcholine exhange protein. Measurement of Inside--Outside Transitions. Biochemistry 1975;14:2809–2816.

[71] Johnson, L.W., Hughes, M.E., & Zilversmit, D.B. Use of phospholipid exchange protein to measure inside-outside transposition in phosphatidylcholine liposomes. Biochim Biophys Acta 1975;375:176–185.

[72] Armstrong, V.T., Brzustowicz, M.R., Wassall, S.R, Jenski, L.J., & Stillwell, W. Rapid flip-flop in polyunsaturated (docosahexaenoate) phospholipid membranes. Arch Biochem Biophys 2003;414:74–82.

[73] Smith, M., & Jungalwala, F.B. Reversed-phase high performance liquid chromatography of phosphatidylcholine: a simple method for determining relative hydrophobic interaction of various molecular species. J Lipid Res 1981;22:697–704.

[74] Nakano, M., Fukuda, M., Kudo, T., et al. Flip-flop of phospholipids in vesicles: kinetic analysis with time-resolved small-angle neutron scattering. J Phys Chem B 2009;113:6745–6748.

[75] LeBarron, J., & London, E. Effect of lipid composition and amino acid sequence upon transmembrane peptide-accelerated lipid transleaflet diffusion (flip-flop). Biochim Biophys Acta 2016;1858:1812–1820.

[76] Garg, S., Porcar, L., Woodka, A.C., Butler, P.D., & Perez-Salas, U. Noninvasive neutron scattering measurements reveal slower cholesterol transport in model lipid membranes. Biophys J 2011;101:370–377.

[77] Brown, D., & London, E. Functions of lipid rafts in biological membranes. Annu Rev Cell Dev Biol 1998;14:111–136.

[78] Pathak P, & London E. The effect of membrane lipid composition on the formation of lipid ultrananodomains. Biophys J 2015;109:1630–1638.

[79] Kiessling, V., Yang, S.T., & Tamm, L.K. Supported lipid bilayers as models for studying membrane domains. Curr Top Membr 2015;75:1–23.

[80] Lin, Q., & London, E. Ordered raft domains induced by outer leaflet sphingomyelin in cholesterol-rich asymmetric vesicles. Biophys J 2015;108:2212–2222.

[81] Hussain, N.F., Siegel, A.P., Ge, Y., Jordan, R., & Naumann, C.A. Bilayer asymmetry influences integrin sequestering in raft-mimicking lipid mixtures. Biophys J 2013;104:2212–2221.

[82] Wan, C., Kiessling, V., Cafiso, D.S., & Tamm, L.K. Partitioning of synaptotagmin I C2 domains between liquid-ordered and liquid-disordered inner leaflet lipid phases. Biochemistry 2011;50:2478–2485.

[83] Tweten, R.K., Hotze, E.M., & Wade, K.R. The unique molecular choreography of giant pore formation by the cholesterol-dependent cytolysins of Gram-positive bacteria. Annu Rev Microbiol 2015;69:323–340.

[84] Lin, Q., & London, E. The influence of natural lipid asymmetry upon the conformation of a membrane-inserted protein (perfringolysin O). J Biol Chem 2014;289:5467–5478.

[85] Kainu, V., Hermansson, M., & Somerharju, P. Introduction of phospholipids to cultured cells with cyclodextrin. J Lipid Res 2010;51:3533–3541.

[86] Kainu, V., Hermansson, M., & Somerharju, P. Electrospray ionization mass spectrometry and exogenous heavy isotope-labeled lipid species provide detailed information on aminophospholipid acyl chain remodeling. J Biol Chem 2008;283:3676–3687.

Volker Kiessling, Lukas K. Tamm

15 Application and characterization of asymmetric-supported membranes

Abstract: The plasma membrane of cells is a heterogeneous compartment that consists of a variety of different lipid and protein species. Its structure and dynamics are determined by complex lipid–lipid, protein–protein, and lipid–protein interactions within the membrane and with its environment. Studies of model membranes have played crucial roles and contributed essential new knowledge to our current understanding of the structure and function of cell membranes. Recently, several groups have developed new model membrane systems in order to study consequences of the compositional asymmetry between the two leaflets of the plasma membrane. Our group has focused on the development, characterization, and application of asymmetric-supported planar bilayers. In this chapter, we first summarize methods of preparation and properties of asymmetric supported planar bilayers. We then highlight how we have utilized these membranes to address relevant biological questions in the fields of trans-bilayer lipid domain coupling, soluble N-ethylmaleimide-sensitive factor attachment proteins receptor (SNARE)-mediated membrane fusion, and viral entry into cells.

15.1 Introduction

Many fundamental biological functions of the cell take place at the plasma membrane. This membrane separates the extracellular space from the cytoplasm and it is the site of communication between these spaces. The composition and structure of the plasma membrane is closely linked to its functions. Considering the large variety of very specific processes that take place in parallel, it is not surprising that the plasma membrane is a highly dynamic and heterogeneous place. Part of this heterogeneity is the asymmetry of the membrane relative to its extracellular and cytoplasmic sides. The two leaflets of the plasma membrane differ in lipid composition, protein content, and exposure of specific domains of integral membrane proteins.

In order to gain a molecular understanding of membrane biology, it is desirable to isolate the biochemical reactions that take place at the membrane. This is complicated by the fact that the lipid bilayer of the membrane is not only the site where the reaction takes place but is often an active player in the process. Reconstituting a complete or partial biological reaction in model

---

Volker Kiessling, Lukas K. Tamm, Center for Membrane and Cell Physiology, University of Virginia, Charlottesville, Va.

https://doi.org/10.1515/9783110544657-015

membranes is a very attractive starting point for developing biochemical assays that involve membrane proteins. The goal is to investigate a certain process in a membrane environment that is simpler compared to the complex plasma membrane, but complex enough to reproduce the main aspects of the function of the relevant membrane components. However, often it is not known what the "minimum complexity" is or there is no model system available to mimic the necessary characteristics of the cell membrane. One of the crucial properties that has attracted the attention of many investigators over the last years is the aforementioned asymmetry of the membrane. Several model membrane systems that result in asymmetric lipid compositions have been developed. Here, we focus solely on asymmetric supported planar membranes. Due to their geometry, supported planar membranes are ideally suited to be incorporated into quantitative fluorescent microscopy assays. We describe the preparation and characterization of asymmetric-supported planar membranes and discuss some applications in the fields of transmembrane phase-coupling and membrane fusion. The large majority of the summarized work has been published by our group, and for details, we refer readers to the cited literature, where one can also find references about the specific biological questions asked.

15.2 Preparation and properties of asymmetric-supported membranes

The reconstitution of integral membrane proteins into planar supported membranes is a prerequisite for isolating complex membrane processes in a supported model membrane platform. Unfortunately, the standard vesicle fusion (VF) technique results in supported membranes with immobile and often nonfunctional protein content. Most likely, this is due to protein domains that protrude from the outer vesicle surface and interact strongly with the glass support during the fusion process. To overcome this problem, Kalb et al. in 1992 introduced the Langmuir–Blodgett (LB)/VF technique [1, 2], where the supported membrane is prepared in two steps: (1) a lipid monolayer is transferred from the water/air interface of a Langmuir trough onto a hydrophilic glass substrate, and (2) vesicles or proteoliposomes are fused to the supported monolayer to form a supported bilayer (Figure 15.1). It was hypothesized that the encounter of the proteoliposomes with the hydrophobic acyl-chains of the supported monolayer would better preserve the structure of the membrane proteins during the bilayer formation process. In the following paragraphs, we will discuss the basic preparation method of the LB/VF technique and several variations of this technique. While all of them make use of the two-step preparation principle, they differ in the kind of vesicles that are added in the second step (Figure 15.1).

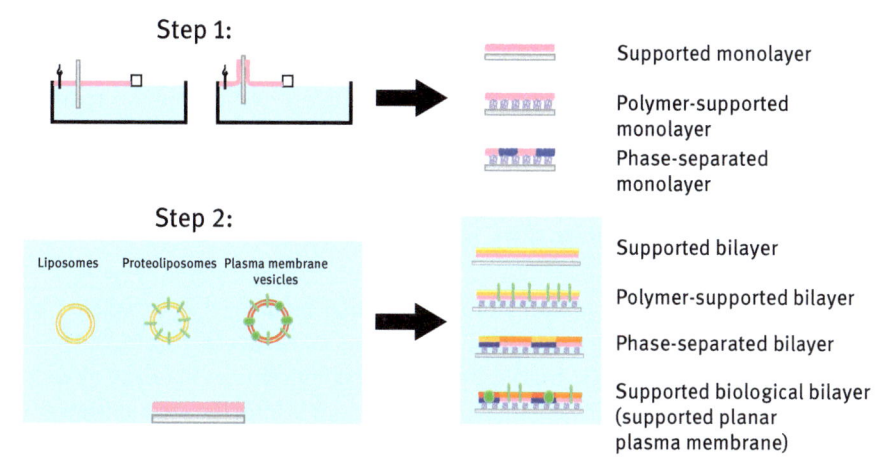

Figure 15.1: Two step Langmuir-Blodgett/Vesicle fusion technique. In the first step a lipid monolayer is transferred from the air-water interface of a Langmuir trough onto a glass substrate. As depicted on the right this monolayer can include polymer to increase the distance from the substrate and to stabilize phase separated lipids. In step 2, the second leaflet of the supported membrane is formed by adding liposomes, proteoliposomes, or extracted natural plasma membrane vesicles. The different types of supported bilayers that have been formed by this method are pictured on the right.

15.2.1 Step 1: The supported monolayer

In order to prepare the supported monolayer, lipids are first spread from organic solvent (usually chloroform) on the water surface of a Langmuir trough. The lipid monolayer at the water/air interface is compressed by a computer-controlled barrier until the surface pressure reaches typically 32 mN/m. This surface pressure corresponds to a packing density of the lipids that is equivalent to their packing density within biological membranes. A clean hydrophilic substrate (e.g., glass or quartz or an oxidized Si wafer) is quickly immersed through the water/air interface into the bulk water of the Langmuir trough and then slowly retracted in order to transfer a monolayer onto the support. During the transfer process, the surface pressure is held constant by controlling the barrier and therefore the available surface area for the lipid film. The change of the available surface area on the trough is equivalent to the area of transferred monolayer area onto the substrate.

Adding a small amount of a polymer-conjugated lipid (we typically use 3 mol% of a PEGylated lipid) to the monolayer stabilizes the transferred monolayer and adds more space between substrate and membrane [3–5]. The latter is important for the reconstitution of laterally mobile integral membrane proteins and for the observation of single fusion events by the transfer of fluorescent fusion proteins discussed below.

Lipid compositions for the monolayer can range from pure synthetic lipid species to complex mixtures of extracted natural lipids that under some conditions may phase separate into coexisting lipid domains. Microscopic domains of lipids in gel-, liquid-ordered (lo), or liquid-disordered (ld) phases can be visualized by the addition of fluorescently labeled lipids or lipid analogs that partition preferentially in one or the other phase. The structure of the lipid domains is preserved during the transfer from the trough onto the substrate [6].

15.2.2 Monolayers + liposomes (asymmetry)

The formation of a supported bilayer from a supported monolayer is completed by rehydrating the monolayer with a solution that contains liposomes (Figure 15.1). These liposomes that may be prepared from different lipid compositions than the already prepared monolayer fuse with the monolayer and their lipids form the second leaflet of the supported bilayer. For fluorescently labeled lipids, the asymmetry of the membrane can be quantified by quenching the fluorescence of the accessible distal leaflet with heavy metal ions like Co^{2+} [7] or by measuring the distance between the fluorescent layer and the substrate by fluorescence interference contrast microscopy [8, 9]. It could be shown that the asymmetry of the fluorophore distribution is maintained for several hours. The distribution of 0.5 mol% rhodamine-labeled phosphatidylethanolamine in an asymmetric supported fluid 1-palmitoyl-2-oleoyl-*sn*-glycero-3-phosphocholine (POPC) membrane decreased from 98:2 (distal-:proximal-leaflet) immediately after preparation to 88:12 after 4 h [9]. This high level of asymmetry that is maintained over a relatively long time mimics the slow uncatalyzed flip-flop rate of lipids in model and biological membranes and allows the performance of the experiments discussed below that require a defined membrane asymmetry.

The first biological problem that was approached with this system was if and how the lipid phase in one leaflet influences the phase in the opposing leaflet. The physical basis for the "raft-hypothesis" of biological membranes [10] is the strong interaction of cholesterol with sphingomyelin, which is located predominantly in the outer leaflet of the plasma membrane. Ternary lipid mixtures consisting of sphingomyelin or saturated phospholipids, cholesterol, and unsaturated phospholipids separate into liquid-ordered lo and liquid-disordered ld phases at room temperature. Model membranes consisting of lipids that are typical for the inner leaflet of the plasma membrane do not phase-separate. However, for "lipid rafts" to be platforms that are important for cell signaling, they would have to span across the whole membrane and the coupling between the two leaflets at the site of lipid rafts would have to be mediated by transmembrane proteins or by the lipids themselves. Using asymmetric supported membranes, we showed that lipid coupling between leaflets is sufficient to induce rafts across the membrane and we characterized molecular determinants of this coupling by preparing asymmetric supported membranes that contain phase-separating ternary lipid mixtures in their proximal leaflet. Indeed, it was shown that monolayers with coexisting lo- and ld phases can induce phase separation in the opposing distal leaflet if it is composed of different lipid species [9, 11]. These studies suggested that the coupling of the inner leaflet mixtures depends more on their intrinsic chain melting temperature than on their specific headgroup classes. Later, it was shown that the alteration of the lipid phases in the proximal (cytoplasmic-mimicking) leaflet modulates the binding of the peripheral C2 protein domains of synaptotagmin-1 [12].

15.2.3 Monolayers + proteoliposomes (oriented and mobile membrane proteins)

Reconstituting integral membrane proteins into the supported membrane is essential for the investigation of many relevant membrane processes. Recombinant membrane proteins can be purified in detergent and reconstituted into liposomes by detergent removal, either by dialysis or size exclusion. The resulting proteoliposomes can then be added to supported monolayers in order to incorporate the protein into a supported membrane with defined lipid compositions in both leaflets [2]. SNARE (soluble N-ethylmaleimide-sensitive factor attachment proteins receptor) proteins are particularly well-suited examples that illustrate the advantages of this reconstitution method. SNAREs such as syntaxin-1a and synaptobrevin-2 are single span integral membrane proteins with large N-terminal cytosolic domains and only a few residues at the C-terminal ends of their transmembrane domains. When reconstituted into supported membranes by the LB/VF method, SNAREs will be oriented predominantly (>80%) with their cytosolic domains pointing away from the substrate [7, 13]. This orientation results in laterally mobile and functionally active proteins [13, 14].

15.2.4 Monolayers + giant plasma membrane vesicles (native proteins in phase-separated bilayers)

Vesicles containing all the lipid and protein material from a native plasma membrane can be directly derived from cultured cells by inducing the blebbing of giant plasma membrane vesicles (GPMVs) by treatment with formaldehyde and disulfide blocking agents [15–17] or by applying hypertonic buffers [18]. After harvesting GPMVs from cell cultures, they may be added to supported monolayers and supported planar plasma membranes (SPPMs) can be prepared by the same procedure as described above. The lipid content of GPMVs that have been derived from HeLa cells separates into lo- and ld phases, and when added to phase separated monolayers, the resulting supported membrane will reflect the domain structure of the original LB monolayer [19]. Specific protein contents and their preference for certain lipid phases can be visualized by fluorescent antibodies. It could be shown that the T-cell receptors CD4 and CCR5, expressed in HeLa cells, partition into the lo phase and in the boundary between lo- and ld phases, respectively [19].

15.3 Membrane fusion assays

Membrane fusion, the process during which two lipid bilayers merge, is essential for many biological processes including synaptic transmission, hormone secretion, fertilization, intracellular trafficking, and viral infection. Reconstituting membrane fusion in model membranes has been proven to be a useful and unique tool to investigate the process at the molecular level. In the following paragraphs, we will discuss how asymmetric supported membranes in combination with total internal reflection fluorescence (TIRF) microscopy were utilized in single fusion assays applied to different biological systems (Figure 15.2).

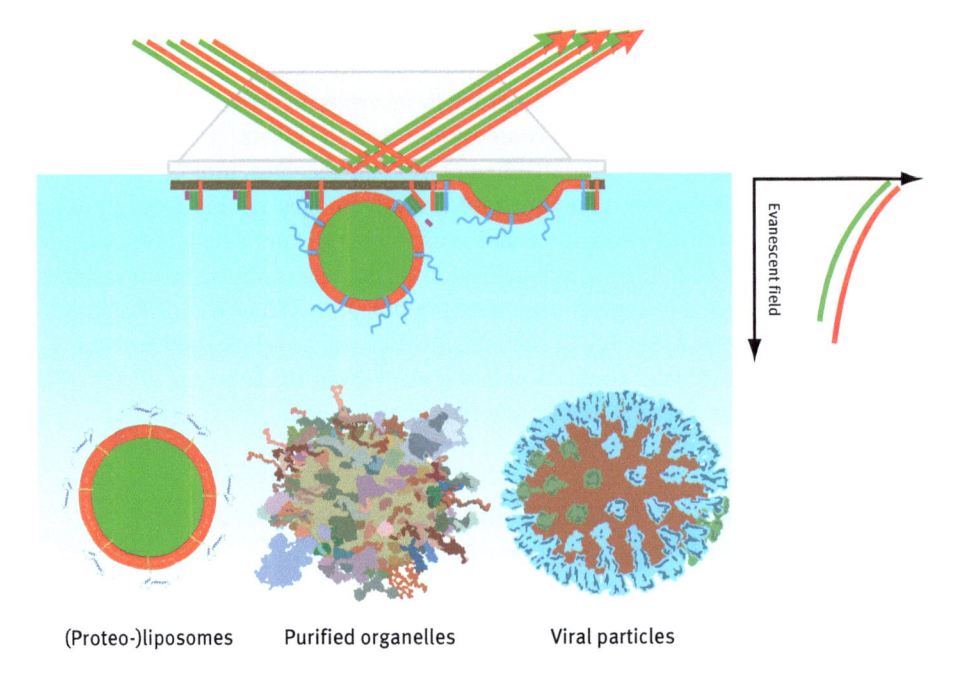

Figure 15.2: Total internal reflection fluorescence (TIRF) microscopy based single vesicle fusion assay. A supported membrane mounted to a TIRF microscope acts as target membrane for vesicles that are content- or membrane-labeled. Vesicles are detected when the fluorophores are excited within the evanescent field of the microscope, typically ranging from ~80 - 150 nm into the sample. Binding-, hemifusion-, and fusion events of single vesicles with the target membrane can be distinguished. Different types of vesicles can be used. Examples shown are a proteo-liposome, a purified synaptic vesicle, and a viral particle.

15.3.1 Lipid membranes + virions/fusion peptide

Viral infection of a cell by enveloped viruses is accomplished by fusion of the viral envelope with the target membrane of the cell which can be the plasma membrane or an endosomal membrane. This fusion process is mediated by specialized fusion proteins that are embedded in the viral membrane. A common property of these proteins is that they contain a mostly hydrophobic domain that can be inserted into a target membrane. This "fusion peptide" is protected within the much larger protein structure and is usually exposed by a conformational change of the whole protein that is triggered by a signal like a change of the environmental pH or the interaction with a host–cell receptor.

The fusion peptide by itself is capable of mediating fusion between lipid bilayers. Many details about viral membrane fusion mechanisms have been derived from fusion assays consisting of liposomes and fusion peptides. A single HIV fusion-peptide mediated fusion assay with phase-separated supported membranes and liposomes revealed that this fusion peptide preferentially inserts at the phase boundaries between lo- and ld phases in the lipid bilayer [20]. The observation of single virions that were exposed to moderate temperature elevation to expose the fusion peptide confirmed the strong interaction of this peptide with lipid phase boundaries. Observing single fusion events by following the fluorescence from labeled lipids in the viral membrane by TIRF microscopy in these assays allows distinguishing full membrane fusion events from hemi-fusion events, where fusion results in only one connected lipid leaflet and no content transfer [21] (Figure 15.2). Thanks to the planar geometry of supported membranes, it is relatively easy to quantify the (hemi-)fusion efficiencies and kinetics in different lipid environments. The targeting of the lipid phase boundary of the fusion peptide depends on the specific fusion peptide. For example, phase boundary targeting as observed with the HIV fusion peptide did not occur with the fusion peptide of influenza hemagglutinin. It was later shown that the promotion of fusion at lipid phase boundaries by the HIV fusion peptide depends strongly on the magnitude of the line tension that exists at these boundaries [22]. An exploitation of line tension might have evolved as a source of energy driving the fusion of HIV particles with the plasma membrane at sites of membrane phase discontinuity.

15.3.2 SPPM + virions

While the supported membranes in the examples above were, with the exception of the added polymer lipids in the proximal leaflet, not asymmetric, they demonstrate the advantages and quantitative nature of these assays. By performing these assays on asymmetric membranes that were prepared from supported phase-separated monolayers and GPMVs that contained the native HIV receptors, these studies could be extended to a system that includes virion membrane binding, exposure of the fusion peptide by the conformational change of the HIV glycoprotein (gp41/gp120), and membrane fusion [19]. HIV binds to CD4 receptors in the plasma membrane which triggers a conformational change that exposes a binding site for the co-receptor CCR5. Binding of the co-receptor in turn exposes the fusion peptide, which inserts into the plasma membrane and mediates fusion. Because of the above-discussed partitioning of the receptors in phase-separated SPPMs, binding and fusion of HIV pseudoviruses happens preferentially at the phase boundaries. It could also be shown that treatment of the target membrane by lineactants (molecules that are preferentially located at the phase boundary) or extraction of cholesterol inhibits HIV membrane fusion [19]. While inhibiting membrane fusion, these treatments also abolish the coexistence of ordered and disordered lipid phases in in the target membranes, demonstrating a clear correlation between the two processes.

15.3.3 SNARE-mediated fusion

The release of neurotransmitter from synaptic vesicles at the presynaptic plasma membrane as well as the release of hormones like insulin by exocytotic events is a highly regulated process and involves the assembly of a variety of proteins and lipids. A common feature of all these membrane fusion processes is the requirement for SNARE proteins which are at the core of the proteinacious membrane fusion machinery. Because SNAREs are capable of catalyzing membrane fusion in vitro, we refer to this process as "SNARE-mediated membrane fusion." Neurotransmitter release after membrane depolarization and influx of Ca^{2+} into cells at synapses is particularly fascinating because of its precision and fast kinetics. Despite many years of research and much progress on the molecular requirement for synaptic vesicle fusion, no consensus has been found on exactly how these molecules orchestrate Ca^{2+}-triggered release. One reason is the lack of quantitative biochemical assays that are able to examine the overall fusion reaction or parts of it in a lipid environment that mimics the essential properties of a native cell membrane. Several fluorescence microscopy assays with supported membranes have been developed recently to target specifically the questions around SNARE-mediated membrane fusion [23]. Here, we focus especially on the single vesicle/liposome fusion assays that are geared toward observing the overall fusion reaction and that make use of the asymmetry of the supported target membrane.

15.3.3.1 Single liposome-supported membrane fusion assay

Reconstituting the plasma membrane SNAREs syntaxin-1a and SNAP-25 into supported membranes and the synaptic vesicle SNARE synaptobrevin-2/VAMP2 into liposomes showed that SNAREs not only are sufficient to promote fusion but also that fusion occurs with fast kinetics. In symmetric membranes, composed of only POPC and cholesterol, the fusion kinetics revealed that eight reactions have to take place between liposome docking and the onset of fusion [14]. One interpretation of this result is that on average, eight SNARE complexes have to be formed for fusion to occur efficiently at this speed. This requirement is flexible and depends on the lipid composition as was shown by adding increasing amounts of phosphatidylserine and phosphatidylethanolamine to the distal leaflet of the supported membrane [24]. When a lipid composition of the distal leaflet of the supported membrane was chosen that better mimics the composition of the inner plasma membrane leaflet by increasing the PE concentration, the number of parallel reactions (number of formed SNARE complexes) was reduced to 3.

With only SNAREs and neutral lipids present in the two bilayers, the fusion reaction is not Ca^{2+}-dependent [14]. Adding the known calcium sensor of evoked neurotransmitter release, synaptotagmin-1 to the synaptic vesicle mimicking liposomes and making the distal leaflet of the supported membrane even more plasma membrane-like by adding 15% PS and 3% phosphatidylinositol-4,5-bisphosphate (PIP2), the assay becomes highly Ca^{2+}-sensitive [25]. Fusion of single proteoliposomes was observed by the transfer of fluorescently labeled lipids from the liposome membrane into the supported membrane and by the transfer of fluorescent content from the lumen of the proteoliposomes into the narrow cleft between substrate and supported membrane. The fusion efficiencies went from ~40% in the absence of Ca^{2+} to ~80% in the presence of Ca^{2+}. While these results showed that a minimal system consisting of SNAREs, synaptotagmin, and anionic lipids in the distal leaflet of the supported membrane is capable of reproducing Ca^{2+} stimulation, the observation that there was still relatively high fusion in the absence of Ca^{2+} also showed that something was missing in this system to fully reproduce the primed vesicle state in which the vesicle is docked but does not fuse before Ca^{2+} arrives.

15.3.3.2 Fusion of purified organelles with asymmetric-supported membranes

An attractive alternative to the reconstitution of synaptobrevin, synaptotagmin, and possibly more secretory vesicle proteins into proteoliposomes is the purification of native synaptic vesicles from neurons. When synaptic vesicles from rat brain are fluorescently labeled by a membrane dye and added to asymmetric supported membranes, they fuse in a SNARE- and Ca^{2+}-dependent manner [25]. Just as synaptobrevin/synaptotagmin containing proteoliposomes, synaptic vesicle fusion is increased with Ca^{2+} but is not completely inhibited in the absence of Ca^{2+}.

While it is in principle possible to genetically manipulate the protein content of tissue-derived vesicles, this is more easily achieved by purifying vesicles from cultured cells. Indeed, dense core vesicles (DCVs) from PC12 cells have been proven to be a very useful tool to further our understanding of the Ca^{2+} sensitivity of release [26]. Docking and fusion of DCVs can be observed on a TIRF microscope by expressing a fluorescent protein that is fused to the secretory neuropeptide Y in the lumen of the vesicle. In a minimal system, which contains only the SNAREs syntaxin and SNAP25 in the supported membrane, the DCV fusion assay confirms the already discussed results, namely, that fusion is SNARE dependent and that it increases when Ca^{2+} and anionic lipids are present in the distal leaflet of the planar membrane. In addition, it could be shown that docking, that is, binding to the supported membrane, is also Ca^{2+}-dependent and that this docking efficiency follows a different Ca^{2+} dependency than the fusion efficiency. Using genetic knockdown approaches, it was revealed that synaptotagmin is indeed responsible for the Ca^{2+} dependence of fusion, while the calcium-dependent activator protein for secretion (CAPS) is responsible for the observed Ca^{2+} dependence of docking. A high population of a primed state where vesicles do not fuse in the absence of Ca^{2+}, but do so with high efficiency in its presence, could be accomplished by adding the soluble proteins complexin and Munc18 to the system. These proteins are known to be essential for regulated exocytosis. The triggering of fusion from the primed state depends very strongly on the anionic lipids PS and PIP2 in the distal leaflet of the supported membrane and on the presence of the CAPS protein in the DCV as could be shown by additional knockdown experiments. Overall, the preparation of the asymmetric supported membrane was crucial to reproduce, for the first time, the physiological requirement for regulated exocytosis and observe the triggering of fusion by Ca^{2+} from a primed state.

15.3.3.3 Modulation of the fusion-pore lifetime by membrane asymmetry

The fusion of content-labeled DCVs in the assay discussed above is observed by a characteristic fluorescent signal in a TIRF microscope [26]. Due to the relatively large size (~200 nm diameter) of the DCVs, the signal is very sensitive to the mean position of fluorophores within the evanescent field of the exciting laser light. Fusion happens in two distinct phases: first, a fusion pore opens and content is released into the cleft between support and membrane while the overall structure of the vesicle stays intact. Second, vesicles collapse into the plane of the supported membrane. This assay, therefore, allows the measurement of a fusion-pore lifetime by determining the time between the onset of fusion and the collapse of the vesicle. It has been proposed that the intrinsic negative curvature of PE lipids could stabilize the formation of highly curved fusion pores. Indeed, by preparing different asymmetric-supported membranes with different PE concentrations in the two leaflets, it was found that PE in the distal leaflet of the supported membrane stabilizes the fusion pore while PE in the proximal leaflet inhibits the formation of a fusion pore [27]. The results from these experiments also implicitly prove that it is possible to prepare supported membranes with asymmetric PE distributions between the two leaflets.

15.4 Conclusion

We have discussed the preparation and application of asymmetric supported membranes. It was our intention to present a model membrane system that provides an environment that is ideally suited to address a variety of fascinating biological membrane problems. Many of the discussed applications wouldn't have been possible in other previously introduced membrane systems. In many examples, the asymmetry of the membrane was a prerequisite for the quality of the assay and not the research topic itself. However, we are confident that supported asymmetric membranes will be applied also to address more specifically biological questions around bilayer asymmetry. While almost all of the studies discussed in this chapter have been performed in our own lab, we are hopeful that many other research groups will utilize and further develop the asymmetric supported planar membrane platform in their own research.

References

[1] Kalb, E., Frey, S., & Tamm, LK. Formation of supported planar bilayers by fusion of vesicles to supported phospholipid monolayers. Biochim Biophys Acta 1992; 1103:307–316.
[2] Kalb, E., & Tamm, LK. Incorporation of cytochrome b5 into supported phospholipid-bilayers by vesicle fusion to supported monolayers. Thin Solid Films 1992; 210:763–765.
[3] Purrucker, O., Fortig, A., Jordan, R., & Tanaka, M. Supported membranes with well-defined polymer tethers-incorporation of cell receptors. Chemphyschem 2004; 5:327–335.
[4] Wagner, ML., & Tamm, LK. Tethered polymer-supported planar lipid bilayers for reconstitution of integral membrane proteins: silane-polyethyleneglycol-lipid as a cushion and covalent linker. Biophys J 2000; 79:1400–1414.
[5] Kiessling, V., & Tamm, LK. Measuring distances in supported bilayers by fluorescence interference-contrast microscopy: polymer supports and SNARE proteins. Biophys J 2003; 84:408–418.
[6] Crane, JM., & Tamm, LK. Role of cholesterol in the formation and nature of lipid rafts in planar and spherical model membranes. Biophys J 2004; 86:2965–2979.
[7] Liang, B., Kiessling, V., & Tamm, LK. Prefusion structure of syntaxin-1A suggests pathway for folding into neuronal trans-SNARE complex fusion intermediate. Proc Natl Acad Sci U S A 2013; 110:19384–19389.
[8] Crane, JM., Kiessling, V., & Tamm, LK. Measuring lipid asymmetry in planar supported bilayers by fluorescence interference contrast microscopy. Langmuir 2005; 21:1377–1388.
[9] Kiessling, V., Crane, JM., & Tamm, LK. Transbilayer effects of raft-like lipid domains in asymmetric planar bilayers measured by single molecule tracking. Biophys J 2006; 91:3313–3326.
[10] Simons, K., & Ikonen, E. Functional rafts in cell membranes. Nature 1997;387:569–572.
[11] Wan, C., Kiessling, V., & Tamm, LK. Coupling of cholesterol-rich lipid phases in asymmetric bilayers. Biochemistry 2008; 47:2190–2198.
[12] Wan, C., Kiessling, V., Cafiso, DS., & Tamm, LK. Partitioning of synaptotagmin I C2 domains between liquid-ordered and liquid-disordered inner leaflet lipid phases. Biochemistry 2011; 50:2478–2485.
[13] Wagner, ML., & Tamm, LK. Reconstituted syntaxin1A/SNAP25 interacts with negatively charged lipids as measured by lateral diffusion in planar supported bilayers. Biophys J 2001; 81:266–275.
[14] Domanska, MK., Kiessling, V., Stein, A., Fasshauer, D., & Tamm, LK. Single vesicle millisecond fusion kinetics reveals number of SNARE complexes optimal for fast SNARE-mediated membrane fusion. J Biol Chem 2009; 284:32158–32166.
[15] Scott, RE. Plasma membrane vesiculation: a new technique for isolation of plasma membranes. Science 1976; 194:743–745.

[16] Baumgart, T., Hammond, AT., Sengupta, P., et al. Large-scale fluid/fluid phase separation of proteins and lipids in giant plasma membrane vesicles. Proc Natl Acad Sci U S A 2007; 104:3165–3170.
[17] Levental, I., Lingwood, D., Grzybek, M., Coskun, U., & Simons, K. Palmitoylation regulates raft affinity for the majority of integral raft proteins. Proc Natl Acad Sci U S A 2010; 107:22050–22054.
[18] Del Piccolo, N., Placone, J., He, L., Agudelo, SC., & Hristova, K. Production of plasma membrane vesicles with chloride salts and their utility as a cell membrane mimetic for biophysical characterization of membrane protein interactions. Anal Chem 2012; 84: 8650–8655.
[19] Yang, ST., Kreutzberger, AJB., Kiessling, V., Ganser-Pornillos, BK., White, JM., & Tamm, LK. HIV virions sense plasma membrane heterogeneity for cell entry. Sci Adv 2017; 3: e1700338.
[20] Yang, ST., Kiessling, V., Simmons, JA., White, JM., & Tamm, LK. HIV gp41-mediated membrane fusion occurs at edges of cholesterol-rich lipid domains. Nat Chem Biol 2015; 11: 424–431.
[21] Kiessling, V., Liang, B., & Tamm, LK. Reconstituting SNARE-mediated membrane fusion at the single liposome level. Methods Cell Biol 2015; 128: 339–363.
[22] Yang, ST., Kiessling, V., & Tamm, LK. Line tension at lipid phase boundaries as driving force for HIV fusion peptide-mediated fusion. Nat Commun 2016; 7: 11401.
[23] Kiessling, V., Liang, B., Kreutzberger, AJ., & Tamm, LK. Planar supported membranes with mobile SNARE proteins and quantitative fluorescence microscopy assays to study synaptic vesicle fusion. Front Mol Neurosci 2017; 10: 72.
[24] Domanska, MK., Kiessling, V., & Tamm, LK. Docking and fast fusion of synaptobrevin vesicles depends on the lipid compositions of the vesicle and the acceptor SNARE complex-containing target membrane. Biophys J 2010; 99: 2936–2946.
[25] Kiessling, V., Ahmed, S., Domanska, MK., Holt, MG., Jahn, R., & Tamm, LK. Rapid fusion of synaptic vesicles with reconstituted target SNARE membranes. Biophys J 2013; 104: 1950–1958.
[26] Kreutzberger, AJB., Kiessling, V., Liang, B., et al. Reconstitution of calcium-mediated exocytosis of dense-core vesicles. Sci Adv 2017; 3: e1603208.
[27] Kreutzberger, AJB., Kiessling, V., Liang, B., Yang, ST., Castle, JD., & Tamm, LK. Asymmetric phosphatidylethanolamine distribution controls fusion pore lifetime and probability. Biophys J 2017; 113: 1912–1915.

Andrew F. Craig, Indra D. Sahu, Carole Dabney-Smith, Dominik Konkolewicz, Gary A. Lorigan

16 Styrene-maleic acid copolymers: a new tool for membrane biophysics

Abstract: One of the most difficult aspects to studying membrane proteins is mimicking the native lipid bilayer while maintaining the structure and function of the protein. A promising new structural biology tool exists to mimic the native lipid bilayer while maintaining protein integrity using styrene-maleic acid (SMA) copolymers. SMA polymers have shown the ability to interact in detergent-free systems to solubilize large heterogeneous vesicles into homogenous disc-like structures known as SMA lipid nanoparticles (SMALPS). SMALPS have been used in an array of biophysical techniques for many systems. In this book chapter, we will highlight both the chemical properties of SMA polymers and the multiple biophysical applications of the polymers. The future applications of the polymers and new potential methodologies will be discussed as well.

Keywords: Styrene-maleic acid, SMALPs, Membrane protein, Membrane mimetic system, Biophysical techniques

16.1 Introduction

Mimicking the native lipid bilayer is one of the most difficult aspects of studying membrane proteins. Membrane proteins account for approximately one-third of overall proteins in the human genome but make up less than 1% of the known protein structures [1]. Membrane proteins have a variety of functions across the cell including protein transport, enzymatic activities, cellular signaling, and structural support for cell walls [1, 2]. Membrane proteins are pharmacological targets of approximately 50% of current drug targets [1, 3]. The lack of structural knowledge from membrane bound proteins in comparison with water soluble proteins is not due to lack of biological relevance or biological abundance, but due to the difficulty in studying membrane proteins. Membrane proteins by nature tend to be hydrophobic so that these proteins can exist natively in a membrane bilayer. Studying membrane proteins directly in the lipid bilayer is not possible because the proteins must be extracted from the native bilayer [4, 5]. Furthermore, membrane protein extraction requires the native bilayer to be solubilized for purification making artificial mimetics required to

---

Andrew F. Craig, Indra D. Sahu, Carole Dabney-Smith, Dominik Konkolewicz, Gary A. Lorigan, Department of Chemistry and Biochemistry, Miami University, Oxford, OH, USA

https://doi.org/10.1515/9783110544657-016

study membrane proteins. It is difficult to mimic this bilayer without affecting the folding or activity of the protein. Membrane proteins tend to exist in the native bilayer in anisotropic conditions with amphipathic lipids surrounding them with polar head groups and a surrounding core of aliphatic lipid chains. In order to study the structures of membrane proteins, the proteins must be extracted and purified from the native lipid bilayer. Purification of membrane proteins has proven much more labor intensive when compared to aqueous soluble proteins because of the requirement to maintain a functional membrane environment [2]. Many techniques exist to help solubilize membrane proteins, with each method having distinct advantages and disadvantages, but the styrene-maleic acid (SMA) copolymers have emerged as potential new lipid mimetics. SMAs are polymers that when added to intact lipids form polymer-lipid discs known as SMA lipid nanoparticles (SMALPs) [6–11].

In this chapter, we will discuss different methods for solubilization and purification of membrane proteins with a focus on SMA polymers. We will focus on method-based applications rather than theoretical uses. A history of SMA polymers synthesis and its earlier applications will be discussed. The characterization of SMA polymers will be described and applications of SMA polymers for biophysical studies of membrane protein system will be summarized. Furthermore, we will discuss several potential future directions as well as possible alternative polymer-based lipid mimetic systems.

16.2 Membrane protein solubilization techniques

Solubilizing the native membrane to obtain well-defined lipid particles is one of the largest challenges in membrane protein research. The goal of membrane protein purification is to find a mimetic that stabilizes the protein, allows the protein to be purified at high concentrations and to maintain the protein's activity and natively folded state. Several different systems exist to study these proteins including detergents, lipid vesicles, bicelles, and nanodiscs [4, 12–14]. Each technique has distinct advantages as well as disadvantages. A more recently developed system is the SMA polymer, which can directly solubilize membranes into a disc-like structure (Figure 16.1).

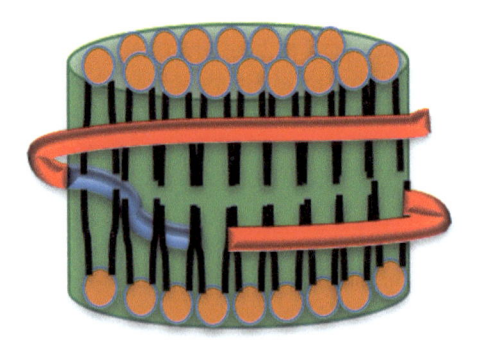

Figure 16.1: Schematic diagram of the shape of a SMALP. The *red* color represents a section of polymer that has an alternative structure of maleic acid and styrene. *Blue* represents a section of styrene units.

16.3 Detergents

Detergents are the most widely used membrane mimetic used to study membrane proteins in a variety of techniques including solution NMR and X-ray crystallography [5, 13, 15–17]. Detergents are advantageous because they can solubilize in a similar manner as a native lipid bilayer and form spherical detergent micelles [13]. Micelles exist in a variety of classes including positively charged, negatively charged, and zwitterionic that can be used for a variety of different membranes [4]. Detergents have several disadvantages that make them potentially problematic to use. The primary disadvantage of detergents is that the detergent micelles formed do not accurately mimic the native lipid bilayer. Detergent micelles exist as primarily spherical structures rather than the extended flat structure of native lipids [17]. The lack of a bilayer environment may influence the folding as well as the activity of membrane proteins [2, 17, 18]. Another issue of detergents includes the empirical understanding required to choose the correct detergent. In order to optimize membrane protein solubilization, a variety of detergents are screened in order to find the ideal detergent. The ideal detergent/s are chosen using indirect methods such as choosing a detergent that gives the narrowest line-widths in solution NMR spectroscopy [17, 19]. A final issue of detergents exists in the difference in water permeability and lateral pressure profiles of micelles and bilayer systems. As a result, the stability of membrane proteins in detergents is lower when compared to bilayers which can lead to protein aggregation, inactivation or adoption of nonnative conformations [13]. Therefore, detergent-based micelles can be a useful structural tool to study membrane proteins but due to the poor mimic of the native lipid bilayer, more suitable lipid mimics are sought.

16.4 Lipid vesicles or liposomes

The most native mimic of the native lipid bilayer exists in lipid bilayer vesicles or liposomes. Liposomes are advantageous to be used as a lipid mimic because the structure most accurately mimics the native bilayer of membrane proteins [4, 13]. However, liposomes are large, heterogeneous bilayer systems which have solvent inaccessible inner cores which make cytoplasmic studies difficult [20]. Incorporation of a physiologically high concentration of membrane proteins into liposomes is often challenging which makes biophysical studies difficult [4, 12, 21–25]. This leads to significant challenges for solid-state NMR studies which typically require large amounts of protein (~mg scale).

16.5 Bicelles

Bicelles are another lipid bilayer mimetic that forms homogeneous discoidal-like structures that are composed of a mixture of long-chain phospholipids (e.g., 1,2-dimyristoyl-sn-glycero-3-phosphocholine (DMPC)) and short-chain phospholipids (e.g., 1,2-dihexanoyl-*sn*-glycero-3-phosphocholine (DHPC)) [4, 12, 23–25]. Due to the disc-like structure of bicelles, studies can be performed on both the extracellular and cytoplasmic domain of membrane proteins [20]. The alignment of bicelles in a magnetic field can provide valuable structural and dynamic information for NMR and EPR spectroscopic studies [26]. A disadvantage of bicelles is that only certain lipid compositions form bicelles and the protein stability can be an issue.

16.6 Nanodiscs

Protein-lipid nanodiscs are another class of a lipid mimetic system. Nanodiscs consist of lipids encapsulated by a membrane scaffold protein, such as an apolipoprotein, to form a homogenous discoidal bilayer [27–29]. The disc-like structure allows for membrane proteins to be studied in a bilayer-like system similar to the bicelles [14, 28, 29]. The diameter of nanodiscs is most commonly ~10 nm but can range in size from ~6 to 17 nm [29, 30]. The ability to control the size of the nanoparticle of nanodiscs is a significant advantage over most lipid mimetics. Nanodiscs can be formed from a wide range of lipids, which is a significant advantage over bicelles since many proteins require specific lipid compositions for proper folding and function [14, 28, 29]. A drawback of the nanodisc is that the protein to be studied must be purified first into a detergent prior to incorporation into nanodiscs. In addition, the membrane scaffold protein's spectroscopic properties can interfere with the spectral properties of membrane protein to be studied [28, 29]. Nanodiscs can be difficult to form and often require extensive optimization to get the scaffolding proteins to fold correctly [31]. The drawbacks of the described bilayer mimics left a desire for another lipid mimetic that combines the advantages of all models without the disadvantages.

16.7 Styrene-maleic acid copolymers formed lipid nanoparticles

SMA copolymers have been used to create a new bilayer mimic for both biochemical and biophysical studies. These amphiphilic polymers form stable-homogenous nanoparticles upon addition to a lipid bilayer. A common example of the SMA being used is the commercially available lipodisq™ [6, 23, 31–34]. SMA polymers have been used in a variety of systems including in detergent-free systems [31, 33]. These applications will be discussed in greater detail later. SMA polymers have been used for years in a variety of systems.

16.8 The styrene-maleic acid copolymer

SMA is formed from the hydrolyzed form of the styrene-maleic anhydride (SMAn) copolymer. SMAn is synthesized by the copolymerization of styrene and maleic anhydride monomers (Figure 16.2). Both forms of the polymer are used commercially and have a variety of applications. SMAn is commonly used as a thermal stabilizer in plastic blends. Several suppliers sell this polymer but the most prominent are TOTAL Cray Valley (Beaufort, TX) and Polyscope (Geleen, NL).

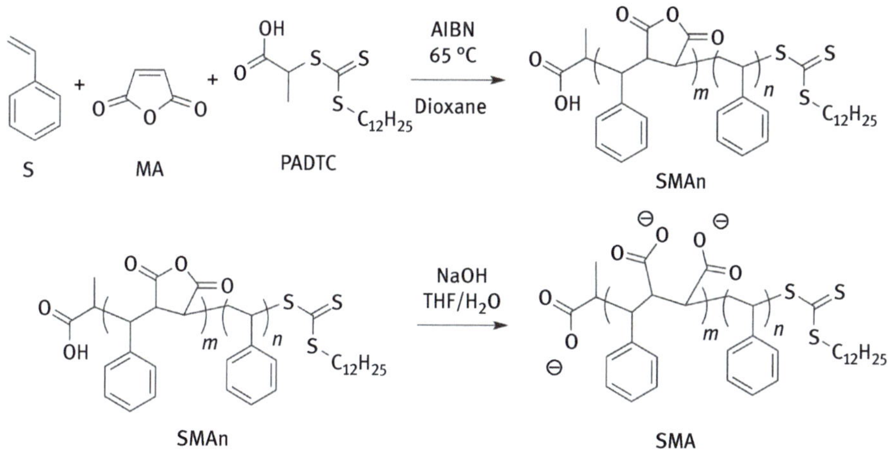

Figure 16.2: Synthesis of the poly(styrene-alt-maleic anhydride-b-styrene) block copolymer (SMAn), followed by the hydrolysis of this polymer to give the water soluble poly(styrene-alt-maleate-b-styrene) block copolymer (SMA).

SMA polymers have been used previously as drug conjugates in cancer therapies [35] SMA polymers have been more recently developed into a male contraceptive currently in phase III clinical trials [36]. These polymers have shown promise as a drug delivery system [37]. The newest application of SMA polymers exists in using them to form disc-like structures around a lipid bilayer. The original polymerization was patented by Malvern Cosmeceutics (Worcester, UK) in 2006 [37]. The use of SMA copolymers has greatly expanded and the complete list of commercial sources is

Table 16.1: This is the list of commercially available SMA polymers obtained from smalp.net.

Product	SMA ratio	M_w (g/mol)	M_n (g/mol)	Source
Xiran SZ20010	4.3:1	7,300	2,500	Polyscope
Xiran SZ23110	3.3:1	110,000	44,000	Polyscope
Xiran SZ25010	3:1	10,000	4,000	Polyscope
Xiran SZ26080	2.8:1	80,000	32,000	Polyscope
Xiran SZ26120	2.8:1	120,000	48,000	Polyscope
Xiran SZ28110	2.6:1	110,000	44,000	Polyscope
Xiran SZ28130	2.6:1	130,000	n/a	Polyscope
Xiran SZ30010	2.3:1	6,500	2,500	Polyscope
Xiran SZ33030	2:1	30,000	9,000	Polyscope
Xiran SZ40005	1.2:1	5,000	2,000	Polyscope
SMA1000	1:1	5,500	2,000	Cray Valley
SMA2000	2:1	7,500	3,000	Cray Valley
SMA2021	2:1	21,000	12,000	Cray Valley
SMA3000	3:1	9,500	3,800	Cray Valley
SMA4000	4:1	n/a	3,800	Cray Valley
Lipodisq L8920	2:1	n/a	n/a	Sigma
Lipodisq L9045	3:1	n/a	n/a	Sigma
Lipodisq L9170	3:1	n/a	n/a	Sigma

Notes: not all of these polymers are designed for membrane biophysics.

shown in Table 16.1. SMA copolymers can be obtained in several ratios and can be synthesized using new methods or purchased commercially [11].

16.9 Synthesis of styrene-maleic anhydride copolymers

The polymerization of styrene and maleic anhydride (MAn) monomers (Figure 16.2) is a radical chain growth reaction that leads to the formation SMAn copolymers with varying molecular weights. Until recently, all SMAn copolymers were synthesized by conventional-free radical polymerization [37]. Although this polymerization method is highly efficient, it results in a broad distribution of molecular weight, as characterized calculated using the polydispersity index (PDI), typically determined from size exclusion chromatography data [38]. PDI is calculated as the ratio of the weight-average molecular weight and the number-average molecular weight [38]. The closer the value to 1.0, the narrower is the weight distribution of polymers. The PDI of polymers synthesized by radical polymerization typically range from about 2.0 to 2.5 [34].

Molecular weight dispersity is an issue in radical polymer synthetic chemistry. Therefore, recent methods have been sought after to control the radical synthesis.

Recent studies have shown the macromolecular architecture can be fine-tuned by controlling the radical process [39–41]. The name for techniques that allow for this control of the polymer structure through radical methods are called reversible deactivation radical polymerization (RDRP) methods [42]. An example of RDRP is reversible addition-fragmentation chain transfer (RAFT) polymerization [43–46]. RAFT polymerization is one of the most versatile RDRP methods since it creates well-defined polymers from very wide variety of monomers, including both styrene and maleic anhydride [45]. RAFT is advantageous when compared to traditional radical polymerization including polymers with relative low PDI, facile control over the molecular weight of the polymer, and controlling the unique microstructure of individual polymers [43, 46, 47]. This is evidenced by the fact that RAFT polymerization controls the polymer molecular weight giving materials with PDI values between of 1.05 and 1.4 [11].

In the synthesis of lipid mimetics, RAFT was used for the copolymerization of styrene and maleic anhydride with an excess of styrene. The result was a block architecture created in a one-pot synthesis [11]. This created a block copolymer with a nearly perfect architecture of styrene alternating with maleic anhydride, and a second block of a homopolymer of styrene. This polymer's unique architecture results from the near-zero reactivity ratios of styrene and maleic anhydride, which promotes this alternating structure in the first block [48]. Additionally, this RAFT polymerization method offers a simple method to synthesize a set of polymers at varying molecular weights and SMA ratios.

16.10 Hydrolysis of SMA polymers

In order for dissolution of lipid to occur via the addition of SMA polymers, the anhydride form must be hydrolyzed to the maleate form. Hydrolysis is required as the maleic anhydride form is insoluble in water. This hydrolysis can be seen in Figure 16.2. SMAn polymer hydrolysis occurs when mixed with water or an alkaline solution to the acid form in which the two carboxyl groups become partly deprotonated to yield water-soluble polymers. Hydrolysis in water is often slow and so basic solutions are used to accelerate the hydrolysis reaction to overnight [11, 34, 37].

After polymer hydrolysis, the SMA solution can be purified and isolated in several ways. If an excess of base has been used, the solution can be brought into an SMA solution using either a dialysis method [49] or an acid-based precipitation reaction [11, 50]. SMA polymers are insoluble at pH values below ~6 due to the protonation of the carboxyl groups [34, 36]. The exact pH that the polymers are soluble can vary depending on both the SMA ratio of the polymer as well as the system it is solubilized in. After precipitation in acids, often dilute hydrochloric acid, the polymers can be dried and isolated by lyophilization. This lyophilization yields a fully protonated polymer that can be stored at room temperature for months. These polymers can be brought up in water and the pH can be readily adjusted by either KOH or NaOH.

16.11 Utilization of SMA polymers in membrane protein research

So far, the most common SMA variant used in membrane protein research is the commercially available lipodisq™ nanoparticle. This is a well-characterized polymer that is used in many different systems [6, 23, 31–34, 51, 52]. Until recently, the copolymers used had SMA ratios of 2:1 or 3:1 with an average polymer weight of approximately 7.5–10 kD. The typical pH range used for membrane solubilization ranged from ~7 to 8. It has been previously reported that at this range, the polymers adopt a random coil conformation and the balance between the hydrophobic effect and electrostatic interaction will be at the optimal range [34]. Recently, the Lorigan lab used the RAFT technique to not only vary the molecular weight of the polymers but also used SMA ratios not previously characterized [11].

16.12 Use of SMA polymers for membrane mimetics

SMAs have been used for the last decade as a potential new mimetic. Multiple studies have been performed to study the effects of lipid membranes and SMA polymers. Most of these techniques have used model membranes as the lipid bilayer. These model membranes are useful for biophysical studies because they allow a step-wise variation in lipid parameters and because they can promote the formation of SMALPS. In this section, we will give an overview and description of SMA with model membrane with and without membrane proteins.

SMALPS have been studied and characterized in a variety of systems. Size analysis of the SMALPS formed is the most commonly performed by transmission electron microscopy (TEM) [6, 10, 15, 23, 50], size exclusion chromatography (SEC) [10, 50], and dynamic light scattering (DLS) [6, 11, 23, 31, 33, 49]. With the use of the commercially available lipodisq™, the size of the resulting nanoparticles is around 10 nm. The RAFT synthesized polymers with a 3:1 SMA ratio form nanoparticles of diameter ~10 nm regardless of the molecular weight of the polymer. It should be noted the RAFT polymers can create larger particle sizes that the commercially available polymers cannot [11]. This discrepancy in size, independent of lipid composition, is likely a result of the difference in microstructure of the RAFT-SMALPS when compared to the lipodisq form SMALPS. It has been reported that relatively low SMA to lipid ratios can affect the average particle size [6, 15, 23]. Using lower SMA to lipid ratios tends to result in larger and more heterogeneous particle sizes likely due to the incomplete dissolution of the lipid bilayer [6, 15, 23]. Qualitatively, it is easy to visually see when SMALPS form as the opaque lipid solutions scatters less upon addition of more polymer (Figure 16.3), consistent with the formation of smaller average particle sizes upon addition of SMA polymer [11, 23]. Most experiments have been performed using excess SMA in order to fully solubilize the native lipid bilayer.

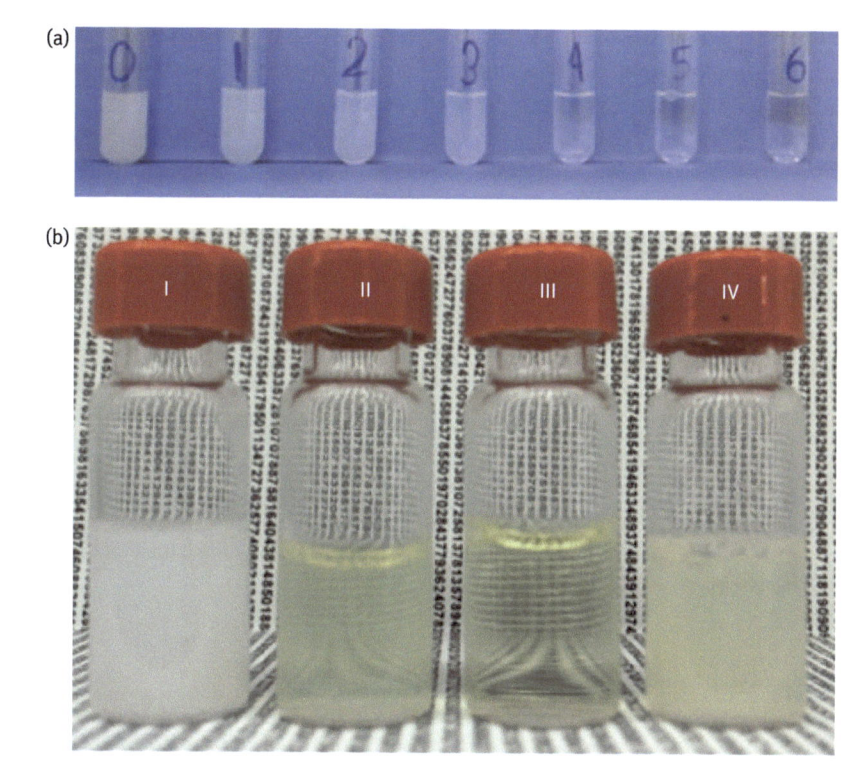

Figure 16.3: **(a)** Visible observation of lipodisq nanoparticles formation through the combination of POPC/POPG vesicles and SMA polymers at different lipid to polymer weight ratios and normal POPC/POPG vesicles as a control: (0) control POPC/POPG vesicles, (1) lipodisq nanoparticles (1/0.25), (2) lipodisq nanoparticles (1/0.5), (3) lipodisq nanoparticles (1/0.75), (4) lipodisq nanoparticles (1/1.25), (5) lipodisq nanoparticles (1/1.75), (6) lipodisq nanoparticles (1/2.25) (adapted from Zhang et al. [23] with permission). **(b)** Vial I is a control of just POPC/POPG vesicles, vial II is POPC/POPG with the addition of a 2:1 ratio SMA polymer, vial III is the addition of a 3:1 ratio SMA polymer and vial IV is a POPC lipid with a 4:1 SMA ratio polymer (adapted from Craig et al. [11] with permission).

The interactions between lipids and SMA polymers have been studied using both Fourier transform infrared (FTIR) spectroscopy and nuclear magnetic resonance (NMR) spectroscopy [10, 11, 31]. These interactions showed that the styrene portion of SMA polymers directly interacts with the acyl groups of the lipids and the carboxylate groups interact with the polar head groups. This proved the prediction that the styrene groups were located inside the bilayer with the carboxylate groups facing the bulk aqueous layer. Electron paramagnetic resonance (EPR) was used to show that the acyl groups in certain acyl chains are restricted [31]. Similarly, EPR showed that spin-labeled proteins inside the SMALPs also had restricted motion from the styrene which can lead to narrower distance distribution in double electron–electron resonance (DEER) experiments [32].

These findings show that the SMA copolymers not only interact with lipid bilayers but also promote disc-like nanoparticles, similar to nanodiscs. A schematic of how this interaction is proposed to occur is shown in Figure 16 4.

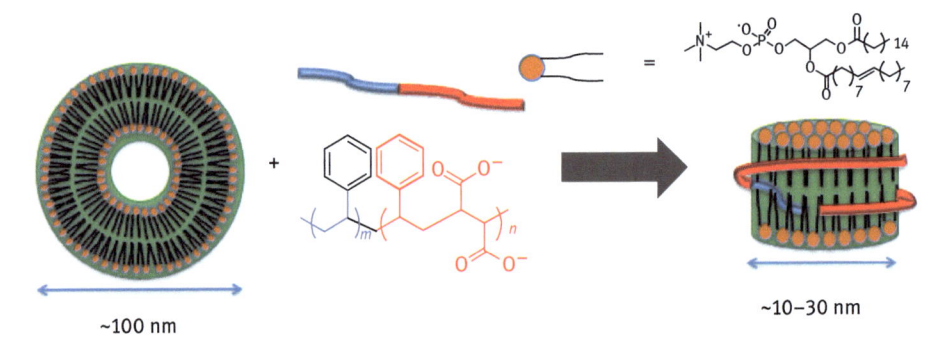

Figure 16.4: Cartoon representation of the addition of a SMA polymer to an intact vesicle and the structural implications of the addition to form the membrane RAFT SMALP. (Adapted from ref Craig et al. [11] with permission)

16.13 Kinetics of formation of SMALPs

In order for SMALPs to form, the polymer must dissolve the lipid bilayer. The kinetics of the dissolution has been studied by turbidimetry [50]. Liposomes are large heterogeneous complexes that scatter UV light, whereas SMALPs are much smaller and are largely translucent to light. Thus, the solubilization process could be followed by monitoring the decrease in light scattering, and commensurate increase in transmittance, using a spectrophotometer. Using this technique, the effect of lipid composition, salt, and SMA concentration was studied. Furthermore, a model for solubilization was proposed. The first step consisted of binding of SMA to the lipid bilayer. This binding can be modulated using alternative head groups such as anionic lipids as the negative charge causes electrostatic repulsion. During the second step, the SMA polymer inserts into the hydrophobic core of the membrane. The final step is the actual solubilization of the bilayer and the formation of SMALPs [50]. Additionally, the rate limiting step appears to be the polymer insertion into the hydrophobic core of the membrane.

16.14 Membrane protein incorporation in SMALPs

The above studies were conducted with the lipids in the absence of membrane proteins. However, many studies have been conducted where SMALPs have been formed with membrane proteins in the lipids. Initially, membrane proteins PagP and bacteriorhodopsin were incorporated into SMALPS [33, 49], but later functional proteins such as KCNE1 and multisubunit Complex IV holoenzyme were studied in SMALPs [32, 51].

Two main strategies have been utilized to incorporate membrane proteins into SMALPs. The first consists of proteins being solubilized in detergents and incorporated into vesicles using conventional methods [53] and then supplemented with SMA polymers after lipid incorporation to form SMALPs [6, 32, 34]. Importantly, the activity and structure of these proteins were found to be unaffected by the incorporation of the protein into SMALPS. The other main strategy to incorporate membrane proteins into SMALPs is through detergent-free methods [33, 51]. This occurs by isolating either native chloroplasts or inclusion bodies and directly purifying the protein from the native lipids. The fact the SMA polymers can purify proteins directly from cells is advantageous that no other current membrane mimetic can achieve. Some membrane proteins are difficult to solubilize and purify from native membranes and the efficiency can be improved by the addition of lipids such as 1,2-dimyristoyl-sn-glycero-3-phosphocholine (DMPC) [50].

16.15 Purification of SMALPs

Once SMALPs have been formed, a further purification may be required in certain cases to obtain the protein of interest. These purification procedures often use standard biochemical purification procedures. In order to purify SMALPs from the small amount of undissolved vesicles or cell lysates, size exclusion chromatography is often utilized [51]. Alternative purification procedure utilizes chromatography with respect to Ni^{2+} affinity to His-tagged proteins [54–56] or binding to antibodies [55]. Both of these methods give good yields and purity when compared to detergent-based methods. A potential issue arises with the use of Ni^{2+}-based protocols. It is possible that the SMA polymers may bind themselves to the affinity tag due to the negative charge on the maleic acid carboxylate groups and the resulting electrostatic interactions. This binding of SMA can immobilize the Ni^{2+} and prevent protein binding. A previous report proposed several alternatives and solution to this immobilization in removal of excess SMA by filtration, decreasing the amount of SMA used, increasing the amount of Ni^{2+}, increasing the number of histidines or by simply diluting SMALPs samples prior to addition of Ni^{2+} containing samples [34]. There is currently no data on SMALP compatibility with other purification systems (i.e., FLAG tag, maltose-binding protein etc.), but the same issues faced while using Ni^{2+} tags could inhibit binding as well. However, conditions could likely be optimized to improve yields to levels similar to detergents using these alternative affinity tags.

16.16 Protein stability in SMALPs

Protein stability is a constant issue when working with complicated membrane protein complexes. Detergents offer less stability when compared to other lipid mimics but are often a necessity for purification. The result of using detergents makes it difficult to obtain high concentration of proteins required for techniques such as NMR spectroscopy. Therefore, having an alternative to be used for purification than detergents remains a promising opportunity for proteins that are inherently unstable such as G-proteins [57]. SMA polymers have been used to successfully purify and isolate a G-protein [56]. Several other reports exist where an increased stability is reported in SMA polymers when compared to detergents [49, 54, 55]. In addition to greater stability with protein, SMA polymers have the ability to be stored at room temperature for months with no degradation in a powder form, compared to some detergents used in membrane protein purification that need to be stored at −20 °C otherwise degradation can occur [34].

16.17 Protein–protein interactions in SMALPs

A new desire is to use SMALPs to study either larger protein complexes or to study protein–protein interactions within an individual SMALP. Larger functional proteins have been incorporated in SMALPs [51, 58] but the small particle size of commercially available SMA polymers has prevented multiple protein complexes to be incorporated simultaneously. The creation of SMALP particles with larger particle sizes gives the potential for not only larger protein complexes to be incorporated but also the possibility for multiple complexes to be incorporated. By having multiple complexes, systems could be studied such as energy transfer in photosynthetic systems. These studies could occur because of the increase from empty SMALP particle sizes of approximately ~10 nm to sizes of ~30 nm. This threefold increase in diameter correlated with a ninefold increase in the area of the particle. Additionally, this larger particle size could aid in SMALPs being used in cryo-EM to study large protein complexes in a homogenous environment when compared to heterogeneous vesicles.

16.18 Structural studies involving SMALPs

SMALPs have been utilized for many structural studies using a variety of biophysical techniques. Optical techniques have been successfully utilized because of the small size of the discs created allowing for a high degree of light scattering. This has allowed the use of circular dichroism (CD) [33], fluorescence spectroscopy [54, 55] among other techniques. An advantage of SMA polymers when compared to nanodiscs is that nanodiscs have a high α-helical that interferes with studying the secondary structure of the protein of interest. It is important to note that even though styrene absorbs in the far-UV, the signal from styrene is low enough to be subtracted out with a baseline correction [54]. After baseline correction, the CD is still sensitive enough to obtain the complex structural data resulting from proteins secondary structures [54].

SMALPs in conjunction with EPR spectroscopy have been used to study the structure and dynamic properties of membrane protein complexes [31, 32, 59]. Results showed not only that the average distance measurements remained the same with SMA polymers compared to the vesicles [33], but also gave narrower and more precise line shapes when compared to liposomes [32]. Studies on protein structures utilizing solution NMR are currently lacking for SMALPs. Using NMR to study SMALPs for solution, NMR structures are problematic for several reasons. The first is that the size of the protein-lipid polymer complex may be close to exceeding the resolving power of NMR. Additionally, NMR is traditionally performed at slightly acidic pH values to maximize the number of protons for analysis. SMA polymer effects are optimal at a slightly basic pH values and the biophysics required for both to work optimally may prevent its usage. Bicelles have been used for magnetic alignment [26] and this would be desirable for SMALPs. However, all studies using solid state NMR have revealed isotropic peaks in ^{31}P which suggests fast reorientation in all directions [11, 15, 23].

16.19 Functional analysis of membrane-incorporated SMALPs

All previously described studies on membrane proteins in SMALPs have been biophysical in nature rather than probing the protein's functionally. Biophysics takes advantage of the fact that soluble domains of the protein are accessible to the solvent unlike what occurs in nature or in vesicles. Therefore, it is desirable to use SMA polymers in a functionally relevant system. SMA polymers do have several potential limitations for their functional studies. The higher order of lipids from model membranes can cause rigidity in the lipid movement giving results that are unlikely to occur in a biological system [31]. Rigidity from the polymers is what does give rise to greater stability in SMALPs, but it could hinder conformational dynamics from helical movement or interfere with function or protein binding [34]. The charge of SMA polymers could affect the binding of molecules and more specifically negatively charged molecules. Promising results have been obtained using KcsA when placed into solution electrophysiological in which functionality preserved [54]. This is an encouraging result that the polymers can be used for functional studies as well as biophysical studies.

16.20 General protocol

The general protocol for preparing SMALPs is described in this section. The polymer of interest should be either synthesized or purchased and brought up in the solution of interest. A workflow chart describing the general protocol is shown in Figure 16.5. Typically, the membrane protein is incorporated using traditional biophysical methods into liposomes prior to the formation of SMALPs using previously described methods [6, 32–34, 53]. Once SMALPs have been formed, there is currently no evidence of membrane proteins being inserted. Therefore, the membrane protein should be inserted into the vesicles prior to SMALPs formation either by traditional biophysical methods [6, 32, 34, 59] or detergent-free methods [7, 31, 33, 51, 55]. Ensure that the polymer is dissolved in the buffer of interest prior to addition to lipids. Add the SMA in excess, typically anywhere from 2.5 to 5 mol%, to the solution and mix thoroughly. If the SMA is not fully dissolved, it may be necessary to add a small amount of base, typically 1M KOH or 1M NaOH, to fully solubilize polymers. The procedure needs to be optimized for detergent-free systems based as discussed in the literature [8, 31, 33, 51, 55–56]. A qualitative proof that SMALPs have formed is the solution turns from opaque to clear. Freeze/thawing can make this process occur faster but can also introduce stability issues especially when membrane proteins are present. Mix the solution anywhere from 5 min to overnight depending on the sample. If the sample requires additional purification including affinity chromatography or size exclusion chromatography to remove non-specific proteins or the small amount of undissolved vesicles perform the purification. The sample should be ready for analysis using techniques relevant to the system being studied.

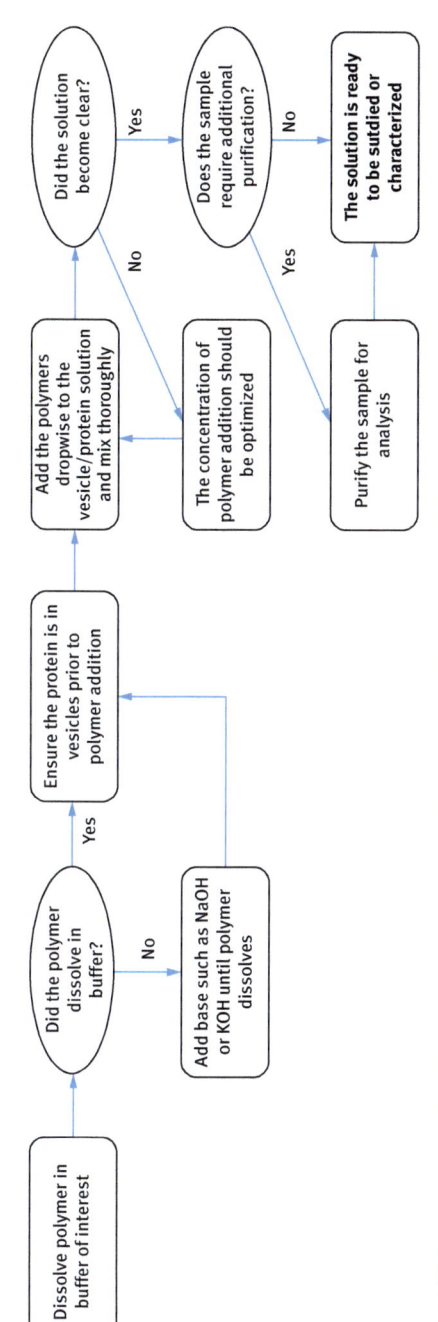

Figure 16.5: Generic workflow that can be used as a guideline for using SMA to form SMALPS.

16.21 Conclusion

SMA polymers are a useful and promising new structural biology tool. These polymers have been used for a variety of systems and have shown great promise in all types of cells. SMAs have many advantages over other current membrane mimics. Although the polymers already show great promise, the potential applications for them are numerous including functional studies, cryo-em, protein–protein interactions, and energy transfer studies. As such, SMA polymers will likely have a large influence on membrane protein studies for years to come.

References

[1] Hemminga, M.A., & Berliner, L.J. ESR spectroscopy in membrane biophysics, Springer, New York, 2007.
[2] Sachs, J.N., & Engelman, D.M. Introduction to the membrane protein reviews: the interplay of structure, dynamics, and environment in membrane protein function, Annu Rev Biochem 75 (2006) 707–712.
[3] Overington, J.P., Al-Lazikani, B., & Hopkins, A.L. How many drug targets are there?, Nat Rev Drug Discov 5 (12) (2006) 993–996.
[4] Seddon, A.M., Curnow, P., & Booth, P.J. Membrane proteins, lipids and detergents: not just a soap opera, Bba-Biomembranes 1666 (1-2) (2004) 105–117.
[5] Rigaud, J.L., & Levy, D. Reconstitution of membrane proteins into liposomes, Methods Enzymol 372 (2003) 65–86.
[6] Zhang, R., Sahu, I.D., Bali, A.P., Dabney-Smith, C., & Lorigan, G.A. Characterization of the structure of lipodisq nanoparticles in the presence of KCNE1 by dynamic light scattering and transmission electron microscopy, Chem Phys Lipids 203 (2017) 19–23.
[7] Postis, V., Rawson, S., Mitchell, J.K., Lee, S.C., Parslow, R.A., Dafforn, T.R., Baldwin, S.A., & Muench, S.P. The use of SMALPs as a novel membrane protein scaffold for structure study by negative stain electron microscopy, Bba-Biomembranes 1848 (2) (2015) 496–501.
[8] Lee, S.C., Knowles, T.J., Postis, V.L.G., Jamshad, M., Parslow, R.A., Lin, Y.P., Goldman, A., Sridhar, P., Overduin, M., Muench, S.P., & Dafforn, T.R. A method for detergent-free isolation of membrane proteins in their local lipid environment, Nat Protoc 11 (7) (2016) 1149–1162.
[9] Lee, S.C., Khalid, S., Pollock, N.L., Knowles, T.J., Edler, K., Rothnie, A.J., T. O, R.T., & Dafforn, T.R. Encapsulated membrane proteins: a simplified system for molecular simulation, Biochim Biophys Acta 38 (2016) 137–144.
[10] Jamshad, M., Grimard, V., Idini, I., Knowles, T.J., Dowle, M.R., Schofield, N., Sridhar, P., Lin, Y.P., Finka, R., Wheatley, M., Thomas, O.R.T., Palmer, R.E., Overduin, M., Govaerts, C., Ruysschaert, J.M., Edler, K.J., & Dafforn, T.R. Structural analysis of a nanoparticle containing a lipid bilayer used for detergent-free extraction of membrane proteins, Nano Res 8 (3) (2015) 774–789.
[11] Craig AF, C.E., Sahu, I.D., Zhang, R., Frantz, N.D., Al-Abdul, W.M.S., Dabney, S.C., Konkolewicz, D., & Lorigan, G.A. Tuning the size of styrene-maleic acid copolymer-lipid nanoparticles (SMALPs) using RAFT polymerization for biophysical studies, Biochimica et Biophysica Acta (BBA) - Biomembranes 1858 (11) (2016) 2931–2939.
[12] Sanders, C.R., & Prosser, R.S. Bicelles: a model membrane system for all seasons?, Structure 6 (10) (1998) 1227–1234.

[13] Zhou, H.X., & Cross, T.A., Influences of membrane mimetic environments on membrane protein structures, Annu Rev Biophys 42 (2013) 361–392.
[14] Denisov, I.G., Grinkova, Y.V., Lazarides, A.A., & Sligar, S.G. Directed self-assembly of monodisperse phospholipid bilayer Nanodiscs with controlled size, J Am Chem Soc 126 (11) (2004) 3477–3487.
[15] Vargas, C., Arenas, R.C., Frotscher, E., & Keller, S. Nanoparticle self-assembly in mixtures of phospholipids with styrene/maleic acid copolymers or fluorinated surfactants, Nanoscale 7 (48) (2015) 20685–20696.
[16] Tate, C.G. Practical considerations of membrane protein instability during purification and crystallisation, Methods Mol Biol 601 (2010) 187–203.
[17] Prive, G.G. Detergents for the stabilization and crystallization of membrane proteins, Methods 41 (4) (2007) 388–397.
[18] Garavito, R.M., & Ferguson-Miller, S. Detergents as tools in membrane biochemistry, J Biol Chem 276 (35) (2001) 32403–32406.
[19] Page, R.C., Moore, J.D., Nguyen, H.B., Sharma, M., Chase, R., Gao, F.P., Mobley, C.K., Sanders, C.R., Ma, L., Sonnichsen, F.D., Lee, S., Howell, S.C., Opella, S.J., & Cross, T.A. Comprehensive evaluation of solution nuclear magnetic resonance spectroscopy sample preparation for helical integral membrane proteins, J Struct Funct Genomics 7 (1) (2006) 51–64.
[20] Raschle, T., Hiller, S., Etzkorn, M., & Wagner, G. Nonmicellar systems for solution NMR spectroscopy of membrane proteins, Curr Opin Struc Biol 20 (4) (2010) 471–479.
[21] Geertsma, E.R., Nik Mahmood, N.A., Schuurman-Wolters, G.K., & Poolman, B. Membrane reconstitution of ABC transporters and assays of translocator function, Nat Protoc 3 (2) (2008) 256–266.
[22] Fang, G., Friesen, R., Lanfermeijer, F., Hagting, A., Poolman, B., & Konings, W.N., Manipulation of activity and orientation of membrane-reconstituted di-tripeptide transport protein DtpT of Lactococcus lactis, Mol Membr Biol 16 (4) (1999) 297–304.
[23] Zhang, R.F., Sahu, I.D., Liu, L.S., Osatuke, A., Corner, R.G., Dabney-Smith, C., & Lorigan, G.A. Characterizing the structure of lipodisq nanoparticles for membrane protein spectroscopic studies, Bba-Biomembranes 1848 (1) (2015) 329–333.
[24] Vold, R.R., & Prosser, R.S. Magnetically oriented phospholipid bilayered micelles for structural studies of polypeptides. Does the ideal bicelle exist?, J Magn Reson Ser B 113 (3) (1996) 267–271.
[25] Lee, D., Walter, K.F.A., Bruckner, A.K., Hilty, C., Becker, S., & Griesinger, C., Bilayer in small bicelles revealed by lipid-protein interactions using NMR spectroscopy, J Am Chem Soc 130 (42) (2008) 13822–13823.
[26] Howard, K.P., & Opella, S.J. High-resolution solid-state NMR spectra of integral membrane proteins reconstituted into magnetically oriented phospholipid bilayers, J Magn Reson B 112 (1) (1996) 91–94.
[27] Borch, J., & Hamann, T. The nanodisc: a novel tool for membrane protein studies, Biol Chem 390 (8) (2009) 805–814.
[28] Bayburt, T.H., Vishnivetskiy, S.A., McLean, M.A., Morizumi, T., Huang, C.C., Tesmer, J.J., Ernst, O.P., Sligar, S.G., & Gurevich, V.V. Monomeric rhodopsin is sufficient for normal rhodopsin kinase (GRK1) phosphorylation and arrestin-1 binding, J Biol Chem 286 (2) (2011) 1420–1428.
[29] Bayburt, T.H., & Sligar, S.G. Membrane protein assembly into Nanodiscs, FEBS Lett 584 (9) (2010) 1721–1727.
[30] Hagn, F., Etzkorn, M., Raschle, T., & Wagner, G. Optimized phospholipid bilayer nanodiscs facilitate high-resolution structure determination of membrane proteins, J Am Chem Soc 135 (5) (2013) 1919–1925.

[31] Orwick, M.C., Judge, P.J., Procek, J., Lindholm, L., Graziadei, A., Engel, A., Grobner, G., Watts, A. Detergent-free formation and physicochemical characterization of nanosized lipid-polymer complexes: lipodisq, Angew Chem Int Edit 51 (19) (2012) 4653–4657.

[32] Sahu, I.D., McCarrick, R.M., Troxel, K.R., Zhang, R., Smith, H.J., Dunagan, M.M., Swartz, M.S., Rajan, P.V., Kroncke, B.M., Sanders, C.R., & Lorigan, G.A. DEER EPR measurements for membrane protein structures via bifunctional spin labels and lipodisq nanoparticles, Biochemistry 52 (38) (2013) 6627–6632.

[33] Orwick-Rydmark, M., Lovett, J.E., Graziadei, A., Lindholm, L., Hicks, M.R., & Watts, A. Detergent-free incorporation of a seven-transmembrane receptor protein into nanosized bilayer lipodisq particles for functional and biophysical studies, Nano Letters 12 (9) (2012) 4687–4692.

[34] Dorr, J.M., Scheidelaar, S., Koorengevel, M.C., Dominguez, J.J., Schafer, M., van Walree, C.A., & Killian, J.A. The styrene-maleic acid copolymer: a versatile tool in membrane research, Eur Biophys J 45 (1) (2016) 3–21.

[35] Maeda, H. SMANCS and polymer-conjugated macromolecular drugs: advantages in cancer chemotherapy, Adv Drug Deliv Rev 46 (1-3) (2001) 169–185.

[36] Banerjee, S., Pal, T.K., & Guha, S.K. Probing molecular interactions of poly(styrene-co-maleic acid) with lipid matrix models to interpret the therapeutic potential of the co-polymer, Biochim Biophys Acta 1818 (3) (2012) 537–550.

[37] Tonge, S.R. Compositions Comprising a Lipid and Copolymer of Styrene and Maleic Acid, Malvern Cosmeceutics Limited, United Kingdom, 2006.

[38] Gaborieau, M., Gilbert, R.G., Gray-Weale, A., & Hernandez, J.M., C. P Theory of multiple-detection size-exclusion chromatography of complex branched polymers, Macromol Theory Simul 16 (1) (2007) 13–28.

[39] Matyjaszewski, K., & Tsarevsky, N.V. Macromolecular engineering by atom transfer radical polymerization, J Am Chem Soc 136 (18) (2014) 6513–6533.

[40] Konkolewicz, D., Krys, P., & Matyjaszewski, K. Explaining unexpected data via competitive equilibria and processes in radical reactions with reversible deactivation, Acc Chem Res 47 (10) (2014) 3028–3036.

[41] Moad, G., Rizzardo, E., & Thang, S.H. Polymerization, RAFT. and some of its applications, Chem-Asian J 8 (8) (2013) 1634–1644.

[42] Braunecker, W.A., & Matyjaszewski, K. Controlled/living radical polymerization: features, developments, and perspectives, Prog Polym Sci 32 (1) (2007) 93–146.

[43] Chiefari, J., Chong, Y.K., Ercole, F., Krstina, J., Jeffery, J., Le, T.P.T., Mayadunne, R.T.A., Meijs, G.F., Moad, C.L., Moad, G., Rizzardo, E., & Thang, S.H. Living free-radical polymerization by reversible addition-fragmentation chain transfer: the RAFT process, Macromolecules 31 (16) (1998) 5559–5562.

[44] Moad, G., Rizzardo, E., & Thang, S.H. Living radical polymerization by the RAFT Process - A third update, Aust J Chem 65 (8) (2012) 985–1076.

[45] Hill, M.R., Carmean, R.N., & Sumerlin, B.S. Expanding the scope of RAFT polymerization: recent advances and new horizons, Macromolecules 48 (16) (2015) 5459–5469.

[46] Gregory, A., & Stenzel, M.H. Complex polymer architectures via RAFT polymerization: From fundamental process to extending the scope using click chemistry and nature's building blocks, Prog Polym Sci 37 (1) (2012) 38–105.

[47] McCormack, C.L., & Lowe, A.B. Aqueous RAFT polymerization: Recent developments in synthesis of functional water-soluble (Co)polymers with controlled structures, Accounts Chem Res 37 (5) (2004) 312–325.

[48] Hill, D.J.T., Odonnell, J.H., & Osullivan, P.W. Analysis of the mechanism of copolymerization of styrene and maleic-anhydride, Macromolecules 18 (1) (1985) 9–17.

[49] Knowles, T.J., Finka, R., Smith, C., Lin, Y.P., Dafforn, T., & Overduin, M. Membrane proteins solubilized intact in lipid containing nanoparticles bounded by styrene maleic acid copolymer, J Am Chem Soc 131 (22) (2009) 7484-+.

[50] Scheidelaar, S., Koorengevel, M.C., Pardo, J.D., Meeldijk, J.D., Breukink, E., & Killian, J.A. Molecular model for the solubilization of membranes into nanodisks by styrene maleic acid copolymers, Biophys J 108 (2) (2015) 279–290.

[51] Long, A.R., O'Brien, C.C., Malhotra, K., Schwall, C.T., Albert, A.D., Watts, A., & Alder, N.N. A detergent-free strategy for the reconstitution of active enzyme complexes from native biological membranes into nanoscale discs, Bmc Biotechnol 13 (2013) 41.

[52] Bell, A.J., Frankel, L.K., & Bricker, T.M. high yield non-detergent isolation of photosystem I-lightharvesting chlorophyll II membranes from spinach thylakoids implications for the organization of the PS lantennae in higher plants, J Biol Chem 290 (30) (2015) 18429–18437.

[53] Coey, A.T., Sahu, I.D., Gunasekera, T.S., Troxel, K.R., Hawn, J.M., Swartz, M.S., Wickenheiser, M.R., Reid, R.J., Welch, R.C., Vanoye, C.G., Kang, C.B., Sanders, C.R., & Lorigan, G.A. Reconstitution of KCNE1 into lipid bilayers: comparing the structural, dynamic, and activity differences in micelle and vesicle environments, Biochemistry 50 (50) (2011) 10851–10859.

[54] Dorr, J.M., Koorengevel, M.C., Schafer, M., Prokofyev, A.V., Scheidelaar, S., van der Cruijsen, E.A.W., Dafforn, T.R., Baldus, M., & Killian, J.A. Detergent-free isolation, characterization, and functional reconstitution of a tetrameric K+ channel: The power of native nanodiscs, P Natl Acad Sci USA 111 (52) (2014) 18607–18612.

[55] Gulati, S., Jamshad, M., Knowles, T.J., Morrison, K.A., Downing, R., Cant, N., Collins, R., Koenderink, J.B., Ford, R.C., Overduin, M., Kerr, I.D., Dafforn, T.R., & Rothnie, A.J. Detergent-free purification of ABC (ATP-binding-cassette) transporters, Biochem J 461 (2) (2014) 269–278.

[56] Jamshad, M., Charlton, J., Lin, Y.P., Routledge, S.J., Bawa, Z., Knowles, T.J., Overduin, M., Dekker, N., Dafforn, T.R., Bill, R.M., Poyner, D.R., & Wheatley, M. G-protein coupled receptor solubilization and purification for biophysical analysis and functional studies, in the total absence of detergent, Biosci Rep 35 (2) (2015) e00188.

[57] Rasmussen, S.G., Choi, H.J., Rosenbaum, D.M., Kobilka, T.S., Thian, F.S., Edwards, P.C., Burghammer, M., Ratnala, V.R., Sanishvili, R., Fischetti, R.F., Schertler, G.F., Weis, W.I., & Kobilka, B.K. Crystal structure of the human beta2 adrenergic G-protein-coupled receptor, Nature 450 (7168) (2007) 383–387.

[58] Swainsbury, D.J., Scheidelaar, S., van Grondelle, R., Killian, J.A., & Jones, M.R. Bacterial reaction centers purified with styrene maleic acid copolymer retain native membrane functional properties and display enhanced stability, Angew Chem Int Ed Engl 53 (44) (2014) 11803–11807.

[59] Sahu, I.D., Kroncke, B.M., Zhang, R., Dunagan, M.M., Smith, H.J., Craig, A., McCarrick, R.M., Sanders, C.R., & Lorigan, G.A. Structural investigation of the transmembrane domain of KCNE1 in proteoliposomes, Biochemistry 53 (40) (2014) 6392–6401.

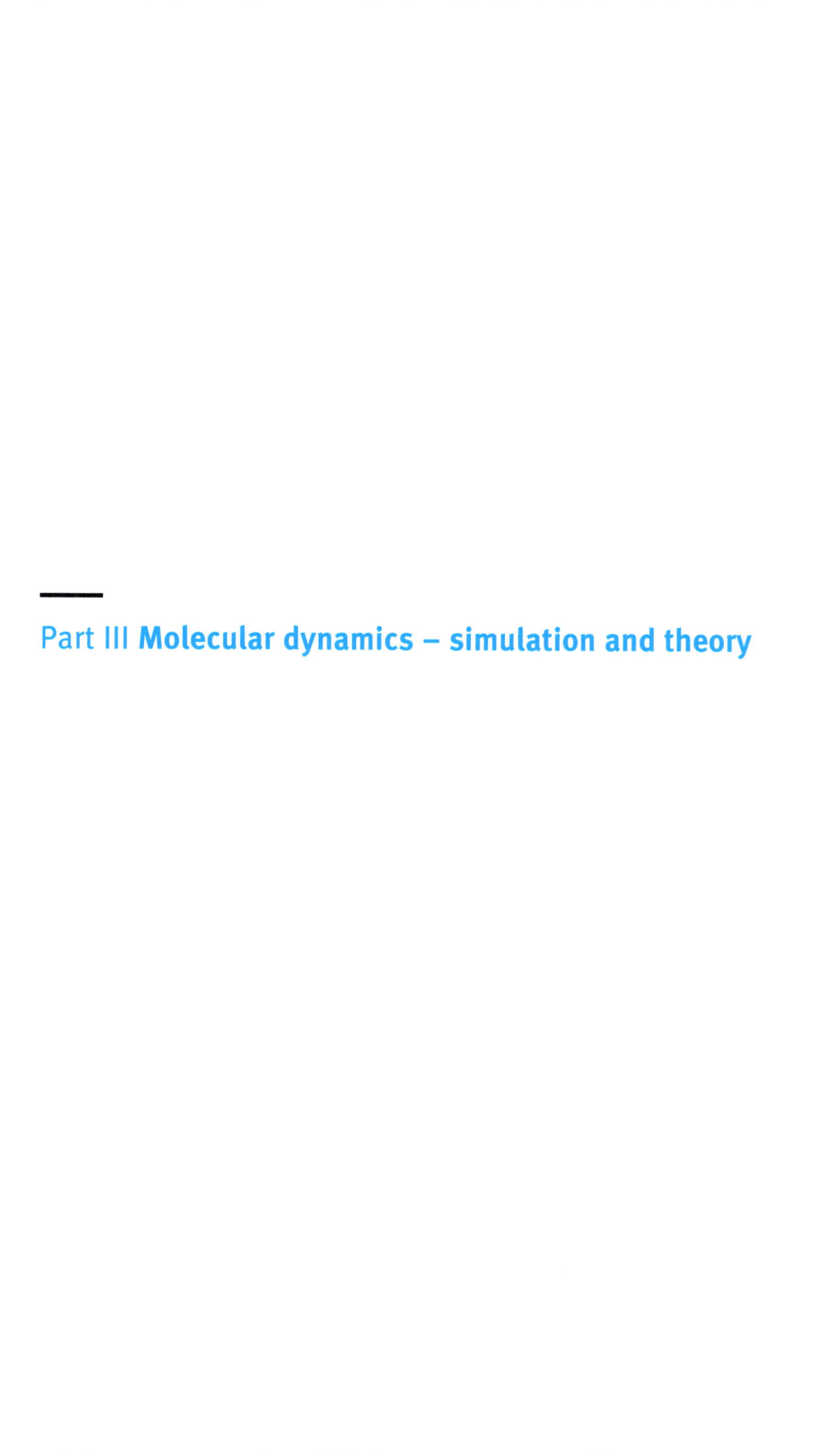

Part III Molecular dynamics – simulation and theory

M. Schick

17 On the origin of "Rafts": The plasma membrane as a microemulsion

Abstract: The hypothesis that the plasma membrane is heterogeneous characterized by regions of saturated lipids and cholesterol which float, like rafts, in a sea of unsaturated lipids seems to be widely accepted. There is no agreement, however, on the physical mechanism that gives rise to these inhomogeneities and accounts for their characteristic size, which is on the order of a 100 nm. In this chapter, one model with a clear physical basis is reviewed. In posits that the inhomogeneities are those characteristic of a microemulsion, one brought about by the coupling of fluctuations in the height of the membrane to fluctuations in the membrane composition.

Keywords: rafts, microemulsion, spontaneous curvature

17.1 Introduction

The hypothesis that the plasma membrane is not homogeneous, but rather is characterized by "rafts", of the order of 100 nm, that are enriched in sphingomyelin (SM) and cholesterol, and that float in a "sea" of phosphatidylcholine (PC) and other phospholipids [1], is an intriguing one. Because they are enriched in saturated SM and cholesterol, the rafts have a greater areal density than the sea. Transmembrane proteins, and the anchors of proteins attached to the plasma membrane from either side, are sensitive to the density difference between rafts and sea and will prefer one environment to the other. Hence, instead of being uniformly distributed throughout the plasma membrane, proteins will be found at a greater density in one or the other of these environs, and thus will work more efficiently. Hence, physical organization leads to functional organization. This is indeed a very attractive idea. There is only one catch: why do lipids organize into regions on the order of 100 nm?

There have been a few answers proposed to this question. The most popular is that rafts are brought about by phase separation, that is, that the rafts and sea are two coexisting phases. This idea is bolstered by the many observations of ternary mixtures of saturated lipids, like SM, unsaturated lipids, like dioleoylphosphatidylcholine (DOPC), and cholesterol. These mixtures do, indeed, undergo phase separation at biological temperatures [2]. There are several difficulties, however, with this proposal. The most obvious one is that it cannot explain the origin of the characteristic size of 100 nm. In phase separation, the size of the phases which coexist is as large as the system itself.

M. Schick, Department of Physics, University of Washington, Seattle, WA, USA

https://doi.org/10.1515/9783110544657-017

Thus, one has to invoke something which will give rise to this length, such as the cytoskeleton, and assert that it prevents the nanometer size regions from coarsening into macroscopic ones [3]. Another difficulty arises from the fact that membranes with the composition of the cytoplasmic leaf of the plasma membrane have no tendency to undergo phase separation [4]. This is due to the fact that the separation in the ternary mixtures noted above is driven by the poor chain packing of the saturated chains of SM and the unsaturated chains of PC. However, almost all of the SM in the plasma membrane is in the outer leaf, and there is very little of it in the inner leaf. So membranes with a composition like that of the outer leaf separate readily while those with compositions resembling the inner leaf do not. Coupling between leaves can induce a transition in the inner leaf if its composition is not far from a phase transition [5], but the composition of the inner leaf of the plasma membrane seems to be far from that needed for phase separation. Furthermore, even if the inner leaf did undergo separation due to a coupling to the outer leaf, the difference in density between the coexisting phases would be small, again due to the lack of SM. As this density difference is what makes a raft useful for protein separation, a raft formed in the inner leaflet would not be a useful one. A final objection to this proposal is that phase separation has never been observed in the plasma membrane at biological temperatures.

Separation has been observed in blebs at much lower temperatures [6]. As a consequence, the hypothesis has been proposed that phase separation could occur in the plasma membrane at temperature lower than biological and that rafts are the critical fluctuations of this unseen transition. My objections to this idea are twofold. First, for the fluctuation to have a characteristic size of 100 nm, both the temperature and the composition of the membrane must be controlled such that the actual system is just a certain distance from the critical transition. Second, and more severe, is the effect of the lack of SM in the inner leaf. Even were there critical fluctuations, the density difference between the inside and outside of a droplet would be so small that it would be ineffective as a raft.

The other explanation for the origin of rafts is that the plasma membrane is a microemulsion, that is, a fluid with structure of a characteristic size. Microemulsions are well known in bulk three-dimensional systems of oil, water, and a surfactant, which solubilizes the oil and water into a single disordered fluid, but one which is characterized not only by the usual correlation length but also by an additional length, one of the size of the regions of oil and water [7]. The mechanisms that bring about the microemulsion in the two-dimensional plasma membrane vary and have been nicely reviewed by Schmid [8]. There are several difficulties to overcome in thinking of the plasma membrane as a microemulsion. Among them are the following: what is the nature of the different regions which are solubilized; what is the mechanism that reduces the line tension between these regions so that they do solubilize? My answer to these questions is that the two different regions which are solubilized are those with a large spontaneous curvature and those with a small spontaneous curvature. In the exoplasmic leaf of the plasma membrane, the former would include regions of SM, while the latter would include those of PC. In the cytoplasmic leaflet, regions of large spontaneous curvature are those with appreciable concentrations of phosphatidylethanolamine (PE), while regions of small spontaneous curvature are those rich in PC and phosphatidylserine (PS). The mechanism which reduces the line tension between regions rich in large or small spontaneous curvatures is that proposed by Liebler and Andelman [9] and illustrated in Figure 17.1. The basic idea is that the free energy of the membrane can be reduced if the composition and height fluctuations of the membrane are coupled such that components with negative spontaneous curvature go to regions which are locally concave in the outer leaf while those with positive spontaneous curvature go to regions which are locally convex. A similar statement applies to the inner leaf. In other words, a flat bilayer has an internal tension because its components prefer regions of non-zero curvature. By allowing the different components to segregate locally at parts of the membrane which have a non-zero curvature like their own spontaneous curvature, this internal tension can be relieved, and the free energy decreased. This idea will be incorporated in a simple calculation in the next section. It will be sufficient to display the defining property of the regions being separated, their characteristic size, and how this size varies with temperature, surface tension, and other variables under experimental control.

17.2 The model

17.2.1 The order parameter

I will consider the outer leaf of the plasma membrane to be composed primarily of PC, SM, and cholesterol, and the inner leaf to be composed of PS, PE, and cholesterol. Let the local mol fractions of the components in the outer leaf be denoted by $\phi_{PC}(r)$, $\phi_{SM}(r)$, and $\phi_{C,o}(r)$, and those of the inner leaf by $\phi_{PS(r)}$, $\phi_{PE}(r)$, and $\phi_{C,i}(r)$. To incorporate the idea that the regions to be separated differ in the spontaneous curvature, I define order parameters in the outer leaf and in the inner leaf to be

$$\phi_o(r) \equiv \phi_{SM}(r) + \phi_{C,o}(r) - \phi_{PC}(r) \tag{17.1}$$

$$\phi_i(r) \equiv \phi_{PE}(r) + \phi_{C,i}(r) - \phi_{PS}(r) \tag{17.2}$$

Further I define

$$\begin{aligned}\phi(r) &\equiv \frac{\phi_o(r) - \phi_i(r)}{2} \\ &= \frac{[\phi_{SM}(r) + \phi_{C,o}(r) - \phi_{PC}(r)] - [\phi_{PE}(r) + \phi_{C,i}(r) - \phi_{PS}(r)]}{2}\end{aligned} \tag{17.3}$$

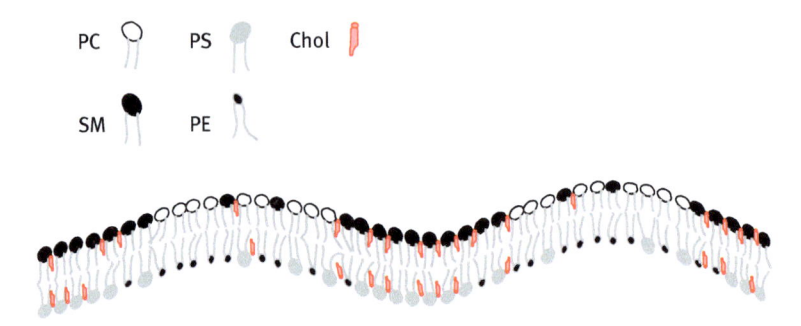

Figure 17.1: Regions rich in SM and cholesterol in the outer leaf and of PS in the inner leaf floating in a sea of PC in the outer leaflet and PE and cholesterol in the inner leaf.

The first bracket contains the difference in mole fractions between the components in the outer leaf with large spontaneous curvatures, SM, and cholesterol, and that with a small spontaneous curvature, PC. Similarly, the second bracket is the difference between the components in the inner leaf with large spontaneous curvatures, PE, and cholesterol, and that with a small one, PS [10]. Figure 17.1 shows that one expects this order parameter to have a large variation from positive to negative values as a function of position in the membrane. For completeness, I also define the other combination

$$\psi(r) \equiv \frac{\phi_o(r) + \phi_i(r)}{2}$$

$$= \frac{[\phi_{SM}(r) + \phi_{C,o}(r) - \phi_{PC}(r)] + [\phi_{PE}(r) + \phi_{C,i}(r) - \phi_{PS}(r)]}{2} \quad (17.4)$$

The variation in this order parameter is expected to be much less than the variation in $\phi(r)$.

17.2.2 The free energy

The free energy of the system can be written as the sum of several terms. First, there is the free energy of the bilayer if it were planar. This can be written, to second order in the local parameters $\phi(r)$ and $\psi(r)$,

$$F_{\text{planar}} = \int d^2r \left[\frac{b_\phi}{2}(\nabla\phi)^2 + a_\phi\phi^2 + \frac{b_\psi}{2}(\nabla\psi)^2 + a_\psi\psi^2 \right] \quad (17.5)$$

Then, there is the additional free energy because the membrane is not flat. There is the cost of adding surface area to the membrane and of bending it. These energies are written here in terms of the height, $h(r)$, above some reference plane and in the Monge representation:

$$F_{\text{curv}} = \int d^2 r \frac{1}{2} \{\gamma[\nabla h(r)]^2 + \kappa[(\nabla^2 h) - H_0[\phi]]^2\} \tag{17.6}$$

where γ is the surface tension of the membrane, κ is its bending modulus, and $H_0[\phi]$ is the spontaneous curvature of the membrane, which is a functional of the order parameter ϕ. I expand this contribution in a power series in ϕ and keep terms in the free energy up to second order. I ignore terms which contribute a constant to the free energy, or renormalize the coefficient a_ϕ in eq. (17.5), or renormalize the bending modulus. What remains is

$$F_{\text{curv}} = \int d^2 r \frac{1}{2} [\gamma(\nabla h)^2 + \kappa(\nabla^2 h)^2 - 2\kappa H'_0 \phi(r) \nabla^2 h(r)] \tag{17.7}$$

where $H'_o = \delta H_0[\phi]/\delta\phi$, the first functional derivative of the spontaneous curvature with respect to the order parameter. One clearly sees that the local composition, in the order parameter $\phi(r)$, couples to the local curvature, $\nabla^2 h(r)$ [9].

It is convenient to write the total free energy, $F_{\text{tot}} = F_{\text{planar}} + F_{\text{curv}}$ in terms of $\tilde\phi(k)$, $\tilde\psi(k)$, and $\tilde h(k)$, the Fourier transforms of $\phi(r)$, $\psi(r)$, and $h(r)$;

$$F_{\text{tot}}[\phi, h] = \int d^2 k \left[\left(a_\phi + \frac{b_\phi}{2} k^2\right) \tilde\phi(k)\tilde\phi(-k) + \left(a_\psi + \frac{b_\psi}{2} k^2\right) \tilde\psi(k)\tilde\psi(-k) \right.$$

$$\left. + \frac{1}{2}(\kappa k^4 + \gamma k^2)\tilde h(k)\tilde h(-k) - \kappa H'_o k^2 \tilde h(-k)\tilde\phi(k) \right] \tag{17.8}$$

Because the free energy is only quadratic in the $\tilde h(k)$, they can be integrated out. Equivalently, one can minimize the free energy with respect to them: $\delta F_{\text{tot}}/\delta \tilde h(k) = 0$ with the result

$$\tilde h(k) = \left(\frac{\kappa H'_o}{\gamma}\right) \frac{1}{[1 + \kappa k^2/\gamma]} \tilde\phi(k) \tag{17.9}$$

This shows explicitly that the free energy can be decreased when the height and composition variations are correlated. Upon substitution of this result into the free energy, eq. (17.8), one obtains

$$F_{\text{tot}}[\phi] = \int d^2 k \left(\left\{ a_\phi + \frac{b_\phi}{2}\left[1 - \left(\frac{(\kappa H'_o)^2}{b_\phi \gamma}\right) \frac{1}{(1 + \kappa k^2/\gamma)} \right] k^2 \right\} \tilde\phi(k)\tilde\phi(-k) \right.$$

$$\left. + \left[a_\psi + \frac{b_\psi}{2} k^2 \right] \tilde\psi(k)\tilde\psi(-k) \right) \tag{17.10}$$

17.2.3 The phase diagram

There are several phases of interest in this model. First there are two condensed phases characterized by $<\tilde{\phi}(0)> \neq 0$, where the brackets denote an ensemble average. These two phases can coexist at temperatures below T_c given by $a_\phi(T_c) = 0$. There are also modulated phases which are characterized by $<\tilde{\phi}(k)> \neq 0$ for some non-zero values of k. These are either characterized by alternating stripes of regions rich in large spontaneous curvature components and small spontaneous curvature components, or hexagonal arrays of one of these within a background of the other. Finally, there is the disordered phase characterized by $<\tilde{\phi}(k)> = 0$ for all values of k. It is this phase which is of the most interest to us. Some of its properties can best be understood by examining the structure function $S(k) \equiv <\tilde{\phi}(k)\tilde{\phi}(-k)>$. It is essentially a response function, or susceptibility, that displays the sensitivity of the order parameter to external perturbations that couple to it. It has a peak at the wavevector of the fluctuation in the system that is most responsive to external perturbations. The inverse of the structure factor is directly proportional to the coefficient of $\tilde{\phi}(k)\tilde{\phi}(-k)$ in eq. (17.10). From this structure, factor one finds that its peak occurs at the wavevector k^* where

$$k^* = 0, \quad \frac{\kappa|H'_0|}{(b_\phi\gamma)^{1/2}} \leq 1$$

$$= \left(\frac{\gamma}{\kappa}\right)^{1/2} \left[\frac{\kappa|H'_0|}{(b_\phi\gamma)^{1/2}} - 1\right]^{1/2}, \quad \frac{\kappa|H'_0|}{(b_\phi\gamma)^{1/2}} \geq 1 \qquad (17.11)$$

This behavior divides the disordered phase into two regions; one which displays the peak at $k^* = 0$ which I denote as an ordinary fluid, and the other in which $k^* \neq 0$ and which I denote a microemulsion. In the ordinary fluid, the real space correlation function $<\phi(0)\phi(r)>$ is characterized by its large-distance exponential decay to zero on the scale of the correlation length ξ. In the microemulsion, the correlation function behaves at large distances like an oscillatory function of wavevector k^* whose amplitude decays exponentially with correlation length ξ. Thus, the microemulsion is a fluid, because the correlation function decays exponentially, but it is a fluid with structure that is apparent in the oscillations of the correlation function, or equivalently in the peak in the structure function at $k^* \neq 0$.

A phase diagram as obtained from mean-field theory, in which the free energy $F_{tot}[\phi]$ is simply minimized with respect to ϕ, is shown in Figure 17.2. It is plotted in the plane of a_ϕ, which is essentially the temperature, and τ where

$$\tau = 1 - \frac{(\kappa H'_0)^2}{b_\phi\gamma} \qquad (17.12)$$

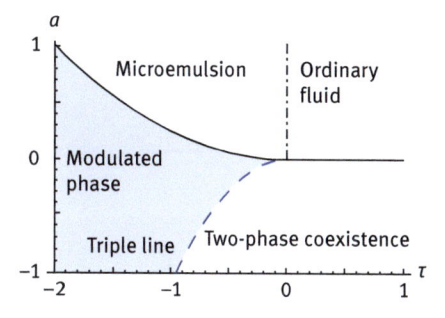

Figure 17.2: Phase diagram of the model calculated within the mean-field approximation, as a function of the two parameters a_ϕ and τ. Dashed lines denote first-order transitions; solid lines, continuous transitions. The region of macroscopic phase separation is denoted as a two-phase coexistence. A modulated phase appears for $\tau \leq 0$. The dash-dot line is the Lifshitz line. To the right of it, the fluid is an ordinary one; to the left of it, the fluid is a microemulsion.

The system has been taken such that the mol fractions of components with large spontaneous curvatures and small spontaneous curvatures are equal; that is, the global average of the order parameter $\phi(r)$ vanishes.

Within mean-field theory, there is a line of continuous transitions between the ordinary fluid and the two phases which are in coexistence. There is also a line of continuous transitions between the microemulsion and the modulated phases. A triple line separates the two coexisting phases from the modulated phase. Note that there is no direct path between the two coexisting phases and the microemulsion.

I emphasize that there is no phase transition between the ordinary fluid and the microemulsion; that is, there is no singularity in the free energy as one passes from one to the other. Hence, the distinction between phases is arbitrary. The locus of points at which the peak in the structure function moves off of $k^* = 0$ to a some non-zero value, a locus denoted the Lifshitz line, is a convenient marker because the structure function can be determined from scattering experiments. It is this marker which is adopted in Figure 17.2. At the Lifshitz line, the correlation length, ξ, is much less than the characteristic wavelength $2\pi/k^*$. As a consequence, no oscillations are discernible in the real space correlation function. Another possible marker for the distinction between ordinary fluid and microemulsion is the locus at which ξ is just equal to $2\pi/k^*$, which is denoted by the disorder line [11]. At this point, oscillations in the real-space correlation function are visible.

Numerical simulations have been carried out on almost the same model [12] (They differ only at large wavevectors, or small wavelengths, which are not of interest here). In contrast to mean-field theory, such simulations include the effects of fluctuations, which can be large in two-dimensional systems [13]. The resulting phase diagram is shown in Figure 17.3.

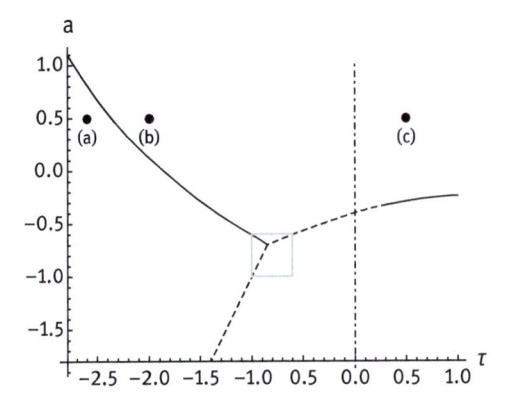

Figure 17.3: Three dots indicate the locus of the systems from which were drawn the snapshots shown in Fig. 17.4.

It can be seen that fluctuations increase the phase space of the microemulsion. The Lifshitz line is shown as a dotted-dashed line. In addition, a portion of the line of continuous transitions from two-phase coexistence has been driven to be first order, and there is now a direct path between the two coexisting phases and the microemulsion. Representative configurations from the modulated phase, the microemulsion, and the ordinary fluid are shown in Figures 17.4 (a), (b), and (c), respectively. The characteristic sizes, $2\pi/k^*$, of the regions of large and small spontaneous curvature in the microemulsion are clearly visible. Additional simulations on related models of microemulsions have been carried out by Schmid and coworkers [14, 15].

17.3 The plasma membrane as a microemulsion

That the plasma membrane might be a microemulsion is an attractive idea because a microemulsion is characterized by fluctuating regions of a well-defined size. But should we expect from the above analysis that the plasma membrane is in a regime of a microemulsion, and if so, what is the characteristic size? Well to be in the region of a microemulsion, eq. (17.11) says that $\kappa |H'_0|/(b_\phi \gamma)^{1/2}$ must be greater than unity. I take $\kappa = 44 k_B T = 188$ pN nm [16], $b_\phi = 5 k_B T = 21.4$ pN nm [17] and for the surface tension, $\gamma = 0.02$ pN/nm [18]. Thus, the plasma membrane would be in the regime of a microemulsion if $|H'_0| > 3 \times 10^{-3}$ nm^{-1}. Recall that $H'_0 \equiv \delta H_0[\phi]/\delta\phi$ so that if we take the spontaneous curvature to be a linear function of the mol fractions of its components, as is usually done, then $H'_0 = H_0$, the mol-fraction weighted difference between the large and small spontaneous curvatures of the components. For the main components of the plasma membrane, the magnitude of H_0 ranges from the small value of

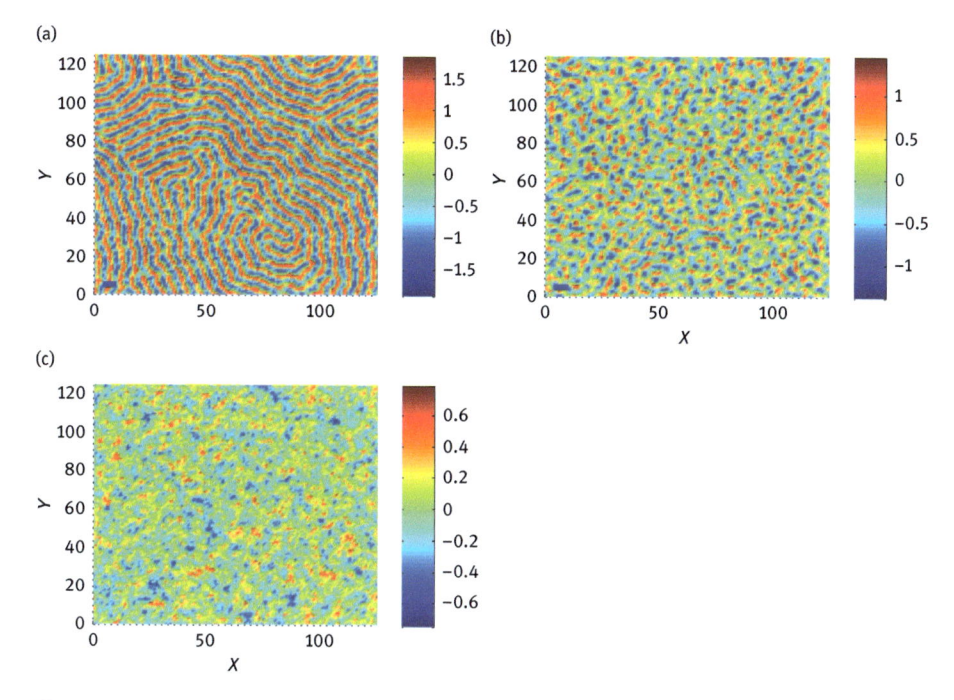

Figure 17.4: Representative configurations from different phases of the system. The parameter b_ϕ is set to 4. (a) The location of the system is $a_\phi = 0.5$ and $\tau = -2.6$, and the system is in the stripe phase. (b) The location of this system is $a_\phi = 0.5$, $\tau = -2.0$. The system is a microemulsion. (c) $a_\phi = 0.5$ and $\tau = 0.5$. The system is an ordinary fluid. (Bar, lower-left corner) characteristic size $2\pi\sqrt{-2/\tau}$.

0.022 nm^{-1} for PC to the large value of 0.32 nm^{-1} for PE [10]. Thus, the condition is certainly fulfilled.

If this is so, then what is the characteristic size of the regions of the microemulsion? With $\kappa|H'_0|(b_\phi\gamma)^{1/2}$ much greater than unity as expected, the expression for k^*, eq. (17.11), simplifies to

$$k^* \approx \left(\frac{\gamma(H'_0)^2}{b_\phi}\right)^{1/4}$$

$$= \left(\frac{\gamma}{b_\phi}\right)^{1/2} \left(\frac{b_\phi(H'_0)^2}{\gamma}\right)^{1/4} \tag{17.13}$$

A typical size of the regions either rich or poor in lipids of large spontaneous curvature would then be

$$\frac{\pi}{k^*} \approx \pi \left(\frac{b_\phi}{\gamma}\right)^{1/2} \left(\frac{\gamma}{b_\phi (H'_0)^2}\right)^{1/4} \tag{17.14}$$

I write the result in this form because $(b_\phi/\gamma)^{1/2}$ is a length. With the values above, it is 33 nm. Multiplication by π yields 104 nm. The particular value one obtains for π/k^* depends on the value of H_0', the effective spontaneous curvature of the bilayer. This in turn depends on which lipids one includes in its evaluation, and the values for their individual spontaneous curvatures. For example, if I ignore all phospholipids except PE, which has the largest spontaneous curvature, then $\phi = \phi_{PE}$, $H_0[\phi] = H_{0,PE}\phi_{PE}$ and $H'_0 = \delta H_0[\phi]/\delta\phi_{PE} = H_{0,PE} = 0.316$ nm^{-1} [10]. Putting this, and the above values for b_ϕ and γ into the above eq. (17.14), I obtain an estimate of 32 nm for the characteristic size in the microemulsion. I have employed slightly different models elsewhere, but they both yield estimates of the same order of magnitude for the size of the inhomogeneities; 121nm in Ref [19], 75nm in Ref [20]. It hardly needs to be said that this is just the estimated size of rafts [21].

17.4 Evidence in vitro and in vivo for microemulsions in membranes

Have microemulsions been observed in model membranes or in biological ones? The answer is definitely affirmative, although the microemulsions have not always been identified as such. There are several examples. The quaternary system composed of dioleoylphosphatidylcholine (DOPC), 1-palmitoyl, 2-oleoylphosphatidylcholine (POPC), distearoylphosphatidylcholine (DSPC), and cholesterol clearly displays a microemulsion, as seen in Figures 17.2B, D, and E of Ref [22] which is reproduced in Figure 17.5 below. The phase was, I believe, misidentified there as being within a region of two-phase coexistence.

Recent examples appear both in a model membrane of diphytanoylphosphatidylcholine (DiPhyPC), dipalmitoylphosphatidylcholine (DDPC), and cholesterol, and in cell-derived giant plasma membrane vesicles [23]. An example of a microemulsion in the model membrane is shown in Figure 17.6.

The advantage of the model system is, of course, that one can readily vary the external parameters. In particular, the model vesicle could be made more taut or more flaccid by varying the osmotic pressure. By decreasing the osmotic pressure and thereby decreasing the surface tension, γ, one can decrease the parameter τ of eq. (17.12) and, as seen, in Figure 17.3, move from a region of two-phase coexistence to a disordered phase. It was observed that this transition was a first-order one and that the disordered phase was a microemulsion [23]. This is in

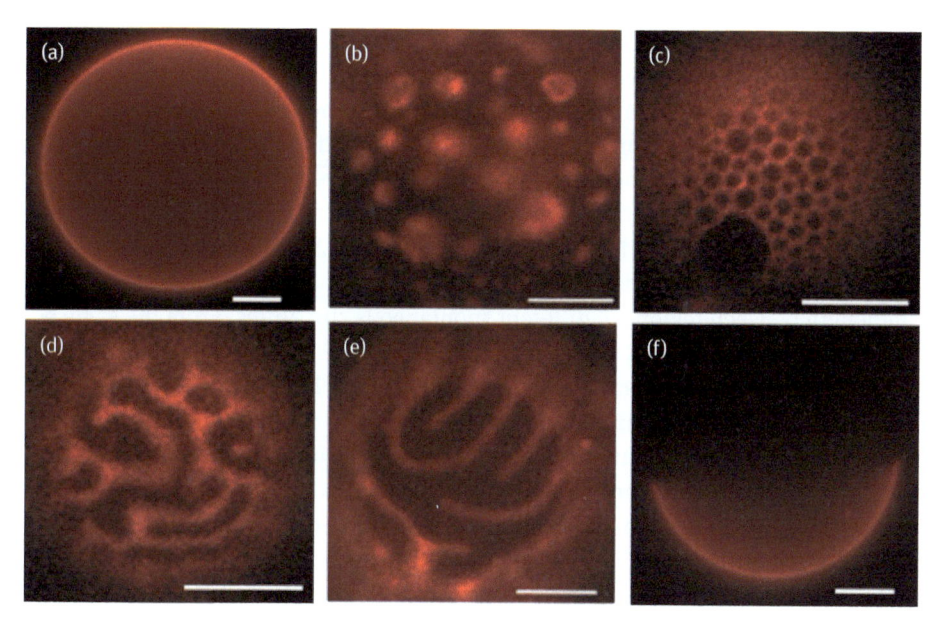

Figure 17.5: GUV patterns from four-component mixtures of DSPC, POPC, DOPC, and cholesterol as POPC is increasingly replaced by DOPC in going from A to F. I believe B, D, and E are microemulsions. Figure reproduced from Ref. [22]

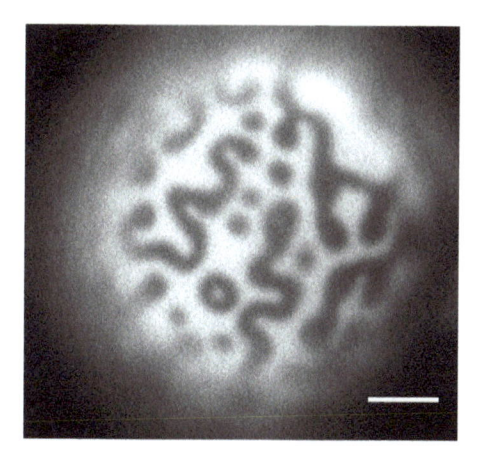

Figure 17.6: Vesicle of 35/35/30 DiphyPC/DPPC/cholesterol at room temperature. Scale bar is 20 μm. Courtesy of C. Cornell and S.L. Keller.

accord with the phase diagram of Figure 17.3. The observation is significant because in the absence of mechanisms that can bring about a microemulsion, first-order transitions from two-phase coexistence leads to an ordinary disordered phase, not to a microemulsion.

Having emerged from two-phase coexistence into a microemulsion, the system is presumably not far from the Lifshitz line, that is the characteristic wavevector of the inhomogeneities, k^* of eq. (17.11), is small, and the characteristic size is large. As the surface tension decreases, one moves further from the Lifshitz line, that is, the factor $[(\kappa|H'_0|)/(b_\phi \gamma)^{1/2} - 1]$ becomes larger, so that the wavevector increases. Thus as the surface tension decreases in this region, the characteristic size of inhomogeneities decreases. This is the behavior observed in experiment [23]. Similarly, an increase of temperature is expected to decrease the parameter b_ϕ that is related to the energy per unit length between regions of different spontaneous curvatures. From the same eq. (17.11), one see that this will increase the characteristic wavevector and decrease the characteristic size. This is also observed in experiment [23].

Of the two observations, the characteristic size of inhomogeneities decreases with an increase in temperature and with a decrease in surface tension, the first does not discriminate between various theories of the origin of the inhomogeneities, while the second does. By this I mean the following. If the energy per unit length between regions of one kind and regions of the other decreases with increasing temperature, as one would expect, then it is less expensive to make such interfaces and the characteristic size will decrease. Any reasonable theory will predict this.

However, that the characteristic size decreases with decreasing surface tension indicates that the inhomogeneities are related in some way to variations in the membrane height. These variations create additional membrane area and the cost of this is related to the surface tension. If this tension decreases, the system can make more such variations in the membrane height with the consequence that the characteristic size of the inhomogeneities decreases. The theory presented here predicts this effect.

17.5 Discussion

I have reviewed the application of a theory, due to Leibler, Andelman, and co-workers [9, 24, 25] which I have applied to the question of the origin of rafts. In this picture, local variations in the membrane shape couple to local variations in its composition in order to relieve tension and lower the system's free energy. The theory provides a characteristic size for the variations in composition in the system's fluid phase, which is a microemulsion. This size, π/k^*, with k^* given in eq. (17.11), depends upon various physical parameters, such as the membrane's bending modulus, surface tension, and spontaneous curvature. Liu et al [26] considered this mechanism to be the origin of rafts, but concluded that the mechanism could not be the cause. I have argued elsewhere [27] that this negative conclusion resulted from the particular values they chose for the physical parameters. A more reasonable choice, as utilized here, results in a characteristic size of the inhomogeneities of the order of 100 nm. It could be objected that the estimate depends upon the spontaneous curvature of the membrane; that the usual assumption that this curvature can be calculated as a linear combination of the spontaneous curvatures of the components weighted by their mol fractions is not correct [28]. I believe that the objection is valid, but I do not believe it will have a large effect on the estimate of the characteristic size. If one considers the estimate given in eq. (17.14), the factor $\pi(b_\phi/\gamma)^{1/2}$ is equal to 104 nm. The value of the spontaneous curvature, H'_0 which would make the second-factor unity is 0.03 nm^{-1}, which is the right order of magnitude for the spontaneous curvature. Further, the characteristic size only depends on $(H'_0)^{-1/2}$, a slow variation. For example, if we take the spontaneous curvature to be as large as that of PE, 0.316 nm^{-1} as I did above, we obtain 32 nm for the characteristic size. As the spontaneous curvatures of most lipids are between 0.03 nm^{-1} and 0.31 nm^{-1}, it is unlikely that the spontaneous curvature of the membrane, no matter how calculated, will change very much the estimate of the characteristic size from the order of 100 nm.

I have, elsewhere, extended the theory presented here so that the two leaves of the plasma membrane are treated independently [20]. I assumed that the membrane is of constant thickness which implies that the local curvatures of the two leaves are equal and opposite. This leads to a strong coupling between them that results in an anti-correlation between regions rich in SM in the outer leaf and regions rich in PE in the inner leaf. Thus, my view of a raft would be a region rich in SM and cholesterol in the outer leaf and a region rich in PS below it in the inner leaf floating in a sea of PC in the outer leaf and PE and cholesterol in the inner leaf. Applied to recent experiments on model membranes [23], one would expect liquid-ordered and liquid-disordered regions in the two leaves to be anti-correlated. The expectation is contrary to what is observed in the experiments on the model membranes [23]. This is not too surprising given that the strength of the coupling between leaves in model membranes is measured to be [29] an order of magnitude less than I predict to occur in the plasma membrane.

It would be of great interest to carry out experiments on model membranes of differing compositions in order to compare the characteristic lengths obtained as a function of those components; in particular, it would be of interest to increase the relative fraction of lipids with a large spontaneous curvature, such as PE. One would expect from the theory presented here that the characteristic size in the microemulsion would decrease. Such experiments would be very helpful in obtaining a definitive answer to the question of the origin of rafts. But at least one has in hand a consistent theory which provides reasonable answers to the questions of the origin of rafts and of their characteristic size.

Acknowlegment

It has been a pleasure to work with so many people on such an interesting topic. Among theorists, I particularly want to thank David Allender, David Andelman, Ha

Giang, Lutz Maibaum, and Roie Shlomovitz; among experimentalists, special thanks go to Caitlin Cornell and Sarah Keller.

References

[1] Simons, K., & Ikonen, E. 1997. Functional rafts in cell membranes. *Nature*. 387:569–572.
[2] Veatch, S.L., & Keller, S.L. 2005. Seeing spots: Complex phase behavior in simple membranes. *Biochim. Biophys. Acta*. 1746:172–185.
[3] Kusumi, A., Sako, Y., & Yamamoto, M. 1993. Confined lateral diffusion of membrane receptors as studied by single particle tracking (nanovid microscopy). effects of calcium-induced differentiation in cultured epithelial cells. *Biophys. J.* 65:2021–2040.
[4] Wang, T.Y., & Silvius, J.R. 2001. Cholesterol does not induce segregation of liquid-ordered domains in bilayers modeling the inner leaflet of the plasma membrane. *Biophys. J.* 81:2762–2773.
[5] Collins, M.D., & Keller, S.L. 2008. Tuning lipid mixtures to induce domains across leaflets of unsupported asymmetric bilayers. *PNAS*. 105:124–128.
[6] Veatch, S.L., Sengupta, P., Honerkamp-Smith, A., Holowka, D., & Baird, B. 2008. Critical fluctuations in plasma membrane vesicles. *ACS Chem. Bio*. 3:287–293.
[7] Gompper, G., & Schick, M. 1994. Self-assembling amphiphilic systems. Academic Press, San Diego.
[8] Schmid, F. 2017. Physical mechanisms of micro- and nanodomain formation in multicomponent lipid membranes. *BBA-Biomembr*. 1859:509–528.
[9] Leibler, S., & Andelman, D. 1987. Ordered and curved meso-structures in membranes and amphiphilic films. *J. Physique*. 48:2013–2018.
[10] Kollmitzer, B., Heftberger, P., Rappolt, M., & Pabst, G. 2013. Monolayer spontaneous curvature of raft-forming membrane lipids. *Soft Matter*. 9:10877–10884.
[11] Fisher, M., & Widom, B. 1969. Decay of correlations in linear systems. *J. Chem. Phys*. 50:3756–3772.
[12] Shlomovitz, R., Maibaum, L., & Schick, M. 2014. Macroscopic phase separation, modulated phases, and microemulsions: A unified picture of rafts. *Biophys. J.* 106:1979–1985.
[13] Toner, J., & Nelson, D. 1981. Smectic, cholesteric, and rayleigh-bernard order in two dimensions. *Phys. Rev. B*. 23:316–334.
[14] Toppozini, L., Meinhardt, S., Armstrong, C., Yamani, Z., Kuvcerka, N., Schmid, F., & Reinhardt, M. 2014. The structure of cholesterol in lipid rafts. *Phys. Rev. Lett*. 113:228101.
[15] Meinhardt, S., Vink, R., & Schmid, F. 2013. Monolayer curvature stabilizes nanoscale raft domains in mixed lipid bilayers. *PNAS*. 110:4476–4481.
[16] Evans, E. 1983. Bending elastic modulus of red blood cell membrane derived from buckling instability in micropipet aspiration tests. *Biophys. J.* 43:27–30.
[17] Lipowsky, R. 1992. Budding of membranes induced by intramembrane domains. *J. Phys. II*. 2:1825–1840.
[18] Dai, J., & Sheetz, M.P. 1999. Membrane tether formation from blebbing cells. *Biophys. J.* 77:3363–3370.
[19] Shlomovitz, R., & Schick, M. 2013. Model of a raft in both leaves of an asymmetric lipid bilayer. *Biophys. J.* 105:1406–1413.
[20] Schick, M. 2018. Strongly correlated rafts in both leaves of an asymmetric bilayer. *J. Phys. Chem. B*, 122:3251–3258, 2018.
[21] Pike, L. 2006. Rafts defined: A report on the keystone symposium on lipid rafts and cell function. *J. Lipid Res*. 47:1597–1598.

[22] Konyakhina, T., Goh, S., Amazon, J., Heberle, F., Wu, J., & Feigenson, G. 2011. Control of a nanscopic-to-macroscopic transition: Modulated phases in four-component dspc/dopc/popc. chol giant unilamellar vesicles. *Biophys. J.* 101:L08–L10.

[23] Cornell, C.E., Skinkle, A.D., He, S., Levental, I., Levental, K.R., & Keller, S.L. 2018. Tuning length scale of small features of cell-derived membranes and synthetic model membranes. *Biophys. J.* 115:690–701

[24] Kawakatsu, T., Andelman, D., Kawasaki, K., & Taniguchi, T. 1993. Phase transitions and shapes of two component membranes and vesicles i: Strong segregation limit. *J. Phys. II. France.* 3:971–997.

[25] Taniguchi, T., Kawasaki, K., Andelman, D., & Kawakatsu, T. 1994. Phase transitions and shapes of two component membranes and vesicles ii: Weak segregation limit. *J. Phys. II. France.* 4:1333–1362.

[26] Liu, J., Qi, S., Groves, J., & Chakraborty, A. 2005. Phase segregation on different length scales in a model cell membrane system. *J. Phys. Chem.* 109:19960–19969.

[27] Schick, M. 2012. Membrane heterogeneity: Manifestation of a curvature-induced microemulsion. *Phys. Rev. E.* 85:031902-1–031902-4.

[28] Sodt, A., Venable, R., Lyman, E., & Pastor, R. 2015. Lipid-lipid interactions determine the membrane spontaneous curvature. *Biophys. J.* 108:181a.

[29] Blosser, M., Honerkamp-Smith, A., Haataja, M., & Keller, S. 2015. Transbilayer colocalization of lipid domains explained via measurements of strong coupling parameters. *Biophys. J.* 109:2317–2327.

Jonathan D. Nickels, John Katsaras

18 Combining experiment and simulation to study complex biomimetic membranes

Abstract: The three-dimensional architecture of biological membranes has functional consequences for the living cell. In the outer leaflet of the plasma membrane, lipids are thought to self-organize into domains enriched in high-melting lipids and cholesterol. Currently, there is much evidence implicating these lipid domains in a variety of membrane processes, including protein sorting, cell signaling, and the maintenance of membrane physical properties. Cells also actively maintain an asymmetric distribution of different lipid types between their plasma membrane inner and outer leaflets, resulting in monolayers with different fluidities and charge densities. Moreover, it is an open question how these and other bilayer properties are coupled, if at all. A variety of biophysical techniques are being employed to interrogate both lateral and transverse bilayer structures with sub-nanometer resolution. Molecular dynamic simulations are also used to probe the membrane's finer details, providing a glimpse of membrane organization with atomic resolution. In this chapter, we present a discussion of these two approaches to investigate the organization of complex biomimetic membranes and present recent examples of how they have been used to complement each other.

Keywords: Lipid, Raft, Asymmetry, Molecular Dynamics, Coarse Grained, NMR, Fluorescence, Neutron Scattering, X-ray Scattering

18.1 The structure of biomembranes

Lipid membranes are a vital part of every organism, viruses being somewhat of an exception. Biological membranes control key life processes; e.g., defining the boundaries of cells and their cellular components, maintaining the crucial transmembrane potential, and regulating the exchange of small molecules. Because the plasma membrane is at the nexus of so many biological processes, it has increasingly become the subject of study in a number of scientific disciplines. With passing time and an increasing number of studies, layers of structural and dynamic complexity have been recognized in the plasma membranes of living cells [1].

Jonathan D. Nickels, Department of Chemical and Environmental Engineering, University of Cincinnati, Cincinnati, OH, USA
John Katsaras, Neutron Scattering Directorate, Oak Ridge National Laboratory, Oak Ridge, TN, USA; Shull Wollan Center, Oak Ridge National Laboratory, Oak Ridge, TN, USA

https://doi.org/10.1515/9783110544657-018

A key feature of biological membranes is their compositional asymmetry [2]. The chemical differences between the two apposing leaflets affects protein and ion binding and can result in differences in the physical properties of the two lipid leaflets [3–5]. Within 5 years [6–12] of the publication of the fluid mosaic model [13], it became known that the majority of the amino phospholipids resided on the membrane inner leaflet of red blood cells, with sphingomyelin and choline phospholipids primarily populating the outer leaflet [8, 10]. Membrane asymmetry is associated with a number of biological functions, including the signaling of apoptosis [14–17], thrombosis [11, 18], and phagocytosis [14, 19–21], and as an indicator of tumorigenic cells [22, 23]. Bilayer asymmetry is maintained by adenosine triphosphate (ATP)-dependent enzymes that shuttle lipids between the bilayer leaflets [24, 25]. Until quite recently, however, the generation of asymmetric model systems has not been practical [26–29].

Lateral organization of lipids within the plane of the membrane is the other feature of cell membrane complexity [30] that will be discussed in this chapter. Membrane proteins are thought to selectively partition in regions of different lipid composition. For example, the ordered regions, often called lipid rafts, control protein–protein interactions, enhancing certain associations while suppressing others [31, 32]. Rafts are thought to mediate a range of cellular processes [33], including cell adhesion and migration [34], cell recognition [35], protein sorting [32, 36], synaptic transmission [37], cytoskeletal organization [38], signal transduction [39], and apoptosis [40]. Recently, it has also been suggested that lipid rafts function as buffers of membrane physical properties [41]. Needless to say, lateral organization has been studied for many years [41] and raft structures have been studied extensively since the raft hypothesis was first postulated [42]. Rafts are now thought to be heterogeneous in size and shape [43], ranging between 10 and 100 nm in size [44], are highly dynamic [45], and composed of lipids, sterols, carbohydrates, and proteins [46]. Because of their physiological importance rafts have been studied extensively in animal cells [8, 47] and microbes [48–50], but are also thought to be important in bacterial and viral pathogenesis [49, 51, 52], and to the organization and function of plant membranes [53].

These complex structural features in biological membranes have been studied using a range of computational and experimental approaches. Here, we will discuss some of the ways that experiments and simulations have and will continue to complement each other, to better understand the cell membrane. Specifically, we will discuss important experimental methods that have been used to study lipid asymmetry and lipid rafts, with a focus on data that can be directly compared to molecular dynamics (MD) simulations. We will then discuss some of the simulation approaches which have been used for lipid systems and consider a few recent examples where simulations and experiments have provided us with new insights into biological membranes.

18.2 Experimental methods

New discoveries about the structure of biological membranes have been predominantly driven by experimental observation. Experimental studies of bilayer asymmetry started in the early 1970s [6, 8] and combined lipid degrading enzymes with freeze fracture electron microscopy to identify the different lipid compositions of the cell membrane leaflets. Other methods were also used to gain insight into lipid asymmetry such as radiochemical assays [6, 12], nuclear magnetic resonance (NMR) [12], specifically with lanthanide shift reagents that do not penetrate lipid membranes [54], electron spin resonance (ESR) [24, 55], atomic force microscopy (AFM) [56], and versions of fluorescence microscopies [57, 58] (e.g., fluorescence correlation spectroscopy (FCS) [59], fluorescence interference contrast microscopy (FLIC) [60], total internal reflection fluorescence microscopy [61, 62]). Biochemists, rather than structural biologists or biophysicists, have dominated the lipid asymmetry field [12] because it was quickly recognized that transverse lipid diffusion or flip-flop was a slow process. This implied, as was later shown to be the case, that specific enzymes are responsible for establishing and maintaining lipid asymmetry – accomplished via a series of ATP-dependent enzymes known as flippases, floppases, and scramblases [24, 27, 63–65].

Natural membrane systems are typically noncrystalline and rapidly lose asymmetry if the bilayer is removed from its native biological environment. Model systems derived from natural sources, such as purple membranes, show signs of asymmetric orientation of their liposaccharides [66] when the bilayer is kept in a crystalline state, a nonbiologically relevant condition. It is therefore advantageous to develop and fabricate model systems, which contain some degree of bilayer asymmetry if we want to study its effects in a controlled manner. Supported lipid bilayers [67–70] offer a path to asymmetry via the sequential deposition of lipid monolayers using the Langmuir–Blodgett/Langmuir–Schafer or the Langmuir–Blodgett/vesicle fusion method [71, 72]. Although supported bilayers present unique analytical advantages, unsupported bilayers, such as unilamellar lipid vesicles (ULVs), are generally thought to be better mimics of biological membranes. Microfluidic methods are also useful for the production of large, monodisperse asymmetric lipid vesicles [73, 74]. In addition, the water-in-organic-in-water double-emulsion method offers the chance to support each interface with a distinct lipid mixture and has been used in a number of recent studies [75–79]. Large ULVs can also be produced using electroswelling methods, and smaller ULVs can be prepared via sonication or extrusion through polycarbonate filters of defined size. ULVs are compositionally symmetric when prepared using these approaches, with the exception of curvature-induced asymmetry observed in some ULV lipid compositions [80]. A cyclodextrin-mediated lipid exchange method was developed by London and coworkers to generate bilayer asymmetry [3] in ULVs. Using this method, multilamellar vesicles composed of the desired outer leaflet composition are incubated with cyclodextrin and ULVs containing the desired inner leaflet composition. Specifically, methyl-β-cyclodextrin is used for pure lipid systems [3, 4, 81, 82] and hydroxypropyl-α-cyclodextrin-based lipid exchange is used when cholesterol is present in the acceptor ULVs [82–85].

Lateral lipid organization presents a different analytical challenge. Lipid rafts are generally thought to be small, short lived, and possess very little inherent contrast between the raft and surrounding "sea" of fluid lipids. Early detection approaches were based on detergent resistance [86, 87] and cyclodextrin depletion of cholesterol [88, 89]. In the detergent resistance approach, Triton-X 100, or some similar surfactant, is used to dissolve the cell membrane, leaving behind so-called detergent resistant membrane fragments. A great deal about lipid rafts was learned by determining the lipid and protein compositions of the detergent-resistant and detergent-soluble fractions by mass spectrometry and other biochemical assays [86]. In the case of cholesterol depletion studies, they largely used functional assays to induce changes in cellular behavior or activity [90] through the removal of cholesterol, presumably from lipid rafts. Despite the problems associated with both methods, they have proven vital in our understanding of lipid domains. Other methods which more directly probe the structure, composition, and other properties of lipids within the raft phase include mass spectrometry [91], NMR [92], various fluorescence spectroscopies [93, 94] and microscopies [95–100], AFM (Figure 18.1) [101–104], and x-ray [105] and neutron scattering [50, 106–108].

Lipid rafts are thought to be an important organizational motif in biological lipid membranes. Because of this, it is most useful to study these rafts in their native environment within living cells. Much evidence has emerged showing changes in cell function and organization of cellular proteins in living cells [38, 90, 98, 109] associated with lipid rafts, despite the fact that they were never directly observed. Recently, direct observations of lipid domains in living bacterial cells on the order of 40 nm in size were made using neutron scattering [50], providing perhaps the most direct evidence to date of their existence. It is important to remember, however, that rafts, or lipid domains, are heterogeneous both within a single cell and across many types of biological membranes; the sizes observed in *Bacillus subtilis* cell membranes almost certainly do not reflect a universal lipid raft size. Because of the difficulties in resolving lipid domains in vivo and the complex composition of biological membranes, researchers rely on model lipid mixtures and those extracted from cells to study the structural and dynamical properties of lipid rafts. Giant plasma membrane vesicles derived from living cells through chemical treatment show clear indications of large-scale liquid–liquid phase separation [110]. It has also been noted that detergent resistance was a useful way to isolate and identify raft lipids and their cargoes [111]. The introduction of synthetic model lipid mixtures that exhibit microscopic liquid–liquid phase coexistence [112–115] has facilitated detailed studies of the physical properties of lipid rafts [116].

Among all of the methods used for observing lateral separation, fluorescence-based approaches are possibly the most common. Although limited by the diffraction limit, fluorescence microscopy [97] is able to interrogate domain size, shape, and physical properties [100] based on the partitioning of dye molecules into the different

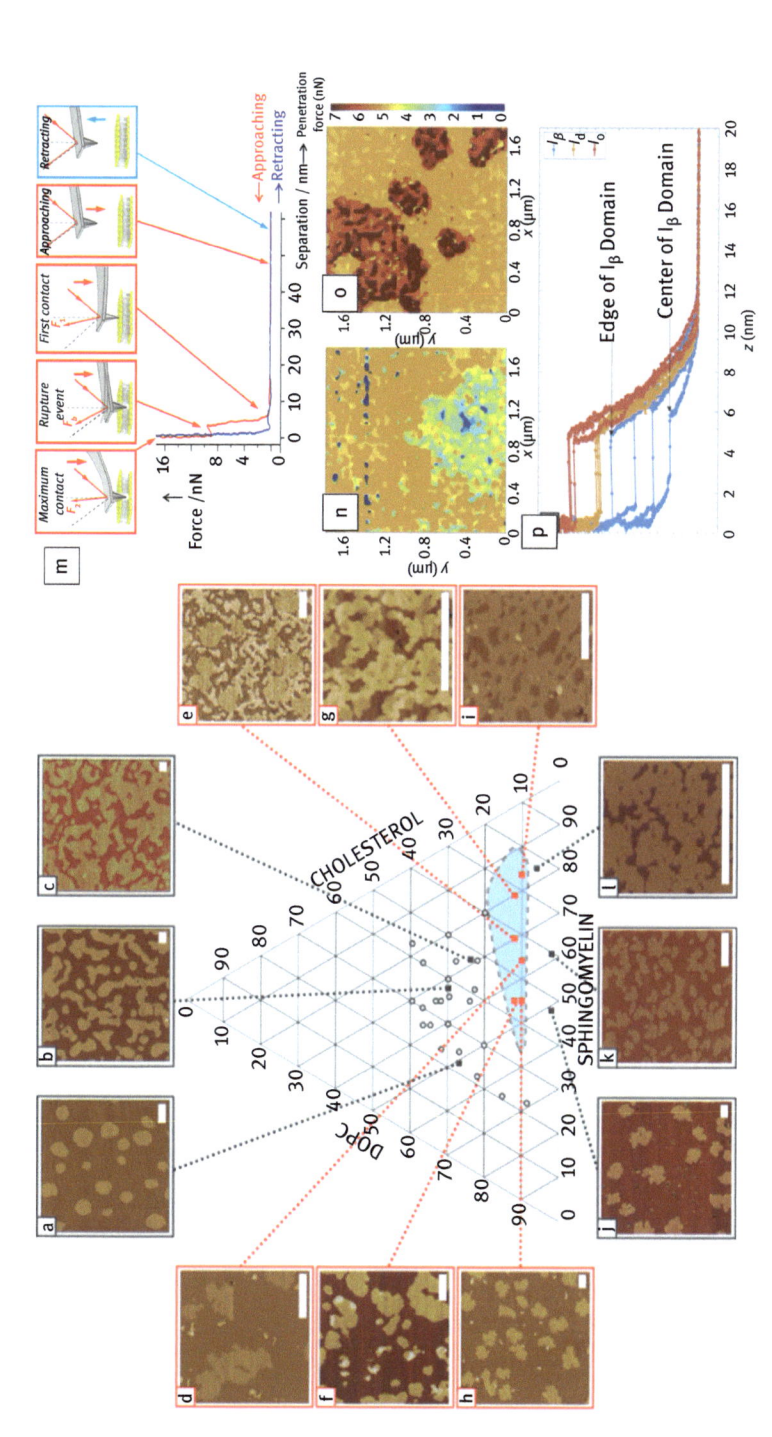

18 Combining experiment and simulation to study complex biomimetic membranes — 519

Figure 18.1: Another way to study lipid domains is to "feel" them. AFM is a scanning probe method that detects topological differences based on the deflection of a scanning probe, a technique used to study lipid bilayers since the early 2000s [101–104]. Here we show the example of a DOPC/Sphingomyelin/Cholesterol mixture, demonstrating topological changes at different points in the phase diagram (A-L) (from Aufderhorst-Roberts et al. [271]). (M) AFM can also be used to probe local mechanical properties using the force to break through the membrane (from Gumi-Audenis et al. [272]). (N-P) Bilayer penetration measurements by Aufderhorst-Roberts et al. [271] for a DOPC/Sphingomyelin/Cholesterol mixture, clearly showing systematic differences in the breakthrough force for the different lipid phases.

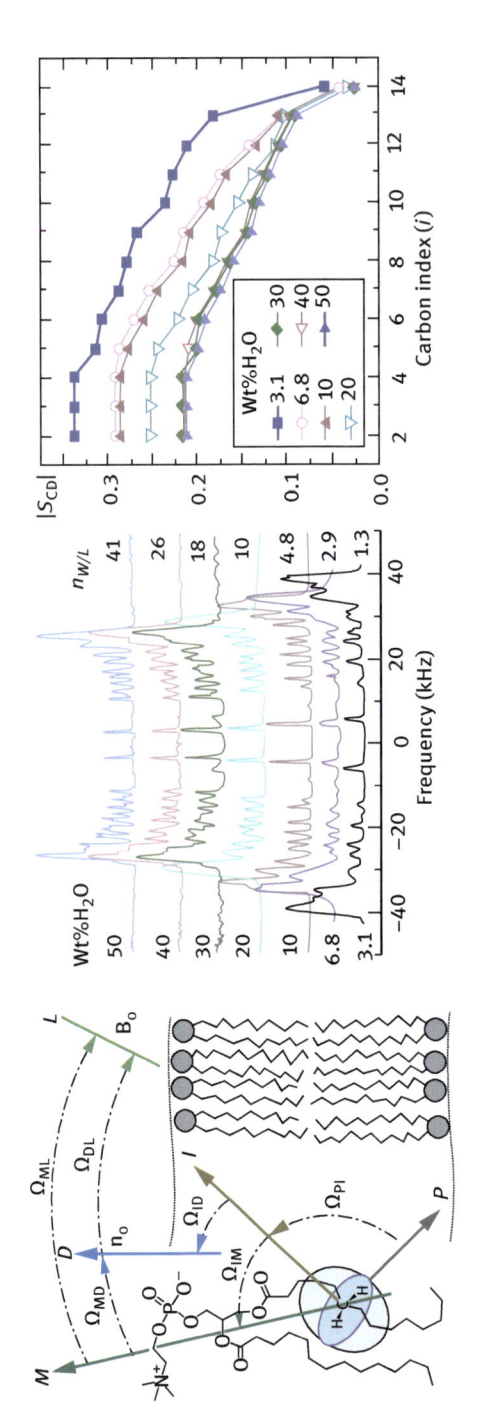

Figure 18.2: (left) The carbon–deuterium order parameter reflects the orientation of the carbon–deuterium bond in the methylene group of the lipid tail versus the bilayer normal. (middle, right) Solid state ^{2}H spectra and resulting order parameters, S_{CD}, as a function of hydration and carbon index for d54-DMPC at 30°C [273]. This segmental description of lipid structure can be directly computed from MD simulations, where the angle distribution of a given pair of atoms is easily obtained. Figures adapted from Kinnun et al. [273].

lipid phases [117], and through the diffusion and ligand binding of raft lipid analogs in model and cellular plasma membranes [118]. Fluorescence recovery after photo-bleaching [119] and FCS [93] are particularly useful approaches for studying lateral diffusion in bilayers [120] and distinguishing the liquid-ordered phase in raft forming lipid mixtures [121, 122]. These techniques work by photo-bleaching a defined membrane area and measuring the rate at which fluorophores diffuse back into the bleached spot. Confocal and two-photon fluorescence [109] microscopies have been used to directly image large domains, and Förster resonance energy transfer [123] has proven extremely useful for studying domain formation and phase behavior [124–126], as well as identifying raft associating molecules [127]. Similarly, other multiphoton methods [128], super-resolution microscopies [129], and single molecule [130] approaches have been developed to overcome the diffraction limit of optical techniques.

Vibrational spectroscopy is another method used to address lipid organization [131]. The vibrational frequencies of specific functional groups are sensitive to the local environment of lipids and proteins. Given sufficient signal-to-noise ratios, shifts in vibrational frequency and relative amplitudes can be descriptive of lipid order. The most widespread vibrational spectroscopy method, namely infrared spectroscopy, has been used in several studies to detect raft-like domains in model membrane systems [132, 133]. Over the years, Raman scattering has also been used to study lipid systems [134] with recent technical developments being applied to the study of asymmetric and laterally organized lipid bilayers. Tip-enhanced Raman spectroscopy is an attractive way to enhance the observable signal, especially when combined with isotopic substitutions [135]. Multiphoton Raman methods such as coherent anti-Stokes Raman spectroscopy [136, 137] have also been used to observe lateral demixing in the plane of model lipid membranes. In addition, the multiphoton Raman method may provide some advantages over standard Raman in terms of a lower excitation power, resulting in less potential damage to the sample. Nonlinear methods such as sum frequency vibrational spectroscopy have been used to study bilayer asymmetry [138] and lipid phase behavior [139]. This method is sensitive to spectral features associated with bilayer asymmetry, but the resulting flip-flop rates obtained by this technique appear to be faster than those from other methods [54]. This may be due to the presence of defects in the supported lipid bilayers needed for this method. For example, it has been known for several decades that pore formation dramatically increases the rate of "lipid flip-flop" [140]. In fact, lipid diffusion through such defects would be a competing mechanism, where one might confuse lateral diffusion with transmembrane flip-flop.

NMR and scattering methods (i.e., X-ray and neutron) are two techniques whose data are well suited for comparison with MD simulations because their time and length scales overlap with those currently accessible by simulations. NMR, specifically solid-state NMR, provides detailed information about lipid speciation, local structure, and molecular motions [141, 142]. Several nuclei [141, 143, 144] are commonly used for NMR in the study of lipids, namely ^{1}H, ^{2}H, ^{13}C, and ^{31}P. NMR is also well suited to the determination of bilayer asymmetry through the use of membrane insoluble paramagnetic compounds [145, 146]. For example, if a paramagnetic shift reagent is added to the bulk solution outside of the vesicle, such as a lanthanide ion, there will be a detectable shift in the NMR signal of outer bilayer leaflet [145–147]. This approach is also used to track asymmetric membrane protein insertion [147] and has recently been used to track lipid flip-flop in asymmetric vesicles [54].

For the case of lateral organization, ^{2}H NMR has played an important role in quantifying the compositional changes within phase separating model lipid mixtures [92]. It should be noted, however, that the authors were not comfortable in ascribing this result to strict phase coexistence, citing the possibility of critical fluctuations within a single phase; though subsequent observations of nanoscopic heterogeneities in similar model lipid compositions by scattering methods indicate that there is a population with ensemble properties of nanoscopic heterogeneities consistent with the presence of distinct lipid phases. Carbon–deuterium (C–D) order parameters are a tremendously useful reporter of structural order in lipid acyl chains that reside in the bilayer's hydrophobic region. Moreover, because an order parameter reflects the angle between the C–D bond and the bilayer normal, it is an obvious measure that connects NMR experimental results with MD simulations (Figure 18.3). Similar data can be obtained by fluorescence depolarization [148], however, the need of a fluorescent probe does introduce a potentially perturbing element into the system; NMR has the benefit of using an isotopic label. Accessing the T1 relaxation time reveals similar information about lipid

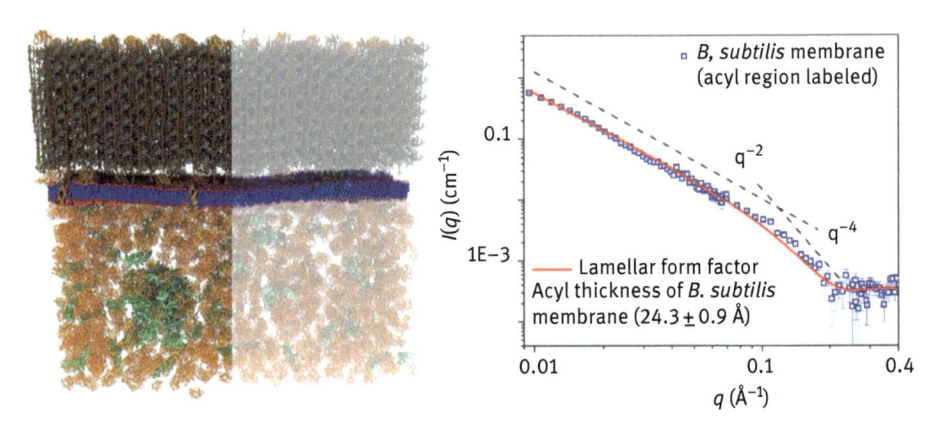

Figure 18.3: Small-angle neutron scattering is an important technique for the study of biomembranes. Neutrons scatter strongly from hydrogen nuclei that are abundant in lipids and scatter differently for the isotopes of hydrogen, ^{1}H and ^{2}H. This is the basis of what is known as the contrast-matching technique, where substitution of the hydrogen isotopes can be used to create null-scattering conditions that allow for the observation of one component within a complex, multicomponent system. Recently, this has been used to observe the structure of the cell membrane in a living *Bacillus subtilis* cell [50]. This figure, from Nickels et al., depicts (left) the contrast enhancement achieved using specific deuteration and (right) the resulting scattering observation of a lamellar form factor.

ordering [149]. Heteronuclear or cross-polarization methods, where polarization is transferred between two atoms, is another strategy for obtaining information about the ordering of lipid acyl chains [150]. The use of magic angle spinning conditions should also be mentioned as a way of dramatically enhancing the signal of the system, avoiding the need for oriented samples [151]. MD simulation and NMR are also well matched for the description of local motions. The pulsed field gradient method has been used to measure the lateral diffusion rate of lipids and cholesterol in the plane of the bilayer [152, 153] and has been applied to study the dynamical properties of lipids in model raft forming mixtures [154, 155]. In MD simulations, the diffusion rates are directly determined from the atomic displacements in their trajectories.

Neutron and X-ray scattering are powerful structural biology techniques that have been instrumental in the determination of lipid membrane structure. Structurally, they probe important molecular length scales (angstrom to hundreds of nanometers), and neutrons are sensitive to molecular motions from hundreds of nanoseconds to fractions of a picosecond. The coherent elastic scattering of both X-rays and neutrons reflect the pair distribution function of atoms in the sample. Such pair distribution functions reflect the molecular structure and can be directly computed from MD simulations. Indeed, the lamellar structure of certain lipid phases [156] and the liquid-like conformations of the lipid tails in the bilayer hydrophobic region [156–158] have been observed by X-ray diffraction. Moreover, X-ray diffraction has played a critical role in quantifying the linear relationship between acyl chain length and bilayer thickness in lipid bilayers [159], and remains a critical structural technique for biomembranes [160]. The technique is also sensitive to the lateral organization of lipids [161], even for complex lipid mixtures [162]. Through these observations, and the presence of multiple d-spacings at lower scattering wave vectors, phase coexistence can be inferred from X-ray scattering observations [163].

Neutron scattering is another diffraction technique, similar to X-ray scattering, in its ability to probe length scales from hundreds of nanometers to angstroms. However, there are some important differences between the two probes. First, neutrons scatter from atomic nuclei whose scattering power is denoted by the nuclear scattering length, b; X-rays scatter from the electrons surrounding the nucleus. Second, because cold and thermal neutrons have energies on the order of meVs, the observed changes in momentum (differences in the velocity of the scattered neutron) can be detected and directly related to the collective and self-motions of the atoms in the sample [164]. Note, that the inelastic scattering of X-rays is also used to probe atomic motions, but at very high frequencies and longer wave vectors than neutrons, making inelastic scattering of X-rays less useful for the study of lipids.

The neutron scattering length, b, does not vary with atomic number, as does the electron density in the case of X-rays, but varies widely between elements and even between isotopes of the same element [165]. This has two important implications with regard to the study of lipids and complex lipid structures. First, this means that low atomic number elements, such as hydrogen, can scatter equally well as high atomic number elements, such as lead. As lipids and water are rich in hydrogen, neutrons have proven a vital tool in studying the structural details of biomembranes [166–168], specifically the location of water around biointerfaces [169–174], including bilayer surfaces [175].

Isotopic substitution, specifically the isotopes of hydrogen, (^{1}H) and deuterium (^{2}H), is used to systematically vary the scattering contrast of biological systems [176]. This so called contrast-matching approach has proven very useful for the study of lateral and transverse lipid organization. Contrast-matching refers to the generation of a null scattering condition between structures within the same sample that might ordinarily scatter when unlabeled. This makes neutrons uniquely useful for the study of complex biomembrane systems, where one can introduce deuterated lipids into a specific lipid phase to match the solvent [107] or the theoretical match point of the bilayer when perfectly mixed [50, 106, 108]. This approach has been used to study proteins [176], lipids [177], nucleic acids [178], polymers [179], and other materials. This matching approach is precisely the strategy recently used to directly observe lateral demixing in the cell membrane of a living bacterium, confirming the presence of nanoscopic heterogeneities and consistent with the concept of lipid rafts [50].

The inelastic scattering of neutrons provides additional information in the time domain about the motions of lipid molecules, ranging from bilayer collective undulations [180, 181] and thickness fluctuations [182], to the lateral diffusion of lipids [183], and the local motions of individual lipids [174, 183, 184], even into the vibrational regime [185]. These motions are correlated with important physical properties of lipid membranes, such as the bending modulus [181]. Recent work has shown that the modulus of nanoscopic lipid domains is distinct from the "sea" of lipids surrounding them [107]. This experiment also showed that phase separation can be influenced by the local distribution of elastic moduli. These dynamical features are directly connected to the time-dependent self and pair correlation functions that are accessible by MD simulations. By defining these correlation functions for specific modes, one can extract dynamical terms that are analogous to the ones observed by experiment.

18.3 MD simulation

Simulations are an essential tool for the investigation of complex lipid membranes. MD methods provide access to molecular scale information, complementing experimental results and enhancing our molecular understanding. Visualization software for simulation results, such as VMD [186], have become critical tools for communicating not only MD simulations, but also for the molecular implications of experimental studies. The earliest MD lipid simulations used highly simplified models (water/fatty acid/alcohol) [187] and hydrated DLPE (1,2-didodecanoyl-sn-glycero-3-phosphoethanolamine) membranes [188], with a focus on understanding the structure of water near the membrane surface. Later simulations extending to hundreds of picoseconds allowed for the study of lipid structure and dynamics [189–192]. A number of excellent review articles [193–205] have chronicled the progress of MD simulations for lipid membranes over the last 30 years. All-atom simulations on the order of 1 μs are now somewhat common, enabling one to ask questions about complex lipid structures that better mimic biological membranes.

Recent advances in efficiency and computing power have enabled longer and larger simulations. Overcoming the computational cost of a simulation is a prime consideration when it comes to evaluating the utility of MD approaches for a given scientific problem. Increases in simulation size (number of particles, N), time step, and duration increase accuracy at the cost of computational complexity. This may or may not be acceptable. Take for example simulation size (i.e., number of particles, N). The size must be sufficient to reproduce the phenomena of interest, keeping in mind, however, that the computational cost scales as N^2 if one explicitly evaluates the interparticle interactions across the entire system. Although methods such as Ewald summation [206] and the imposition of reasonable cut-off distances can reduce these costs to $N \log N$, there is a very real computational cost when larger and more realistic conditions are simulated.

Even though many strategies have been developed to mitigate the above-mentioned scaling effects, they must still be weighed carefully when considering all-atom versus coarse-grained (CG) approaches (Figure 18.4). CG approaches are useful

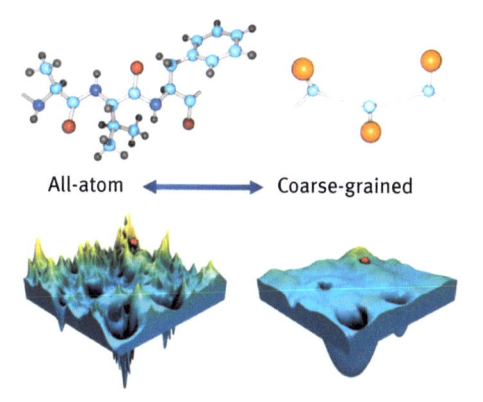

Figure 18.4: All-atom models seek to reproduce molecular structure and dynamics, in full detail, based on the precise chemical structure at the expense of sample size, simulation time, and computational resources. On the other hand, CG simulations capture the majority of a system's important features over longer simulation times, sample sizes, and at reduced computational costs, at the expense of full atomic detail. Image from Kmieck et al. [207].

because they reduce the number of particles in a simulation by grouping several atoms into pseudoparticles. One can readily appreciate how this reduces the computational cost and increases accessible time and length scales. Different force fields are needed for these pseudoparticles, and extensive development is required. They are, however, an excellent approach to scale a simulation down to a computationally manageable size. Sometimes coarse graining is used in combination with atomistic simulations in what is sometimes called multiscale modeling, an approach commonly used in protein simulations [207]. It is therefore safe to say that currently there are a number of good options for constructing simulations that complement experiment from the atomic to the microscale.

There are several "varieties" of simulation used to study lipid membranes and their complexity. Here, we will focus on all-atom and CG MD approaches. To place these approaches in context, one might first describe Monte Carlo (MC) simulations, which preceded the above-mentioned MD approaches. Although MC methods do not generally give information about the dynamics of a system they remain extremely useful to this day. MC simulations, named by Nicholas Metropolis [208] in dubious honor of Stanislaw Ulam's uncle, refer to a group of methods using sampling from a defined probability distribution and assigning an outcome from that selection, say a move within a conformation space. Based on that outcome, another step is taken and so on, until a sufficient number of steps are taken for an "accurate" average value to be obtained. This method was pioneered at Los Alamos National Laboratory in the late 1940s [209, 210] by Ulam, Metropolis, John von Neuman, and co-workers. Naming this approach was not Metropolis' only contribution; the Metropolis–Hastings algorithm [211] is still widely used today. When one introduces a probability distribution for the underlying variables, and there is no memory effect (i.e., the current and future state are not influenced by a prior state), these systems can be considered as using the Markov approach.

18.4 All-atom simulation approaches

All-atom simulations evolved rapidly from the early MC simulations. By the mid-1950s, the classic study by Fermi, Pasta, and Ulam describing simulations of elastic collisions between spheres [212] had been published. Less than ten years later, the field had progressed to simulations of liquid argon using the Lennard-Jones (LJ) potential in the classic work of Rahman [213]. Simulations of lipid systems did not appear until the late 1980s [187], as was discussed earlier. Recently, there are examples of membrane lateral organization studies using all-atom simulations [107]. However, the domains are preformed because the time needed for lipid domain formation is beyond the capability of current simulation approaches. Similarly, asymmetric bilayers are also the subject of atomistic MD simulations [214] and are beginning to interrogate how transverse and lateral features are coupled [215].

MD simulations proceed by evaluating Newton's equation of motion for each atom (or pseudoparticle) within the simulation box and updating its position and velocity at each timestep, which is typically on the order of 1–2 fs. Initial coordinates for the simulations must be obtained or generated to begin the simulation. It is often challenging to assemble complex, multicomponent membrane models, a task which requires specialized software. One such tool is the CHARMM-GUI [216]. It is a flexible platform that is commonly used for lipid bilayer assembly and which supports a number of different force fields. Another assembly software is the so-called INSANE CG building tool [217] maintained by the Martini team. Once the initial assembly is complete, models then typically undergo rapid energy minimization using software such as Tinker [218] (tinkertools.org), followed by some form of simulated annealing.

Once the system is set up, the performance of the simulations is achieved using a program which can efficiently compute the Newtonian equations of motion for systems with hundreds to millions of particles. There are many software capable of this and we will mention a few commonly used programs, but this list is by no means exhaustive. For example, the Chemistry at HARvard Macromolecular Mechanics (CHARMM) [219] program is targeted at biological systems such as peptides and lipids. This software was developed initially by the group of Nobel laureate Martin Karplus and is currently maintained by an active community. Features include a comprehensive set of energy functions, a variety of enhanced sampling methods, support for multiscale techniques, including QM/MM and MM/CG, and a range of implicit solvent models. GROMACS [220], or GROningen MAchine for Chemical Simulations, is a popular, free and open-source package able to run on both CPUs and GPUs. It was initially developed by the Berendensen Group at the University of Groningen and is now led out of the Science for Life Library in Stockholm. NAMD [221], also known as "Nanoscale MD" or "Not Another MD program", was developed at the University of Illinois at Urbana-Champaign by the Theoretical and Computational Biophysics Group in the Beckman Institute for Advanced Science and Technology. It is noted for its efficiency in executing the simulation in parallel over hundreds to thousands of cores, making it a popular program for large systems containing millions of atoms.

All of these programs require information about atom–atom interactions, which is provided by the force field. A few popular force fields used for lipid simulations include CHARMM, AMBER, and GROMOS, although there are numerous others that have been used in the literature to simulate lipid bilayers. As can probably be gathered from the names of the selected force fields below, they were developed in parallel with the simulation program by a number of user communities. These programs have, for the most part, now matured to the point that force fields are compatible across simulation platforms, a feature which is convenient for new users. AMBER [222, 223], or Assisted Model Building with Energy Refinement, is a group of force fields developed at the University of California at San Francisco. The standard AMBER force field results showed some deviations from experimental results and were subsequently modified to be used with lipid membrane systems [224, 225]. This refinement, however, is not unique to AMBER. The Chemistry at HARvard Macromolecular Mechanics (CHARMM) force field family includes CHARMM for lipids [226], a version optimized for lipid bilayers by changing a number of interaction parameters to better reproduce observed lipid features [227, 228]. More generic force fields, such as OPLS are also available, though less frequently applied to lipid systems [229]. We will not discuss the use of polarizable force fields [230, 231] or those that explicitly account for quantum effects [232], although such considerations can be included at the expense of computational complexity.

A more exhaustive description of the process of MD simulations can be found in the book by Frenkel and Smit [233], or in shorter form, in the book chapter by Allen [234]. Additional details that should be considered are: what kind of interaction constraints are being used (SHAKE [235], LINCS [220], etc.) and the implementation of interaction cut-off distances, which greatly improve computational efficiency at the expense of truncating long-range interactions. Another vital component of lipid simulations is water [236]. Water surrounds biological membranes and is typically crowded with proteins, ions, and other solutes. Explicit hydration is computationally expensive, but more accurately describes a given system. Water force fields are available with increasing levels of complexity. Three site models (TIP3P [237] or SPC/E [238]) are most commonly used, although four (TIP4P [237] and other), five, and other interaction site models are also available. Polarizable options for the water force field [230, 239] exist and are more likely to capture the anomalous physical properties of water at the cost of computational time.

18.5 CG MD simulation

Currently, the time and length scales accessible to all-atom lipid bilayer simulations are limited to a few thousand lipid molecules for a duration on the order of a microsecond. Although a great many studies can be successfully carried out with our current technical simulation capabilities, larger and longer simulations are better able to capture long wavelength phenomena and to bridge molecular motions to the time and length scales that parallel most experiments.

While progress in processor capacity and algorithm efficiencies are helping solve problems using the brute force approach, an alternative route is the simplification of the system using CG simulations. In the case of CG simulations the investigator trades atomic-level detail for a larger or longer simulation in which the atoms are combined into pseudoparticles, or coarse grains. This has the effect of reducing the number of calculations needed, enabling more molecules to be included with fewer calculations per timestep. Because of the simplifications, however, it is important to understand the type of information one wants to obtain, the necessary time and length scales to obtain it, and the degree to which a chosen CG model changes the outcome of the simulation when considering to use CG methods. It should be noted that there is some potential to "recover" some level of atomic detail using reverse transform tools that map all-atom structure back onto CG models [240].

Numerous strategies have been developed for mapping the all-atom structure to a configuration of pseudoparticles, each of which must be accompanied by a specific force field. Early CG simulations of lipid systems [241] were extremely coarse, resembling what we today refer to as a 3-bead model (Figure 18.5) [242]. However, various groups have optimized the united atom force field parameters in recent years [243, 244]. Currently, the most popular CG approach uses the Martini force field [245] that was developed by Marrink and co-workers. Since its initial development – simply described as a CG model for semiquantitative lipid simulations [246] – it has been expanded to include other molecule types. Martini is implemented in many of the most popular simulation programs, including GROMACS and NAMD.

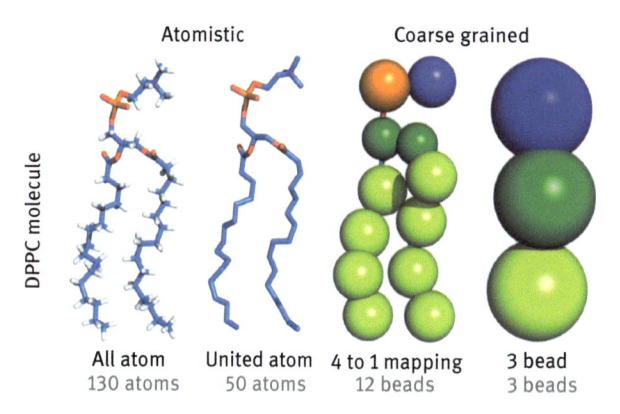

Figure 18.5: CG simulations reduce the total number of particles. Whereas all-atom simulations include individual particles representing individual atoms, various CG simulations adopt different strategies to decrease the number of particles in a simulation. For example, in the united atom approach, hydrogen is implied, while a 4:1 mapping is used in the Martini model and three beads in the Cooke model [242]. Image modified from the review of Pluhackova and Böckmann [274].

As mentioned, CG simulations are currently being used to study phase behavior [247] and lateral phase separation in lipid membranes [248–250]. The figure from Risselada and Marrink [249] shows an example of how large-scale simulations that include raft-like structures are now within the reach of MD simulations, including model systems like DPPC:Chol (Figure 18.6) [251] and biomimetic systems replicating the plasma membrane composition (Figure 18.7) [252]. CG simulations have also been used to understand how lateral organization relates to lipid asymmetry [253], where it was shown that membrane curvature can result in domain anti-registration.

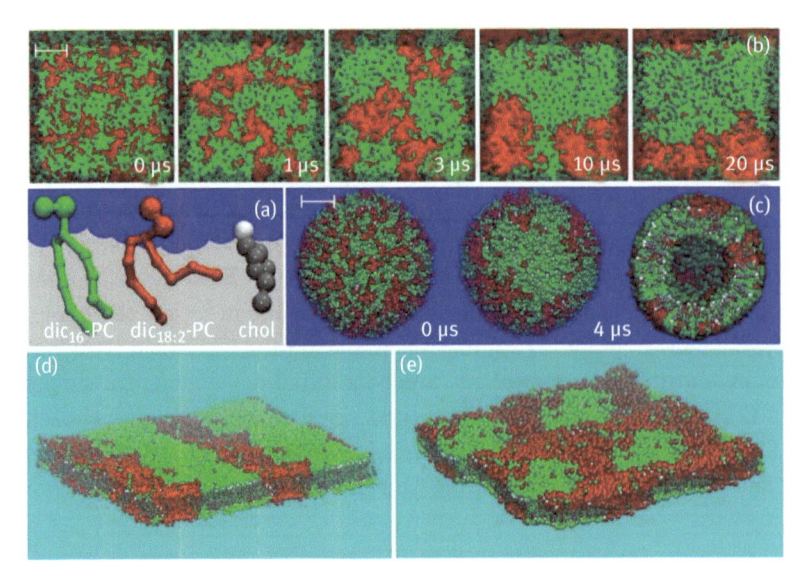

Figure 18.6: The Martini [245] force field is currently the most popular CG simulation approach for lipid membrane simulations. There are now many examples using it to simulate large-scale systems, such as those which exhibit lipid phase separation. Figure from Risselada and Marrink [249].

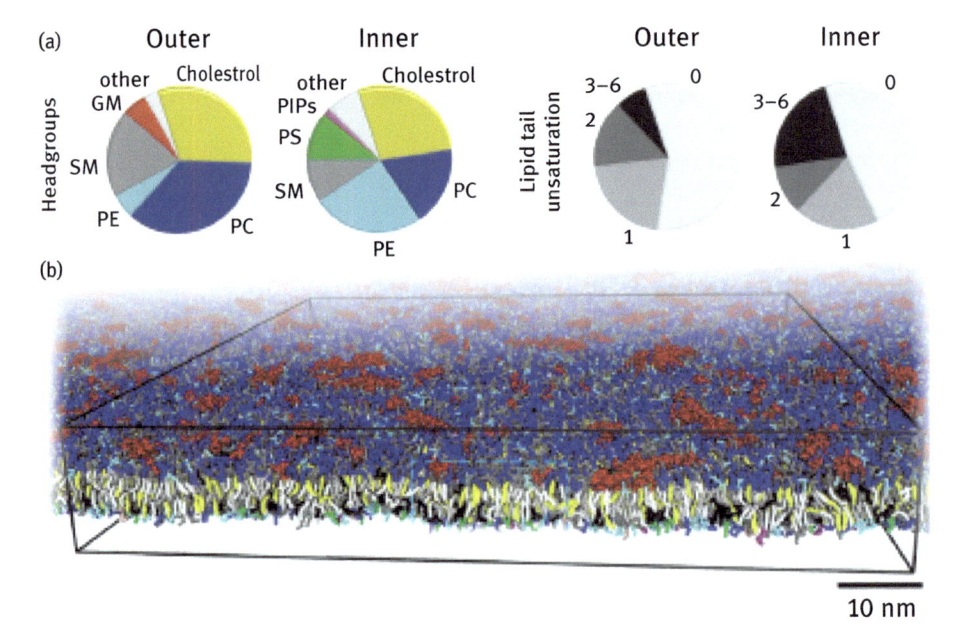

Figure 18.7: Large-scale simulations aim to reproduce the complex structures of the mammalian plasma membrane, including lateral organization and bilayer asymmetry. Figure used with permission from Ingolfsson et al. [252].

18.6 Integrating experiment and simulation

It should be pointed out that MD simulations are only as good as the force fields used. Because of this, the first and most obvious way that experiment and simulations were combined was in the optimization of the force fields against experimental results [254]. White and co-workers proposed using neutron and X-ray scattering [255] data for the validation of simulations using both the all-atom CHARMM22/27 force field and the united-atom GROMACS approach. Scattering continues to provide quality information for lipid force field optimization, driving studies such as Kucerka's and Katsaras's work [256, 257]. Similarly, orientational order parameters were used to check changes in the LJ potentials in the simulation force field [258]. NMR order parameters were also used to tune other parameters such as the LJ hydrocarbon and torsional parameters, and the partial atomic charges and torsional parameters of the phosphate moiety [227]. These new data motivated a full update of the CHARMM force field for lipids, which was validated using both scattering and NMR observations [226]. The united-atom force field has also been updated using CD order parameters from NMR [259].

These, and other efforts, have resulted in a number of well-parameterized force fields, which are now producing increasingly accurate results and enabling one to address scientific questions that cannot be answered by experiment alone [195]. An example of NMR observations motivating subsequent MD simulations can be found in the 2002 account of Pastor et al. [260] regarding simulations of the molecular motions of DPPC. There, they describe how experimental observations of the ^{13}C NMR T_1 relaxation [149] motivated a series of simulations that demonstrated the concept that after ~100 ps, lipids adopt a cylindrical shape, but that they "wobble" into a cone-like configuration on the nanosecond timescale.

Another way that simulations are being deployed in concert with experimental results is to refine the structures obtained from intrinsically lower resolution techniques, such as SAXS or SANS. Typically, one determines the scattering density profile from these methods, and these profiles are related to the pair distribution function for atoms in a lipid bilayer; as weighted by their respective contrast terms. Lipid bilayers, even simple ones, have a rich atomic scale structure that is averaged out by these methods. Yet, we can compute the pair distribution function from MD simulation trajectories rather trivially. Even the structure of the seemingly simple, fully hydrated liquid crystalline phase of dimyristoylphosphatidylcholine (DMPC) has benefited from advanced simulations. By creating a series of simulation boxes at different area per lipid, the simulated electron density profile can be computed and compared, yielding the most optimal structure [261]. The benefit of this becomes clearer when looking at complex bilayer compositions, such as those from natural extracts [262] or canonical phase separating model lipid mixtures (Figure 18.8) [107, 108]. In the case of lipid extracts, where the compositional space is large, simulations are well suited at capturing the system's complexity by computing the total neutron scattering

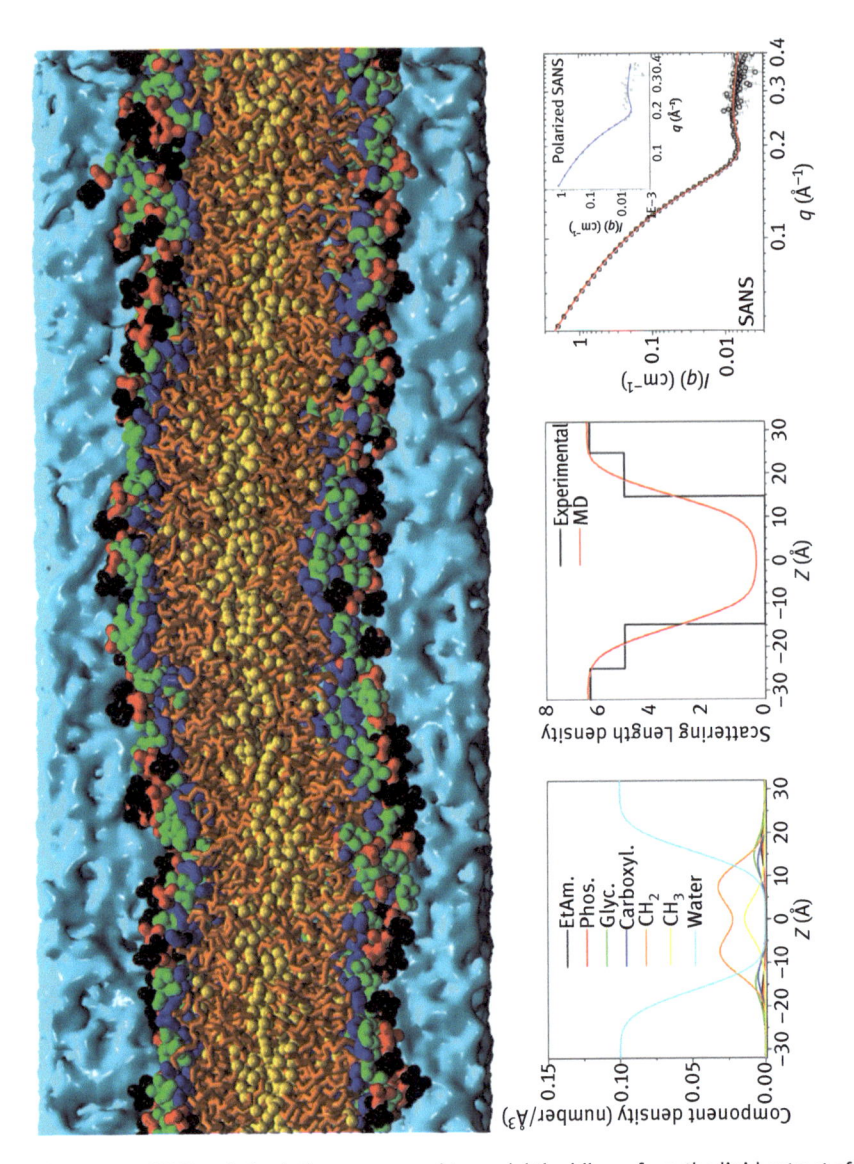

Figure 18.8: SANS and simulations were used to model the bilayer from the lipid extract of Bacillus subtilis [262]. When compositional space is large, such as that from natural extracts which contain many lipid species, simulations are a great tool to interpret experimentally observed scattering length density profiles. By comparing to simulation, one can extract meaningful average parameters that are shared across the many different lipid species, like the membrane hydrophobic thickness, area per lipid, or average number of water molecules in the head group region. Figure adapted from Nickels et al. [262].

length density (SLD) profile and connecting it to meaningful average bilayer properties, such as area per lipid or water molecules per lipid.

Simulations have been used to extract meaningful information from experiments studying lipid compositions that phase separate. One example is POPC (1-palmitoyl-2-oleoyl-glycero-3-phosphocholine) / DSPC (1,2-distearoyl-sn-glycero-3-phosphocholine) / Cholesterol in a 22:39:39 ratio [107]. This composition phase separates into coexisting Ld and Lo phases and has been studied in detail. Based on a published ternary phase diagram [108, 126], it is possible to tune the neutron SLD of each phase through judicious hydrogen/deuterium substitution of the constituent lipids. These phases can then be matched to the solvent SLD by varying the D_2O/H_2O ratio, rendering one lipid phase effectively "invisible" to neutrons. This enables neutron techniques to isolate the scattering signal from only the noncontrast-matched phase. In doing so, one can determine the size and composition of the nanoscopic domains but also enables subsequent experiments to determine the bending modulus of the two phases. Ultimately, we hope that all of this information, when combined, will lead to insights into the mechanism that is responsible for nanoscopic phase separation.

Using such a phase separating lipid composition, Nickels et al. performed atomistic MD simulations that brought to light molecular arrangements, such as the alignment of the lipids relative to the domain boundary, which could not be easily obtained by experiment. For example, Figure 18.9 demonstrates how it is possible to evaluate the orientation of the *sn*-2 position fatty acid chain in a lipid with respect to the Lo/Ld interface for a given frame of the simulation trajectory. This is repeated many times for sufficient statistical sampling, and the orientational preferences can thus be probed. This result demonstrates that the lipid head group points toward the Ld/Lo interface and may be affecting line tension. In the case of POPC, the saturated chain (*sn*-1 palmitoyl in this case) is hypothesized to associate with the Lo phase, while the unsaturated chain (*sn*-2 oleoyl in this case) prefers the Ld phase. In this way, the acyl chains can pack more

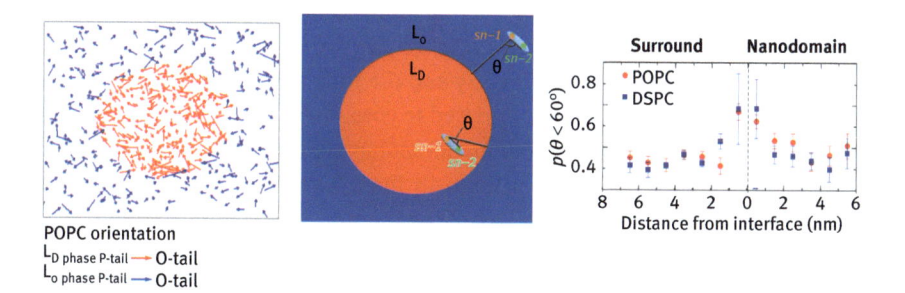

Figure 18.9: POPC/DSPC/Chol. forms nanoscopic domains in 60 nm diameter unilamellar vesicles. Simulations inspired by these structural measurements can provide a level of structural detail nominally inaccessible by experiment [107]. One such molecular detail is how lipids orient the *sn*-2 chain preferentially toward the domain interface to within 1 nm of the boundary.

favorably, thus reducing the interfacial energy and stabilizing the domain interface. However, both the fully saturated acyl chain DSPC and the mixed chain POPC molecules – considered as a potential linactant molecules – show this orientational preference.

Order parameters can also be computed to better understand the nanodomain interface at the molecular scale. An order parameter describes the state of the system, such as the CD order parameter that we mentioned in a previous section. In that case, the angle of the CD bond (generically the CH bond) is considered relative to the bilayer normal and is used to describe the packing of the lipid hydrocarbon chains – this parameter is commonly used in many combined experimental and simulation studies because it is precisely measured by NMR [263]. However, order parameters can be determined for many relevant physical properties and typically have values between one and zero. Often they are used to distinguish between two phases, but in simulations they can also be used to map molecular scale deviations from the average properties of a system. In the POPC/DSPC/Chol. system, interesting trends emerge in the order parameters near the domain interface. Order parameters describing the lipid splay, lipid tilt, and average CD bond angle were computed and are shown in Figure 18.10 as a function of distance from the Lo/Ld interface. There are clear minima near the interface for all parameters. This disordering is understood to be a compensatory overshoot, as predicted by numerous theories. The minima in the tilt and splay parameters, along with the thinning of the bilayer at the interface, point to local bilayer flexibility.

Mechanical properties, more precisely, those describing membrane flexibility are of great interest and simulations provide us with a number of ways to address them [264]. Evaluation of the tensile force in cylindrical bilayer systems is one convenient way [265]. Another brute force approach is to run simulations of sufficient size and duration that capture some of the low-frequency bilayer undulations influenced by the bending modulus [266]. This is somewhat analogous to the use of coherent neutron scattering to observe the thermal fluctuations of a bilayer using the neutron spin echo (NSE) technique. In that case, the decay of the intermediate scattering function can be related to the undulational dynamics of the bilayer bending modulus. The model of Zilman and Granek treats the bilayer as a flat sheet, a condition considered valid for systems with $qR \gg 1$. Alternatively, lipid splay and tilt are intimately connected to the bending of the bilayer [267] and this can be used to compute an approximate bending modulus from an MD simulation [268]. Simulations have the benefit of being able to compute these motions for regions of interest in the bilayer, such as a single lipid domain. Experimentally, determining the bending moduli of individual lipid phases, especially within coexisting systems and for nanoscopic phases, is extremely challenging. For the case of micron scale domains, individual phase moduli can be determined by fluorescent microscopy techniques (Figure 18.11) [269], but for nanoscopic systems, NSE is currently the only method capable of doing so [107]. Using the POPC/DSPC/Chol. system, it was recently shown that nanoscopic lipid domains exhibit distinct mechanical properties from the surrounding lipid phase, but do not exhibit emergent

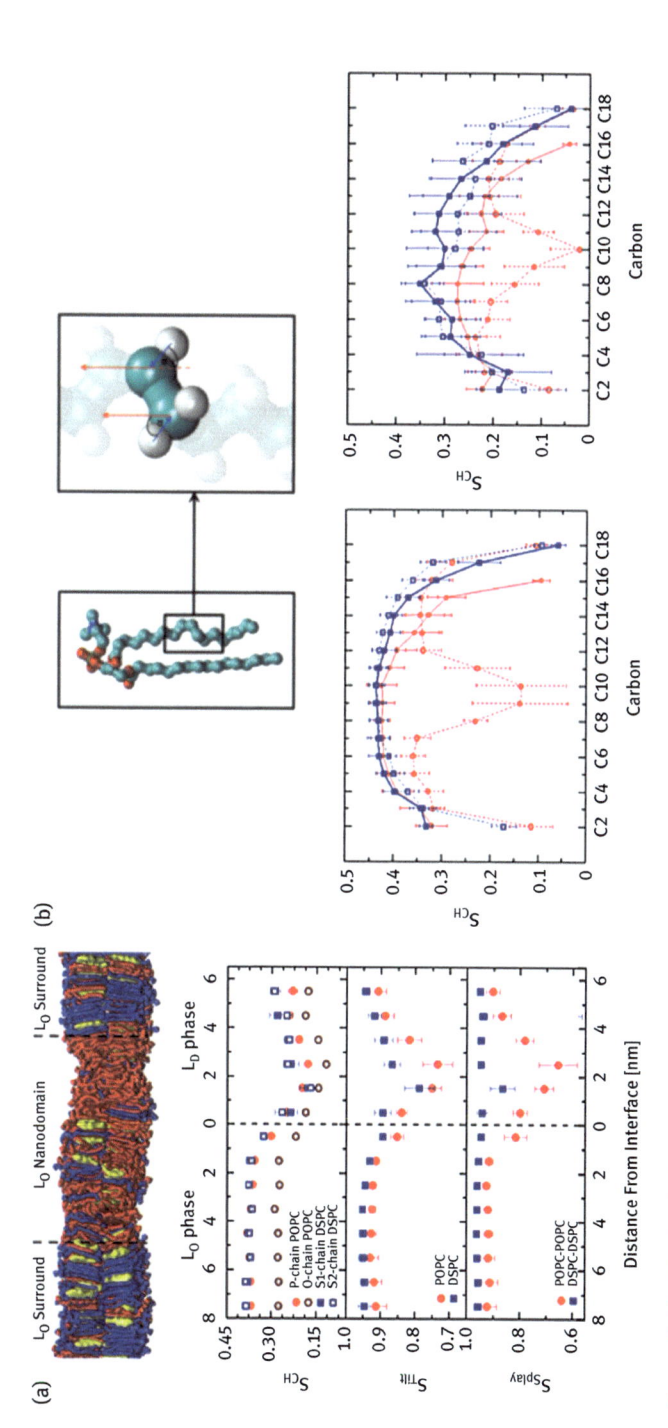

Figure 18.10: MD simulations reveal many additional details of the domain interface. (a) Cross-section of simulated Ld/Lo boundary after 150 ns of simulation (DSPC, blue; POPC, red; and cholesterol, yellow). A series of order parameters can be computed, such as SCH, reflecting the angle between the CH bond and the bilayer normal, STilt describing the average tilt angle of a lipid relative to the bilayer normal, and SSplay reflecting the divergence of local lipid tilt. These parameters describe the disorder associated with the interfacial region. (b) Representation of how the carbon-hydrogen (or carbon-deuterium) order, SCH, parameter is computed. It is based on the angle α between the bond and the bilayer normal. This can be computed for each carbon atom in a lipid, and show a dependence illustrating that the chains are less ordered at the ends and in the region of unsaturations (solid red circle – P-chain of POPC, open red circle – O-chain of POPC, solid blue square – sn-1 chain DSPC, open blue square – sn-2 chain DSPC). Figure adapted from Nickels et al. ref. [107].

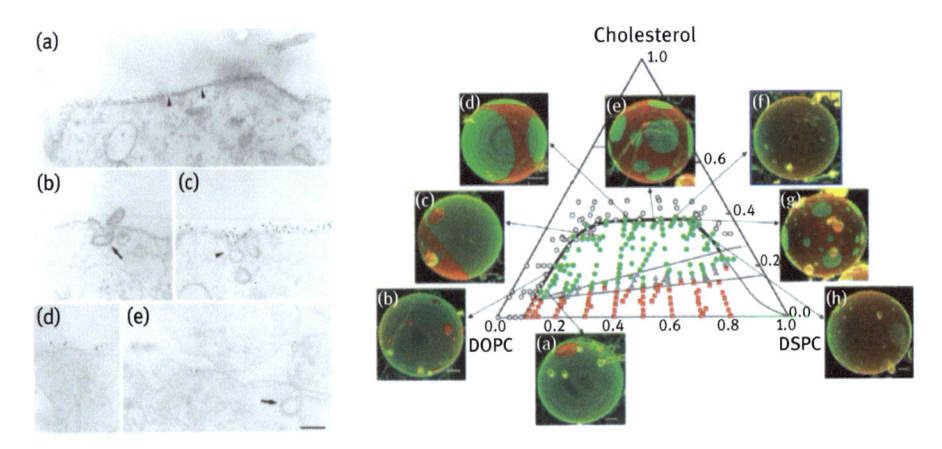

Figure 18.11: Visualization of lipid domains provides some of the most compelling pieces of evidence for the existence of lipid rafts. Examples of electron [270] (left) and fluorescence microscopies [124] (right) clearly show lipid organization. While providing insights into raft formation, and being intuitive and visually compelling evidence, these approaches do not tend to overlap with the time and length scales probed by MD simulation.

physical properties due to their nanoscopic size. This conclusion was arrived through NSE experimentation and MD simulations using the splay/tilt calculation, and the direct calculation of bilayer undulations for the lipid domains. However, due to limitations imposed by the simulation box size and the duration of the simulation, the long wavelength undulation modes observed experimentally were not fully captured by the simulations, even though they were carried out for several hundred nanoseconds. As such, it was not possible to determine the absolute bending moduli. Nonetheless, a qualitative agreement between all the different methods was observed, and there was a clear indication that the bending properties of nanoscopic lipid domains differ from the phase in which they reside in.

18.7 Concluding remarks

In this chapter we have described complementary experimental and computational methods used to describe the structure and dynamics of complex biomimetic membranes. Current understanding is built predominantly upon experimental data, but MD simulations are steadily becoming an accepted tool for providing a molecular understanding of these complex systems. With the increasing sophistication of simulation tools and steady improvements in computational power, the possibility that simulations will revolutionize our understanding of membrane biophysics is becoming a reality. This is especially true for complex biomembranes exhibiting compositional asymmetry and lateral organization.

References

[1] Gorter, E., & Grendel, F. On bimolecular layers of lipoids on the chromocytes of the blood. The Journal of Experimental Medicine 1925;41:439–443.
[2] Van Meer, G., Voelker, D.R., & Feigenson, G.W. Membrane lipids: where they are and how they behave. Nature Reviews Molecular Cell Biology 2008;9:112–124.
[3] Cheng, H.T. & London, E. Preparation and properties of asymmetric vesicles that mimic cell membranes: effect upon lipid raft formation and transmembrane helix orientation. J Biol Chem 2009;284:6079–92.
[4] Chiantia, S., & London, E. Acyl chain length and saturation modulate interleaflet coupling in asymmetric bilayers: effects on dynamics and structural order. Biophysical Journal 2012;103:2311–2319.
[5] Devaux, P.F. Static and dynamic lipid asymmetry in cell membranes. Biochemistry 1991;30:1163–1173.
[6] Bretscher, M.S. Asymmetrical lipid bilayer structure for biological membranes. Nature 1972;236:11–12.
[7] Bretscher, M.S. Membrane structure: some general principles. Science 1973;181:622–629.
[8] Verkleij, A., Zwaal, R., Roelofsen, B., Comfurius, P., Kastelijn, D., & Van Deenen, L. The asymmetric distribution of phospholipids in the human red cell membrane. A combined study using phospholipases and freeze-etch electron microscopy. Biochimica et Biophysica Acta (BBA)-Biomembranes 1973;323:178–193.
[9] Steck, T.L., & Dawson, G. Topographical distribution of complex carbohydrates in the erythrocyte membrane. Journal of Biological Chemistry 1974;249:2135–2142.
[10] Rothman, J., & Lenard, J. Membrane asymmetry. Science 1977;195:743–753.
[11] Zwaal, R.F.A., Comfurius, P., & Van Deenen, L.L.M. Membrane asymmetry and blood coagulation. Nature 1977;268:358–360.
[12] dO penKamp, J.A. Lipid asymmetry in membranes. Annual Review of Biochemistry 1979;48:47–71.
[13] Singer, S, Nicolson, G.L. The fluid mosaic model of the structure of cell membranes. Science 1972;175:720–731.
[14] Fadok, V.A., Voelker, D.R., Campbell, P.A., Cohen, J.J., Bratton, D.L., & Henson, P.M. Exposure of phosphatidylserine on the surface of apoptotic lymphocytes triggers specific recognition and removal by macrophages. The Journal of Immunology 1992;148:2207–2216.
[15] Martin, S., Reutelingsperger, C., McGahon, A.J., et al. Early redistribution of plasma membrane phosphatidylserine is a general feature of apoptosis regardless of the initiating stimulus: inhibition by overexpression of Bcl-2 and Abl. The Journal of Experimental Medicine 1995;182:1545–1556.
[16] Bennett, M., Gibson, D., Schwartz, S., & Tait, J. Binding and phagocytosis of apoptotic vascular smooth muscle cells is mediated in part by exposure of phosphatidylserine. Circulation Research 1995;77:1136–1142.
[17] Casciola-Rosen, L., Rosen, A., Petri, M., & Schlissel, M. Surface blebs on apoptotic cells are sites of enhanced procoagulant activity: implications for coagulation events and antigenic spread in systemic lupus erythematosus. Proceedings of the National Academy of Sciences 1996;93:1624–1629.
[18] Bevers, E.M., Comfurius, P., Vanrijn, J., Hemker, H.C., & Zwaal, R.F.A. Generation of prothrombin-converting activity and the exposure of phosphatidylserine at the outer surface of platelets. European Journal of Biochemistry 1982;122:429–436.
[19] Tanaka, Y., & Schroit, A. Insertion of fluorescent phosphatidylserine into the plasma membrane of red blood cells. Recognition by autologous macrophages. Journal of Biological Chemistry 1983;258:11335–11343.

[20] Schroit, A.J., Madsen, J.W., & Tanaka, Y. In vivo recognition and clearance of red blood cells containing phosphatidylserine in their plasma membranes. Journal of Biological Chemistry 1985;260:5131–5138.
[21] Fadok, V.A., Laszlo, D.J., Noble, P.W., Weinstein, L., Riches, D., & Henson, P. Particle digestibility is required for induction of the phosphatidylserine recognition mechanism used by murine macrophages to phagocytose apoptotic cells. The Journal of Immunology 1993;151:4274–4285.
[22] Connor, J., Bucana, C., Fidler, I.J., & Schroit, A.J. Differentiation-dependent expression of phosphatidylserine in mammalian plasma membranes: quantitative assessment of outer-leaflet lipid by prothrombinase complex formation. Proceedings of the National Academy of Sciences 1989;86:3184–3188.
[23] Utsugi, T., Schroit, A.J., Connor, J., Bucana, C.D., & Fidler, I.J. Elevated expression of phosphatidylserine in the outer membrane leaflet of human tumor cells and recognition by activated human blood monocytes. Cancer Research 1991;51:3062–3066.
[24] Seigneuret, M., & Devaux, P.F. ATP-dependent asymmetric distribution of spin-labeled phospholipids in the erythrocyte membrane: relation to shape changes. Proceedings of the National Academy of Sciences 1984;81:3751–3755.
[25] Bevers, E.M., Comfurius, P., Dekkers, D.W., & Zwaal, R.F. Lipid translocation across the plasma membrane of mammalian cells. Biochimica et Biophysica Acta (BBA)-Molecular and Cell Biology of Lipids 1999;1439:317–330.
[26] Kiessling, V., Wan, C., & Tamm, L.K. Domain coupling in asymmetric lipid bilayers. Biochimica at Biophysica Acta 2009;1788:64–71.
[27] Devaux, P.F., & Morris, R. Transmembrane asymmetry and lateral domains in biological membranes. Traffic 2004;5:241–246.
[28] May, S. Trans-monolayer coupling of fluid domains in lipid bilayers. Soft Matter 2009;5:3148–3156.
[29] Marquardt, D., Geier, B., & Pabst, G. Asymmetric lipid membranes: towards more realistic model systems. Membranes 2015;5:180.
[30] Pike, L.J. Rafts defined: a report on the keystone symposium on lipid rafts and cell function. Journal of Lipid Research 2006;47:1597–1598.
[31] Sezgin, E., Levental, I., Mayor, S., & Eggeling, C. The mystery of membrane organization: composition, regulation and roles of lipid rafts. Nature Reviews Molecular Cell Biology 2017;18:361.
[32] Lingwood, D., & Simons, K. Lipid rafts as a membrane-organizing principle. Science 2010;327:46–50.
[33] Brown, D., & London, E. Functions of lipid rafts in biological membranes. Annual Review of Cell And Developmental Biology 1998;14:111–136.
[34] Del Pozo, M.A., Alderson, N.B., Kiosses, W.B., Chiang, -H.-H., Anderson, R.G., & Schwartz, M.A. Integrins regulate Rac targeting by internalization of membrane domains. Science 2004;303:839–842.
[35] Pierce, S.K. Lipid rafts and B-cell activation. Nature Reviews Immunology 2002;2:96–105.
[36] Jacobson, K., Mouritsen, O.G., & Anderson, R.G. Lipid rafts: at a crossroad between cell biology and physics. Nature Cell Biology 2007;9:7–14.
[37] Hering, H., Lin, C.-C., & Sheng, M.. Lipid rafts in the maintenance of synapses, dendritic spines, and surface AMPA receptor stability. The Journal of Neuroscience 2003;23:3262–3271.
[38] Villalba, M., Bi, K., Rodriguez, F., Tanaka, Y., Schoenberger, S., & Altman, A. Vav1/Rac-dependent actin cytoskeleton reorganization is required for lipid raft clustering in T cells. The Journal of Cell Biology 2001;155:331–338.
[39] Simons, K., & Toomre, D. Lipid rafts and signal transduction. Nature Reviews Molecular Cell Biology 2000;1:31–39.

[40] Gajate, C., & Mollinedo, F. The antitumor ether lipid ET-18-OCH3 induces apoptosis through translocation and capping of Fas/CD95 into membrane rafts in human leukemic cells. Blood 2001;98:3860–3863.
[41] Nickels, J.D., Smith, M.D., Alsop, R.J., et al. The Journal of Physical Chemistry B DOI: 10.1021/acs.jpcb.8b12126.
[42] Shimshick, E.J., & McConnell, H.M. Lateral phase separation in phospholipid membranes. Biochemistry 1973;12:2351–2360.
[43] Nickels JD, Smith MD, Alsop RJ, et al. The Journal of Physical Chemistry B DOI: 10.1021/acs.jpcb.8b12126.
[44] Simons, K., & Ikonen, E. Functional rafts in cell membranes. Nature 1997;387:569–572.
[45] Pike, L. Lipid rafts: heterogeneity on the high seas. Biochemical Journal 2004;378:281–292.
[46] Pralle, A., Keller, P., Florin, E.-L., Simons, K., & Hörber, J. Sphingolipid–cholesterol rafts diffuse as small entities in the plasma membrane of mammalian cells. The Journal of Cell Biology 2000;148:997–1008.
[47] Samsonov, A.V., Mihalyov, I., & Cohen, F.S. Characterization of cholesterol-sphingomyelin domains and their dynamics in bilayer membranes. Biophysical Journal 2001;81:1486–1500.
[48] Brown, D.A., & London, E. Structure and function of sphingolipid-and cholesterol-rich membrane rafts. Journal of Biological Chemistry 2000;275:17221–17224.
[49] Allen, J.A., Halverson-Tamboli, R.A., & Rasenick, M.M. Lipid raft microdomains and neurotransmitter signalling. Nature Reviews Neuroscience 2007;8:128–140.
[50] López, D., & Kolter, R. Functional microdomains in bacterial membranes. Genes & Development 2010;24:1893–1902.
[51] García-Fernández, E., Koch, G., Wagner, R.M., et al. Membrane microdomain disassembly inhibits MRSA antibiotic resistance. Cell 2017;171:1354–1367e20.
[52] Nickels, J.D., Chatterjee, S., Stanley, C.B., et al. The in vivo structure of biological membranes and evidence for lipid domains. PLOS Biology 2017;15:e2002214.
[53] van der Goot FG, Harder T. Raft membrane domains: from a liquid-ordered membrane phase to a site of pathogen attack. Seminars in immunology; Academic Press. 2001;13:89-97
[54] Dick, R.A., Goh, S.L., Feigenson, G.W., & Vogt, V.M. HIV-1 Gag protein can sense the cholesterol and acyl chain environment in model membranes. Proceedings of the National Academy of Sciences 2012;109:18761–18766.
[55] Mongrand, S., Morel, J., Laroche, J., et al. Lipid rafts in higher plant cells purification and characterization of triton X-100-insoluble microdomains from tobacco plasma membrane. Journal of Biological Chemistry 2004;279:36277–36286.
[56] Heberle F.A, Marquardt, D, Doktorova, M, et al. Sub-nanometer Structure of an Asymmetric Model Membrane: Interleaflet Coupling Influences Domain Properties. Langmuir 2016;32:5195-5200.
[57] Marsh, D. Electron spin resonance: spin labels. Membrane spectroscopy: Springer; 1981:51–142.
[58] Lin, W.-C., Blanchette, C.D., Ratto, T.V., & Longo, M.L. Lipid asymmetry in DLPC/DSPC-supported lipid bilayers: a combined AFM and fluorescence microscopy study. Biophysical Journal 2006;90:228–237.
[59] McIntyre, J.C., & Sleight, R.G. Fluorescence assay for phospholipid membrane asymmetry. Biochemistry 1991;30:11819–11827.
[60] Kinosita, K., Kawato, S., & Ikegami, A. A theory of fluorescence polarization decay in membranes. Biophysical Journal 1977;20:289–305.
[61] Chiantia, S., Schwille, P., Klymchenko, A.S., & London, E. Asymmetric GUVs prepared by MβCD-mediated lipid exchange: an FCS study. Biophysical Journal 2011;100:L1–L3.
[62] Crane, J.M., Kiessling, V., & Tamm, L.K. Measuring lipid asymmetry in planar supported bilayers by fluorescence interference contrast microscopy. Langmuir 2005;21:1377–1388.

[63] Yuan, J., Hao, C., Chen, M., Berini, P., & Zou, S. Lipid reassembly in asymmetric Langmuir–Blodgett/Langmuir–Schaeffer bilayers. Langmuir 2012;29:221–227.
[64] Wan, C., Kiessling, V., & Tamm, L.K. Coupling of cholesterol-rich lipid phases in asymmetric bilayers. Biochemistry 2008;47:2190–2198.
[65] van Meer, G. Dynamic transbilayer lipid asymmetry. Cold Spring Harbor Perspectives In Biology 2011;3:a004671.
[66] Lhermusier, T., Chap, H., & Payrastre, B. Platelet membrane phospholipid asymmetry: from the characterization of a scramblase activity to the identification of an essential protein mutated in Scott syndrome. Journal of Thrombosis and Haemostasis 2011;9:1883–1891.
[67] Daleke, D.L. Regulation of transbilayer plasma membrane phospholipid asymmetry. Journal of Lipid Research 2003;44:233–242.
[68] Weik, M., Patzelt, H., Zaccai, G., & Oesterhelt, D. Localization of glycolipids in membranes by in vivo labeling and neutron diffraction. Molecular Cell 1998;1:411–419.
[69] Tamm, L.K., & McConnell, H.M. Supported phospholipid bilayers. Biophysical Journal 1985;47:105–113.
[70] Richter, R.P., Bérat, R., & Brisson, A.R. Formation of solid-supported lipid bilayers: an integrated view. Langmuir 2006;22:3497–3505.
[71] Wagner, M.L., & Tamm, L.K. Tethered polymer-supported planar lipid bilayers for reconstitution of integral membrane proteins: silane-polyethylene glycol-lipid as a cushion and covalent linker. Biophysical Journal 2000;79:1400–1414.
[72] Castellana, E.T., & Cremer, P.S. Solid supported lipid bilayers: From biophysical studies to sensor design. Surface Science Reports 2006;61:429–444.
[73] Zasadzinski, J., Viswanathan, R., Madsen, L., Garnaes, J., & Schwartz, D. Langmuir-Blodgett films. Science 1994;263:1726–1733.
[74] Kalb, E., Frey, S., & Tamm, L.K. Formation of supported planar bilayers by fusion of vesicles to supported phospholipid monolayers. Biochimica et Biophysica Acta (BBA)-Biomembranes 1992;1103:307–316.
[75] Pautot, S., Frisken, B.J., & Weitz, D. Engineering asymmetric vesicles. Proceedings of the National Academy of Sciences 2003;100:10718–10721.
[76] Shum, H.C., Lee, D., Yoon, I., Kodger, T., & Weitz, D.A. Double emulsion templated monodisperse phospholipid vesicles. Langmuir 2008;24:7651–7653.
[77] Funakoshi, K., Suzuki, H., & Takeuchi, S. Lipid bilayer formation by contacting monolayers in a microfluidic device for membrane protein analysis. Analytical Chemistry 2006;78:8169–8174.
[78] Funakoshi, K., Suzuki, H., & Takeuchi, S. Formation of giant lipid vesiclelike compartments from a planar lipid membrane by a pulsed jet flow. Journal of the American Chemical Society 2007;129:12608–12609.
[79] Hu, P.C., Li, S., & Malmstadt, N. Microfluidic fabrication of asymmetric giant lipid vesicles. ACS Applied Materials & Interfaces 2011;3:1434–1440.
[80] Richmond, D.L., Schmid, E.M., Martens, S., Stachowiak, J.C., Liska, N., & Fletcher, D.A. Forming giant vesicles with controlled membrane composition, asymmetry, and contents. Proceedings of the National Academy of Sciences 2011;108:9431–9436.
[81] Hamada, T., Miura, Y., Komatsu, Y., Kishimoto, Y., Vestergaard, M., & Takagi, M. Construction of asymmetric cell-sized lipid vesicles from lipid-coated water-in-oil microdroplets. The Journal of Physical Chemistry B 2008;112:14678–14681.
[82] Różycki, B., & Lipowsky, R. Spontaneous curvature of bilayer membranes from molecular simulations: Asymmetric lipid densities and asymmetric adsorption. The Journal of Chemical Physics 2015;142:054101.

[83] Cheng, H.T., & London, E. Preparation and properties of asymmetric large unilamellar vesicles: interleaflet coupling in asymmetric vesicles is dependent on temperature but not curvature. Biophysical Journal 2011;100:2671–2678.

[84] Huang, Z., & London, E. Effect of cyclodextrin and membrane lipid structure upon cyclodextrin-lipid interaction. Langmuir 2013;29:14631–14638.

[85] Lin, Q., & London, E. Preparation of artificial plasma membrane mimicking vesicles with lipid asymmetry. PLoS One 2014;9:e87903.

[86] Son, M., & London, E. The dependence of lipid asymmetry upon polar headgroup structure. Journal of Lipid Research 2013;54:3385–3393.

[87] Son, M., & London, E. The dependence of lipid asymmetry upon phosphatidylcholine acyl chain structure. Journal of Lipid Research 2013;54:223–231.

[88] Lingwood, D., & Simons, K. Detergent resistance as a tool in membrane research. Nature Protocols 2007;2:2159–2165.

[89] London, E., & Brown, D.A. Insolubility of lipids in triton X-100: physical origin and relationship to sphingolipid/cholesterol membrane domains (rafts). Biochimica et Biophysica Acta (BBA)-Biomembranes 2000;1508:182–195.

[90] Zidovetzki, R., & Levitan, I. Use of cyclodextrins to manipulate plasma membrane cholesterol content: evidence, misconceptions and control strategies. Biochimica et Biophysica Acta (BBA)-Biomembranes 2007;1768:1311–1324.

[91] Lawrence, J.C., Saslowsky, D.E., Edwardson, J.M., & Henderson, R.M. Real-time analysis of the effects of cholesterol on lipid raft behavior using atomic force microscopy. Biophysical Journal 2003;84:1827–1832.

[92] Kabouridis, P.S., Janzen, J., Magee, A.L., & Ley, S.C. Cholesterol depletion disrupts lipid rafts and modulates the activity of multiple signaling pathways in T lymphocytes. European Journal of Immunology 2000;30:954–963.

[93] Kraft, M.L., Weber, P.K., Longo, M.L., Hutcheon, I.D., & Boxer, S.G. Phase separation of lipid membranes analyzed with high-resolution secondary ion mass spectrometry. Science 2006;313:1948–1951.

[94] Veatch, S., Polozov, I., Gawrisch, K., & Keller, S. Liquid domains in vesicles investigated by NMR and fluorescence microscopy. Biophysical Journal 2004;86:2910–2922.

[95] Bacia, K., Scherfeld, D., Kahya, N., & Schwille, P. Fluorescence correlation spectroscopy relates rafts in model and native membranes. Biophysical Journal 2004;87:1034–1043.

[96] Schwille, P., Korlach, J., & Webb, W.W. Fluorescence correlation spectroscopy with single-molecule sensitivity on cell and model membranes. Cytometry Part A 1999;36:176–182.

[97] Zipfel, W.R., Williams, R.M., Christie, R., Nikitin, A.Y., Hyman, B.T., & Webb, W.W. Live tissue intrinsic emission microscopy using multiphoton-excited native fluorescence and second harmonic generation. Proceedings of the National Academy of Sciences 2003;100:7075–7080.

[98] Baumgart, T., Hunt, G., Farkas, E.R., Webb, W.W., & Feigenson, G.W. Fluorescence probe partitioning between L o/L d phases in lipid membranes. Biochimica et Biophysica Acta (BBA)-Biomembranes 2007;1768:2182–2194.

[99] Stöckl, M.T., & Herrmann, A. Detection of lipid domains in model and cell membranes by fluorescence lifetime imaging microscopy. Biochim Biophys Acta 2010;1798:1444–1456.

[100] Zacharias, D.A., Violin, J.D., Newton, A.C., & Tsien, R.Y. Partitioning of lipid-modified monomeric GFPs into membrane microdomains of live cells. Science 2002;296:913–916.

[101] Korlach, J., Schwille, P., Webb, W.W., & Feigenson, G.W. Characterization of lipid bilayer phases by confocal microscopy and fluorescence correlation spectroscopy. Proceedings of the National Academy of Sciences 1999;96:8461–8466.

[102] Baumgart, T., Hess, S.T., & Webb, W.W. Imaging coexisting fluid domains in biomembrane models coupling curvature and line tension. Nature 2003;425:821–824.

[103] Yuan, C., Furlong, J., Burgos, P., & Johnston, L.J. The size of lipid rafts: an atomic force microscopy study of ganglioside GM1 domains in sphingomyelin/DOPC/cholesterol membranes. Biophysical Journal 2002;82:2526–2535.

[104] Johnston, L.J. Nanoscale imaging of domains in supported lipid membranes. Langmuir 2007;23:5886–5895.

[105] Tokumasu, F., Jin, A.J., Feigenson, G.W., & Dvorak, J.A. Nanoscopic lipid domain dynamics revealed by atomic force microscopy. Biophysical Journal 2003;84:2609–2618.

[106] Goksu, E.I., Vanegas, J.M., Blanchette, C.D., Lin, W.C., & Longo, M.L. AFM for structure and dynamics of biomembranes. Biochimica at Biophysica Acta 2009;1788:254–266.

[107] Mills, T.T., Toombes, G.E., Tristram-Nagle, S., Smilgies, D.-M., Feigenson, G.W., & Nagle, J.F. Order parameters and areas in fluid-phase oriented lipid membranes using wide angle X-ray scattering. Biophysical Journal 2008;95:669–681.

[108] Pencer, J., Mills, T., Anghel, V., Krueger, S., Epand, R.M., & Katsaras, J. Detection of submicron-sized raft-like domains in membranes by small-angle neutron scattering. The European Physical Journal E 2005;18:447–458.

[109] Nickels, J.D., Cheng, X., Mostofian, B., et al. Mechanical properties of nanoscopic lipid domains. Journal of the American Chemical Society 2015;137:15772–15780.

[110] Heberle, F.A., Petruzielo, R.S., Pan, J., et al. Bilayer thickness mismatch controls domain size in model membranes. Journal of the American Chemical Society 2013;135:6853–6859.

[111] Gaus, K., Gratton, E., Kable, E.P., et al. Visualizing lipid structure and raft domains in living cells with two-photon microscopy. Proceedings of the National Academy of Sciences 2003;100:15554–15559.

[112] Baumgart, T., Hammond, A.T., Sengupta, P., et al. Large-scale fluid/fluid phase separation of proteins and lipids in giant plasma membrane vesicles. Proceedings of the National Academy of Sciences 2007;104:3165–3170.

[113] Ahmed, S.N., Brown, D.A., & London, E. On the origin of sphingolipid/cholesterol-rich detergent-insoluble cell membranes: physiological concentrations of cholesterol and sphingolipid induce formation of a detergent-insoluble, liquid-ordered lipid phase in model membranes. Biochemistry 1997;36:10944–10953.

[114] Veatch, S.L., & Keller, S.L. Separation of liquid phases in giant vesicles of ternary mixtures of phospholipids and cholesterol. Biophysical Journal 2003;85:3074–3083.

[115] Veatch, S.L., & Keller, S.L. Organization in lipid membranes containing cholesterol. Physical Review Letters 2002;89:268101.

[116] Dietrich, C., Bagatolli, L., Volovyk, Z., et al. Lipid rafts reconstituted in model membranes. Biophysical Journal 2001;80:1417–1428.

[117] Edidin, M. The state of lipid rafts: from model membranes to cells. Annual Review of Biophysics and Biomolecular Structure 2003;32:257–283.

[118] Jacobson, K., Mouritsen, O.G., & Anderson, R.G. Lipid rafts: at a crossroad between cell biology and physics. Nature Cell Biology 2007;9:7.

[119] Klymchenko, A.S., & Kreder, R. Fluorescent probes for lipid rafts: from model membranes to living cells. Chemistry & Biology 2014;21:97–113.

[120] Sezgin, E., Levental, I., Grzybek, M., et al. Partitioning, diffusion, and ligand binding of raft lipid analogs in model and cellular plasma membranes. Biochimica et Biophysica Acta 2012;1818:1777–1784.

[121] Wu, E., Jacobson, K., & Papahadjopoulos, D. Lateral diffusion in phospholipid multibilayers measured by fluorescence recovery after photobleaching. Biochemistry 1977;16:3936–3941.

[122] Derzko, Z., & Jacobson, K. Comparative lateral diffusion of fluorescent lipid analogs in phospholipid multibilayers. Biochemistry 1980;19:6050–6057.

[123] Mouritsen, O.G., & Jørgensen, K. Dynamical order and disorder in lipid bilayers. Chemistry and Physics of Lipids 1994;73:3–25.
[124] Almeida, P.F., Vaz, W.L., & Thompson, T. Lateral diffusion and percolation in two-phase, two-component lipid bilayers. Topology of the solid-phase domains in-plane and across the lipid bilayer. Biochemistry 1992;31:7198–7210.
[125] Jares-Erijman, E.A., & Jovin, T.M. FRET imaging. Nature Biotechnology 2003;21:1387–1395.
[126] Zhao, J., Wu, J., Heberle, F.A., et al. Phase studies of model biomembranes: complex behavior of DSPC/DOPC/cholesterol. Biochimica et Biophysica Acta (BBA)-Biomembranes 2007;1768:2764–2776.
[127] de Almeida, R.F., Loura, L.M., Fedorov, A., & Prieto, M. Lipid rafts have different sizes depending on membrane composition: a time-resolved fluorescence resonance energy transfer study. Journal of Molecular Biology 2005;346:1109–1120.
[128] Feigenson, G.W. Phase diagrams and lipid domains in multicomponent lipid bilayer mixtures. Biochimica et Biophysica Acta (BBA)-Biomembranes 2009;1788:47–52.
[129] Kenworthy, A.K., Petranova, N., & Edidin, M. High-resolution FRET microscopy of cholera toxin B-subunit and GPI-anchored proteins in cell plasma membranes. Molecular Biology of the Cell 2000;11:1645–1655.
[130] Bagatolli, L., & Gratton, E. Two photon fluorescence microscopy of coexisting lipid domains in giant unilamellar vesicles of binary phospholipid mixtures. Biophysical Journal 2000;78:290–305.
[131] Simons, K., & Gerl, M.J. Revitalizing membrane rafts: new tools and insights. Nature Reviews Molecular Cell Biology 2010;11:688–699.
[132] Schütz, G.J., Kada, G., Pastushenko, V.P., & Schindler, H. Properties of lipid microdomains in a muscle cell membrane visualized by single molecule microscopy. The EMBO Journal 2000;19:892–901.
[133] Mendelsohn, R., & Moore, D.J. Vibrational spectroscopic studies of lipid domains in biomembranes and model systems. Chemistry and Physics of Lipids 1998;96:141–157.
[134] Schultz, Z.D., & Levin, I.W. Lipid microdomain formation: characterization by infrared spectroscopy and ultrasonic velocimetry. Biophysical Journal 2008;94:3104–3114.
[135] Lewis, R.N., & McElhaney, R.N. Membrane lipid phase transitions and phase organization studied by Fourier transform infrared spectroscopy. Biochimica et Biophysica Acta (BBA)-Biomembranes 2013;1828:2347–2358.
[136] Czamara, K., Majzner, K., Pacia, M.Z., Kochan, K., Kaczor, A., & Baranska, M. Raman spectroscopy of lipids: a review. Journal of Raman Spectroscopy 2015;46:4–20.
[137] Opilik, L., Bauer, T., Schmid, T., Stadler, J., & Zenobi, R. Nanoscale chemical imaging of segregated lipid domains using tip-enhanced Raman spectroscopy. Physical Chemistry Chemical Physics 2011;13:9978–9981.
[138] Potma, E.O., & Xie, X.S. Direct visualization of lipid phase segregation in single lipid bilayers with coherent anti-stokes Raman scattering microscopy. ChemPhysChem 2005;6:77–79.
[139] Li, L., Wang, H., & Cheng, J.-X. Quantitative coherent anti-stokes Raman scattering imaging of lipid distribution in coexisting domains. Biophysical Journal 2005;89:3480–3490.
[140] Brown, K.L., & Conboy, J.C. Lipid flip-flop in binary membranes composed of phosphatidylserine and phosphatidylcholine. The Journal of Physical Chemistry B 2013;117:15041–15050.
[141] Liu, J., & Conboy, J.C. Phase transition of a single lipid bilayer measured by sum-frequency vibrational spectroscopy. Journal of the American Chemical Society 2004;126:8894–8895.
[142] Fattal, E., Nir, S., Parente, R.A., & Szoka, J.F.C. Pore-forming peptides induce rapid phospholipid flip-flop in membranes. Biochemistry 1994;33:6721–6731.

[143] Davis, J.H. The description of membrane lipid conformation, order and dynamics by 2H-NMR. Biochimica et Biophysica Acta (BBA)-Reviews on Biomembranes 1983;737:117–171.
[144] Seelig, A., & Seelig, J. Dynamic structure of fatty acyl chains in a phospholipid bilayer measured by deuterium magnetic resonance. Biochemistry 1974;13:4839–4845.
[145] Henderson, T.O., Glonek, T., & Myers, T.C. Phosphorus-31 nuclear magnetic resonance spectroscopy of phospholipids. Biochemistry 1974;13:623–628.
[146] Petrache, H.I., Dodd, S.W., & Brown, M.F. Area per lipid and acyl length distributions in fluid phosphatidylcholines determined by 2H NMR spectroscopy. Biophysical Journal 2000;79:3172–3192.
[147] Solomon, I. Relaxation processes in a system of two spins. Physical Review 1955;99:559.
[148] Bloembergen, N. Proton relaxation times in paramagnetic solutions. The Journal of Chemical Physics 1957;27:572–573.
[149] Su, Y., Mani, R., & Hong, M. Asymmetric insertion of membrane proteins in lipid bilayers by solid-state NMR paramagnetic relaxation enhancement: a cell-penetrating peptide example. Journal of the American Chemical Society 2008;130:8856–8864.
[150] Heyn, M.P. Determination of lipid order parameters and rotational correlation times from fluorescence depolarization experiments. FEBS Letters 1979;108:359–364.
[151] Brown, M.F., Ribeiro, A.A., & Williams, G.D. New view of lipid bilayer dynamics from 2H and 13C NMR relaxation time measurements. Proceedings of the National Academy of Sciences 1983;80:4325–4329.
[152] Hong, M., Schmidt-Rohr, K., & Pines, A. NMR measurement of signs and magnitudes of CH dipolar couplings in lecithin. Journal of the American Chemical Society 1995;117:3310–3311.
[153] Gross, J.D., Warschawski, D.E., & Griffin, R.G. Dipolar recoupling in MAS NMR: a probe for segmental order in lipid bilayers. Journal of the American Chemical Society 1997;119:796–802.
[154] Orädd, G., & Lindblom, G. Lateral diffusion studied by pulsed field gradient NMR on oriented lipid membranes. Magnetic Resonance in Chemistry 2004;42:123–131.
[155] Scheidt, H.A., Huster, D., & Gawrisch, K. Diffusion of cholesterol and its precursors in lipid membranes studied by 1H pulsed field gradient magic angle spinning NMR. Biophysical Journal 2005;89:2504–2512.
[156] Orädd, G., Westerman, P.W., & Lindblom, G. Lateral diffusion coefficients of separate lipid species in a ternary raft-forming bilayer: a Pfg-NMR multinuclear study. Biophysical Journal 2005;89:315–320.
[157] Lindblom, G., Orädd, G., & Filippov, A. Lipid lateral diffusion in bilayers with phosphatidylcholine, sphingomyelin and cholesterol: An NMR study of dynamics and lateral phase separation. Chemistry and Physics of Lipids 2006;141:179–184.
[158] Luzzati, V., & Husson, F. The structure of the liquid-crystalline phases of lipid-water systems. The Journal of Cell Biology 1962;12:207–219.
[159] Tardieu, A., Luzzati, V., & Reman, F. Structure and polymorphism of the hydrocarbon chains of lipids: a study of lecithin-water phases. Journal of Molecular Biology 1973;75:711–733.
[160] Engelman, D.M. Lipid bilayer structure in the membrane of Mycoplasma laidlawii. Journal of Molecular Biology 1971;58:153–165.
[161] Lewis, B.A., & Engelman, D.M. Lipid bilayer thickness varies linearly with acyl chain length in fluid phosphatidylcholine vesicles. Journal of Molecular Biology 1983;166:211–217.
[162] Nagle, J.F., & Tristram-Nagle, S. Structure of lipid bilayers. Biochimica et Biophysica Acta (BBA)-Reviews on Biomembranes 2000;1469:159–195.
[163] Koenig, B.W., Strey, H.H., & Gawrisch, K. Membrane lateral compressibility determined by NMR and x-ray diffraction: effect of acyl chain polyunsaturation. Biophysical Journal 1997;73:1954–1966.

[164] Majewski, J., Kuhl, T., Kjaer, K., & Smith, G. Packing of ganglioside-phospholipid monolayers: an x-ray diffraction and reflectivity study. Biophysical Journal 2001;81:2707–2715.

[165] Heftberger, P., Kollmitzer, B., Rieder, A.A., Amenitsch, H., & Pabst, G. In situ determination of structure and fluctuations of coexisting fluid membrane domains. Biophysical Journal 2015;108:854–862.

[166] Bee, M. Quasielastic neutron scattering: principles and applications in solid state chemistry, biology, and materials science. Bristol: Adam Hilger; 1988.

[167] Sears, V.F. Neutron scattering lengths and cross sections. Neutron News 1992;3:26–37.

[168] Büldt, G., Gally, H., Seelig, A., Seelig, J., & Zaccai, G. Neutron diffraction studies on selectively deuterated phospholipid bilayers. Nature 1978;271:182.

[169] Büldt, G., Gally, H., Seelig, J., & Zaccai, G. Neutron diffraction studies on phosphatidylcholine model membranes: I. Head group conformation. Journal of Molecular Biology 1979;134:673–691.

[170] Zaccai, G., Büldt, G., Seelig, A., & Seelig, J. Neutron diffraction studies on phosphatidylcholine model membranes: II. Chain conformation and segmental disorder. Journal of Molecular Biology 1979;134:693–706.

[171] Toppozini, L., Roosen-Runge, F., I. Bewley, R., et al. Anomalous and anisotropic nanoscale diffusion of hydration water molecules in fluid lipid membranes. Soft Matter 2015;11:8354–8371.

[172] Nickels, J.D., O'Neill, H., Hong, L., et al. Dynamics of protein and its hydration water: neutron scattering studies on fully deuterated GFP. Biophysical Journal 2012;103:1566–1575.

[173] Perticaroli, S., Ehlers, G., Stanley, C.B., et al. Description of hydration water in protein (green fluorescent protein) solution. Journal of the American Chemical Society 2017;139:1098–1105.

[174] König, S., Sackmann, E., Richter, D., Zorn, R., Carlile, C., & Bayerl, T. Molecular dynamics of water in oriented DPPC multilayers studied by quasielastic neutron scattering and deuterium-nuclear magnetic resonance relaxation. The Journal of Chemical Physics 1994;100:3307.

[175] Settles, M., & Doster, W. Anomalous diffusion of adsorbed water: a neutron scattering study of hydrated myoglobin. Faraday Discussions 1996;103:269–279.

[176] Swenson, J., Kargl, F., Berntsen, P., & Svanberg, C. Solvent and lipid dynamics of hydrated lipid bilayers by incoherent quasielastic neutron scattering. The Journal of Chemical Physics 2008;129:045101.

[177] Zaccai, G., Blasie, J., & Schoenborn, B. Neutron diffraction studies on the location of water in lecithin bilayer model membranes. Proceedings of the National Academy of Sciences 1975;72:376–380.

[178] Jacrot, B. The study of biological structures by neutron scattering from solution. Reports on Progress in Physics 1976;39:911.

[179] Knoll, W., Ibel, K., & Sackmann, E. Small-angle neutron scattering study of lipid phase diagrams by the contrast variation method. Biochemistry 1981;20:6379–6383.

[180] Pardon, J., Worcester, D., Wooley, J., Tatchell, K., Van Holde, K., & Richards, B. Low-angle neutron scattering from chromatin subunit particles. Nucleic Acids Research 1975;2:2163–2176.

[181] Nickels, J.D., Atkinson, J., Papp-Szabo, E., et al. Structure and hydration of highly-branched, monodisperse phytoglycogen nanoparticles. Biomacromolecules 2016;17:735–743.

[182] Watson, M.C., & Brown, F.L. Interpreting membrane scattering experiments at the mesoscale: the contribution of dissipation within the bilayer. Biophysical Journal 2010;98:L9–L11.

[183] Zilman, A., & Granek, R. Undulations and dynamic structure factor of membranes. Physical Review Letters 1996;77:4788.

[184] Woodka, A.C., Butler, P.D., Porcar, L., Farago, B., & Nagao, M. Lipid bilayers and membrane dynamics: insight into thickness fluctuations. Physical Review Letters 2012;109:058102.

[185] Pfeiffer, W., Henkel, T., Sackmann, E., Knoll, W., & Richter, D. Local dynamics of lipid bilayers studied by incoherent quasi-elastic neutron scattering. EPL (Europhysics Letters) 1989;8:201.
[186] König, S., Pfeiffer, W., Bayerl, T., Richter, D., & Sackmann, E. Molecular dynamics of lipid bilayers studied by incoherent quasi-elastic neutron scattering. Journal de Physique II 1992;2:1589–1615.
[187] Rheinstädter, M., Ollinger, C., Fragneto, G., Demmel, F., & Salditt, T. Collective dynamics of lipid membranes studied by inelastic neutron scattering. Physical Review Letters 2004;93:108107.
[188] Humphrey, W., Dalke, A., & Schulten, K. VMD: visual molecular dynamics. Journal of Molecular Graphics 1996;14:33–38.
[189] Egberts, E., & Berendsen, H. Molecular dynamics simulation of a smectic liquid crystal with atomic detail. The Journal of Chemical Physics 1988;89:3718–3732.
[190] Berkowitz, M.L., & Raghavan, K. Computer simulation of a water/membrane interface. Langmuir 1991;7:1042–1044.
[191] Damodaran, K., Merz, J.K.M., & Gaber, B.P. Structure and dynamics of the dilauroylphosphatidylethanolamine lipid bilayer. Biochemistry 1992;31:7656–7664.
[192] Egberts, E., Marrink, S.-J., & Berendsen, H.J. Molecular dynamics simulation of a phospholipid membrane. European Biophysics Journal 1994;22:423–436.
[193] Heller, H., Schaefer, M., & Schulten, K. Molecular dynamics simulation of a bilayer of 200 lipids in the gel and in the liquid crystal phase. The Journal of Physical Chemistry 1993;97:8343–8360.
[194] Venable, R.M., Zhang, Y., Hardy, B.J., & Pastor, R.W. Molecular dynamics simulations of a lipid bilayer and of hexadecane: an investigation of membrane fluidity. Science 1993;262:223–226.
[195] Tobias, D.J., Tu, K., & Klein, M.L. Atomic-scale molecular dynamics simulations of lipid membranes. Current Opinion in Colloid & Interface Science 1997;2:15–26.
[196] Pastor, R.W. Molecular dynamics and Monte Carlo simulations of lipid bilayers. Current Opinion in Structural Biology 1994;4:486–492.
[197] Feller, S.E. Molecular dynamics simulations of lipid bilayers. Current Opinion in Colloid & Interface Science 2000;5:217–223.
[198] Ingólfsson, H.I., Lopez, C.A., Uusitalo, J.J., et al. The power of coarse graining in biomolecular simulations. Wiley Interdisciplinary Reviews: Computational Molecular Science 2014;4:225–248.
[199] Venable, R.M., Brown, F.L., & Pastor, R.W. Mechanical properties of lipid bilayers from molecular dynamics simulation. Chemistry and Physics of Lipids 2015;192:60–74.
[200] Pabst G, Kučerka N, Nieh M-P, & Katsaras J. Liposomes, lipid bilayers and model membranes: from basic research to application: CRC Press, Boca Raton; 2014.
[201] Lyubartsev, A.P., & Rabinovich, A.L. Recent development in computer simulations of lipid bilayers. Soft Matter 2011;7:25–39.
[202] Nickels, J.D., Smith, J.C., & Cheng, X. Lateral organization, bilayer asymmetry, and inter-leaflet coupling of biological membranes. Chemistry and Physics of Lipids 2015;192:87–99.
[203] Rabinovich, A., & Lyubartsev, A.P. Computer simulation of lipid membranes: Methodology and achievements. Polymer Science Series C 2013;55:162–180.
[204] Berkowitz, M.L., & Kindt, J.T. Molecular detailed simulations of lipid bilayers. Reviews in Computational Chemistry, Volume 27 2010:253–286.
[205] Mouritsen, O.G., & Jørgensen, K. Small-scale lipid-membrane structure: simulation versus experiment. Current Opinion in Structural Biology 1997;7:518–527.
[206] Scott, H.L. Modeling the lipid component of membranes. Current Opinion in Structural Biology 2002;12:495–502.

[207] Khakbaz P, Monje-Galvan V, Zhuang X, Klauda JB. Modeling lipid membranes. Biogenesis of Fatty Acids, Lipids and Membranes Handbook of Hydrocarbon and Lipid Microbiology. Springer, Cham. 2017:1–19.
[208] Essmann, U., Perera, L., Berkowitz, M.L., Darden, T., Lee, H., & Pedersen, L.G. A smooth particle mesh Ewald method. The Journal of Chemical Physics 1995;103:8577–8593.
[209] Kmiecik, S., Gront, D., Kolinski, M., Wieteska, L., Dawid, A.E., & Kolinski, A. Coarse-grained protein models and their applications. Chemical Reviews 2016;116:7898–7936.
[210] Metropolis N. Los Alamos science. Special Issue 1987;125-130.
[211] Metropolis, N., & Ulam, S. The monte carlo method. Journal of the American Statistical Association 1949;44:335–341.
[212] Metropolis, N., Rosenbluth, A.W., Rosenbluth, M.N., Teller, A.H., & Teller, E. Equation of state calculations by fast computing machines. The Journal of Chemical Physics 1953;21:1087–1092.
[213] Hastings, W.K. Monte carlo sampling methods using Markov chains and their applications. Biometrika 1970;57:97–109.
[214] Fermi I, Pasta, P, Ulam, S, Tsingou M. Studies of the nonlinear problems. Los Alamos Scientific Laboratory, New Mexico; 1955.
[215] Rahman, A. Correlations in the motion of atoms in liquid argon. Physical Review 1964;136: A405–A11.
[216] López Cascales, J., Otero, T., Smith, B.D., Gonzalez, C., & Marquez, M. Model of an asymmetric DPPC/DPPS membrane: effect of asymmetry on the lipid properties. A molecular dynamics simulation study. The Journal of Physical Chemistry B 2006;110:2358–2363.
[217] Tian J, Nickels, JD, Katsaras J, Cheng X. The Behavior of Bilayer Leaflets in Asymmetric Model Membranes: Atomistic Simulation Studies. The Journal of Physical Chemistry B 2016;120:8438-8448.
[218] Jo, S., Kim, T., Iyer, V.G., & Im, W. CHARMM-GUI: a web-based graphical user interface for CHARMM. Journal of Computational Chemistry 2008;29:1859–1865.
[219] Wassenaar, T.A., Ingólfsson, H.I., Böckmann, R.A., Tieleman, D.P., & Marrink, S.J. Computational lipidomics with insane: a versatile tool for generating custom membranes for molecular simulations. Journal of Chemical Theory and Computation 2015;11:2144–2155.
[220] Ponder, J.W., & Richards, F.M. An efficient newton-like method for molecular mechanics energy minimization of large molecules. Journal of Computational Chemistry 1987;8:1016–1024.
[221] Brooks, B.R., Bruccoleri, R.E., Olafson, B.D., States, D.J., Swaminathan, S., & Karplus, M. CHARMM: a program for macromolecular energy, minimization, and dynamics calculations. Journal of Computational Chemistry 1983;4:187–217.
[222] Hess, B., Bekker, H., Berendsen, H.J., & Fraaije, J.G. LINCS: a linear constraint solver for molecular simulations. Journal of Computational Chemistry 1997;18:1463–1472.
[223] Phillips, J.C., Braun, R., Wang, W., et al. Scalable molecular dynamics with NAMD. Journal of Computational Chemistry 2005;26:1781–1802.
[224] Wang, J., Wolf, R.M., Caldwell, J.W., Kollman, P.A., & Case, D.A. Development and testing of a general amber force field. Journal of Computational Chemistry 2004;25:1157–1174.
[225] Cornell, W.D., Cieplak, P., Bayly, C.I., et al. A second generation force field for the simulation of proteins, nucleic acids, and organic molecules. Journal of the American Chemical Society 1995;117:5179–5197.
[226] Dickson, C.J., Rosso, L., Betz, R.M., Walker, R.C., & Gould, I.R. GAFFlipid: a general amber force field for the accurate molecular dynamics simulation of phospholipid. Soft Matter 2012;8:9617–9627.
[227] Dickson, C.J., Madej, B.D., Skjevik, Å.A., et al. Lipid14: the amber lipid force field. Journal of Chemical Theory and Computation 2014;10:865–879.

[228] Klauda, J.B., Venable, R.M., Freites, J.A., et al. Update of the CHARMM all-atom additive force field for lipids: validation on six lipid types. The Journal of Physical Chemistry B 2010;114:7830–7843.

[229] Feller, S.E., & MacKerell, A.D. An improved empirical potential energy function for molecular simulations of phospholipids. The Journal of Physical Chemistry B 2000;104:7510–7515.

[230] Feller, S.E., Gawrisch, K., & MacKerell, A.D. Polyunsaturated fatty acids in lipid bilayers: intrinsic and environmental contributions to their unique physical properties. Journal of the American Chemical Society 2002;124:318–326.

[231] Jorgensen, W.L., Maxwell, D.S., & Tirado-Rives, J. Development and testing of the OPLS all-atom force field on conformational energetics and properties of organic liquids. Journal of the American Chemical Society 1996;118:11225–11236.

[232] Ponder, J.W., Wu, C., Ren, P., et al. Current status of the AMOEBA polarizable force field. The Journal of Physical Chemistry B 2010;114:2549–2564.

[233] Lagardère L, Jolly, L-H, Lipparini F, et al. Tinker-HP: a massively parallel molecular dynamics package for multiscale simulations of large complex systems with advanced point dipole polarizable force fields. Chemical Science 2018;9:956-972.

[234] Shao, Y., Gan, Z., Epifanovsky, E., et al. Advances in molecular quantum chemistry contained in the Q-Chem 4 program package. Molecular Physics 2015;113:184–215.

[235] Frenkel D, Smit, B. Understanding molecular simulation: from algorithms to applications: Academic Press, San Diengo; 2001.

[236] Allen, M.P. Introduction to molecular dynamics simulation. Computational Soft Matter: From Synthetic Polymers to Proteins 2004;23:1–28.

[237] Ryckaert, J.-P., Ciccotti, G., & Berendsen, H.J. Numerical integration of the cartesian equations of motion of a system with constraints: molecular dynamics of n-alkanes. Journal of Computational Physics 1977;23:327–341.

[238] Nickels, J.D., & Katsaras, J. Water and Lipid Bilayers. Membrane Hydration: Springer International Publishing; 2015:45–67.

[239] Jorgensen, W.L., Chandrasekhar, J., Madura, J.D., Impey, R.W., & Klein, M.L. Comparison of simple potential functions for simulating liquid water. The Journal of Chemical Physics 1983;79:926–935.

[240] Berendsen, H., Grigera, J., & Straatsma, T. The missing term in effective pair potentials. Journal of Physical Chemistry 1987;91:6269–6271.

[241] Ren, P., & Ponder, J.W. Polarizable atomic multipole water model for molecular mechanics simulation. The Journal of Physical Chemistry B 2003;107:5933–5947.

[242] Wassenaar, T.A., Pluhackova, K., BöCkmann, R.A., Marrink, S.J., & Tieleman, D.P. Going backward: a flexible geometric approach to reverse transformation from coarse grained to atomistic models. Journal of Chemical Theory and Computation 2014;10:676–690.

[243] Smit, B., Hilbers, P., Esselink, K., Rupert, L., Van Os, N., & Schlijper, A. Computer simulations of a water/oil interface in the presence of micelles. Nature 1990;348:624.

[244] Cooke, I.R., Kremer, K., & Deserno, M. Tunable generic model for fluid bilayer membranes. Physical Review E 2005;72:011506.

[245] Ulmschneider, J.P., & Ulmschneider, M.B. United atom lipid parameters for combination with the optimized potentials for liquid simulations all-atom force field. Journal of Chemical Theory and Computation 2009;5:1803–1813.

[246] Chiu, S.-W., Pandit, S.A., Scott, H., & Jakobsson, E. An improved united atom force field for simulation of mixed lipid bilayers. The Journal of Physical Chemistry B 2009;113:2748–2763.

[247] Marrink, S.J., Risselada, H.J., Yefimov, S., Tieleman, D.P., & De Vries, A.H. The MARTINI force field: coarse grained model for biomolecular simulations. The Journal of Physical Chemistry B 2007;111:7812–7824.

[248] Marrink, S.J., De Vries, A.H., & Mark, A.E. Coarse grained model for semiquantitative lipid simulations. The Journal of Physical Chemistry B 2004;108:750–760.

[249] Rodgers, J.M., Sørensen, J., de Meyer, F.-M., Schiøtt, B., & Smit, B.. Understanding the phase behavior of coarse-grained model lipid bilayers through computational calorimetry. The Journal of Physical Chemistry B 2012;116:1551–1569.

[250] Bennett, W.D., & Tieleman, D.P. Computer simulations of lipid membrane domains. Biochimica et Biophysica Acta (BBA)-Biomembranes 2013;1828:1765–1776.

[251] Risselada, H.J., & Marrink, S.J. The molecular face of lipid rafts in model membranes. Proceedings of the National Academy of Sciences of the United States of America 2008;105:17367–17372.

[252] Carpenter, T.S., Lopez, C.A., Neale, C., et al. Accurate phase separation of complex lipid mixtures (DPPC/DOPC/CHOL) with a refined coarse grained martini model. Biophysical Journal 2018;114:102a.

[253] Wang, Y., Gkeka, P., Fuchs, J.E., Liedl, K.R., & Cournia, Z. DPPC-cholesterol phase diagram using coarse-grained molecular dynamics simulations. Biochimica et Biophysica Acta (BBA) - Biomembranes 2016;1858:2846–2857.

[254] Ingólfsson, H.I., Melo, M.N., Van Eerden, F.J., et al. Lipid organization of the plasma membrane. Journal of the American Chemical Society 2014;136:14554–14559.

[255] Perlmutter, J.D., & Sachs, J.N. Interleaflet interaction and asymmetry in phase separated lipid bilayers: molecular dynamics simulations. Journal of the American Chemical Society 2011;133:6563–6577.

[256] Poger, D., Caron, B., & Mark, A.E. Validating lipid force fields against experimental data: Progress, challenges and perspectives. Biochimica et Biophysica Acta (BBA)-Biomembranes 2016;1858:1556–1565.

[257] Benz, R.W., Castro-Román, F., Tobias, D.J., & White, S.H. Experimental validation of molecular dynamics simulations of lipid bilayers: a new approach. Biophysical Journal 2005;88:805–817.

[258] Kučerka, N., Nagle, J.F., Sachs, J.N., et al. Lipid bilayer structure determined by the simultaneous analysis of neutron and X-Ray scattering data. Biophysical Journal 2008;95:2356–2367.

[259] Kučerka, N., M-P, N., & Katsaras, J.. Fluid phase lipid areas and bilayer thicknesses of commonly used phosphatidylcholines as a function of temperature. Biochimica et Biophysica Acta (BBA)-Biomembranes 2011;1808:2761–2771.

[260] Berger, O., Edholm, O., & Jähnig, F. Molecular dynamics simulations of a fluid bilayer of dipalmitoylphosphatidylcholine at full hydration, constant pressure, and constant temperature. Biophysical Journal 1997;72:2002–2013.

[261] Smondyrev, A.M., & Berkowitz, M.L. United atom force field for phospholipid membranes: constant pressure molecular dynamics simulation of dipalmitoylphosphatidicholine/water system. Journal of Computational Chemistry 1999;20:531–545.

[262] Pastor, R.W., Venable, R.M., & Feller, S.E. Lipid bilayers, NMR relaxation, and computer simulations. Accounts of Chemical Research 2002;35:438–446.

[263] Klauda, J.B., Kučerka, N., Brooks, B.R., Pastor, R.W., & Nagle, J.F. Simulation-based methods for interpreting x-ray data from lipid bilayers. Biophysical Journal 2006;90:2796–2807.

[264] Nickels, J.D., Chatterjee, S., Mostofian, B., et al. Bacillus subtilis lipid extract, a branched-chain fatty acid model membrane. The Journal of Physical Chemistry Letters 2017;8:4214–4217.

[265] Ferreira, T.M., Coreta-Gomes, F., Ollila, O.S., Moreno, M.J., Vaz, W.L., & Topgaard, D. Cholesterol and POPC segmental order parameters in lipid membranes: solid state 1 H–13 C NMR and MD simulation studies. Physical Chemistry Chemical Physics 2013;15, 1976–1989.

[266] Venable, R.M., Brown, F.L.H., & Pastor, R.W. Mechanical properties of lipid bilayers from molecular dynamics simulation. Chemistry and Physics of Lipids 2015;192:60–74.

[267] Harmandaris, V.A., & Deserno, M. A novel method for measuring the bending rigidity of model lipid membranes by simulating tethers. The Journal of Chemical Physics 2006;125:204905.

[268] Watson, M.C., Brandt, E.G., Welch, P.M., & Brown, F.L.H. Determining biomembrane bending rigidities from simulations of modest size. Physical Review Letters 2012;109:028102.

[269] Kuzmin, P.I., Akimov, S.A., Chizmadzhev, Y.A., Zimmerberg, J., & Cohen, F.S. Line tension and interaction energies of membrane rafts calculated from lipid splay and tilt. Biophysical Journal 2005;88:1120–1133.

[270] Khelashvili, G., Kollmitzer, B., Heftberger, P., Pabst, G., & Harries, D. Calculating the bending modulus for multicomponent lipid membranes in different thermodynamic phases. Journal of Chemical Theory and Computation 2013;9:3866–3871.

[271] Baumgart, T., Das, S., Webb, W.W., & Jenkins, J.T. Membrane elasticity in giant vesicles with fluid phase coexistence. Biophysical Journal 2005;89:1067–1080.

[272] Harder, T., Scheiffele,, P., Verkade, P., & Simons, K. Lipid domain structure of the plasma membrane revealed by patching of membrane components. The Journal of Cell Biology 1998;141:929–942.

[273] Aufderhorst-Roberts, A., Chandra, U., & Connell, S.D. Three-phase coexistence in lipid membranes. Biophysical Journal 2017;112:313–324.

[274] Gumí-Audenis, B., Costa, L., Carlá, F., Comin, F., Sanz, F., & Giannotti, M.I. Structure and nanomechanics of model membranes by atomic force microscopy and spectroscopy: insights into the role of cholesterol and sphingolipids. Membranes 2016;6:58.

[275] Kinnun, J.J., Mallikarjunaiah, K.J., Petrache, H.I., & Brown, M.F. Elastic deformation and area per lipid of membranes: Atomistic view from solid-state deuterium NMR spectroscopy. Biochimica et Biophysica Acta (BBA) - Biomembranes 2015;1848:246–259.

[276] Pluhackova, K., & Böckmann, R.A. Biomembranes in atomistic and coarse-grained simulations. Journal of Physics: Condensed Matter 2015;27:323103.

Yevhen Cherniavskyi, D. Peter Tieleman
19 Simulations of biological membranes with the Martini model

Abstract: Molecular dynamics (MD) simulations are an extremely useful class of methods for characterization of biological membranes. Constant increase of computer power and simulation software efficiency during the past few decades significantly pushed the boundaries in the field. With modern all-atom force fields, one can easily simulate small membrane system using desktop computer and obtain atomistic level of structural detail. But despite this progress, all-atom force fields introduce pose practical limits on the system sizes and timescales accessible for MD. Less detailed, coarse-grained force fields significantly reduce the amount of computer power required for simulations, but this speedup comes at a cost of structural detail. The Martini model is the most widely used coarse-grain force field for MD simulations of biological membranes and membrane proteins. While retaining significant chemical detail of the lipids, Martini reduces the computational cost of simulations up to 10^3 times compared to all-atom force fields. This enables simulations of complex behavior of mixed membranes by MD. Here, we begin with an overview of the Martini model, its advantages and limitations in the context of MD simulations of biological membranes, followed by recent examples of large-scale simulations of mixed membrane systems.

Keywords: Lipid bilayer, molecular dynamics, MARTINI, coarsegrain force field, mixed membranes

19.1 Introduction

Theoretical studies of biological membranes can be an extremely challenging task. The level of details provided by analytically tractable models is rather limited compared to the data obtained by various experimental techniques, but the interpretation of many experimental results relies on model assumptions about structure and dynamics of processes that take place in the membrane. Molecular dynamics (MD) simulation is a middle ground between experiment and analytical theory.

In a nutshell, MD can be described as a broad class of computational methods that are based on classical statistical mechanics and (semi-)atomistic description of the system of interest. First, one has to describe the system in terms of interaction potentials between atoms that make up the molecules. In classical MD, all interactions between atoms can be divided into two classes: bonded and nonbonded interactions. Bonded interactions correspond to the potentials that describe covalent bonds between atoms within a molecule. Nonbonded interactions are the potentials that take into account electrostatic and van der Waals interactions. The set of parameters that define interaction potentials for all the components of the system is called force field. Different force fields can have different functional forms that implement the interaction potentials and can have a different level of detail used in the modeling of the system. Once all the interactions in the system are adequately described, we can simulate the behavior of the system. MD utilizes classical mechanics to get a statistical ensemble of the system and relies on the ergodicity assumption to get the thermodynamic properties of the system.

A general approach is rather simple: setup initial positions for all the atoms in the system and propagate the state of the system through time by numerically integrating classical equations of motion with interaction potentials defined by the force field. If the length of the trajectory obtained by this procedure is long enough (much larger than the characteristic time of the process of interest) and the ergodicity assumption is satisfied, we effectively obtain enough information on the partition function of our system to calculate thermodynamic properties of interest. Now we can substitute the ensemble average in the calculation of an average of the property of interest by the average over the trajectory obtained in our simulation. This approach eliminates the need for direct calculation of partition function, which is an impossible task even numerically given the number of degrees of freedom in a typical biomolecular system. Another advantage of MD simulations is that one can obtain information not only about average values of different properties of the system but also have a detailed look at non-equilibrium transition states of the system.

There are two main factors which define the quality of the resulting simulation: accuracy of the force field and the length of the simulation. The question of the quality of the force field can be subdivided into two parts – level of detail that can be captured by the force field and the accuracy of the interaction parameters. Different force fields use a different level of detail to describe the structure of the molecules used in the simulation. Some force fields use separate atoms as the smallest building block of the molecule (atomistic force fields [1–3]), while other force fields describe the molecules as groups of atoms combined into single interaction site (united atom [4–6] and coarse-grain force fields [7–9]).

The choice of an appropriate force field is an important step of the simulation process. The questions that need to be answered before choosing a force field for your simulation are "What type of effects I am interested in?," "What is the driving force behind these effects?," "Is the force field that I want to use can reproduce these effects?," "Can I reproduce the same effects with a less computationally expensive force field?". Based on the answers to these questions, one should be able to choose the force field that is the best trade-off between simulation accuracy and speed. One can choose the most detailed force field that can reproduce very fine details of molecular structure (like the polarizable Drude force field [2]) and look at the structure of the system that is not significantly influenced by the polarizability effects. The Drude force field is designed to increase the accuracy of the atomistic force field by partially taking into account the redistribution of the charge of the molecule in the presence of external electric field. For example, the proximity of an ion can induce temporary polarization of the molecule. But if the effect of such induced charges is negligibly small for the system in hand, one should consider using less computationally expensive force field (e.g., simple atomistic or united atom force field) as it will increase the simulation speed and will not have significant influence on the accuracy of the simulation.

The accuracy of the interaction parameters, while being an important factor when choosing the force field to use, is much harder to assess in general. Additionally, improved parameters of the same force field are released regularly. Multiple comparative studies of the performance of different force fields when applied to a particular problem exist in the literature and one should be aware of the possible limitations of the force field before performing a simulation.

19.2 Molecular dynamics and lipid membranes

In MD simulations of biological membranes atomistic, united atom and coarse-grain force fields are all commonly used. Atomistic and united-atom force fields are a good choice when one is interested in the fine details of the bilayer structure and simulations on a microsecond timescale are adequate to sample the timescales of interest. For example, if one wants to simulate bilayer self-assembly from a randomly distributed lipid molecules in water and then study the density distribution of different components of the lipid molecules, atomistic or united-atom force fields are the reasonable approaches. Modern software for MD simulations is rather efficient and one can perform such simulation on a desktop machine.

One of the advantages of the atomistic force fields is that it is easy to compare the results of such simulations with experimental data. If one wants to assess how well a given atomistic force field reproduces the fine details of the bilayer structure, it is straightforward to compare simulation results with small-angle X-ray or neutron scattering form factors [10]. In addition, acyl chain order parameters of lipid tails can be directly compared to deuterium NMR experiments [11]. All these quantities depend on the positions of separate atoms which are readily available from the trajectory for the case of atomistic force fields.

But the simulations with atomistic force fields are limited in terms of the simulation length. Timescales that are accessible for such simulations are few orders of magnitude shorter if compared to coarse-grained simulations. As the result, it is very hard to use atomistic force fields to study slow processes (e.g., lateral diffusion of lipids in lipid bilayers) or large systems. The time required for a proper equilibration of the bilayer which consists of a complex mixture of multiple lipid types can exceed hundreds of microseconds. The need of long simulations is especially pronounced if one is interested in the processes that take place in phase-separating lipid mixtures. Lateral diffusion of lipids in the regions of the gel phase can be extremely slow, making the time required for the equilibration far beyond what is accessible for atomistic force fields.

The problem of limited size and timescale available for the atomistic force fields can be partially alleviated by the use of united-atom force fields. This class of the force fields represents heavy atoms with their adjacent hydrogen atoms as a single interaction site. For example, CH_3 group, in this case, can be represented as a single particle, instead of four separate atoms. In most of the cases, such approximation has a minor effect on the overall accuracy but reduces the total number of particles that have to be simulated. But the overall gain in the simulation speed is still far beyond what is required to simulate slow processes which take place in lipid membranes. One of the reasons of a relatively small speedup compared to atomistic force fields is the algorithms that are used to calculate nonbonded interactions. There are two types of nonbonded interactions that are taken into account – electrostatic and van der Waals interactions. The strength of van der Waals interactions decreases as r^{-6} with the distance between the interacting particles. As the result, the contribution to the total force that comes from van der Waals interactions can be neglected if the distance between interacting particles is larger than a certain cutoff. The value of this distance cutoff for van der Waals interactions is specific for a given force field and often is of the order of 1 nm. Unfortunately, the contribution which comes from electrostatic forces decrease as $1/r$ as a function of distance, so we cannot follow the same approach as with van der Waals interactions. But the calculation of particle–particle electrostatic interaction for all the atom pairs in the system is an extremely computationally heavy process, which significantly slows down the simulation. Most of the atomistic and united-atom force fields use optimized methods (e.g., PME [11, 12] or other) for the treatment of electrostatic interactions. These methods decrease the overhead of proper calculation of electrostatic interactions, but still require a significant amount of resources, which makes this part of calculation a bottleneck for an overall performance of the simulation.

19.3 MARTINI force field

Coarse-grained force fields are force fields in which multiple heavy atoms are represented as one interaction site. MARTINI is one of the most widely used coarse-grained force field in the field of biomolecular simulations. This force field has a medium level of granularity with an average four heavy atoms (with adjacent hydrogen atoms) represented as one bead (Figure 19.1). For example, four water molecules are represented as a single bead, which corresponds to the widely observed tetrahedral arrangement of water molecules in the bulk [13]. The MARTINI force field version 2.0 was released in 2007 [9] and constantly improved and expanded to include new types of molecules [14–16]. The model designed to be as simple as possible, while retaining significant

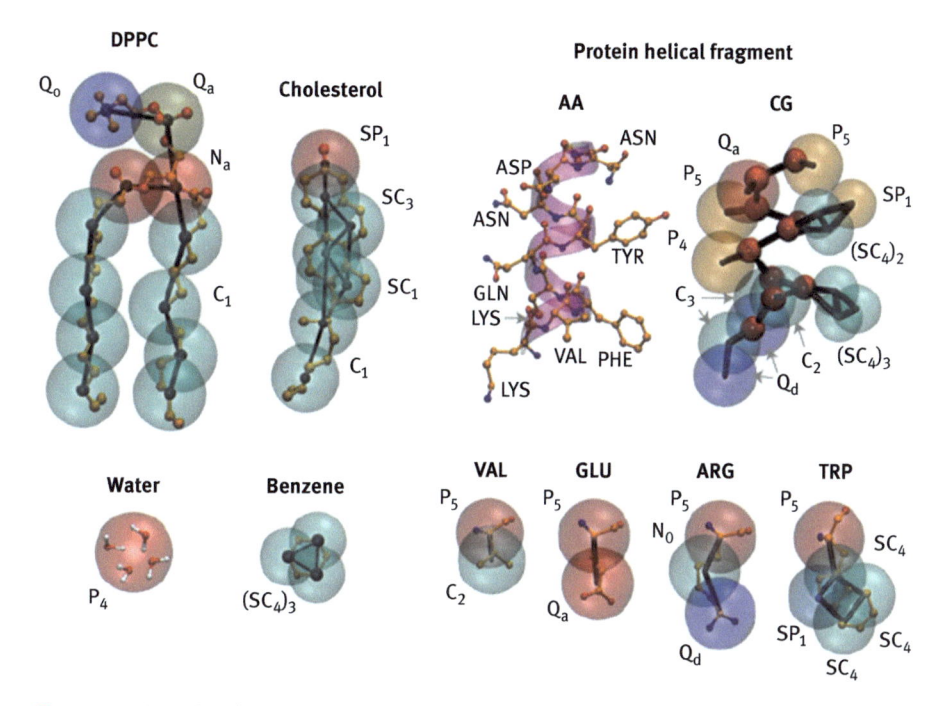

Figure 19.1: Mapping the atomistic structure to coarse-grained MARTINI beads.

chemical detail of lipids and other molecules. All the beads in the model have one out of four interaction types: charged, polar, non-polar, and apolar. These interaction types can be subdivided into subtypes based on their ability to form hydrogen bonds and the degree of polarity with the total of 18 different subtypes. Van der Waals interaction in MARTINI force field is modeled with Lenard-Jones 12-6 potential:

$$U_{LJ} = 4\epsilon_{ij}\left[\left(\frac{\sigma_{ij}}{r}\right)^{12} - \left(\frac{\sigma_{ij}}{r}\right)^{6}\right] \quad (19.1)$$

where r is a distance between the beads and σ_{ij} represents the effective size of the bead and is equal to 0.47 nm for all the bead types, except ring particles which are used for a small ring compounds. The strength of interaction with Lenard-Jones potential, defined by ϵ_{ij}, is divided into 10 interaction levels and defined by the types of interacting particles. The beads with the charged bead type carry a non-zero charge q and interact via Coulombic potential function:

$$U_{el} = \frac{q_i q_j}{4\pi\epsilon_0\epsilon_r r} \quad (19.2)$$

where ϵ_0 is dielectric permeability of vacuum, ϵ_r is relative dielectric constant. Lenard-Jones interactions are taken into account only for the pairs of beads within the distance cutoff of 1.1 nm. The potential is gradually shifted from 0.9 to 1.1 nm to ensure zero energy and force on the cutoff distance. Electrostatic interactions are normally treated with reaction-field approach [17] with relative dielectric constant of 15 and cutoff distance of 1.1 nm. As the result, electrostatic interactions within 1.1 nm cutoff are treated explicitly and outside of this distance, the medium is assumed to have uniform dielectric constant. Thus, the charge of the molecule is assumed to induce polarization of the media, which implicitly screens the charge. Such approach significantly increases simulation speed as one does not need to explicitly calculate electrostatic interactions between all the bead pairs.

Nonbonded interactions in the MARTINI force field are parametrized based on thermodynamic properties of different coarse-grained particles. The free energies of hydration, vaporization, and partitioning between water and organic phases were obtained from MD simulations of different coarse-grained beads. These energies are then compared to experimental data for small molecules that correspond to a given bead [9]. For example, free energy of vaporization can be obtained from the equilibrium densities of corresponding beads in water and vapor. With MARTINI force fields, it is possible to obtain a multi-microsecond simulation of water slab surrounded by vacuum layer, which mimics water/vapor interface. The length of such simulation is big enough to obtain equilibrium partitioning of the beads between water and vapor, thus providing an easy way to calculate hydration-free energies. A similar approach can be used to calculate partitioning-free energies for an organic phase/water type of systems (Figure 19.2). In general, MARTINI model tends to overestimate absolute values for free energies of hydration and vaporization, but the overall trend is correct. MARTINI model shows much better results in the case of oil/water partitioning free energies, which is a much more relevant parameter for the

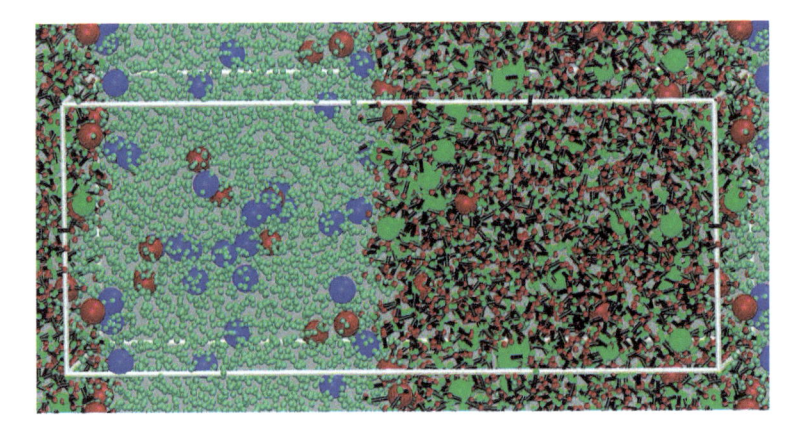

Figure 19.2: Equilibrium configuration of a water/octanol system used to determine partitioning free energies. Small beads denote the water phase (P_4) in cyan, and the octanol phase consisting of dimers of hydrocarbon (C_1) in green and alcohol (P_1) in red. The larger beads represent solutes: butane (C_1) in green, propanol (P_1) in red, and sodium ions (Q_d) in blue. The simulation box is indicated by thick gray lines.

Table 19.1: Free energies of vaporization ΔG^{vap}, hydration ΔG^{hydr}, and partitioning G^{part} between water (W) and organic phases (H, hexadecane; C, chloroform; E, ether; O, octanol) are compared to experimental values.

type	building block	examples	ΔG^{vap} exp	ΔG^{vap} CG	ΔG^{hyd} exp	ΔG^{hyd} CG	ΔG^{part}_{HW} Exp	ΔG^{part}_{HW} CG	ΔG^{part}_{CW} Exp	ΔG^{part}_{CW} CG	ΔG^{part}_{EW} exp	ΔG^{part}_{EW} CG	ΔG^{part}_{OW} exp	ΔG^{part}_{OW} CG
Qda	H$_3$N$^+$–C$_2$–OH	ethanolamine (protonated)				−25		<−30		−18		−13		−18
Qd	H$_3$N$^+$–C$_3$	1-propylamine (protonated)				−25		<−30		−18		−13		−18
Qa	NA$^+$OH	sodium (hydrated)				−25		<−30		−18		−13		−18
Qa	PO$_4^-$	phosphate				−25		<−30		−18		−13		−18
Qa	CL$^-$HO	chloride (hydrated)				−25		<−30		−18		−13		−18
Qo	C$_3$N$^+$	choline				−25		<−30		−18		−13		−18
P5	H$_2$N–C$_2$=O	acetamide	sol	sol	−40	−25	−27	−28	(−20)	−18	−15	−13	−8	−10
P4	HOH (× 4)	water	−27	−18	−27	−18	−25	−23		−14	−10	−7	−8	−9
P4	HO–C$_2$–OH	ethanediol	−35	−18	−33	−18	−21	−23		−14		−7	−8	−9
P3	HO–C$_2$=O	acetic acid	−31	−18	−29	−18	−19	−21	−9	−10	−2	−6	−1	−7
	C–NH– C=O	methylformamide	−35	−18		−18		−21		−10		−6	−5	−7
P2	C$_2$–OH	ethanol	−22	−16	−21	−14	−13	−17	−5	−2	−3	1	−2	−2
P1	C$_3$–OH	1-propanol	−23	−16	−21	−14	−9	−11	−2	−2	0	1	1	−1
P1	C$_3$–OH	2-propanol	−22	−16	−20	−14	−10	−11	−2	−2	−1	1	0	−1

(continued)

Table 19.1 (continued)

N_{da}	Group	Name	ΔG^{vap}		ΔG^{hyd}		ΔG^{part}_{HW}		ΔG^{part}_{CW}		ΔG^{part}_{EW}		ΔG^{part}_{OW}	
N_d	C_4-OH	1-butanol	−25	−16	−20	−9	−5	−7	2	0	4	2	4	3
N_d	H_2N-C_3	1-propylamine	−17	−13	−18	−9	(−6)	−7	(1)	0	(−3)	2	(3)	3
N_a	$C_3=O$	2-propanone	−17	−13	−16	−9	−6	−7	1	0	−1	2	−1	3
	$C-NO_2$	nitromethane	−23	−13	−17	−9	−6	−7		0		2	−2	3
	$C_3\equiv N$	proprionitrile	−22	−13	−17	−9	−5	−7		0		2	1	3
	$C-O-$, $C=O$	methylformate	−16	−13	−12	−9	(−6)	−7	(4)	0	(−1)	2	(0)	3
N_o	$C_2HC=O$	propanol		−13	−15	−9	−4	−7		0	2	2	3	3
	$C-O-C_2$	methoxyethane	−13	−10	(−8)	−2	(1)	−2		6	(3)	6	(3)	5
C_5	C_3-SH	1-propanethiol	−17	−10	1	1		5		10		10		6
	$C-S-C_2$	methyl ethyl sulfide	−17	−10	−6	1	(7)	5		10		10	(9)	6
C_4	$C_2=C_2$	2-butyne	−15	−10	−1	5		9		13		13	9	9
	$C=C$, $C=C$	1,3-butadiene		−10	2	5	11	9		13		13	11	9
C_3	$C-X_4$	chloroform	−18	−10	−4	5	(7)	9	14	13		13	11	9
	$C_2=C_2$	2-butene		−10		5		13		13		13	13	14
	C_3-X	1-chloropropane	−16	−10	−1	5	12	13		13		13	12	14
		2-bromopropane	−16	−10	−2	5		13		13		13	12	14
C_2	C_3	propane	gas	−10	8	10		16		15		14	14	16
C_1	C_4	butane	−11^b	−10	9	14	18	18		18		14	16	17
		isopropane	gas	−10	10	14		18		18		14	16	17

simulations of biological membranes. Table 19.1 summarizes the free energy values obtained from MD with MARTINI force field and experimental data.

Bonded interactions between covalently connected beads are modeled with a set of harmonic-like energy functions:

$$U_{bonded} = \frac{1}{2}K_{bond}(R-R_{bond})^2 + \frac{1}{2}K_{angle}(\cos(\theta)-\cos(\theta_0))^2 + K_d(1+\cos(\alpha-\alpha_d)) \\ + K_{id}(\phi-\phi_{id})^2 \quad (19.3)$$

The first term of eq. (19.3) is used to preserve the distance R between covalently bound beads. The second term is responsible for maintaining the angle θ between three covalently bound beads. The third term controls dihedral angles. The forth term ring particles and is added to prevent out of plane distortions of such compounds. Lenard-Jones interactions between nearest neighbors are excluded. Equilibrium parameters for the bound potentials are picked to mimic the distribution of the center of mass of the group of atoms that corresponds to a given bead. Atomistic models are used for such parameterization.

19.3.1 Advantages and limitations of the MARTINI model

The use of coarse-grained MARTINI force field can significantly increase the system size and timescale accessible for the simulation. This speedup is achieved through a combination of factors. First, the number of particles is reduced by a factor of approximately 10 compared to atomistic force fields. Second, with MARTINI force field, the use of computationally expensive algorithms (e.g., PME) to calculate non-bonded interactions is not needed any more. Explicit treatment of the nonbonded interactions only for the particles within a cutoff distance provides sufficient level of accuracy for the model. This also reduces the computational overhead which comes from the communication between compute nodes if the simulation is performed on distributed cluster-like systems. Third, coarse-graining of the system leads to the reduction of the total number of degrees of freedom of the system. This leads to a number of consequences, such as significantly smoother energy landscape of the system. As the result, we can use increased integration time step, the simulations with MARTINI force field. Typical integration time steps in atomistic simulations are 1 or 2 fs. The smoother energy landscape with MARTINI allows time steps of up to 20–40 fs, contributing one order of magnitude to the total speedup. The combination of these factors gives a total speedup of up to 10^3 times compared to all-atom force fields. For example, the simulation of a large DPPC bilayer (c.a. 180000 coarse-grain beads) with the MARTINI force field on 250 cores can give the performance of approximately 6 microseconds per day [18].

Reduced number of degrees of freedom has some other consequences besides the use of increased integration step. First, the simulation time no longer corresponds to the real time. This means that if some process takes 2 microseconds of simulation time then that does not necessarily imply that it will take the same 2 microseconds in reality. For example, if one compare the speed of lateral diffusion of lipids in the lipid membrane, the diffusion will occur approximately 4–5 times faster than in real membrane [9, 19–21]. This is a direct consequence of decreased "roughness" of the energy landscape, which allows faster diffusion. But the problem is that for different processes the speedup can be different. As the result, timescales with MARTINI force field require additional effort to be interpreted in a rigorous manner. The second limitation is the shift between entropic and enthalpic contributions to the free energy of the system. Decreased number of degrees of freedom leads to reduced contribution from the entropic term in the total free energy. This leads to increased contribution from the enthalpic term, as a total free energy stays the same compared to the atomistic representation of the system. As the result, correct behavior of the different properties of the system as a function of temperature can be hard to reproduce, because the entropic term in the free energy is significantly underestimated in the MARTINI force field. Nevertheless, if the simulation is performed at approximately physiological temperatures, this effect should not significantly influence the simulation results.

Another limitation of the MARTINI model, which is not very relevant for its application to biological membranes, is fixed secondary structure of proteins. The secondary structure of a protein is defined by the network of hydrogen bonds that can be formed by the backbone atoms. The level of granularity in the MARTINI force field does not explicitly capture these effects. As the result, the secondary structure of proteins should be fixed by introduction of additional potentials to avoid spontaneous unfolding of the protein. This is commonly done by creating the network of week harmonic potentials (called ELNeDyn [22]), which keeps secondary structure fixed. Consequently, simulations with the MARTINI model cannot explicitly capture the effects which involve significant changes in the secondary structure of the protein.

Despite the fact that the MARTINI was parameterized to reproduce thermodynamic properties of the system, many structural properties of lipid bilayers still can be obtained with a high level of accuracy. For example, areas per lipid molecule in the lipid bilayer in a fluid phase agree well with experimental data. Surprisingly, in certain cases, the MARTINI even outcompete atomistic force fields – DOPC area per lipid with the MARTINI is approximately 67.8 A^2 at 303K, experimentally measured area per lipid is 67.4 A^2 [23] and all-atom CHARMM36m force field produce area per lipid of 69.3 A^2. But despite rather good agreement between experimental data and MARTINI simulations for the structural properties of many lipid types, the MARTINI model cannot discriminate between the lipid types which have acyl chain length that differs less than by four carbon atoms. Thus, DLPC and DMPC lipids will have exactly the same structure in the MARTINI model.

19.3.2 Applications of the MARTINI model

The combination of the speed and accuracy of the properties of the lipid bilayers makes the MARTINI force field a very powerful tool, which can be used for the simulations of biological membranes. This becomes evident if one is interested in the behavior of the membranes which are composed of multiple lipid types and can form different phases, depending on the composition and temperature. A recent study by Baoukina et al. [24] is focused on the properties of binary and ternary lipid mixtures which are known to phase separate and form domains enriched in liquid disordered, liquid ordered, or gel phases at the different temperatures. In this study, authors used DPPC/DUPC and DPPC/DUPC/Chol mixtures simulated in a temperature range of 270–340 K. Such composition and the temperature range allowed them to directly look at lipid mixing and demixing which is thought to be the driving force behind lipid rafts formation. All the simulations in this study started as a random mixture of approximately 4.6 thousands of lipid molecules in bilayer phase and were simulated for 10–30 microseconds. Such a long simulation length and big system size are essential for the study of phase-separating lipid bilayers. Slow lateral diffusion of lipids imposes very strict limitations on the minimum length of the simulation, which is required for the equilibration of the system. The size of the system is also important in this case, as if the system is too small it will artificially limit the growth of the lipid domains. This study provides a detailed look at different structural properties of phase-separating lipid mixtures at a different temperature (Figure 19.3).

The systems that can be simulated with the MARTINI force field are not limited to model membranes. Recent work by Ingólfsson et al. [25] provides an interesting insight into the organization of lipids in the mammalian plasma membrane. In this work, realistic model of the mammalian plasma membrane was constructed and simulated for 40 microseconds. The system was composed of 63 different lipid types and has had an asymmetric distribution of lipids in inner and outer leaflets (Figure 19.4). The total number of lipid molecules was approximately 20,000, which corresponds to the system size of $71 \times 71 \times 11$ nm. Such system provides a detailed look how lipids interact with each other, form domains, and redistribute in the real plasma membrane. This type of simulations is an exclusive source of near-atomic details of plasma membrane organization, which is extremely hard to achieve modern experimental techniques. It is likely that in the nearest future it will be possible to explicitly simulate whole organelles with coarse-grained force fields like MARTINI.

On a length scale that exceeds bilayer thickness, the energetics of membrane is dictated mainly by its elastic properties. While some of the elastic properties of lipid bilayers can be measured experimentally, other important quantities remain hard to prob. One of such properties is Gaussian curvature elasticity. It is hard to design a measurement, which will be sensitive to the Gaussian curvature modulus. Hu et al. in their paper [26] propose a robust way of Gaussian curvature modulus measurement. The method is based on the measurement of the probability of the spontaneous

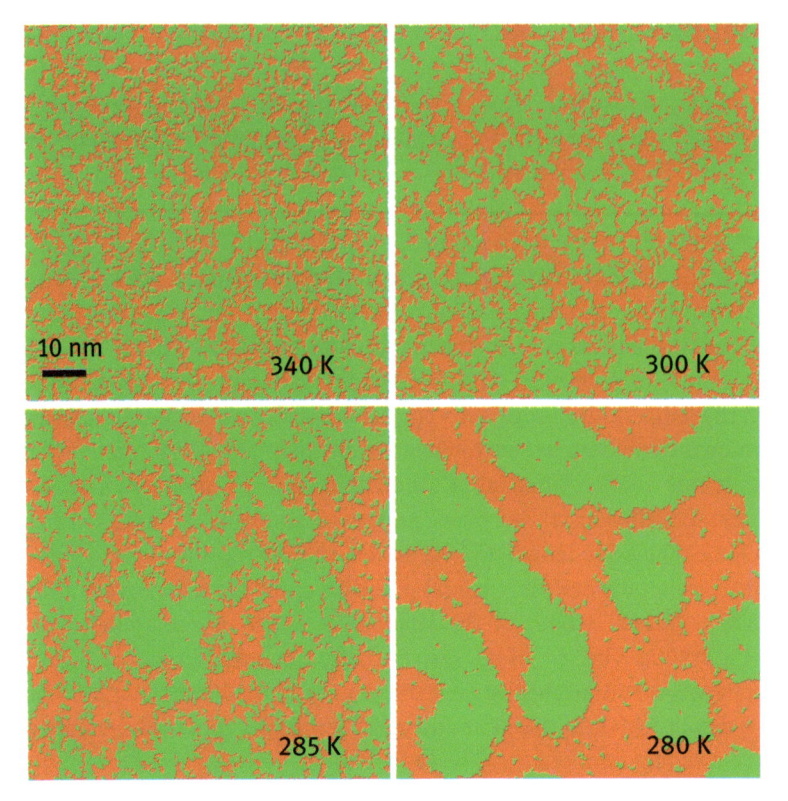

Figure 19.3: Phase behavior of the DPPC/DUPC 3:2 large bilayers at selected temperatures. View from top on the upper leaflet is given. DPPC is shown in green, DUPC in orange; water not shown.

transition of circular membrane patch closing up into a vesicle as a function of initial membrane curvature. Hu et al. performed six sets of simulations of curved membrane patches with 1200 DMPC lipids per patch using the Martini model. Each set was composed of 60 independent simulations in order to get a reliable value of closure probabilities, which can be converted into the Gaussian curvature modulus. This method was compared with a stress profile method, which provides seemingly more direct way to measure the Gaussian curvature modulus and is based on the integration of the bilayer stress profile. Stress profile can be used to measure different elastic properties of the bilayer – surface tension, lipid curvature, and Gaussian curvature modulus. Additionally, it requires much simpler simulation protocol and one does not need to perform multiple sets of simulations. But, as shown by Hu et al., this method can produce incorrect result as it is significantly influenced by the regions of the stress profile located at the water/lipid interface [27], which are hard to measure with a high accuracy in simulations. The patch-closure method, while being more demanding for the number and complexity of the simulations that has to be performed, gives much more reliable way to measure the Gaussian curvature modulus. This study

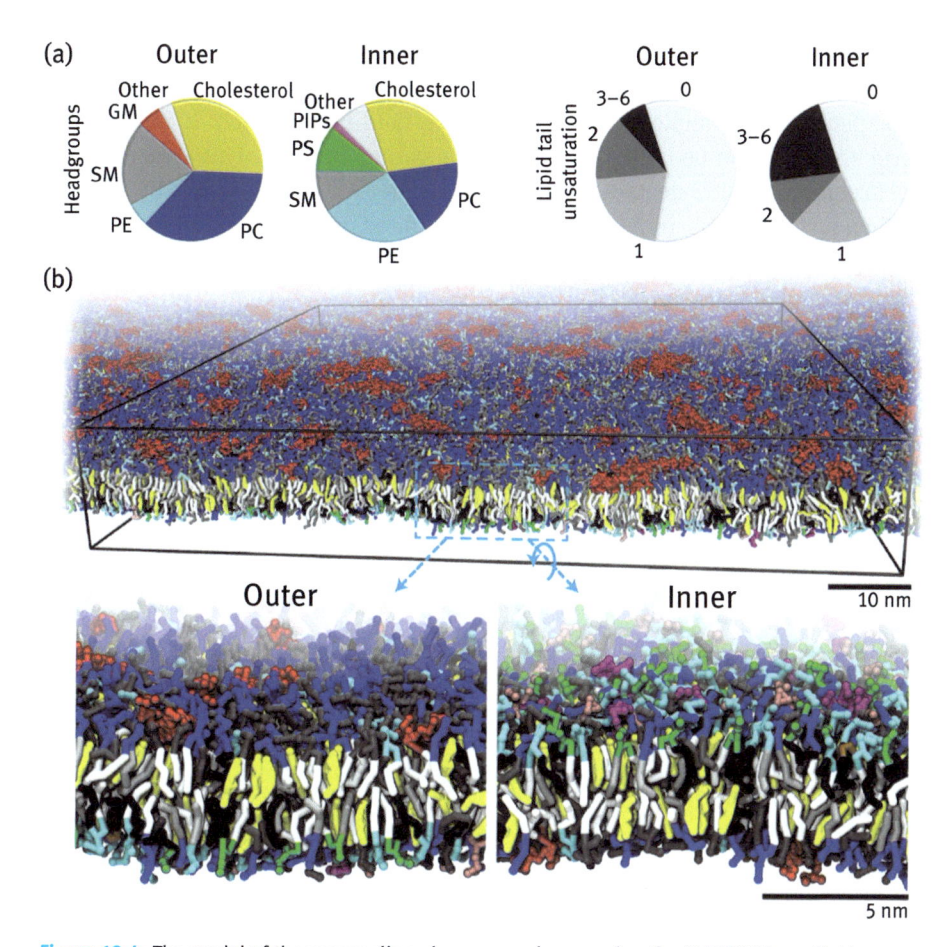

Figure 19.4: The model of the mammalian plasma membrane using the MARTINI force field.

illustrates how the advantages of the Martini model (computationally cheep, but still accurate) can be used for robust and reliable methods to study elastic properties of the lipid membranes. Also, it is always important to keep in mind the limitations of the Martini force field. For example, another study of elastic properties of lipid bilayers by Wang and Deserno [28] concluded that the energy required for a lipid molecule to tilt in the Martini model is two to three times higher in comparison with more highly resolved united atom Berger force field. Additionally, they have found that DMPC lipids in Berger force field [29] have different stiffness of the tail and headgroup regions, the feature which is not captured by coarse-grain Martini force field [28].

Coarse-grain Martini force field is a valuable model which can provide insight into the processes which are hard to simulate with atomistic or united-atom force fields due to the limitations on accessible size and timescale that can be efficiently sampled with a limited amount of computational resources. However, in certain cases, the use of only coarse-grained model is not enough. For many applications, the resolution provided my the Martini model is too coarse to resolve a subtle details lipid–lipid or lipid–protein interactions, but the system in hand is very computationally expensive to simulate with atomistic force field from scratch. In such cases, hybrid approach can be used – the simulation can be performed on a coarse-grain level to equilibrate slow degrees of freedom (e.g., lateral lipid diffusion in the bilayer) and then reverse coarse-grained to atomistic or united-atom representation to get an insight into the fine details of molecular interactions. For such a purpose, the tool called Backwards was developed [30]. This tool allows user to change the resolution of existing system by backmapping a coarse-grained structure to atomistic one. For example, in the paper by Tahir et al. [31], the authors were studying the effect of lipid membrane composition on the frequency of lipid tail protrusions into the solvent region. The Martini force field does not capture very well the flexibility of lipid tails, but for the mixed bilayers, the time required for a proper mixing of different components is rather high. The authors decided to perform coarse-grain Martini simulation of lipid bilayers composed of DPPC/DOPC and DPPC/DLPC mixtures, 512 lipid molecules per system, and then backmap prequilibrated structure to atomistic resolution (Figure 19.5). By performing the coarse-grain simulation for 3 microseconds, proper

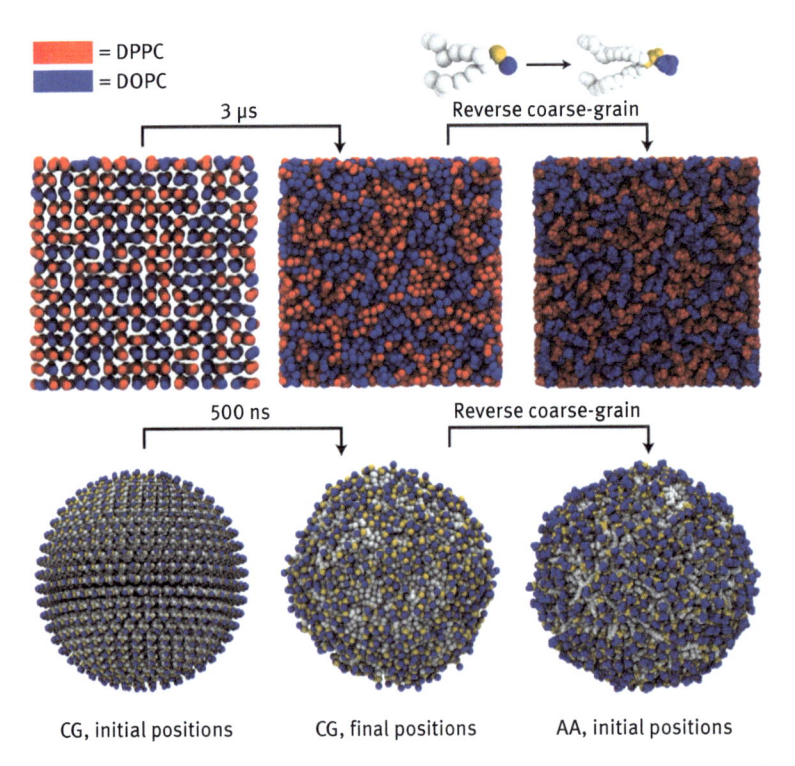

Figure 19.5: The preparation of pre-equilibrated atomistic DPPC/DOPC bilayer (top) and DOPC vesicle using the Martini coarse-grained model.

mixing of different lipid types was achieved. Subsequent simulation with the atomistic resolution allowed the authors to get an insight into the fine-grain details of the bilayer defects that originate from lipid tail protrusions. The backmapping approach is not limited to the cases where it is hard to prepare properly equilibrated initial system setup. Herzog et al. used the Martini force field to study the interaction and conformational dynamics of focal adhesion kinase protein which binds peripherally to the lipid membranes [32]. The use of the Martini force field allowed them to simulate multiple systems for a 5 microseconds each and sample different binding modes of the protein as a function of membrane composition. To get a better understanding of lipid–protein interactions, the authors backmapped the selected structures from a coarse-grain trajectory and used them as a starting point for the atomistic simulation. Similar approach was used by Cheng et al. to get an insight on membrane–protein interactions and preferred position of a model $A\beta^{42}$ peptide [33].

19.4 Conclusions

MD simulations can be a very valuable tool in the study of biological membranes. This class of methods is a middle ground between theory and experiments. The combination of all-atom and coarse-grained simulations can provide a detailed look at the fine structure and general properties of complex lipid mixtures. Recent advances in the hardware and software efficiency significantly push the limits for the simulations toward the state where explicit simulation of realistic cell membranes and organelles becomes possible. Some authors even refer to MD as "computational microscopy" [25]. While simulations will never be a substitution for the real experiment, it is hard to overestimate the significance of computational methods as a source of the structural and dynamic data for biological membranes.

In this chapter, we gave a brief overview of MD as a method, with special attention paid to coarse-grained MARTINI force field, its advantages and limitations and its applications in the field of membrane structure.

References

[1] Klauda, J.B., Venable, R.M., Freites, J.A., O'Connor, J.W., Tobias, D.J. Mondragon-Ramirez, C., Vorobyov, I., MacKerell, A.D., & Pastor, R.W. "Update of the CHARMM all-atom additive force field for lipids: Validation on six lipid types," *J. Phys. Chem. B*, vol. 114, no. 23, pp. 7830–7843, 2010.

[2] Li, H., Chowdhary, J., Huang, L., He, X., MacKerell, A.D., & Roux, B. "Drude polarizable force field for molecular dynamics simulations of saturated and unsaturated zwitterionic lipids," *J. Chem. Theory Comput.*, vol. 13, no. 9, pp. 4535–4552, Sep. 2017.

[3] Jorgensen, W.L., Maxwell, D.S., & Tirado-Rives, J. "Development and testing of the OPLS all-atom force field on conformational energetics and properties of organic liquids," *J. Am. Chem. Soc.*, vol. 118, no. 45, pp. 11225–11236, Jan. 1996.

[4] Schmid, N., Eichenberger, A.P., Choutko, A., Riniker, S., Winger, M., Mark, A.E., & Gunsteren, W.F. "Definition and testing of the GROMOS force-field versions 54A7 and 54B7," *Eur. Biophys. J.*, vol. 40, no. 7, pp. 843–856, 2011.

[5] Wang, L., Wu, Y., Deng, Y., Kim, B., Pierce, L., Krilov, G., Lupyan, D., Robinson, S., Dahlgren, M.K., Greenwood, J., Romero, D.L., Masse, C., Knight, J.L., Steinbrecher, T., Beuming, T., Damm, W., Harder, E., Sherman, W., Brewer, M., Wester, R., Murcko, M., Frye, L., Farid, R., Lin, T., Mobley, D.L., Jorgensen, W.L., Berne, B.J., Friesner, R.A., & Abel, R. "Accurate and reliable prediction of relative ligand binding potency in prospective drug discovery by way of a modern free-energy calculation protocol and force field," *J. Am. Chem. Soc.*, vol. 137, no. 7, pp. 2695–2703, Feb. 2015.

[6] Jorgensen, W.L., & Tirado-Rives, J. "The OPLS [optimized potentials for liquid simulations] potential functions for proteins, energy minimizations for crystals of cyclic peptides and crambin," *J. Am. Chem. Soc.*, vol. 110, no. 6, pp. 1657–1666, Mar. 1988.

[7] Cooke, I.R., Kremer, K., & Deserno, M. "Tunable generic model for fluid bilayer membranes," *Phys. Rev. E*, vol. 72, no. 1, p. 11506, Jul. 2005.

[8] Sodt, A.J., & Head-Gordon, T. "An implicit solvent coarse-grained lipid model with correct stress profile," *J. Chem. Phys.*, vol. 132, no. 20, p. 205103, May 2010.

[9] Marrink, S.J., Risselada, H.J., Yefimov, S., Tieleman, D.P., & de Vries, A.H. "The MARTINI force field: Coarse grained model for biomolecular simulations," *J. Phys. Chem. B.*, vol. 111, no. 27, pp. 7812–7824, Jul. 2007.

[10] Kučerka, N., Katsaras, J., & Nagle, J.F. "Comparing membrane simulations to scattering experiments: Introducing the SIMtoEXP software," *J. Membr. Biol.*, vol. 235, no. 1, pp. 43–50, May 2010.

[11] Singer, P.M., Asthagiri, D., Chapman, W.G., & Hirasaki, G.J. "Molecular dynamics simulations of NMR relaxation and diffusion of bulk hydrocarbons and water," *J. Magn. Reson.*, vol. 277, pp. 15–24, 2017.

[12] Essmann, U., Perera, L., Berkowitz, M.L., Darden, T., Lee, H., & Pedersen, L.G. "A smooth particle mesh ewald method," *J. Chem. Phys.*, vol. 103, no. 19, pp. 8577–8593, Nov. 1995.

[13] Voloshin, V.P., & Naberukhin, Y.I. "Proper and improper hydrogen bonds in liquid water," *J. Struct. Chem.*, vol. 57, no. 3, pp. 497–506, 2016.

[14] Monticelli, L., Kandasamy, S.K., Periole, X., Larson, R.G., Tieleman, D.P., & Marrink, S.-J. "The MARTINI coarse-grained force field: Extension to proteins," *J. Chem. Theory Comput.*, vol. 4, no. 5, pp. 819–834, May 2008.

[15] Uusitalo, J.J., Ingólfsson, H.I., Akhshi, P., Tieleman, D.P., & Marrink, S.J. "Martini coarse-grained force field: Extension to DNA," *J. Chem. Theory Comput.*, vol. 11, no. 8, pp. 3932–3945, Aug. 2015.

[16] López, C.A., de Vries, A.H., & Marrink, S.J. "Computational microscopy of cyclodextrin mediated cholesterol extraction from lipid model membranes," *Sci. Rep.*, vol. 3, p. 2071, Jun. 2013.

[17] Barker, J.A., & Watts, R.O. "Monte Carlo studies of the dielectric properties of water-like models," *Mol. Phys.*, vol. 26, no. 3, pp. 789–792, Sep. 1973.

[18] de Jong, D.H., Baoukina, S., Ingólfsson, H.I., & Marrink, S.J. "Martini straight: Boosting performance using a shorter cutoff and GPUs," *Comput. Phys. Commun.*, vol. 199, pp. 1–7, 2016.

[19] Bennett, W.F.D., & Tieleman, D.P. "Computer simulations of lipid membrane domains," *Biochim. Biophys. Acta – Biomembr.*, vol. 1828, no. 8, pp. 1765–1776, 2013.

[20] Tocanne, J.-F., Dupou-Cézanne, L., & Lopez, A. "Lateral diffusion of lipids in model and natural membranes," *Prog. Lipid Res.*, vol. 33, no. 3, pp. 203–237, 1994.

[21] Vaz, W.L., & Almeida, P.F. "Microscopic versus macroscopic diffusion in one-component fluid phase lipid bilayer membranes," *Biophys. J.*, vol. 60, no. 6, pp. 1553–1554, 1991.
[22] Periole, X., Cavalli, M., Marrink, S.-J., & Ceruso, M.A. "Combining an elastic network with a coarse-grained molecular force field: Structure, dynamics, and intermolecular recognition," *J. Chem. Theory Comput.*, vol. 5, no. 9, pp. 2531–2543, Sep. 2009.
[23] Kučerka, N., Nagle, J.F., Sachs, J.N., Feller, S.E., Pencer, J., Jackson, A., & Katsaras, J., "Lipid bilayer structure determined by the simultaneous analysis of neutron and X-ray scattering data," *Biophys. J.*, vol. 95, no. 5, pp. 2356–2367, Sep. 2008.
[24] Baoukina, S., Rozmanov, D., & Tieleman, D.P. "Composition fluctuations in lipid bilayers," *Biophys. J.*, vol. 113, no. 12, pp. 2750–2761, 2017.
[25] Ingólfsson, H.I., Arnarez, C., Periole, X., & Marrink, S.J., "Computational 'microscopy' of cellular membranes," *J. Cell Sci.*, vol. 129, no. 2, p. 257 LP-268, Jan. 2016.
[26] Hu, M., Briguglio, J.J., & Deserno, M. "Determining the gaussian curvature modulus of lipid membranes in simulations," *Biophys. J.*, vol. 102, no. 6, pp. 1403–1410, Mar. 2012.
[27] Hu, M., de Jong, D.H., Marrink, S.J., & Deserno, M. "Gaussian curvature elasticity determined from global shape transformations and local stress distributions: A comparative study using the MARTINI model," *Faraday Discuss.*, vol. 161, no. 0, pp. 365–382, 2013.
[28] Wang, X., & Deserno, M. "Determining the lipid tilt modulus by simulating membrane buckles," *J. Phys. Chem. B.*, vol. 120, no. 26, pp. 6061–6073, Jul. 2016.
[29] Berger, O., Edholm, O., & Jähnig, F. "Molecular dynamics simulations of a fluid bilayer of dipalmitoylphosphatidylcholine at full hydration, constant pressure, and constant temperature," *Biophys. J.*, vol. 72, no. 5, pp. 2002–2013, 1997.
[30] Wassenaar, T.A., Pluhackova, K., B??ckmann, R.A., Marrink, S.J., & Tieleman, D.P. "Going backward: A flexible geometric approach to reverse transformation from coarse grained to atomistic models," *J. Chem. Theory Comput.*, vol. 10, no. 2, pp. 676–690, 2014.
[31] Tahir, M.A., Van Lehn, R.C., Choi, S.H., & Alexander-Katz, A. "Solvent-exposed lipid tail protrusions depend on lipid membrane composition and curvature," *Biochim. Biophys. Acta – Biomembr.*, vol. 1858, no. 6, pp. 1207–1215, 2016.
[32] Herzog, F.A., Braun, L., Schoen, I., & Vogel, V. "Structural insights how PIP2 imposes preferred binding orientations of FAK at lipid membranes," *J. Phys. Chem. B.*, vol. 121, no. 15, pp. 3523–3535, Apr. 2017.
[33] Cheng, K.H., Qiu, L., Cheng, S.Y., & Vaughn, M.W. "Lipid insertion domain unfolding regulates protein orientational transition behavior in a lipid bilayer," *Biophys. Chem.*, vol. 206, pp. 22–39, 2015.

Zhiqiang Shen, Alessandro Fisher, Huilin Ye, Ying Li
20 Multiscale modeling of lipid membrane

Abstract: Cell membranes that consist of lipids, glycolipids, sterols, and membrane proteins are of great importance for biological functions. Due to the complexity of cell membranes, it is preferable to investigate the model lipid bilayers that consist of several different lipid species, natural proteins or artificial peptides to provide a fundamental understanding on their behaviors. Molecular dynamics (MD) simulation for lipid membranes has been gained increasing attentions due its ability to explore the structural and dynamics details at molecular scale. MD simulation serves as an indispensable tool for the experimental and theoretical researches. In simulations, the development of multiscale lipid models is crucial to explore the multiscale phenomena occurred in cell membranes. In this chapter, we will introduce these multiscale lipid models in simulations, ranging from the all-atom lipid model to the triangulated membrane model. We hope that our chapter can serve as a tutorial for beginners in this field and provide an overview about the computational simulations for cell membranes.

Keywords: cell membrane, molecular dynamic simulation, multiscale modeling

20.1 Introduction

Cell membranes that provide a functional barrier between cells and their surroundings are critical in many biological processes. As shown in the schematic of a cell membrane in Figure 20.1, the cell membrane is composed of lipids, glycolipids, sterols, and membrane proteins that are imbedded or translocated throughout the membrane [1–3]. These membrane-relevant molecules contain a large variety of species. The lipid molecules in cell membranes can be classified into eight main classes. And over 30,000 different lipid structures have been discovered [4]. The membrane proteins are also in great diversity to take in charge of various kinds of functionalities. Take a simple prokaryotic, *Escherichia coli*, for example. Its outer membrane contains lots of pore-like proteins. And the inner membrane encloses large numbers of specific transport systems, such as lactose permease. Apart from that, the inner membrane is composed of three main phospholipids [5]. This great complexity of the compositions and structures of cell membranes determines their complex physical properties and functionalities. For instance, phosphatic acid acts

---

Zhiqiang Shen, Department of Mechanical Engineering, University of Connecticut, Storrs, USA
Alessandro Fisher, Department of Biomedical Engineering, University of Connecticut, Storrs, USA
Huilin Ye, Department of Mechanical Engineering, University of Connecticut, Storrs, USA
Ying Li, Department of Mechanical Engineering, University of Connecticut, Storrs, USA; Institute of Materials Science, University of Connecticut, Storrs, USA

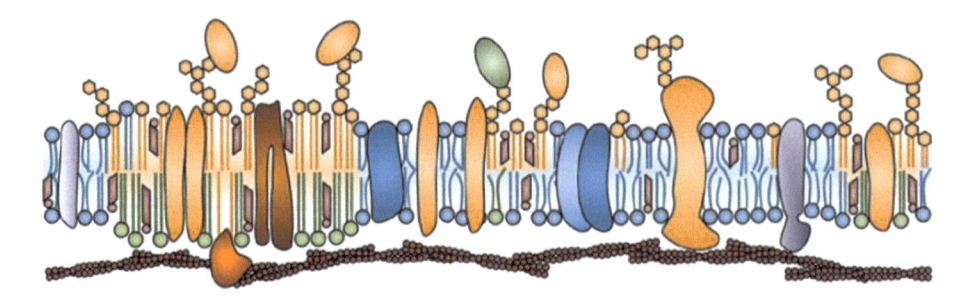

Figure 20.1: Schematic illustration of a cell membrane. The blue background indicates the fluid region of membrane. The orange background represents the liquid-order phase that contains various lipids, glycolipids and cholesterol. The brown rods represent the cytoskeleton that supports the membrane. The figure is adopted from Ref. [16] with permission.

as a pH sensor, which connects the environmental pH value and intercellular signaling [6]. Phosphatidylinositol in the inner leaflet behaves like a second messenger [7]. Furthermore, lipids and membrane proteins interact with each other to determine membrane structure. For instance, the asymmetrical lipid distributed between the two leaflets of the membrane is affected by the proteins, flippases, and floppases. These two proteins are in charge of the transport of lipids across two leaflets through an adenosine triphosphate cost process [8]. All of these diversities in compositions, structures, and functionalities in cell membrane are crucial to maintain the cell life.

Due to the complexity of cell membranes, it is preferable to investigate simplified membrane systems. Such reconstituted lipid bilayers consist of several different lipid species, natural proteins, or artificial peptides that can be treated as a model system to mimic the cell membrane. Research focusing on reconstituted lipid bilayers may provide fundamental understanding of the relevant biological processes. Experimental and theoretical approaches on reconstituted lipid bilayers are of great importance for the investigation of model systems. However, due to the limitation in experimental measurements, it is inherently difficult to provide the dynamic and structural insight at molecular scales. Molecular dynamics (MD) simulation has attracted more attentions for its ability to observe dynamic processes [9]. MD simulations not only act as a method to prove and improve the theories, but also serve as an explanation and guideline for experimental works [10, 11].

The behaviors of lipid membranes occur at different time and length scales [12]. For example, the orientation change of the lipid head group with respect to the membrane planar surface may happen within a nanosecond. But the movement of the whole lipid membrane is much slower. The exchange of positions between two neighboring lipids might take a few nanoseconds. Furthermore, the exchanging of lipid molecules between two leaflets needs to take from minutes to several hours [13]. On the other hand, the length of a lipid molecule and the thickness of lipid bilayer are on the order of ten Angstroms. The lipid raft in the membrane is typically on the order from 10 to 100 nm [14]. The protein-rich region in the plasma membrane is around 1 μm [15], while the size of the cell is on the order of several micrometers.

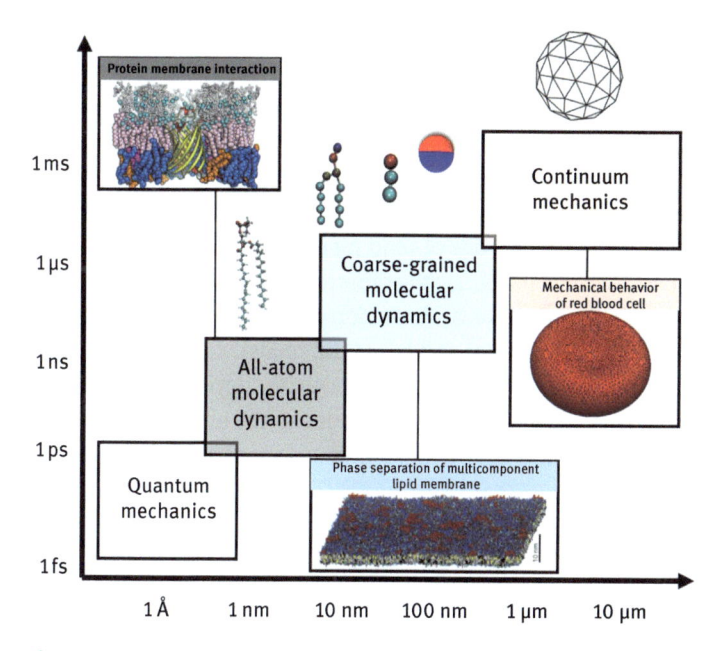

Figure 20.2: A schematic diagram of the length and time scales for lipid membranes accessible by computational simulations.

To understand the lipid membrane in different time and length scales, multiscale lipid models are necessary to explore the relevant phenomena (Figure 20.2). For example, the all-atom MD simulation can provide both a high time resolution for the dynamics of a membrane as well as its sub-Angstrom spatial resolution. But it is difficult to be used to investigate the phase separation of lipid membrane on long time scale. In some situations, instead of focusing on the details of each atom, we are more interested in the overall lipids or membrane properties. Therefore, coarse-grained lipid is a widely used method that reduces the computational degree of freedom but preserves the necessary information for the problems in the specific time and length scales. In the coarse-grained MD lipid models, a group of atoms in the lipid is usually considered as a single bead according to different mapping methods. These coarse-grained lipid models can be classified into two categories: explicit solvent lipid models and implicit solvent lipid models. For instance, Martini [17, 18] and dissipative particle dynamics (DPD) [19] lipids belong to the explicit solvent lipid models, in which several water molecules are represented by a bead as they interact with the coarse-grained lipids. Nevertheless, to simulate the lipid membrane with explicit solvent lipid models, we need to take far more water beads than actual lipids. Therefore, the interaction between water beads accounts for the most of the computational cost. To solve this problem, the implicit solvent lipid models have been proposed, where the water molecules are not considered in the simulation. In the simulation with implicit solvent lipid models, the membrane can be on the order of 100 nm. Despite the success of coarse-grained lipid models in MD simulation, in some situations, we need to consider the whole cell, such as a red blood cell (RBC). To realize this, the cell surfaces are usually modeled by a triangulated membrane surface [20–22].

In this chapter, we will introduce these multiscale lipid models in simulations, ranging from the all-atom lipid model to the one-bead coarse-grained lipid model, as shown in Figure 20.3. The following parts of this chapter are arranged as described below: First, we will introduce the properties of lipid membranes and their measurements in simulations. Second, we will introduce the different lipid models. Particularly, we will talk about the Chemistry at HARvard Macromolecular Mechanics (CHARMM) lipid model for all-atom lipid model, Martini and DPD lipid models for explicit solvent lipid model, and one-bead lipid and three-bead lipid models for implicit lipid models. We will also briefly cover the triangulated membrane model. In the end of this chapter, we will go through those widely used simulation packages for simulations.

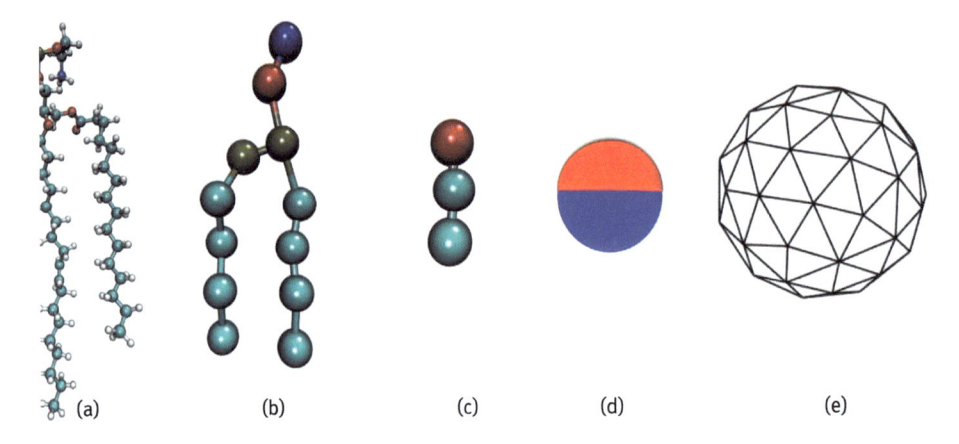

Figure 20.3: Lipid models in simulation. **(a)** All-atom model of 1,2-dioleoyl-*sn*-glycero-3-phosphoethanolamine (DOPE) [42]. **(b)** A 12-particle coarse-grained lipid model [17, 18]. **(c)** A three-particle coarse-grained model [68]. **(d)** A one-particle coarse-grained lipid model [70]. **(e)** Triangulated vesicle surface. In the triangulated membrane model each triangle represents a patch of membrane.

20.2 Membrane properties and their measurements in simulation

The properties of membranes or lipids are of great importance when considering the interactions between membrane and other materials. In this part, we will introduce these well-defined and widely used properties, including diffusion constant, order parameter, membrane tension, stretch modulus, and bending modulus, to capture the characteristics of membrane. We will give a brief review about these properties and their measurement in the computational simulation.

20.2.1 Diffusion constant

The diffusion constant of a lipid molecule in a membrane is a parameter to measure its dynamic property. It is also highly related to the state of the lipid membrane. For example, the lipid diffusion constant of a fluid-state membrane is larger than that of a gel-state membrane. In simulations, we can measure the lipid diffusion constant by monitoring the mean squared displacement (MSD) of lipid molecules in a membrane. The relation between MSD and the lipid diffusion constant is given below:

$$D_{\text{lipid}} = \frac{1}{4\Delta t} \left\langle [r_{\|,i}(t+\Delta t) - r_{\|,i}(t)]^2 \right\rangle \tag{20.1}$$

where Δt is the period over which diffusivity is measured. The $r_{\|,i}(t)$ is the position vector of the i th lipid into the bilayer plane at time t. $[r_{\|,i}(t+\Delta t) - r_{\|,i}(t)]^2$ is the MSD for ith lipid. The angular bracket represents an average over all the lipids in the membrane. Because of this linear relationship between the MSD and diffusion constant, we can obtain the diffusion constant by fitting the MSD over a certain period.

20.2.2 Order parameter

The order parameter is used to understand the molecular structure of individual lipid molecules. In the simulation, the order parameter S can be measured by the equation as follows:

$$S = \left\langle \frac{3}{2} \cos^2 \theta_i - \frac{1}{2} \right\rangle \tag{20.2}$$

where θ is the angle between bond vectors to the bilayer normal. In the all-atom simulation, the bond is referred to the C–H bond. While in the coarse-grained lipid model, the bond is usually taken as the bond between tail beads. The angular bracket means that the equation is applied and averaged throughout the entire bilayer. $S = 1$ corresponds to the direction that aligns with the bilayer normal. $S = 0$ represents randomly orientated lipid molecules.

20.2.3 Membrane tension

Membrane tension is an important state parameter for the whole membrane. In biology, cells can adjust their membrane tensions according to their surrounding environment [23]. In computational simulations, the membrane tension is an important parameter to investigate the membrane response under different conditions. The membrane tension Σ can be directly measured in the simulation as follows [9, 24]:

$$\Sigma = \int dz \left[p_{zz}(z) - 0.5(p_{xx}(z) + p_{yy}(z)) \right] \quad (20.3)$$

where the z-axis is the normal direction of the planar bilayer interface. $p_{zz}(z)$, $p_{xx}(z)$, and $p_{yy}(z)$ are the pressure components along the z, y, and x directions, respectively.

20.2.4 Stretch modulus

The stretch modulus is an overall mechanical property of membranes, which is used to measure the energy penalty for expanding the membrane per unit area. There are two methods in simulations to measure the stretch modulus of a lipid membrane: (1) the fluctuation method and (2) the stretching method. In the fluctuation method, we will control the membrane tension at a constant value in simulation and then measure the fluctuation of the membrane area. The stretch modulus can be obtained as [25]:

$$K_A = k_B T \frac{\langle A \rangle}{N \langle (A - A_0)^2 \rangle} \quad (20.4)$$

where k_B and T are the Boltzmann constant and absolute temperature, respectively. N is the number of lipids in each leaflet. A is the area per lipid and A_0 is its equilibrium area. Usually, the membrane tension is set as zero during the testing. In the stretching method, the stretching modulus can be obtained by measuring the relation between membrane tension and its projected area. The relationship between the stretch modulus and membrane project is given as [26]

$$\Sigma = K_A \frac{A - A_0}{A_0} \quad (20.5)$$

where A is the lipid area, which is defined as projected in-plane area of each lipid molecule. A_0 is the lipid area at zero surface tension.

20.2.5 Bending rigidity

The membrane bending rigidity is another important mechanical property to determine the energy penalty required to deform the membrane from its spontaneous curvature to a different curvature. The bending rigidity of the planar membrane can be extracted from the membrane fluctuation spectrum [27]. According to the membrane profile function $h(x,y)$ of a planar membrane, its Fourier transform could be expressed as [27]:

$$h(q) = \frac{l}{L}\sum_n h(r)\exp(i\boldsymbol{q} \cdot \boldsymbol{r}) \tag{20.6}$$

where L is the lateral side length of the planar membrane. l is the membrane patch length, when we divide the membrane into small patches to capture profile function $h(x,y)$. $\boldsymbol{q} = 2\pi(n_x, n_y)/l$ is the wave vector. q is the norm of wave vector. Based on the equipartition theorem, the power spectrum is given as [27, 28]:

$$\left\langle |h(q)|^2 \right\rangle = \frac{k_B T}{l^2[\kappa q^4 + \Sigma q^2]} \tag{20.7}$$

where κ and Σ are the bending rigidity and membrane tension, respectively. Then, the bending rigidity of a membrane can be obtained by fitting the measured fluctuation spectrum to Eq. (20.7). For simplicity, the membrane tension is usually set as zero, when testing the bending rigidity. Apart from this method, according to the elastic theory, the value of the bending rigidity can be estimated by the following formula [26, 29]:

$$\kappa = \frac{K_A l_{me}^2}{48} \tag{20.8}$$

where l_{me} is the bilayer thickness.

20.2.6 Energy of membrane

With the membrane tension, stretch modulus, and bending rigidity at hand, the analysis of membrane energy change during a certain biological process is of great importance to reveal the potential underlying governing mechanisms. For a homogenous lipid membrane, its energy state can be captured by the classical Canham–Helfrich's fluid membrane mode [28, 30]:

$$H = \int_S ds \left\{\frac{\kappa}{2}M^2 + \Sigma + \kappa_G M_G\right\} + \int_V \Delta p dV \tag{20.9}$$

where H represents the energy of a lipid membrane. M and M_G are the mean curvature and the Gaussian curvature, respectively. κ_G is the Gaussian rigidity. According to the Gauss–Bonnet theory [31, 32], the term $\kappa_G M_G$ is zero if the membrane keeps its topology under the deformation. The term of $\Delta p dV$ exists only if an enclosed structure, such as vesicle, is considered. Δp represents the pressure difference inside and outside the enclosed structure. V is its total volume.

20.3 Lipid models

20.3.1 All-atom simulation model – CHARMM lipid model

The atomic-level or near-atomic-level lipid models can be classified into all-atom and united-atom models. The all-atom force fields for lipids molecules include: CHARMM [33] and AMBER [34]. While the united-atom force fields include BERGER [35] force field developed based on the OPLS united-atom force field [36] and GROMOS-based force field [37]. Among these different force fields, the CHARMM force field is one of the most used force fields to model the lipid membranes. Here, we will use the CHARMM force field as an example to introduce the lipid model in the atomic level. Please refer to the corresponding references for other all-atom lipid models.

20.3.1.1 Model and potential

The CHARMM force field is designed to simulate the biomolecules such as proteins, DNA, RNA, lipid, and drug-like molecules. The development of the CHARMM force field for lipid molecules is traced from the early 1990s, C22 [38]. The C22 force field was later optimized as the C27 [39] and the C27r [40, 41] force fields. The most recent CHARMM force field is C36 [42], which was published in 2010. The potential in the CHARMM force field includes non-bonded pair-wise, bond, angle, and dihedral interactive potentials. The formulas of the potentials are given as follows:

$$V(\hat{R}) = \sum_{bonds} K_b(b-b_0)^2 + \sum_{angles} K_\theta(\theta-\theta_2)^2 + \sum_{Dihedrals}\left[\sum_j K_{\varphi,j}(1+\cos(n_j\varphi-\delta_j))\right]$$
$$+ \sum_{\substack{nobonded \\ pair\, i,j}} \varepsilon_{ij}\left[\left(\frac{R_{min,ij}}{r_{ij}}\right)^{12} - 2\left(\frac{R_{min,ij}}{r_{ij}}\right)^{6}\right] + \sum_{\substack{nobonded \\ pair\, i,j}} \frac{q_i q_j}{\varepsilon_D r_{ij}}$$

(20.10)

where harmonic bond and angle potentials are applied with force constants K_b and K_θ, respectively. The corresponding equilibrium bond length and angle values are b_0 and θ_0. The dihedral potential is a sum of sinusoids with force constant $K_{\varphi,j}$, multiplicity n, and offset δ. In the dihedral potential j can range from 1 to 6 depending on the involved atomic properties. The non-bonded pair-wise interactions consist of two parts. The first non-bonded interaction term of Lennard–Jones (LJ) potential is used to model the Van der Waals interactions between atoms, where ε_{ij} is the depth of potential well, $R_{min,ij}$ is the distance at which the LJ potential reaches its minimum value $-\varepsilon_{ij}$. The second term is the Columbic interactive potential, where q_i and q_j are the partial atomic charges. ε_D is the dielectric constant, which is taken as 1 with explicit water molecules. The CHARMM force field works well with different lipid molecules species, such as 1,2-dipalmitoyl-sn-phosphatidylcholine (DPPC), 1-palmitoyl-2-oleoyl-sn-phosphatidylcholine (POPC), cholesterol, and 1,2-dioleoyl-sn-glycero-3-phosphoethanolamine (DOPE). The specific interactive parameters can be found in the CHARMM website: http://mackerell.umaryland.edu/charmm_ff.shtml. Additionally, the rigid TIP3P water model [42] is used for the water molecule under the framework of CHARMM force field.

20.3.1.2 Membrane properties

For simulations on atomic level, the dynamics of each lipid are of great importance when investigating the interaction between lipids and proteins, or other biomolecules. To capture the dynamic structure of lipid molecules, the order parameters of lipid molecules should be reproduced in the simulation. Taking the lipids of DPPC as an example, shown in Figure 20.4, the order parameters tested in the simulation

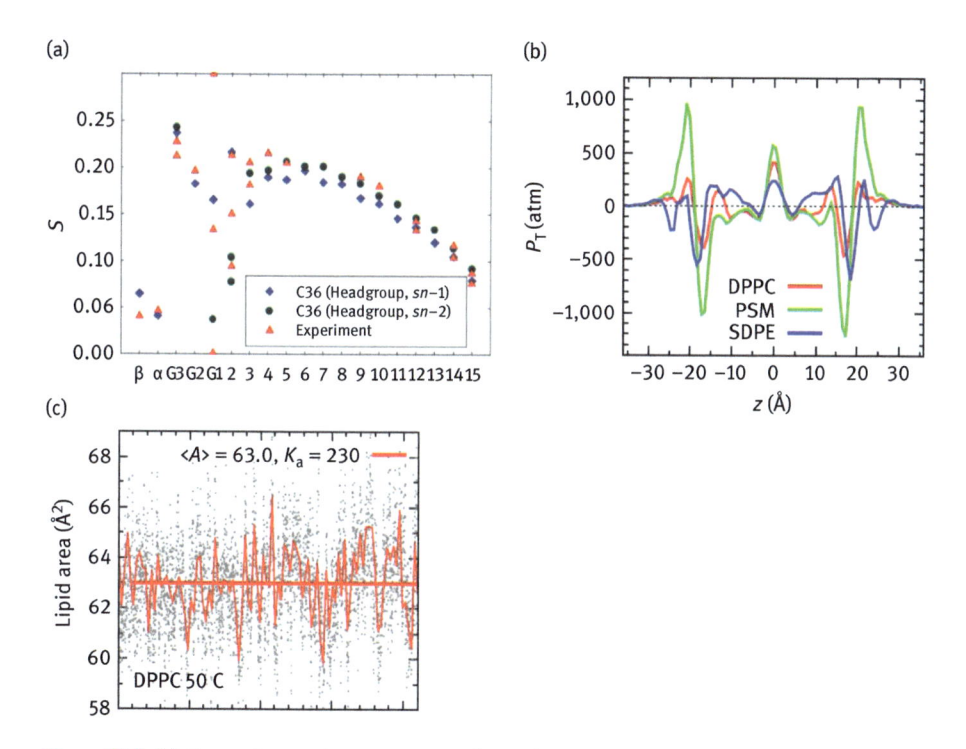

Figure 20.4: (a) Comparison order parameters of DPPC between experimental results and simulation with CHARMM C36 potential. (b) Stress distribution along the bilayer thickness directions of bilayers with components of DPPC, PSM, and SDPE, respectively. (c) Time series of area for DPPC small systems. The gray symbols are the instant values at 100 ps intervals. The fluctuation lines are Bezier smoothed 5 ps averages; the horizontal lines are the averages over 20–420 ns. Figure (a) is adopted from Ref. [42] with permission. Figures (b) and (c) are adopted from Ref [43] with permission.

under the C36 potential agree well with the experimental results. Particularly, the slight difference for the order parameters between the two chains, due to the quadrupole splitting in some parts of theses chains, is also captured. Additionally, the order parameters in the two chains decrease in the center of the bilayer, which indicates that there is more disorder in the middle plane of the lipid bilayer. The order parameters of other lipid types, such as 1-palmitoyl-2-oleoyl-sn-glycero-3-phosphocholine (POPC) and 1-palmitoyl-2-oleoyl phosphatidylethanolamine (POPE), can also be reproduced with the C36 potential.

Though the order parameter is a good property to reflect the behavior of an individual lipid, it is not enough to describe the overall bilayer characteristics. The stress profiles of DPPC, N-palmitoyl sphingomyelin (PSM), and 1-stearoyl-2-docosahexaenoyl-phosphatidylethanolamine (SDPE) lipid bilayer are given in Figure 20.4b. The stress profile is symmetric with the middle plane of the bilayer. Additionally, the stress near the water/hydrocarbon interface experiences large oscillation. Testing the area fluctuation of the bilayer and its corresponding average membrane area for the 72 DPPC lipid bilayer is shown in Figure 20.4c. We can see that the stretch modulus of the DPPC bilayer is around 215 dyn/cm, which is consistent with the experimental result [44]. Similarly, the bending rigidity of the DPPC bilayer has been measured according to eq. (20.7) in simulations and is around $9.6 \pm 0.1 \times 10^{-20}$ J.

20.3.1.3 Application

One of the advantages of the all-atom lipid model is that we can explore atomic details of the interactions between molecules in simulations. For instance, the interactions between proteins and membranes are of great importance for the biological processes. As shown in Figure 20.5, the authors adopted the C36 lipid model to investigate the *Escherichia coli* (*E. coli*) outer membrane and its interaction with outer membrane phospholipase A (OmpLA) [45]. The inner leaflet of *E. coli* is composed of phospholipids. While its outer membrane is made up of serialized amphipathic molecules denoted as lipopolysaccharide that consists of lipid A, a core oligosaccharide, and an O-antigen polysaccharide. The protein OmpLA is buried in the membrane and across the two leaflets. In the simulation, the behaviors of lipids have been explored. As shown in Figures 20.5b and 20.5c, the two-dimensional z-position distributions of the C2 and C4 atoms of lipid A (Figure 20.5b) and the acyl chain C2 atoms of phospholipids (Figure 20.5c) are given. Because of the hydrophobic parts between proteins and lipid membrane, the local thickening of lipids is founded near proteins. Furthermore, the dynamic behaviors about the structure variation of the protein can be observed in simulations. Please refer to Ref [45] for details about the simulation.

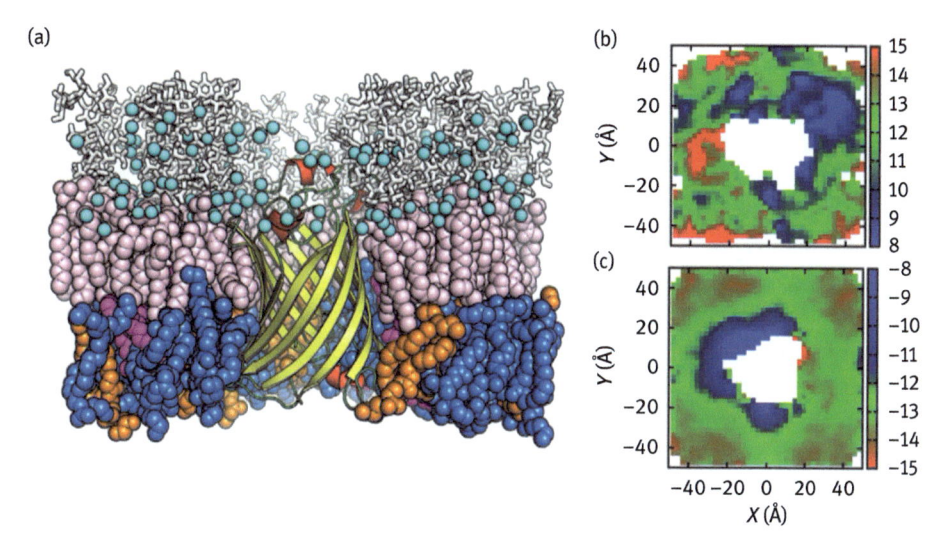

Figure 20.5: (a) A snapshot of asymmetric lipid membrane and the embedded protein. For the configuration of protein, the yellow part denotes the barrel. The red part represents the helix structure. The green part denotes the loop and turns. For the lipid membrane, Lipid A, R1 core (i.e., without O-antigen polysaccharide), 1-palmitoyl(16:0)-2-palmitoleoyl (16:1 cis-9)-phosphatidylethanolamine (PPPE), 1-palmitoyl(16:0)-2-vacenoyl(18:1 cis-11)-phosphatidylglycerol (PVPG), and 1,10-palmitoyl-2,20-vacenoyl cardiolipin with a net charge of −2e (PVCL2) are pink, white (sticks), and blue, orange, and magenta spheres, respectively. Calcium ions are cyan spheres. (B–C) Two-dimensional z-position distributions of C2 and C5 atoms of lipid A (b) and the acyl chain C2 atoms of PLS(c). Figures (a–c) are adopted from Ref. [45] with permission.

20.3.1.4 Limitation

C36 CHARMM also has its limitations. The first problem of C36 is its inconsistency of the bilayer versus the monolayers. For example, for the current LJ truncation of 12 Å, the surface tension of the DPPC monolayer at the lipid area of 64 Å^2/lipid is underestimated by 15 dyn/cm^3. A similar problem is the underestimation for the surface tension of the alkane/area interfaces. Another limitation of C36 is the overestimation of dipole potential [46]. For instance, the dipole potential for DPPC bilayers is around 700 mV, which is much larger than the experimental value of 250 mV.

20.3.2 Coarse-grained explicit solvent lipid model – Martini lipid model

20.3.2.1 Model and potential

The Martini lipid model is one of the most widely used and standard coarse-grained lipid models [47–49]. The Martini lipid model is computational efficient comparing with the all-atom lipid model by mapping four or three heavy atoms into a single bead. And it is better to capture the chemical difference between different lipid species compared to other coarse-grained lipid models. An interactive bead in the Martini model represents a group of atoms. Particularly, an interactive bead represents four heavy atoms for linear structure. And it represents three heavy atoms for ring structure in lipid molecules. The interactive beads in Martini model are divided into four main types: polar (P), nonpolar (N), apolar (C), and charged (Q). To more precisely reflect the chemical nature of the atomic structure of the interactive bead, each main type contains several subtypes. These subtypes are denoted by a subscript to represent the hydrogen-bonding capability (d = donor, a = acceptor, da = both, 0 = none) or the degree of polarity (from 1, low polarity, to 5 high polarity) [17, 18].

The non-bonded interactive beads interact with each other through a shifted LJ 12-6 potential:

$$U_{LJ}(r) = 4\varepsilon_{ij}\left[\left(\frac{\sigma_{ij}}{r}\right)^{12} - \left(\frac{\sigma_{ij}}{r}\right)^{6}\right] \quad (20.11)$$

where σ_{ij} is the distance between two beads when approaching each other and ε_{ij} is the interactive strength. Importantly, the different bead types are reflected on the interaction strength ε_{ij} and distance σ_{ij}.

Additionally, a shift Coulombic potential exists between charged beads (Q) to reflect the electronic interaction:

$$U_{el}(r) = \frac{q_i q_j}{4\pi\varepsilon_0 \varepsilon_r r} \quad (20.12)$$

where the relative dielectric constant ε_r is set to 15 in the simulations. For the non-bonded interactions of LJ and Coulombic potentials, the cutoff distance is taken at $r_{cut} = 1.2$ nm. The LJ potential is shifted from 0.9 nm to r_{cut} and the Coulombic potential is shifted from 0.0 to r_{cut}.

For beads within a lipid molecule, a harmonic potential is applied between chemically connected interactive beads:

$$V_{bond}(R) = \frac{1}{2}K_{bond}(R - R_{bond})^2 \quad (20.13)$$

where the bond strength is $K_{bond} = 1,250$ kJ/mol/nm^2 and the equilibrium distance is set as $R_{bond} = 0.47$ nm. Note that the LJ interaction is excluded between bonded beads. To ensure the stiffness of the lipid tail, a harmonic angle potential is used:

$$V_{angle}(\theta) = \frac{1}{2}K_{angle}[\cos(\theta) - \cos(\theta_0)]^2 \quad (20.14)$$

Three different sets of angle strength K_{angle} and equilibrium angle θ_0 are utilized according to the chemical conditions: (1) for aliphatic chain, $K_{angle} = 25$ kJ/mol,

$\theta_0 = 180°$; (2) for cis-unsaturated bonds, $K_{\text{angle}} = 45\text{kJ/mol}$, $\theta_0 = 120°$; (3) for trans-unsaturated bonds, $K_{\text{angle}} = 45 \text{ kJ/mol}$, $\theta_0 = 180°$.

In addition to the interactive bead types and interactive potentials mentioned above, it is also noteworthy that to correctly model the ring structure in a lipid molecule such as cholesterol. Spherical ring particles are introduced by reducing the LJ interactive strength ε_{ij} and length σ_{ij}. Inside the ring structure, a set of constraints is applied to avoid fast oscillation. Furthermore, in the Martini model, the coarse-grained water molecules tend to freeze between 280 and 300 K. To prevent the freezing of water bead in the simulation, antifreeze particles with adjusted LJ interactive length are introduced. Please refer to Refs [17, 18] about the details of the interaction parameters for the ring particles and antifreeze water particles. The information about different molecules can be found at the website: http://cgmartini.nl/index.php/martini

20.3.2.2 Membrane properties

Under control of the potentials mentioned above, the Martini lipid model can properly describe membrane properties. For example, under the temperature of 323 K, the DPPC lipid bilayer in the Martini model can reproduce a lipid area 0.64 nm^2, which matches the experimental value 0.64 nm^2 at the same temperature [50]. Additionally, when compared to the atomic simulation, the Martini model can also reproduce the order parameter and stress distribution across the membrane. As shown in Figure 20.6a, the order parameter in the Martini model agrees well with the ones in atomistic simulations. In Figure 20.6, the center of mass of four methylene groups in atomic simulations was taken to get the position corresponding to the coarse-grained bead in the Martini model. The glycerol linkage and phosphate-choline bonds prefer being parallel to the bilayer surface, while other bonds are more likely to be perpendicular to the bilayer surface. Furthermore, comparing the stress distribution in all-atom lipid model in Figure 20.6b, the Martini lipid model can capture the characteristic of stress distribution.

20.3.2.3 Application

Using the framework of the Martini method, one can capture the chemistry and physical properties of the lipid membrane. Martini is a particularly good computational tool to investigate the lipid membrane with different lipid components and explore details about the cooperative behavior of lipids within the multicomponent lipid bilayer. For example, it is well-known in the experiments that the order parameter of lipid tails around cholesterol will increase [52], which is known as condensing effect of cholesterol in the lipid membrane. As shown in Figure 20.7a, in the cholesterol/DPPC bilayer, the area per lipid in the bilayer is decreasing as the fraction

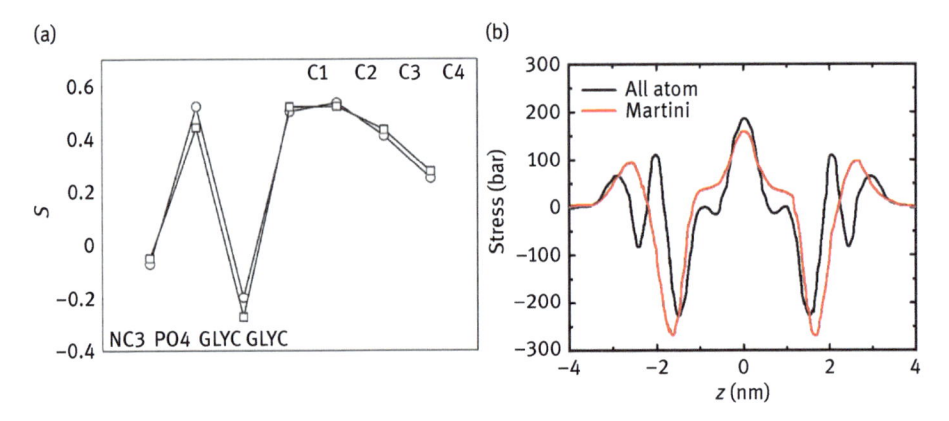

Figure 20.6: **(a)** Comparison of the order parameter of the consecutive bonds between martini DPPC lipid model and those of all-atom lipid models. **(b)** Comparison of the stress distribution along with the bilayer thickness direction of DPPC bilayer between Martini model and all-atom model. (a) is adopted from Ref. [17] with permission. (b) is a replot from the data in Ref. [18].

cholesterol in the bilayer increases. This kind of trend is consistent with the atomic and experimental results. Furthermore, in Ref. [51], a multicomponent lipid membrane with more than 60 different lipid types is investigated by the Martini model to explore the lipid organization in the real cellular membrane. It has been found in the simulation that a general nonideal mixing of different lipids in the two leaflets and each lateral planar monolayer occurs. For example, the cholesterol is enriched in the outer leaflet.

20.3.3 Coarse-grained explicit solvent lipid model – DPD model

20.3.3.1 DPD method

The DPD method was originally introduced by Hoogerbrugge and Koelman [53]. It can capture both hydrodynamics and thermodynamics. The beads in the DPD method interact with each other through three types of different forces: (1) a conservative force, (2) a dissipative force, and (3) a random force. Specifically, the conservative force is a soft repulsion [54]:

$$\boldsymbol{F}_{ij}^C = \begin{cases} a_{ij}(1-r_{ij})\boldsymbol{e}_{ij}, & |r_{ij}| < r_0 \\ 0, & |r_{ij}| \geq r_0 \end{cases} \quad (20.15)$$

where a_{ij} represents the maximum repulsive force. r_{ij} denotes the distance between the two beads i and j, and e_{ij} is the unit vector pointing from i to j. Here, r_0 is the cutoff distance for pairwise interactions. The dissipative force $\boldsymbol{F}_{ij}^D$ and random $\boldsymbol{F}_{ij}^R$ force can

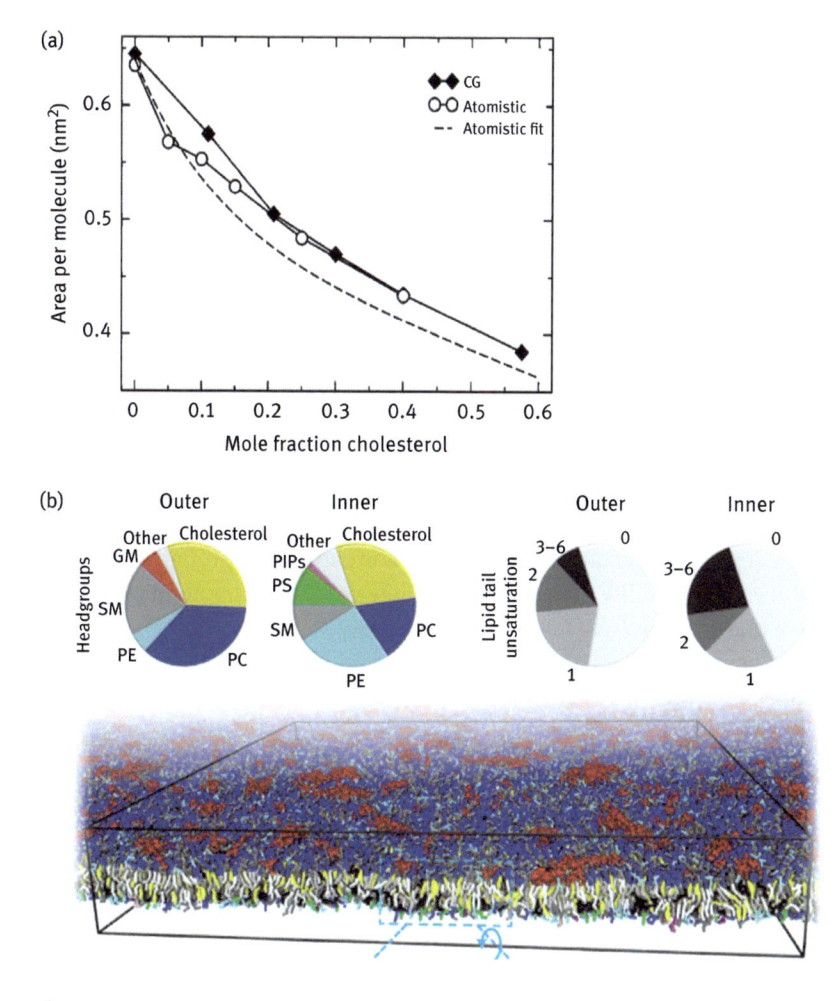

Figure 20.7: (a) Condensation effect of cholesterol on lipid area; (b) The pie charts on the upper panel represent the distribution of the main lipid head group and the level of tail unsaturation in the inner and outer leaflets. The figure in the bottom panel is the snapshot of membrane. Cholesterols are colored in yellow. The lipid head groups are colored based on their types. The tails are colored based on the number of unsaturated bond (0, white;1, light gray;2,dark gray;3–6,black). (a) and (b) are adopted from Refs [18] and [51], respectively.

be considered as "Pairwise Brownian dashpot" to represent the viscous forces and thermal noise between interactive beads. The formulas of them are given by:

$$F_{ij}^D = -\gamma \omega^D(r_{ij})(e_{ij} \cdot v_{ij})e_{ij}, \quad F_{ij}^R = \sigma \omega^R(r_{ij})\zeta_{ij}e_{ij}/\sqrt{\Delta t} \quad (20.16)$$

where v_{ij} is the relative velocity vector between beads i and j. ζ_{ij} is a random number with zero mean and unit variance. Δt is the timestep used in the simulation. ω^D and

w^R are the distance-dependent weight functions. γ and σ are the friction coefficient and the amplitude of noise, respectively. To satisfy the steady state under the canonical ensemble, relation between the dissipative and random forces needs to be satisfied [55]:

$$w^D(r) = [w^R(r)]^2, \sigma^2 = 2\gamma k_B T \tag{20.17}$$

Usually $w^D(r)$ will be taken as

$$w^D(r) = \begin{cases} (1 - r/r_0)^2, & |r < r_0 \\ 0, & |r \geq r_0 \end{cases} \tag{20.18}$$

Because of the soft repulsive force, a larger timestep can be used in the DPD simulation than in the hard-core LJ pairwise interaction. Therefore, DPD is able to explore larger simulation system around 100 nm. Additionally, the dissipative and random forces serve as a thermostat, which can conserve the linear and angular momentum in the simulated system. To correctly describe the thermodynamic state of a liquid, the compressibility of the system should be correctly captured. In the DPD simulation, the compressibility of system κ^{-1} is determined by the number density ρ of the system and the repulsive force amplitude a [54]:

$$\kappa^{-1} \approx 1 + 0.2a\rho/k_B T \tag{20.19}$$

Considering the compressibility of water $\kappa^{-1} \approx 16$, we should have the relation of $a \approx 75 k_B T/\rho$. Therefore, to model the membrane in the solvent, the repulsive force between water is usually taken as $a = 25\ k_B T$ and the total bead density in the simulated system is usually set as $\rho = 3$.

To model the liquid–liquid interface with the DPD method, such as the membrane–water interface, we need to capture the beads mutual solubility, which is usually described by the Forly–Hugins χ-parameter in polymer chemistry. The χ-parameter represents the excess free energy of mixing different materials in the Flory–Huggins mode. Taking A and B, two components, as an example, if the parameter χ is positive, then A and B do not favor each other because of the large free energy increment when mixing. They prefer to separate into two different phases. If the parameter χ is negative, then A and B prefer to stay with each to minimize the totally free energy. Following this procedure, the parameter χ between the hydrophobic lipid tail and water will be positive. In the DPD method, it has been proved that the parameter χ is linearly related to with the excess AB repulsion over the AA repulsion [54]. If we take $\rho = 3$, $a_{AA} = a_{BB} = 25\ k_B T$, this relation will be expressed as:

$$\chi = (0.286 \pm 0.002)\Delta a \tag{20.20}$$

where $\Delta a = a_{AB} - a_{AA}$. Following this protocol, a large repulsion is usually set between lipid tails and water to capture the hydrophobic effect in the membrane with DPD method.

20.3.3.2 Different DPD model and membrane properties

To model the lipid molecule, the interactive beads within a lipid molecule are connected by spring bond to form different lipid structures. With the DPD method, both single tail and double tails lipid architectures are widely used in the simulation based on the specific problem [56, 57]. Additionally, an angle-dependent potential is usually applied on the tail beads to ensure the tail stiffness. To form a stable lipid bilayer in the simulation, relatively larger repulsive force is usually applied for the tail-head and tail-water bead interactions [24]. By specifically calculating the parameter χ of different groups in the lipid molecules, the obtained DPD lipid model can also be used to investigate the effect of cholesterol on the lipid bilayer and its phase transition under different temperatures [58–60]. Here, we will take the lipid model developed by Lipowsky's group as an example to explain the potentials and the corresponding membrane properties [19, 56, 61, 62].

The architecture of the lipid molecule is shown in Figure 20.8a. The spring bond potential is given as:

$$V_b(r) = \frac{1}{2} k_b (r_{ij} - l_0)^2 \tag{20.21}$$

where the spring constant is $k_b = 128\ k_B T/r_0^2$ and the equilibrium length $l_0 = 0.5 r_0$. The angular potential is given as

$$V_a(\varphi) = \frac{1}{2} k_s [1 - \cos(\varphi)] \tag{20.22}$$

where $k_s = 15\ k_B T$, φ is the angle formed by the three sequentially connected beads in the tail. Note that this angle potential is also applied on the angle of head–tail–tail.

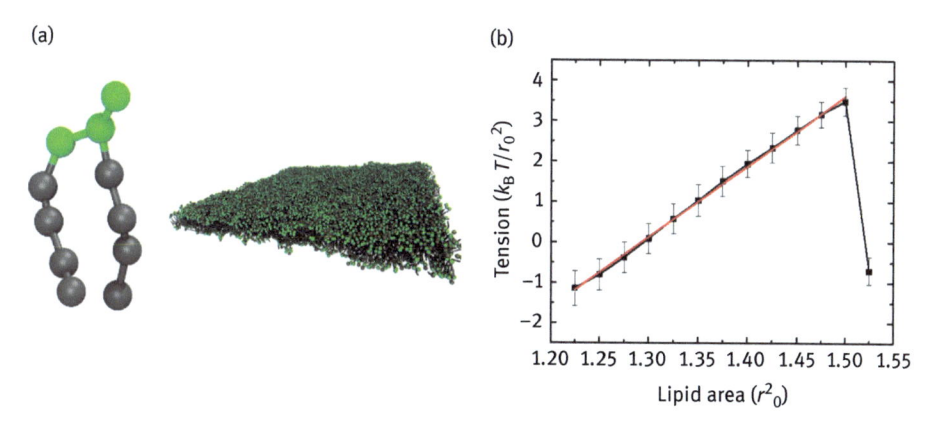

Figure 20.8: (a) Lipid molecular model and the assembled lipid bilayer. The heads in the lipids are colored in green. The tails are colored in gray. (b) The membrane tension as a function of lipid area.

However, the head–head–tail angle is not constrained by this angle potential. Please refer to Ref. [56] for the details about the pairwise interaction.

Under the control of these parameters, the lipid molecules can self-assemble into a stable lipid bilayer. And this set of parameters can well capture the fusion process between vesicle and lipid bilayer. Particularly, as shown in Figure 20.8b, the membrane tension is linearly increasing with the lipid area. The stretch modulus of the lipid bilayer can be calculated by fitting the linear relation, which is around 18 $k_B T/r_0$. The bilayer thickness of the bilayer is around 3.5 r_0. The bending rigidity of the membrane can be obtained according to the relation in eq. (20.8), which is around 6 $k_B T$.

20.3.3.3 Application

One of the advantages of the DPD method is that it can both capture the thermodynamics and hydrodynamics. For instance, DPD lipid model can be used to investigate the behavior of liposome under the shear flow [63]. The shear flow can be activated by creating two layers of rigid wall in the top and bottom of the simulation box and assigning a constant velocity for one of them. As shown in Figure 20.9a, the deformation and rupture process of the liposome under the shear force can be clearly observed in the simulation. Furthermore, we can estimate the membrane tension distribution on the liposome surface by calculating the lipid area. As given in Figure 20.9b, the large membrane tension at the tip of liposome is the reason that induces the rupture of liposome. Furthermore, the DPD method is also widely applied to investigate the endocytosis of nanoparticles, where the size of nanoparticle is around 10–60 nm [64–67]. As shown in Figure 20.9c, during the endocytosis process, we can investigate the deformation of the membrane and the nanoparticles. Additionally, the configurations in the simulations can be used in the theory to analyze the free energy change during this process and explore the underlying physical mechanisms [67].

20.3.4 Coarse-grained implicit solvent lipid model: three-bead model

20.3.4.1 Model and potentials

A three-bead implicit solvent lipid molecule developed by Deserno et al. is to capture the physical properties of lipid membrane [68]. In this model, a lipid molecule is represented by three beads: a lipid head and two lipid tail beads. Pairwise interactions are applied between the beads to reflect the hydrophobic and hydrophilic properties of lipid tail and head, respectively. Harmonic bond interactions connect the lipid beads as a molecule and ensure molecular stiffness.

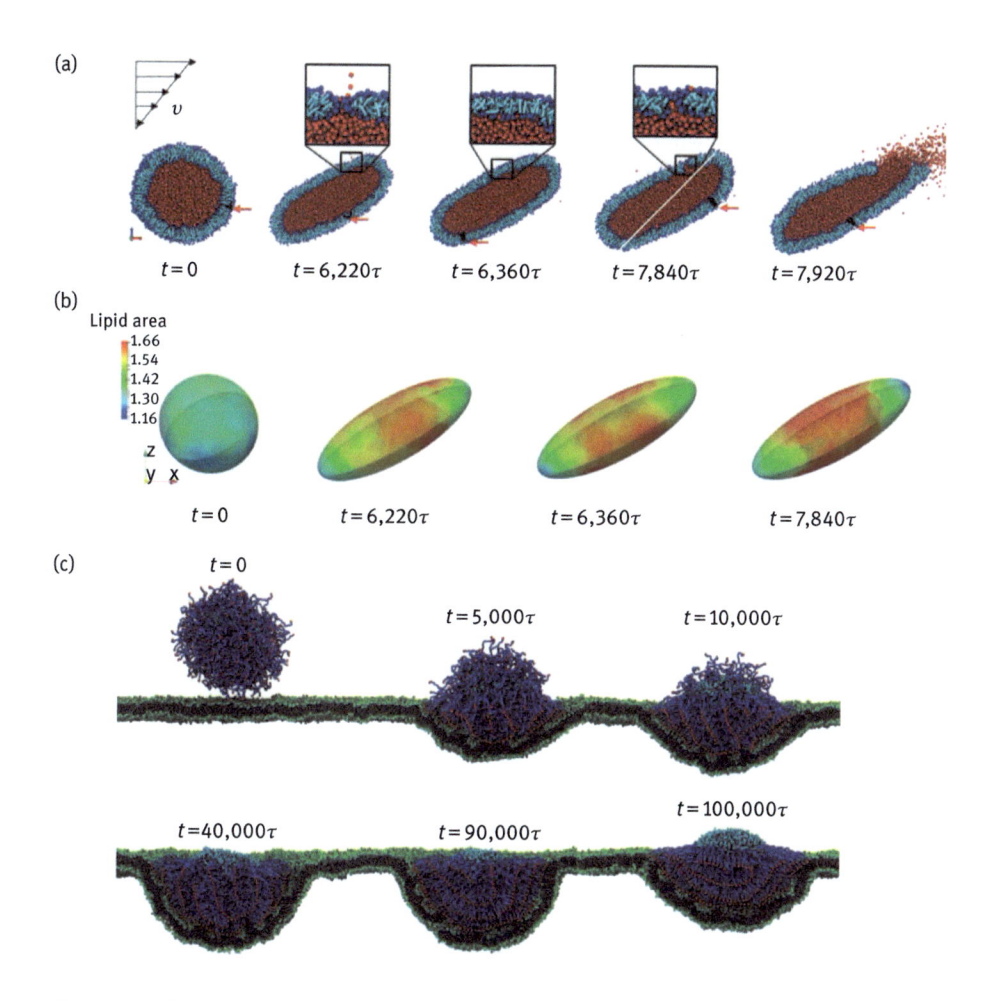

Figure 20.9: (a) Deformation of a vesicle at different timesteps under the simple shear flow. (b) The distribution of lipid area on the surface of liposome. The large lipid area is corresponding to the large membrane tension. (c) The endocytosis process of liposome with decorated polyethylene glycol (PEG) polymers. The lipid membrane is colored green for head group and gray for tail group, respectively. The PEG polymers are colored in blue. The liposome is colored in cyan.

A Weeks–Chandler–Andersen (WCA) potential is applied between interactive beads to define repulsive interactions that determines the size of the phospholipids [68]:

$$V_{\text{rep}}(r) = \begin{cases} 4\varepsilon\left[\left(\frac{b}{r}\right)^{12} - \left(\frac{b}{r}\right)^{6} + \frac{1}{4}\right], & |r \leq r_c \\ 0, & |r > r_c \end{cases} \quad (20.23)$$

where ε is the unit of energy. Additionally, $b_{\text{head, head}} = b_{\text{head, head}} = 0.95\sigma$ and $b_{\text{head, tail}} = 1.0\sigma$ (σ is the length unit). These variables ensure the cylindrical shape of each lipid molecule, where $r_c = 2^{1/6}b$ is the cutoff distance. An attractive potential is needed between the tail beads to capture the hydrophobicity of tails group in solvent. This kind of attraction must be strong enough to allow the hydrophobic tails to aggregate, but not too strong. Otherwise, the aggregated lipid bilayer will become too solid. A proper, weaker potential, allows the phospholipids mobility, producing a fluid lipid bilayer phase. The attractive potential applied in this solvent free model is given as

$$V_{\cos}(r) = \begin{cases} -\varepsilon, & |r < r_c \\ -\varepsilon \cos^2[\pi(r - r_c)/2w_c], & |r_c \le r \le r_c + w_c \\ 0, & |r > r_c + w_c \end{cases} \quad (20.24)$$

This potential smoothly connects the WCA potential at r_c. Furthermore, it acts as an attractive force between r_c and $r_c + w_c$. And w_c is the key parameter to regulate the attractive range, which ultimately determines the state of the lipid bilayer.

Apart from the pairwise potentials above, bond potentials are used to connect the head/tail beads within a lipid molecule and ensure the stiffness of each molecule. In a lipid molecule, each bead is connected with its nearest neighbor through a finite extensible nonlinear elastic bond potential:

$$V_{\text{bond}}(r) = -\frac{1}{2}k_{\text{bond}}r_\infty^2 \log\left[1 - (r/r_\infty)^2\right] \quad (20.25)$$

where the bond stiffness is $k_{\text{bond}} = 30\varepsilon/\sigma^2$ and the divergence length is $r_\infty = 1.5\sigma$. Additionally, a harmonic potential is applied between the head bead and the second tail bead to ensure the molecular stiffness:

$$V_{\text{bend}}(r) = \frac{1}{2}k_{\text{bend}}(r - 4\sigma)^2 \quad (20.26)$$

where the constant, $k_{\text{bend}} = 10\varepsilon/\sigma^2$.

20.3.4.2 Membrane properties

Under the potentials described above, these coarse-grained lipid molecules can self-assemble into a lipid bilayer. After tuning the attractive range w_c, three different states exist for a lipid bilayer. The phase diagram of lipid bilayers against w_c and temperature is given in Figure 20.10. At values $w_c < 0.8\,\sigma$, the attractive range between lipid molecules is too small resulting in instability at high temperatures, and gel phases at low temperature. When $w_c > 0.8\,\sigma$, a stable fluid-state bilayer emerges under a suitable temperature range as the attractive range increases. This fluid state

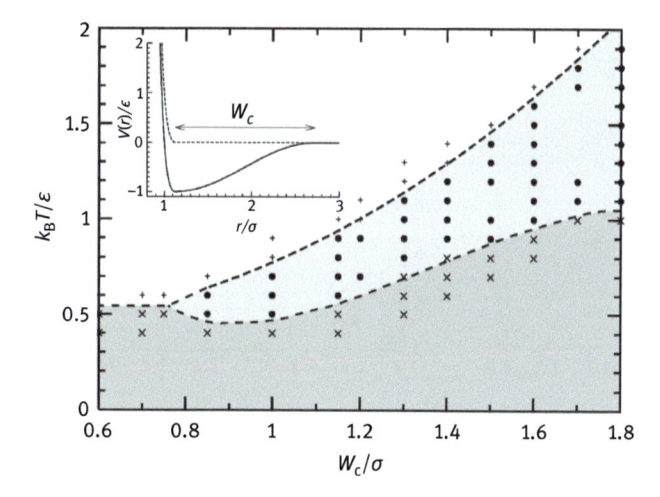

Figure 20.10: Phase diagram of lipid membrane in the space of attraction range w_c and temperature at zero membrane tension. (X) represents gel state; filled dot represents fluid state; (+) represents unstable membrane in the simulation. The figure is adopted from Ref. [68] with permission.

lipid bilayer in the simulation corresponds to the fluid state lipid membrane in biology, which could be applied to investigate the physical properties of the lipid membrane and membrane-related interfacial interactions. Additionally, under the range of $w_c > 0.8\ \sigma$, two temperature boundaries appear for the lipid bilayer in the simulation. When the temperature is higher than the upper boundary, the thermal fluctuation of lipids is too large. And the lipid bilayer becomes unstable. While the temperature is less than that of the lower boundary, the smaller thermal fluctuation facilitates the tight packing of lipids in the bilayer, resulting in a gel-state bilayer. It is also interesting to note in the simulation that as w_c gets larger, the temperature range for the fluid state becomes wider. In the following part of this model, we will discuss the properties of the fluid state bilayer. Particularly, we will introduce how the attractive range w_c affects these properties in this model.

As the attractive range of lipid interactions increases, diffusivity decreases. As seen from Figure 20.11a, an increase in the attractive potential by less than 0.5 results in a reduction of diffusivity by a factor of 3. This relation between D_b and w_c is expected. As by increasing the attractive range w_c, a lipid molecule within a bilayer will attract more neighboring lipids, leading to a more tightly lipid packing with much smaller diffusivity. The order parameter S is measured in Figure 20.11b. For a highly coarse-grained three bead model, the θ in eq. (20.2) is defined as the angle between the whole lipid and bilayer normal. As the attractive range w_c is increased, order parameter S also increases from $S = 0.35$ at $w_c = 1.3\sigma$ to $S = 0.85$ at $w_c = 1.8\sigma$, which indicates that at the long attractive range, the lipid molecules are highly ordered in the bilayer.

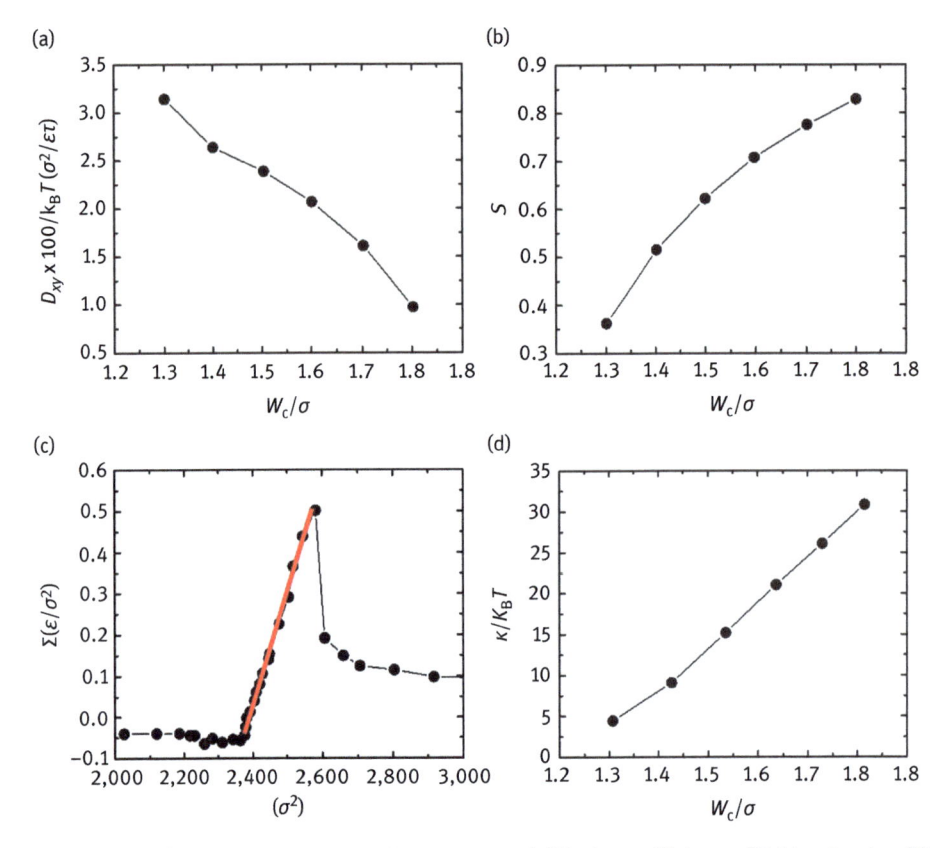

Figure 20.11: (a) Relation between attractive range w_c and diffusion coefficients of lipid molecules. (b) The relation between attractive range w_c and order parameter. (c) The relation between membrane area and membrane tension; The red line is obtained by the linear fitting to calculate the stretch modulus. (d) The relation between attractive range w_c and bending rigidity of lipid membrane. The temperature during all of these tests is controlled at $T = 1.1\,\varepsilon/k_B$. The figures are replotted from the data in Ref. [68].

The mechanical properties of the lipid bilayers can also be captured by the three-bead model. The stretch modulus can be measured with the stretching method in eq. (20.4). As shown in Figure 20.11c, the relation between membrane tension and total area of membrane is obtained by stretching a bilayer with 4,000 lipids. The temperature and attractive range are controlled at $T = 1.1\,\varepsilon/k_B$ and $w_c = 1.6\sigma$, respectively. With the small membrane area, the membrane tension is not sensitive to the variation of membrane area because of the buckling that occurs at the membrane. After the membrane is flatten, the membrane tension linearly increases along with the membrane area. After reaching a critical value, the membrane ruptures to release its tension, which results in the drop of membrane tension. Fitting the linear region, we can get the membrane stretch molecules as $K_A = 6.4\,\varepsilon/\sigma^2$.

The bending modulus of membrane can be calculated with the fluctuation method as mentioned above. As shown in Figure 20.11d, at the temperature $T = 1.1\varepsilon/k_B$, the bending modulus of membrane increases linearly with the increment of attractive range. Its value can be tuned from $\kappa = 5k_BT$ at $w_c = 1.3\sigma$ to $\kappa = 32k_BT$ at $w_c = 1.8\sigma$.

20.3.4.3 Application

Without the water molecules, the advantage of the solvent-free three-bead lipid model can be applied to investigate the problems on the order of 100 nm length and millisecond timescales. For example, Deserno et al. [69] investigated the cooperation of the nanoparticles/proteins during their interaction with lipid membrane. The side length of lipid membrane in the simulation is 160 nm. As shown in Figure 20.12a, it has been

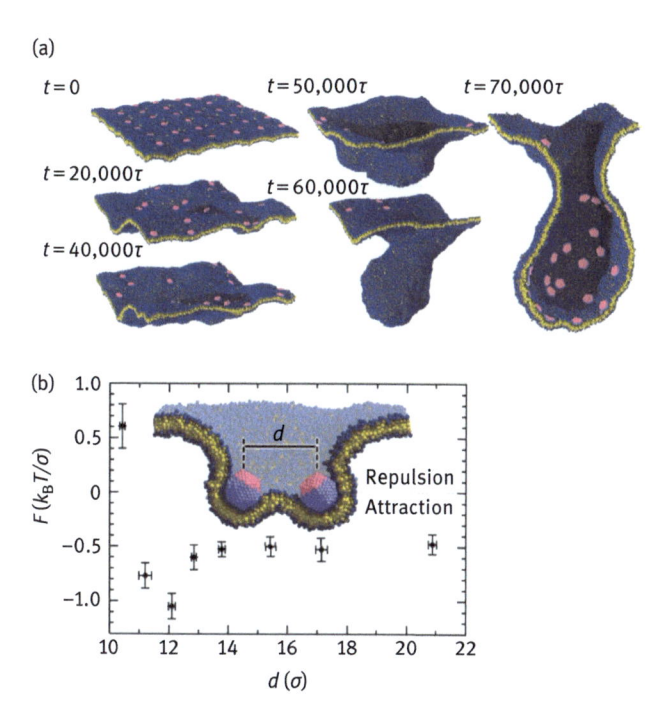

Figure 20.12: (a) Snapshots of membrane and nanoparticles during the interactions. The lipid head is colored in dark blue. And the lipid tail is colored in yellow. The nanoparticles are colored in pink and light blue. The light blue part of the nanoparticle has an attractive force between lipid head in membrane. (b) The relation between force experienced by two nanoparticles and their relatively distance. To get the magnitude of the force at each distance, the relative distance between two nanoparticles are fixed during the testing for each point. The figures are adopted from Ref. [69] with permission.

found in the simulation that these nanoparticles prefer to aggregate and finally are internalized in a cooperative way. To explore the details, the authors hold the distance between two nanoparticles and monitor the force between them. It has been discovered that there is an attractive force existing between these two nanoparticles because of the curvature-mediated interaction, as shown in Figure 20.11b. Please refer to Ref. [69] for the details about the simulation methods and results.

20.3.5 Coarse-grained implicit lipid model – one-bead lipid model

20.3.5.1 Model and interactive potential

Under the framework of particle-based membrane, the one-bead lipid model pushes the time and length scale to the limit by considering each lipid molecule as a particle. To realize the self-assembly of one-bead lipid molecules and get the corresponding physical bilayer properties, the interactive particles are considered as spherical beads. As shown in Figure 20.13a, the axes of symmetry for each bead represent the longitudinal directions. Correspondingly, the interactive force between beads depends on both their relative distance and orientation. In Figure 20.13a, $r_{ij} = r_i - r_j$ is the relative distance vector between two beads; n_i and n_j represent the axes of symmetry of particles, by which we can get their relative orientation.

The properties of the lipid molecule are governed by the pairwise interaction. The potential function between beads is given as [70, 71]:

$$U(r_{ij}, n_i, n_j) = \begin{cases} u_R(r) + [1 - \phi(\hat{r}_{ij}, n_i, n_j)], & |r < r_{min} \\ u_A(r)\phi(\hat{r}_{ij}, n_i, n_j), & |r_{min} < r < r_c \end{cases} \quad (20.27)$$

where $u_R(r)$ and $u_A(r)$ are the distance-dependent repulsive and attractive potentials, respectively. r_{min} is the repulsive force range. And r_c is the cutoff distance of the pairwise interaction. $\phi(\hat{r}_{ij}, n_i, n_j)$ is a weight function to tune the interaction force depending on their relative orientations. Specifically, the distance-dependent function is taken as:

$$U(r) = \begin{cases} U_R(r) = \varepsilon\left[\left(\frac{r_{min}}{r}\right)^4 - 2\left(\frac{r_{min}}{r}\right)^2\right], & |r < r_{min} \\ U_A(r) = -\varepsilon\cos^{2\zeta}\left[\frac{\pi}{2}\frac{(r - r_{min})}{(r_c - r_{min})}\right], & |r_{min} < r < r_c \end{cases} \quad (20.28)$$

where ε and σ are energy and length units, respectively, which are taken as one in the simulation. The constant $r_{min} = \sqrt[6]{2}\sigma$ and $r_c = 2.6\sigma$. The repulsive part is the LJ-42 potential. The attractive part is a cosine function, which smoothly connects the repulsive part at r_{min} and decays to zero at the cutoff distance r_c. ζ is used to adjust

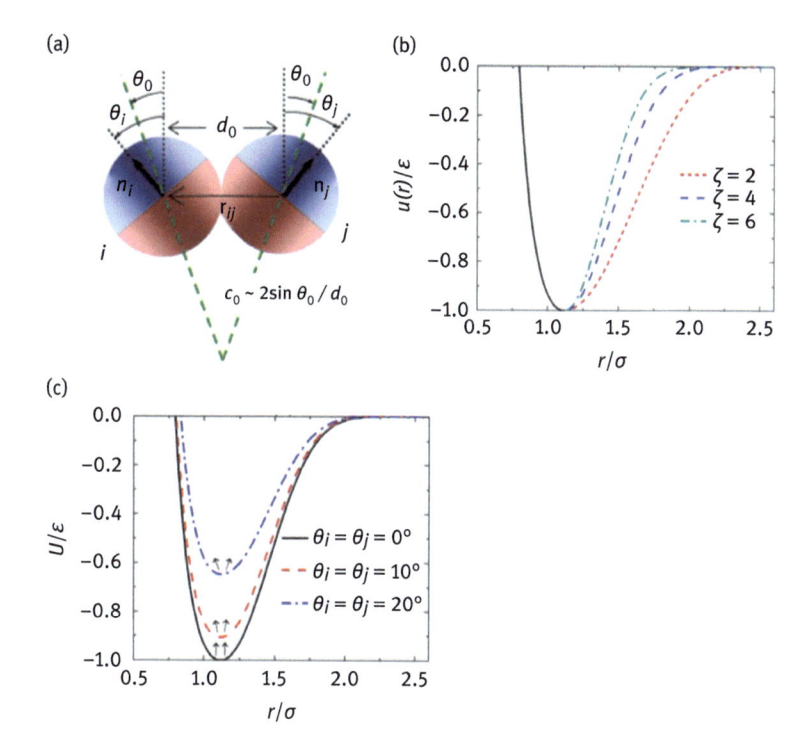

Figure 20.13: (a) Schematic of the orientation-dependent interparticle interaction; The color scheme used here is to facilitate visualization. (b) The distance-dependent function and its variation with the changing of exponent ζ. (c) The inter-particle interaction and its dependence on the distance and relative orientation angle θ_j. The figures are adopted from Ref. [70] with permission.

the slope of the attractive branch of the potential, as shown in Figure 20.13b. The orientation-dependent function is given as

$$\phi = 1 + \mu\left[a(\widehat{r}_{ij}, n_i, n_j) - 1\right] \quad (20.29)$$

$$a = (n_i \times \widehat{r}_{ij}) \cdot (n_j \times \widehat{r}_{ij}) + \sin\theta_0 (n_j - n_i) \cdot \widehat{r}_{ij} - \sin^2\theta_0 \quad (20.30)$$

where ϕ reaches its maximum of 1 when the angle between the two lipid molecules is θ_0. Otherwise, ϕ is less than 1. Therefore, θ_0 is the most energetically favorable angle between two lipids. In this perspective, θ_0 can be considered to determine the spontaneous orientation of lipid, which is directly related to the spontaneous curvature $c_0 \sim 2\sin\theta_0/d_0$ (c_0 and d_0 are the spontaneous curvature and average interparticle distance, respectively). The parameter μ is a weight constant to determine the energy penalty when the angle between two lipids deviates from θ_0. It means that μ is correlated with the bending rigidity of the lipid membrane.

20.3.5.2 Membrane properties

Under control of the pairwise interactions above, the lipid beads can form vesicle or planar membrane through the self-assembly process with suitable parameters. For the listed parameters above, ζ and μ are the most important control parameters to change the properties of the membrane. Additionally, we can also investigate the spontaneous curvature of membrane by tuning the parameter θ_0. In the following part, we will introduce how the membrane properties can change along with these parameters. For simplicity, θ_0 is fixed as 0.

In this one-bead model, the gel-fluid states of the membrane are mainly determined by the temperature and ζ as given in Figure 20.14a, where μ is fixed as 3. As we can see, three different states of the membrane can be obtained, namely, fluid, gel, and gas states. The gas state in the phase diagram is defined for the situation that at least one lipid bead flies away from the membrane. The gel state is defined when the diffusion coefficient of lipid is smaller than 0.01 σ^2/τ. Importantly, we can get a broad range of fluid-state membrane, under which state the lipid diffusion coefficient is on the order of 0.1 σ^2/τ. It is also the fluid-state membrane, which is widely used to

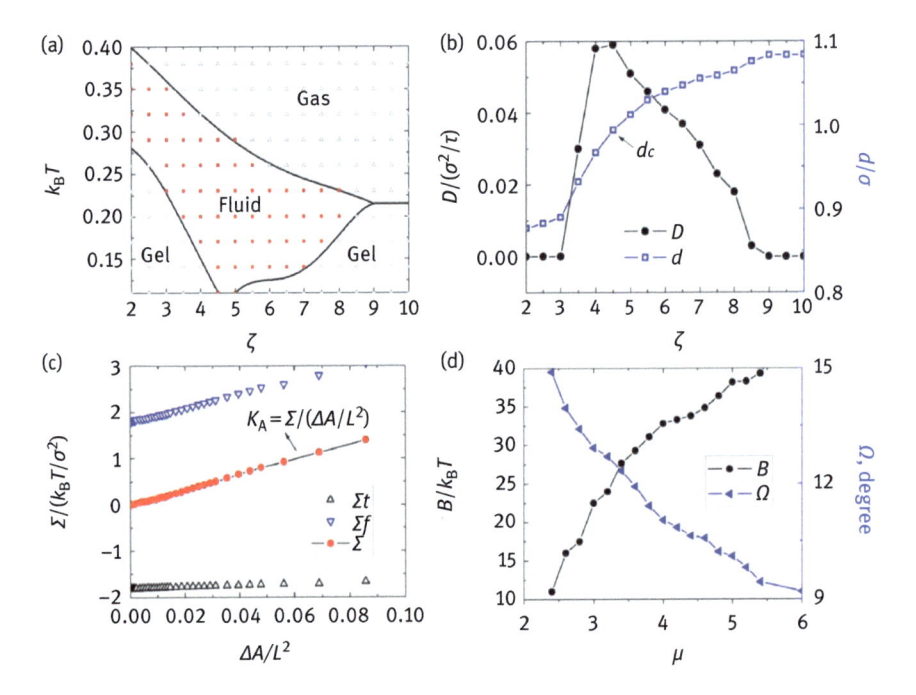

Figure 20.14: (a) Phase diagram of lipid membrane in the space of temperature and exponent ζ. (b) Diffusion constant D and inter-bead distance d as a function of exponent ζ. (c) Membrane tension as a function of area strain. The temperature is controlled as $k_B T = 0.23\ \varepsilon$. $\zeta = 4$, and $\mu = 3$. (d) The bending rigidity κ as a function of exponent ζ. The figures are adopted from Ref. [70] with permission.

investigate the membrane properties or the interaction between the membrane and other materials.

As we can see in the phase diagram, the state of the membrane can be effectively controlled by ζ. And the state of membrane is highly related to the diffusion constant of lipids. The ζ value in the function is used to control the attraction force under certain distance, which can affect the equilibrium inter-particle distance between lipids and finally determine the lipid diffusion constant. As shown in Figure 20.14b, as the increment of the ζ, the equilibrium interparticle distance d monotonically increases. While the lipid diffusion constant first increases and then decreases, after it reaches the maximum values of around $0.06\,\sigma^2/\tau$. When $\zeta < 4$, the equilibrium inter-bead distance between particle is smaller than $1.0\,\sigma$. Therefore, the tightly packed lipid molecules move relatively slower. When the $\zeta > 4$, equilibrium inter-bead distance between particle is larger than $1.0\,\sigma$, under which region the inter-bead interactive force is attractive. Therefore, the diffusion coefficient decreases along with ζ because of the stronger attractive force.

After understanding the dynamic properties of the lipid membrane with this model, we will discuss about its mechanical properties. Similar to the other models, the stretch modulus of membrane can be measured by testing the membrane tension under a series of different area strains. As shown in Figure 20.14c, under the temperature of $k_B T = 0.23\,\varepsilon$, with parameters of $\zeta = 4$ and $\mu = 3$, the stretch modulus of the membrane is around $18 k_B T/\sigma^2$. Furthermore, the bending rigidity of membrane can be obtained by the fluctuation method. As given in Figure 20.14d, a relationship can be found between the bending rigidity and μ. When we fix the $\zeta = 4$, $k_B T = 0.23\,\varepsilon$, and change μ from 2.4 to 6, the bending rigidity of membrane will change from $12\,k_B T$ to $40\,k_B T$, and this range falls within that of experimental values.

20.3.5.3 Application

The side length of lipid membrane built by the one-bead lipid model can be up to 500 nm in simulations. Therefore, the one-bead model is widely applied for problems related to vesicle-shape transformation and the endocytosis process. And one of the featured applications of this model is that it can be used to investigate the membrane properties of RBCs, which consist of both a lipid bilayer and a spectrin network. The spectrin network in the RBCs is of great importance to maintain RBC configurations under the shear stress. As shown in Figure 20.14, by combining a lipid membrane with a spectrin network in the simulated model, we can investigate the shear response of the composite membrane. As we can observe from Figure 20.15c, the shear resistance in RBC is mainly contributed by the spectrin network. Please refer to Ref. [72] for details about the model and simulation results.

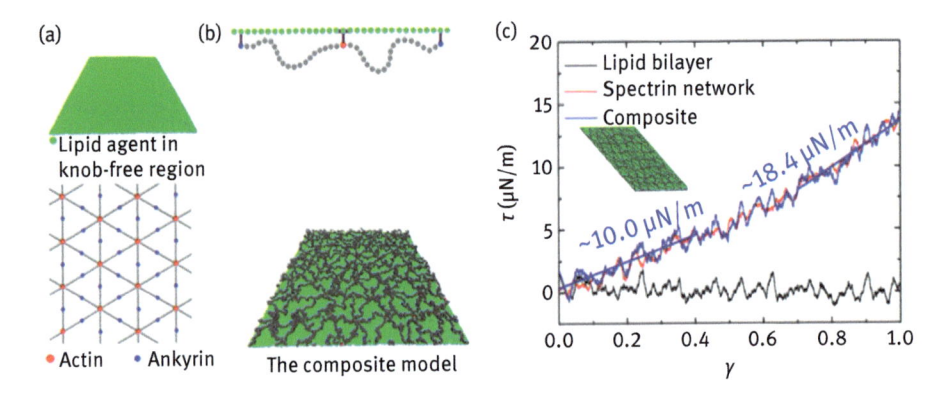

Figure 20.15: **(a)** Molecular models of lipid bilayer and spectrin network. **(b)** The side and top views of RBC membrane with both lipid membrane and spectrin network. **(C)** Responses of lipid bilayer, spectrin network and composite membrane under shear strain. The figures are adopted from Ref. [72] with permission.

20.3.6 Triangulated membrane model

Apart from the particle-based lipid membrane model, the triangulated surface membrane is also a widely used approach to investigate the behavior of lipid vesicles or RBCs [21, 74]. As shown in Figure 20.16, in this method, the vesicle is discretized into a set of triangular plaquettes. Each of plaquettes represents a flat patch of membrane. And these plaquettes are interconnected by links, which are intersected at vertices points. The elastic energy of the discretized membrane can be expressed as [73]

$$H = \lambda_b \sum_v \sum_{\{e\}_v} \left\{ 1 - \hat{N}[f_1(e)] \times \hat{N}[f_2(e)] \right\} + \Delta p V \qquad (20.31)$$

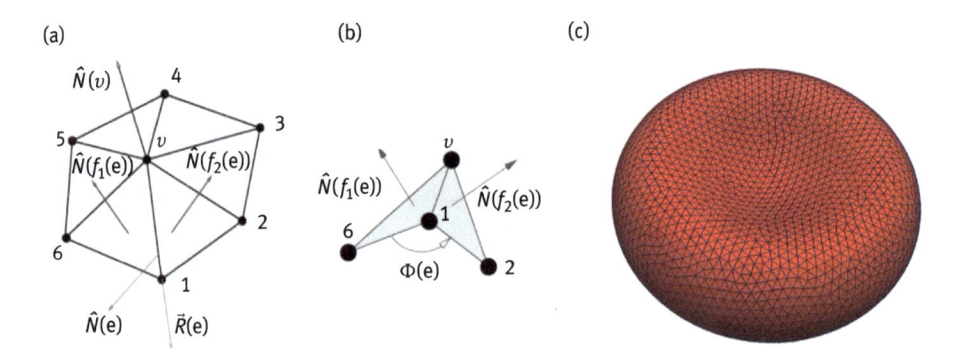

Figure 20.16: **(a)** Surface patches of triangulated membrane around vertex v. **(b)** The single dihedral angle between faces f_1 and f_2. **(c)** Triangulated red blood surface. **(a)** and **(b)** are adopted from Ref. [73] with permission.

where $\{e\}_v$ and v are the numbers of linkers and vertex points, respectively. $\hat{N}[f_1(e)]$ and $\hat{N}[f_2(e)]$ are the normal vectors of faces f_1 and f_2, respectively. f_1 and f_2 share an edge e. V is the volume of the vesicle. λ_b is directly related to the bending rigidity as $\lambda_b = \sqrt{3}\kappa$ for a spherical vesicle. The evolution of triangulated mesh can be directly tracked through an energy minimization process. Due to the simplicity of the triangulated membrane model, the size of membrane can be under the length scale of a micrometer. Details about the implementation of the membrane models can be founded in the Ref. [75]. In addition, please refer to these review papers for the application of the triangulated membrane [76].

20.4 Simulation packages for membrane modeling

There are lots of programs, which could be used during the modeling of lipid membrane and their interaction with other materials. Here, we just list the most widely used programs. For example, NAMD [77], GROMACS [78], LAMMPS [79], CHARMM [80], and Amber [81] are good MD packages for these simulations. VMD [82] is usually used as a visualization software for analyzing the simulation results. Moltemplate [83] and Packmol [84] are those programs used to build the initial configuration of lipid systems. For further information about these programs, please refer the corresponding references and manuals.

20.5 Conclusion

In this chapter, we provide an overview about the computation simulations on cell membrane. Multiscale lipid models, ranging from the all-atom lipid model to triangulated membrane model, are introduced. Particularly, the all-atom lipid model of CHARMM and coarse-grained lipid model of Martini, DPD, three-bead and one-bead are introduced in detail. For each model, we cover their interactive potentials, membrane properties, and the typical applications. We hope our chapter provides a comprehensive tutorial on multiscale modeling of lipids and membranes, which will be of interest to readers newly entering this field.

References

[1] Dupuy, A.D., & Engelman, D.M. *Protein area occupancy at the center of the red blood cell membrane*. Proceedings of the national academy of sciences. 2008. **105**(8):p. 2848–2852.
[2] McGuffee, S.R., & Elcock, A.H. *Diffusion, crowding & protein stability in a dynamic molecular model of the bacterial cytoplasm*. PLoS Computational Biology. 2010. **6**(3):p. e1000694.

[3] Takamori, S., Holt, M., Stenius, K., Lemke, E.A., Grønborg, M., Riedel, D., Urlaub, H., Schenck, S., Brügger, B., & Ringler, P. *Molecular anatomy of a trafficking organelle.* Cell. 2006. **127**(4): p. 831–846.

[4] Barrera, N.P., Zhou, M., & Robinson, C.V. *The role of lipids in defining membrane protein interactions: insights from mass spectrometry.* Trends in cell biology. 2013. **23**(1):p. 1–8.

[5] Dowhan, W. *Molecular basis for membrane phospholipid diversity: why are there so many lipids?.* Annual review of biochemistry. 1997. **66**(1):p. 199–232.

[6] Shin, J.J. & Loewen, C.J. *Putting the pH into phosphatidic acid signaling.* BMC biology. 2011. **9**(1):p. 85.

[7] Van Meer, G., Voelker, D.R., & Feigenson, G.W. *Membrane lipids: where they are and how they behave.* Nature reviews molecular cell biology. 2008. **9**(2):p. 112.

[8] Menon, I., Huber, T., Sanyal, S., Banerjee, S., Barré, P., Canis, S., Warren, J.D., Hwa, J., Sakmar, T.P., & Menon, A.K. *Opsin is a phospholipid flippase.* Current biology. 2011. **21**(2): p. 149–153.

[9] Venturoli, M., Sperotto, M.M., Kranenburg, M., & Smit, B. *Mesoscopic models of biological membranes.* Physics reports. 2006. **437**(1–2):p. 1–54.

[10] Li, Y., Abberton, B.C., Kröger, M., & Liu, W.K. *Challenges in multiscale modeling of polymer dynamics.* Polymers. 2013. **5**(2):p. 751–832.

[11] Greene, M.S., Li, Y., Chen, W., & Liu, W.K. *The archetype-genome exemplar in molecular dynamics and continuum mechanics.* Computational mechanics. 2014. **53**(4):p. 687–737.

[12] König, S. & Sackmann, E. *Molecular and collective dynamics of lipid bilayers.* Current opinion in colloid & interface science. 1996. **1**(1):p. 78–82.

[13] den Kamp, J.A.O., Roelofsen, B., & van Deenen, L.L. *Structural and dynamic aspects of phosphatidylcholine in the human erythrocyte membrane.* Trends in biochemical sciences. 1985. **10**(8):p. 320–323.

[14] Jacobson, K., Mouritsen, O.G., & Anderson, R.G. *Lipid rafts: at a crossroad between cell biology and physics.* Nature cell biology. 2007. **9**(1):p. 7.

[15] Lodish, H., A. Berk, S.L. Zipursky, P. Matsudaira, D. Baltimore, & Darnell, J. *Molecular cell biology.* Vol. 3. 1995: WH Freeman New York.

[16] Pluhackova, K. & Böckmann, R.A. *Biomembranes in atomistic and coarse-grained simulations.* Journal of physics: condensed matter. 2015. **27**(32):p. 323103.

[17] Marrink, S.J., De Vries, A.H., & Mark, A.E. *Coarse grained model for semiquantitative lipid simulations.* The journal of physical chemistry B, 2004. **108**(2):p. 750–760.

[18] Marrink, S.J., Risselada, H.J., Yefimov, S., Tieleman, D.P., & De Vries, A.H. *The MARTINI force field: coarse grained model for biomolecular simulations.* The journal of physical chemistry B. 2007. **111**(27):p. 7812–7824.

[19] Shillcock, J.C. & Lipowsky, R. *Equilibrium structure and lateral stress distribution of amphiphilic bilayers from dissipative particle dynamics simulations.* The journal of chemical physics. 2002. **117**(10):p. 5048–5061.

[20] Fedosov, D.A., Caswell, B., & Karniadakis, G.E. *A multiscale red blood cell model with accurate mechanics, rheology, and dynamics.* Biophysical journal. 2010. **98**(10): p. 2215–2225.

[21] Ye, H., Shen, Z., Yu, L., Wei, M., & Li, Y. *Anomalous vascular dynamics of nanoworms within blood flow.* ACS biomaterials science & engineering. 2018, 4 (1): pp. 66–77.

[22] Artemieva, A.B., Schwille, P., & Petrov, E.P. *Lattice-based Monte Carlo simulations of lipid membranes: correspondence between triangular and square lattices.* Biophysical journal. 2014. **106**(2):p. 290a–291a.

[23] Kosmalska, A.J., Casares, L., Elosegui-Artola, A., Thottacherry, J.J., Moreno-Vicente, R., González-Tarragó, V., Del Pozo, M.Á., Mayor, S., Arroyo, M., & Navajas, D. *Physical principles of*

membrane remodelling during cell mechanoadaptation. Nature communications. 2015. **6:** p. 7292.

[24] Groot, R.D. & Rabone, K. *Mesoscopic simulation of cell membrane damage, morphology change and rupture by nonionic surfactants.* Biophysical journal. 2001. **81**(2):p. 725–736.

[25] Feller, S.E. & Pastor, R.W. *Constant surface tension simulations of lipid bilayers: the sensitivity of surface areas and compressibilities.* The journal of chemical physics. 1999. **111**(3): p. 1281–1287.

[26] Goetz, R., Gompper, G., & Lipowsky, R. *Mobility and elasticity of self-assembled membranes.* Physical review letters. 1999. **82**(1):p. 221.

[27] Boal, D. & Boal, D.H. *Mechanics of the Cell.* 2012: Cambridge University Press. Cambridge, UK.

[28] Helfrich, W. *Elastic properties of lipid bilayers: theory and possible experiments.* Zeitschrift für Naturforschung C. 1973. **28**(11–12):p. 693–703.

[29] Mutz, M. & Helfrich., W. *Bending rigidities of some biological model membranes as obtained from the Fourier analysis of contour sections.* Journal de physique. 1990. **51**(10):p. 991–1001.

[30] Canham, P.B. *The minimum energy of bending as a possible explanation of the biconcave shape of the human red blood cell.* Journal of theoretical biology. 1970. **26**(1):p. 61–81.

[31] Seifert, U. *Configurations of fluid membranes and vesicles.* Advances in physics. 1997. **46**(1): p. 13–137.

[32] Toponogov,V.A. *Differential geometry of curves and surfaces.* 2006: Springer. New York City, USA.

[33] Best, R.B., Zhu, X., Shim, J., Lopes, P.E., Mittal, J., Feig, M., & MacKerell, A.D. Jr. *Optimization of the additive CHARMM all-atom protein force field targeting improved sampling of the backbone ϕ, ψ and side-chain $\chi1$ and $\chi2$ dihedral angles.* Journal of chemical theory and computation. 2012. **8**(9):p. 3257–3273.

[34] Duan, Y., Wu, C., Chowdhury, S., Lee, M.C., Xiong, G., Zhang, W., Yang, R., Cieplak, P., Luo, R., & Lee., T. *A point-charge force field for molecular mechanics simulations of proteins based on condensed-phase quantum mechanical calculations.* Journal of computational chemistry. 2003. **24**(16):p. 1999–2012.

[35] Berger, O., Edholm, O., & Jähnig, F. *Molecular dynamics simulations of a fluid bilayer of dipalmitoylphosphatidylcholine at full hydration, constant pressure, and constant temperature.* Biophysical journal. 1997. **72**(5):p. 2002–2013.

[36] Jorgensen, W.L., Maxwell, D.S., & Tirado-Rives., J. *Development and testing of the OPLS all-atom force field on conformational energetics and properties of organic liquids.* Journal of the American Chemical Society. 1996. **118**(45):p. 11225–11236.

[37] Schmid, N., Eichenberger, A.P., Choutko, A., Riniker, S., Winger, M., Mark, A.E., & van Gunsteren., W.F. *Definition and testing of the GROMOS force-field versions 54A7 and 54B7.* European Biophysics Journal. 2011. **40**(7):p. 843.

[38] Schlenkrich, M., Brickmann, J., MacKerell, A.D., & Karplus, M. *An empirical potential energy function for phospholipids: criteria for parameter optimization and applications,* in Biological Membranes. 1996, Springer. New York City, USA. p. 31–81.

[39] Feller, S.E., Yin, D., Pastor, R.W., & MacKerell, A.D. Jr *Molecular dynamics simulation of unsaturated lipid bilayers at low hydration: parameterization and comparison with diffraction studies.* Biophysical journal. 1997. **73**(5):p. 2269–2279.

[40] Klauda, J.B., Brooks, B.R., MacKerell, A.D., Venable, R.M., & Pastor, R.W. *An ab initio study on the torsional surface of alkanes and its effect on molecular simulations of alkanes and a DPPC bilayer.* The journal of physical chemistry B. 2005. **109**(11):p. 5300–5311.

[41] Klauda, J.B., Pastor, R.W., & Brooks, B.R. *Adjacent gauche stabilization in linear alkanes: implications for polymer models and conformational analysis.* The journal of physical chemistry B. 2005. **109**(33):p. 15684–15686.

[42] Klauda, J.B., Venable, R.M., Freites, J.A., Connor, J.W. O., Tobias, D.J., Mondragon-Ramirez, C., Vorobyov, I., MacKerell, A.D. Jr., & Pastor, R.W. *Update of the CHARMM all-atom additive force field for lipids: validation on six lipid types.* The journal of physical chemistry B. 2010. **114**(23): p. 7830–7843.

[43] Venable, R.M., Brown, F.L., & Pastor, R.W. *Mechanical properties of lipid bilayers from molecular dynamics simulation.* Chemistry and physics of lipids. 2015. **192**:p. 60–74.

[44] Rawicz, W., Olbrich, K., McIntosh, T., Needham, D., & Evans., E. *Effect of chain length and unsaturation on elasticity of lipid bilayers.* Biophysical journal. 2000. **79**(1): p. 328–339.

[45] Wu, E.L., Fleming, P.J., Yeom, M.S., Widmalm, G., Klauda, J.B., Fleming, K.G., & Im, W. *E. coli outer membrane and interactions with OmpLA.* Biophysical journal. 2014. **106**(11): p. 2493–2502.

[46] Pastor, R. & MacKerell, A. Jr. *Development of the CHARMM force field for lipids.* The journal of physical chemistry letters. 2011. **2**(13):p. 1526–1532.

[47] López, C.s.A., Sovova, Z., van Eerden, F.J., de Vries, A.H., & Marrink, S.J. *Martini force field parameters for glycolipids.* Journal of chemical theory and computation. 2013. **9**(3): p.1694–1708.

[48] Marrink, S.J. & Tieleman, D.P. *Perspective on the martini model.* Chemical society reviews. 2013. **42**(16):p. 6801–6822.

[49] Periole, X. & Marrink, S.-J. *The Martini coarse-grained force field,* in *Biomolecular Simulations.* 2013, Springer. New York City, USA. p. 533–565.

[50] Nagle, J.F. & Tristram-Nagle, S. *Structure of lipid bilayers.* Biochimica et Biophysica Acta (BBA)- Reviews on Biomembranes. 2000. **1469**(3):p. 159–195.

[51] Ingólfsson, H.I., Melo, M.N., Van Eerden, F.J., Arnarez, C.m., Lopez, C.A., Wassenaar, T.A., Periole, X., De Vries, A.H., Tieleman, D.P., & Marrink, S.J. *Lipid organization of the plasma membrane.* Journal of the American Chemical Society. 2014. **136**(41): p. 14554–14559.

[52] McMullen, T.P., Lewis, R.N., & McElhaney, R.N. *Cholesterol–phospholipid interactions, the liquid-ordered phase and lipid rafts in model and biological membranes.* Current opinion in colloid & interface science. 2004. **8**(6):p. 459–468.

[53] Hoogerbrugge, P. & Koelman, J. *Simulating microscopic hydrodynamic phenomena with dissipative particle dynamics.* EPL (Europhysics Letters). 1992. **19**(3):p. 155.

[54] Groot, R.D. & Warren, P.B. *Dissipative particle dynamics: bridging the gap between atomistic and mesoscopic simulation.* The journal of chemical physics. 1997. **107**(11): p. 4423–4435.

[55] Espanol, P. & Warren, P. *Statistical mechanics of dissipative particle dynamics.* EPL (Europhysics Letters). 1995. **30**(4):p. 191.

[56] Grafmüller, A., Shillcock, J., & Lipowsky, R. *The fusion of membranes and vesicles: pathway and energy barriers from dissipative particle dynamics.* Biophysical journal. 2009. **96**(7): p. 2658–2675.

[57] Laradji, M. & Kumar, P.S. *Dynamics of domain growth in self-assembled fluid vesicles.* Physical review letters. 2004. **93**(19):p. 198105.

[58] de Meyer, F. & Smit, B. *Effect of cholesterol on the structure of a phospholipid bilayer.* Proceedings of the national academy of sciences. 2009. **106**(10):p. 3654–3658.

[59] de Meyer, J.-M, F.d.r., Benjamini, A., Rodgers, J.M., Misteli, Y., & Smit, B. *Molecular simulation of the DMPC-cholesterol phase diagram.* The Journal of Physical Chemistry B. 2010. **114**(32):p. 10451–10461.

[60] Kranenburg, M. & Smit, B. *Simulating the effect of alcohol on the structure of a membrane.* FEBS letters. 2004. **568**(1–3):p. 15–18.

[61] Hu, J., Lipowsky, R., & Weikl, T.R. *Binding constants of membrane-anchored receptors and ligands depend strongly on the nanoscale roughness of membranes.* Proceedings of the national academy of sciences. 2013. **110**(38):p. 15283–15288.

[62] Shillcock, J.C. & Lipowsky, R. *Tension-induced fusion of bilayer membranes and vesicles.* Nature materials. 2005. **4**(3):p. 225.

[63] Shen, Z., Ye, H., Kröger, M., & Li, Y. *Self-assembled core–polyethylene glycol–lipid shell nanoparticles demonstrate high stability in shear flow.* Physical chemistry chemical physics. 2017. **19**(20):p. 13294–13306.

[64] Li, Y., Kröger, M., & Liu, W.K. *Endocytosis of PEGylated nanoparticles accompanied by structural and free energy changes of the grafted polyethylene glycol.* Biomaterials. 2014. **35**(30):p. 8467–8478.

[65] Li, Y., Kröger, M., & Liu, W.K. *Shape effect in cellular uptake of PEGylated nanoparticles: comparison between sphere, rod, cube and disk.* Nanoscale. 2015. **7**(40):p. 16631–16646.

[66] Shen, Z., Loe, D.T., Awino, J.K., Kröger, M., Rouge, J.L., & Li, Y. *Self-assembly of core-polyethylene glycol-lipid shell (CPLS) nanoparticles and their potential as drug delivery vehicles.* Nanoscale. 2016. **8**(31):p. 14821–14835.

[67] Shen, Z., Ye, H., Kröger, M., & Li, Y. *Aggregation of polyethylene glycol polymers suppresses receptor-mediated endocytosis of PEGylated liposomes.* Nanoscale. 2018. **10**(9):p. 4545–4560.

[68] Cooke, I.R. & Deserno, M. *Solvent-free model for self-assembling fluid bilayer membranes: stabilization of the fluid phase based on broad attractive tail potentials.* The journal of chemical physics. 2005. **123**(22):p. 224710.

[69] Reynwar, B.J., Illya, G., Harmandaris, V.A., Müller, M.M., Kremer, K., & Deserno, M. *Aggregation and vesiculation of membrane proteins by curvature-mediated interactions.* Nature. 2007. **447**(7143):p. 461.

[70] Yuan, H., Huang, C., Li, J., Lykotrafitis, G., & Zhang, S. *One-particle-thick, solvent-free, coarse-grained model for biological and biomimetic fluid membranes.* Physical review E. 2010. **82**(1):p. 011905.

[71] Fu, S.-P., Peng, Z., Yuan, H., Kfoury, R., & Young, Y.-N. *Lennard-Jones type pair-potential method for coarse-grained lipid bilayer membrane simulations in LAMMPS.* Computer physics communications. 2017. **210**:p. 193–203.

[72] Zhang, Y., Huang, C., Kim, S., Golkaram, M., Dixon, M.W., Tilley, L., Li, J., Zhang, S., & Suresh, S. *Multiple stiffening effects of nanoscale knobs on human red blood cells infected with Plasmodium falciparum malaria parasite.* Proceedings of the national academy of sciences. 2015. **112**(19):p. 6068–6073.

[73] Ramakrishnan, N., Kumar, P., & Ipsen, J.H. *Modeling anisotropic elasticity of fluid membranes.* Macromolecular theory and simulations. 2011. **20**(7):p. 446–450.

[74] Ye, H., Shen, Z., & Li, Y. *Computational modeling of magnetic particle margination within blood flow through LAMMPS.* Computational mechanics. 2017:p. 1–20.

[75] Ramakrishnan, N., Kumar, P.S., & Radhakrishnan, R. *Mesoscale computational studies of membrane bilayer remodeling by curvature-inducing proteins.* Physics reports. 2014. **543**(1):p. 1–60.

[76] David, N., Tsvi, P., & Steven, W. *Statistical mechanics of membranes and surfaces.* 2004: World Scientific. Singapore.

[77] Phillips, J.C., Braun, R., Wang, W., Gumbart, J., Tajkhorshid, E., Villa, E., Chipot, C., Skeel, R.D., Kale, L., & Schulten, K. *Scalable molecular dynamics with NAMD.* Journal of computational chemistry. 2005. **26**(16):p. 1781–1802.

[78] Lindahl, E., Hess, B., & Van Der Spoel, D. *GROMACS 3.0: a package for molecular simulation and trajectory analysis.* Molecular modeling annual. 2001. **7**(8):p. 306–317.

[79] Plimpton, S. *Fast parallel algorithms for short-range molecular dynamics*. Journal of computational physics. 1995. **117**(1):p. 1–19.
[80] Brooks, B.R., Brooks, C.L., MacKerell, A.D., Nilsson, L., Petrella, R.J., Roux, B., Won, Y., Archontis, G., Bartels, C., & Boresch, S. *CHARMM: the biomolecular simulation program*. Journal of computational chemistry. 2009. **30**(10):p. 1545–1614.
[81] Case, D., Darden, T., Cheatham, T. Simmerling, III, C., Wang, J., Duke, R., Luo, R., Walker, R., Zhang, W., & Merz, K. *AMBER 12; University of California: San Francisco, 2012*. There is no corresponding record for this reference, 2010:p. 1–826.
[82] Humphrey, W., Dalke, A., & Schulten, K. *VMD: visual molecular dynamics*. Journal of molecular graphics. 1996. **14**(1):p. 33–38.
[83] Jewett, A. *Moltemplate Manual*. 2017, University of California, Santa Barbara Shea Lab (August 2015).
[84] Martínez, L., Andrade, R., Birgin, E.G., & Martínez, J.M. *PACKMOL: a package for building initial configurations for molecular dynamics simulations*. Journal of computational chemistry. 2009. **30**(13):p. 2157–2164.

Chun Chan, Xiaolin Cheng

21 Molecular dynamics simulation studies of small molecules interacting with cell membranes

Abstract: Biomembranes are an indispensable element of all cells, whether prokaryotic or eukaryotic. They function as the barriers that allow for cellular compartmentalization, as a first line of defense against harmful foreign substances, and as a medium for cellular communication. Membranes are composed of a mélange of lipids, sterols, proteins, and even exogenous molecules. As such, the interaction of small molecules with lipid membranes is of great importance for membrane functions. In this review, we will give account of the recent progress in molecular simulations of small molecule–lipid bilayer interactions. We will first discuss how the chemical and physical properties (e.g., hydrophobic thickness mismatch and phase behavior) of the lipid bilayers affect the distribution and orientation of small molecules within the membranes. This will be followed by an account of simulation studies on how the presence of small molecules modulates the structure and dynamics of lipid membranes. Finally, we will review emerging mechanisms by which small molecules (e.g., α-tocopherol) traverse a cell membrane as revealed by molecular simulations using advanced sampling techniques. Detailed knowledge of these mechanisms will have significant implication for understanding physiological distribution, pharmacokinetics, and resistance of drug molecules.

Keywords: Molecular Dynamics Simulation, Small Molecule, Membrane Partitioning, Membrane Properties, Membrane Dynamics

21.1 Introduction

Biological membranes, formed by amphiphilic lipid molecules, are highly organized structures. Membranes function as the barriers that allow for cellular compartmentalization, as a first line of defense against harmful foreign substances, and as a medium for cellular communication. Complex compositions of lipids with different melting temperatures lead to interesting phenomena, for instance membrane phases, lipid raft formation, and cross-leaflet coupling. As such, the interaction of small

---

Chun Chan, College of Pharmacy, Medicinal Chemistry & Pharmacognosy, The Ohio State University, Columbus
Xiaolin Cheng, College of Pharmacy, Medicinal Chemistry & Pharmacognosy, The Ohio State University, Columbus; Biophysics Graduate Program, The Ohio State University, Columbus; Translational Data Analytics Institute, The Ohio State University, Columbus

https://doi.org/10.1515/9783110544657-021

molecules with lipid membranes is of great importance for membrane functions. A variety of experimental techniques have been used to explore these interesting phenomena of membranes, including neutron scattering [1–4], NMR [5, 6], and X-ray diffraction techniques [1]. All-atom molecular dynamics (MD) simulations have the capability of providing structural and dynamic information at the atomic level which is not directly accessible by any of experimental techniques. Thus, combination of MD simulation with various experiments has proven to be extremely useful in membrane research [1]. By calculation of forces acting on each individual atoms according to the force fields parameterized from *ab initio* calculations and/or experimental data, MD simulation is able to visualize the time evolution of complex biological systems, providing atomistic details of physical movements of the systems across considerable spatial and temporal scales [7, 8].

In this chapter, we will give account of the recent progress in molecular simulations of small molecule–lipid bilayer interactions. We will first discuss how the chemical and physical properties (e.g., hydrophobic thickness mismatch and phase behavior) of the lipid bilayers affect the distribution and orientation of small molecules within the membranes. This will be followed by an account of simulation studies on how the presence of small molecules modulates the structure and dynamics of lipid membranes. Given that most physical properties are intimately related to the lateral and transversal organization in membranes, whether and how the addition of small molecules will reinforce or disrupt these couplings, which has drawn much attention in recent years, will be discussed. Finally, we will review emerging mechanisms by which small molecules traverse a cell membrane as revealed by molecular simulations using advanced sampling techniques. Detailed knowledge of these mechanisms will have significant implication for understanding physiological distribution, pharmacokinetics, and resistance of drug molecules.

21.2 Small molecule–lipid interactions

The polar head groups of lipid molecules expose to the exterior of biological membranes and form an intriguing membrane–water interface. At this interface, polar and/or charged groups generate strong electrostatic fields as well as short-ranged interactions such as hydrogen bonds (H-bonds) and water bridges. Water molecules play a key role in heavily charged regions by not only directly interacting with the polar groups (hydration) but also indirectly bridging lipids and/or small molecules (water bridges).

H-bond is usually defined by measuring the distance between a hydrogen donor (D) and a hydrogen acceptor (A) to be less than 0.325 nm and the angle between the H-D bond and the vector connecting D and A to be less than 35° [9, 10]. This choice of distance and angle is supported by radial distribution functions and crystallographic data [9]. A charge pair (or salt bridge) is formed between a positively charged choline group (N-CH$_3$) and a negatively charged lipid oxygen atom within a separation of 0.4 nm in phosphatidylcholine (PC) lipid bilayers [11]. Strong interactions between choline and phosphate groups were found to be crucial for the properties of biological membranes [10–12]. A water molecule has simultaneously a H-bond to two polar groups either within the same molecule or between two adjacent molecules, thereby creating a water bridge. Water bridges stabilize membrane structure and/or small molecules at the membrane–water interfacial regions in a similar way that the structural water does in proteins.

The stability of polar interactions can be quantified by their lifetimes. The lifetime is defined as a period between the first and the last occurrences of the given interaction. Allowance for short breaks during the period varies between calculations [9], resulting in a wide range of lifetimes from a fraction of a picosecond [13, 14] to over tens of nanoseconds [9, 11]. The distinct glucolipid head group structures dictate their lateral segregation in membrane microdomains, where hydrogen bonding interactions between head group sugars increase the microdomain rigidity [15, 16]. Such structural differences of constituent lipids have enabled the usage of electron microscopy techniques together with uranyl acetate staining to visualize functional membrane microdomains [17].

Polar groups from small molecules can form strong electrostatic interactions with both lipids and water molecules at the membrane–water interface. Taking cholesterol as an example, the cholesterol hydroxyl group can act as both H-bond donor and an acceptor, or a charged group. Simulations and experiments have shown conflicting results on which part of PC lipids a cholesterol hydroxyl group interacts with. While most results suggested that the carbonyl oxygen atoms were the partner [18–23], the phosphate oxygen atoms were observed to preferentially interact with the cholesterol hydroxyl group in more recent MD studies [24]. In these cases, MD studies were able to provide valuable atomistic information on how small molecules behave in the interfacial regions.

Compared to polar interactions, nonpolar interactions are weaker and generally nonspecific. They are dominated by van der Waals interactions and are closely related to the lipid packing [25–27]. Radial distribution function (RDF) is often used to quantify the distribution of lipids around a chosen lipid, which can be calculated through the formula:

$$RDF_{ij} = \frac{1}{\rho}\left\langle \frac{n_j(r)}{4\pi r^2 dr}\right\rangle \quad (21.1)$$

where $n_j(r)$ is the number of particles j found inside a spherical shell of radius r and width dr centered at atom i, $\langle\rangle$ denotes the ensemble average. RDF for acyl chains is quite conserved for commonly seen PC bilayers such as dimyristoylphosphatidylcholine (DMPC), dioleoylphosphatidylcholine (DOPC), 1-palmitoyl-2-oleoylphosphatidylcholine (POPC) or sphingolipid bilayers (Figure 21.1). Typical features of RDFs include two maxima at ~0.5 and ~0.9 nm, and a minimum at ~0.7 nm. The first maximum corresponds to the sum of the van der Waals radii of two CH$_2$ groups,

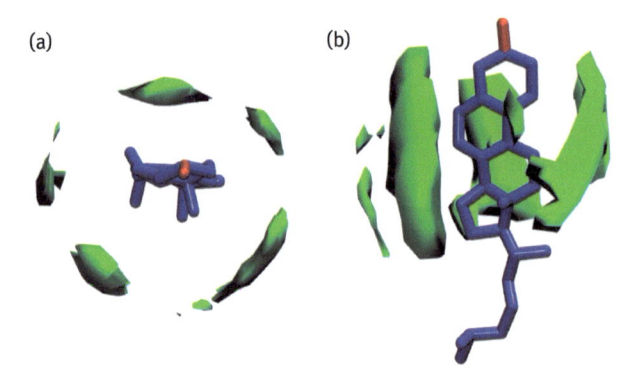

Figure 21.1: 3D spatial density distribution of lipid hydrocarbon chains in the vicinity of cholesterol.

indicating a close contact between methylene groups in the hydrophobic region of the bilayers. By adding small molecules to the membrane, RDF of acyl chains can be severely altered, reflecting disruption to the packing of lipids [28].

21.3 How membrane affects the small molecule partitioning

21.3.1 Transverse partitioning

Considering various interactions between lipid and small molecules, it is not surprising that different small molecules can be found at different locations with different orientations in biological membranes. Studying the partition and orientation of small molecules is crucial to understand their effects on membrane structures and dynamics. Chemical structure of a small molecule determines, to a large extent, how it partitions inside lipid bilayers. If the molecule is overall hydrophobic, it is more likely to be found near the center region of the membrane. Otherwise, if the molecule is less hydrophobic, it can then be found near the membrane–water interface. Such molecules are usually small in size, such as inorganic gases or highly fluorinated organic molecules used as general anesthetics (GA). Molecules with similar chemical properties can still vary in their detailed partition and orientation within membranes. For example, xenon was observed to reside at the membrane center and aggregate upon increase of external pressure [29]. Halothane, on the other hand, was found to stay across the lipid tail region and slightly concentrate in the glycerol region [30].

Interactions between human sex hormones (progesterone and testosterone) and DMPC bilayers were studied using coarse-grained MD simulations [31]. The steroid-based hormones were found to locate deeply in the membrane due to their hydrophobic nature. Hypericin is a large polyphenolic molecule derived from plants with eight rings in a planar geometry. Individual hypericin molecules were reported to adsorb the membrane surface, forming multiple hydrogen bonds with water. Hypericin molecules aggregate spontaneously in water [32]. In addition, brominated hypericin can spontaneously translocate across a cholesterol-containing lipid bilayer and reach the interior of the membrane [33].

An amphipathic molecule is both hydrophobic and hydrophilic, which allows a wide range of possibilities by altering the degree of polarity and hydrophobicity. Depending on the carried charges, the membrane–water interface and the lipid head group–lipid tails interface are the two preferred binding sites for most amphipathic membrane-binding molecules. For instance, local anesthetics (LA) share similar chemical structures, containing an aromatic ring and an ionizable amino group. The charge on a molecule can vary with the local environment, such as the pH values, the charged or uncharged LA molecule may then prefer different locations inside the membrane. Most LA molecules are found in neutral and protonated forms at physiological pH. Uncharged species were often reported to bind to membranes more strongly [34]. Using MD simulations, charged lidocaine molecules were found to reside near the membrane–water interface, farther away from the membrane center than the uncharged form [35]. In comparison, uncharged lidocaine molecules are more free to move laterally in the hydrophobic region as well as flip-flop between the two leaflets. Uncharged articaine (another LA molecule) also favors the membrane center instead of the membrane–water interface but with a much localized state [36]. Structurally speaking, different from lidocaine, articaine has a more lipophilic thiophene aromatic ring and an additional ester group. Despite this difference, the behavior of articaine inside the membrane is similar to that of lidocaine. However, the relationship between protonation state and membrane partition does not hold among all the LA molecules. Benzocaine, which is found only in the uncharged form under physiological conditions, behaves similarly to other charged LAs. That is, instead of favoring the hydrophobic center region, benzocaine was found to locate near the lipid head region [37]. A possible reason for this is that the amino group in benzocaine has a low pK_a due to its bonding to an electron-withdrawing benzene ring [38, 39].

The penetration depth of anesthetic molecules in membranes is crucial for the binding to their membrane-bound receptors. Some potent intravenous GAs, such as propofol [40] and ketamine [41], were found to reside in the same depth of the binding sites in the transmembrane region of the protein. This partition of propofol also explains why only acyl carbon atoms near the interface were observed to become ordered in the presence of propofol. Different protonation states of drugs also have implication for their location inside the lipid bilayers. For instance, uncharged form of a neuroleptic drug, chlorpromazine, tends to reside in the lipid tail regions [42, 43]. Although the lipid surface density does not influence the penetration depth of chlorpromazine in a dipalmitoylphosphatidylcholine (DPPC) monolayer, it affects the orientation of the molecule and thus its lipid ordering ability, which will be discussed in later sections.

Other examples of amphipathic molecules include β-blockers, which were studied using a dual-resolution computational method and were found to partition near the lipid head groups in DMPC bilayers [31]. Dioscin, a sterol-like molecule extracted from plants, adopts a cholesterol-like orientation in the membrane. Moreover, dioscin was found to diffuse toward the cholesterol-rich fraction of the membrane and bind to cholesterol even more strongly than cholesterol does to itself [44]. Such dioscin-cholesterol aggregates can bend the membrane drastically, which was suggested to have physiological significance such as hemolysis of red blood cells or cytotoxicity caused by saponin molecules, to which dioscin belongs [44, 45].

Ceramide (CER) is another example of amphipathic molecules that has important physiological roles. CER has been found to strongly interact with lipid molecules in binary mixtures of PC lipids and CER at various CER concentrations [46]. MD simulations have shown that CER has considerable effects on membrane properties, often mediated through hydrogen bonding, which implies the location of CER molecules should facilitate their hydrogen bonding. Interestingly, while the length of hydrophobic tails was found to have little effects on lateral organization of surrounding lipid molecules, they do affect bilayer thickness and free volume in the hydrophobic regions, modulating the membrane permeability [47].

Tamoxifen, an anticancer drug, was found to insert into lipid bilayers with flexible orientations, oscillating between 30° and 150° with respect to the bilayer normal [48]. As tamoxifen is shorter in length compared to most lipid acyl chains, it was found to reside in apolar chain region of bilayers in a wide range of depth. MD simulations also revealed that tamoxifen's orientation vary as a function of its depth within the bilayer. Tamoxifen's hydrophilic amine group is, on the other hand, preferentially located either at the bilayer center or at the head group-tail interface.

Besides straightforward MD simulations, advanced sampling techniques such as umbrella sampling have been widely used to compute the free energies of membrane-small molecule interactions. The large hydrophobic moiety of steroids tends to facilitate penetrating deeply into the lipid bilayers; the penetration depth of small molecules is found to increase with the number of aromatic rings in the molecules. Unlike molecules with large hydrophobic cores, for example, steroids with a rigid cyclic skeleton, the presence of charges in the hydrophilic group in nonsteroidal anti-inflammatory drugs (NSAIDs) has a much drastic effect on their interactions with lipid molecules. In a study of NSAIDs, Boggara and Krishnamoorti [49] quantified the free energy differences in crossing the membrane between neutral and charged forms of ibuprofen and aspirin. The charged drug molecules experience a large energy barrier near the membrane center, which is, however, significantly flattened for the neutral forms. Moreover, the free energy well, determining the preferential location of the molecules, shifts from the membrane center to the membrane–water interfacial region upon charging the molecules. Study of the distributions of charged side chains in lipid bilayers [50] has reached similar conclusion as protonated drug molecules – charged side chains have higher energy barriers at the membrane center than their neutral counterparts.

MD-based free energy calculations are one of the most rigorous methods to determine the location and orientation of small molecules in bilayers. Kessel et al. [51] concluded that, using free energy calculations, cholesterol prefers embedding its backbone in the hydrocarbon core and the OH group in the polar head group region. This optimal location was thought to be determined by the desolvation of cholesterol, whereas spatial fluctuations around this state are governed by the elastic response of neighboring lipids trying to adapt to the shape of cholesterol. This result is consistent with the condensing effect of cholesterol on membranes, where lipid molecules are tightly packed in the presence of cholesterol.

21.3.2 Tilting angle and orientation

Tilting angle and orientation of a small molecule are defined specifically to the molecular configuration. Taking cholesterol as an example [52], the orientation of cholesterol molecules in the membrane is described using the vector joining C3 and C17 atoms (or C15 in some literatures [53]) on the cholesterol ring and the bilayer normal. As usual for polar coordinates, the tilting angle θ is defined in the range [0°; 180°]. In this definition, $\theta = 0°$ represents a cholesterol orientation where the ring plane is parallel to the bilayer normal z-axis. One can then construct the probability distribution function of cholesterol tilt angle by dividing [0°; 180°] into small bins and histograming θ obtained from MD simulations. The histogram is finally normalized by the total number of samples to yield the normalized probability density function $P(\theta)$. Given a $P(\theta)$, the averaged cholesterol tilt angle can be obtained by

$$\langle \theta \rangle = \int_0^{180} \theta P(\theta) d\theta \tag{21.2}$$

For symmetric bilayers, $P(\theta)$ is usually constructed to be in the range [0°; 90°].

Furthermore, the orientational angle histogram can be used to derive a potential of mean force ($PMF(\theta)$), providing a measure of the free energy change associated with the tilting of a small molecule in membranes

$$PMF(\theta) = -k_B T \ln \frac{P(\theta)}{P_0(\theta)} \tag{21.3}$$

where T is the absolute temperature. By Taylor expansion of the PMF function around its minimum value up to the quadratic order in θ, one can obtain the tilt modulus χ_t [51]

$$\Delta PMF(\theta) = PMF(0) + \frac{1}{2}\chi_t \theta^2 \qquad (21.4)$$

In the studies of mixtures of cholesterol and DMPC lipids [52], increase of cholesterol concentration was found to lead to a significant shift in cholesterol orientation angle distributions. The tilt modulus was also found to increase upon addition of cholesterol, indicating an increase in rigidity of orientational degree of freedom in the bilayer.

21.3.3 Lateral partitioning

Lateral diffusion and phase separation have been known to exist in functional cell membranes since early 1973 [54]. The concept of "lipid rafts" was originally proposed to explain the self-associative properties unique to sphingolipid and cholesterol [55]. Regional lipid phase transition has been the mostly accepted physical explanation for the formation of lipid rafts [55, 56]. Although formation of lipid rafts was found to be crucially important for various cellular functions, such as signal transduction [57], plasma membrane microorganization [58], and membrane traffic [59], the molecular details of lipid rafts remain elusive. Given the highly complex and dynamic nature of cell membranes, model membranes have been widely used to provide valuable insights into molecular mechanisms behind cellular processes [60]. For example, coexisting liquid phases, known as "liquid ordered" (L_o) and "liquid disordered" (L_d) in cholesterol-containing lipid mixtures have been studied extensively, and substructures within the phases have also been revealed by MD simulations [7, 61]. It has been demonstrated in model membrane systems that a protein-mediated cross-linking of gangliosides could induce the formation of coexisting liquid phases from a single-phase membrane [62]. More recently, owing to the advances in experimental techniques as well as close cooperation with molecular simulations, the physical properties of individual phases could be isolated from the bulk properties, such as bending modulus or thickness [2].

On the other hand, the localized membrane phases may foster laterally differential partitioning of small molecules. Small molecules such as cholesterol have been found crucial for membrane-phase behavior, owing to their lipid ordering or disrupting abilities. As a mutual effect, properties of the lipid molecules, such as tail lengths or degree of saturation, which contribute to the membrane phase formation, also affect the lateral partitioning of cholesterol [63, 64]. Generally, the unsaturated bilayers are less sensitive to rigid rings of the cholesterol compared to the saturated ones [65, 66]. Carotenoids, considered as a cholesterol "surrogate" for organisms that do not contain cholesterol in their membranes, not only laterally partition in mixed-phase membranes but also shift its role of a membrane stiffener or a membrane softener in different phases [67]. In the presence of carotenoids, the L_o bilayer is thinned by bending the lipid tails, while the L_d bilayer is thinned by way of interdigitation of the lipid tails. As a result, carotenoids tend to decrease the order of the L_o bilayer but increase the order of the L_d one. Simulations of structured membranes that combine ordered and disordered regions show that chloroform molecules accumulate preferentially in the lipid phases that are more disordered [68]. It is thus suggested that the anesthetic action of chloroform arises from its unique partitioning behavior, both transversally and laterally, in cell membranes.

Vitamin E (α-tocopherol) has long been recognized as the major antioxidant in biological membranes and is found to partition into domains that are enriched in polyunsaturated phospholipids, amplifying the concentration of the vitamin in the place where it is most needed [69]. Polyphosphoinositides are also found to play an important role in the structure or function of detergent-insoluble membrane domains, connecting the functional role of lipid rafts and the regulation of exocytosis [70, 71]. Microdomains of plasma membrane enriched in cholesterol and sphingolipids also affect the binding of membrane-associated proteins, such as Ras, which is a lipidated membrane-bound GTPase mediating numerous growth-factor-related signaling pathways [72–74]. Using time-lapse tapping-mode atomic force microscopy, it has been shown that partitioning of N-Ras occurs preferentially in L_d lipid domains, independent of the membrane-localization motifs [75].

21.4 How small molecules affect membranes

21.4.1 Membrane dipole potential

Partitioning of polar or charged small molecules can alter the electrostatic properties of the bilayer. In the study of lidocaine [76], influence of the charged and uncharged drug molecule on electrostatic potential, charge distribution, and headgroup orientation was shown to account for the lidocaine anesthetic action. The dipole electrostatic potential was affected by the presence of either form of lidocaine. Interestingly, the electrostatic potential in the lipid tail region turned out to be almost the same for both charged and uncharged lidocaine at equal concentrations, though the underlying mechanisms were different. The charged lidocaine has a strong influence on the lipid headgroups, leading to a decreased tilting of the phosphorus-nitrogen dipole vector. This alignment of dipoles increases the total electrostatic potential inside the membrane. The uncharged lidocaine, on the other hand, retains the lipid structure and causes almost no change to the lipid charge distribution. The total electrostatic potential in the middle of the membrane increases because of partial charges on the lidocaine itself, with the main contribution from the dipole moment of the carbonyl group. In such ways, the presence of lidocaine modulates the electrostatic properties and ultimately functions of membranes. Similar phenomena have also been observed for other molecules such as anesthetic chloroform [68] and cationic lipids [77].

21.4.2 Ordering effects

The molecular order parameter, S_{mol}, can be used to quantitatively evaluate the order of lipid chains:

$$S_{mol} = \frac{1}{2}\left(3\langle\cos^2\theta_n\rangle - 1\right) \quad (21.5)$$

where θ_n is the angle between the vector linking two consecutive carbon atoms in the hydrocarbon chain and the bilayer normal. The angular bracket $\langle\rangle$ denotes the ensemble average. S_{mol} can be obtained from both molecular simulations and various experimental techniques, including NMR, EPR, and fluorescence spectroscopy [78]. The deuterium order parameter (S_{CD}) measured in NMR is related to the molecular order parameter as $S_{mol} = -2S_{CD}$ [79]. Alternatively, S_{CD} can be calculated from MD trajectories by redefining θ_n in eq. 21.5 to the angle between the C-H bond and the bilayer normal [80–82]. MD simulations of DMPC and DPPC bilayers containing cholesterol have shown an increase in the order parameters of the PC chains compared to the respective cholesterol-free PC bilayers [19, 83–87]. Other small molecules such as trehalose, which is a food addictive, were also studied in a similar fashion as cholesterol [86].

Other parameters used to quantify the order of the PC chains are the probability distributions of the *gauche* rotamers or the average tilt angles. The *gauche* rotamer distribution function can be obtained from MD trajectories by measuring the torsion angles of the hydrocarbon chains. The *trans* conformation corresponds to the torsion angle of 180° ± 30°, while the *gauche* conformation corresponds to 60° ± 30° and 300° ± 30°. Note that for single bonds adjacent to a double bond in the acyl chain, neither *trans* nor *gauche* is the most stable conformation [88]. In experiments, the probability of *gauche* rotamers is difficult but possible to be measured with infrared spectroscopies [89]. Cholesterol has been shown to reduce the rate of *trans-gauche* isomerization in bilayers [89, 90].

The tilt angle measures the rigid-body angular fluctuation of a lipid chain. It can be calculated as the angle between the vector linking the first and the last carbon atoms in the chain and the bilayer normal. The angle can also be experimentally measured by fluorescence or EPR spectroscopy [91]. The difference is within 4° between simulation and experimentally measured values [53, 90, 91]. Generally, introduction of cholesterol into PC lipid bilayers will interfere with the first torsion angles of the acyl chains, reflected by the decreased *gauche* conformations, and reduced tilt angles, making the lipids more ordered.

Cholesterol is one of the best-studied membrane-interacting small molecules. It has long been considered as an integral constituent of biological membranes, owing to its structure very compatible with lipid molecules. Cholesterol is abundant in animal cell membranes and plays important roles in modulating mechanical properties and phase behaviors of the biological membranes. While typically around 20–30 mol percentage, the cholesterol concentration can be as high as 70 mol percentage in ocular lens membranes [92]. Structurally, cholesterol consists of a steroid ring, a β-hydroxyl group, and a short hydrocarbon chain attached to the steroid ring, extending the whole structure in a linear and planar manner. One generic effect that cholesterol has on lipid bilayers in the physiologically relevant fluid phase is the increase of acyl chain order. And due to this ordering ability, cholesterol is thought to be responsible for the liquid–liquid coexistence. The L_o phase is driven by direct interactions between cholesterol and saturated lipids; structure and thermodynamics of the L_o phase have been heavily studied by mixing cholesterol and saturated phospholipids [93–97].

In contrast to ordering, small molecules can also disrupt the packing of lipid molecules. Menthol, an extract from traditional Chinese medicine mint oils, was studied using MD simulations on its penetration enhancement effects [98]. It was found that menthol perturbs ordered lipids, making them floppier, thereby leading to enhanced permeability through the bilayers. Such a disordering effect also results in an increase in area per lipid and a decrease in membrane thickness. Amantadine, rimantadine, and memantine are tricycle-decane derivatives used in the treatment of influenza. These molecules disrupt membranes even more than charged LAs and NSAIDs do, likely owing to their molecular shapes incompatible for lipid packing.

21.4.3 Condensing effects

While the ordering ability of small molecules mainly affects local lipid acyl chains, small molecules can also have an impact on the lateral organization of membranes by altering the membrane surface area. Cholesterol is known to increase the membrane surface density of mixed PC bilayers, thus reducing the area per PC molecule [99, 100]. Quantifying this condensing effect can be tricky. For a single component lipid bilayer, the average area per lipid can be calculated by dividing the total surface area by the number of lipids in one leaflet. For a mixed lipid bilayer doped with small molecules, however, one has to independently obtain the areas for lipid A_L and for small molecule A_{SM}.

Taking cholesterol as an example, one classical approach is to assume A_{SM} of cholesterol, A_{chol}, is constant, which can be subtracted from the total surface area of the bilayer [101–103]. The assumption was backed up by cholesterol monolayer studies, in which A_{chol} was measured to be $0.39 - 0.41 \, \text{nm}^2$ both in simulations [25] and experiments [104–106]. However, such an assumption is too simple and not fully justified for certain cases. The cross-sectional area of the steroid ring is much larger than that of the acyl tail, therefore the distribution of free areas would greatly depend on the position of cholesterol in the bilayer; it has been shown that the volume of free space varies from high to low along the axis from the bilayer center to a distance of 1–1.5 nm from the center [107]. As a result, tight packing of lipid acyl chains around the cholesterol reduces its surrounding free area, leading to an overestimation of A_{chol} if taken directly from a cholesterol monolayer.

Another approach taking water molecules into account was also applied initially to the cholesterol cases. A_L is calculated from the formula [108]:

$$A_L = \frac{2A_{sys}}{N_L}\left(1 - \frac{N_{chol}V_{chol}}{V_{sys} - N_w V_w}\right) \tag{21.6}$$

where A_{sys} and V_{sys} are the area and volume of the simulation box, N_L is the number of lipid molecules, N_w and V_w are the number and the volume of water molecules, and N_{chol} and V_{chol} are the number and the volume of cholesterol molecules. By assuming that V_{chol} is constant, A_{chol} is calculated to be $0.27 - 0.29 \text{ nm}^2$ with a slight dependence on the cholesterol concentration.

The dependence of surface area on small molecule concentration, however, provides a more general approach to determining the molecular areas. Assuming a linear dependence, one can obtain both A_L and A_{SM} by plotting average area per lipid, A_{avg}, against small molecule concentration x [109],

$$A_{avg} = xA_{SM} + (1-x)A_L \tag{21.7}$$

A decrease in areas occupied by lipid molecules is often observed upon adding small molecules into the membrane. For instance, DMPC bilayers were reported to have areas comparable to that in its gel phase ($0.42 - 0.47 \text{ nm}^2$) at 35% cholesterol concentration or above in both MD simulations [109] and X-ray diffraction experiments [106, 110, 111]. Free areas of molecules are one of the experimentally accessible observables, which can also be calculated from molecular simulations, thus providing means for cross-validation and complementary studies of membranes.

Using MD simulations [67], carotenoids were shown to decrease the order of L_o bilayer and increase the order of the L_d one. Overall, the bilayers were found to become thinner upon carotenoid addition. The L_o bilayer was thinned via the compression of the individual monolayers, while the L_d bilayer was thinned by greater interdigitation of the lipid tails from the two leaflets.

Biological membranes are in fact composed of a few hundreds of different lipid species. Unsaturated PCs, together with sphingolipids and cholesterol, are the main components of the outer leaflet of eukaryotic cell membranes. Overall, the induced ordering and condensing effects are weaker for unsaturated lipids. Replacing the saturated lipids in previous simulations with monounsaturated POPC lipids revealed that cholesterol has weaker interactions with the POPC lipid tails [63]. Similar effects were observed in experiments for di-unsaturated DOPC compared to saturated DSPC even though the acyl chains are of equal length [64]. MD simulations provided a molecular explanation for this weakened interaction. As the double bond on the acyl chain increases the lipid tail tilting, the small molecule interacting with the lipid tails tends to tilt more accordingly, reducing its membrane ordering and condensing ability. For example, the tilt angle of cholesterol was found to increase significantly from DMPC to POPC bilayers [53].

21.4.4 Lateral pressure

The concept of lateral pressure is well established for phospholipid monolayers at the oil–water interface. However, biological membranes are naturally in a tension-free state. Therefore, lateral pressure should instead be considered to account for the modulation of molecular components inside the membrane. This internal lateral pressure in biological membranes cannot be determined directly due to the lack of a "resultant" lateral pressure (i.e., the tension-free state). Therefore, in many cases while molecules partition into the membrane, it is the lateral compressibility (i.e., the derivative of the internal lateral pressure with respect to area) rather than the lateral pressure itself that is relevant. Internal lateral pressure is composed of the components of the interactions between the membrane constituents, specifically the derivatives of their free energy with respect to area. Nevertheless, a lateral pressure profile is a useful quantity accounting for many important biological processes such as anesthesia [68, 112, 113]. While it is straightforward to suggest that the general anesthetic molecules act by binding to certain membrane-bound proteins, for example, ion channels, it is difficult to explain the wide chemical variety of these molecules and very few proteins identified so far that are capable of anesthetics binding. As a result, Cantor proposed that anesthetic molecules may alter the bilayer lateral pressure properties, which in turn shift the conformational equilibrium of certain membrane-bound ion channels, eventually leading to anesthetic action [112].

Computing lateral pressure profiles from MD simulations can be challenging [114]. First, the local pressure tensor is calculated in a similar fashion as the bulk pressure tensor by summing up the kinetic and virial (or configurational) contributions [115, 116]:

$$P_{\alpha\beta}(z) = \frac{1}{V}\left(\sum_{i \in slab} m_i v_{i\alpha} v_{i\beta} + \sum_{i \in slab}\sum_{j>i} F_{ij\alpha} r_{ij\beta} f(z, z_i, z_j)\right) \quad (21.8)$$

where $P(z)$ is an element in the pressure tensor, α and β are the Cartesian coordinates x, y, or z, V is the volume, m_i is the mass of particle i, $v_{i\alpha}$ is its velocity in the α direction, $F_{ij\alpha}$ is the α component of the total force on particle i due to particle j, and $r_{ij\beta}$ is the β component of the vector $r_i - r_j$. Using the components of the local pressure tensor, the lateral pressure is given by

$$P_{lateral} = \frac{P_{xx}(z) + p_{yy}(z)}{2} \tag{21.9}$$

While the kinetic term can be trivially calculated from the velocities of the particles, the virial term however involves computing an ensemble average of the interaction force contributions to the pressure. In order to compute the lateral pressure profile, which is actually a spatial variation of the bulk pressure, one must integrate the individual contributions over some contours. Following Lindahl and Edholm [117], the simulation space is partitioned into slabs along the z axis normal to the membrane surface, and the contributions due to particles interacting within or through each slab are then computed. This can be done mathematically by substituting α and β by x and y in eq. 21.8. A weighting function $f(z, z_i, z_j)$ has to be introduced to the virial term depending on which slab the contributing particle resides. The weight is 1.0 if both particles i and j are in the same slab, $dz/|z_i - z_j|$ if one particle is in the slab, and $\Delta z/|z_i - z_j|$ if neither particle is in the slab, but the slab lies between the two particles. Here, dz is the distance to the boundary of the slab (taking periodic boundary conditions into account), and Δz is the slab thickness. Such a weighting function distributes the virial evenly on the connecting line between two interacting particles and therefore smoothens the contribution to the local pressure profile through the slabs. This choice of contour is referred as the Irving-Kirkwood (IK) method [118, 119] and has been used in most bilayer studies [117, 120–122]. Another choice of contour is the Harasima (H) expression [123], and the significance of that is its compatibility with calculating the viral contribution from the reciprocal space part of the Ewald sum. However, as observed by Sonne et al. [114], the invariance of the local pressure to the slab width actually favors the use of the IK expression over the H expression, leading to the conclusion that it is not possible to rule in favor of either expressions.

Specific examples of using lateral pressure profile to explain the molecular mechanism of anesthetics include the study of ketamine [41], in which no significant changes to the membrane thickness and lateral area per lipid were found, instead a significant change of lateral pressure and a pressure shift toward the center of the bilayer were induced by insertion of ketamine at the membrane–water interface. Further estimation confirmed that the effect is large enough to subsequently modulate the activity of membrane-bound ion channels by affecting their opening probabilities. Characteristic changes in lateral pressure also occurred for cholesterol in DPPC bilayers. When the cholesterol concentration was increased to above 20%, the pressure profile transformed from a rather flat shape into an oscillating pattern with large positive and negative values. This might have an implication for relating the abundance of cholesterol in eukaryotic cell membranes to diverse protein activities [124]. By replacing cholesterol with other sterol molecules, notable differences could be induced in lateral pressure profiles [125], potentially altering other molecules' function inside the membrane.

The lateral pressure profile also has implications for the partition of small molecules into bilayers. The accumulation of drugs (β-blockers) near the lipid head groups can be explained by the large attractive lateral pressure troughs in that region [31]. And the repulsive forces present above and below the head group-tail region prevent free movement of these drug molecules.

21.4.5 Curvature and bending modulus

Bending modulus, K_c, is a macroscopic membrane property that can be determined by both experiments [126–128] and simulations [129–131]. K_c can be obtained from MD simulations of membranes using the Helfrich Canham (HC) approach [132, 133], which postulated that the membrane is a thin, structureless, and homogeneous fluid sheet. Throughout the simulation, overall shape of the membrane is described through the "height field," $h(x, y)$, which indicates the z-direction displacement of the bilayer from the reference. The power spectrum of the height fluctuation is predicted to be

$$\langle |h_q|^2 \rangle = \frac{k_B T}{K_c q^4} \tag{21.10}$$

where q is the wave vector, T is temperature, and k_B is Boltzmann's constant. Fitting the simulated power spectrum to the equation yields K_c. Note that this approach fails for small systems when bilayer thickness is comparable to the box size so is only valid in the limit of $q \to 0$. Such a limitation originates from molecular fluctuations of lipids such as lipid protrusion [134] and lipid tilting [135–137], which contributes to the smaller wavelength fluctuations and is not captured by the HC model. An improvement to the theory is to decouple the effects of lipid tilting from the lipid orientation fluctuations associated with K_c [137, 138]. The modified height fluctuation is written as

$$\langle |h_q|^2 \rangle = k_B T \left(\frac{1}{K_c q^4} + \frac{1}{K_\theta q^2} \right) \tag{21.11}$$

where K_θ is the lipid tilt modulus and therefore explicitly includes the contribution of lipid tilting to the height spectra at large q.

To deal with multicomponent membranes, the bending modulus can be more conveniently derived from the splay modulus χ. The splay modulus is associated with the free energy cost for splaying one lipid molecule with respect to another [131]. To obtain χ, one first defines a local lipid director vector $\vec{t}$. For PC lipids, $\vec{t}$ is the vector that connects the midpoint between the phosphate and backbone C2 atoms to the center of the three terminal carbons in the two lipid chains [136]. Although the definition of $\vec{t}$ may not be universal among lipids and small molecules, it has been shown that different definitions of $\vec{t}$ vectors could keep the calculated bending moduli within an acceptable uncertainty threshold [131]. The probability distributions $P(\alpha)$ of the splay angle α between all possible lipid–lipid pairs are then computed. Note that $P(\alpha)$ contains only those pairs for which at least one of the participant molecules is titled by no more than 10°. Finally, by performing a quadratic fit of the potential of mean force $PMF(\alpha) = -k_B T \ln[p(\alpha)/\sin\alpha]$ in the interval of small α angles, χ can be obtained as the coefficient yielding the best fit [52, 130].

For membranes of multiple components, one has to calculate all possible pairs of lipids and weigh the corresponding splay contributions appropriately. Following the analogous weighting for the bending rigidity of mixed membranes, each splay component is weighted by the occurrence probability of the corresponding molecular pair in the simulation [139]. Taking cholesterol-containing membrane as an example, the effective bending modulus can be expressed as

$$\frac{1}{K_c^{eff}} = \frac{1}{\varphi_{CC} + \varphi_{LL} + \varphi_{CL}} \left(\frac{\varphi_{CC}}{\chi_{CC}} + \frac{\varphi_{LL}}{\chi_{LL}} + \frac{\varphi_{CL}}{\chi_{CL}} \right) \qquad (21.12)$$

where χ is the splay moduli, φ denotes the number of pairs, and the subscript CC, LL, and CL denote the Chol–Chol, lipid–lipid, and Chol–lipid pairs, respectively [130]. Note that the number of pairs φ is obtained by counting pairs within 10 Å of each other.

Other approaches of calculating the bending modulus involve simulating tethers, that is, actively imposing a deformation on the membrane and then measuring the force required to hold it [140]. Such a method has been applied to large areas of curvature or even tubulation induced by peripheral proteins [141], but is not compatible with the relatively small and local undulational fluctuations caused by inserted small molecules.

21.4.6 Membrane dynamics

Many biological processes, including cellular signaling and energy transduction, are controlled to a great extent by the dynamics of lipids in the cell membranes. Lipid molecules exhibit complex dynamical behavior ranging from dynamics of individual molecules to collective motions of the entire membrane, such as vibrational, torsional and rotational motions, membrane bending and undulation, lateral diffusion, and flip-flop, etc. They cover a broad range of length and time scales: from angstroms for local lipid molecule motions to a few tens of nanometers for curvature generation of the bilayer; and from femtoseconds for molecular vibrations to a few hundred seconds for the trans-leaflet flip flop.

Diffusion of individual lipids in a membrane is quantified by the diffusion coefficient D, which can be calculated from the long-time mean square displacement (MSD):

$$D = \lim_{t \to \infty} \frac{\langle |\mathbf{r}(t+t_0) - \mathbf{r}(t_0)|^2 \rangle}{4t} \tag{21.13}$$

where r is the vector denoting the position of a lipid molecule. The averaging indicated by the brackets is performed over all lipid molecules and all initial time origins t_0. In pure membrane system, lipids diffuse in both lateral and membrane normal directions. However, the motion in the direction perpendicular to the bilayer plane is restricted and after some time it reaches a limiting value that is insignificant compared to the dimension of the lipid molecule, corresponding to a small intrinsic fluctuation of membrane thickness. This membrane thickness fluctuation decreases upon addition of cholesterol molecules [108]. The lateral diffusion is instead of more relevance to planar objects like membranes. A decrease in the diffusion coefficients of both cholesterol and DPPC molecules with increasing cholesterol concentrations was observed in MD simulations [108]. Experimental results agree on the result of cholesterol slowing down the diffusion in the liquid crystalline phase, albeit with a much smaller magnitude [142]. FRAP (fluorescence recovery after photobleaching) measurements reported a decrease from between 1.6 and 0.4×10^{-7} cm^2/s (depending on temperature) to between 1.3 and 0.2×10^{-7} cm^2/s going from 0% to 40% cholesterol in a DMPC bilayer [143]. Similar figures have also been reported for a DPPC/cholesterol system based on pulsed NMR measurements [144]. The effect of cholesterol on lateral diffusion is tightly coupled with the lipid ordering and condensation, which all together manifests the phase behavior changes upon addition of cholesterol. The diffusion of DPPC is slower than that of cholesterol, but the difference becomes smaller at higher cholesterol concentrations. The slower diffusion of DPPC is likely due to the stronger DPPC headgroup interactions.

Although motions of individual lipid molecules have successfully been investigated using both aforementioned macroscopic or microscopic experimental techniques, there are still large discrepancies between the lipid diffusion coefficients measured by macroscopic and microscopic techniques. The lateral diffusion of the entire lipid molecule is Fickian in nature and was extensively verified by macroscopic experiments. Recently, internal motions of micelles or lipid molecules have been reported using high-resolution neutron scattering techniques [145, 146]. In the ordered solid gel (L_o) phase, the lipids are more ordered and undergo uniaxial rotational motion. However, in the disordered fluid (L_d) phase, the hydrogen atoms of the lipid tails undergo spatially localized translation diffusion which is a manifestation of the flexibility of the chains in the fluid phase. The local diffusivity and the confinement volume increase in the linear fashion from near the lipid's polar headgroup to the end of its hydrophobic tail.

21.4.7 Membrane permeability

Due to the slow process of membrane translocation, direct MD simulation of membrane permeation of small molecules is hardly achievable, instead insights into the membrane permeation process are usually obtained indirectly via computation of the free energy and local diffusion rate profiles of individual small molecules across the membrane. These calculations must be handled with care as the quantities are to be reconstructed from biased ensembles and/or nonequilibrium processes. Initially, this was done for water molecules [147] and an inhomogeneous model was proposed against the simple solubility-diffusion model. Later, it was applied to gas molecules such as oxygen and ammonia [148] and recently has been extended to even more complex membrane systems involving embedded proteins [149].

The experimentally accessible permeability coefficient P can be related to the free energy profile ΔG and local diffusion D of the molecule:

$$\frac{1}{P} = \int_{z_1}^{z_2} \frac{\exp(\Delta G(z)/k_B T)}{D(z)} dz \qquad (21.14)$$

where T is the temperature, and z is along the membrane normal direction. $\Delta G(z)$ can be obtained by a geometrical free energy calculation approach using constrained MD simulations and the relation [147]:

$$\Delta G(z) = -k_B T \ln P(z) \qquad (21.15)$$

where $P(z)$ is the probability of finding the molecule in each discretized layer along the z-axis sampled by simulations. The local diffusion constant D can be calculated from the slope of the mean square displacement (MSD) against time:

$$D = \frac{\langle (z(t+\Delta t) - z(t))^2 \rangle}{2\Delta t} \qquad (21.16)$$

where z is the z-coordinates of the molecule, and Δt is the chosen time interval. Δt is here limited to small values (normally less than 5 ps), so that D best represents the local diffusion coefficient in the region it is calculated for. Note that computing MSD in short time interval limit may fail due to large free energy barriers [147]. In such cases, D can be calculated from a relation between diffusion constant and autocorrelation function of the random forces $\Delta F(t)$ acting on the molecule through the fluctuation-dissipation theorem [150]:

$$D(z) = \frac{(k_BT)^2}{\int_0^\infty \langle \Delta F(z, t)\Delta F(z, 0)\rangle dt} \quad (21.17)$$

For the above relationship to hold, one has to assume that during the decay time of the friction coefficient, which is related to the local diffusion coefficient via Einstein's relation, the molecule remains in a region of constant free energy.

Computing local diffusion coefficients of small molecules with respect to z-coordinates reveals distinct interaction regions inside the membrane. In addition to their differential partitioning (location and orientation) in the membrane, small molecules may also have different dynamic behaviors in different positions of the membrane. In studies of benzene molecules penetrating bilayers [151], it was found that the translational diffusion of benzene is faster in the center near than the zwitterionic head groups of DMPC lipids, which is in agreement with experimental results. Also, a "hopping" mechanism, which was initially proposed for the diffusion of gases through soft polymers, was shown to be the mechanism of diffusion in lipid bilayers. Jumps of up to 8 Å can occur within as little as 5 ps, whereas average motions for that time period are only ~1.5 Å. Torsional changes in the lipid hydrocarbon chains moderate these jumps and serve as "gates" between voids through which the benzene molecules move.

Sampling small molecules spontaneously passing through biological membranes is still beyond current computational limits of conventional MD simulations. Therefore, enhanced sampling methods have been applied with MD simulations, usually combined with free energy calculations, to study the translocation processes. Lv et al. [152] have computed the transmembrane free energy landscapes of hydrogen sulfide (H_2S) and its structural analogue water (H_2O) using orthogonal space tempering (OST) method. As H_2S is partially amphipathic, compared to the evidently more polar water molecule, it resides at the interface between the polar head-group and nonpolar acyl chain regions. Moreover, the moderate polarity of the molecule determines the free energy barrier height between the interfacial region and membrane center; and its small hydrophobic moiety leads to a relatively high binding affinity for the bilayer hydrophobic core. Taken together, when H_2S translocates from the bulk solution to the membrane center, the free energy difference is negligible, in good agreement with experimental water-to-hexadecane transfer free energy and measured permeability ratio.

Comer and his colleagues have performed a series of studies on membrane permeability of short-chain alcohols using importance-sampling MD simulations [153–155]. The membrane permeability of alcohols was found to be correlated with the length of the aliphatic tail in a positive manner within the range from 0.15 cm/s to 7.30 cm/s. These results agree well with experimental measurement, in which water was significantly slower (13.0×10^{-3} cm/s) than methanol in a POPC bilayer.

21.5 Concluding remarks

We have reviewed a handful of studies concerning the interplay between cell membranes and small molecules from a perspective of molecular simulations. We started by describing the molecular interactions between lipid and small molecules. These interactions are at least partially responsible for how membranes may affect the interacting small molecules in terms of their transversal location and orientation, and lateral partitioning in the membranes. Vice versa, governed by the same molecular interactions, the small molecules may have significant modulating effects on the membranes, including ordering/disrupting membrane structure, and altering membrane electrostatic potentials, lateral pressures, rigidity and permeability, which in turn affect the thermodynamic and dynamical properties (i.e., both local and global dynamics) of the small molecules inside the membranes. The membrane–small molecule interactions are truly a mutual effect between membranes and small molecules. The chemical and structural properties of the small molecules cannot solely account for their interactions with (partition in) the membranes. The composition, structural, and phase properties of the membranes also play important roles in shaping their interactions with (accommodation of) the small molecules, which may in turn induce ordering and condensation effects on the membranes.

Although interactions between cell membranes and small molecules have been studied for decades, a molecular-level understanding of these interactions in complex biological systems is still lacking. MD simulations, which can provide atomistic information of the dynamic and heterogeneous systems, are narrowing down the gap between experiment and simulation at both temporal and spatial scales. In combination with state-of-the-art experimental techniques, molecular simulations have been the new driving force in clarifying how molecular interactions lead to unique membrane structures and organizations, ultimately giving rise to membrane's complex biological functions.

References

[1] Pan, J., Cheng, X., Monticelli, L., Heberle, F.A., Kuerka, N., Peter Tieleman, D., & Katsaras, J. The molecular structure of a phosphatidylserine bilayer determined by scattering and molecular dynamics simulations. *Soft Matter*, 10(21):3716–10, 2014.

[2] Nickels, J.D., Cheng, X., Mostofian, B., Stanley, C., Lindner, B., Heberle, F.A., Perticaroli, S., Feygenson, M., Egami, T., Standaert, R.F., Smith, J.C., Myles, D.A.A., Ohl, M., & Katsaras, J. Mechanical properties of nanoscopic lipid domains. *Journal of the American Chemical Society*, 137(50):15772–15780, 2015.

[3] Pan, J., Cheng, X., Sharp, M., Ho, C.S., Khadka, N., & Katsaras, J. Structural and mechanical properties of cardiolipin lipid bilayers determined using neutron spin echo, small angle

neutron and X-ray scattering, and molecular dynamics simulations. *Soft Matter*, 11(1):130–138, 2015.

[4] Nickels, J.D., Chatterjee, S., Stanley, C.B., Qian, S., Cheng, X., Myles, D.A.A., Standaert, R.F., Elkins, J.G., & Katsaras, J. The in vivo structure of biological membranes and evidence for lipid domains. *PLoS Biol*, 15(5):e2002214, 2017.

[5] Brown, M.F., Ribeiro, A.A., & Williams, G.D. New view of lipid bilayer dynamics from 2H and 13C NMR relaxation time measurements. *Proceedings of the National Academy of Sciences*, 80 (14):4325–4329, 1983.

[6] Brown, M.F. Theory of spin-lattice relaxation in lipid bilayers and biological membranes. Dipolar relaxation. *The Journal of Chemical Physics*, 80(6):2808–2831, 1984.

[7] Sodt, A.J., Sandar, M.L., Gawrisch, K., Pastor, R.W., & Lyman, E. The molecular structure of the liquid-ordered phase of lipid bilayers. *Journal of the American Chemical Society*, 136(2): 725–732, 2014.

[8] Needham, S.R., Roberts, S.K., Arkhipov, A., Mysore, V.P., Tynan, C.J., Laura C, Z.-D., Kim, E.T., Losasso, V., Korovesis, D., Hirsch, M., Rolfe, D.J., Clarke, D.T., Winn, M.D., Lajevardipour, A., Clayton, A.H.A., Pike, L.J., Perani, M., Parker, P.J., Shan, Y., Shaw, D.E., & Marisa L, M.-F. EGFR oligomerization organizes kinase-active dimers into competent signalling platforms. *Nature Communications*, 7:13307, 2016.

[9] Pasenkiewicz-Gierula, M., Takaoka, Y., Miyagawa, H., Kitamura, K., & Kusumi, A. Hydrogen bonding of water to phosphatidylcholine in the membrane as studied by a molecular dynamics simulation: location, geometry, and lipid-lipid bridging via hydrogen-bonded water. *The Journal of Physical Chemistry A*, 101(20):3677–3691, 1997.

[10] Pasenkiewicz-Gierula, M., Róg, T., Kitamura, K., & Kusumi, A. Cholesterol effects on the phosphatidylcholine bilayer polar region: a molecular simulation study. *Biophysical Journal*, 78(3):1376–1389, 2000.

[11] Pasenkiewicz-Gierula, M., Takaoka, Y., Miyagawa, H., Kitamura, K., & Kusumi, A. Charge pairing of headgroups in phosphatidylcholine membranes: a molecular dynamics simulation study. *Biophysical Journal*, 76(3):1228–1240, 1999.

[12] Yeagle, P.L., Hutton, W.C., Huang, C.H., & Martin, R.B. Phospholipid head-group conformations – intermolecular interactions and cholesterol effects. *Biochemistry*, 16(20):4343–4349, 1977.

[13] Róg, T., Murzyn, K., & Pasenkiewicz-Gierula, M. The dynamics of water at the phospholipid bilayer surface: a molecular dynamics simulation study. *Chemical Physics Letters*, 352(5–6):323–327, 2002.

[14] Murzyn, K., Zhao, W., Karttunen, M., Kurdziel, M., & Tomasz, R. Dynamics of water at membrane surfaces: effect of headgroup structure. *Biointerphases*, 1(3):98–105, 2006.

[15] Thompson, T.E., & Tillack, T.W. Organization of glycosphingolipids in bilayers and plasma membranes of mammalian cells. *Annual Review of Biophysics and Biophysical Chemistry*, 14(1):361–386, 1985.

[16] Rock, P., Allietta, M., Young, W.W., Thompson, T.E., & Tillack, T.W. Organization of glycosphingolipids in phosphatidylcholine bilayers: use of antibody molecules and Fab fragments as morphologic markers. *Biochemistry*, 29(36):8484–8490, 1990.

[17] Garca-Fernández, E., Koch, G., Wagner, R.M., Fekete, A., Stengel, S.T., Schneider, J., Mielich-Süss, B., Geibel, S., Markert, S.M., Stigloher, C., & Lopez, D. Membrane microdomain disassembly inhibits MRSA antibiotic resistance. *Cell*, 171(6):1354–1367. e20, 2017.

[18] Kechuan, T., Klein, M.L., & Tobias, D.J. Constant-pressure molecular dynamics investigation of cholesterol effects in a dipalmitoylphosphatidylcholine bilayer. *Biophysical Journal*, 75 (5):2147–2156, 1998.

[19] Smondyrev, A.M., & Berkowitz, M.L. Structure of dipalmitoylphosphatidylcholine/cholesterol bilayer at low and high cholesterol concentrations: molecular dynamics simulation. *Biophysical Journal*, 77(4):2075–2089, 1999.

[20] Soubias, O., Jolibois, F., Milon, A., & Reat, V. High-resolution C-13 NMR of sterols in model membrane. *Comptes Rendus Chimie*, 9(3–4):393–400, 2006.

[21] Tantipolphan, R., Rades, T., Strachan, C.J., Gordon, K.C., & Medlicott, N.J. Analysis of lecithin-cholesterol mixtures using Raman spectroscopy. *Journal of Pharmaceutical and Biomedical Analysis*, 41(2):476–484, 2006.

[22] Arsov, Z., & Quaroni, L. Direct interaction between cholesterol and phosphatidylcholines in hydrated membranes revealed by ATR-FTIR spectroscopy. *Chemistry and Physics of Lipids*, 150(1):35–48, 2007.

[23] Mondal, S., & Mukhopadhyay, C. Molecular insight of specific cholesterol interactions: a molecular dynamics simulation study. *Chemical Physics Letters*, 439(1–3):166–170, 2007.

[24] Hénin, J., & Chipot, C. Hydrogen-bonding patterns of cholesterol in lipid membranes. *Chemical Physics Letters*, 425(4–6):329–335, 2006.

[25] Róg, T., & Pasenkiewicz-Gierula, M. Cholesterol effects on the phospholipid condensation and packing in the bilayer: a molecular simulation study. *FEBS Letters*, 502(1–2):68–71, 2001.

[26] Cao, H., Tokutake, N., & Regen, S.L. Unraveling the mystery surrounding cholesterol's condensing effect. *Journal of the American Chemical Society*, 125(52):16182–16183, 2003.

[27] Bonn, M., Roke, S., Berg, O., Juurlink, L.B.F., Stamouli, A., & Muller, M. A molecular view of cholesterol-induced condensation in a lipid monolayer. *Journal of Physical Chemistry B*, 108(50):19083–19085, 2004.

[28] Pöyry, S., Róg, T., Karttunen, M., & Vattulainen, I. Significance of cholesterol methyl groups. *Journal of Physical Chemistry B*, 112(10):2922–2929, 2008.

[29] Yamamoto, E., Akimoto, T., Shimizu, H., Hirano, Y., Yasui, M., & Yasuoka, K. Diffusive nature of xenon anesthetic changes properties of a lipid bilayer: molecular dynamics simulations. *Journal of Physical Chemistry B*, 116(30):8989–8995, 2012.

[30] Tu, K.M., Matubayasi, N., Liang, K.K., Todorov, I.T., Chan, S.L., & Chau, P.L. A possible molecular mechanism for the pressure reversal of general anaesthetics: aggregation of halothane in POPC bilayers at high pressure. *Chemical Physics Letters*, 543:148–154, 2012.

[31] Orsi, M., & Essex, J.W. Permeability of drugs and hormones through a lipid bilayer: insights from dual-resolution molecular dynamics. *Soft Matter*, 6(16):3797–12, 2010.

[32] Eriksson, E.S.E., Dos Santos, D.J.V.A., Guedes, R.C., & Eriksson, L.A. Properties and permeability of hypericin and brominated hypericin in lipid membranes. *Journal of Chemical Theory and Computation*, 5(12):3139–3149, 2009.

[33] Eriksson, E.S.E., & Eriksson, L.A. The influence of cholesterol on the properties and permeability of hypericin derivatives in lipid membranes. *Journal of Chemical Theory and Computation*, 7(3):560–574, 2011.

[34] Butterworth, J.F., & Strichartz, G.R. Molecular mechanisms of local anesthesia: a review. *Anesthesiology*, 72(4):711–734, 1990.

[35] Högberg, C.-J., Maliniak, A., & Lyubartsev, A.P. Dynamical and structural properties of charged and uncharged lidocaine in a lipid bilayer. *Biophysical Chemistry*, 125(2):416–424, 2007.

[36] Mojumdar, E.H., & Lyubartsev, A.P. Molecular dynamics simulations of local anesthetic articaine in a lipid bilayer. *Biophysical Chemistry*, 153(1):27–35, 2010.

[37] Bernardi, R.C., Gomes, D.E.B., Gobato, R., Taft, C.A., Ota, A.T., & Pascutti, P.G. Molecular dynamics study of biomembrane/local anesthetics interactions. *Molecular Physics*, 107(14):1437–1443, 2009.

[38] Porasso, R.D., Drew Bennett, W.F., Oliveira-Costa, S.D., & Cascales, J.J.L. Study of the benzocaine transfer from aqueous solution to the interior of a biological membrane. *Journal of Physical Chemistry B*, 113(29):9988–9994, 2009.

[39] López Cascales, J.J., Oliveira Costa, S.D., & Porasso, R.D. Thermodynamic study of benzocaine insertion into different lipid bilayers. *Journal of Chemical Physics*, 135(13):10B604, 2011.

[40] Nury, H., Van Renterghem, C., Weng, Y., Tran, A., Baaden, M., Dufresne, V., Changeux, J.-P., Sonner, J.M., Delarue, M., & Corringer, P.-J. X-ray structures of general anaesthetics bound to a pentameric ligand-gated ion channel. *Nature*, 469(7330):428–431, 2011.

[41] Jerabek, H., Pabst, G., Rappolt, M., & Stockner, T. Membrane-mediated effect on ion channels induced by the anesthetic drug ketamine. *Journal of the American Chemical Society*, 132(23):7990–7997, 2010.

[42] Pickholz, M., Oliveira, O.N., & Skaf, M.S. Molecular dynamics simulations of neutral chlorpromazine in zwitterionic phospholipid monolayers. *Journal of Physical Chemistry B*, 110(17):8804–8814, 2006.

[43] Pickholz, M., Oliveira Jr, O.N., & Skaf, M.S. Interactions of chlorpromazine with phospholipid monolayers: effects of the ionization state of the drug. *Biophysical Chemistry*, 125(2–3):425–434, 2007.

[44] Lin, F., & Wang, R. Hemolytic mechanism of dioscin proposed by molecular dynamics simulations. *Journal of Molecular Modeling*, 16(1):107–118, 2010.

[45] Osbourn, A., Goss, R.J.M., & Field, R.A. The saponins – polar isoprenoids with important and diverse biological activities. *Natural Product Reports*, 28(7):1261–1268, 2011.

[46] Wang, E., & Klauda, J.B. Molecular dynamics simulations of ceramide and ceramide-phosphatidylcholine bilayers. *The Journal of Physical Chemistry B*, 121(43):10091–10104, 2017.

[47] Gupta, R., Dwadasi, B.S., & Rai, B. Molecular dynamics simulation of skin lipids: effect of ceramide chain lengths on bilayer properties. *The Journal of Physical Chemistry B*, 120(49):12536–12546, 2016.

[48] Khadka, N.K., Cheng, X., Ho, C.S., Katsaras, J., & Pan, J. Interactions of the anticancer drug tamoxifen with lipid membranes. *Biophysical Journal*, 108(10):2492–2501, 2015.

[49] Boggara, M.B., & Krishnamoorti, R. Partitioning of nonsteroidal antiinflammatory drugs in lipid membranes: a molecular dynamics simulation study. *Biophysical Journal*, 98(4):586–595, 2010.

[50] Libo, L., Vorobyov, I., Dorairaj, S., & Allen, T.W. Chapter 15 charged protein side chain movement in lipid bilayers explored with free energy simulation. *Current Topics in Membranes*, 60:405–459, 2008.

[51] Kessel, A., Ben-Tal, N., & May, S. Interactions of cholesterol with lipid bilayers: the preferred configuration and fluctuations. *Biophysical Journal*, 81(2):643–658, 2001.

[52] Khelashvili, G., Pabst, G., & Harries, D. Cholesterol orientation and tilt modulus in DMPC bilayers. *Journal of Physical Chemistry B*, 114(22):7524–7534, 2010.

[53] Róg, T., & Pasenkiewicz-Gierula, M. Cholesterol effects on the phosphatidylcholine bilayer nonpolar region: a molecular simulation study. *Biophysical Journal*, 81(4):2190–2202, 2001.

[54] Shimshick, E.J., & McConnell, H.M. Lateral phase separation in phospholipid membranes. *Biochemistry*, 12(12):2351–2360, 1973.

[55] Simons, K., & Ikonen, E. Functional rafts in cell membranes. *Nature*, 387(6633):569–572, 1997.

[56] Brown, D.A., & London, E. Structure and origin of ordered lipid domains in biological membranes. *The Journal of Membrane Biology*, 164(2):103–114, 1998.

[57] Simons, K., & Toomre, D. Lipid rafts and signal transduction. *Nature Reviews Molecular Cell Biology*, 1(1):31–39, 2000.

[58] Parton, R.G., & Hancock, J.F. Lipid rafts and plasma membrane microorganization: insights from Ras. *Trends in Cell Biology*, 14(3):141–147, 2004.

[59] Hanzal-Bayer, M.F., & Hancock, J.F. Lipid rafts and membrane traffic. *FEBS Letters*, 581(11):2098–2104, 2007.
[60] Fan, J., Sammalkorpi, M., & Haataja, M. Formation and regulation of lipid microdomains in cell membranes: theory, modeling, and speculation. *FEBS Letters*, 584(9):1678–1684, 2010.
[61] Pan, J., Cheng, X., Heberle, F.A., Mostofian, B., Kuerka, N., Drazba, P., & Katsaras, J. Interactions between ether phospholipids and cholesterol as determined by scattering and molecular dynamics simulations. *The Journal of Physical Chemistry B*, 116(51):14829–14838, 2012.
[62] Hammond, A.T., Heberle, F.A., Baumgart, T., Holowka, D., & Baird, B., and G W Feigenson. Crosslinking a lipid raft component triggers liquid ordered-liquid disordered phase separation in model plasma membranes. *Proceedings of the National Academy of Sciences*, 102(18):6320–6325, 2005.
[63] Róg, T., & Pasenkiewicz-Gierula, M. Cholesterol effects on a mixed-chain phosphatidylcholine bilayer: a molecular dynamics simulation study. *Biochimie*, 88(5):449–460, 2006.
[64] Martinez-Seara, H., Róg, T., Pasenkiewicz-Gierula, M., Vattulainen, I., Karttunen, M., & Reigada, R. Interplay of unsaturated phospholipids and cholesterol in membranes: effect of the double-bond position. *Biophysical Journal*, 95(7):3295–3305, 2008.
[65] Aittoniemi, J., Róg, T., Niemelä, P., Pasenkiewicz-Gierula, M., Karttunen, M., & Vattulainen, I. Tilt: major factor in sterols' ordering capability in membranes. *Journal of Physical Chemistry B*, 110(51):25562–25564, 2006.
[66] Róg, T., Vattulainen, I., Jansen, M., Ikonen, E., & Karttunen, M. Comparison of cholesterol and its direct precursors along the biosynthetic pathway: effects of cholesterol, desmosterol and 7-dehydrocholesterol on saturated and unsaturated lipid bilayers. *Journal of Chemical Physics*, 129(15):154508, 2008.
[67] Johnson, Q.R., Mostofian, B., Gomez, G.F., Smith, J.C., & Cheng, X. Effects of carotenoids on lipid bilayers. *Physical Chemistry Chemical Physics*, 41:409, 2018.
[68] Reigada, R. Atomistic study of lipid membranes containing chloroform: looking for a lipid-mediated mechanism of anesthesia. *PLoS ONE*, 8(1):e52631–10, 2013.
[69] Atkinson, J., Harroun, T., Wassall, S.R., Stillwell, W., & Katsaras, J. The location and behavior of α-tocopherol in membranes. *Molecular Nutrition & Food Research*, 54(5):641–651, 2010.
[70] Hope, H.R., & Pike, L.J. Phosphoinositides and phosphoinositide-utilizing enzymes in detergent-insoluble lipid domains. *Molecular Biology of the Cell*, 7(6):843–851, 1996.
[71] Salaün, C., James, D.J., & Chamberlain, L.H. Lipid rafts and the regulation of exocytosis. *Traffic (Copenhagen, Denmark)*, 5(4):255–264, 2004.
[72] Wittinghofer, A., & Pal, E.F. The structure of Ras protein: a model for a universal molecular switch. *Trends in Biochemical Sciences*, 16:382–387, 1991.
[73] Chiu, V.K., Bivona, T., Hach, A., Bernard Sajous, J., Silletti, J., Wiener, H., Johnson, R.L., Cox, A.D., & Philips, M.R. Ras signalling on the endoplasmic reticulum and the Golgi. *Nature Cell Biology*, 4(5):343–350, 2002.
[74] Hancock, J.F., & Parton, R.G. Ras plasma membrane signalling platforms. *Biochemical Journal*, 389(Pt 1):1–11, 2005.
[75] Weise, K., Triola, G., Brunsveld, L., Waldmann, H., & Winter, R. Influence of the lipidation motif on the partitioning and association of N-Ras in model membrane subdomains. *Journal of the American Chemical Society*, 131(4):1557–1564, 2009.
[76] Högberg, C.-J., & Lyubartsev, A.P. Effect of local anesthetic lidocaine on electrostatic properties of a lipid bilayer. *Biophysical Journal*, 94(2):525–531, 2008.
[77] Gurtovenko, A.A., Patra, M., Karttunen, M., & Vattulainen, I. Cationic DMPC/DMTAP lipid bilayers: molecular dynamics study. *Biophysical Journal*, 86(6):3461–3472, 2004.

[78] Seelig, A., & Seelig, J. Dynamic structure of fatty acyl chains in a phospholipid bilayer measured by deuterium magnetic resonance. *Biochemistry*, 13(23):4839–4845, 1974.
[79] Vermeer, L.S., de Groot, B.L., Réat, V., Milon, A., & Czaplicki, J. Acyl chain order parameter profiles in phospholipid bilayers: computation from molecular dynamics simulations and comparison with 2H NMR experiments. *European Biophysics Journal*, 36(8):919–931, 2007.
[80] Davis, J.H. The description of membrane lipid conformation, order and dynamics by H-2-nmr. *Biochimica et Biophysica Acta – Molecular Basis of Disease*, 737(1):117–171, 1983.
[81] Douliez, J.P., Léonard, A., & Dufourc, E.J. Restatement of order parameters in biomembranes: calculation of C-C bond order parameters from C-D quadrupolar splittings. *Biophysical Journal*, 68(5):1727–1739, 1995.
[82] Petrache, H.I., Tu, K., & Nagle, J.F. Analysis of simulated NMR order parameters for lipid bilayer structure determination. *Biophysical Journal*, 76(5):2479–2487, 1999.
[83] Gabdoulline, R.R.R., Vanderkooi, G., & Zheng, C. Comparison of the structures of dimyristoylphosphatidylcholine in the presence and absence of cholesterol by molecular dynamics simulations. *The Journal of Physical Chemistry*, 100(39):15942–15946, 1996.
[84] Chiu, S.W., Jakobsson, E., Scott, H.L., & Monte, C. Carlo and molecular dynamics simulation of hydrated lipid-cholesterol lipid bilayers at low cholesterol concentration. *Biophysical Journal*, 80(3):1104–1114, 2001.
[85] Jedlovszky, P., & Mezei, M. Effect of cholesterol on the properties of phospholipid membranes. 1. Structural features. *Journal of Physical Chemistry B*, 107(22):5311–5321, 2003.
[86] Doxastakis, M., Sum, A.K., & de Pablo, J.J. Modulating membrane properties: the effect of trehalose and cholesterol on a phospholipid bilayer. *Journal of Physical Chemistry B*, 109(50):24173–24181, 2005.
[87] de Meyer, F., & Smit, B. Effect of cholesterol on the structure of a phospholipid bilayer. *Proceedings of the National Academy of Sciences*, 106(10):3654–3658, 2009.
[88] Martinez-Seara, H., Róg, T., Karttunen, M., Reigada, R., & Vattulainen, I. Influence of cis double-bond parametrization on lipid membrane properties: how seemingly insignificant details in force-field change even qualitative trends. *Journal of Chemical Physics*, 129(10), 2008, 105103.
[89] Mendelsohn, R., Davies, M.A., Brauner, J.W., Schuster, H.F., & Dluhy, R.A. Quantitative determination of conformational disorder in the acyl chains of phospholipid bilayers by infrared spectroscopy. *Biochemistry*, 28(22):8934–8939, 1989.
[90] Róg, T., Pasenkiewicz-Gierula, M., Vattulainen, I., & Karttunen, M. What happens if cholesterol is made smoother: importance of methyl substituents in cholesterol ring structure on phosphatidylcholine-sterol interaction. *Biophysical Journal*, 92(10):3346–3357, 2007.
[91] Murari, R., Murari, M.P., & Baumann, W.J. Sterol orientations in phosphatidylcholine liposomes as determined by deuterium NMR. *Biochemistry*, 25(5):1062–1067, 1986.
[92] Li, L.K., So, L., & Spector, A. Membrane cholesterol and phospholipid in consecutive concentric sections of human lenses. *Journal of Lipid Research*, 26(5):600–609, 1985.
[93] Ohvo-Rekila, H., Ramstedt, B., Leppimaki, P., & Slotte, J.P. Cholesterol interactions with phospholipids in membranes. *Progress in Lipid Research*, 41(1):66–97, 2002.
[94] Harden M, M., & Radhakrishnan, A. Condensed complexes of cholesterol and phospholipids. *BBA – Biomembranes*, 1610(2):159–173, 2003.
[95] McMullen, T.P.W., Lewis, R.N.A.H., & McElhaney, R.N. Cholesterol-phospholipid interactions, the liquid-ordered phase and lipid rafts in model and biological membranes. *Current Opinion in Colloid & Interface Science*, 8(6):459–468, 2004.
[96] Róg, T., Pasenkiewicz-Gierula, M., Vattulainen, I., & Karttunen, M. Ordering effects of cholesterol and its analogues. *Biochimica et Biophysica Acta – Molecular Basis of Disease*, 1788(1):97–121, 2009.

[97] Chong, P.L.-G., Zhu, W., & Venegas, B. On the lateral structure of model membranes containing cholesterol. *BBA – Biomembranes*, 1788(1):2–11, 2009.

[98] Wang, H., & Meng, F. The permeability enhancing mechanism of menthol on skin lipids: a molecular dynamics simulation study. *Journal of Molecular Modeling*, 23(10):279, 2017.

[99] Marsh, D., & Smith, I.C. An interacting spin label study of the fluidizing and condensing effects of cholesterol on lecithin bilayers. *Biochimica et Biophysica Acta – Molecular Basis of Disease*, 298(2):133–144, 1973.

[100] Yeagle, P.L. Cholesterol and the cell membrane. *Biochimica et Biophysica Acta – Molecular Basis of Disease*, 822(3–4):267–287, 1985.

[101] Smaby, J.M., Brockman, H.L., & Brown, R.E. Cholesterol's interfacial interactions with sphingomyelins and phosphatidylcholines: hydrocarbon chain structure determines the magnitude of condensation. *Biochemistry*, 33(31):9135–9142, 1994.

[102] Smaby, J.M., Momsen, M., Kulkarni, V.S., & Brown, R.E. Cholesterol-induced interfacial area condensations of galactosylceramides and sphingomyelins with identical acyl chains. *Biochemistry*, 35(18):5696–5704, 1996.

[103] Smaby, J.M., Momsen, M.M., Brockman, H.L., & Brown, R.E. Phosphatidylcholine acyl unsaturation modulates the decrease in interfacial elasticity induced by cholesterol. *Biophysical Journal*, 73(3):1492–1505, 1997.

[104] Hyslop, P.A., Morel, B., & Sauerheber, R.D. Organization and interaction of cholesterol and phosphatidylcholine in model bilayer membranes. *Biochemistry*, 29(4):1025–1038, 1990.

[105] Brzozowska, I., & Figaszewski, Z.A. The equilibrium of phosphatidylcholine–cholesterol in monolayers at the air/water interface. *Colloids and Surfaces B: Biointerfaces*, 23(1):51–58, 2002.

[106] Hung, W.-C., Lee, M.-T., Chen, F.-Y., & Huang, H.W. The condensing effect of cholesterol in lipid bilayers. *Biophysical Journal*, 92(11):3960–3967, 2007.

[107] Falck, E., Patra, M., Karttunen, M., Hyvönen, M.T., & Vattulainen, I. Lessons of slicing membranes: interplay of packing, free area, and lateral diffusion in phospholipid/cholesterol bilayers. *Biophysical Journal*, 87(2):1076–1091, 2004.

[108] Hofsäß, C., Lindahl, E., & Edholm, O. Molecular dynamics simulations of phospholipid bilayers with cholesterol. *Biophysical Journal*, 84(4):2192–2206, 2003.

[109] Chiu, S.W., Jakobsson, E., Jay Mashl, R., & Scott, H.L. Cholesterol-induced modifications in lipid bilayers: a simulation study. *Biophysical Journal*, 83(4):1842–1853, 2002.

[110] Koberl, M., Hinz, H.J., & Rapp, G. Temperature scanning simultaneous small- and wide-angle X-ray scattering studies on glycolipid vesicles: areas, expansion coefficients and hydration. *Chemistry and Physics of Lipids*, 91(1):13–37, 1998.

[111] Tristram-Nagle, S., Liu, Y., Legleiter, J., & Nagle, J.F. Structure of gel phase DMPC determined by X-ray diffraction. *Biophysical Journal*, 83(6):3324–3335, 2002.

[112] Cantor, R.S. The lateral pressure profile in membranes: a physical mechanism of general anesthesia. *Biochemistry*, 36(9):2339–2344, 1997.

[113] Fábián, B., Sega, M., Voloshin, V.P., Medvedev, N.N., & Jedlovszky, P. Lateral pressure rofile and free volume properties in phospholipid membranes containing anesthetics. *Journal of Physical Chemistry B*, 121(13):2814–2824, 2017.

[114] Sonne, J., Hansen, F.Y., & Peters, G.H. Methodological problems in pressure profile calculations for lipid bilayers. *Journal of Chemical Physics*, 122(12):124903–124910, 2005.

[115] Walton, J.P.R.B., Tildesley, D.J., Rowlinson, J.S., & Henderson, J.R. The pressure tensor at the planar surface of a liquid. *Molecular Physics*, 48(6):1357–1368, 1983.

[116] Aldert R, V.B., Marrink, S.-J., & Berendsen, H.J.C. A molecular dynamics study of the decane/water interface. *The Journal of Physical Chemistry*, 97(36):9206–9212, 1993.

[117] Lindahl, E., & Edholm, O. Spatial and energetic-entropic decomposition of surface tension in lipid bilayers from molecular dynamics simulations. *Journal of Chemical Physics*, 113(9): 3882–3893, 2000.

[118] Kirkwood, J.G., & Buff, F.P. The statistical mechanical theory of surface tension. *The Journal of Chemical Physics*, 17(3):338–343, 1949.

[119] Irving, J.H., & Kirkwood, J.G. The Statistical mechanical theory of transport processes. IV. The equations of hydrodynamics. *The Journal of Chemical Physics*, 18(6):817–829, 1950.

[120] Goetz, R., & Lipowsky, R. Computer simulations of bilayer membranes: self-assembly and interfacial tension. *Journal of Chemical Physics*, 108(17):7397–7409, 1998.

[121] Gullingsrud, J., & Schulten, K. Gating of MscL studied by steered molecular dynamics. *Biophysical Journal*, 85(4):2087–2099, 2003.

[122] Gullingsrud, J., & Schulten, K. Lipid bilayer pressure profiles and mechanosensitive channel gating. *Biophysical Journal*, 86(6):3496–3509, 2004.

[123] Harasima, A. Molecular theory of surface tension. *Advances in Chemical Physics*, 1:203–237, 1958.

[124] Patra, M. Lateral pressure profiles in cholesterol–DPPC bilayers. *European Biophysics Journal*, 35(1):79–88, 2005.

[125] Samuli Ollila, O.H., Róg, T., Karttunen, M., & Vattulainen, I. Role of sterol type on lateral pressure profiles of lipid membranes affecting membrane protein functionality: comparison between cholesterol, desmosterol, 7-dehydrocholesterol and ketosterol. *Journal of Structural Biology*, 159(2):311–323, 2007.

[126] Evans, E., & Rawicz, W. Entropy-driven tension and bending elasticity in condensed-fluid membranes. *Physical Review Letters*, 64:2094–2097, 1990.

[127] Simson, R., Wallraff, E., Faix, J., Niewöhner, J., Gerisch, G., & Sackmann, E. Membrane bending modulus and adhesion energy of wild-type and mutant cells of dictyostelium lacking talin or cortexillins. *Biophysical Journal*, 74(1):514–522, 1998.

[128] Cuvelier, D., Derényi, I., Bassereau, P., & Nassoy, P. Coalescence of membrane tethers: experiments, theory, and applications. *Biophysical Journal*, 88(4):2714–2726, 2005.

[129] Lindahl, E., & Edholm, O. Mesoscopic undulations and thickness fluctuations in lipid bilayers from molecular dynamics simulations. *Biophysical Journal*, 79(1):426–433, 2000.

[130] Khelashvili, G., & Harries, D. How cholesterol tilt modulates the mechanical properties of saturated and unsaturated lipid membranes. *Journal of Physical Chemistry B*, 117(8): 2411–2421, 2013.

[131] Khelashvili, G., Kollmitzer, B., Heftberger, P., Pabst, G., & Harries, D. Calculating the bending modulus for multicomponent lipid membranes in different thermodynamic phases. *Journal of Chemical Theory and Computation*, 9(9):3866–3871, 2013.

[132] Canham, P.B. The minimum energy of bending as a possible explanation of the biconcave shape of the human red blood cell. *Journal of Theoretical Biology*, 26(1):61–81, 1970.

[133] Helfrich, W. Elastic properties of lipid bilayers: theory and possible experiments. *Zeitschrift für Naturforschung C*, 28:693, 1973.

[134] Goetz, R., Gompper, G., & Lipowsky, R. Mobility and elasticity of self-assembled membranes. *Physical Review Letters*, 82(1):221–224, 1999.

[135] Hamm, M., & Kozlov, M.M. Elastic energy of tilt and bending of fluid membranes. *The European Physical Journal E*, 3(4):323–335, 2000.

[136] May, E.R., Narang, A., & Kopelevich, D.I. Role of molecular tilt in thermal fluctuations of lipid membranes. *Physical Review. E*, 76:021913, 2007.

[137] Watson, M.C., Penev, E.S., Welch, P.M., & Brown, F.L.H. Thermal fluctuations in shape, thickness, and molecular orientation in lipid bilayers. *The Journal of Chemical Physics*, 135 (24):244701, 2011.

[138] Watson, M.C., Brandt, E.G., Welch, P.M., & Brown, F.L.H. Determining biomembrane bending rigidities from simulations of modest size. *Physical Review Letters*, 109 (2):028102EP–, 2012.

[139] Kozlov, M.M., & Helfrich, W. Effects of a cosurfactant on the stretching and bending elasticities of a surfactant monolayer. *Langmuir*, 8(11):2792–2797, 1992.

[140] Harmandaris, V.A., & Deserno, M. A novel method for measuring the bending rigidity of model lipid membranes by simulating tethers. *The Journal of Chemical Physics*, 125(20):204905, 2006.

[141] Arkhipov, A., Yin, Y., & Schulten, K. Four-scale description of membrane sculpting by BAR domains. *Biophysical Journal*, 95(6):2806–2821, 2008.

[142] Almeida, P.F.F., Vaz, W.L.C., & Thompson, T.E. Lateral diffusion in the liquid phases of dimyristoylphosphatidylcholine/cholesterol lipid bilayers: a free volume analysis. *Biochemistry*, 31(29):6739–6747, 1992.

[143] Vaz, W.L.C., Clegg, R.M., & Hallmann, D. Translational diffusion of lipids in liquid crystalline phase phosphatidylcholine multibilayers. A comparison of experiment with theory. *Biochemistry*, 24(3):781–786, 1985.

[144] Kuo, A.-L., & Wade, C.G. Lipid lateral diffusion by pulsed nuclear magnetic resonance. *Biochemistry*, 18(11):2300–2308, 1979.

[145] Sharma, V.K., Mitra, S., Verma, G., Hassan, P.A., Garcia Sakai, V., & Mukhopadhyay, R. Internal dynamics in SDS micelles: neutron scattering study. *Journal of Physical Chemistry B*, 114 (51):17049–17056, 2010.

[146] Sharma, V.K., Mamontov, E., Anunciado, D.B., O'Neill, H., & Urban, V. Nanoscopic dynamics of phospholipid in unilamellar vesicles: effect of gel to fluid phase transition. *Journal of Physical Chemistry B*, 119(12):4460–4470, 2015.

[147] Marrink, S.-J., & Berendsen, H.J.C. Simulation of water transport through a lipid membrane. *The Journal of Physical Chemistry*, 98(15):4155–4168, 1994.

[148] Marrink, S.J., & Berendsen, H.J.C. Permeation process of small molecules across lipid membranes studied by molecular dynamics simulations. *The Journal of Physical Chemistry*, 100(41):16729–16738, 1996.

[149] Wang, Y., Cohen, J., Boron, W.F., Schulten, K., & Tajkhorshid, E. Exploring gas permeability of cellular membranes and membrane channels with molecular dynamics. *Journal of Structural Biology*, 157(3):534–544, 2007.

[150] Kubo, R. The fluctuation-dissipation theorem. *Reports on Progress in Physics*, 29(1):255–284, 1966.

[151] Bassolino-Klimas, D., Alper, H.E., & Stouch, T.R. Solute diffusion in lipid bilayer membranes: an atomic level study by molecular dynamics simulation. *Biochemistry*, 32(47):12624–12637, 1993.

[152] Chao, L., Aitchison, E.W., Dongsheng, W., Zheng, L., Cheng, X., & Yang, W. Comparative exploration of hydrogen sulfide and water transmembrane free energy surfaces via orthogonal space tempering free energy sampling. *Journal of Computational Chemistry*, 37(6):567–574, 2016.

[153] Comer, J., Schulten, K., & Chipot, C. Diffusive Models of Membrane Permeation with explicit orientational freedom. *Journal of Chemical Theory and Computation*, 10(7):2710–2718, 2014.

[154] Comer, J., Schulten, K., & Chipot, C. Calculation of lipid-bilayer permeabilities using an average force. *Journal of Chemical Theory and Computation*, 10(2):554–564, 2014.

[155] Comer, J., Schulten, K., & Chipot, C. Permeability of a fluid Lipid bilayer to short-chain alcohols from first principles. *Journal of Chemical Theory and Computation*, 13(6): 2523–2532, pages acs.jctc.7b00264–10, 2017.